Arrangements of Hyperplanes—Sapporo 2009

ADVANCED STUDIES
IN PURE MATHEMATICS 62

Chief Editor of the Series: Sadayoshi Kojima (Tokyo Institute of Technology)

Arrangements of Hyperplanes —Sapporo 2009

Edited by

Hiroaki Terao (Hokkaido University)
Sergey Yuzvinsky (University of Oregon)

Mathematical Society of Japan

This book was typeset by $\mathcal{A}_\mathcal{M}\mathcal{S}$-TEX and $\mathcal{A}_\mathcal{M}\mathcal{S}$-LATEX, the TEX macro systems of the American Mathematical Society, together with the style files aspm.sty *and* aspmfm.sty *for $\mathcal{A}_\mathcal{M}\mathcal{S}$-TEX written by Dr. Chiaki Tsukamoto and* aspmproc.sty *for $\mathcal{A}_\mathcal{M}\mathcal{S}$-LATEX written by Dr. Akihiro Munemasa and* aspm.cls *for $\mathcal{A}_\mathcal{M}\mathcal{S}$-LATEX provided by Livretech Co., Ltd.*

TEX is a trademark of the American Mathematical Society.

Edited by the Mathematical Society of Japan.

Published by the Mathematical Society of Japan.

Distributed by the Mathematical Society of Japan, the American Mathematical Society and the World Scientific Publishing Co., Ltd.
Distributed exclusively in North America by the American Mathematical Society.

Advanced Studies in Pure Mathematics 62
ISBN 978-4-931469-67-9

PRINTED IN JAPAN
by Livretech Co., Ltd.

2010 Mathematics Subject Classification.
Primary 14N20.
Secondary 32S22, 52C35.

Preface

This volume of Advanced Studies in Pure Mathematics is the proceedings of the Second MSJ-SI (Mathematical Society of Japan-Seasonal Institute)

"Arrangements of Hyperplanes"
August 1–13, 2009
Hokkaido University, Sapporo, Japan.

The MSJ-SI is a major annual conference series of the Mathematical Society of Japan whose subject is selected by the Scientific Committee of the MSJ.

The subject of the Second MSJ-SI was arrangements of hyperplanes. The conference featured the following eight series of lectures by world's-leading experts:

Arrangements and Topology: (3 lectures)
by Toshitake Kohno (University of Tokyo, Japan)

Arrangements and Singularities: (3 lectures)
by Kyoji Saito (Institute for the Physics and Mathematics of the Universe, Japan)

Arrangements and Computations: (3 lectures)
by Hal Schenck (University of Illinois, USA)

Arrangements and Combinatorics: (3 lectures)
by Richard Stanley (MIT, USA)

Arrangements and Statistics: (3 lectures)
by Akimichi Takemura (University of Tokyo, Japan)

Arrangements, Hypergeometric Integrals and Physics: (3 lectures)
by Alexander Varchenko (University of North Carolina at Chapel Hill, USA)

Arrangements and Algebraic Geometry: (3 lectures)
by Masahiko Yoshinaga (Kyoto University, Japan)

Arrangements (Introduction): (2 lectures)
by Sergey Yuzvinsky (University of Oregon, USA).

The first week of the meeting was devoted to these survey lectures. We believe that the carefully-prepared lectures gave the audience an excellent introduction to the theory of arrangements of hyperplanes. In the second week, twenty-four active researchers gave invited talks on the themes related to arrangements of hyperplanes. In addition, nine relatively young mathematicians talked with vigor about their recent results as short communications. Thus we had fifty-six lectures in total which were presented to the 179 participants, sixty-two of whom were from abroad. The overseas participants included three mathematicians from the Korean Mathematical Society and three from the Mathematical Society of Republic of China whom the MSJ invited to promote mathematical activities in the East Asia.

The organizing committee members were Hiroaki Terao (chair, Hokkaido University), Takuro Abe (Kyoto University), Masahiko Yoshinaga (Kyoto University) and Sergey Yuzvinsky (University of Oregon). The conference was made possible by generous financial supports provided by the MSJ, Hokkaido University, the Sapporo International Communication Plaza Foundation and Professor Akimichi Takemura's Grant-in-Aid for Scientific Research. The staff of MSJ, secretaries of the Mathematics Department, especially Ms. Akiko Yamashita, volunteer interpreters, graduate students successfully helped the conference to go on smoothly through the entire period. Ms. Akiko Taniguchi helped us to edit the files for this volume with her expertise on TEX. We express our gratitude to all of them. Lastly we would like to express our sincere gratitude to the contributors to this volume as well as the referees of the papers.

Editors:
Hiroaki Terao (Hokkaido University)
Sergey Yuzvinsky (University of Oregon)

All papers in this volume have been refereed and are in final form. No version of any of them will be submitted for publication elsewhere.

Program Outline

Date: August 1–13, 2009
Venue: Conference Hall, 2nd floor, Hokkaido University
Sapporo, Japan
Organizing committee: Takuro Abe
Hiroaki Terao (chair)
Masahiko Yoshinaga
Sergey Yuzvinsky
Contents: The meeting will be divided into two parts. In the first part, we will have a series of lectures. Each series is composed of 3 one-hour lectures. In the second part, we will have several single talks.

August 1st (Sat.)

9:45–10:45	Sergey Yuzvinsky (University of Oregon, USA) Arrangements (Introduction)
11:15–12:15	Sergey Yuzvinsky (University of Oregon, USA) Arrangements (Introduction)
14:15–15:15	Hal Schenck (University of Illinois, USA) Arrangements and Computations
15:45–16:45	Alexander Varchenko (University of North Carolina at Chapel Hill, USA) Arrangements, Hypergeometric Integrals and Physics
17:00–18:00	Toshitake Kohno (University of Tokyo, Japan) Arrangements and Topology

August 2nd (Sun.)

9:45–10:45	Toshitake Kohno (University of Tokyo, Japan) Arrangements and Topology
11:15–12:15	Richard Stanley (MIT, USA) Arrangements and Combinatorics
14:15–15:15	Hal Schenck (University of Illinois, USA) Arrangements and Computations
15:45–16:45	Alexander Varchenko (University of North Carolina at Chapel Hill, USA) Arrangements, Hypergeometric Integrals and Physics
17:00–18:00	Toshitake Kohno (University of Tokyo, Japan) Arrangements and Topology

August 3rd (Mon.)

9:45–10:45	Masahiko Yoshinaga (Kyoto University, Japan) Arrangements and Algebraic Geometry
11:15–12:15	Richard Stanley (MIT, USA) Arrangements and Combinatorics
14:15–15:15	Hal Schenck (University of Illinois, USA) Arrangements and Computations
15:45–16:45	Alexander Varchenko (University of North Carolina at Chapel Hill, USA) Arrangements, Hypergeometric Integrals and Physics
17:00–18:00	Masahiko Yoshinaga (Kyoto University, Japan) Arrangements and Algebraic Geometry

August 4th (Tue.)

9:45–10:45	Masahiko Yoshinaga (Kyoto University, Japan) Arrangements and Algebraic Geometry
11:15–12:15	Richard Stanley (MIT, USA) Arrangements and Combinatorics
14:15–15:15	Kyoji Saito (Institute for the Physics and Mathematics of the Universe, Japan) Arrangements and Singularities
15:45–16:45	Akimichi Takemura (University of Tokyo, Japan) Arrangements and Statistics
18:00–20:00	Reception at the Aspen Hotel

August 5th (Wed.)

9:45–10:45	Akimichi Takemura (University of Tokyo, Japan) Arrangements and Statistics
11:15–12:15	Akimichi Takemura (University of Tokyo, Japan) Arrangements and Statistics
14:15–15:15	Kyoji Saito (Institute for the Physics and Mathematics of the Universe, Japan) Arrangements and Singularities
15:45–16:45	Kyoji Saito (Institute for the Physics and Mathematics of the Universe, Japan) Arrangements and Singularities

August 8th (Sat.)

9:30–	Takashi Tsuboi (President of the Mathematical Society of Japan) Welcome speech
9:45–10:45	Kazuhiko Aomoto (Nagoya University, Japan) Hypersphere arrangement and imaginary cycles for hypergeometric integrals
11:15–12:15	Christos Athanasiadis (University of Athens, Greece) Hyperplane arrangements, random walks and eigenvalues
14:15–15:15	Anatoly Libgober (University of Illinois at Chicago, USA) Characteristic varieties of arrangements with isolated non normal crossings
15:45–16:45	Eva Maria Feichtner (University of Bremen, Germany) On tropical arrangement compactifications
17:00–18:00	Richard Rimanyi (University of North Carolina at Chapel Hill, USA) Equivariant classes of matroid realization spaces

August 9th (Sun.)

9:45–10:45	Uli Walther (Purdue University, USA) Bockstein maps in local cohomology Joint with Anurag Singh (Utah)
11:15–12:15	Claudio Procesi (Università di Roma, Italy) From splines to the index theorem
14:15–15:15	Alexandru Suciu (Northeastern University, USA) On the geometry and topology of cohomology jump loci
15:45–16:45	Igor Dolgachev (University of Michigan, USA) Invariant theory of arrangements of hyperplanes
17:00–17:30	[short communication I] Simona Settepanella (Scuola Superiore Sant'Anna, Italy) Arrangements and political economics
17:30–18:00	[short communication II] Luca Moci (Università di Roma, Italy) An introduction to toric arrangements
18:00–18:30	[short communication III] Yoshio Sano (RIMS, Japan) Graphical Arrangements and Braid Arrangements – Graph orientations and linear extensions –

August 10th (Mon.)

9:45–10:45	Anders Björner (Mittag-Leffler Institute, Sweden) (presented by Axel Hultman (Royal Institute of Technology, Sweden)) Random walks on complex hyperplane arrangements and self-organizing libraries
11:15–12:15	Katsuhisa Mimachi (Tokyo Institute of Technology, Japan) Arrangements of hyperplanes and solutions of the Fuchsian diferential equations free from accessary parameters
14:15–15:15	Dai Tamaki (Shinshu University, Japan) Iterated loop spaces and oriented matroids
15:45–16:45	Mario Salvetti (Università di Pisa, Italy) Some applications of discrete Morse theory to arrangements
17:00–18:00	Antoine Deza (McMaster University, Canada) Hyperplane arrangements with large average diameter

August 11th (Tue.)

9:45–10:15	[short communication IV] Meital Eliyahu (Bar Ilan University, Israel) A characterization of line arrangement whose fundamental group is a direct sum of free groups
10:15–10:45	[short communication V] David Garber (Holon Institute of Technology, Israel) A conjugation-free geometric presentation of fundamental groups of arrangements
11:15–12:15	Yoshishige Haraoka (Kumamoto University, Japan) Arrangements of hyperplanes arising from sections of rigid local systems
14:15–15:15	Hironobu Kimura (Kumamoto University, Japan) On a problem of arrangements related to the hypergeometric integrals of confluent type
15:45–16:45	Eduard Looijenga (Universiteit Utrecht, Netherlands) Representations out of polydifferentials and the KZ-system
17:00–17:30	[short communication VI] Amanda Hager (University of Iowa, USA) Freeness of arrangements and arrangement bundles

17:30–18:00 [short communication VII]
Miguel Angel Marco-Buzunariz (Instituto de Ciencias Matematicas Consejo Superior de Investigaciones Cientificas, Spain)
On Rybnikov's example

18:00–18:30 [short communication VIII]
Suho Oh (MIT, USA)
Bruhat order and hyperplane arrangements

18:30–19:00 [short communication IX]
Max Wakefield (US Naval Academy, USA)
Formality of subspace arrangements

August 12th (Wed.)

9:45–10:45 Michael Falk (Northern Arizona University, USA)
Vanishing products in Orlik–Solomon algebras

11:15–12:15 David Mond (University of Warwick, UK)
Some new free divisors associated with versal deformations and stable maps

14:15–15:15 Ichiro Shimada (Hiroshima University, Japan)
On lattice-theoretic invariants of curves on a surface

15:45–16:45 Jorge Vitório Pereira (IMPA, Brazil)
Webs and arrangements

17:00–18:00 Graham Denham (University of Western Ontario, Canada)
Critical sets of master functions: resonance and log derivations

August 13th (Thu.)

9:45–10:45 Daniel Cohen (Louisiana State University, USA)
Collision-free motion planning on surfaces

11:15–12:15 Tomohide Terasoma (The University of Tokyo, Japan)
Lines in Fermat hypersurfaces and $M_{0,n}$

14:15–15:15 Ivan Cherednik (The University of North Carolina at Chapel Hill, USA)
Affine root systems and harmonic analysis

CONTENTS

Advanced Studies in Pure Mathematics 62, 2012
Arrangements of Hyperplanes—Sapporo 2009
pp. 1–26

Hypersphere arrangement and imaginary cycles for hypergeometric integrals

Kazuhiko Aomoto

Abstract.

We construct imaginary cycles as Lefschetz cycles for hypergeometric integrals associated with a hypersphere arrangement and discuss the relation between the twisted rational de Rham cohomology. We also discuss it in degenerate cases where several hyperspheres contact with each other. We pose two geometric problems involved in it.

§1. Introduction

First we want to illustrate in one-dimensional case the main objective discussed in this article. Let Q be the complex circle: $\{\xi = (\xi_1, \xi_2); \xi_1^2 + \xi_2^2 = 1\}$ in the complex affine plane $\mathbf{C}^2$. Q is isomorphic to $\mathbf{C}^*$ by taking $\xi_1 + \sqrt{-1}\xi_2 = \zeta$. Consider a family of m complex lines $H_j : f_j = 0$ where

$$f_j = u_{j,0} + u_{j,1}\xi_1 + u_{j,2}\xi_2$$

such that $u_{j,0}, u_{j,1}, u_{j,2} \in \mathbf{R}$ and that $u_{j,1}^2 + u_{j,2}^2 - u_{j,0}^2 = 1$. Denote

$$a_{i,j} = u_{i,1}u_{j,1} + u_{i,2}u_{j,2} - u_{i.0}u_{j,0},\ a_{i,0} = a_{0,i} = u_{i,0}, a_{0,0} = -1.$$

The intersection of Q and H_j consisits of two different points which we denote by ζ_j, ζ_j^* such that $|\zeta_j| = |\zeta_j^*| = 1$. Let R be the $\mathbf{C}[\xi_1, \xi_2]$

Received December 21, 2009.
2010 *Mathematics Subject Classification.* Primary 32S22; Secondary 14F40, 33C70, 52C35.
Key words and phrases. Hypersphere arrangement, configuration matrix, hypergeometric integrals, twisted de Rham cohomology, real and imaginary cycles.

module

$$R = \sum_{\nu_1 \geq 0, \ldots, \nu_m \geq 0} \mathbf{C}[\xi_1, \xi_2] \prod_{j=1}^{m} f_j^{-\nu_j}$$
$$= \sum_{\nu_1 \geq 0, \ldots, \nu_m \geq 0} \mathbf{C}[\zeta, \zeta^{-1}] \prod_{j=1}^{m} \{(\zeta - \zeta_j)(\zeta - \zeta_j^*)\}^{-\nu_j}$$

because f_j can be written as

$$f_j = \sqrt{-1}\frac{(\zeta - \zeta_j)(\zeta - \zeta_j^*)}{(\zeta_j^* - \zeta_j)\zeta}.$$

Consider the multiplicative function

$$\Phi_0(\xi) = \prod_{j=1}^{m} f_j^{\lambda_j} \quad (\lambda_j \in \mathbf{R}_{>\mathbf{0}})$$

and the associated rational de Rham cohomology on $Y = Q - \cup_{j=1}^{m} H_j$

$$H^1(Y, \nabla_0) \cong R\tau_Q / \nabla_0(R)$$

defined by the covariant differential $\nabla_0(\psi) = d\psi + d\log \Phi_0 \psi$, where we denote

$$\tau_Q = -\xi_1 d\xi_2 + \xi_2 d\xi_1 = \sqrt{-1}\frac{d\zeta}{\zeta}.$$

Suppose that $\zeta_1, \zeta_1^*, \ldots, \zeta_m, \zeta_m^*$ are different from each other. Then one can prove that for generic λ_j

$$H^1(Y, \nabla_0) \cong \mathbf{C}^{2m}$$

and it is spanned by

$$\varphi_Q(\emptyset)\tau_Q = \tau_Q, \varphi_Q(j)\tau_Q = \frac{\tau_Q}{f_j} = d\log\frac{\zeta - \zeta_j}{\zeta - \zeta_j^*}\ (1 \leq j \leq m),$$
$$\varphi_Q(j,k)\tau_Q = \frac{\tau_Q}{f_j f_k}\ (1 \leq j < k \leq m).$$

These one-forms are not linearly independent on Y. For any different i, j, k there exists the fundamental linear relation

$$(1.1) \quad c_i\varphi_Q(i) + c_j\varphi_Q(j) + c_k\varphi_Q(k) + c_{j,k}\varphi_Q(j,k) + c_{k,i}\varphi_Q(k,i) + c_{i,j}\varphi_Q(i,j) = 0$$

where $c_i, c_j, c_k, c_{j,k}, c_{k,i}, c_{i,j}$ can be written in terms of $a_{i,j}, a_{k,0}$ as

$$c_i = -\frac{A(0,i)}{A\begin{pmatrix} 0, & i, & k \\ 0, & i, & j \end{pmatrix}}, \quad c_{j,k} = \frac{A(j,k)}{A\begin{pmatrix} i, & j, & k \\ 0, & j, & k \end{pmatrix}}$$

$c_j, c_k, c_{k,i}, c_{i,j}$ being defined in the same way cyclically. Moreover $A(0,i) = -1 - a_{i,0}^2$, $A(j,k) = 1 - a_{j,k}^2$ and $A\begin{pmatrix} i, & j, & k \\ i', & j', & k' \end{pmatrix}$ denotes the determinant of the matrix whose components are $a_{p,q}$ $(p = i,j,k; q = i', j', k')$.

The twisted homology $H_1(Y, \hat{\mathcal{L}}_0)$ dual to $H^1(Y, \nabla_0)$ is spanned by the linearly independent cycles which are expressed by the closures (arcs) of the connected components of $\Re Y$.

Suppose now that for a fixed pair i, j, one of ζ_i or ζ_i^* coincides with one of ζ_j or ζ_j^*. This occurs if and only if $A(i,j) = 0$, i.e.,$a_{i,j} = \pm 1$. If ζ_j tends to the point ζ_i^*, then the arc connecting the points ζ_i^*, ζ_j in $\Re Q$ reduces to a point. Hence if $\zeta_i^* = \zeta_j$, the dimension of $H_1(Y, \hat{\mathcal{L}}_0)$ decreases by one. On the other hand one can show that $\varphi_Q(i,j)$ can be described cohomologically as a linear combination of $\varphi_Q(k,i), \varphi_Q(k,j), \varphi_Q(k), \varphi_Q(\emptyset)$:

$$(1.2)\quad 2(\lambda_i + \lambda_j - 1)\varphi_Q(i,j) \sim -\sum_{k \neq i,j} \lambda_k \{\frac{A(k,i)}{a_{k,i} + a_{k,j}}\varphi_Q(k,i) + \frac{A(k,j)}{a_{k,i} + a_{k,j}}\varphi_Q(k,j)\} + \sum_{k=0}^{m} \lambda_k a_{k,0}\varphi_Q(k) - (\lambda_\infty - 1)\{\frac{A(0,i)}{a_{i,0} + a_{j,0}}\varphi_Q(i) + \frac{A(0,j)}{a_{i,0} + a_{j,0}}\varphi_Q(j)\} - \lambda_\infty \varphi_Q(\emptyset),$$

where we denote $\lambda_\infty = \sum_{j=1}^m \lambda_j$.

In particular, consider the case where $m = 3$ and $A(1,2) = A(1,3) = A(2,3) = 0$, i.e., $\zeta_1^* = \zeta_2, \zeta_2^* = \zeta_3, \zeta_3^* = \zeta_1$. Then (1.1) reduces to the only one identity:

$$\varphi_Q(1) + \varphi_Q(2) + \varphi_Q(3) = 0$$

and there are three identities of type (1.2):

$$2(\lambda_i + \lambda_j - 1)\varphi_Q(i,j) \sim \sum_{k=1}^{3} \lambda_k a_{k,0}\varphi_Q(k) + (\lambda_\infty - 1)\left\{\frac{1 + a_{i,0}^2}{a_{i,0} + a_{j,0}}\varphi_Q(i) + \frac{1 + a_{j,0}^2}{a_{i,0} + a_{j,0}}\varphi_Q(j)\right\} - \lambda_\infty \varphi_Q(\emptyset).$$

Hence $H^1(Y, \nabla_0)$ is of dimension three and is spanned by a basis of representatives $\varphi_Q(\emptyset), \varphi_Q(1), \varphi_Q(2)$.

In this article we want to extend the above observation to the n-dimensional cases ($n \geq 1$). The twisted de Rham cohomology $H^n(Y, \nabla_0)$ can be formulated in the space Y, the complement of a union of $(n-1)$-dimensional hyperspheres in the n-dimensional fundamental complex hypersphere Q. There arise two problems.

In the first place, based on my preceding articles (see [2], [5]), we formulate the twisted de Rham cohomology presenting explicitly its basis in terms of the invariants $a_{i,j}$ obtained by the Lorentz inner products of the coefficients of two linear functions defining hyperspheres, together with $a_{k,0}$ which are the constant terms of a linear function. The differential structures of the hypergeometric integrals are described by the invariants $a_{i,j}, a_{k,0}$ under Lorentz groups or orthogonal groups. In the final section we show how they should be modified in some degenerate cases.

In the second place the basis of the twisted homology $H_n(Y, \hat{\mathcal{L}}_0)$ cannot always be realized by real domains in $\Re Y$ except in special domain of parameters. We must construct some of them as imaginary cycles. We want to show that this can be done by deforming real Lefschetz cycles from a special domain of parameters involved where all the cycles can be realized by real domains (see Theorem 3.11).

§2. Basic properties

Let $\mathcal{A}$ be an arrangement of m hyperplanes H_j ($1 \leq j \leq m$) defined over the real field of coefficients in the $(n+1)$-dimensional complex affine space $\mathbf{C}^{n+1}$. Each hyperplane can be described as

$$H_j : u_{j,0} + \sum_{\nu=1}^{n+1} u_{j,\nu}\xi_\nu = 0$$

for $\xi = (\xi_1, \ldots, \xi_{n+1}) \in \mathbf{C}^{n+1}$.

We denote by $N(\mathcal{A})$ the union of hyperplanes: $= \bigcup_{H_j \in \mathcal{A}} H_j$ and by $X = M(\mathcal{A})$ the complement of $N(\mathcal{A}) := \mathbf{C}^{n+1} - N(\mathcal{A})$. Let Q be the complex hypersphere: $f_0 = 0$ defined by the quadratic polynomial$f_0 = 1 - \sum_{\nu=1}^{n+1} \xi_\nu^2$. Each intersection $H_j \cap Q$ defines a hypersphere in Q provided it does not reduce to a point i.e., $-u_{j,0}^2 + \sum_{\nu=1}^{n+1} u_{j,\nu}^2 \neq 0$. Throughout this article we shall assume this condition and so may assume that

$$-u_{j,0}^2 + \sum_{\nu=1}^{n+1} u_{j,\nu}^2 = 1.$$

The family $\mathcal{A}' = \{H_j \cap Q\}_{1\le j\le m}$ defines a hypersphere arrangement in Q. We denote by Y the intersection of X and $Q : Y = Q - \cup_{j=1}^m H_j \cap Q$. We denote by $\Re Q$ the real part of Q which is identified with the n-dimensional real hypersphere. The real part $S_j = H_j \cap \Re Q$ is a real $(n-1)$-dimensional hypersphere in $\Re Q$.

We define the $(m+1)\times(m+1)$ configuration matrix $A = (a_{i,j})_{0\le i,j\le m}$ associated with $\mathcal{A}'$, whose components are Lorentz inner products

$$a_{i,j} = -u_{i0}u_{j0} + \sum_{\nu=1}^{n+1} u_{i,\nu}u_{j,\nu} \ (1 \le i, j \le m); \ a_{i,0} = a_{0,i} = u_{i0}; \ a_{0,0} = -1$$

so that $a_{i,i} = 1$ for $1 \le i \le m$.

For a set of indices $I = \{i_1, \dots, i_p\} \subset \{0, 1, 2, \dots, m\}$ the size p will be denoted by $|I|$. We say that I is admissible if $I \subset \{1, 2, \dots, m\}$. For two sets of indices $I = \{i_1, i_2, \dots, i_p\}$ and $J = \{j_1, \dots, j_p\}$ we define the subdeterminant

$$A\begin{pmatrix} I \\ J \end{pmatrix} = \begin{vmatrix} a_{i_1,j_1} & a_{i_1,j_2} & \cdots & a_{i_1,j_p} \\ a_{i_2,j_1} & a_{i_2,j_2} & \cdots & a_{i_2,j_p} \\ \vdots & \vdots & & \vdots \\ a_{i_p,j_1} & a_{i_p,j_2} & \cdots & a_{i_p,j_p} \end{vmatrix}.$$

We abbreviate $A\begin{pmatrix} I \\ I \end{pmatrix}$ by $A(I)$.

For an admissible set $I = \{i_1, i_2, \dots, i_p\}$ and a set $J = \{j_1, j_2, \dots, j_p\} \subset \{0, 1, 2, \dots, n+1\}$ $(p \le n+2)$ we denote the subdeterminant

$$U\begin{pmatrix} i_1, i_2, \dots, i_p \\ j_1, \dots, j_p \end{pmatrix} = \begin{vmatrix} u_{i_1,j_1} & u_{i_1,j_2} & \cdots & u_{i_1,j_p} \\ u_{i_2,j_1} & u_{i_2,j_2} & \cdots & u_{i_2,j_p} \\ \vdots & \vdots & & \vdots \\ u_{i_p,j_1} & u_{i_p,j_2} & \cdots & u_{i_p,j_p} \end{vmatrix}.$$

Remark. *The matrix A has at most rank $n+2$. Assume that I is admissible. Then $A(I) = 0$ for $|I| \ge n+3$, and*

$$A(I) = -U^2\begin{pmatrix} i_1, i_2, \dots, i_{n+2} \\ 0, 1, \dots, n+1 \end{pmatrix}$$

for $|I| = n+2$. On the other hand $A(0, I) = 0$ for $|I| \ge n+2$ and

$$A(0, I) = -U^2\begin{pmatrix} i_1, i_2, \dots, i_{n+1} \\ 1, \dots, n+1 \end{pmatrix}$$

for $|I| = n+1$.

Lemma 2.1. *Fix an admissible set I and consider the intersection subspace $V = \cap_{j\in I} H_j$ in $\mathbf{C}^{n+1}$.*

(i) *In case where $|I| \leq n$, $A(I) = 0$ if and only if V has contact with Q at one point.*

(ii) *In case where $|I| = n+1$, $A(I) = 0$ if and only if V has a common point with Q.*

(iii) *In case where $|I| = n+2$, $A(I) = 0$ if and only if V is not empty.*

Assume that

$(\mathcal{H}1)$: $A(0, I) < 0$ for an arbitrary admissible set I such that $|I| \leq n+1$, i.e., the homogeneous parts of $|I|$ linear functions f_j $(j \in I)$ are linearly independent.

Assume further that

$(\mathcal{H}2)$: $A(I) \neq 0$ for an arbitrary admissible I $(2 \leq |I| \leq n+2)$.

Then for any $I \subset \{1, 2, \ldots, m\}$ with $|I| = n+2$ the $(n+2) \times (n+2)$ symmetric submatrix $(a_{i,j})_{i,j\in I}$ has the signature of $n+1$ $(+)$sign and one $(-)$sign so that $A(I) < 0$ for $|I| = n+2$. This is equivalent to say that for any sequence of increasing admissible sets of indices

$$I_1 \subset I_2 \subset \cdots \subset I_{n+1} \subset I_{n+2}$$

such that $|I_r| = r$, the signs of $A(I_r)A(I_{r+1})$ $(1 \leq r \leq n+1)$ are positive except for one. Hence $A(I) < 0$ implies $A(J) < 0$ if $I \subset J$, $|J| \leq n+2$ (see [8]). In particular the following two cases are interesting:

$(\mathcal{H}2\text{a})$: $A(I) > 0$ for all admissible I with $2 \leq |I| \leq n+1$ and $A(I) < 0$ for all admissible I with $|I| = n+2$.

$(\mathcal{H}2\text{b})$: $A(I) < 0$ for all admissible I with $2 \leq |I| \leq n+2$.

Let τ be the $(n+1)$-form $d\xi_1 \wedge \cdots \wedge d\xi_{n+1.}$ on $\mathbf{C}^{n+1}$. We denote the n-form $-\tau_Q$ on $\mathbf{C}^{n+1}$ such that its restriction to Q is the standard volume form on $\Re Q$:

$$-\tau_Q = \sum_{\nu=1}^{n+1} (-1)^{\nu-1} \xi_\nu d\xi_1 \wedge \cdots \wedge d\xi_{\nu-1} \wedge d\xi_{\nu+1} \wedge \cdots \wedge d\xi_{n+1}$$

such that $df_0 \wedge \tau_Q \equiv \tau \mod (f_0)$. We consider the multiplicative function on X

$$\Phi_0(\xi) = \prod_{j=1}^{m} f_j(\xi)^{\lambda_j}$$

where we assume that every $\lambda_j \in \mathbf{R}$ is positive and generic. We denote by $H^r(X - Y, \nabla_0)$ and $H^r(Y, \nabla_0)$ the r-dimensional twisted rational de Rham cohomologies on $X - Y$ and Y associated with the covariant differentiation ∇_0 respectively:

$$\nabla_0(\psi) = d\psi + d\log\Phi_0 \wedge \psi.$$

These cohomologies are defined in a standard way by using differential forms ψ on $X - Y$ or Y with values in the $\mathbf{C}[\xi_1, \ldots, \xi_{n+1}]$-module

$$R = \sum_{\nu_1 \geq 0, \ldots, \nu_m \geq 0} \mathbf{C}[\xi_1, \ldots, \xi_{n+1}] \prod_{k=1}^{m} f_k(\xi)^{-\nu_k}$$

$\mathcal{L}_0$ be the local systems on $X - Y$ and Y defined by $\Phi_0(\xi)$ respectively, and $\hat{\mathcal{L}}_0$ be their duals defined by $\Phi_0(\xi)^{-1}$. Then the $(n+1)$- and n-dimensional homologies $H_{n+1}(X - Y, \hat{\mathcal{L}}_0)$ and $H_n(Y, \hat{\mathcal{L}}_0)$ represented by twisted cycles are dual to the twisted rational de Rham cohomologies $H^{n+1}(X - Y, \nabla_0)$ and $H^n(Y, \nabla_0)$ through the pairs of integrals respectively

(2.1)
$$H^{n+1}(X - Y, \nabla_0) \times H_{n+1}(X - Y, \hat{\mathcal{L}}_0) \ni (\varphi, \mathfrak{c}) \longrightarrow \langle \varphi, \mathfrak{c} \rangle = \int_{\mathfrak{c}} \Phi_0 \varphi \tau,$$

(2.2)
$$H^n(Y, \nabla_0) \times H_n(Y, \hat{\mathcal{L}}_0) \ni (\varphi, \mathfrak{c}) \longrightarrow \langle \varphi, \mathfrak{c} \rangle = \int_{\mathfrak{c}} \Phi_0 \varphi \tau_Q.$$

The following two Propositions have been proved in [2] and [3] (see Proposition 3.2_p, 3.3_p and Lemma 4.2 in [2, I], and also [5]).

Proposition 2.2. *Under the conditions* $(\mathcal{H}1), (\mathcal{H}2)$ *we have the isomorphism*

$$H^n(Y, \nabla_0) \cong \mathbf{C}^{\kappa_n}$$

where $\kappa_n = \sum_{\nu=0}^{n} \binom{m}{\nu} + \binom{m-1}{n}$. $H^n(Y, \nabla_0)$ *has a basis represented by the differential* n*-forms*

$$\varphi_Q(I)\tau_Q = \frac{\tau_Q}{f_{i_1} \cdots f_{i_p}}$$

where I *moves over the admissible sets* I *of indices such that* $0 \leq |I| \leq n+1$. *We denote* $\varphi_Q(\emptyset) = 1$ *for* $|I| = 0$. *There exist the fundamental relations among them of the following type. For an arbitrary admissible* *set of indices J with $|J| = n+2$ there exists the identity:*

$$\frac{1}{2}\sum_{\mu\neq\nu}(-1)^{\mu+\nu}\varphi_Q(\partial_\mu\partial_\nu J)\frac{A(0,\partial_\mu\partial_\nu J)}{A\begin{pmatrix}0,\partial_\mu J\\0,\partial_\nu J\end{pmatrix}} + \sum_{\mu=1}^{n+2}(-1)^{\mu-1}\varphi_Q(\partial_\mu J)\frac{A(\partial_\mu J)}{A\begin{pmatrix}0,\partial_\mu J\\J\end{pmatrix}} = 0 \tag{2.3}$$

where $\partial_\mu J$ denotes the subset of J deleted by the μth index j_μ. Further for $|I| = n+2$ a partial fraction gives

$$U\begin{pmatrix}I\\0,1,\dots,n+1\end{pmatrix}\varphi_Q(I) = \sum_{\mu=1}^{n+2}(-1)^{\mu-1}U\begin{pmatrix}\partial_\mu I\\1,\dots,n+1\end{pmatrix}\varphi_Q(\partial_\mu I). \tag{2.4}$$

We denote by $\mathcal{B}$ a linear space spanned by the representatives $\varphi_Q(I)$, $0 \le |I| \le n+1$.

Proposition 2.3. *Under the condition $(\mathcal{H}1), (\mathcal{H}2a)$ $H_n(Y, \hat{\mathcal{L}}_0)$ has a basis represented by the closures of all the connected components of $\Re Y = \Re Q \cap Y$. Their number is equal to κ_n. In other words, $H_n(Y, \hat{\mathcal{L}}_0)$ is spanned by only real twisted cycles defined by connected components of $\Re Y$.*

For example we have $\kappa_1 = 2m$, $\kappa_2 = m^2 - m + 2$, $\kappa_3 = \frac{1}{3}m^3 - m^2 + \frac{8}{3}m$.

Remark. *The number κ_n is also equal to the number of non-compact connected components of $\mathbf{R}^{n+1} - N(\mathcal{A})$.*

§3. Twisted imaginary cycles

We may assume without losing generality

$$u_{j,0} \le 0 \quad \text{for all } j, 1 \le j \le m. \tag{3.1}$$

Define the set

$$S_{j,+} := \{\xi \in \Re Q; f_j(\xi) > 0\}$$

as the inside of the real hypersphere $S_j = \Re Q \cap H_j$. We denote by $\boldsymbol{\nu}_j$ the unit normal of $\Re H_j$:

$$\boldsymbol{\nu}_j = \frac{(u_{j,1}, u_{j,2}, \dots, u_{j,n+1})}{\sqrt{\sum_{\nu=1}^{n+1} u_{j,\nu}^2}}. \tag{3.2}$$

Remark that $\boldsymbol{\nu}_j \in S_{j,+}$.

First notice the following:

Lemma 3.1. *Suppose I is admissible. The real affine subspace $\bigcap_{j\in I} \Re H_j$ is disjoint with $\Re Q$ if and only if $A(I) < 0$.*

Proof. In fact the square of the distance between the subspace $\bigcap_{j\in I} \Re H_j$ and the origin is equal to $\{A(I) + A(0,I)\}/A(0,I)$. It is bigger than 1 if and only if $A(I) < 0$ because $A(0,I) < 0$. Q.E.D.

Corollary 3.2. *Suppose that $A(i,j) < 0$, i.e., $a_{i,j}^2 > 1$ for every pair $i,j \in \{1,2,\ldots,m\}, i \neq j$ then every S_j is disjoint with each other. In this case, $S_{i,+}, S_{j,+}$ are disjoint, or the one is included in the other, according as $a_{i,j} < -1$ or $a_{i,j} > 1$.*

Proposition 3.3. *Under the condition $(\mathcal{H}2)$ consider an admissible set I such that $2 \leq |I| \leq n+1$. Suppose further $A(I)$ is a positive number near 0 and that $A(J) < 0$ for any admissible $J \supset I$, $|J| > |I|$. Then the compact domain*

$$\mathfrak{l}(I) := \{\xi \in \Re Q; f_j(\xi) \geq 0 \quad (j \in I)\}$$

gives a twisted real cycle representing an element $H_n(Y, \hat{\mathcal{L}}_0)$. This cycle vanishes if it is deformed in an isotopic way by the matrix A as $A(I)$ tends to 0, any other $A(K)$ being never equal to 0.

Proof. Since $A(I) > 0$ and near 0, $\mathfrak{l}(I)$ is one of the compact components of $\Re Y$. This reduces to a point for $A(I) \to 0$ as is seen from Lemma 3.1. Q.E.D.

Definition 3.4. *The cycle $\mathfrak{l}(I)$ mentioned in* Proposition 3.3 *is called* **the twisted vanishing cycle (Lefschetz cycle) at the singularity** $A(I) = 0$.

Assume now the conditions $(\mathcal{H}2b)$ together with $(\mathcal{H}1)$. Then each $(n-1)$-dimensional hypersphere $\Re Q \cap H_j$ is disjoint with each other. This means that $\Re Y$ has only $m+1$ connected components which make only a part of the basis of $H_n(Y, \hat{\mathcal{L}}_0)$. We want to construct a basis of $H_n(Y, \hat{\mathcal{L}}_0)$ consisting of imaginary cycles in addition to real ones, using special cells $\Delta^*(I), \Delta^*_+(I;J)$ as follows.

Let I be an admissible set such that $|I| = p, 2 \leq p \leq n$. We have $A(I) < 0$. We may assume without losing generality $I = \{1,2,\ldots,p\}$. Choose an orthonormal basis $\{e_1, e_2, \ldots, e_{n+1}\}$ of $\mathbf{R}^{n+1}$ such that $\boldsymbol{\nu}_1 = e_1$ and the j-dimensional subspace $\langle \boldsymbol{\nu}_1, \ldots, \boldsymbol{\nu}_j \rangle$ is spanned by $e_1, \ldots, e_j$

$(1 \leq j \leq p)$ where the coefficient of $\boldsymbol{\nu}_j$ with respect to e_j is positive. Hence $f_1, \ldots, f_p$ can be represented as

$$(3.3)\quad f_j(\xi) = f_j(\xi_1, \ldots, \xi_j) = u_{j,0} + \sum_{\nu=1}^{j} u_{j,\nu}\xi_\nu,\ (u_{j,j} > 0) \quad (1 \leq j \leq p).$$

We say that the basis $\{e_1, e_2, \ldots, e_{n+1}\}$ is attached to I.

We denote by $\Delta^*(I)$ the n-dimensional cell consisting of all points

$$\xi = \sum_{k=1}^{p} \xi_k e_k + \sqrt{-1} \sum_{k=p+1}^{n+1} \xi_k^* e_k \in \{\mathbf{R}^p \times (\sqrt{-1}\mathbf{R})^{n+1-p}\} \cap Q$$

which is a piece of an ultra hyperboloid

$$(3.4)\qquad \xi_1^2 + \cdots + \xi_p^2 - {\xi_{p+1}^*}^2 - \cdots - {\xi_{n+1}^*}^2 = 1,$$

$$(3.5)\qquad f_j(\xi_1, \ldots, \xi_j) \leq 0 \quad (1 \leq j \leq p).$$

Denote by $\Delta_k^*(\partial_k I)$ the $(n-1)$-dimensional face $\Delta^*(\partial_k I) \cap \Delta^*(I)$ (To define $\Delta_k^*(\partial_k I)$ one should take another orthonormal basis attached to $\partial_k I$). In particular one sees $\Delta_p^*(\partial_p I) = \Delta^*(I) \cap \{\xi_p = 0\}$.

Furthermore consider a pair of admissible sets of indices I, J ($|I| = p, |J| = q$) such that $I \supset J$ and $2 \leq q \leq p-1$. For simplicity assume that $I = \{1, 2, \ldots, p\}$ and $J = \{1, 2, \ldots, q\}$. There exist a basis $\{e_1, \ldots, e_q, e_{q+1}, \ldots, e_p, e_{p+1}, \ldots, e_{n+1}\}$ of $\mathbf{R}^{n+1}$ attached to the pair (I, J) such that $||e_k|| = 1$ $(1 \leq k \leq n+1)$ ($||\ ||$ denotes the length), which satisfies the following properties $\mathcal{P}$:

(i) $\{e_1, \ldots, e_q\}$ is an orthonormal system such that the subspace $\langle e_1, \ldots, e_j \rangle$ spanned by $e_1, \ldots, e_j$ coincides with the subspace $\langle \boldsymbol{\nu}_1, \ldots, \boldsymbol{\nu}_j \rangle$ $(1 \leq j \leq q)$.

(ii) The systems $\{e_1, \ldots, e_q\}$ and $\{e_{q+1}, \ldots, e_p\}$ are orthogonal to each other.

(iii) $\boldsymbol{\nu}_j$ lies in the subspace spanned by $e_1, \ldots, e_q, e_j$ $(q+1 \leq j \leq p)$.

(iv) $\{e_{p+1}, \ldots, e_{n+1}\}$ is an orthonormal system such that it is orthogonal to the system $\{e_1, \ldots, e_p\}$.

We may assume that each coefficient of $\boldsymbol{\nu}_j$ with respect to e_j is positive $(1 \leq j \leq p)$.

Then in terms of the coordinates $\xi = \sum_{k=1}^{n+1} \xi_k e_k$ the fundamental hypersphere Q can be represented

$$(3.6)\qquad \sum_{k=1}^{n+1} \xi_k^2 + 2 \sum_{q+1 \leq j < k \leq p} \gamma_{j,k}\xi_j\xi_k = 1$$

where $\gamma_{j,k}$ denotes the inner product (e_j, e_k). Remark that the geodesic $(p-1)$-simplex $\langle \boldsymbol{\nu}_1, \ldots, \boldsymbol{\nu}_p \rangle$ with vertices $\boldsymbol{\nu}_1, \ldots \boldsymbol{\nu}_p$ is contained in the real oriented cone

$$C(I;J) : \{\xi_{q+1} \geq 0, \ldots, \xi_p \geq 0\} \subset \mathbf{R}^{n+1}$$

and the hyperplane $\xi_k = 0$ $(p \geq k \geq q+1)$ contains the totally geodesic face $\langle \boldsymbol{\nu}_1, \ldots, \boldsymbol{\nu}_{k-1}, \boldsymbol{\nu}_{k+1}, \ldots, \boldsymbol{\nu}_p \rangle$. Denote by $\Delta_+^*(I;J)$ the cell consisting of the points $\xi = \sum_{k=1}^q \xi_k e_k + \sqrt{-1} \sum_{k=q+1}^{n+1} \xi_k^* e_k \in \{\mathbf{R}^q \times (\sqrt{-1}\mathbf{R})^{n+1-q}\} \cap Q$ satisfying

$$\sum_{k=1}^{q} \xi_k^2 - \sum_{k=q+1}^{n+1} \xi_k^{*2} - 2 \sum_{q+1 \leq j < k \leq p} \gamma_{j,k} \xi_j^* \xi_k^* = 1 \tag{3.7}$$

$$f_1(\xi_1) \leq 0, \ldots, f_q(\xi_1, \ldots, \xi_q) \leq 0, \xi_{q+1}^* \geq 0, \ldots, \xi_p^* \geq 0. \tag{3.8}$$

In particular we put $\Delta_+^*(I;I) = \Delta^*(I)$ and $\Delta_+^*(I;J)$ for $q = 1$ to be an empty set. The above construction is also possible for the pairs $(\partial_k I; \partial_k J)$ or $(\partial_k I; J)$: $\Delta_+^*(\partial_k I; \partial_k J) \cap \Delta_+^*(I;J)$ and $\Delta_+^*(I;J) \cap \{\xi_k^* = 0\} = \Delta_+^*(\partial_k I; J) \cap \{\xi_k^* = 0\}$, being the $(n-1)$-dimensional k-th faces of $\Delta_+^*(I;J)$, are denoted by $\Delta_{k,+}^*(\partial_k I; \partial_k J)$ $(q \geq k \geq 1, q \geq 3)$ and by $\Delta_{k,+}^*(\partial_k I; J)$ $(p \geq k \geq q+1, q \geq 2)$ respectively. This is compatible because the coordinates $(\xi_1, \ldots, \xi_q, \xi_{q+1}^*, \ldots, \xi_{n+1}^*)$ attached to $(I;J)$ can be identified with the one attached to $(\partial_k I; J)$ or $(\partial_k I; \partial_k J)$ provided $\xi_k^* = 0$.

Remark 3.5. *$\Delta_+^*(I;J)$ is included in the oriented cone transformed from $C(I;J)$ by the rotations $(\xi_{q+1}, \cdots, \xi_{n+1}) \to e^{\sqrt{-1}\theta}(\xi_{q+1}, \ldots, \xi_{n+1})$ $(0 \leq \theta \leq \frac{\pi}{2})$ and corresponds one-to-one to the real cell $\Delta_+(I;J) = C(I;J) \cap \Delta(I)$.*

Then

Lemma 3.6. *For $2 \leq p \leq n+1$ we have the boundary formulae modulo $\cup_{k \in I} H_k \cap Q$:*

$$\partial \Delta^*(I) \equiv 0 \quad (p = 2),$$

$$\partial \Delta^*(I) \equiv \sum_{k=1}^{p} (-1)^{k-1} \Delta_k^*(\partial_k I) \quad (p \geq 3),$$

$$\begin{aligned}\partial\Delta_+^*(I;\partial_K I) &\equiv \sum_{m\in I-K}(-1)^{m-1}\Delta_{m,+}^*(\partial_m I;\partial_m\partial_K I)\\ &\quad+\sum_{m\in K}(-1)^{m-1}\Delta_{m,+}^*(\partial_m I;\partial_K I)\quad (l\geq 2,\ p-l\geq 3),\\ &\equiv \sum_{m\in K}(-1)^{m-1}\Delta_{m,+}^*(\partial_m I;\partial_K I)\quad (l\geq 2, p-l=2),\\ &\equiv \sum_{m\in I-K}(-1)^{m-1}\Delta_{m,+}^*(\partial_m I;\partial_m\partial_K I)+(-1)^{k-1}\Delta_k^*(\partial_k I)\\ &\quad (l=1, K=\{k\}, p\geq 3)\end{aligned}$$

where K, $|K|=l$, denotes a subset of ordered indices $K=\{k_1,\dots,k_l\}\subset I$ and ∂_K denotes $\partial_{k_1}\dots\partial_{k_l}$.

The following Lemma follows immediately from the preceding Lemma:

Lemma 3.7. *Suppose $2\leq p\leq n+1$. Then the n-chain*

$$\mathfrak{l}^*(I)=\Delta^*(I)+\sum_{l=1}^{p-2}\sum_{K,|K|=l}\Delta_+^*(I;\partial_K I)(-1)^l \tag{3.9}$$

defines an n-cycle in $H_n(Y,\hat{\mathcal{L}}_0)$. Here K moves over the family of subsets of ordered indices $K=\{k_1,\dots,k_l\}\subset I$.

Definition 3.8. $\mathfrak{l}^*(I)$ *is called* **twisted Lefschetz cycle** *associated with I. If $A(I)$ is near 0, this is a deformation of $\mathfrak{l}(I)$ as the matrix A moves from the part $A(I)>0$ to the one $A(I)<0$ being detoured from the singularity $A(I)=0$.*

Note that $\mathfrak{l}^*(I)$ coincides with $\Delta^*(I)$ if $|I|=2$ and with $\Delta^*(I)-\sum_{k=1}^3\Delta_+^*(I;\partial_k I)$ if $|I|=3$.

Assume in particular that $p=n+1$ in case $m\geq n+1$. First consider the case where $n=1$, $m\geq 2$. By hypothesis every $a_{i,j}^2>1$. The $2m$ points $\cup_{j=1}^m\Re Q\cap H_j$ are different from each other. Therefore $\Re Y$ consists of $2m$ connected components which make a basis of $H_1(Y,\hat{\mathcal{L}}_0)$. None of imaginary cycles occur as stated in the Introduction. Next consider the case where $n=2$, $m\geq 3$. Suppose $I=\{1,2,3\}$ is admissible with $|I|=3$. Let $\Delta(I)$ be the geodesic triangle in $\Re Q$ with vertices $\boldsymbol{\nu}_1,\boldsymbol{\nu}_2,\boldsymbol{\nu}_3$. Then

$$\mathfrak{l}^*(I)=\Delta^*(I)-\sum_{k=1}^3\Delta_+^*(I;\partial_k I)$$

where $\Delta^*(I)=\overline{\Delta(I)-S_{1,+}-S_{2,+}-S_{3,+}}$ (the overline $\overline{\Delta-\cdots}$ denotes the closure).

In general there exist $\binom{m}{n+1}$ cycles $\mathfrak{l}^*(I)$ with $|I| = n+1$. All the cycles thus constructed are not necessarily linearly independent in $H_n(Y, \hat{\mathcal{L}}_0)$. In fact we have

Lemma 3.9. *For an admissible I with $|I| = n+2$, the following identity holds modulo $\cup_{k\in I} H_k \cap Q$:*

$$\sum_{j=1}^{n+2} (-1)^{j-1} \mathfrak{l}^*(\partial_j I) \equiv 0. \tag{3.10}$$

We may choose as linearly independent ones the cycles $\mathfrak{l}^*(I), |I| = n+1$ such that $1 \in I$ so that their number is equal to $\binom{m-1}{n}$.

Proof. We may assume that $I = \{1, 2, \ldots, n+2\}$. Then

$$\begin{aligned} \sum_{j=1}^{n+2} (-1)^{j-1} \mathfrak{l}^*(\partial_j I) &= \sum_{j=1}^{n+2} (-1)^{j-1} \Delta^*(\partial_j I) \\ &+ \sum_{j=1}^{n+2} (-1)^{j-1} \sum_{l=1}^{n-1} \sum_{K \subset \partial_j I, |K|=l} \Delta_+^*(\partial_j I; \partial_j \partial_K I). \end{aligned} \tag{3.11}$$

On the other hand we see

$$\sum_{j=1}^{n+2} (-1)^{j-1} \Delta^*(\partial_j I) \equiv 0, \ \sum_{j\in L} (-1)^{j-1} \Delta_+^*(\partial_j I; \partial_L I) \equiv 0$$

for $L \subset I$, $|L| = l+1 \geq 2$ being fixed. In view of Remark 3.5 the last identity is derived from the one

$$\sum_{j\in L} (-1)^{j-1} C(\partial_j I; \partial_L I) = 0.$$

Hence the RHS of (3.11) vanishes modulo $\cup_{k\in I} H_k \cap Q$. Q.E.D.

As an immediate consequence we have

Proposition 3.10. *The number of Lefschetz cycles is equal to κ_n.*

Summing up the above we have proved the following:

Theorem 3.11. *Under the conditions ($\mathcal{H}1$) and ($\mathcal{H}2b$), as a basis of $H_n(Y, \hat{\mathcal{L}}_0)$, one can choose the representatives of twisted cycles of the following kinds:*

(i) *Real cycles. This can be realized by the real chambers which are the closures of the connected components of $\Re Y$. Their number is equal to $1 + m$.*

(ii) *Imaginary Lefschetz cycles* $\mathfrak{l}^*(I)$ *such that* $2 \leq |I| \leq n$. *Their number is equal to* $\sum_{\nu=2}^{n}\binom{m}{\nu}$.

(iii) *Imaginary Lefschetz cycles* $\mathfrak{l}^*(I), |I| = n+1$ *such that* $1 \in I$. *Their number is equal to* $\binom{m-1}{n}$.

§4. Stereographic projection

The cycles defined in the previous section can also be described in the n-dimensional Euclidean space as below. The complement of the south pole, $\Re Q - \{(-1, 0, \ldots, 0)\}$, is isomorphic to $\mathbf{R}^n$ through the stereopgraphic projection

$$\eta_1 = \frac{\xi_2}{1+\xi_1}, \ldots, \eta_n = \frac{\xi_{n+1}}{1+\xi_1} \tag{4.1}$$

which is a conformal transformation. Then a hypersphere S in $\Re Q$ corresponds to a hypersphere or a hyperplane $\tilde{S}$ in $\mathbf{R}^n$:

$$\sum_{\nu=1}^{n}(\eta_\nu - v_\nu)^2 = r^2 \quad (r > 0)$$

where a hyperplane can be regarded as a limiting case for $r = \infty$. Denote the center of $\tilde{S}$ by $\mathbf{v} = (v_1, \ldots, v_n)$ and its length by $||\mathbf{v}|| = \sqrt{\sum_{\nu=1}^{n} v_\nu^2}$. Then we have

$$S : u_0 + \sum_{\nu=0}^{n+1} u_\nu \xi_\nu = 0, \ (u_0 \leq 0)$$

$$u_0 = \frac{r^2 - 1 - ||\mathbf{v}||^2}{2r}, u_1 = \frac{r^2 + 1 - ||\mathbf{v}||^2}{2r}, u_{\nu+1} = \frac{v_\nu}{r} \ (1 \leq \nu \leq n)$$

or

$$-u_0 = \frac{r^2 - 1 - ||\mathbf{v}||^2}{2r}, -u_1 = \frac{r^2 + 1 - ||\mathbf{v}||^2}{2r}, -u_{\nu+1} = \frac{v_\nu}{r} \ (1 \leq \nu \leq n)$$

according as $r^2 - 1 - ||\mathbf{v}||^2 \leq 0$ or > 0, namely

Lemma 4.1. S_+ *corresponds to the inside or the outside of* $\tilde{S}$ *according as* $r^2 - 1 - ||\mathbf{v}||^2 < 0$ *or* > 0.

As for $a_{i,j}$

$$a_{i,j} = \frac{r_i^2 + r_j^2 - ||\mathbf{v}^{(i)} - \mathbf{v}^{(j)}||^2}{2r_i r_j}$$

where $r_i, r_j, \mathbf{v}^{(i)}, \mathbf{v}^{(j)}$ denote the radii and the centers of $\tilde{S}_i, \tilde{S}_j$ respectively. Hence

Lemma 4.2. *We have*

$$A(i,j) = \frac{(r_i - r_j + a)(-r_i + r_j + a)(r_i + r_j + a)(r_i + r_j - a)}{4r_i^2 r_j^2}$$

where we put $a = ||\mathbf{v}^{(i)} - \mathbf{v}^{(j)}||$. *This implies* $a_{i,j} > 1$ *if and only if* $|r_i - r_j| > a$. $a_{i,j} < -1$ *if and only if* $r_i + r_j < a$.

In the same way

Lemma 4.3. *Suppose that* $|a_{i,j}| < 1$ *for an admissible* $I = \{i, j, k\}$ *and put* $-\cos \alpha_{i,j} = a_{i,j}$ *such that* $0 < \alpha_{i,j} < \pi$. *Then*

$$A(i,j,k) = -4 \cos \frac{\alpha_{i,j} + \alpha_{j,k} + \alpha_{i,k}}{2} \cdot \cos \frac{-\alpha_{i,j} + \alpha_{j,k} + \alpha_{i,k}}{2}$$
$$\cdot \cos \frac{\alpha_{i,j} - \alpha_{j,k} + \alpha_{i,k}}{2} \cdot \cos \frac{\alpha_{i,j} + \alpha_{j,k} - \alpha_{i,k}}{2}.$$

The three hyperspheres $\tilde{S}_i, \tilde{S}_j, \tilde{S}_k$ *intersect each other and* $\pi - \alpha_{i,j}$ *is equal to the angle subtended by the tangents of* $\tilde{S}_i, \tilde{S}_j$ *at an intersection point of* $\tilde{S}_i \cap \tilde{S}_j$. $A(1,2,3) = 0$ *if and only if* $\alpha_{i,j} + \alpha_{j,k} + \alpha_{i,k} = \pi$, *or* $-\alpha_{i,j} + \alpha_{j,k} + \alpha_{i,k} = \pi$, *or* $\alpha_{i,j} - \alpha_{j,k} + \alpha_{i,k} = \pi$, *or* $\alpha_{i,j} + \alpha_{j,k} - \alpha_{i,k} = \pi$.

Lemma 4.4. *For an arbitrary admissible* $I, |I| \leq n + 1$ *there exist the new coordinates* $\eta = (\eta_1, \ldots, \eta_n)$ *of* $\mathbf{R}^n$ *such that* $\mathbf{v}^{(i_1)} = 0$ *and* $\mathbf{v}^{(i_{l+1})}$ *lies in the* $\eta_1, \ldots, \eta_l$*-subspace* $2 \leq l \leq |I| - 1$.

Proof. We may assume that $I = \{1, 2, \ldots, p\}$. By the change of coordinates (4.1), there exist the coordinates $\xi_1, \ldots, \xi_{n+1}$ such that $\boldsymbol{\nu}_{l+1}$ lies in the $\xi_1, \ldots, \xi_{l+1}$-subspace $(1 \leq l \leq p)$. Since $u_{1,\nu} = 0$ for $\nu \geq 2$, $\mathbf{v}^{(1)} = 0$. And $u_{l,\nu} = 0$ for $\nu \geq l+1$, $\mathbf{v}^{(l+1)}$ lies in the $\eta_1, \ldots, \eta_l$-subspace.

The cycles equivalent to the one constructed in Section 3 are described as follows:

Consider the case where $m = 2$. Let $\mathbf{v}^{(1)}, \mathbf{v}^{(2)}$ be the centers of $\tilde{S}_1, \tilde{S}_2$ and the insides of $\tilde{S}_1, \tilde{S}_2$ be denoted by $\tilde{S}_{1,+}, \tilde{S}_{2,+}$ respectively. Suppose first that $|a_{1,2}| < 1$. Then $\tilde{S}_{1,+} \cap \tilde{S}_{2,+}$ is a non-empty domain so that $\mathbf{R}^n - \tilde{S}_1 \cup \tilde{S}_2$ consists of 4 connected components:

$$\mathbf{R}^n - \tilde{S}_{1+} \cup \tilde{S}_{2,+}, \tilde{S}_{1,+} - \tilde{S}_{2,+}, \tilde{S}_{2,+} - \tilde{S}_{1,+}, \tilde{S}_{1,+} \cap \tilde{S}_{2,+}.$$

Their closures make the representatives of a basis of $H_n(Y, \hat{\mathcal{L}}_0)$.

Suppose that $a_{1,2} < -1$. Then $\tilde{S}_{1,+}$ is disjoint with $\tilde{S}_{2,+}$. We have three real domains $\tilde{S}_{1,+}, \tilde{S}_{2,+}, \mathbf{R}^n - \tilde{S}_{1+} \cup \tilde{S}_{2,+}$. On the other hand suppose that $a_{1,2} > 1$. Then $\tilde{S}_{1,+}$ includes or is included in $\tilde{S}_{2,+}$. Assume for example that $\tilde{S}_{1,+} \supset \tilde{S}_{2,+}$. Then there are three real domains $\mathbf{R}^n - \tilde{S}_{1,+}, \tilde{S}_{1,+} - \tilde{S}_{2+}, \tilde{S}_{2,+}$.

There is the Lefschetz cycle enclosed by two pieces of hyperboloids

$$\tilde{\mathfrak{l}}(\{1,2\}) : \{\eta = (\eta_1, \sqrt{-1}\eta_2^*, \ldots, \sqrt{-1}\eta_n^*) \in \mathbf{R} \times (\sqrt{-1}\mathbf{R})^{n-1};$$
$$\eta_1^2 - \sum_{\nu=2}^{n} \eta_\nu^{*2} \geq r_1^2, (\eta_1 - v_1^{(2)})^2 - \sum_{\nu=2}^{n} \eta_\nu^{*2} \geq r_2^2\}.$$

More generally suppose that $A(I) < 0$ for $|I| = p\,(2 \leq p \leq n)$. We may assume that $I = \{1, 2, \ldots, p\}$.

There exists an orthonormal basis $(\tilde{e}_k)_k$ of $\mathbf{R}^n$ such that the new coordinates $\eta = \sum_{k=1}^{n} \eta_k \tilde{e}_k \in \mathbf{R}^n$ define $\tilde{S}_j\,(1 \leq j \leq p)$ by

$$\sum_{\nu=1}^{j-1} (\eta_\nu - v_\nu^{(j)})^2 + \sum_{\nu=j}^{n} \eta_\nu^2 = r_j^2.$$

Corresponding to the $(p-1)$-simplex $\tilde{\Delta}(I) = \langle \mathbf{v}^{(1)}, \ldots, \mathbf{v}^{(p)} \rangle$ with vertices $\mathbf{v}^{(1)}, \ldots, \mathbf{v}^{(p)}$, define the cell $\tilde{\Delta}^*(I)$ enclosed by p pieces of ultra hyperboloids

$$\tilde{\Delta}^*(I) = \{\eta = (\eta_1, \ldots, \eta_{p-1}, \sqrt{-1}\eta_p^*, \ldots, \sqrt{-1}\eta_n^*) \in \mathbf{R}^{p-1}$$
$$\times (\sqrt{-1}\mathbf{R})^{n-p+1}; \sum_{\nu=1}^{j-1} (\eta_\nu - v_\nu^{(j)})^2 + \sum_{\nu=j}^{p-1} \eta_\nu^2 - \sum_{\nu=p}^{n} \eta_\nu^{*2} \geq r_j^2$$
$$(1 \leq j \leq p)\}.$$

Further we put

$$\tilde{\Delta}_+^*(I; \partial_p I) = \{\eta = (\eta_1, \ldots, \eta_{p-2}, \sqrt{-1}\eta_{p-1}^*, \ldots, \sqrt{-1}\eta_n^*) \in \mathbf{R}^{p-2}$$
$$\times (\sqrt{-1}\mathbf{R})^{n-p+2}; \eta_{p-1}^* \geq 0, \sum_{\nu=1}^{j-1} (\eta_\nu - v_\nu^{(j)})^2 + \sum_{\nu=j}^{p-2} \eta_\nu^2 - \sum_{\nu=p-1}^{n} \eta_\nu^{*2}$$
$$\geq r_j^2, (1 \leq j \leq p-1)\}.$$

which is the n-cell enclosed by $p-1$ pieces of ultra hyperboloids and the hyperplane $\eta_{p-1}^* = 0$. By exchange of coordinates one can similarly define the cells $\tilde{\Delta}_+^*(I; \partial_k I)\,(1 \leq k \leq p-1)$. More generally for a pair $I, J\,(I \supset J,\ |I| = p, |J| = q)$ such that $2 \leq q \leq p-1$ there exists a basis $\tilde{e}_k\,(1 \leq k \leq n)$ of $\mathbf{R}^n$ attached to the pair satisfying the properties $\mathcal{P}$

in §3, q, p being replaced by $q-1, p-1$ respectively, such that one can define the n-cell

$$\tilde{\Delta}_+^*(I;J) := \{\eta = \sum_{k=1}^{q-1} \eta_k e_k + \sqrt{-1}\sum_{k=q}^{n} \eta_k^* e_k \in \mathbf{R}^{q-1} \times (\sqrt{-1}\mathbf{R})^{n-q+1};$$

$$\sum_{k=1}^{j-1}(\eta_k - v_k^{(j)})^2 + \sum_{k=j}^{q-1}\eta_k^2 - \sum_{k=q}^{n}\eta_k^{*2} - 2\sum_{q\le j<k\le p-1}\tilde{\gamma}_{j,k}\eta_j^*\eta_k^* \ge r_j^2,$$

$$(1 \le j \le q), \eta_q^* \ge 0, \ldots, \eta_{p-1}^* \ge 0\}$$

where $\tilde{\gamma}_{j,k}$ denotes the inner product $(\tilde{e}_j, \tilde{e}_k)$. Then according to (3.9) the Lefschetz cycle $\tilde{\mathfrak{l}}^*(I)$ $(2 \le p \le n+1)$ is defined to be

$$\tilde{\mathfrak{l}}^*(I) = \tilde{\Lambda}^*(I) + \sum_{j=1}^{p-2}(-1)^j \sum_{K\subset I, |K|=j} \tilde{\Delta}_+^*(I;\partial_K I).$$

In the same way as Lemma 3.9 we have

Lemma 4.5. *For any admissible I such that $|I| = n+2$, the identity holds*

$$\sum_{j=1}^{n+2}(-1)^{j-1}\tilde{\mathfrak{l}}^*(\partial_j I) \equiv 0$$

modulo the complexification of $\cup_{j\in I}\tilde{S}_j$ in $\mathbf{C}^n$.

In conclusion we may choose admissible I with $|I| = n+1$, such that $1 \in I$ so that any other can be a linear combination of them. We have the same conclusion as Theorem 3.11.

§5. Degenerate cases

In Section 3, and 4 we have assumed $(\mathcal{H}1)$ and $(\mathcal{H}2)$. In this section we discuss the cases where these conditions are not necessarily satisfied.

First note the following (for example, see [2, I], Lemma 4.2):

Lemma 5.1. *We have the commutative diagram:*

$$\begin{array}{ccc} H^{n+1}(X-Y;\nabla_0) & \xrightarrow{\text{Res}} & H^n(Y;\nabla_0) \\ \updownarrow & & \updownarrow \\ H_{n+1}(X-Y;\hat{\mathcal{L}}_0) & \xleftarrow{\delta} & H_n(Y;\hat{\mathcal{L}}_0) \end{array}$$

where Res *denotes the Residue along Y, and δ means the boundary operation (Leray map) into a tubular neighborhood of Y in $X-Y$.*

For an arbitrary $\varphi(\xi)\tau \in R\tau$ such that its representative $\in H^{n+1}(X-Y, \nabla_0)$, denote

$$\varphi^{(1)}\tau_Q = \mathrm{Res}\left(\frac{\varphi}{f_0}\tau\right) = \left[\frac{\varphi\tau}{df_0}\right]_Y, \quad \varphi^{(2)}\tau_Q = \mathrm{Res}\left(\frac{\varphi}{f_0^2}\tau\right).$$

Then $\varphi^{(1)}(I)$ is equal to the restriction of

$$\varphi(I) = \frac{1}{f_{i_1}\cdots f_{i_p}}$$

to Q. As for $\varphi^{(2)}(I)$ the following two recurrence relations play an important role in the sequel:

Lemma 5.2. *For an admissible I with $|I| = p\,(0 \leq p \leq n+1)$*

$$(5.1)\quad A(I)\varphi^{(2)}(I) \sim \sum_{k\notin I} \lambda_k A\begin{pmatrix} 0, & I \\ k, & I \end{pmatrix} \varphi_Q(k, I)$$
$$+(\lambda_\infty + n - p - 1)A(0, I)\varphi_Q(I) - \sum_{\nu=1}^{p}(-1)^{\nu-1} A\begin{pmatrix} & I \\ 0, & \partial_\nu I \end{pmatrix} \varphi^{(2)}(\partial_\nu I).$$

In particular

$$\varphi^{(2)}(\emptyset) \sim \sum_{k=1}^{m} \lambda_k a_{k,0}\varphi_Q(k) - (\lambda_\infty + n - 1)\varphi(\emptyset),$$
$$\varphi^{(2)}(j) \sim \sum_{k\neq j} \lambda_k A\begin{pmatrix} 0, & j \\ k, & j \end{pmatrix} \varphi_Q(k, j) - \sum_{k=1}^{m} \lambda_k a_{j,0}a_{k,0}\varphi_Q(k)$$
$$+(\lambda_\infty + n - 2)A(0, j)\varphi_Q(j) + (\lambda_\infty + n - 1)a_{j,0}\varphi_Q(\emptyset).$$

Therefore $\varphi^{(2)}(I)$ can be described as a linear combination of $\varphi_Q(J)$ such that $|J - J\cap I| \leq 1$ with the coefficients of rational functions of $a_{i,j}$, $a_{k,0}$ whose denominators are products of $A(K)$ for $K \subset I$.

For the proof see [2, I], Proposition 4.2.

Lemma 5.3. *Fix an admissible I with $p = |I| \leq n+1$. Then an arbitrary $\mu, 1 \leq \mu \leq p$*

$$(5.2)\quad (-1)^{\mu-1}(\lambda_{i_\mu} - 1)A(I)\frac{\varphi_Q(I)}{f_{i_\mu}} \sim -\sum_{k \notin I} \lambda_k A\begin{pmatrix} I \\ k, \partial_\mu I \end{pmatrix} \varphi_Q(k, I)$$
$$- (\lambda_\infty + n - p - 1)A\begin{pmatrix} I \\ 0, \partial_\mu I \end{pmatrix} \varphi_Q(I)$$
$$+ \sum_{\nu=1}^{p} (-1)^{\nu-1} A\begin{pmatrix} \partial_\mu I \\ \partial_\nu I \end{pmatrix} \varphi^{(2)}(\partial_\nu I).$$

For the proof see [2, I], Proposition 4.2.

Owing to Lemma 5.2 and 5.3 an arbitray form $\frac{\tau_Q}{\prod_{k=1}^m f_k^{\nu_k}}$ $(\nu_k \geq 0)$ can be described explicitly as a linear combination of the representatives of admissible forms $\varphi_Q(I)\tau_Q$.

Proposition 5.4. *In addition to $(\mathcal{H}1)$ suppose the following condition:*

$(\mathcal{H}IV(p))$ *For a fixed admissible I with $p = |I| \leq n$, $A(I) = 0$. But for any other admissible J such that $|J| \leq n+2$ $A(J) \neq 0$.*

Then $\mathfrak{l}^*(I)$ vanishes. The dimension of $H_n(Y, \hat{\mathcal{L}}_0)$ decreases by one and is equal to $\kappa_n - 1$. On the other hand the representatives $\varphi_Q(I)$ in Proposition 2.2 *does not make a basis of $H^n(Y, \nabla_0)$. We have a linear relation*

(5.3)

$$(\sum_{j \in I} \lambda_j + \lambda_\infty + n - p - 1)A(0, I)\varphi_Q(I) + \sum_{k \notin I} \lambda_k A\begin{pmatrix} 0, & I \\ k, & I \end{pmatrix} \varphi_Q(k, I)$$
$$- \sum_{k \notin I} \lambda_k \sum_{\nu=1}^{p} (-1)^{\nu-1} \frac{A\begin{pmatrix} I \\ 0, & \partial_\nu I \end{pmatrix} A\begin{pmatrix} k, & \partial_\nu I \\ 0, & \partial_\nu I \end{pmatrix}}{A(\partial_\nu I)} \varphi_Q(k, \partial_\nu I)$$
$$\equiv 0 \mod \mathcal{B}(I)$$

where $\mathcal{B}(I)$ denotes a linear space spanned by $\varphi_Q(J)$ such that $|J - J \cap I| \leq 1$ and $|J| < |I|$. $H^n(Y, \nabla_0)$ is of dimension $\kappa_n - 1$ and is spanned by $\varphi_Q(J)$ such that $J \neq I$ and $|J| \leq n+1$ with the fundamental relations (2.3), (2.4).

Proof. In fact since $A(I) = 0$, the LHS of (5.1) vanishes. A repeated application of (5.1) to $\varphi^{(2)}(\partial_\nu I)$ shows the RHS of (5.1) equals the RHS of (5.3) in view of the Jacobi identities

$$A^2\begin{pmatrix} I & \\ 0, & \partial_\nu I \end{pmatrix} = -A(0,I)A(\partial_\nu I).$$

Q.E.D.

Proposition 5.5. *In addition to* $(\mathcal{H}1)$ *suppose the following condition:*

$\mathcal{H}IV(n+1)$ *For a fixed admissible* I *with* $|I| = n+1$, $A(I) = 0$. *But for any other admissible* J *such that* $|J| \le n+2$ $A(J) \ne 0$. *Then* $\mathfrak{l}^*(I)$ *vanishes and* $\dim H_n(Y, \hat{L}_0) = \kappa_n - 1$. *We have a linear relation*

$$(5.4) \qquad 2(\sum_{j\in I}\lambda_j - 1)A(0,I)\varphi_Q(I) \equiv \sum_{k\notin I}\sum_{\nu=1}^{n+1}(-1)^{\nu-1}\lambda_k \cdot \frac{A(0,I)A\begin{pmatrix} I \\ k, \partial_\nu I \end{pmatrix}A(k,\partial_\nu I)}{A(k,I)A(\partial_\nu I)}\varphi_Q(k,\partial_\nu I) \mod \mathcal{B}(I).$$

$H^n(Y, \nabla_0)$ *is of dimension* $\kappa_n - 1$ *and is spanned by* $\varphi_Q(J)$ *such that* $J \ne I$ *and* $|J| \le n+1$ *with the fundamental relations* (2.3), (2.4).

Proof. Since $A(I) = 0$ the LHS of (5.1) vanishes. Applying repeatedly (2.4) to $\varphi_Q(k, I)$ and (5.1) to $\varphi^{(2)}(\partial_\nu I)$ one sees that the RHS of (5.1) equals

$$-2(\sum_{j\in I}\lambda_j - 1)A(0,I)\varphi_Q(I)$$
$$+\sum_{k\notin I}\sum_{\nu=1}^{n+1}(-1)^{\nu-1}\lambda_k\frac{A\begin{pmatrix} 0,I \\ k,I \end{pmatrix}A\begin{pmatrix} I \\ 0,\partial_\nu I \end{pmatrix}A(k,\partial_\nu I)}{A(k,I)A(\partial_\nu I)}\varphi_Q(k,\partial_\nu I)$$
$$\mod \mathcal{B}(I).$$

Hence (5.4) follows owing to the identities

$$A(k,I) = -U^2\begin{pmatrix} k, I \\ 0,1,\dots,n+1 \end{pmatrix},$$
$$A\begin{pmatrix} 0,I \\ k,I \end{pmatrix}A\begin{pmatrix} 0,\partial_\nu I \\ I \end{pmatrix} = A(0,I)A\begin{pmatrix} I \\ k,\partial_\nu I \end{pmatrix}.$$

Q.E.D.

Corollary 5.6. *Suppose that $m = n+2, n \geq 1$ and that $A(I) = 0$ for all admissible I with $|I| = n+1$. Then $H^n(Y, \nabla_0)$ is of dimension $\kappa_n - (n+2) = 2^{n+2} - n - 4$ and is spanned by the representatives $\varphi_Q(I)$ with $|I| \leq n$ with the one fundamental relation: For $J = \{1, 2, \ldots, n+2\}$.*

$$(5.5) \qquad \sum_{\mu \neq \nu} (-1)^{\mu+\nu} \varphi_Q(\partial_\nu \partial_\nu J) \frac{A(0, \partial_\mu \partial_\nu J)}{A\begin{pmatrix} 0, & \partial_\mu J \\ 0, & \partial_\nu J \end{pmatrix}} = 0.$$

$\varphi_Q(I)$ ($|I| = n+1$) can be expressed as

$$(5.6) \qquad 2(\sum_{j \in I} \lambda_j - 1) A(0, I) \varphi_Q(I) \equiv 0 \mod \mathcal{B}(I)$$

over the coefficients of rational functions of $a_{i,j}, u_{k,0}$ with the denominators $A(K)$ ($|K| \leq n$). This identity is just an n-dimensional version of (1.2).

Proof. (5.5) is a special case of (2.3) since $A(\partial_\mu J) = 0$. On the other hand (5.6) is a special case of (5.4) since $A(k, \partial_\nu I) = 0$. Q.E.D.

Proposition 5.7. *In addition to ($\mathcal{H}1$) suppose the following condition:*

($\mathcal{H}IV(n+2)$) *For a fixed admissible I with $|I| = n+2$, $A(I) = 0$. But for any other admissible J such that $|J| \leq n+2$ $A(J) \neq 0$. Then there is no vanishing of Lefschetz cycles and* $\dim H_n(Y, \hat{\mathcal{L}}_0) = \kappa_n$. *On the other hand, for any fixed μ*

$$(5.7) \quad (\sum_{\nu \in I} \lambda_\nu - 1) U\begin{pmatrix} \partial_\mu I \\ 1, \ldots, n+1 \end{pmatrix} \varphi_Q(I) \equiv 0 \mod \mathcal{B}(I).$$

$H^n(Y, \nabla_0)$ is of dimension κ_n and is spanned by $\varphi_Q(J)$ such that $|J| \leq n+1$ with the fundamental relations (2.3), (2.4).

Proof. By hypothesis the LHS of (2.4) vanishes. By a multiplication by f_{i_μ} of both sides of (2.4) one sees that $\varphi_Q(I)$ is linearly dependent on $\frac{\varphi_Q(\partial_\nu I)}{f_{i_\mu}}$ $(\nu \neq \mu)$. On the other hand due to (5.2) each $\frac{\varphi_Q(\partial_\nu I)}{f_{i_\mu}}$ is linearly dependent on admissible $\varphi_Q(J)$ with $|J| \leq n+1$. Hence the Proposition follows. Q.E.D.

Finally we consider the special case where $n \geq 2, m = n+2$ and $A(i,j) = 0$, i.e., $a_{i,j} = \pm 1$ for all $i, j \in \{1, 2, \ldots, n+2\}$ $(i \neq j)$. Since the signature of A is of type $(n+1, 1)$, we have $A(1,2) = 0, A(1, 2, \ldots, p) < 0$ if $3 \leq p \leq n+2$.

By a suitable Lorentz transformation we may assume that $a_{i,j} = -1$ for all i, j $(i \neq j)$. In fact

Lemma 5.8. *There exist a diagonal matrix P with diagonal elements equal to ± 1 such that $B = P \cdot A \cdot {}^tP$ is the matrix with diagonal elements* 1 *and off-diagonal elements* -1.

Proof. Denote by B_r the matrix of size $r+2$ with diagonal elements 1 and off-diagonal elements -1. Let A_r be the matrix with the (i,j) elements $a_{i,j}$ $(1 \le i,j \le r+2)$. For $r=0$ the Lemma is trivial. Suppose that the Lemma is true for A_{r-1}. There exists a diagonal matrix P_{r-1} with diagonal elements ± 1 such that $B_{r-1} = P_{r-1} \cdot A_{r-1} \cdot {}^tP_{r-1}$. Let $\tilde{P}_r$ be the diagonal matrix of size $r+2$ such that the first $r+1$ diagonal elements coincides with the ones of P_{r-1} and the last one equal to 1. Then $\tilde{P}_r \cdot A_r \cdot {}^t\tilde{P}_r$ has the same components as B_r except for the off-diagonal components in the last column or row. Denote these components by $\varepsilon_1, \dots, \varepsilon_{r+1}$. Then we have

$$\det(\tilde{P}_r \cdot A_r \cdot{}^t \tilde{P}_r) = (1-r)2^r + (r-2)2^{r-1}\sum_{k=1}^{r+1}\varepsilon_k^2 \\ -2^r \sum_{1\le i<j\le r+1} \varepsilon_i\varepsilon_j < 0.$$

But this inequality goes to a contradiction except for the case where all ε_j equal 1 or all ε_j equal -1. One sees that the first case is equivalent to B_r, while the last one coincides with B_r. Q.E.D.

Lemma 5.9. *Suppose $I = \{1,2,\dots,n+2\}$. The matrix A for all off diagonal elements $a_{i,j} = -1$ defines the hypersphere arrangement $\mathcal{A}'$ if and only if $\{a_{j,0}\}_j$ satisfy the quadratic relation*

$$(n-1)\sum_{j=1}^{n+2} a_{j,0}^2 - 2\sum_{1\le j<k\le n+2} a_{j,0}a_{k,0} + 2n = 0.$$

Proof. First remark that if $A(0,I) \le 0$ then $A(0,J) < 0$ for $J \subset I, J \ne I$. In fact it is sufficient to show this in case $J = \{1,\dots,r\}$ $(3 \le r \le n+1)$. This follows by lowering induction from the identity

$$A(0,1,\dots,r)A(1,\dots,r+1) - A^2\begin{pmatrix} 0,1,\dots,r \\ r+1,1,\dots,r\end{pmatrix} \\ = A(0,1,\dots,r+1)A(1,\dots,r)$$

because $A(1,\dots,r), A(1,\dots,r+1)$ are both negative. On the other hand

$$A(0,I) = 2^n\{(n-1)\sum_{j=1}^{n+2} a_{j,0}^2 - 2\sum_{1\le j<k\le n+2} a_{j,0}a_{k,0} + 2n\}.$$

Hence the Lemma follows. Q.E.D.

We now apply to it the formula (5.3) for $p=2$.
For $I=\{i,j\}$ we have

$$A(0,i,j)=-(a_{i,0}+a_{j,0})^2,\ A\begin{pmatrix}0,i,j\\k,i,j\end{pmatrix}=2(a_{i,0}+a_{j,0}).$$

Hence (5.3) and Lemma 5.2 give

$$\begin{aligned}(5.8)\quad &(\lambda_\infty+n-3+\lambda_i+\lambda_j)(a_{i,0}+a_{j,0})\varphi_Q(i,j)\\ &+\sum_{k\neq i,j}\lambda_k(a_{k,0}+a_{i,0})\varphi_Q(k,i)+\sum_{k\neq i,j}\lambda_k(a_{k,0}+a_{j,0})\varphi_Q(k,j)\sim w_{i,j}\end{aligned}$$

where we put

$$\begin{aligned}w_{i,j}=2\sum_{k\neq i,j}\lambda_k\varphi_Q(k,i,j)&+(a_{i,0}+a_{j,0})\sum_{k=1}^{n+2}\lambda_k a_{k,0}\varphi_Q(k)\\ &-(\lambda_\infty+n-1)\cdot(a_{i,0}+a_{j,0})\varphi_Q(\emptyset)\\ &-(\lambda_\infty+n-2)\{A(0,i)\varphi_Q(i)+A(0,j)\varphi_Q(j)\}.\end{aligned}$$

To solve (5.4) with respect to $\varphi_Q(i,j)$ we put $v_{i,j}=(a_{i,0}+a_{j,0})\varphi_Q(i,j)$ and

$$\begin{aligned}V_i&=\sum_{k\neq i}\lambda_k v_{k,i},\ V_\infty=\sum_{i\neq j}\lambda_i\lambda_j v_{i,j},\\ W_i&=\sum_{k\neq i}\lambda_k w_{k,i},\ W_\infty=\sum_{i\neq j}\lambda_i\lambda_j w_{i,j}.\end{aligned}$$

Then (5.8) is equivalent to

$$(5.9)\qquad (\lambda_\infty+n-3)v_{i,j}+V_i+V_j\sim w_{i,j}.$$

(5.9) can be uniquely solved for $v_{i,j}$:

$$\begin{aligned}(5.10)\qquad &(\lambda_\infty+n-3)v_{i,j}\sim w_{i,j}\\ &+V_\infty\left(\frac{1}{2\lambda_\infty+n-3-2\lambda_i}+\frac{1}{2\lambda_\infty+n-3-2\lambda_j}\right)\\ &-\left(\frac{W_i}{2\lambda_\infty+n-3-2\lambda_i}+\frac{W_j}{2\lambda_\infty+n-3-2\lambda_j}\right)\end{aligned}$$

where V_i and V_∞ are uniquely determined by

$$(2\lambda_\infty + n - 3 - 2\lambda_i)V_i \sim W_i - V_\infty,$$
$$\left(1 + \sum_{k=1}^{n+2} \frac{\lambda_k}{2\lambda_\infty + n - 3 - 2\lambda_k}\right) V_\infty \sim \sum_{k=1}^{n+2} \frac{\lambda_k W_k}{2\lambda_\infty + n - 3 - 2\lambda_k}$$

provided none of $2\lambda_\infty + n - 3 - 2\lambda_k$ or the symmetric polynomial

$$G(\lambda) = \prod_{k=1}^{n+2}(2\lambda_\infty + n - 3 - 2\lambda_k) + \sum_{k=1}^{n+2} \lambda_k \prod_{j\neq k}(2\lambda_\infty + n - 3 - 2\lambda_j)$$

vanishes. In this way we can conclude

Proposition 5.10. *For $m = n+2, n \geq 2$, suppose that in addition to $(\mathcal{H}1)$, $a_{i,j} = -1$ for all i, j $(i \neq j)$, and $A(I) < 0$ for all admissible I with $3 \leq |I| \leq n+2$. Suppose further that neither of $2\lambda_\infty + n - 3 - 2\lambda_k$ or $\lambda_\infty + n - 3$ or $G(\lambda)$ vanish. Then all the Lefschetz cycles $\mathfrak{l}^*(I)$ $(|I| = 2)$ vanish. $H^n(Y, \nabla_0)$ is of dimension $\kappa_n - \binom{n+2}{2}$ and has a basis of representatives $\varphi_Q(I)$ with $0 \leq |I| \leq n+1, |I| \neq 2$ satisfying the fundamental relations (2.3),(2.4). $\varphi_Q(i,j)$ can be described as a linear combination of these representatives as in (5.10).*

§6. Two problems

As is seen from Propositon 2.2, $H^n(Y, \nabla_0)$ is spanned by the representatives $\varphi_Q(I)$. The result due to Orlik–Terao (see [9]) suggests that this fact still holds in general in the following sense:

Conjecture 6.1. *Without any of the hypotheses $(\mathcal{H}1)$ or $(\mathcal{H}2)$, $H^n(Y, \nabla_0)$ is spanned by the representatives $\varphi_Q(I)$, $I \subset \{1, 2, \ldots, m\}$ including $\varphi_Q(\emptyset)$.*

The complex hypersphere Q has the Kähler metric

$$ds^2 = \sum_{\nu=1}^{n+1} |d\xi_\nu|^2 = \sum_{\mu,\nu=1}^{n} g_{\mu,\overline{\nu}} d\zeta^\mu d\overline{\zeta^\nu}$$

with respect to local coordinates $\zeta = (\zeta^\nu)_{1\leq\nu\leq n}$. We put

$$\lambda_j = Nl_j + \lambda'_j \quad (N \in \mathbf{Z}_{>0})$$

for fixed $l = (l_j)_j \in (\mathbf{Z}_{>0})^m, \lambda' = (\lambda'_j)_j \in \mathbf{R}^m$. For a large N the asymptotic behavior of the integral (2.2) can be explicitly evaluated if the cycle **c** is a stable cycle defined by the gradient vector field on Y:

$$d\zeta^{\mu} = \sum_{\nu=1}^{n} \frac{\partial}{\partial \overline{\zeta_{\nu}}} (\sum_{j=1}^{m} l_j \log |f_j|) g^{\mu,\overline{\nu}} dt$$

where $g^{\mu,\overline{\nu}}$ denotes the inverse matrix of the metric tensor $g_{\mu,\overline{\nu}}$. The critical points are determined by the equations on Y (see [1], [10], [14])

$$\sum_{j=1}^{m} l_j \, d\log f_j = 0.$$

Every cycle mentioned in Theorem 3.11 seems to have one-to-one relation with a stable cycle corresponding to these critical points.

Question 6.2. *How can be found the critical points of the gradient vector field in Y? When are they real or imaginary? Are the stable cycles corresponding to them homological to the ones mentioned in* Theorem *3.11?*

Under the conditions $(\mathcal{H}1), (\mathcal{H}2a)$ every critical point is real and lies one-to-one in a connected component of $\Re Y$. On the other hand under the conditions $(\mathcal{H}1), (\mathcal{H}2b)$ all critical points are not real. It can be proved by a perturbative method that all of the critical points lie in $\Re Y$, at least in the case where $m \leq n+1$ and all admissible $A(I)$, $(2 \leq |I| \leq n+1)$ are negative and near 0.

References

[1] K. Aomoto, On vanishing of cohomology attached to certain many valued meromorphic functions, J. Math. Soc. Japan, **27** (1975), 248–255.

[2] K. Aomoto, Configuration and invariant Gauss–Manin connections of integrals I, Tokyo J. Math., **5** (1982), 249–287; II, *ibid*, **6** (1983), 1–24.

[3] K. Aomoto, Errata to “Configuration and invariant Gauss–Manin connections of integrals I, II”, Tokyo J. Math., **22** (1999), 511–512.

[4] K. Aomoto, Vanishing of certain 1-form attached to a configuration, Tokyo J. Math., **9** (1986), 453–454.

[5] K. Aomoto, Gauss–Manin connections of Schläfli type for hypersphere arrangements, Ann. Inst. Fourier, **53** (2003), 977–995.

[6] K. Aomoto and P. Forrester, On a Jacobian identity associated with real hyperplane arrangements, Compositio Math., **121** (2000), 263–295.

[7] K. Aomoto, M. Kita, P. Orlik and H. Terao, Twisted de Rham cohomology groups of logarithmic forms, Adv. Math., **128** (1997), 119–152.

[8] F. Gantmacher, The Theory of Matrices I, Chelsea, 1959.

[9] P. Orlik and H. Terao, Commutative algebra for arrangements, Nagoya Math. J., **134** (1994), 65–73.
[10] P. Orlik and H. Terao, The number of critical points of a product of powers of linear functions, Invent. Math., **120** (1995), 1–14.
[11] P. Orlik and H. Terao, Arrangements and Hypergeometric Integrals, MSJ Memoirs, **9**, Math. Soc. Japan, 2001; 2nd ed., 2007.
[12] F. Pham, Introduction à L'étude Topologique des Singularités de Landau, Gauthiers Villars, 1967.
[13] C. A. Rogers, Packing and Covering, Cambridge, 1964.
[14] A. N. Varchenko, Critical points of the product of powers of linear functions and families of bases of singular vectors, Compositio Math., **97** (1995), 385–401.

Kazuhiko Aomoto
Department of Mathematics, Nagoya University,
Furo-cho 1, Chikusa-ku, Nagoya, 464-8602, Japan
E-mail address: kazuhiko@aba.ne.jp

Advanced Studies in Pure Mathematics 62, 2012
Arrangements of Hyperplanes—Sapporo 2009
pp. 27–57

Complexes, duality and Chern classes of logarithmic forms along hyperplane arrangements

Graham Denham and Mathias Schulze

Abstract.

We describe dualities and complexes of logarithmic forms and differentials for central affine and corresponding projective arrangements. We generalize the Borel–Serre formula from vector bundles to sheaves on $\mathbb{P}^d$ with locally free resolutions of length one. Combining these results we present a generalization of a formula due to Mustaţă and Schenck, relating the Poincaré polynomial of an arrangement in $\mathbb{P}^3$ (or a locally tame arrangement in $\mathbb{P}^d$ with zero-dimensional non-free locus) to the total Chern polynomial of its sheaf of logarithmic 1-forms.

CONTENTS

§1. Introduction

The study of logarithmic differentials and differential forms in the context of hyperplane arrangement singularities now has a thirty-year history that begins with Saito [Sai80]. For a comprehensive survey, we refer to the forthcoming book [CDF+]. This paper collects together

Received April 30, 2010.
Revised August 3, 2010.
2010 *Mathematics Subject Classification.* 32S22, 52C35, 16W25.
Key words and phrases. Hyperplane arrangement, logarithmic differential form, Poincaré polynomial, Chern class.
GD was partially supported by a grant from NSERC of Canada.

some results about complexes of logarithmic forms: some known, others folklore, and some new. In particular, we attempt to clarify the relationship between forms on the complements of central, non-central and projective arrangements, respectively.

In order to be more precise, let $\Omega^p(\mathscr{A})$ denote the module of logarithmic p-forms (defined in (2.2)), for a central arrangement $\mathscr{A}$ of rank ℓ. Following [Dol82, Dol07, Dim92], we consider the submodule $\Omega^p_0(\mathscr{A})$ (Definition 2.4) of forms which vanish when contracted along the Euler differential. We show that the coherent sheaf on this submodule coincides with the usual sheaf of logarithmic forms on the projectivization of $\mathscr{A}$, Definition 2.10. We also consider a *quotient* of $\Omega^p(\mathscr{A})$ which we call the module of *relative* logarithmic forms, denoted $\Omega^p_\sigma(\mathscr{A})$ and defined in Definition 2.7. This was introduced in slightly different terms by Terao and Yuzvinsky [TY95], and discussed more generally in [dGMS09]. We note that this module is isomorphic to $\Omega^p_0(\mathscr{A})$, which gives a noncanonical splitting of the inclusion $\Omega^p_0(\mathscr{A}) \hookrightarrow \Omega^p(\mathscr{A})$. We note that the choice involved amounts to choosing an affine chart. In this way, we understand logarithmic forms for affine, non-central arrangements in terms of their cone.

The modules of logarithmic forms are reflexive, which was observed first in [Sai80, (1.7) Cor.] for $\Omega^1(\mathscr{A})$. The exterior product gives a map $j\colon \bigwedge^p \Omega^1(\mathscr{A}) \to \Omega^p(\mathscr{A})$, which we see is an isomorphism exactly when $\bigwedge^p \Omega^1(\mathscr{A})$ is also reflexive (Proposition 2.2.) We note that, if $\mathscr{A}$ is tame, (Definition 2.3), then j is an isomorphism for values of p less than the codimension of the non-free locus of $\mathscr{A}$ (Proposition 2.9). In the case where $\mathscr{A}$ is free, this is part of Saito's criterion for freeness, from [Sai80, (1.8) Thm.]; in the case where $\mathscr{A}$ is locally free, it was noted by Mustaţă and Schenck [MS01]. The same is true for the variations above.

In §3, we examine the duals of the modules of logarithmic forms, which are modules (or sheaves) of multilinear logarithmic differentials. In the dual setting, the natural construction is a quotient of the module of logarithmic forms $D_p(\mathscr{A})$, denoted by $D^0_p(\mathscr{A})$ and defined in Definition (3.4). Our work with forms allows us to replace the quotient with a submodule, $D^\sigma_p(\mathscr{A})$, again by choosing a chart. The modules of forms are also self-dual, which gives some useful symmetry. We note some equivalent formulations of the homological notions of free and tame arrangements.

Multiplication by a degree-0 logarithmic 1-form ω_λ gives a cochain complex $(\Omega^\bullet(\mathscr{A}), \omega_\lambda)$, as well as for the projective constructions. Orlik and Terao [OT95b] show that this complex is exact, and $(\Omega^\bullet_\sigma(\mathscr{A}), \omega_\lambda)$ has a single non-zero cohomology group, both for suitably "generic" choices of λ, which we discuss below. Their main application is to show that, under the same hypotheses on λ, the function

$$\Phi_\lambda = \prod_{H \in \mathscr{A}} \alpha_H^{\lambda_H} \tag{1.1}$$

has isolated, nondegenerate critical points. (Here, $\mathscr{A}$ is a non-central arrangement defined by affine-linear forms $\{\alpha_H : H \in \mathscr{A}\}$.) Recent generalizations of this result in [CDFV] make use of a parameterized version of this complex, which we discuss in §4.1. We clarify the relationship between the various complexes and, in doing so, improve slightly on a main result of [CDFV].

As further applications, we give some formulas for the Chern classes of the sheaf of logarithmic 1-forms in §5. Mustaţă and Schenck [MS01] showed that, for locally free arrangements, the total Chern polynomial is given by the arrangement's characteristic polynomial. We see in Proposition 5.18 that the coefficients of the two polynomials always agree up to degree k, for any arrangement, if its non-free locus has codimension $> k$. For certain projective arrangements with zero-dimensional non-free locus (including all arrangements of rank ≤ 4), we compute the total Chern polynomial in Theorem 5.13. Our expression is combinatorial if the non-free locus consists of generic closed subarrangements; however, we exhibit two arrangements with the same matroid and different Chern polynomials in Example 5.20.

§2. Log forms

2.1. Normal and reflexive sheaves

Let $\mathscr{A}$ be a central (simple) arrangement of n hyperplanes in an ℓ-dimensional $\mathbb{C}$-vector space $V = \mathbb{A}^\ell$.

We denote by $L(\mathscr{A})$ the intersection lattice of $\mathscr{A}$, and by $L_c(\mathscr{A}) \subseteq L(\mathscr{A})$ the subset of codimension-c flats. For $X \in L(\mathscr{A})$, let $\mathscr{A}_X = \{H \in \mathscr{A} \mid X \subseteq H\}$ be the localization of $\mathscr{A}$ at X; for $H \in \mathscr{A}$ let $\mathscr{A}^H = \{H' \cap H \mid H' \in \mathscr{A} \setminus \{H\}\}$ be the restriction of $\mathscr{A}$ to H.

We abbreviate $\mathbb{P}^d = \mathbb{P}V$, $d := \ell - 1$, and denote by $\mathbb{P}\mathscr{A} := \pi(\mathscr{A})$ the corresponding projective arrangement, where

$$\pi \colon V \setminus \{0\} \to \mathbb{P}V \tag{2.1}$$

is the canonical projection. For $H \in \mathscr{A}$ we denote by $\alpha_H \in V^*$ its defining equation. We can assume that for some $\{H_1, \ldots, H_\ell\} \subset \mathscr{A}$, $x_i := \alpha_{H_i}$ are coordinates on V. Then $f = \prod_{H \in \mathscr{A}} \alpha_H$ is the (reduced) defining equation of $\mathscr{A}$ in the coordinate ring $R = \operatorname{Sym}(V^*)$ of V.

We shall readily identify R-modules with the associated sheaves on the affine space V, and denote $-^\vee = \operatorname{Hom}_R(-, R)$. We identify $\mathscr{A}$ with $\operatorname{Spec}(R/f) \subset V$ and denote by $i\colon U := V \setminus \operatorname{Sing}\mathscr{A} \hookrightarrow V$ the inclusion of the complement of the singular locus of $\mathscr{A}$ in V. Following Barth [Bar77], we say an R-module M is $\mathscr{A}$-normal if $M \to i_* i^* M$ is an isomorphism. Note that R itself is normal. The following easy facts are particularly useful for our purpose.

Lemma 2.1.

(1) Any reflexive module has depth ≥ 2.
(2) Two $\mathscr{A}$-normal modules are dual if they are dual on U.
(3) A torsion-free $\mathscr{A}$-normal module that is reflexive on U is reflexive.
(4) Moreover, $\mathscr{A}$-normality is a consequence of reflexivity.

Proof. The non-obvious statements follow from [Har80, Props. 1.1, 1.6]. Q.E.D.

2.2. Relative logarithmic forms

The modules Ω_V^p form a graded R-module $\Omega_V^\bullet$ closed under exterior product such that the natural map $R \to \Omega_V^1$ has degree 0. In particular, the localization $\Omega_{V,f}^\bullet$ has a natural $\mathbb{Z}$-grading. The module of *logarithmic differential p-forms along* $\mathscr{A}$ is the graded R-module

$$\Omega^p(\mathscr{A}) = \left\{ \omega \in \frac{1}{f}\Omega_V^p \mid \forall H \in \mathscr{A}\colon \frac{d\alpha_H}{\alpha_H} \wedge \omega \in \frac{1}{f}\Omega_V^{p+1} \right\} \subset \Omega_{V,f}^\bullet. \tag{2.2}$$

It is easy to check that $\Omega^\bullet(\mathscr{A})$ is closed under exterior product.

As the modules $\frac{1}{f}\Omega_V^p$ and $\frac{1}{f}\Omega_V^{p+1}$ are free, $\Omega^p(\mathscr{A})$ is $\mathscr{A}$-normal, and

$$\Omega^p(\mathscr{A})_{\mathfrak{p}} = \bigwedge^p \Omega^1(\mathscr{A})_{\mathfrak{p}} \tag{2.3}$$

for primes $\mathfrak{p}$ supported on the non-singular locus U. Both sides are free over $R_{\mathfrak{p}}$, so $\Omega^p(\mathscr{A})$ is reflexive by Lemma 2.1.(3).

The exterior product gives a map $j_p\colon \bigwedge^p \Omega^1(\mathscr{A}) \to \Omega^p(\mathscr{A})$, which is easily seen to be a monomorphism. Under some hypotheses, j is an isomorphism, as we will see in Proposition 2.9. However, exterior powers of a reflexive module need not be reflexive, and in general one has:

Proposition 2.2. *For any $\mathscr{A}$, we have $(\bigwedge^p \Omega^1(\mathscr{A}))^{\vee\vee} \cong \Omega^p(\mathscr{A})$ for $0 \leq p \leq \ell$.*

Proof. Let $E^p = E^p(\mathscr{A})$ denote the cokernel of j_p. By (2.3), the module E^p is supported on $\operatorname{Sing}\mathscr{A}$, which has codimension ≥ 2. Therefore $\operatorname{Ext}^i_R(E^p, R) = 0$ for $i = 0, 1$. It follows that

$$j_p^\vee : \Omega^p(\mathscr{A})^\vee \to (\bigwedge^p \Omega^1(\mathscr{A}))^\vee \tag{2.4}$$

is an isomorphism. We apply $-^\vee$ again and note $\Omega^p(\mathscr{A})$ is reflexive. Q.E.D.

By Lemma 2.1.(1), $\operatorname{pd}_R \Omega^p(\mathscr{A}) \leq \ell - 2$ for all p. The following definition first appeared in [OT95a]:

Definition 2.3. An arrangement $\mathscr{A}$ is *tame* if $\operatorname{pd}_R \Omega^p(\mathscr{A}) \leq p$, for $0 \leq p \leq \ell$.

Since $\Omega^0(\mathscr{A}) = R$ is free, the condition is vacuous except for $1 \leq p \leq \ell - 3$. Accordingly, all arrangements of rank ≤ 3 are tame.

The graded R-dual of $\Omega^1(\mathscr{A})$ is the graded R-module of *logarithmic differentials along* $\mathscr{A}$

$$D(\mathscr{A}) = D_1(\mathscr{A}) = \{\delta \in D_V \mid \forall H \in \mathscr{A} \colon \delta(\alpha_H) \in \alpha_H R\}$$

where $D_V = \operatorname{Der}_{\mathbb{C}}(R, R)$ is the module of polynomial vector fields on V.

Note that the standard Euler differential $\chi = \sum_{i=1}^{\ell} x_i \partial_i$ is an element of $D(\mathscr{A})$. Recall from [OT92, Prop. 4.86] that contraction ι_δ with $\delta \in D(\mathscr{A})$ defines a graded map

$$\iota_\delta \colon \Omega^p(\mathscr{A}) \to \Omega^{p-1}(\mathscr{A}).$$

By $\mathscr{A}$-normality of $\Omega^\bullet(\mathscr{A})$, it suffices to check this on U, where it is clear.

Definition 2.4. We call $\Omega^p_0(\mathscr{A}) := \ker \iota_\chi \subset \Omega^p(\mathscr{A})$ be the module of *relative logarithmic differential p-forms along* $\mathscr{A}$.

Clearly, $\Omega^\bullet_0(\mathscr{A})$ is closed under exterior product, and locally free on U. As above, freeness of $\frac{1}{f}\Omega^p_V$ and $\frac{1}{f}\Omega^{p-1}_V$ implies that $\Omega^p_0(\mathscr{A})$ is $\mathscr{A}$-normal and torsion free. Then Lemma 2.1.(3) and the same argument as Proposition 2.2 gives the following.

Proposition 2.5. *$\Omega^\bullet_0(\mathscr{A})$ is closed under exterior product and reflexive. Moreover,* $\Omega^\bullet_0(\mathscr{A}) \cong (\bigwedge^\bullet \Omega^1_0(\mathscr{A}))^{\vee\vee}$.

We note also the following simple fact (cf. [OT92, Exa. 4.122]).

Proposition 2.6. *$\mathscr{A} \mapsto \Omega^p_0(\mathscr{A})$ is a local functor. That is, for any graded prime ideal $\mathfrak{p}$ of R, $\Omega^p_0(\mathscr{A})_\mathfrak{p} \cong \Omega^p_0(\mathscr{A}_{X(\mathfrak{p})})_\mathfrak{p}$, where $X(\mathfrak{p})$ denotes the set-theoretically smallest subspace in the intersection lattice $L(\mathscr{A})$ containing the zeroes of $\mathfrak{p}$.*

The term “relative” in Definition 2.4 does not refer to a specific map here. It turns out that the so defined differential forms are “relative” to many maps simultaneously. This can be seen as follows: As in [CDFV], we denote

$$\omega_{\mathbf{a}} = d\log(\alpha^{\mathbf{a}}) = \sum_{H\in\mathscr{A}} a_H \frac{d\alpha_H}{\alpha_H} \tag{2.5}$$

for independent variables $\mathbf{a} = (a_H)_{H\in\mathscr{A}} \in (\mathbb{C}^{\mathscr{A}})^\vee$. We write

$$|\mathbf{a}| = \sum_{H\in\mathscr{A}} a_H = \iota_\chi(\omega_{\mathbf{a}}). \tag{2.6}$$

Now for any $\sigma \in \mathbb{C}^{\mathscr{A}}$ with $|\sigma| \neq 0$, $(\Omega^\bullet(\mathscr{A}), \omega_\sigma)$ is split exact (see [OT92, Prop. 4.86]), since

$$\iota_\chi(\omega_{\mathbf{a}} \wedge \omega) = |\mathbf{a}|\omega - \omega_{\mathbf{a}} \wedge \iota_\chi(\omega).$$

Definition 2.7. We define the module of *logarithmic differential p-forms relative to α^σ* to be the graded R-module

$$\Omega^p_\sigma(\mathscr{A}) = \Omega^p(\mathscr{A})/\omega_\sigma \wedge \Omega^{p-1}.$$

For any $\lambda \in \mathbb{C}^{\mathscr{A}}$ with $|\lambda| = 0$, we can identify the complexes

$$(\Omega^\bullet_\sigma(\mathscr{A}), \omega_\lambda) = (\Omega^\bullet_0(\mathscr{A}), \omega_\lambda) \tag{2.7}$$

via the map $[\omega] \mapsto \omega - \omega_\sigma \wedge \iota_\chi(\omega)$. In particular, we can identify

$$\Omega^\bullet(\mathscr{A}) = \omega_\sigma \wedge \Omega_0^{\bullet-1}(\mathscr{A}) \oplus \Omega_0^\bullet(\mathscr{A}) \cong \Omega_0^{\bullet-1}(\mathscr{A}) \oplus \Omega_0^\bullet(\mathscr{A}). \tag{2.8}$$

Then the notion of tameness, Definition 2.3, is unaffected by working with relative logarithmic forms:

Proposition 2.8. *$\mathscr{A}$ is tame if and only if $\operatorname{pd}_R \Omega^p_0(\mathscr{A}) \leq p$ for $1 \leq p \leq \ell - 3$.*

Proof. If $\operatorname{pd}_R \Omega^p_0(\mathscr{A}) \leq p$ for all p, then the same is true for $\Omega^p(\mathscr{A})$ by (2.8). Conversely, if $\mathscr{A}$ is tame, argue by induction on p using (2.8) that $\operatorname{Ext}^q_R(\Omega^p_0(\mathscr{A}), R) = 0$ for all $q > p$. Q.E.D.

The following generalizes a result of Mustaţă–Schenck [MS01, Thm. 5.3].

Proposition 2.9. *If $\mathscr{A}$ is a tame arrangement whose non-free locus has codimension $> k$, then $\bigwedge^p \Omega^1_0(\mathscr{A}) = \Omega^p_0(\mathscr{A})$, and hence also $\bigwedge^p \Omega^1(\mathscr{A}) = \Omega^p(\mathscr{A})$, for $p < k$.*

Proof. By the argument from the proof of [MS01, Lem. 5.2], the hypotheses imply that $\Omega_0^1(\mathscr{A})$ is a $(k-1)$-syzygy. Then, by [Leb77, Satz 3.1], $\mathrm{pd}_R \bigwedge^p \Omega_0^1(\mathscr{A}) = p$ for $p \leq k$. Applying $\mathrm{Ext}_R^q(-, R)$ to the exact sequence

$$0 \to \bigwedge^p \Omega_0^1(\mathscr{A}) \to \Omega_0^p(\mathscr{A}) \to E_0^p \to 0$$

shows that $\mathrm{Ext}_R^q(E_0^p, R) = 0$ for $q > p+1$ and hence for $q > k > p$. But by assumption the E_0^p are supported in codimension $> k$ only, so $\mathrm{Ext}_R^q(E_0^p, R) = 0$ for $q \leq k$ as well, so $E_0^p = 0$ for all $p < k$. Q.E.D.

2.3. Projective logarithmic forms

Geometrically, $\Omega_0^\bullet(\mathscr{A})$ can be considered as an affine version of the sheaf $\Omega^\bullet(\mathbb{P}\mathscr{A})$ defined following the original definition of K. Saito [Sai80], as we will see in Proposition 2.12. Recall that $\mathbb{P}\mathscr{A}$ is the image of $\mathscr{A}$ under the natural projection (2.1).

Definition 2.10. Consider $\mathbb{P}\mathscr{A}$ as a principal divisor $D = (f)$ on $\mathbb{P}V$. Then we call

$$\Omega^p(\mathbb{P}\mathscr{A}) := \Omega_{\mathbb{P}V}^p(D) \cap d^{-1}\Omega_{\mathbb{P}V}^{p+1}(D)$$

is the sheaf of *logarithmic differential p-forms along* $\mathbb{P}\mathscr{A}$.

In other words, $\Omega^\bullet(\mathbb{P}\mathscr{A})$ is the sheaf of rational differential forms ω on $\mathbb{P}V$ for which both ω itself and $d\omega$ have at most a simple pole along $\mathbb{P}\mathscr{A}$. Note that in any chart $\{x_i = 1\}$, $\Omega^\bullet(\mathbb{P}\mathscr{A})$ restricts to $\Omega^\bullet(\mathscr{A}^{\{x_i=1\}})$.

In order to see the claimed relation with $\Omega_0^\bullet(\mathscr{A})$, first fix a chart with index $i \in \{1, \dots, \ell\}$. Then $\Omega_{V,f}^\bullet/\omega_{e_i} \wedge \Omega_{V,f}^\bullet$ is the module of differential forms with poles along $\mathscr{A} \setminus \{H_i\}$ relative to the map x_i on $\{x_i \neq 0\}$. Using the definition of $\Omega^\bullet(\mathscr{A})$, one checks that the composition of canonical maps

$$\Omega^\bullet(\mathscr{A})_{x_i} \hookrightarrow \Omega_{V,f}^\bullet \twoheadrightarrow \Omega_{V,f}^\bullet/\omega_{e_i} \wedge \Omega_{V,f}^\bullet$$

factors through an inclusion

$$\Omega_{e_i}^\bullet(\mathscr{A})_{x_i} \hookrightarrow \Omega_{V,f}^\bullet/\omega_{e_i} \wedge \Omega_{V,f}^\bullet,$$

and it follows that we can identify

$$\begin{aligned} \Omega_{e_i}^\bullet(\mathscr{A}) \otimes_R R/\langle x_i - 1\rangle &= \Omega^\bullet(\mathscr{A}^{\{x_i=1\}}), \text{ and} \\ \Omega_{e_i}^\bullet(\mathscr{A})_{x_i} &= \Omega^\bullet(\mathscr{A}^{\{x_i=1\}}) \otimes_{\mathbb{C}} \mathbb{C}[x_i^{\pm 1}]. \end{aligned} \tag{2.9}$$

This immediately implies the following (see [TY95]):

Proposition 2.11. *This correspondence* (2.9) *combined with* (2.7) *identifies* $(\Omega_0^\bullet(\mathscr{A})_{x_i}, \omega_\lambda)$ *and* $(\Omega^\bullet(\mathscr{A}^{\{x_i=1\}}), \omega_{\lambda_{\hat{i}}})$ *where* $\lambda_{\hat{i}}$ *is obtained by deleting the ith component from* λ.

In order to understand the global relation of $\Omega_0^\bullet(\mathscr{A})$ and $\Omega^\bullet(\mathbb{P}\mathscr{A})$, we follow the approach in the proof of [Har77, Ch. II, Thm. 8.13]. We obtain a logarithmic version of the well-known description of logarithmic differential forms on projective space as the kernel of contraction with the Euler differential, cf. [Dim92, Ch. 6, §1] and [Dol82, §2.1]. For 1-forms along a generic $\mathscr{A}$, this result can be found in [MS01, p. 702–703].

Proposition 2.12. *As sheaves on* $\mathbb{P}V$,

$$\widetilde{\Omega_0^p(\mathscr{A})} = \Omega^p(\mathbb{P}\mathscr{A}).$$

Proof. Analogous to the logarithmic version in Definition 2.4, define

$$\Omega_{V,0}^p := \ker(\iota_\chi\colon \Omega_V^p \to \Omega_V^{p-1}). \tag{2.10}$$

Consider the sequence

$$0 \longrightarrow \Omega_{V,0}^1 \longrightarrow \Omega_V^1 \xrightarrow{\iota_\chi} \Omega_V^0 \longrightarrow 0,$$

which corresponds to the sequence $0 \to M \to E \to S$ in the proof of [Har77, Ch. II, Thm. 8.13]. This sequence is exact away from the origin. Applying $-^\vee$, $\bigwedge^p$, and then $-^\vee$ again, yields a sequence

$$0 \to \bigwedge^p \Omega_{V,0}^1 \longrightarrow \Omega_V^p \xrightarrow{\iota_\chi} \Omega_V^{p-1}, \tag{2.11}$$

which is exact away from the origin. By [Har77, Thm. 8.4], $\widetilde{\Omega}_{V,0}^1 = \Omega_{\mathbb{P}V}^1$, and hence

$$\widetilde{\bigwedge^p \Omega_{V,0}^1} = \Omega_{\mathbb{P}V}^p. \tag{2.12}$$

It follows from (2.10), (2.11), and (2.12) that $\widetilde{\Omega}_0^p = \Omega_{\mathbb{P}V}^p$, and then

$$\widetilde{\Omega_0^p(*\mathscr{A})} = \Omega_{\mathbb{P}V}^p(*\mathbb{P}\mathscr{A}).$$

Comparison of the subsheaves $\widetilde{\Omega_0^p(\mathscr{A})} \subset \widetilde{\Omega_0^p(*\mathscr{A})}$ and $\Omega^p(\mathbb{P}\mathscr{A}) \subset \Omega_{\mathbb{P}V}^p(*\mathbb{P}\mathscr{A})$ can now be done in charts. So the claim follows from Proposition 2.11. Q.E.D.

Definition 2.13. We call $\mathbb{P}\mathscr{A}$ *locally tame* if $\Omega^p(\mathbb{P}\mathscr{A})$ has a locally free resolution of length p, for $0 \leq p \leq \ell - 1$.

All arrangements in $\mathbb{P}^3$ are locally tame by reflexivity of $\Omega^1(\mathbb{P}\mathscr{A})$ and [Har80, Prop. 1.3]. Proposition 2.9 applied in charts gives the following:

Proposition 2.14. *If $\mathbb{P}\mathscr{A}$ is a locally tame arrangement whose non-free locus has codimension $> k$, then $\bigwedge^p \Omega^1(\mathbb{P}\mathscr{A}) = \Omega^p(\mathbb{P}\mathscr{A})$ for $p < k$.*

Recall that $\mathscr{A}$ is called locally free if $\Omega^1(\mathscr{A}_X)$ is free for all $X \in L_{<\ell}(\mathscr{A})$.

Definition 2.15. We call $\mathbb{P}\mathscr{A}$ locally free if the sheaf $\Omega^1(\mathbb{P}\mathscr{A})$ is a vector bundle.

From Proposition 2.12, using that $\mathscr{A} \mapsto \Omega^1(\mathscr{A})$ is a local functor, we deduce the following equivalence.

Lemma 2.16. *$\mathbb{P}\mathscr{A}$ is locally free if and only if $\mathscr{A}$ is locally free. More precisely, $\Omega^1(\mathbb{P}\mathscr{A})$ is free on an open set $U \subseteq \mathbb{P}V$ if and only if, whenever $X \subseteq U$ for some $X \in L(\mathbb{P}\mathscr{A})$, the closed subarrangement $\mathscr{A}_X$ is free.*

§3. Dualities

3.1. Duality with relative log differentials

We define the module of *logarithmic differentials relative to α^σ* to be the graded R-module

(3.1) $$D^\sigma(\mathscr{A}) := \{\delta \in D(\mathscr{A}) \mid \delta(\alpha^\sigma) = 0\}.$$

Then clearly

(3.2) $$D(\mathscr{A}) = R\chi \oplus D^\sigma(\mathscr{A}).$$

Applying $-^\vee$ to the exact sequence

$$0 \longrightarrow D^\sigma(\mathscr{A}) \longrightarrow D(\mathscr{A}) \xrightarrow{\ \phi\ } R \longrightarrow 0,$$

where $\phi(\delta) = \frac{\delta(\alpha^\sigma)}{\alpha^\sigma} = \langle \omega_\sigma, \delta \rangle$, shows that $\phi^\vee = \omega_\sigma$ hence

Lemma 3.1. $D^\sigma(\mathscr{A})^\vee = \Omega^1_\sigma(\mathscr{A})$.

The higher R-modules of logarithmic differentials are defined by

(3.3) $$D_p(\mathscr{A}) = \Big\{\delta \in \bigwedge^p D_V \mid \forall H \in \mathscr{A}, g_2, \dots, g_p \in R \\ : \delta(\alpha_H, g_2, \dots, g_p) \in \alpha_H R\Big\}.$$

As in (2.3), we have

$$D_p(\mathscr{A})_\mathfrak{p} = \bigwedge^p D_1(\mathscr{A})_\mathfrak{p} \tag{3.4}$$

for primes $\mathfrak{p}$ supported on U Generalizing (3.1), we introduce higher relative logarithmic differentials.

Definition 3.2. We define the module of *logarithmic p-differentials relative to* α^σ to be the graded R-module

$$D_p^\sigma(\mathscr{A}) = \{\delta \in D_p(\mathscr{A}) \mid \forall g_2, \dots, g_p \in R \colon \delta(\alpha^\sigma, g_2, \dots, g_p) = 0\}.$$

Both $D_\bullet(\mathscr{A})$ and $D_\bullet^\sigma(\mathscr{A})$ are clearly closed under exterior product. As in the case of forms, both are $\mathscr{A}$-normal and torsion free, and using Lemma 2.1 we obtain

Proposition 3.3. *$D_\bullet^\sigma(\mathscr{A})$ is closed under exterior product and reflexive.*

By [MS01, Prop. 2.2], there is a non-degenerate pairing

$$\langle -, - \rangle \colon \Omega^p(\mathscr{A}) \times D_p(\mathscr{A}) \to R. \tag{3.5}$$

By $\mathscr{A}$-normality of the three modules involved, it is sufficient to check this on U where it is clear.

Under duality such as in (3.5), subspaces of one factor correspond to quotient spaces of the other factor. Therefore we need also a quotient representation of $D_p^\sigma(\mathscr{A})$ independent of σ.

Definition 3.4. We define the module of *relative logarithmic p-differentials along* $\mathscr{A}$ to be the graded R-module

$$D_p^0(\mathscr{A}) := D_p(\mathscr{A}) / \chi \wedge D_{p-1}(\mathscr{A}).$$

As we can check on U, $(D_\bullet(\mathscr{A}), \chi)$ is exact and is splits by ι_{ω_σ}. We can thus, as in (2.7) and (2.8), identify

$$D_p^\sigma(\mathscr{A}) = D_p^0(\mathscr{A}),$$

$$D_p(\mathscr{A}) = \chi \wedge D_{p-1}^0(\mathscr{A}) \oplus D_p^0(\mathscr{A}) \cong D_{p-1}^0(\mathscr{A}) \oplus D_p^0(\mathscr{A}).$$

Proposition 3.5. *The pairing* (3.5) *induces a non-degenerate pairing*

$$\Omega_0^p(\mathscr{A}) \times D_p^0(\mathscr{A}) = \Omega_\sigma^p(\mathscr{A}) \times D_p^\sigma(\mathscr{A}) \to R. \tag{3.6}$$

Proof. To show (3.6) is well-defined, we need to show that $\big(\omega_\sigma \wedge \Omega^{p-1}(\mathscr{A})\big) \times D_p^\sigma(\mathscr{A})$ and $\Omega_0^p(\mathscr{A}) \times \big(\chi \wedge D_{p-1}^0(\mathscr{A})\big)$ are mapped to zero by (3.5), using the sub and quotient representations the two factors. This can be checked on U, and we show only the first statement. By (2.3) and (3.4), locally at points in U, $\Omega^{p-1}(\mathscr{A})$ and $D_p(\mathscr{A})$ are generated by $\omega = \frac{d\alpha_{H_2}}{\alpha_{H_2}} \wedge \cdots \wedge \frac{d\alpha_{H_p}}{\alpha_{H_p}}$ and $\delta = \delta_1 \wedge \ldots, \wedge \delta_p$, $\delta_i \in D(\mathscr{A})$, respectively. Using (2.5),

$$\langle \omega_\sigma \wedge \omega, \delta \rangle = \frac{\det(\delta_i(\alpha^\sigma)|\delta_i(\alpha_{H_j}))}{\alpha^\sigma \alpha_{H_2} \cdots \alpha_{H_p}} = \frac{\delta(\alpha^\sigma, \alpha_{H_1}, \ldots, \alpha_{H_p})}{\alpha^\sigma \alpha_{H_2} \cdots \alpha_{H_p}},$$

so the pairing is well-defined.

Conversely, we need to verify that any $\omega \in \Omega^p(\mathscr{A})$ with $\big\langle \omega, D_p^\sigma(\mathscr{A})\big\rangle = 0$ must be in $\omega_\sigma \wedge \Omega^{p-1}(\mathscr{A})$, and that any $\delta \in D_p(\mathscr{A})$ with $\langle \Omega_0^p(\mathscr{A}), \delta \rangle = 0$ must be in $\chi \wedge D_{p-1}(\mathscr{A})$. Again we show only the first statement, and we can restrict ourselves to local considerations on U: By exactness of $(\Omega^\bullet(\mathscr{A}), \omega_\sigma)$, it suffices to show that $\omega_\sigma \wedge \omega = 0$. From (3.2) and (3.4) we derive that

$$D_{p+1}(\mathscr{A}) = (R\chi \oplus D^\sigma(\mathscr{A})) \wedge \bigwedge^p D^\sigma(\mathscr{A}).$$

As $\Big\langle \omega_\sigma \wedge \omega, \bigwedge^{p+1} D^\sigma(\mathscr{A}) \Big\rangle = 0$ and, for $\delta \in D_p^\sigma(\mathscr{A})$,

$$\langle \omega_\sigma \wedge \omega, \chi \wedge \delta \rangle = |\sigma| \langle \omega, \delta \rangle = 0$$

by hypothesis, we find that $\langle \omega_\sigma \wedge \omega, D_{p+1}(\mathscr{A}) \rangle = 0$. Using (3.4) and local coordinates one shows that $\omega_\sigma \wedge \omega = 0$ and hence that $\omega \in \omega_\sigma \wedge \Omega^{p-1}(\mathscr{A}) = \Omega_0^{p-1}(\mathscr{A})$ using (2.8). The claim follows by $\mathscr{A}$-normality of the latter module. Q.E.D.

From Propositions 2.12 and 3.5 we deduce the following dual version of Propositions 2.12. As in Definition 2.10, we define the sheaf $D_\bullet(\mathbb{P}\mathscr{A})$ on $\mathbb{P}V$ by (3.3) in charts.

Proposition 3.6. *As sheaves on* $\mathbb{P}V$,

$$\widetilde{D_p^0(\mathscr{A})} = D_p(\mathbb{P}\mathscr{A}).$$

3.2. Self-duality of the relative log complex

Let $dx = dx_1 \wedge \cdots \wedge dx_\ell$. Then $\Omega^\ell(\mathscr{A}) = R\frac{1}{f}dx \cong R(n-\ell)$ as a graded R-module. On the other hand, contraction on dx/f gives a graded map

$$D(\mathscr{A}) \to \Omega^{\ell-1}(\mathscr{A})(n-\ell), \quad \delta \mapsto \iota_\delta(dx/f), \tag{3.7}$$

and this can easily be seen to be an isomorphism. Let ν denote the image of the Euler differential:

$$\nu = \iota_\chi(dx/f) = \ell dx/df \in \Omega_\sigma^{\ell-1}(\mathscr{A}). \tag{3.8}$$

By split exactness of $(\Omega^\bullet(\mathscr{A}), \omega_\sigma)$, or by (2.7) and (2.8),

$$\omega_\sigma : \Omega_\sigma^{\ell-1}(\mathscr{A}) \to \Omega^\ell(\mathscr{A}) \tag{3.9}$$

is an isomorphism with inverse ι_χ, and hence $\Omega_\sigma^{\ell-1}(\mathscr{A})$ is a free rank-one R-module generated by ν. For the same reason,

$$\Omega^{\ell-1}(\mathscr{A}) = \omega_\sigma \wedge \Omega_\sigma^{\ell-2}(\mathscr{A}) \oplus \Omega_\sigma^{\ell-1}(\mathscr{A}) \cong \Omega_\sigma^{\ell-2}(\mathscr{A}) \oplus \Omega_\sigma^{\ell-1}(\mathscr{A}). \tag{3.10}$$

For $\delta \in D^\sigma(\mathscr{A})$, $\iota_\delta(\omega_\sigma) = \delta(\alpha^\sigma)/\alpha^\sigma = 0$ and hence

$$\omega_\sigma \wedge \iota_\delta(\Omega^\ell(\mathscr{A})) = -\iota_\delta\left(\omega_\sigma \wedge \Omega^\ell(\mathscr{A})\right) = 0.$$

Thus, $\iota_\delta(\Omega^\ell(\mathscr{A}))$ lies in the first summand of (3.10). By definition of the generator ν of $\Omega_\sigma^{\ell-1}(\mathscr{A})$, $\iota_\chi(\Omega^\ell(\mathscr{A}))$ lies in the second summand of (3.10). Thus, composing (3.9) with ι_δ, the isomorphism (3.7), induces an isomorphism

$$D^\sigma(\mathscr{A}) \to \Omega_\sigma^{\ell-2}(\mathscr{A})(\ell-n), \quad \delta \mapsto \frac{1}{\omega_\sigma}\iota_\delta(\omega_\sigma \wedge \nu) = \iota_\delta(\nu). \tag{3.11}$$

This proves, using Proposition 3.5 for the last part,

Proposition 3.7.

(1) $\Omega_\sigma^\ell(\mathscr{A}) = 0$
(2) $\Omega_\sigma^{\ell-1}(\mathscr{A}) = R\nu \cong R(n-\ell)$
(3) $\Omega_\sigma^{\ell-2}(\mathscr{A}) \cong D^\sigma(\mathscr{A})(n-\ell) \cong (\Omega_\sigma^1(\mathscr{A})(\ell-n))^\vee$

Generalizing this result, consider the non-degenerate pairing

$$-\wedge - : \Omega_{V,f}^p \times \Omega_{V,f}^q \to \Omega_{V,f}^\ell = R_f, \quad p+q=\ell,$$

defined by the exterior product. It induces a non-degenerate pairing

$$-\wedge - : \Omega^p(\mathscr{A}) \times \Omega^q(\mathscr{A}) \to \Omega^\ell(\mathscr{A}) \cong R(n-\ell), \quad p+q=\ell. \tag{3.12}$$

On U this is easy to check in local coordinates using (2.3), then it follows on V by reflexivity. It is immediate from the definition of the pairing that it turns $(\Omega^\bullet(\mathscr{A}), \omega)$ into a self-dual complex for any $\omega \in \Omega^1(\mathscr{A})$.

Proposition 3.8. *The pairing* (3.12) *induces a non-degenerate pairing*

$$(3.13)\quad -\wedge-\colon \Omega_0^p(\mathscr{A})\times\Omega_0^q(\mathscr{A})\to\Omega_0^{\ell-1}(\mathscr{A})\cong R(n-\ell),\quad p+q=\ell-1$$

which makes the complex $(\Omega_0^\bullet(\mathscr{A}),\omega_\lambda)$ *self-dual.*

Proof. Similar to (3.11), (3.13) is induced by (3.12) by identifying $\Omega_0^i(\mathscr{A})=\omega_\sigma\wedge\Omega^i(\mathscr{A})\subset\Omega^{i+1}(\mathscr{A})$ for $i=q,\ell$, and $\Omega_0^p(\mathscr{A})=\Omega_\sigma^p(\mathscr{A})$. Then the pairing is clearly well-defined. We have to show that for $\omega\in\Omega_0^q(\mathscr{A})$ with $\Omega^p(\mathscr{A})\wedge\omega_\sigma\wedge\omega=0$ we have $\omega_\sigma\wedge\omega=0$, or equivalently $\omega\in\omega_\sigma\wedge\Omega^{p-1}(\mathscr{A})$. But this hypothesis implies $\omega_\sigma\wedge\omega=0$ on U and hence on V by $\mathscr{A}$-normality of $\Omega^{q+1}(\mathscr{A})$, so the claim follows. Q.E.D.

§4. Complexes

4.1. The parametric relative log complex

The variety

$$(4.1)\quad \Sigma(\mathscr{A})=\left\{(x,\lambda)\in\big(V\setminus\bigcup_{H\in\mathscr{A}}H\big)\times\mathbb{C}^{\mathscr{A}}\mid\omega_\lambda(x)=0\right\}\subset V\times\mathbb{C}^{\mathscr{A}},$$

was used in [CDFV] to study critical sets of master functions associated with the arrangement $\mathscr{A}$. In [CDFV, Thm 2.9], it was shown that its closure

$$(4.2)\qquad \overline{\Sigma}(\mathscr{A})=\operatorname{Spec}(S/I),$$

where the ideal

$$I=I(\mathscr{A}):=\langle D(\mathscr{A}),\omega_{\mathbf{a}}\rangle\subset S:=R\otimes_{\mathbb{C}}C,\quad C:=\operatorname{Sym}((\mathbb{C}^{\mathscr{A}})^\vee)=\mathbb{C}[\mathbf{a}],$$

was defined in [CDFV, §2.5] using the pairing (3.5). For the moment, we ignore the grading in C and continue to refer only to the $\mathbb{Z}$-grading on R. Note that

$$(4.3)\qquad H^\ell(\Omega_{S/C}^\bullet(\mathscr{A}),\omega_{\mathbf{a}})=(S/I(\mathscr{A}))dx/f\cong(S/I(\mathscr{A}))(\ell-n),$$

where $\Omega_{S/C}^\bullet(\mathscr{A})=\Omega^\bullet(\mathscr{A})\otimes_{\mathbb{C}}C$. In the spirit of this paper, let

$$I_\sigma=I_\sigma(\mathscr{A}):=\langle D^\sigma(\mathscr{A}),\omega_{\mathbf{a}}\rangle\subset S,$$
$$(4.4)\quad I_0=I_0(\mathscr{A}):=\big\langle D^0(\mathscr{A}),\omega_{\mathbf{a}}\big\rangle\subset S_0:=R\otimes_{\mathbb{C}}C_0,\quad C_0:=C/\langle|\mathbf{a}|\rangle.$$

Note that, by (2.6),

$$(4.5)\qquad S/I=S/(I_\sigma+\langle|\mathbf{a}|\rangle)\cong S_0/I_0.$$

To obtain the analogue of (4.3) for I_0 satisfying the parametric version of (2.7), we need to work with parameters in C_0. We use the notations

$$\Omega^\bullet_{S/C,\sigma}(\mathscr{A}) = \Omega^\bullet_\sigma(\mathscr{A}) \otimes_{\mathbb{C}} C, \quad \Omega^\bullet_{S_0/C_0,0}(\mathscr{A}) = \Omega^\bullet_0(\mathscr{A}) \otimes_{\mathbb{C}} C_0.$$

Since $\omega_{\mathbf{a}} \wedge \iota_\delta(\nu) = \iota_\delta(\omega_{\mathbf{a}})\nu$, it follows from (3.11) that

$$H^{\ell-1}(\Omega^\bullet_{S/C,\sigma}(\mathscr{A}), \omega_{\mathbf{a}}) = (S/I_\sigma)\nu \cong (S/I_\sigma)(n-\ell), \tag{4.6}$$

$$H^{\ell-1}(\Omega^\bullet_{S_0/C_0,0}(\mathscr{A}), \omega_{\mathbf{a}}) = (S_0/I_0)\nu \cong (S_0/I_0)(n-\ell). \tag{4.7}$$

Now assume that $\mathscr{A}$ is tame, so that $(\Omega^\bullet_{S/C}(\mathscr{A}), \omega_{\mathbf{a}})$ has cohomology concentrated in degree ℓ by [CDFV, Thm. 3.5]. For the next argument, we need a *second grading* for which the natural map $R \to \Omega^1_V$ has degree -1 and the coefficients $\mathbf{a}$ have degree 1. Degree shifts with respect to this grading are denoted by square brackets $[-]$. While $\omega_{\mathbf{a}}$ is still of degree 0, ω_σ has degree -1 for the second grading. For the remainder of this paragraph, the differential in each complex is given by multiplication by $\omega_{\mathbf{a}}$. Consider now the exact sequence of second-graded complexes,

$$0 \longrightarrow \omega_\sigma \wedge \Omega^{\bullet-1}_{S/C}(\mathscr{A}) \longrightarrow \Omega^\bullet_{S/C}(\mathscr{A}) \xrightarrow{\omega_\sigma} \omega_\sigma \wedge \Omega^\bullet_{S/C}(\mathscr{A})[-1] \longrightarrow 0,$$

and that the second-degrees of $\omega_\sigma \wedge \Omega^\bullet_{S/C}(\mathscr{A})$ have a lower bound, one sees that the cohomology of $\omega_\sigma \wedge \Omega^\bullet_{S/C}(\mathscr{A})$ is concentrated in degree $\ell - 1$. From the exact sequences of S-modules

$$0 \longrightarrow \omega_\sigma \wedge \Omega^{\bullet-1}_{S/C}(\mathscr{A}) \longrightarrow \Omega^\bullet_{S/C}(\mathscr{A}) \longrightarrow \Omega^\bullet_{S/C,\sigma}(\mathscr{A}) \longrightarrow 0,$$

$$0 \longrightarrow \Omega^\bullet_{S/C,\sigma}(\mathscr{A}) \xrightarrow{|\mathbf{a}|} \Omega^\bullet_{S/C,\sigma}(\mathscr{A}) \longrightarrow \Omega^\bullet_{S_0/C_0,0}(\mathscr{A}) \longrightarrow 0,$$

we see $\Omega^\bullet_{S_0/C_0,0}(\mathscr{A})$ has cohomology at most in degrees $\ell-2$ and $\ell-1$. In fact there is no cohomology in degree $\ell - 2$ since $|\mathbf{a}|$ is a non-zero divisor on (4.6). To see this, choose $\sigma = e_1$, so that I_σ has generators independent of a_1. By a linear change of coordinates in C, we may hence replace $|\mathbf{a}|$ by the coordinate a_1, which is then clearly a non-zero divisor on S/I_σ. Together with (4.7), this proves the following:

Proposition 4.1. *For tame $\mathscr{A}$, the complex $(\Omega^\bullet_{S_0/C_0,0}(\mathscr{A}), \omega_{\mathbf{a}})$ resolves $(S_0/I_0(\mathscr{A}))(n-\ell)$.*

In order to give this result a projective interpretation, let $\Gamma := \operatorname{Spec} C_0$, consider the image $\mathbb{P}_\Gamma\mathscr{A}$ of $\mathbb{P}\mathscr{A}$ under the projection

$$\pi_\Gamma \colon \mathbb{P}_\Gamma V := \operatorname{Proj} S_0 \twoheadrightarrow \mathbb{P}V,$$

and use Proposition 2.12 to identify

$$\widetilde{\Omega^p_{S_0/C_0,0}}(\mathscr{A}) = \Omega^p_{\mathbb{P}_\Gamma V/\Gamma}(\mathbb{P}_\Gamma\mathscr{A}) = \pi_\Gamma^*\Omega^p(\mathbb{P}\mathscr{A}).$$

Consider also the preimage $\mathscr{A}_\Gamma$ of $\mathscr{A}$ under affine projection $V_\Gamma := V \times \Gamma \to V$. As in Proposition 2.11, we can now identify $(\widetilde{\Omega^\bullet_{S_0/C_0,0}}(\mathscr{A}), \omega_{\mathbf{a}})$ and $(\Omega^\bullet_{V_\Gamma/\Gamma}(\mathscr{A}_\Gamma^{\{x_i=1\}}), \omega_{\mathbf{a}_{\hat{i}}})$ where $\mathbf{a}_{\hat{i}}$ is obtained by deleting the ith component from $\mathbf{a}$. Setting $C_{\hat{i}} := \mathbb{C}[a_j \mid j \neq i] \cong C_0$ and $\Gamma_{\hat{i}} := \operatorname{Spec} C_{\hat{i}}$, the latter can then be identified with $(\Omega^\bullet_{V_{\Gamma_{\hat{i}}}/\Gamma_{\hat{i}}}(\mathscr{A}_\Gamma^{\{x_i=1\}}), \omega_{\mathbf{a}_{\hat{i}}})$.

Similarly, I_0 can be related to

$$I(\mathbb{P}\mathscr{A}) := \langle D(\mathbb{P}\mathscr{A}), \omega_{\mathbf{a}}\rangle \subset \mathscr{O}_{\mathbb{P}_\Gamma V}. \tag{4.8}$$

Namely, by Proposition 3.6, we have that

$$\widetilde{I_0(\mathscr{A})} = I(\mathbb{P}\mathscr{A}). \tag{4.9}$$

A dual version of (2.9) serves to identify $I_0(\mathscr{A})_{x_i} = I_{e_i}(\mathscr{A})_{x_i}$ and $I(\mathscr{A}^{\{x_i=1\}})$, where the latter is the restriction of the ideal sheaf $I(\mathbb{P}\mathscr{A})$ to the chart $\{x_i = 1\}$. Then we can also relate $\overline{\Sigma}(\mathscr{A})$ to

$$\overline{\Sigma}(\mathbb{P}\mathscr{A}) := \operatorname{Proj}(S_0/I(\mathbb{P}\mathscr{A})).$$

Recall from (4.1) that $\overline{\Sigma}(\mathscr{A})$ is a subvariety of $V \times \mathbb{C}^{\mathscr{A}}$; let $\mathbb{P}\overline{\Sigma}(\mathscr{A})$ denote its projectivization in the first factor. By (4.2), (4.5), and (4.9), $\overline{\Sigma}(\mathbb{P}\mathscr{A}) = \mathbb{P}\overline{\Sigma}(\mathscr{A})$ via the natural inclusion $\mathbb{P}_\Gamma V \hookrightarrow \mathbb{P}V \times \mathbb{C}^{\mathscr{A}}$.

Finally, the projective version of Proposition 4.1 reads

Proposition 4.2. *For tame $\mathscr{A}$, the complex $(\pi_\Gamma^*\Omega^\bullet(\mathbb{P}\mathscr{A}), \omega_{\mathbf{a}})$ resolves $\mathscr{O}_{\overline{\Sigma}(\mathbb{P}\mathscr{A})}(n-\ell)$.*

4.2. The log complex in tame and generic cases

Specializing (4.4) and (4.8) to $\mathbf{a} = \lambda \neq 0$, $|\lambda| = 0$, we set (as in [OT95b, Prop. 2.7]):

$$\begin{aligned} I_{0,\lambda} = I_{0,\lambda}(\mathscr{A}) &:= \langle D^0(\mathscr{A}), \omega_\lambda\rangle \subset R, \\ I_\lambda(\mathbb{P}\mathscr{A}) &:= \langle D(\mathbb{P}\mathscr{A}), \omega_\lambda\rangle \subset \mathscr{O}_{\mathbb{P}V}, \end{aligned}$$

such that (4.9) gives

$$\widetilde{I_{0,\lambda}(\mathscr{A})} = I_\lambda(\mathbb{P}\mathscr{A}). \tag{4.10}$$

We denote by $\overline{\Sigma}_\lambda(\mathscr{A})$ and $\overline{\Sigma}_\lambda(\mathbb{P}\mathscr{A})$ the specializations of $\overline{\Sigma}(\mathscr{A})$ and $\overline{\Sigma}(\mathbb{P}\mathscr{A})$ to $\mathbf{a} = \lambda$ respectively.

Proposition 4.1 has the following application, which improves [CDFV, Prop. 3.9]:

Proposition 4.3. *If $\mathscr{A}$ is tame, then $H^p(\Omega_0^\bullet(\mathscr{A}), \omega_\lambda) \neq 0$ implies that the codimension of $\overline{\Sigma}_\lambda(\mathscr{A})$ is at most p, provided that either $\mathscr{A}$ has rank at most 4, $\mathscr{A}$ is free, or $p \leq 2$.*

Proof. For $\lambda \in \mathbb{C}^{\mathscr{A}}$, let $R_\lambda = S/\langle\{a_i - \lambda_i\}\rangle$. Arguing as in [CDFV, Prop. 3.9], Proposition 4.1 implies the hyper-Ext spectral sequence

$$E_1^{p,q} = \operatorname{Ext}^q_{S_0}(\Omega^{\ell-1-p}_{S_0/C_0,0}(\mathscr{A}), R_\lambda) \cong \operatorname{Ext}^q_R(\Omega_0^{\ell-1-p}(\mathscr{A}), R) \tag{4.11}$$

converges to $\operatorname{Ext}^{p+q}_{S_0}(S_0/I_0(\mathscr{A}), R_\lambda)$, and $E_1^{p,0} \cong \Omega_0^p(\mathscr{A})$ by Proposition 3.8. Since E_2^{p0} receives no nonzero differentials for $p \leq 2$, we see that, if $H^p(\Omega^\bullet_{S_0/C_0,0}(\mathscr{A}), \omega_\lambda)$ is nonzero, then so is $\operatorname{Ext}^p_{S_0}(S_0/I_0(\mathscr{A}), R_\lambda)$ for $p \leq 2$, as well as for $p = 3$ when $\ell - 1 = 3$. This implies that the codimension of $\overline{\Sigma}_\lambda(\mathscr{A})$ is at most p. Q.E.D.

The projective version of this result based on Proposition 4.2 reads:

Proposition 4.4. *If $\mathscr{A}$ is a tame arrangement and $H^p(\pi_\Gamma^*\Omega^\bullet(\mathbb{P}\mathscr{A}), \omega_\lambda)$ is nonzero, then the codimension of $\overline{\Sigma}_\lambda(\mathbb{P}\mathscr{A})$ is at most p, provided that either $\mathscr{A}$ has rank at most 4, $\mathscr{A}$ is free, or $p \leq 2$.*

In [OT95b], Orlik and Terao show that there is a Zariski-open subset $Y \subseteq \mathbb{C}^{\mathscr{A}}$ with the property that, for $\lambda \in Y$, the function (1.1) has non-degenerate, isolated critical points. We will call such λ *generic.* Part of their argument shows that, for a noncentral, affine arrangement $\mathscr{A}$, the cohomology of the complex $(\Omega^\bullet(\mathscr{A}), \omega_\lambda)$ is concentrated in top dimension, for $\lambda \in Y$: see [OT95b, Prop 4.6]. We can prove a projective version of this result, which is a slightly stronger statement, but requires a tameness hypothesis.

Definition 4.5. We call $\mathscr{A}$ *almost tame* if $\operatorname{pd}\Omega^p(\mathscr{A}) \leq p+1$ for $p = 1, \ldots, \ell$.

As in Definition 2.3, the condition is vacuous except for $1 \leq p \leq \ell - 4$. For the sake of stating some results as generally as possible, we have now introduced two new homological boundedness conditions on an arrangement. For the reader's convenience, we summarize their relative strength as follows:

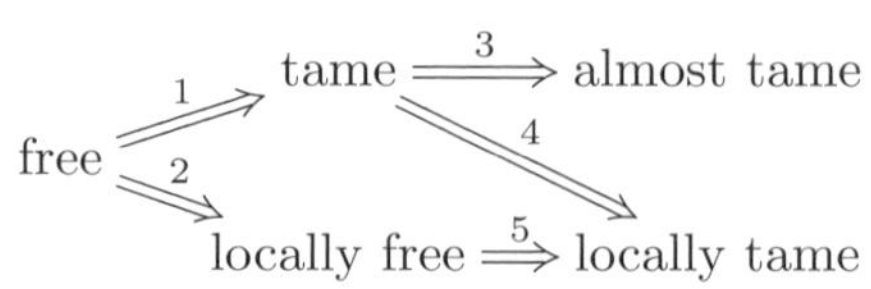

2 3 4

The properties in the column labeled ℓ hold for all arrangements of rank $\leq \ell$. [CDFV, Example 5.3] gives an arrangement of rank 4 which is locally free but not tame, showing implications 2, 3 and 4 cannot be reversed in general. The rank-4 arrangements of Example 5.20 are tame but not locally free, so implications 1 and 5 also cannot be reversed.

Proposition 4.6. *If $\mathscr{A}$ is almost tame and λ is generic (with $|\lambda| = 0$), then*

$$(4.12)\qquad H^p(\Omega_0^\bullet(\mathscr{A}), \omega_\lambda) \cong \begin{cases} 0 & \text{for } p \neq \ell - 1, \\ (R/I_{0,\lambda}(\mathscr{A}))(n-\ell) & \text{for } p = \ell - 1. \end{cases}$$

Proof. Denote by $\hat{\Omega}_0^\bullet(\mathscr{A})$ the complex obtained from $(\Omega_0^\bullet(\mathscr{A}), \omega_\lambda)$ by replacing $\Omega_0^{\ell-1}(\mathscr{A})$ by $\omega_\lambda \wedge \Omega_0^{\ell-2}(\mathscr{A})$. Consider the two hyperhomology spectral sequences of the complex $\hat{\Omega}_0^\bullet(\mathscr{A})$:

$$(4.13)\quad {}^{I}E^2_{p,q} = \operatorname{Ext}_R{}^q(H^p(\hat{\Omega}_0^\bullet(\mathscr{A}), R)), \qquad {}^{II}E^1_{p,q} = \operatorname{Ext}_R^{-q}(\hat{\Omega}_0^p(\mathscr{A}), R).$$

By Proposition 2.11 and [OT95b], it follows that $\hat{\Omega}_0^\bullet(\mathscr{A})$ is exact away from the origin. So ${}^{I}E^2_{p,-q}$ is non-zero only for $q = \ell$ and $p < \ell - 1$, contributing to degrees $p - q < -1$ in the abutment. But by hypothesis and (2.8), ${}^{II}E^1_{p,-q}$ is non-zero only if $q \leq p+1$, contributing to degrees complementary to those of ${}^{I}E^2_{p,q}$. This shows that both sequences converge to zero. But the first sequence degenerates on the E^2-page, and hence ${}^{I}E^2_{p,q} = 0$. This proves the first claim; the second follows by specializing (4.7) to $\mathbf{a} = \lambda$. Q.E.D.

The projective version of this result does not require tameness, as we can apply [OT95b] in charts.

Proposition 4.7. *If λ is generic (with $|\lambda| = 0$), then*

$$(4.14)\qquad H^p(\Omega^\bullet(\mathbb{P}\mathscr{A}), \omega_\lambda) \cong \begin{cases} 0, & p \neq \ell - 1, \\ (\mathscr{O}_{\mathbb{P}V}/I_\lambda(\mathbb{P}\mathscr{A}))(n-\ell), & p = \ell - 1. \end{cases}$$

In special cases, a more detailed understanding is possible. For example, from [CHKS06, Thm. 5] we see that if $\mathscr{A}$ is a generic arrangement, the critical set of the master function (1.1) in $\mathbb{P}V$ is zero-dimensional, for all nonzero λ with $|\lambda| = 0$. From this we note the following:

Proposition 4.8. *If $\mathscr{A}$ is a generic arrangement, then* (4.12) *holds for all $\lambda \neq 0$ (with $|\lambda| = 0$).*

Proof. In [DSS+10], we show for generic $\mathscr{A}$ that $\mathrm{Ext}^q_R(\Omega^p_0(\mathscr{A}), R) = 0$ except for $q = 0$ and $q = p$. Then the E_2-page of the spectral sequence (4.11) is zero except for $p + q = \ell - 1$ and $q > 0$, while $E_2^{p,0} = H^p(\Omega^p_0(\mathscr{A}), \omega_\lambda)$ by Proposition 3.8.

However, by [CHKS06, Thm. 5], the codimension of $\overline{\Sigma}_\lambda(\mathbb{P}\mathscr{A})$, and hence of $\overline{\Sigma}_\lambda(\mathscr{A})$, is $\ell - 1$, so $E_\infty^{p,q} = 0$ for $p + q \leq \ell - 1$. It follows that $E_2^{p,0} = 0$ for $0 \leq p \leq \ell - 2$. Q.E.D.

Passing to coherent sheaves, we obtain the projective analogue of Proposition 4.7 as well.

Corollary 4.9. *If $\mathscr{A}$ is a generic arrangement, then* (4.14) *holds for all $\lambda \neq 0$ (with $|\lambda| = 0$).*

§5. Chern classes

In this section, we prove an analogue of the Borel–Serre formula for sheaves on $\mathbb{P}^d$ with projective resolution of length one (see Theorem 5.7). Then we apply this formula and the theory developed in the preceding sections to prove a generalized Mustaţă–Schenck formula for tame arrangements with zero-dimensional non-free locus (see Theorem 5.13).

5.1. Polynomial identities

We begin with some technical preparations for the following sections. First, we work in the ring $\mathbb{Q}[[u, t]]$. Consider the power series

$$F_\gamma(t, u) = \prod_{i=1}^{n}(1 + ue^{\gamma_i t}) \in \mathbb{Q}[[u, t]], \tag{5.1}$$

with parameters in the ring of symmetric functions in a set of variables γ, for which we refer to [Mac95]. Let $\alpha = \{\alpha_1, \ldots, \alpha_n\}$ and $\beta = \{\beta_1, \ldots, \beta_{n-r}\}$ denote two sets of variables, and let

$$C(t, u) = F_\alpha(t, u)/F_\beta(t, u) \in \mathbb{Q}[[u, t]]. \tag{5.2}$$

Let

$$E_\gamma(t) = \prod_{i=1}^{n}(1 + \gamma_i t) \in \mathbb{Q}[t],$$

$$H_\gamma(t) = \prod_{i=1}^{n}(1 - \gamma_i t)^{-1} \in \mathbb{Q}[[t]]$$

denote the generating function for the elementary symmetric and complete symmetric functions respectively, where the variables are $\gamma = \alpha$ or $\gamma = \beta$.

Denote by $\mathscr{S} \subset \mathbb{Q}[[u,t]]$ the subset of power series in u and t, for which the coefficient of t^k is a polynomial in u of degree at most k. It is easy to see that $\mathscr{S}$ is closed under taking products and multiplicative inverses (whenever defined).

Lemma 5.1. *$\mathscr{F}_\gamma((1+u)t,u)/(1+u)^n \in \mathscr{S}$ for any variables $\gamma = \{\gamma_1,\dots,\gamma_n\}$, and*

$$\left.\frac{F_\gamma((1+u)t,u)}{(1+u)^n}\right|_{u=-1} = E_\gamma(-t).$$

Proof. We expand $F_\gamma((1+u)t,u)$ as a power series in t. Note that, for any i,

$$\begin{aligned} 1+ue^{\gamma_i(1+u)t} &= 1+u+\gamma_i u(1+u)t + u(1+u)^2t^2P_i((1+u)t) \\ &= (1+u)(1+\gamma_i ut + u(1+u)t^2P_i((1+u)t)), \end{aligned}$$

for some power series $P_i(t)$, and hence

$$F_\gamma((1+u)t,u)/(1+u)^n = \prod_{i=1}^n (1+\gamma_i ut + u(u+1)t^2P_i((u+1)t)),$$

from which we see the coefficient of t^k is a polynomial of degree k in u, so we may evaluate to obtain

$$F_\gamma((1+u)t,u)/(1+u)^n|_{u=-1} = E_\gamma(-t),$$

as required. Q.E.D.

Proposition 5.2. *One can write*

$$C(t,u) = \sum_{k\geq 0}(1+u)^{r-k}a_k(u)t^k,$$

where $a_k(u)$ are polynomials in u of degree at most k, such that

$$\sum_{k\geq 0} a_k(-1)t^k = E_\alpha(-t)H_\beta(t). \tag{5.3}$$

Proof. By Lemma 5.1, $F_\alpha((1+u)t,u)/(1+u)^n$ and $F_\beta((1+u)t, u)/(1+u)^{n-r}$ are both in $\mathscr{S}$, so $C((1+u)t,u)/(1+u)^r$ is too. Moreover,

$$\left.\frac{C((1+u)t,u)}{(1+u)^r}\right|_{u=-1} = E_\alpha(-t)/E_\beta(-t) = E_\alpha(-t)H_\beta(t).$$

The claim follows. Q.E.D.

Now we switch to the ring $A[[u]]$, where

$$A := \mathbb{Q}[t]/\langle t^{d+1}\rangle,$$

and consider the image $\overline{C}(t,u)$ in $A[[u]]$ of $C(t,u)$ from (5.2). If $r \geq d$, then, by Proposition 5.2, $\overline{C}(t,u)$ becomes a polynomial in u of degree r, which motivates the following.

Definition 5.3. For $0 \leq p \leq r \geq d$, we call $L^p_{\alpha,\beta}(t) \in \mathbb{Q}[t]/\langle t^{d+1}\rangle$ defined by

$$\sum_{p=0}^{r} L^p_{\alpha,\beta}(t)u^p := \overline{C}(t,u). \tag{5.4}$$

the *pth Lebelt polynomial.*

For any variables γ, let $|\gamma| = \sum_{i=1}^n \gamma_i$.

Lemma 5.4. *For $r = d$,*

$$L^r_{\alpha,\beta} = e^{(|\alpha|-|\beta|)t}, \quad \textit{and}$$

$$L^{r-1}_{\alpha,\beta} = e^{(|\alpha|-|\beta|)t}\left(\sum_{i=1}^{n} e^{-\alpha_i t} - \sum_{i=1}^{n-r} e^{-\beta_i t}\right).$$

Proof. As $\overline{C}(t,u)$ is a polynomial in u of degree r, we may change variables to get

$$L^r_{\alpha,\beta}\overline{C}(t,u) = u^r\overline{C}(t,u^{-1})\big|_{u=0} = \left.\frac{\prod_{i=1}^n (u+e^{\alpha_i t})}{\prod_{i=1}^{n-r}(u+e^{\beta_i t})}\right|_{u=0} = \frac{\prod_{i=1}^n e^{\alpha_i t}}{\prod_{i=1}^{n-r} e^{\beta_i t}},$$

as required. For the second claim,

$$L^{r-1}_{\alpha,\beta} = \frac{d}{du}\left.\frac{\prod_{i=1}^n (u+e^{\alpha_i t})}{\prod_{i=1}^{n-r}(u+e^{\beta_i t})}\right|_{u=0} = e^{(|\alpha|-|\beta|)t}\left(\sum_{i=1}^{n} e^{-\alpha_i t} - \sum_{i=1}^{n-r} e^{-\beta_i t}\right).$$

Q.E.D.

5.2. Lebelt resolutions

We shall now consider a coherent sheaf of rank r on $\mathbb{P}^d$ having a projective resolution of length one,

$$0 \longrightarrow \mathscr{F}_1 \longrightarrow \mathscr{F}_0 \longrightarrow \mathscr{M} \longrightarrow 0. \tag{5.5}$$

For such sheaves we shall prove a Borel–Serre formula in Theorem 5.7. We will compute in the extended rational Chow ring

$$A(\mathbb{P}^d)_{\mathbb{Q}}[[u]] \cong A[[u]].$$

Let α and β denote by the Chern roots of $\mathscr{F}_0$ and $\mathscr{F}_1$.

Proposition 5.5. *Assume that $\mathscr{M}$ is locally a $(k-1)$st syzygy. Then*

$$\mathrm{ch}_t \bigwedge^p \mathscr{M} = L^p_{\alpha,\beta}$$

for $0 \le p \le k$. In particular, this holds vacuously for $p = 1$.

Proof. By the hypothesis on $\mathscr{M}$, we may use the Lebelt resolution [Leb77]

$$0 \to S^p\mathscr{F}_1 \to S^{p-1}\mathscr{F}_1 \otimes \bigwedge^1 \mathscr{F}_0 \to \cdots$$

$$\cdots \to S^1\mathscr{F}_1 \otimes \bigwedge^{p-1} \mathscr{F}_0 \to \bigwedge^p \mathscr{F}_0 \to \bigwedge^p \mathscr{M} \to 0$$

to resolve $\bigwedge^p \mathscr{M}$, for all $p \le k$. Here we are using the fact that local resolutions glue, by the uniqueness of the Lebelt's differential proved in [Leb77, (2.2) Satz]. Then

$$\begin{aligned}
\mathrm{ch}_t(\bigwedge^s \mathscr{M})u^s &= \sum_{p+q=s} (-1)^q \,\mathrm{ch}_t(S^q\mathscr{F}_1)\,\mathrm{ch}_t(\bigwedge^p \mathscr{F}_0)u^2 \\
&= \sum_{p+q=s} (-1)^q \sum_{i_1<\cdots<i_p} ue^{\alpha_{i_1}t}\cdots ue^{\alpha_{i_p}t} \\
&\quad \cdots \sum_{j_1\le\cdots\le j_q} ue^{\beta_{j_1}t}\cdots ue^{\beta_{j_q}t},
\end{aligned}$$

using the splitting principle for symmetric and exterior powers of vector bundles. The claim now follows by expanding the expression (5.2) as a power series in u, and noting that the expression above is the u^s-term, which is $L^s_{\alpha,\beta}$ by Definition 5.3. Q.E.D.

The equation (5.3) gives the Chern polynomial of $\mathscr{M}$:

Lemma 5.6.

$$c_t(\mathscr{M}) = \sum_{k=0}^{d} a_k(-1)(-t)^k.$$

Proof. By (5.5) and the multiplicativity of Chern polynomials,

$$c_t(\mathscr{M}) = c_t(\mathscr{F}_0)/c_t(\mathscr{F}_1) = E_\alpha(t)/E_\beta(t) = E_\alpha(t)H_\beta(-t)$$

and the result follows by (5.3). Q.E.D.

For $r = d$, we can prove an analogue of the Borel–Serre formula for vector bundles, by expressing the top Chern class of $\mathscr{M}$ in terms of Lebelt polynomials.

Theorem 5.7. *For $\mathscr{M}$ of rank $r = d$,*

$$c_r(\mathscr{M}) = (-1)^r \sum_{p=0}^{r} (-1)^p L^p_{\alpha,\beta}.$$

Proof. Since $\overline{C}(t,u)$ is a polynomial in u, we may set $u = -1$ in (5.4) to get

$$\sum_{p=0}^{d} (-1)^p L^p_{\alpha,\beta} = \overline{C}(t,-1) = a_d(-1).$$

Now by Lemma 5.6, our Euler characteristic is obtained as the coefficient of t^d. Q.E.D.

We can mimic another formula for vector bundles: If $\mathscr{M}$ was locally free of rank $r = d$, then we would have an isomorphism $\bigwedge^{r-1}\mathscr{M} \cong \bigwedge^r \mathscr{M} \oplus \mathscr{M}^\vee$, and hence $\mathrm{ch}_t(\bigwedge^r \mathscr{M}) \cdot \mathrm{ch}_{-t}\mathscr{M} = \mathrm{ch}_t(\bigwedge^{r-1}\mathscr{M})$. Lemma 5.4 proves the following analogue of this formula:

Proposition 5.8. *For $\mathscr{M}$ of rank $r = d$,*

$$e^{c_1(\mathscr{M})t} = L^r_{\alpha,\beta},$$
$$e^{c_1(\mathscr{M})t}\,\mathrm{ch}_{-t}(\mathscr{M}) = L^{r-1}_{\alpha,\beta}.$$

5.3. Mustaţă–Schenck formulas

We recall a result of Mustaţă and Schenck [MS01, Thm. 4.1], which we formulate projectively here. Due to a different grading convention, $\Omega^1(\mathscr{A})$ in loc. cit. translates to

$$\Omega^1(\mathscr{A})(1) \cong \Omega^1_0(\mathscr{A})(1) \oplus S(1)$$

and its associated total Chern polynomial in the Chow ring $A(\mathbb{P}^{\ell-1}) \cong \mathbb{Z}[t]/\langle t^\ell \rangle$ to

$$c_t(\Omega^1(\mathscr{A})(1)) = (1+t)c_t(\Omega^1(\mathbb{P}\mathscr{A})(1))$$

using Proposition 2.12 and multiplicativity of c_t. Denote by $\pi(\mathscr{A}, t)$ and $\pi(\mathbb{P}\mathscr{A}, t)$ the Poincaré polynomials of the complements $V \setminus \mathscr{A}$ and $\mathbb{P}V \setminus \mathbb{P}\mathscr{A}$ respectively. By [OT92, Prop. 2.51] (see also [OT92, Prop. 5.1, Thm. 5.90]) they are related by

$$\pi(\mathscr{A}, t) = (1+t)\pi(\mathbb{P}\mathscr{A}, t).$$

Now the projective version of [MS01, Thm. 4.1] reads

Theorem 5.9 ([MS01]). *Let $\mathbb{P}\mathscr{A}$ be a locally free arrangement in $\mathbb{P}V$. Then*

$$c_t(\Omega^1(\mathbb{P}\mathscr{A})(1)) = \pi(\mathbb{P}\mathscr{A}, t) \tag{5.6}$$

in the Chow ring $A(\mathbb{P}V) \cong \mathbb{Z}[t]/\langle t^\ell \rangle$ where $\pi(\mathbb{P}\mathscr{A}, t)$ is the Poincaré polynomial of the complement $\mathbb{P}V \setminus \mathbb{P}\mathscr{A}$, and c_t denotes the total Chern polynomial.

In the following, we generalize this result for locally tame arrangements in $\mathbb{P}^{\ell-1}$ with zero-dimensional non-free locus. Using that $\mathscr{A} \mapsto \Omega^1(\mathscr{A})$ is a local functor, we have

$$\mathscr{E}\mathrm{xt}^p_{\mathscr{O}_{\mathbb{P}V}}(\Omega^1(\mathbb{P}\mathscr{A}), \mathscr{O}_{\mathbb{P}^d}) = \begin{cases} \bigoplus\limits_{X \in L_{\ell-1}(\mathscr{A})} \mathrm{Ext}^1_{R_X}(\Omega^1(\mathscr{A}_X), R_X), & \text{if } p = 1, \\ 0, & \text{if } p \geq 2, \end{cases} \tag{5.7}$$

where we consider $\mathscr{A}_X$ as an arrangement in an affine chart $\mathbb{A}^{\ell-1}$ of $\mathbb{P}V$ with origin X and coordinate ring R_X.

Definition 5.10. For a central arrangement $\mathscr{A}$ in V with zero-dimensional non-free locus, let

$$N(\mathscr{A}) = \mathrm{length}(\mathrm{Ext}^1_R(\Omega^1(\mathscr{A}), R)),$$

a non-negative integer. For an arrangement $\mathbb{P}\mathscr{A}$ in $\mathbb{P}V$ with zero-dimensional non-free locus, let

$$N(\mathbb{P}\mathscr{A}) = h^0(\mathscr{E}\mathrm{xt}^1_{\mathscr{O}_{\mathbb{P}V}}(\Omega^1(\mathbb{P}\mathscr{A}), \mathscr{O}_{\mathbb{P}V})) = \sum_{X \in L_{\ell-1}(\mathscr{A})} N(\mathscr{A}_X), \tag{5.8}$$

using (5.7).

Note that $N(\mathscr{A}) = 0$ if and only if $\mathscr{A}$ is free, and that $N(\mathbb{P}\mathscr{A}) = 0$ if and only if $\mathbb{P}\mathscr{A}$ is locally free.

Recall that a rank-ℓ arrangement $\mathscr{A}$ is called k*-generic* if $\#\mathscr{A}_X = k$ for all $X \in L_k(\mathscr{A})$, and $\mathscr{A}$ is called generic if it is $(\ell - 1)$*-generic.*

By [DSS+10], $N(\mathscr{A}) = \binom{n-1}{\ell}$ for a generic, rank-ℓ arrangement $\mathscr{A}$ of n hyperplanes. This gives the following explicit combinatorial formula:

Theorem 5.11. *If* $\mathbb{P}\mathscr{A}$ *is a* $\ell-2$*-generic arrangement in* $\mathbb{P}^{\ell-1}$*, then*

$$N(\mathbb{P}\mathscr{A}) = \sum_{X \in L_\ell(\mathscr{A})} \binom{\#\mathscr{A}_X - 1}{\ell}.$$

If $\mathscr{A}$ is tame, one can compute $N(\mathscr{A})$ by comparing the Hilbert series of logarithmic forms and differentials:

Proposition 5.12. *For a central arrangement* $\mathscr{A}$ *with a zero-dimensional non-free locus,*

$$N(\mathscr{A}) = \lim_{t \to 1} h(D_1(\mathscr{A}), t) - h(\Omega^1(\mathscr{A}), t^{-1})/(-t)^\ell, \tag{5.9}$$

where $h(M, t)$ *denotes the Hilbert series of a graded module* M*.*

Proof. By the tame hypothesis, $\operatorname{pd} \Omega^1(\mathscr{A}) \leq 1$, so we have a graded free resolution of the form

$$0 \longleftarrow \Omega^1(\mathscr{A}) \longleftarrow F_0 \longleftarrow F_1 \longleftarrow 0.$$

Dualizing gives

$$0 \longrightarrow D_1(\mathscr{A}) \longrightarrow F_0^\vee \longrightarrow F_1^\vee \longrightarrow \operatorname{Ext}^1_R(\Omega^1(\mathscr{A}), R) \longrightarrow 0.$$

Then $h(F^\vee, t) = h(F, t^{-1})/(-t)^\ell$ if F is a free module over a graded ring of dimension ℓ, so

$$\begin{aligned} h(\operatorname{Ext}^1_R(\Omega^1(\mathscr{A}), R), t) &= h(F_1^\vee, t) - h(F_0^\vee, t) + h(D_1(\mathscr{A}), t) \\ &= -h(\Omega^1(\mathscr{A}), t^{-1})/(-t)^\ell + h(D_1(\mathscr{A}), t). \end{aligned}$$

The right-hand side, then, reduces to a polynomial in t, so we may compute $N(\mathscr{A})$ by evaluating at $t = 1$, giving (5.9). Q.E.D.

We can now measure the extent to which the formula of Theorem 5.9 fails for locally tame arrangements with zero-dimensional non-free locus.

Theorem 5.13. *If* $\mathbb{P}\mathscr{A}$ *is a locally tame arrangement in* $\mathbb{P}V$ *with zero-dimensional non-free locus, then*

$$c_t(\Omega^1(\mathbb{P}\mathscr{A})(1)) = \pi(\mathbb{P}\mathscr{A}, t) + N(\mathbb{P}\mathscr{A}) t^{\ell-1}. \tag{5.10}$$

For $\ell = 4$, both hypotheses of Theorem 5.13 are trivially fulfilled by reflexivity of $\Omega^1(\mathbb{P}\mathscr{A})$ and [Har80, Prop. 1.3, Cor. 1.4].

Corollary 5.14. *If $\mathbb{P}\mathscr{A}$ is an arrangement in $\mathbb{P}^3$, then*

$$c_t(\Omega^1(\mathbb{P}\mathscr{A})(1)) = \pi(\mathbb{P}\mathscr{A}, t) + N(\mathbb{P}\mathscr{A})t^3.$$

Corollary 5.15. *If $\mathbb{P}\mathscr{A}$ is an arrangement in $\mathbb{P}^3$, or a locally tame arrangement in $\mathbb{P}V$ with zero-dimensional non-free locus, the Mustaţă–Schenck formula* (5.6) *holds if and only if $\mathscr{A}$ is locally free.*

For an arrangement $\mathbb{P}\mathscr{A}$ in $\mathbb{P}V$, we denote

$$c_t(\Omega^1(\mathbb{P}\mathscr{A})(1)) = 1 + c_1 t + c_2 t^2 + \cdots + c_{\ell-1}t^{\ell-1}, \quad \text{and} \tag{5.11}$$

$$\pi(\mathbb{P}\mathscr{A}, t) = 1 + b_1 t + b_2 t^2 + \cdots + b_{\ell-1}t^{\ell-1}. \tag{5.12}$$

Since $c_1 c_2 \equiv c_3 \mod 2$ for Chern classes of coherent sheaves on $\mathbb{P}^3$ (see, e.g., [Har80, Cor. 2.4]), we observe:

Corollary 5.16. *If $\mathbb{P}\mathscr{A}$ is an arrangement in $\mathbb{P}^3$, then*

$$N(\mathbb{P}\mathscr{A}) \equiv b_1 b_2 + b_3 \mod 2,$$

where the b_1, b_2, and b_3 are the coefficients of the Poincaré polynomial (5.12). *In particular, if $\mathbb{P}\mathscr{A}$ is locally free, then $b_1 b_2 \equiv b_3 \mod 2$.*

The proof will follow some preliminary observations for general ℓ: Let H_0 be a hyperplane in a rank-ℓ arrangement $\mathscr{A}$, and let $\mathscr{A}' = \mathscr{A} \setminus \{H_0\}$. Recall that $\mathscr{A}^{H_0} = \{H \cap H_0 \mid H \in \mathscr{A}'\}$, an arrangement in H_0. The inclusion of $H_0 \subset V$ gives a map $i\colon \mathbb{P}H_0 \to \mathbb{P}V$. By [Zie89, Cor. 4.5], restriction gives an exact sequence:

$$0 \longrightarrow \Omega^p(\mathscr{A})(-1) \xrightarrow{\alpha_{H_0}} \Omega^p(\mathscr{A}') \xrightarrow{i^*} \Omega^p(\mathscr{A}^{H_0}). \tag{5.13}$$

The hyperplane $H_0 \in \mathscr{A}$ is called *generic* if $\operatorname{codim}_{H_0} H_0 \cap X = \operatorname{codim} X$ for all $X \in L_{<\ell}(\mathscr{A}')$. Ziegler notes in [Zie89, Ex. 8.7.(iii)] that i^* need not be surjective, even when H_0 is generic.

However, this subtlety happens at the origin and therefore disappears if one passes to projective space. Note that taking the kernel of ι_χ in (5.13) and then sheafifying yields, using Proposition 2.12, the sequence (5.14) below, and its exactness except at the right.

Proposition 5.17. *Let $\mathbb{P}\mathscr{A}$ be a projective arrangement, and $H_0 \in \mathscr{A}$ generic. Then there are exact sequences of sheaves on $\mathbb{P}V$ and $\mathbb{P}H_0$*

respectively,

(5.14)

$$0 \longrightarrow \Omega^p(\mathbb{P}\mathscr{A})(-1) \xrightarrow{\alpha_{H_0}} \Omega^p(\mathbb{P}\mathscr{A}') \xrightarrow{i^*} i_*\Omega^p(\mathbb{P}\mathscr{A}^{H_0}) \longrightarrow 0,$$

(5.15)

$$0 \longrightarrow \Omega^{p-1}(\mathbb{P}\mathscr{A}^{H_0})(-1) \xrightarrow{d\alpha_{H_0}} i^*\Omega^p(\mathbb{P}\mathscr{A}') \xrightarrow{i^*} \Omega^p(\mathbb{P}\mathscr{A}^{H_0}) \longrightarrow 0,$$

for $0 \leq p \leq \ell - 1$.

Proof. Genericity of H_0 implies that, for any closed point $x \in \mathbb{P}H_0 \cap \mathbb{P}\mathscr{A}'$, $X = \bigcap_{x \in \mathbb{P}H \in \mathbb{P}\mathscr{A}'} \mathbb{P}H$ is at least 1-dimensional, and $\mathbb{P}\mathscr{A}$ is, locally at x, a product of $\mathbb{P}\mathscr{A}'_X$ with the hyperplane $\mathbb{P}H_0$. Thus,

$$\Omega^p(\mathbb{P}\mathscr{A}')_x = \pi^*\Omega^p(\mathbb{P}\mathscr{A}^{H_0})_x \oplus d\alpha_{H_0} \wedge \pi^*\Omega^{p-1}(\mathbb{P}\mathscr{A}^{H_0})_x,$$
$$\Omega^p(\mathbb{P}\mathscr{A})_x = \pi^*\Omega^p(\mathbb{P}\mathscr{A}^{H_0})_x \oplus \frac{d\alpha_{H_0}}{\alpha_{H_0}} \wedge \pi^*\Omega^{p-1}(\mathbb{P}\mathscr{A}^{H_0})_x,$$

where $\pi\colon \mathbb{P}V \setminus \{\alpha_{H_0} = \infty\} \to \mathbb{P}H_0$ is a projection. The claim follows.
Q.E.D.

Proposition 5.18. *Let* $\mathbb{P}\mathscr{A}$ *be a projective arrangement whose non-free locus has codimension* $> k$. *Then the polynomials* (5.11) *and* (5.12) *agree through degree* k, *that is,* $c_i = b_i$ *for* $i = 1, \ldots, k$. *In particular, this holds for* $k = 2$ *without any hypothesis.*

Proof. We argue by induction on ℓ, and we may assume that $k \geq 2$ by reflexivity of $\Omega^1(\mathbb{P}\mathscr{A})$ and [Har77, Cor. 1.4]. If $\ell \leq k + 1$ then $\mathscr{A}$ is locally free, and the result follows by Theorem 5.9. Otherwise, let $H_0 \subset V$ be a linear hyperplane that meets $\mathscr{A}$ generically. Then $L_{\leq \ell-2}(\mathscr{A}) = L_{\leq \ell-2}(\mathscr{A}^{H_0})$, hence

$$\pi(\mathbb{P}\mathscr{A}^{H_0}, t) = 1 + b_1 t + \cdots + b_{l-2} t^{l-2}.$$

We obtain a short exact sequence from (5.15) of Proposition 5.17:

$$0 \longrightarrow \mathscr{O}_{H_0} \longrightarrow i^*\Omega^1(\mathbb{P}\mathscr{A})(1) \longrightarrow \Omega^1(\mathbb{P}\mathscr{A}^{H_0})(1) \longrightarrow 0,$$

from which we see that $c_t(i^*\Omega^1(\mathbb{P}\mathscr{A})(1)) = c_t(\Omega^1(\mathbb{P}\mathscr{A}^{H_0})(1))$.

Now the codimension of the non-free locus in $\mathscr{A}^{H_0}$ is at least as large as that in $\mathscr{A}$, so by induction,

$$c_t(\Omega^1(\mathbb{P}\mathscr{A}^{H_0})(1)) \equiv 1 + b_1 t + \cdots + b_k t^k \mod t^{k+1}.$$

Now let $i\colon \mathbb{P}H_0 \hookrightarrow \mathbb{P}V$ denote the inclusion map; the induced map on Chow rings agrees with the canonical map

$$i^*\colon \mathbb{Z}[t]/t^\ell \to \mathbb{Z}[t]/t^{\ell-1}.$$

Using the functoriality of Chern classes (see, e.g., [Har77, §A.3]) for the third equality, we deduce

$$\begin{aligned} 1+b_1t+\cdots+b_kt^k &\equiv c_t(\Omega^1(\mathbb{P}\mathscr{A}^{H_0})(1)) \\ &= c_t(i^*\Omega^1(\mathbb{P}\mathscr{A})(1)) \\ &= i^*c_t(\Omega^1(\mathbb{P}\mathscr{A})(1)) \\ &\equiv 1+c_1t+\cdots+c_kt^k \mod t^{k+1}. \end{aligned}$$

Q.E.D.

Lemma 5.19. *The Chern character of a reduced point in $\mathbb{P}^d$ is t^d.*

Proof. The point has a Koszul resolution $\bigwedge^\bullet \mathscr{E}$ where $\mathscr{E} = \mathscr{O}_{\mathbb{P}^d}(-1)^d$. By [Ful98, Ex. 3.2.5], its Chern character equals $c_d(\mathscr{E}^\vee)t^d/\operatorname{td}(\mathscr{E}^\vee) = (1-e^{-t})^d = t^d$ in the rational Chow ring $A(\mathbb{P}^d)_\mathbb{Q} \cong \mathbb{Q}[t]/\langle t^{d+1}\rangle$. Q.E.D.

Proof of Theorem 5.13. Proposition 4.7 gives an exact sequence (5.16)

$$0 \longrightarrow \Omega^0(\mathbb{P}\mathscr{A}) \xrightarrow{\omega_\lambda} \Omega^1(\mathbb{P}\mathscr{A}) \xrightarrow{\omega_\lambda} \cdots \xrightarrow{\omega_\lambda} \Omega^{\ell-1}(\mathbb{P}\mathscr{A}) \xrightarrow{\omega_\lambda} \mathscr{F} \longrightarrow 0$$

for generic λ with $|\lambda| = 0$. By [OT95b] and using (4.10),

$$\mathscr{F} = (\mathscr{O}_{\mathbb{P}V}/I_\lambda(\mathbb{P}\mathscr{A}))(n-\ell)$$

is a coherent sheaf supported on points of length equal to the *β-invariant* of $\mathscr{A}$,

$$B = B(\mathscr{A}) := (-1)^{\ell-1}\pi(\mathbb{P}\mathscr{A},-1) = \sum_{i=0}^{\ell-1}(-1)^{\ell-1-i}b_i. \tag{5.17}$$

Using Lemma 5.19, then

$$\operatorname{ch}(\mathscr{F}) = Bt^{\ell-1} \tag{5.18}$$

in $A(\mathbb{P}V)_\mathbb{Q}$.

Recall from Propositions 3.7.(3), 2.12, and 3.6, that

$$\Omega^{\ell-2}(\mathbb{P}\mathscr{A}) \cong \Omega^1(\mathbb{P}\mathscr{A})^\vee(n-\ell).$$

Since $\mathbb{P}\mathscr{A}$ is assumed to be locally tame, $\Omega^1(\mathbb{P}\mathscr{A})$ has a resolution of the form (5.5). Let α and β be the Chern roots of $\mathscr{F}_0$ and $\mathscr{F}_1$, respectively. Using the argument from the proof of Proposition 5.12 (or from the proof of [Har80, Prop. 2.6]), and Lemma 5.19, we compute the Chern character

$$\begin{aligned} \operatorname{ch}_t(\Omega^{\ell-2}(\mathbb{P}\mathscr{A})) &= (\operatorname{ch}_{-t}(\Omega^1(\mathbb{P}\mathscr{A})) + Nt^{\ell-1})e^{(n-\ell)t} \\ &= \operatorname{ch}_{-t}(\Omega^1(\mathbb{P}\mathscr{A}))e^{(n-\ell)t} + Nt^{\ell-1}, \end{aligned} \tag{5.19}$$

where we set $N := N(\mathbb{P}\mathscr{A})$. Similarly $\Omega^{\ell-1}(\mathbb{P}\mathscr{A}) \cong \mathscr{O}_{\mathbb{P}V}(n-\ell)$ gives

$$\operatorname{ch}_t(\Omega^{\ell-1}(\mathbb{P}\mathscr{A})) = e^{(n-\ell)t}. \tag{5.20}$$

By restriction to a generic hyperplane and [Ful98, Rem. 3.2.3.(c)] this gives

$$c_1(\Omega^1(\mathbb{P}\mathscr{A})) = n - \ell. \tag{5.21}$$

Now the two equalities of Proposition 5.8 give

$$L^{\ell-2}_{\alpha,\beta} = \operatorname{ch}_t \Omega^{\ell-2}(\mathbb{P}\mathscr{A}) - Nt^{\ell-1}, \qquad \text{by (5.19), and} \tag{5.22}$$

$$L^{\ell-1}_{\alpha,\beta} = e^{c_1(\Omega^1(\mathbb{P}\mathscr{A}))t} = \operatorname{ch}_t(\Omega^{\ell-1}(\mathbb{P}\mathscr{A})), \qquad \text{by (5.21), (5.20).} \tag{5.23}$$

By Proposition 2.14 and (5.23),

$$\Omega^p(\mathbb{P}\mathscr{A}) = \bigwedge^p \Omega^1(\mathbb{P}\mathscr{A}), \quad p \neq \ell - 2. \tag{5.24}$$

Now we can compute an Euler characteristic of (5.16), using (5.24). Proposition 5.5 together with Theorem 5.7 yields

$$\begin{aligned} \operatorname{ch}_t(\mathscr{F}) &= (-1)^{\ell-1}\sum_{p=0}^{\ell-1}(-1)^p \operatorname{ch}_t(\Omega^p(\mathbb{P}\mathscr{A})) \\ &= (-1)^{\ell-1}\sum_{p=0}^{\ell-1}(-1)^p L^p_{\alpha,\beta} + L^{\ell-2}_{\alpha,\beta} - \operatorname{ch}_t(\Omega^{\ell-2}(\mathbb{P}\mathscr{A})) \\ &= c_{\ell-1}(\Omega^1(\mathbb{P}\mathscr{A}))t^{\ell-1} - Nt^{\ell-1}, \text{ using (5.22).} \end{aligned} \tag{5.25}$$

Now take the coefficient of $t^{\ell-1}$ in (5.18) and (5.25) and use [Ful98, Exa. 3.2.2] to find

$$B = \sum_{i=0}^{\ell-1}(-1)^{\ell-1-i}c_i - N.$$

Finally, use Proposition 5.18 for $k = \ell - 2$ to see $c_i = b_i$ for $0 \le i \le \ell - 2$. It follows from the definition of B in (5.17) that $c_{\ell-1} = b_{\ell-1} + N$, which completes the proof. Q.E.D.

Example 5.20. Let $\mathscr{Z}_1$ and $\mathscr{Z}_2$ be the two arrangements of 9 lines in $\mathbb{P}^2$ introduced in [Zie89, Ex. 8.7] and independently in [Yuz93, Ex. 2.2]. The two arrangements have isomorphic intersection lattices; however, as Schenck notes in [Sch], they are distinguished by the property that the six triple points in $\mathscr{Z}_2$ lie on a conic in $\mathbb{P}^2$, while the triple points in $\mathscr{Z}_1$ do not. By formula (5.6), we see $c_t(\Omega^1(\mathbb{P}\mathscr{Z}_i)(1)) = 1 + 8t + 22t^2$, the Poincaré polynomial, for $i = 1, 2$. On the other hand, the Hilbert series of $\Omega^1_0(\mathscr{A})$ is not combinatorially determined, for arrangements of rank at least 3: $h(\Omega^1_0(\mathscr{Z}_i), t)$ differs for $i = 1, 2$, as noted in [Zie89].

Add a generic hyperplane to each arrangement above, to obtain two combinatorially equivalent arrangements $\mathscr{Z}_1^+$ and $\mathscr{Z}_2^+$ of 10 planes in $\mathbb{P}^3$. Then $\pi(\mathscr{Z}_i^+, t) = (1 + 8t + 22t^2)(1 + t) = 1 + 9t + 30t^2 + 22t^3$, and $\mathscr{Z}_i^+$ has a single non-free closed subarrangement of rank 3, which is $\mathscr{Z}_i$. Computing with Macaulay 2 [GS] shows $N(\mathscr{Z}_1) = 20$, while $N(\mathscr{Z}_2) = 22$. By Theorem 5.13, we find

$$c_t(\Omega^1(\mathbb{P}\mathscr{Z}_i^+)(1)) = \begin{cases} 1 + 9t + 30t^2 + 42t^3 & \text{for } i = 1, \\ 1 + 9t + 30t^2 + 44t^3 & \text{for } i = 2. \end{cases}$$

So we see that, for arrangements of rank at least 4, the Chern polynomial of $\Omega^1(\mathbb{P}\mathscr{A})$ is not combinatorially determined, either.

Acknowledgments. The authors would like to thank Hal Schenck, Uli Walther and Max Wakefield for helpful conversations, as well as the American Institute of Mathematics SQuaREs program for its hospitality while facilitating this collaboration.

References

[Bar77] W. Barth, Some properties of stable rank-2 vector bundles on $\mathbf{P}_n$, Math. Ann., **226** (1977), 125–150, MR0429896 (55#2905).

[CDF+] D. Cohen, G. Denham, M. Falk, H. Schenck, A. Suciu, H. Terao and S. Yuzvinsky, Complex arrangements: algebra, geometry, topology, in preparation.

[CDFV] D. Cohen, G. Denham, M. Falk and A. Varchenko, Critical points and resonance of hyperplane arrangements, Canad. J. Math., **63** (2011), 1038–1057.

[CHKS06] F. Catanese, S. Hoşten, A. Khetan and B. Sturmfels, The maximum likelihood degree, Amer. J. Math., **128** (2006), 671–697, MR2230921 (2007m:13036).

[dGMS09] I. de Gregorio, D. Mond, and C. Sevenheck, Linear free divisors and Frobenius manifolds, Compos. Math., **145** (2009), 1305–1350, MR2551998.

[Dim92] A. Dimca, Singularities and Topology of Hypersurfaces, Universitext, Springer-Verlag, 1992, MR1194180 (94b:32058).

[Dol82] I. Dolgachev, Weighted projective varieties, Group actions and vector fields, Vancouver, B.C., 1981, Lecture Notes in Math., **956**, Springer-Verlag, 1982, pp. 34–71, MR704986 (85g:14060).

[Dol07] I. Dolgachev, Logarithmic sheaves attached to arrangements of hyperplanes, J. Math. Kyoto Univ., **47** (2007), 35–64, MR2359100 (2008h:14018).

[DSS$^+$10] G. Denham, H. Schenck, M. Schulze, M. Wakefield and U. Walther, Local cohomology of logarithmic forms on hyperplane complements, Ann. Inst. Fourier (Grenoble), to appear.

[Ful98] W. Fulton, Intersection Theory, Second ed., Ergeb. Math. Grenzgeb. (3), **2**, Springer-Verlag, 1998, MR1644323 (99d:14003).

[GS] D. Grayson and M. Stillman, Macaulay2—a software system for algebraic geometry and commutative algebra, available at http://www.math.uiuc.edu/Macaulay2.

[Har77] R. Hartshorne, Algebraic Geometry, Grad. Texts in Math., **52**, Springer-Verlag, 1977, MR0463157 (57 #3116).

[Har80] R. Hartshorne, Stable reflexive sheaves, Math. Ann., **254** (1980), 121–176, MR597077 (82b:14011).

[Leb77] K. Lebelt, Freie Auflösungen äusserer Potenzen, Manuscripta Math., **21** (1977), 341–355, MR0450253 (56 #8549).

[Mac95] I. G. Macdonald, Symmetric Functions and Hall Polynomials, Second ed., With contributions by A. Zelevinsky, Oxford Math. Monogr., The Clarendon Press, Oxford Univ. Press, New York, 1995, Oxford Science Publ., MR1354144 (96h:05207).

[MS01] M. Mustaţă and H. Schenck, The module of logarithmic p-forms of a locally free arrangement, J. Algebra, **241** (2001), 699–719, MR1843320 (2002c:32047).

[OT92] P. Orlik and H. Terao, Arrangements of hyperplanes, Grundlehren Math. Wiss., **300**, Springer-Verlag, 1992, MR1217488 (94e:52014).

[OT95a] P. Orlik and H. Terao, Arrangements and Milnor fibers, Math. Ann., **301** (1995), 211–235, MR1314585 (96f:52014).

[OT95b] P. Orlik and H. Terao, The number of critical points of a product of powers of linear functions, Invent. Math., **120** (1995), 1–14, MR1323980 (96a:32062).

[Sai80] K. Saito, Theory of logarithmic differential forms and logarithmic vector fields, J. Fac. Sci. Univ. Tokyo Sect. IA Math., **27** (1980), 265–291, MR586450 (83h:32023).

[Sch] H. Schenck, Hyperplane arrangements: computations and conjectures, In: this volume, pp. 323–358.

[TY95] H. Terao and S. Yuzvinsky, Logarithmic forms on affine arrangements, Nagoya Math. J., **139** (1995), 129–149, MR1355272 (97k:52015).

[Yuz93] S. Yuzvinsky, The first two obstructions to the freeness of arrangements, Trans. Amer. Math. Soc., **335** (1993), 231–244, MR1089421 (93c:52013).

[Zie89] G. M. Ziegler, Combinatorial construction of logarithmic differential forms, Adv. Math., **76** (1989), 116–154, MR1004488 (90j:32016).

Graham Denham
Department of Mathematics, University of Western Ontario,
London, ON N6A 5B7, Canada
E-mail address: gdenham@uwo.ca

Mathias Schulze
Department of Mathematics, Oklahoma State University,
Stillwater, OK 74078, United States
E-mail address: mschulze@math.okstate.edu

Advanced Studies in Pure Mathematics 62, 2012
Arrangements of Hyperplanes—Sapporo 2009
pp. 59–74

Hyperplane arrangements with large average diameter: a computational approach

Antoine Deza, Hiroyuki Miyata, Sonoko Moriyama and Feng Xie

Abstract.

We consider the average diameter of a bounded cell of a simple arrangement defined by n hyperplanes in dimension d. In particular, we investigate the conjecture stating that the average diameter is no more than the dimension d. Previous results in dimensions 2 and 3 suggested that specific extensions of the cyclic arrangement might achieve the largest average diameter. We show that the suggested arrangements do not always achieve the largest diameter and disprove a related conjecture dealing with the minimum number of facets belonging to exactly one bounded cell. In addition, we computationally determine the largest possible average diameter in dimensions 3 and 4 for arrangements defined by no more than 8 hyperplanes via the associated uniform oriented matroids. These new entries substantiate the hypothesis that the largest average diameter is achieved by an arrangement minimizing the number of facets belonging to exactly one bounded cell. The computational framework to generate specific arrangements, and to compute the average diameter and the number of facets belonging to exactly one bounded cell is presented.

§1. Introduction

Let $\mathcal{A}_{d,n}$ be a simple arrangement formed by n hyperplanes in dimension d. We recall that an arrangement is called simple if $n \geq d+1$ and any d hyperplanes intersect at a unique distinct point. The number of bounded cells (closures of the bounded connected components of

Received April 20, 2010.
Revised September 20, 2010.
2010 *Mathematics Subject Classification.* Primary 52C35; Secondary 90C05.
Key words and phrases. Simple hyperplane arrangements, average diameter, Hirsch conjecture, oriented matroids.

the complement) of $\mathcal{A}_{d,n}$ is $I = \binom{n-1}{d}$. Let $\delta(\mathcal{A}_{d,n})$ denote the average diameter of a bounded cell P_i of $\mathcal{A}_{d,n}$; that is,

$$\delta(\mathcal{A}_{d,n}) = \frac{\sum_{i=1}^{i=I} \delta(P_i)}{I}$$

where $\delta(P_i)$ denotes the diameter of P_i, i.e., the smallest number such that any two vertices of P_i can be connected by a path with at most $\delta(P_i)$ edges. Let $\Delta_{\mathcal{A}}(d,n)$ denote the largest possible average diameter of a bounded cell of a simple arrangement defined by n inequalities in dimension d. We consider the following conjecture stating that the average diameter is no more than the dimension d.

Conjecture 1. [**7**] *The average diameter of a bounded cell of a simple arrangement defined by n inequalities in dimension d is not greater than d, i.e., $\Delta_{\mathcal{A}}(d,n) \leq d$.*

1.1. Conjecture 1 as a discrete analogue of a result of Dedieu, Malajovich, and Shub

Conjecture 1 can be regarded as a discrete analogue of a result of Dedieu, Malajovich, and Shub [**6**] on the average total curvature of the central path associated to a bounded cell of a simple arrangement. We first recall the definitions of the central path and of the total curvature. For a polytope, i.e. bounded polyhedron, $P = \{x : Ax \geq b\}$ with $A \in \Re^{n \times d}$, the central path corresponding to $\min\{c^T x : x \in P\}$ is a set of minimizers of $\min\{c^T x + \mu f(x) : x \in P\}$ for $\mu \in (0, \infty)$ where $f(x) = -\sum_{i=1}^{n} \ln(A_i x - b_i)$—the standard logarithmic barrier function [**19**]. Intuitively, the total curvature [**21**] is a measure of how far off a certain curve is from being a straight line. Let $\psi : [\alpha, \beta] \to \Re^d$ be a $C^2((\alpha-\varepsilon, \beta+\varepsilon))$ map for some $\varepsilon > 0$ with a non-zero derivative in $[\alpha, \beta]$. Denote its arc length by $l(t) = \int_\alpha^t \|\dot{\psi}(\tau)\| d\tau$, its parametrization by the arc length by $\psi_{\text{arc}} = \psi \circ l^{-1} : [0, l(\beta)] \to \Re^d$, and its curvature at the point t by $\kappa(t) = \ddot{\psi}_{\text{arc}}(t)$. The total curvature is defined as $\int_0^{l(\beta)} \|\kappa(t)\| dt$. The requirement $\dot{\psi} \neq 0$ insures that any given segment of the curve is traversed only once and allows to define a curvature at any point on the curve. Let $\lambda^c(\mathcal{A}_{d,n})$ denote the average associated total curvature of a bounded cell P_i of a simple arrangement $\mathcal{A}_{d,n}$; that is,

$$\lambda^c(\mathcal{A}_{d,n}) = \sum_{i=1}^{i=I} \frac{\lambda^c(P_i)}{I}$$

where $\lambda^c(P)$ denotes the total curvature of the central path corresponding to the linear optimization problem $\min\{c^T x : x \in P\}$. Dedieu,

Malajovich and Shub [**6**] demonstrated that $\lambda^c(\mathcal{A}_{d,n}) \leq 2\pi d$ for any fixed c. Keeping the linear optimization approach but replacing central path following interior point methods by simplex methods, Haimovich's probabilistic analysis of the shadow-vertex simplex algorithm, see [**3**, Section 0.7], showed that the expected number of pivots is bounded by d. Note that while Dedieu, Malajovich and Shub consider only the bounded cells (the central path may not be defined over some unbounded ones), Haimovich considers the average over bounded and unbounded cells. The result of Haimovich and Conjecture 1 are similar in nature but differ in some aspects as, while the diameter is a lower bound for the number of pivots in the worst case, the number of pivots could be smaller than the diameter for some cells.

Considering the largest $\lambda^c(P)$ over all possible c, we obtain the quantity $\lambda(P)$, referred to as the curvature of a polytope. Following the approach regarding the curvature $\lambda(P)$ as a continuous analogue of the diameter $\delta(P)$, analogues of the results of Holt and Klee [**14**], and Klee and Walkup [**15**] were given in [**7**, **8**]. Namely, a family of polytopes which attain the conjectured order of the largest curvature was given, and it was proved that if the order of the curvature is less than the dimension d for all polytopes defined by $2d$ inequalities and for all d, then the order of the curvature is less than the number of inequalities for all polytopes.

As pointed out by an anonymous referee, it would be interesting to consider the average diameter over all cells, bounded and unbounded. The following property, see [**12**], illustrates the mathematical appeal of this approach. For an oriented matroid of rank $d+1$, $f_k \leq \binom{d}{k} f_d$ where f_k denotes the number of k-faces. Setting $k = d-1$ yields $f_{d-1} \leq d\, f_d$; that is, since any facet belongs to exactly 2 cells, the average number of facets of a cell $\frac{2f_{d-1}}{f_d}$ is no more than $2d$. Note that the inequality holds without the simplicity assumption.

1.2. Conjecture 1 and the conjecture of Hirsch

The conjecture of Hirsch formulated in 1957 and reported in [**5**] states that the diameter of a polyhedron defined by n inequalities in dimension d is not greater than $n-d$. The conjecture holds for $d \leq 3$ and for $n-d \leq 6$, but it had been speculated to be false for large enough n and d, and was recently disproved by Santos [**20**]. It was noticed in [**7**] that Conjecture 1 is nearly implied by the conjecture of Hirsch using the following straightforward argument. If we assume that the diameter of a bounded cell P_i defined by n_i hyperplanes of $\mathcal{A}_{d,n}$ is bounded by $n_i - d$, then:

$$\delta(\mathcal{A}_{d,n}) \leq \frac{\sum_{i=1}^{i=I}(n_i - d)}{I} = \frac{\sum_{i=1}^{i=I} n_i}{I} - d = \frac{2n\binom{n-2}{d-1} - \phi(\mathcal{A}_{d,n})}{I} - d$$

where $\phi(\mathcal{A}_{d,n})$ denotes the number of *external* facets of $\mathcal{A}_{d,n}$; that is, the number of facets belonging to exactly one bounded cell. This inequality yields $\Delta_{\mathcal{A}}(d,n) \leq d + \frac{2d}{n-1}$ as $\phi(\mathcal{A}_{d,n}) \geq 0$, and suggests that arrangements minimizing $\phi(\mathcal{A}_{d,n})$ might be good candidates for achieving a large $\delta(\mathcal{A}_{d,n})$. Let $\Phi_{\mathcal{A}}(d,n)$ denote the minimum number of external facets for any simple arrangement defined by n hyperplanes in dimension d. It was hypothesized in [**7**] that: $\Phi_{\mathcal{A}}(d,n) \geq d\binom{n-2}{d-1}$. While the conjecture of Hirsch turned out to be false, we speculate that the weaker Conjecture 1 could still hold.

We recall previous results concerning Conjecture 1 in Section 2. In Section 3 we consider specific single element extensions of the cyclic arrangements. The computational framework to generate such arrangements, as well as to compute the average diameter and the number of external facets, is presented in Section 4. Finally, in Section 5 we show that, while providing a promising lower bound, the specific single element extensions of the cyclic arrangements do not always achieve the largest diameter and disprove the hypothesized lower bound for the number of external facets. We also substantiate Conjecture 1 by determining new entries for the largest possible average diameter in dimensions 3 and 4 for arrangements defined by no more than 8 hyperplanes. We provide computational evidence that maximizing the average diameter and minimizing the number external facets might be achieved simultaneously.

We refer to the books of Grünbaum [**13**] and Ziegler [**25**] for polytopes and arrangements, to the books of Renegar [**18**] and Roos et al. [**19**] for linear programming and the central path, and to the books of Bokowski [**1**] and Bjöner et al. [**4**] for oriented matroids.

§2. Previous results

Let $\mathcal{A}^*_{d,n}$ denote a simple arrangement combinatorially equivalent to the cyclic hyperplane arrangement which is dual to the cyclic polytope. Proposition 2 recalls that, since the bounded cells of $\mathcal{A}^*_{d,n}$ are mainly combinatorial cubes, the dimension d is an asymptotic lower bound for $\Delta_{\mathcal{A}}(d,n)$ for fixed d. Additional results for small d and $n-d$ are recalled in Proposition 3. See Figure 1 and Figure 2 for an illustration of arrangements maximizing the average diameter for $(d,n) = (2,7)$ and $(3,6)$.

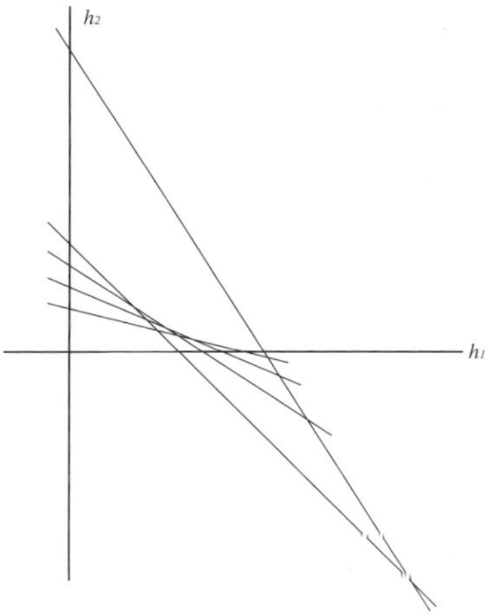

Fig. 1. An arrangement formed by 7 lines maximizing the average diameter

Proposition 2. [9] *For $n \geq 2d$, we have*
$$\Delta_{\mathcal{A}}(d,n) \geq \frac{(d-1)\binom{n-d}{d}+\binom{n-1}{d}+(n-d)(n-d-1)}{\binom{n-1}{d}}.$$

Proposition 3. [9] *We have*

(a) $\Delta_{\mathcal{A}}(d,d+2) = \frac{2d}{d+1}$
(b) $\Delta_{\mathcal{A}}(2,n) = 2 - \frac{2\lceil \frac{n}{2} \rceil}{(n-1)(n-2)}$ *for $n \geq 3$*
(c) $\Delta_{\mathcal{A}}(3,6) = 2$
(d) $3 - \frac{6}{n-1} + \frac{6(\lfloor \frac{n}{2} \rfloor - 2)}{(n-1)(n-2)(n-3)} \leq \Delta_{\mathcal{A}}(3,n) \leq 3 + \frac{4(2n^2-16n+21)}{3(n-1)(n-2)(n-3)}$.

One can easily check that the arrangement resulting from the addition of one hyperplane to $\mathcal{A}^*_{2,n-1}$ such that all the vertices are on one side of the added hyperplane, simultaneously maximizes the average diameter and minimizes the number of external facets. Noticing that the arrangement given in Figure 2 maximizing the average diameter for $(d,n) = (3,6)$ is obtained by adding a hyperplane to $\mathcal{A}^*_{3,5}$ such that all the vertices of $\mathcal{A}^*_{3,5}$ are on one side of the added hyperplane, we investigate extensions of $\mathcal{A}^*_{d,n-1}$ as potential candidates to achieve the largest average diameter $\Delta_{\mathcal{A}}(d,n)$ and minimize the number of external facets $\Phi_{\mathcal{A}}(d,n)$.

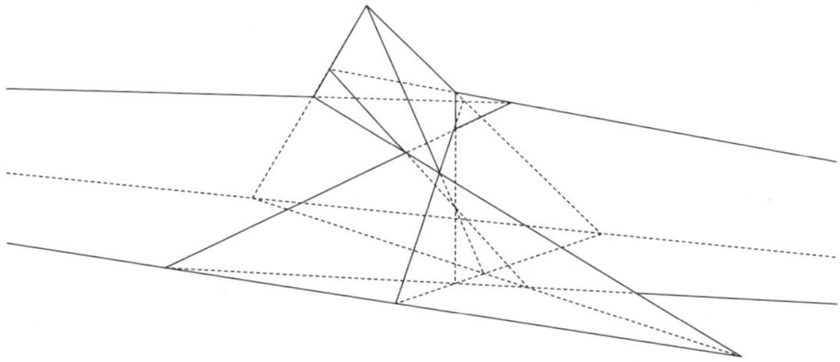

Fig. 2. An arrangement formed by 6 planes maximizing the average diameter

§3. Covering extensions of cyclic arrangements

3.1. Covering extensions of cyclic arrangements

The alternating oriented matroid $\mathcal{M}^*_{d+1,n}$ of rank $d+1$ and with n elements can be realized as a cyclic arrangement $\mathcal{A}^*_{d,n-1}$. Let $(\mathcal{M}^*_{d+1,n}, k)$ be the corresponding affine alternating oriented matroid where k is the infinity element. The combinatorics of the addition of a pseudo-hyperplane to the cyclic hyperplane arrangement, and its relationship with higher Bruhat orders, is studied in detail in [**24**]. Note that since the combinatorial type of an arrangement defined by $d+2$ hyperplanes is unique, all arrangements defined by $d+3$ hyperplanes are extensions of $\mathcal{A}^*_{d,d+2}$. In order to avoid the NP-hard realizability problem [**23**], we focus on single element extensions of $(\mathcal{M}^*_{d+1,n}, k)$ for which all the cocircuits are on one side of the added pseudo-hyperplane. These affine oriented matroids are, by definition, realizable. Let us call *covering extensions* of $\mathcal{A}^*_{d,n-1}$ arrangements whose underlying affine oriented matroid is a single element extension of $(\mathcal{M}^*_{d+1,n}, k)$ for which all the cocircuits are on one side of the added pseudo-hyperplane. In other words, covering extensions of cyclic arrangements are obtained by adding one hyperplane to $\mathcal{A}^*_{d,n-1}$ such that all the vertices are on one side of the added hyperplane. For example, the cyclic arrangement $\mathcal{A}^*_{d,n}$ and the arrangements given in Figures 1 and 2 are covering extensions of a cyclic arrangement.

3.2. Generating covering extensions of cyclic arrangements

One can generate covering extensions of cyclic arrangements via single element extensions of the alternating oriented matroid $\mathcal{M}^*_{d+1,n}$, see [**1**, **4**], and selecting one element as the infinity element.

We recall that the alternating oriented matroid $\mathcal{M}^*_{d+1,n}$ can be represented as an arrangement of n hemispheres on the d-sphere. We generate the covering extensions by adding one hemisphere resulting from a proper perturbation of the infinity hemisphere. Let $\mathcal{C}^*$ be the set of cocircuits of $\mathcal{M}^*_{d+1,n}$. Each cocircuit $X \in \mathcal{C}^*$ corresponds to a vertex on the hemisphere arrangement representing $\mathcal{M}^*_{d+1,n}$. A single element extension of $\mathcal{M}^*_{d+1,n}$ is determined by the cocircuit signature $\sigma : \mathcal{C}^* \to \{+,-,0\}$, see [**4**]. We obtain the cocircuit signature σ_c corresponding to the covering extension via a series of sphere rotations. Let H_k with $1 \leq k \leq n$ be the infinity hemisphere to be perturbed, and H_{n+1} denote the resulting new hemisphere. Regardless of the directions of the rotations, we can keep it small enough so that the vertices which are on the positive, respectively negative, side of H_k are also on the positive, respectively negative, side of H_{n+1}, i.e., we have $\sigma_c(X) = X_k$ if $X_k \neq 0$, where X_k is the kth sign of X. The signature of the remaining cocircuits $\{X \in \mathcal{C}^* : X_k = 0\}$ is determined by the rotations in the following way. Let us choose a pair of antipodal vertices Y and $-Y$ both on the hemisphere H_k, i.e. which have, besides the kth sign, $(d-1)$ zero-signs. Let the ordered index set of zeros in Y, except the one at index k, be $(i_1, i_2, \ldots, i_{d-1})$. We use a sign vector $O \in \{+,-\}^{d-1}$ to represent the rotations, where O_j records the direction of rotation of the d-sphere around the axis defined by the intersection of the hemispheres H_k and H_{i_j} for $j = 1, \ldots, d-1$. The signature of any cocircuit X, except for the antipodal pair Y and $-Y$, is therefore given by:

$$\sigma_c(X) = \begin{cases} X_k, & \text{if } X_k \neq 0; \\ X_{i_j} O_j, & \text{if } X_k = 0,\ j\ : \text{smallest index such that } X_{i_j} \neq 0. \end{cases}$$

Finally, for computational purposes, we set the signature of the antipodal pair Y and $-Y$ by extending the length of the orientation vector O by 1 and use O_d to record the chosen orientation of the antipodal pair. Thus, the signature of any cocircuit X is given by:

$$\sigma_c(X) = \begin{cases} O_d, & \text{if } X = Y; \\ -O_d, & \text{if } X = -Y; \\ X_k, & \text{if } X_k \neq 0; \\ X_{i_j} O_j, & \text{if } X_k = 0,\ j\ : \text{smallest index such that } X_{i_j} \neq 0. \end{cases}$$

See Figure 3 for an illustration of a covering extension of $\mathcal{M}^*_{3,4}$. The covering hemisphere, colored in red, is obtained as a perturbation of hemisphere H_4 corresponding to the choice Y =0++0 and $-Y$ =0--0 as antipodal pair, rotating the sphere around the intersection of hemispheres H_4 and H_1 towards the reader, and then perturbing the antipodal pairs clockwise, that is, $O_d = \{-\}$.

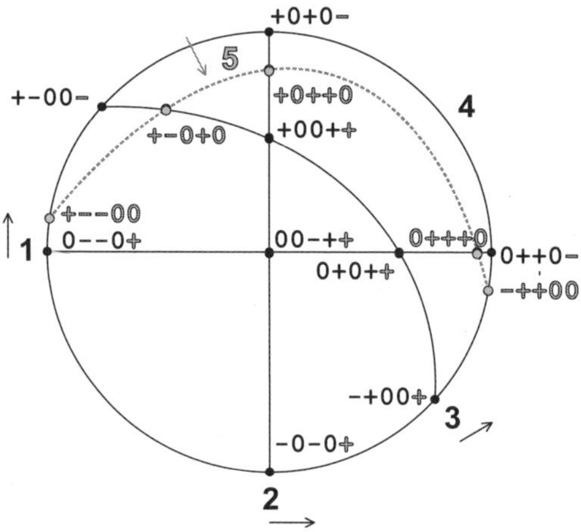

Fig. 3. A covering extension of $\mathcal{M}^*_{3,4}$, the alternating oriented matroid of rank 3 with 4 elements

We have n choices for the infinity hemisphere, $\binom{n-1}{d-1}$ possible pairs of antipodal vertices, $(d-1)!$ different ways of ordering the $(d-1)$ rotations, 2 directions per rotation, and finally 2 choices for the signature of the antipodal pair. Therefore, the total number of potential perturbations is $n\binom{n-1}{d-1}2^{d-1}(d-1)!\,2 = n\binom{n-1}{d-1}2^d(d-1)!$. Note that some affine oriented matroids obtained by the operation described above might belong to the same dissection type [**10**], i.e., the equivalence class of a realizable oriented matroid defined by combinations of relabeling and reorientation that map infinity element to infinity element.

§4. Computational framework

The developed C++ package for computing the average diameter and the number of external facets of an affine oriented matroid can be found at [**22**]. All computations are performed in a combinatorial way which allows us to avoid numerical errors. The code was run on a 32-core server with each core running at a clock rate of 2.3 GHz. Realizability is checked only for the few affine oriented matroids which either maximize the average diameter or minimize the number of external facets. The realizability checking was done on a 64-core server with each core running at a clock rate 2.2 GHz.

4.1. Uniform oriented matroids enumeration

The list of all (uniform) oriented matroids $\mathcal{M}_{d+1,n+1}$ for $d+1 \leq 5$ and $n+1 \leq 8$ as well as for $(d+1, n+1) = (3, 9)$ can be found on the online database [**11**]. The enumeration of all uniform oriented matroids for $(d+1, n+1) = (4, 9)$ and $(5, 9)$ was recently performed by Finschi, Fukuda, and Moriyama and will be uploaded on the online database [**11**] soon.

4.2. Computing the average diameter of a simple arrangement

An affine uniform oriented can be represented as a simple affine pseudo-sphere arrangement. One can easily check if a cell is bounded by verifying that any vertex on the infinity pseudo-sphere is not conformal to the sign vector of the cell. Recall that the sign vector of a cell does not contain any 0. Given the sign vector of a bounded cell, we compute its diameter by constructing its skeleton graph. Two vertices X and Y are adjacent if X can be pivoted to Y, or vice versa, i.e. if there exists a pair of indices (i, j) with $1 \leq i, j \leq n$, $i \neq j$ such that: (i) $X_k = Y_k$ for all $k \neq i, j$, (ii) $X_i = 0$, $Y_i \neq 0$, and (iii) $X_j \neq 0$, $Y_i = 0$. The diameter of the cell is obtained by computing the diameter of the resulting skeleton graph.

4.3. Computing the number of external facets of a simple arrangement

One can check if a facet is bounded by verifying that the sign vector of any vertex at infinity is not conformal to the sign vector of the facet. Recall that the sign vector of a facet contains exactly one 0. A bounded facet is external facet if its sign vector is conformal to the sign vector of some unbounded cell.

4.4. Realizability of uniform oriented matroids

Deciding the realizability of an oriented matroid is known to be NP-hard [**16**] in general but could be tractable for small instances. In particular, following the approach used in [**17**] for $\mathcal{M}_{4,8}$, $\mathcal{M}_{3,9}$, and $\mathcal{M}_{6,9}$, a software was developed to check the realizability for $\mathcal{M}_{4,9}$ and $\mathcal{M}_{5,9}$.

Let $(\{1, \ldots, n+1\}, \chi)$ be a uniform oriented matroid where χ is the associated chirotope, and $(v_1, \ldots, v_{n+1}) \in \mathbb{R}^{(d+1)\times(n+1)}$ the corresponding vector configuration if realizable. The realizability of $(\{1, \ldots, n+$

$1\}, \chi)$ is equivalent to the feasibility of the following polynomial system:

$$\operatorname{sign}(\det(v_{i_1}, \ldots, v_{i_{d+1}})) = \chi(i_1, \ldots, i_{d+1})$$

$$\text{for } 1 \leq i_1 < \cdots < i_{d+1} \leq n+1.$$

This system contains many redundancies which can be exploited to solve it efficiently for small instances. First, since the feasibility of this system is invariant under linear transformations A, with $\det(A) > 0$, of the configuration $(v_1, \ldots, v_{n+1})$ and positive scalar multiplications of $v_1, \ldots v_{n+1}$, we can assume the submatrix $(v_{i_1}, \ldots, v_{i_{d+1}})$ to be an identity matrix for some $(d+1)$-tuple $(i_1, \ldots, i_{d+1})$, and the 1st coordinate of $v_1, \ldots, v_{n+1}$ to be 1 or -1 according to the sign constraints. In addition, square-free variables can be regarded as redundant in the following case:

$$\begin{cases} y < R_i(x_1, \ldots, x_{n+1}) \text{ for } i = 1, \ldots, l \\ y > L_j(x_1, \ldots, x_{n+1}) \text{ for } j = 1, \ldots, m \end{cases}$$

where R_i, L_j are rational functions for $i = 1, \ldots, l, j = 1, \ldots, m$. In this case, we can eliminate the variable y:

$$L_j(x_1, \ldots, x_n) < R_i(x_1, \ldots, x_n) \text{ for } i = 1, \ldots, l, j = 1, \ldots, m.$$

The solvability sequence method applies this rule to special polynomial systems consisting of determinants under the *bipartiteness condition* [**2**]. This elimination rule can be combined with the following *branching rule.*

Proposition 4. *Let $A_1(x_1, \ldots, x_n), \ldots, A_p(x_1, \ldots, x_n), B_1(x_1, \ldots, x_n), \ldots, B_p(x_1, \ldots, x_n)$ be real polynomials. Then the following system*

$$A_k(x_1, \ldots, x_n)y < B_k(x_1, \ldots, x_n) \text{ for } k = 1, \ldots, p$$

is feasible if and only if, for at least one $s : \{1, \ldots, p\} \to \{+, -\}$, the following system is feasible

$$\begin{cases} \operatorname{sign}(A_i(x_1, \ldots, x_n)) = s(i) \text{ for } i = 1, \ldots, p \\ y < \frac{B_i(x_1, \ldots, x_n)}{A_i(x_1, \ldots, x_n)} \text{ for } s(i) = + \\ y > \frac{B_i(x_1, \ldots, x_n)}{A_i(x_1, \ldots, x_n)} \text{ for } s(i) = - \end{cases}$$

We search a square-free variable y, and then apply the branching and elimination rules. If we can eliminate all variables by successive applications of the branching and elimination rules, and obtain consistent inequalities of rationals at some branch of the tree, the oriented matroid is realizable. If we have a polynomial system without a square-free variable at some branch, we try random assignments to the remaining variables. If the random assignments at some branch are successful, the oriented matroid is realizable. Thus, we transform the realizability problem into a tree search problem for which an iterative deepening depth-first search can be used.

§5. Computational results

5.1. Maximal average diameter

For $d \leq 4$ and $n \leq 8$, in order to determine the entries for $\Delta_{\mathcal{A}}(d,n)$, we consider the set $\mathcal{M}_{d+1,n+1}$ of uniform oriented matroids. For each uniform oriented matroid, we consider the $n+1$ choices of setting one element as the infinity element, and compute the average diameter of the resulting affine oriented matroid. Finally, we check the realizability of the oriented matroids maximizing the average diameter. Note that all oriented matroids maximizing the average diameter turned out to be realizable which leads to the following question: *Can an affine non-realizable oriented matroid achieve the maximal average diameter?* The entries for $\Delta_{\mathcal{A}}(d,n)$, including the four new entries for $(d,n) = (3,7), (3,8), (4,7)$ and $(4,8)$, are listed in Table 1. The list of hyperplane arrangements satisfying $\delta(\mathcal{A}_{d,n}) = \Delta_{\mathcal{A}}(d,n)$ can be found in [**22**] where arrangements are represented by the chirotope of its corresponding affine oriented matroid. The signs of the chirotope are ordered reverse lexicographically and the infinity element is always the last one. To avoid redundancy, affine oriented matroids with equivalent dissection types were removed.

(d,n)	$\|\mathcal{M}_{d+1,n+1}\|$	$\Delta_{\mathcal{A}}(d,n)$
(2,5)	4	1.5
(2,6)	11	1.7
(2,7)	135	1.73...
(2,8)	4,382	1.80...
(3,6)	11	2
(3,7)	2,628	2.25
(3,8)	9,276,595	2.42...
(4,7)	135	2.2
(4,8)	9,276,595	2.71...

Table 1. Entries for $\Delta_{\mathcal{A}}(d,n)$ for $d \leq 4$ and $n \leq 8$

One can easily check, by removing a hyperplane and checking if the associated oriented matroid is alternating, if an arrangement satisfying $\delta(\mathcal{A}_{d,n}) = \Delta_{\mathcal{A}}(d,n)$ corresponds to a single element extension of $\mathcal{A}^*_{d,n-1}$. Similarly, one can easily check if an arrangement satisfying $\delta(\mathcal{A}_{d,n}) = \Delta_{\mathcal{A}}(d,n)$ corresponds to a covering extension. As stated in Proposition 5, the computational results disprove the hypothesis that $\Delta_{\mathcal{A}}(d,n)$ is always achieved by an (covering) extension of $\mathcal{A}^*_{d,n-1}$.

Proposition 5. *While a single element extension of $\mathcal{A}^*_{3,6}$ achieves $\Delta_{\mathcal{A}}(3,7)$, none of the arrangements achieving $\Delta_{\mathcal{A}}(3,8)$ and $\Delta_{\mathcal{A}}(4,8)$ is a single element extension of the cyclic arrangement. In addition, the single element extension of $\mathcal{A}^*_{3,6}$ achieving $\Delta_{\mathcal{A}}(3,7)$ is not a covering extension.*

Remark 6. *For line arrangements, $\Delta_{\mathcal{A}}(2,n)$ is always achieved by covering extensions of $\mathcal{A}^*_{2,n-1}$. The largest average diameter $\Delta_{\mathcal{A}}(2,n)$ is achieved only by covering extensions for $n \le 6$, by covering and non-covering extensions as well as non-extensions of $\mathcal{A}^*_{2,6}$ for $n = 7$, and by covering extensions as well as non-extensions of $\mathcal{A}^*_{2,7}$ for $n = 8$.*

While the extensions of $\mathcal{A}^*_{d,n-1}$ fail to always reach $\Delta_{\mathcal{A}}(d,n)$, the covering extensions of $\mathcal{A}^*_{d,n-1}$ could provide a good lower bound. See Table 2 where the value of $\Delta_{\mathcal{A}}(d,n)$ is compared with the maximal average diameter $\Delta^c_{\mathcal{A}}(d,n)$ over all covering extensions of $\mathcal{A}^*_{d,n-1}$ as well as with $\delta(\mathcal{A}^*_{d,n})$, the average diameter of the cyclic arrangement $\mathcal{A}^*_{d,n}$. In particular, we obtain a new lower bound for $\Delta_{\mathcal{A}}(3,9)$. It was showed in [**9**] that $\Delta_{\mathcal{A}}(2,n) = \Delta^c_{\mathcal{A}}(2,n)$ and $\Delta_{\mathcal{A}}(3,6) = \Delta^c_{\mathcal{A}}(3,6)$. The computation shows that $\Delta_{\mathcal{A}}(4,7) = \Delta^c_{\mathcal{A}}(4,7)$, and all arrangements achieving $\Delta_{\mathcal{A}}(3,6)$ and $\Delta_{\mathcal{A}}(4,7)$ are covering arrangements.

(d,n)	$\Delta_{\mathcal{A}}(d,n)$	$\Delta^c_{\mathcal{A}}(d,n)$	$\delta(\mathcal{A}^*_{d,n})$
(3,6)	2	2	1.8
(3,7)	2.25	2.1	2
(3,8)	2.42	2.34...	2.14...
(3,9)	?	2.39...	2.25
(4,7)	2.2	2.2	2
(4,8)	2.71...	2.45...	2.28...

Table 2. Covering extensions of as a lower bound for $\Delta_{\mathcal{A}}(d,n)$

5.2. Minimal number of external facets

While computing entries for $\Delta_{\mathcal{A}}(d,n)$, one can also obtain $\Phi_{\mathcal{A}}(d,n)$, the minimum number of external facets, i.e. facets belonging to exactly one bounced cell, over all arrangements formed by n hyperplanes in dimension d. It was showed in [**9**] that $\Phi_{\mathcal{A}}(2,n) = 2(n-1)$. The computed entries for $\Phi_{\mathcal{A}}(d,n)$ are listed in Table 3. In particular, as stated in Proposition 7, the entry for $(d,n) = (3,8)$ is a counterexample for the hypothesized lower bound. Similarly to Section 5.1, all affine oriented matroids minimizing the number of external facets turned out to be realizable. The list of hyperplane arrangements satisfying $\phi(\mathcal{A}_{d,n}) = \Phi_{\mathcal{A}}(d,n)$ can be found in [**22**].

(d,n)	$\|\|\mathcal{M}_{d+1,n+1}\|\|$	$\Phi_{\mathcal{A}}(d,n)$
(2,5)	4	8
(2,6)	11	10
(2,7)	135	12
(2,8)	4,382	14
(3,6)	11	22
(3,7)	2,628	32
(3,8)	9,276,595	44
(4,7)	135	47
(4,8)	9,276,595	84

Table 3. Entries for $\Phi_{\mathcal{A}}(d,n)$ for $d \leq 4$ and $n \leq 8$

Proposition 7. *The entry for for $(d,n) = (3,8)$ disproves the hypothesized inequality $\Phi_{\mathcal{A}}(d,n) \geq d\binom{n-2}{d-1}$ as we have $\Phi_{\mathcal{A}}(3,8) = 44 < 45 = 3\binom{6}{2}$.*

In addition of providing a promising lower bound for $\Delta_{\mathcal{A}}(d,n)$, the covering extensions of $\mathcal{A}^*_{d,n-1}$ might provide a good upper bound for $\Phi_{\mathcal{A}}(d,n)$. See Table 4 where the value of $\Phi_{\mathcal{A}}(d,n)$ is compared with the minimum number of external facets $\Phi^c_{\mathcal{A}}(d,n)$ over all covering extensions of $\mathcal{A}^*_{d,n-1}$ as well as with $\phi(\mathcal{A}^*_{d,n})$, the number of external facets of the cyclic arrangement $\mathcal{A}^*_{d,n}$, and with the hypothesized lower bound $d\binom{n-2}{d-1}$. In particular, we obtain an upper bound for $\Phi_{\mathcal{A}}(3,9)$. We have $\Phi_{\mathcal{A}}(3,6) = \Phi^c_{\mathcal{A}}(3,6)$ and $\Phi_{\mathcal{A}}(4,7) = \Phi^c_{\mathcal{A}}(4,7)$, and all arrangements achieving $\Phi_{\mathcal{A}}(3,6)$ and $\Phi_{\mathcal{A}}(4,7)$ are covering arrangements.

(d,n)	$\Phi_{\mathcal{A}}(d,n)$	$\Phi^c_{\mathcal{A}}(d,n)$	$\phi(\mathcal{A}^*_{d,n})$	$d\binom{n-2}{d-1}$
(3,6)	22	22	24	18
(3,7)	32	34	40	30
(3,8)	44	48	60	45
(3,9)	?	64	84	63
(4,7)	47	47	50	40
(4,8)	84	88	100	80

Table 4. Covering extensions as an upper bound for $\Phi_{\mathcal{A}}(d,n)$

Remark 8. *While the covering extensions turn out to provide only lower, respectively upper, bound for $\Delta_{\mathcal{A}}(d,n)$, respectively $\Phi_{\mathcal{A}}(d,n)$, the hypothesized relation between maximizing $\Delta_{\mathcal{A}}(d,n)$ and minimizing $\Phi_{\mathcal{A}}(d,n)$ is computationally substantiated by the existence for $d \leq 4$ and $n \leq 8$ of at least one simple arrangement simultaneously maximizing $\Delta_{\mathcal{A}}(d,n)$ and minimizing $\Phi_{\mathcal{A}}(d,n)$.*

The known entries for $\Delta_{\mathcal{A}}(d,n)$ are summarized in Table 5.

	2	3	4	5	6	...	$n-d$	...
2	$\frac{4}{3}$	$\frac{3}{2}$	$\frac{17}{10}$	$\frac{26}{15}$	$\frac{38}{21}$	...	$2-\frac{2\lceil\frac{n}{2}\rceil}{(n-1)(n-2)}$	
3	$\frac{3}{2}$	2	$\frac{9}{4}$	$\frac{17}{7}$	$\geq\frac{67}{28}$			
4	$\frac{8}{5}$	$\frac{11}{5}$	$\frac{19}{7}$					
⋮	⋮							
d	$\frac{2d}{d+1}$							
⋮	⋮							

Table 5. Known entries for $\Delta_{\mathcal{A}}(d,n)$

Acknowledgments. The authors are grateful to the anonymous referees for helpful suggestions. Research supported for the first and last authors by an NSERC Discovery grant, a MITACS grant and the Canada Research Chairs program. The second and third authors were partially supported by KAKENHI.

References

[1] J. G. Bokowski, Computational Oriented Matroids, Cambridge Univ. Press, 2006.

[2] J. G. Bokowski and B. Sturmfels, On the coordinatization of oriented matroids, Discrete Comput. Geom., **1** (1986), 293–306.

[3] K. H. Borgwardt, The Simplex Method, Algorithms Combin. Study Res. Texts, **1**, Springer-Verlag, 1987.

[4] A. Bjöner, M. L. Vergnas, B. Sturmfels, N. White and G. M. Ziegler, Oriented Matroids, Cambridge Univ. Press, 1993.

[5] G. Dantzig, Linear Programming and Extensions, Princeton Univ. Press, 1963.

[6] J.-P. Dedieu, G. Malajovich and M. Shub, On the curvature of the central path of linear programming theory, Found. Comput. Math., **5** (2005), 145–171.

[7] A. Deza, T. Terlaky and Y. Zinchenko, Polytopes and arrangements: diameter and curvature, Oper. Res. Lett., **36** (2008), 215–222.

[8] A. Deza, T. Terlaky and Y. Zinchenko, A continuous d-step conjecture for polytopes, Discrete Comput. Geom., **41** (2009), 318–327.

[9] A. Deza and F. Xie, Hyperplane arrangements with large average diameter, In: Polyhedral Computation, CRM Proc. Lecture Notes, **48**, Amer. Math. Soc., Providence, RI, 2009, pp. 103–114.

[10] L. Finschi and K. Fukuda, Combinatorial generation of small point configurations and hyperplane arrangements, In: Discrete and Computational Geometry, The Goodman–Pollack Festschrift, Algorithms Combin., **25**, Springer-Verlag, 2003, pp. 425–440.

[11] L. Finschi and K. Fukuda, Oriented matroids database, http://www.om.math.ethz.ch.

[12] K. Fukuda, S. Saito and A. Tamura, Combinatorial face enumeration in arrangements and oriented matroids, Discrete Appl. Math., **31** (1991), 141–149.

[13] B. Grünbaum, Convex Polytopes, Grad. Texts in Math., **221**, Springer-Verlag, 2003.

[14] F. Holt and V. Klee, Many polytopes meeting the conjectured Hirsch bound, Discrete Comput. Geom., **20** (1998), 1–17.

[15] V. Klee and D. Walkup, The d-step conjecture for polyhedra of dimension $d < 6$, Acta Math., **133** (1967), 53–78.

[16] M. N. Mnëv, The universality theorems on the classification problem of configuration varieties and convex polytopes varieties, In: Topology and Geometry, Lecture Notes in Math., **1346**, Springer-Verlag, 1988, pp. 527–544.

[17] H. Miyata, S. Moriyama and K. Fukuda, Complete enumeration of small realizable oriented matroids, In: Proceedings of the 22nd Canadian Conference on Computational Geometry, 2010, pp. 143–146.

[18] J. Renegar, A Mathematical View of Interior-Point Methods in Convex Optimization, MPS/SIAM Series on Optimization, 2001.

[19] C. Roos, T. Terlaky and J.-Ph. Vial, Interior Point Methods for Linear Optimization, Springer-Verlag, 2006.
[20] F. Santos, A counter-example to the Hirsch conjecture, Ann. of Math. (2), to appear.
[21] M. Spivak, Comprehensive Introduction to Differential Geometry, Publish or Perish, 1990.
[22] F. Xie, OMC, http://optlab.mcmaster.ca/~feng/omc.
[23] P. Shor, Stretchability of pseudolines is NP-hard, In: Applied Geometry and Discrete Mathematics, DIMACS Ser. Discrete Math. Theoret. Comput. Sci., **4**, 1991, pp. 531–554.
[24] G. M. Ziegler, Higher Bruhat orders and cyclic hyperplane arrangements, Topology, **32** (1993), 259–279.
[25] G. M. Ziegler, Lectures on Polytopes, Grad. Texts in Math., **152**, Springer-Verlag, 1995.

Antoine Deza and Feng Xie
Advanced Optimization Laboratory, Department of Computing and Software, McMaster University, Hamilton, Ontario, L8S 4K1, Canada
E-mail address: deza@mcmaster.ca, xief@mcmaster.ca

Hiroyuki Miyata
Department of Computer Science, The University of Tokyo, 7-3-1 Hongo, Bunkyo-ku, Tokyo, 113-8656, Japan
E-mail address: hmiyata@is.s.u-tokyo.ac.jp

Sonoko Moriyama
Graduate School of Information Sciences, Tohoku University, Aramaki aza Aoba 6-3-09, Aoba-ku, Sendai, 980-8579, Japan
E-mail address: moriso@dais.is.s.tohoku.ac.jp

Advanced Studies in Pure Mathematics 62, 2012
Arrangements of Hyperplanes—Sapporo 2009
pp. 75–107

Vanishing products of one-forms and critical points of master functions

in memory of V. I. Arnol'd

Daniel C. Cohen[1], **Graham Denham**[2], **Michael Falk** and **Alexander Varchenko**[3]

Abstract.

Let $\mathcal{A}$ be an affine hyperplane arrangement in $\mathbb{C}^\ell$ with complement U. Let $f_1, \dots, f_n$ be linear polynomials defining the hyperplanes of $\mathcal{A}$, and $A^\cdot$ the algebra of differential forms generated by the one-forms $d \log f_1, \dots, d \log f_n$. To each $\lambda \in \mathbb{C}^n$ we associate the master function $\Phi = \prod_{i=1}^{n} f_i^{\lambda_i}$ on U and the closed logarithmic one-form $\omega = d \log \Phi$. We assume ω is a general element of a rational linear subspace D of A^1 of dimension $q > 1$ such that the map $\bigwedge^k(D) \to A^k$ given by multiplication in $A^\cdot$ is zero for all $p < k \le q$, and is nonzero for $k = p$. With this assumption, we prove the critical locus $\operatorname{crit}(\Phi)$ of Φ has components of codimension at most p, and these are intersections of level sets of p rational master functions. We give conditions that guarantee $\operatorname{crit}(\Phi)$ is nonempty and every component has codimension equal to p, in terms of syzygies among polynomial master functions.

If $\mathcal{A}$ is p-generic, then D is contained in the degree p resonance variety $\mathcal{R}^p(\mathcal{A})$—in this sense the present work complements previous work on resonance and critical loci of master functions. Any arrangement is 1-generic; we give a precise description of $\operatorname{crit}(\Phi_\lambda)$ in case λ lies in an isotropic subspace D of A^1, using the multinet structure on $\mathcal{A}$ corresponding to $D \subseteq \mathcal{R}^1(\mathcal{A})$. This is carried out in detail for the Hessian arrangement. Finally, for arbitrary p and $\mathcal{A}$, we establish necessary and sufficient conditions for a set of integral one-forms to span such a subspace, in terms of nested sets of $\mathcal{A}$, using tropical implicitization.

Received October 17, 2010.
Revised March 27, 2011.
2010 *Mathematics Subject Classification.* Primary 32S22; Secondary 55N25, 52C35, 14T04.
Key words and phrases. Arrangement of hyperplanes, master function, critical locus, resonance variety, Orlik–Solomon algebra, tropicalization.
[1]Partially supported by National Security Agency grant H98230-05-1-0055.
[2]Partially supported by a grant from NSERC of Canada.
[3]Partially supported by NSF grant DMS-0555327.

§1. Introduction

Let $\mathcal{A} = \{H_1, \ldots, H_n\}$ be an arrangement of distinct affine hyperplanes in $\mathbb{C}^\ell$, with complement $U = \mathbb{C}^\ell - \bigcup_{i=1}^n H_i$. Choose a linear polynomial f_i with zero locus H_i, for each i. Let $\lambda = (\lambda_1 \ldots, \lambda_n) \in \mathbb{C}^n$, and consider the *master function*

$$\Phi_\lambda = \prod_{i=1}^{n} f_i^{\lambda_i}.$$

The multi-valued function Φ_λ has a well-defined critical locus

$$\operatorname{crit}(\Phi_\lambda) = \{x \in U \mid \mathrm{d}\Phi_\lambda(x) = 0\}.$$

Indeed, $\operatorname{crit}(\Phi_\lambda)$ coincides with the zero locus $V(\omega_\lambda)$ of the single-valued closed logarithmic one-form

$$\omega_\lambda = \mathrm{d}\log(\Phi_\lambda) = \sum_{i=1}^{n} \lambda_i \mathrm{d}\log(f_i).$$

In particular, $\operatorname{crit}(\Phi_\lambda)$ is unchanged if λ is multiplied by a non-zero scalar. We are interested in the relation between $\operatorname{crit}(\Phi_\lambda)$ and algebraic properties of the cohomology class represented by ω_λ in $H^1(U, \mathbb{C})$.

For certain arrangements $\mathcal{A}$ and weights λ, the critical points of Φ_λ yield a complete system of eigenfunctions for the commuting hamilitonians of the $\mathfrak{sl}_n(\mathbb{C})$-Gaudin model, via the Bethe Ansatz [23, 25, 17]. That application was the origin of the term master function, introduced in [28]. Much of that theory depends only on combinatorial properties of arrangements, and can be formulated in that general setting—see [29].

Let $A^{\cdot}$ denote the graded $\mathbb{C}$-algebra of holomorphic differential forms on U generated by $\{\mathrm{d}\log(f_i) \mid 1 \le i \le n\}$. By a well-known result of Brieskorn, the inclusion of $A^{\cdot}$ into the de Rham complex of U induces an isomorphism in cohomology, and thus $A^{\cdot} \cong H^{\cdot}(U, \mathbb{C})$, see [1, 4]. In particular $A^1 \cong \mathbb{C}^n$. Since $\omega_\lambda \wedge \omega_\lambda = 0$, left-multiplication by ω_λ makes $A^{\cdot}$ into a cochain complex $(A^{\cdot}, \omega_\lambda)$. For generic λ, $H^p(A^{\cdot}, \omega_\lambda) = 0$ for $p < \ell$, and $\dim H^\ell(A^{\cdot}, \omega_\lambda) = |\chi(U)|$, see [24, 30]. At the same time, for generic λ, Φ_λ has $|\chi(U)|$ isolated, nondegenerate critical points [27, 20, 26]. Here we are concerned with $H^p(A^{\cdot}, \omega_\lambda)$ and $\operatorname{crit}(\Phi_\lambda)$ when λ is not generic.

For $p < \ell$, the ω for which $H^p(A^{\cdot}, \omega) \neq 0$ comprise the pth *resonance variety* $\mathcal{R}^p(\mathcal{A})$ of $\mathcal{A}$, a well-studied invariant of $A^{\cdot}$. On the other hand, precise conditions on λ guaranteeing that Φ_λ has $|\chi(U)|$ isolated critical points are not known. There are also examples where the critical points of Φ_λ are isolated but degenerate.

In some cases $\mathrm{crit}(\Phi_\lambda)$ is positive-dimensional. If $\mathcal{A}$ is a discriminantal arrangement, in the sense of [24], then for certain choices of integral weights λ arising from a simple Lie algebra $\mathfrak{g}$, $\mathrm{crit}(\Phi_\lambda)$ has components of the same positive dimension [25, 16, 18]. In the particular case $\mathfrak{g} = \mathfrak{sl}_2(\mathbb{C})$ of this situation, the codimension of $\mathrm{crit}(\Phi_\lambda)$ is $\ell - 1$. In this case, it was shown in [6] that $\omega_\lambda \in \mathcal{R}^{\ell-1}(\mathcal{A})$ for these λ, with the rank of the skew-symmetric part of $H^{\ell-1}(A^\cdot, \omega_\lambda)$ equal to the number of components of $\mathrm{crit}(\Phi_\lambda)$.

Our work in [5] provides a weak generalization of these results. There we study the *universal critical set*, the set Σ of pairs (x, a) such that $x \in V(\omega_a)$. For fixed λ, $\mathrm{crit}(\Phi_\lambda)$ is the $a = \lambda$ slice of Σ. Let $\overline{\Sigma}$ be the Zariski closure of Σ in $\mathbb{C}^\ell \times \mathbb{C}^n$, and $\overline{\Sigma}_\lambda$ the $a = \lambda$ slice of $\overline{\Sigma}$. In [5] we show, if $\omega_\lambda \in \mathcal{R}^p(\mathcal{A})$, then $\overline{\Sigma}_\lambda$ has codimension at most p, provided $\mathcal{A}$ is tame and either $p \leq 2$ or $\mathcal{A}$ is free. See [5] for definitions of free and tame arrangements; any affine arrangement in $\mathbb{C}^2$ is tame. It is not true in general that $\overline{\Sigma}_\lambda$ is the closure of Σ_λ. Indeed, Σ_λ may be empty under the given hypotheses– that is, $\overline{\Sigma}_\lambda \subseteq \mathbb{C}^\ell \times \mathbb{C}^n$ may lie over $\bigcup_{i=1}^n H_i$.

In this paper, we obtain somewhat more precise information on $\mathrm{crit}(\Phi_\lambda)$, for more general arrangements, but impose a different hypothesis on ω_λ. Namely, we assume that ω_λ has a *decomposable cocycle*, that is, there exists $\psi \in A^p$ such that $\omega_\lambda \wedge \psi = 0$, and ψ is a product of p elements of A^1, whose linear span does not include ω_λ.

We say a subspace D of A^1 is *singular* if the multiplication map $\bigwedge^q(D) \to A^q$ is zero, where $q = \dim D$. Let p be maximal such that $\bigwedge^p(D) \to A^p$ is not the zero map. If $\omega = \mathrm{d}\log(\Phi) \in D - \{0\}$, then ω can be included in a basis of D, and any p-fold product of the other basis elements is a decomposable p-cocycle for ω.

We assume that D is a rational subspace of A^1, that is, D has a basis $\Lambda = \{\omega_{\xi_1}, \ldots, \omega_{\xi_q}\}$ with each ω_{ξ_j} an integer linear combination of $\mathrm{d}\log(f_1), \ldots, \mathrm{d}\log(f_n)$. Associated to Λ is a rational mapping $\Phi_\Lambda = (\Phi_{\xi_1}, \ldots, \Phi_{\xi_q})\colon \mathbb{C}^\ell \rightarrowtail \mathbb{C}^q$, whose image is a quasi-affine subvariety $Y = Y_\Lambda$ of $\mathbb{C}^q$. The dimension of Y is p. If $\omega = \mathrm{d}\log(\Phi) \in D$, then the critical locus $\mathrm{crit}(\Phi)$ is consists of fibers of Φ_Λ and singular points of Φ_Λ. In particular, for generic $\omega \in D$, $\mathrm{crit}(\Phi)$ has codimension at most $\dim(Y) = p$. We obtain more precise conclusions in case the projective closure $\overline{Y}$ is a curve ($p = 1$) or a hypersurface ($p = q - 1$), or $\overline{Y}$ is nonsingular and meets the coordinate hyperplanes transversely. If $\overline{Y}$ is linear, of any codimension, with $Y = \overline{Y} \cap (\mathbb{C}^*)^q$ and Φ_Λ nonsingular, we get a complete description of the $\mathrm{crit}(\Phi)$ for $\mathrm{d}\log(\Phi) \in D$, in terms of critical loci of master functions on the complement of the rank-p arrangement cut out on $\overline{Y}$ by the coordinate hyperplanes.

Every component of $\mathcal{R}^1(\mathcal{A})$ is a rationally-defined and isotropic linear subspace [14], and every element of $\mathcal{R}^1(\mathcal{A})$ has a decomposable cocycle. Moreover, by the theory of multinets and Čeva pencils [10], we can choose Λ so that the variety Y_Λ corresponding to a component of $\mathcal{R}^1(\mathcal{A})$ is linear. We carry out the entire analysis in detail in this case, with special attention to the Hessian arrangement, the one case we know for which Y_Λ is not a hypersurface.

Our approach lends itself to tropicalization, using the main result of [7]. Using the nested set subdivison of the Bergman fan [11], we derive a rank condition for a product of integral one-forms $\omega_{\xi_1} \wedge \cdots \wedge \omega_{\xi_q}$ to vanish. The rank condition can be used in case $\mathcal{A}$ is p-generic to give a combinatorial description of the $(p+1)$-tuples of integral forms in A^1 whose product vanishes, analogous to the description of $\mathcal{R}^1(\mathcal{A})$ in terms of neighborly partitions—see [2].

The outline of this paper is as follows. In Section 2 we introduce Orlik–Solomon algebras and resonance varieties, prove a general result about zero loci of differential forms, and compute critical loci directly for some examples, including the Hessian arrangement. In Section 3 we consider logarithmic one-forms with decomposable p-cocycles satisfying the rationality criterion above, obtaining a precise description of their zero loci, especially in case Φ_Λ is nonsingular and $\overline{Y}_\Lambda$ is a hypersurface meeting the coordinate hyperplanes transversely. We revisit the examples from Section 2. In Section 4 we treat the case where $\overline{Y}_\Lambda$ is linear, returning to the example of the Hessian arrangement. In Section 5 we formulate a test for existence of decomposable cocycles using tropical implicitization.

§2. Resonance, vanishing products, and zeros of one-forms

It will be more convenient for us to consider arrangements of projective hyperplanes in complex projective space $\mathbb{P}^\ell$. Let $[x_0 : \cdots : x_\ell]$ be homogeneous coordinates on $\mathbb{P}^\ell$, and let $\alpha_i \colon \mathbb{C}^{\ell+1} \to \mathbb{C}$ be a nonzero homogeneous linear form, for $0 \leq i \leq n$. Assume without loss that $\alpha_0(x) = x_0$. Let $H_i = \ker(\alpha_i)$, considered as a projective hyperplane in $\mathbb{P}^\ell$, and let $\mathcal{A} = \{H_0, \ldots, H_n\}$. We will denote the corresponding linear hyperplanes in $\mathbb{C}^{\ell+1}$ by cH_i, comprising the central arrangement $c\mathcal{A} = \{cH_0, \ldots, cH_n\}$. Let $U = \mathbb{P}^\ell - \bigcup_{i=0}^n H_i$. We identify $[1 : x_1 : \cdots : x_\ell] \in \mathbb{P}^\ell - H_0$ with $(x_1, \ldots, x_\ell) \in \mathbb{C}^\ell$, and set

$$f_i(x_1, \ldots, x_\ell) = \alpha_i(1, x_1, \ldots, x_\ell)$$

for $1 \leq i \leq n$. Then we recover the affine arrangement $\mathcal{A}$ of the Introduction, with the same complement U. In $\mathbb{P}^\ell$, U is the complement of the singular projective hypersurface defined by

$$Q = \prod_{i=0}^{n} \alpha_i,$$

the (homogeneous) *defining polynomial* of $\mathcal{A}$.

2.1. The projective Orlik–Solomon algebra

Let $\Omega^{\cdot}(U)$ be the complex of holomorphic differential forms on U. The *Orlik–Solomon algebra* of $\mathcal{A}$ is the subalgebra $A^{\cdot}(\mathcal{A})$ of $\Omega^{\cdot}(U)$ generated by $d\log(f_i)$, $1 \leq i \leq n$, as in the Introduction.

We will also study $A^{\cdot}(\mathcal{A})$ in homogeneous coordinates. Let $\omega_i = d\log(\alpha_i)$ for $0 \leq i \leq n$, and let $A^{\cdot}(c\mathcal{A})$ be the algebra of holomorphic forms on $\mathbb{C}^{\ell+1}$ generated by $\omega_0, \dots, \omega_n$. Define $\partial\colon A^{\cdot}(c\mathcal{A}) \to A^{\cdot}(c\mathcal{A})$ by

$$\partial(\omega_{i_1} \wedge \cdots \wedge \omega_{i_k}) = \sum_{j=1}^{k} (-1)^{j-1} \omega_{i_1} \wedge \cdots \wedge \widehat{\omega}_{i_j} \wedge \cdots \wedge \omega_{i_k}$$

and extending linearly. Then ∂ is a graded derivation of degree -1, and

$$\partial(\sum_{i=0}^{n} c_i \omega_i) = \sum_{i=0}^{n} c_i.$$

In general, a holomorphic p-form on $\mathbb{C}^{\ell+1} - \{0\}$ descends to a well-defined form on $\mathbb{P}^{\ell}$ if and only if it is $\mathbb{C}^*$-invariant and its contraction along the Euler vector field $\sum_{i=0}^{n} x_i \frac{\partial}{\partial x_i}$ vanishes, see [8]. This contraction, on $A^{\cdot}(c\mathcal{A})$, is given by ∂, and $A^{\cdot}(c\mathcal{A})$ consists of $\mathbb{C}^*$-invariant forms. Then we may identify $A^{\cdot}(\mathcal{A})$ with the subalgebra $\ker(\partial)$ of $A^{\cdot}(c\mathcal{A})$. This is easily seen to coincide with the subalgebra of $A^{\cdot}(c\mathcal{A})$ generated by $\ker(\partial) \cap A^1(c\mathcal{A})$.

With our choice of coordinates, $d\log(f_i) = \omega_i - \omega_0$ under this identification. $\{\omega_1 - \omega_0, \dots, \omega_n - \omega_0\}$ generates $A^{\cdot}$ by the remark above. Also, $(A^{\cdot}(c\mathcal{A}), \partial)$ is an exact complex, so that $\operatorname{im}(\partial) = \ker(\partial) = A^{\cdot}(\mathcal{A})$, see [19].

There is a well-known presentation of $A^{\cdot}(c\mathcal{A})$ as a quotient of the exterior algebra $E^{\cdot} = \bigwedge^{\cdot}(e_0, \dots, e_n)$. For $C = \{i_1, \dots, i_k\} \subseteq \{0, \dots, n\}$, write $e_C = e_{i_1} \cdots e_{i_k} \in E^k$. Say C is a *circuit* of $c\mathcal{A}$ if C is minimal with the property that

$$\operatorname{codim} \bigcap_{j \in C} H_j < |C|.$$

Then $A^{\cdot}(c\mathcal{A})$ is isomorphic to $E^{\cdot}/I$, where

$$I = \big(\partial e_C \mid C \text{ is a circuit of } c\mathcal{A}\big).$$

2.2. Resonance varieties

Let $\omega = \sum_{i=0}^{n} \lambda_i \omega_i$, where $\lambda = (\lambda_0, \dots, \lambda_n) \in \mathbb{C}^{n+1}$. Assume that $\partial\omega = \sum_{i=0}^{m} \lambda_i = 0$. Since $\mathrm{d}\log(f_i) = \omega_i - \omega_0$, $\omega = \sum_{i=1}^{n} \lambda_i \,\mathrm{d}\log(f_i)$. Then $\omega \in A^1$, and $\omega \wedge \omega = 0$, so we obtain a cochain complex

$$0 \longrightarrow A^0 \xrightarrow{\omega\wedge -} \cdots \xrightarrow{\omega\wedge -} A^p \xrightarrow{\omega\wedge -} \cdots \xrightarrow{\omega\wedge -} A^\ell \longrightarrow 0.$$

Let

$$\begin{aligned} Z^p(\omega) &= \{\psi \in A^p \mid \omega \wedge \psi = 0\}, \\ B^p(\omega) &= \{\psi \in A^p \mid \psi = \omega \wedge \varphi \text{ for some } \varphi \in A^{p-1}\}, \text{ and} \\ H^p(A^{\cdot}, \omega) &= Z^p(\omega)/B^p(\omega). \end{aligned}$$

Then

$$\mathcal{R}^p(\mathcal{A}) = \{\omega \in A^1 \mid H^p(A^{\cdot}, \omega) \neq 0\}$$

is, by definition, the pth *resonance variety* of $\mathcal{A}$.

As observed in the Introduction,

$$\omega = \mathrm{d}\log(\Phi) = \frac{\mathrm{d}\Phi}{\Phi},$$

where $\Phi = \prod_{j=1}^{n} f_j^{\lambda_j}$, and $\mathrm{crit}(\Phi)$ coincides with the zero locus of ω. In homogeneous coordinates, Φ is given by $\prod_{j=0}^{n} \alpha_j^{\lambda_j}$.

2.3. Zeros of forms

We start with an elementary observation about products and zeros of differential forms. If $\psi \in \Omega^k(U)$, for some k, $0 \le k \le \ell$, let

$$V(\psi) = \{x \in U \mid \psi(x) = 0\},$$

a quasi-affine subvariety of $\mathbb{C}^\ell$. Let $U(\psi) = U - V(\psi)$.

Proposition 2.1. *Suppose $\omega \in \Omega^1(U)$ and $\psi \in \Omega^p(U)$ satisfy $\omega \wedge \psi = 0$. Then every component of $V(\omega) - V(\psi)$ has codimension less than or equal to p.*

Proof. We may write $\omega = \sum_{i=1}^{\ell} b_i \mathrm{d}x_i$ for some holomorphic functions $b_1, \dots b_\ell$ on U. Then

$$V(\omega) = \bigcap_{i=1}^{\ell} V(b_i).$$

Similarly, $\psi = \sum_I A_I dx_I$ for some holomorphic functions A_I, where I ranges over all subsets $I = \{i_1, \dots, i_p\}_<$ of $\{1, 2, \dots, \ell\}$, and $dx_I = dx_{i_1} \wedge \cdots \wedge dx_{i_p}$. (The subscript "$<$" is meant to indicate that the elements of I are listed in increasing order.) Set $U_I = U(A_I)$, and let S_I denote the coordinate ring of U_I, i.e., S_I is $\mathbb{C}[x_1, \dots, x_\ell]$, localized at A_I. Then

$$U(\psi) = \bigcup_I U_I.$$

The equation $\omega \wedge \psi = 0$ says, for each subset J of $\{1, \dots, \ell\}$ of size $p+1$,

$$\sum_{i \in J} \sigma(i, J) b_i A_{J - \{i\}} = 0. \tag{2.1}$$

Here $\sigma(i, J) = \pm 1$ depending on the position of i in J.

We have

$$V(\omega) \cap U(\psi) = \bigcup_I V(\omega) \cap U_I.$$

Fix $I = \{i_1, \cdots, i_p\}_<$. For each $i \notin I$, set $J = I \cup \{i\}$ in equation (2.1). Since $A_I \neq 0$ on U_I, one can solve for b_i in terms of $b_{i_1}, \dots, b_{i_p}$. This means b_i lies in the ideal $(b_{i_1}, \dots, b_{i_p})$ of S_I. Since this holds for every $i \notin I$, the defining ideal of $V(\omega) \cap U_I$ in S_I is contained in $(b_{i_1}, \dots, b_{i_p})$. Then each irreducible component of $V(\omega) \cap U_I$ has codimension less than or equal to the codimension of $(b_{i_1}, \dots, b_{i_p})$, which is at most p. Since the U_I cover $U(\psi)$, the result follows. Q.E.D.

Corollary 2.2. *If*

$$\bigcap \{V(\psi) \mid \psi \in \Omega^p(U), \omega \wedge \psi = 0\} = \emptyset,$$

then every component $V(\omega)$ has codimension less than or equal to p.

Corollary 2.3. *Suppose X is a component of $V(\omega)$ of codimension c. If ψ is a p-form satisfying $\omega \wedge \psi = 0$ and $p < c$, then $X \subseteq V(\psi)$.*

Remark 2.4. The preceding results go through without change for any smooth complex analytic variety U, interpreting $x_1, \dots, x_\ell$ as local holomorphic coordinates on U.

A p-form ψ satisfying $\omega \wedge \psi = 0$ will be called a *p-cocycle* for ω. We say ψ is *trivial* if $\psi = \omega \wedge \varphi$ for some $\varphi \in \Omega^{p-1}(U)$. If ψ is a trivial cocycle for ω, then $V(\omega) \subseteq V(\psi)$.

The trivial cocycle condition $\psi = \omega \wedge \varphi$ is generally difficult to characterize. We propose the following conjecture, the converse to the observation above.

Conjecture 2.5. *If $\psi \in A^p$, then $\psi = \omega \wedge \varphi$ for some $\varphi \in A^{p-1}$ if and only if $\omega \wedge \psi = 0$ and $V(\omega) \subseteq V(\psi)$.*

The conjecture is not hard to prove directly in case $p = 1$, and the statement for any p follows from results of [29] if $\omega \in A^1$ is generic.

2.4. Examples

Our first example is linearly equivalent to the rank-three braid arrangement.

Example 2.6. Let $\mathcal{A} = \{H_0, \ldots, H_5\}$ be the arrangement with defining polynomial

$$Q = xyz(x-y)(x-z)(y-z),$$

with the hyperplanes labelled according to the order of factors in Q.

For $a, b, c \in \mathbb{C}$, not all zero, with $a + b + c = 0$, let

$$\Phi_{abc} = [x(y-z)]^a [y(x-z)]^b [z(x-y)]^c.$$

Then $\omega_{abc} := \mathrm{d}\log(\Phi_{abc}) = a(\omega_0 + \omega_5) + b(\omega_1 + \omega_4) + c(\omega_2 + \omega_3)$.

One computes

$$\begin{aligned} \omega_{abc} &= [\mathrm{d}x\,\mathrm{d}y\,\mathrm{d}z] \begin{bmatrix} b_1 \\ b_2 \\ b_3 \end{bmatrix} \\ &= [\mathrm{d}x\,\mathrm{d}y\,\mathrm{d}z] \begin{bmatrix} 1/x & 1/(x-z) & 1/(x-y) \\ 1/(y-z) & 1/y & 1/(y-x) \\ 1/(z-y) & 1/(z-x) & 1/z \end{bmatrix} \begin{bmatrix} a \\ b \\ c \end{bmatrix}. \end{aligned}$$

The zero locus $V(\omega_{abc})$ is given by the vanishing of b_1, b_2, and b_3. The kernel of the matrix is spanned by $(x(y-z), y(z-x), z(x-y))$, so $[x : y : z] \in V(\omega_{abc})$ if and only if $[x(y-z) : y(z-x) : z(x-y)] = [a : b : c]$. Since $a + b + c = 0$, this is equivalent to $[x(y-z) : y(z-x)] = [a : b]$, i.e.,

$$\frac{x(y-z)}{y(z-x)} = \frac{a}{b}$$

or, more symmetrically,

$$\frac{a}{x} + \frac{b}{y} + \frac{c}{z} = 0.$$

If any of a, b, or c are zero, then $\mathrm{crit}(\Phi_{abc}) = V(\omega_{abc})$ is empty. Otherwise $\mathrm{crit}(\Phi_{abc})$ has codimension 1. Moreover, $\mathrm{crit}(\Phi_\lambda)$ is a level set of the master function $\frac{x(y-z)}{y(z-x)}$.

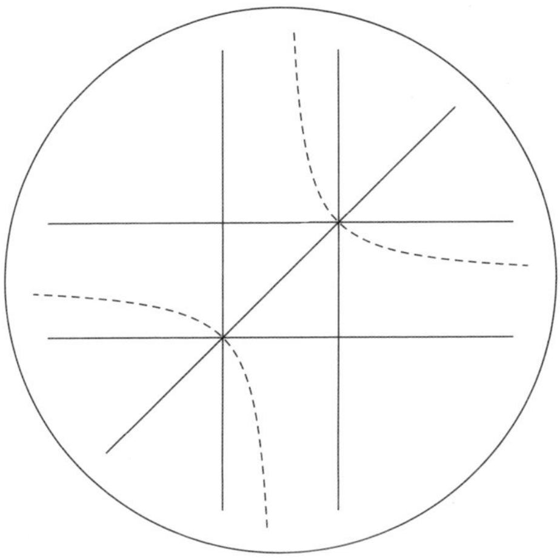

Fig. 1. The braid arrangement with crit($\Phi_{1,1,-2}$)

Let $D = \{\omega_{abc} \mid a+b+c=0\}$. Then $D \subseteq A^1$ and, for any $\omega \in D$, $Z^1(\omega) = D$. Thus $\psi \in Z^1(\omega_{abc})$ if and only if $\psi = d\log \Phi_{a'b'c'}$ with $a'+b'+c'=0$. Then ψ has zero locus given by

$$\frac{a'}{x} + \frac{b'}{y} + \frac{c'}{z} = 0,$$

which one can see is disjoint from $V(\omega_{abc})$ if and only if $\psi \notin B^1(\omega_{abc})$. The Zariski closure of every nonempty critical set contains the four points $[0:0:1]$, $[1:1:1]$, $[1:0:0]$, and $[0:1:0]$—see Figure 1.

Here is a rank-four example.

Example 2.7. Let $\mathcal{A} = \{H_0, \dots, H_7\}$ be the arrangement of eight planes in $\mathbb{P}^3$ with defining polynomial

$$Q = xyzw(x+y+z)(x+y+w)(x+z+w)(y+z+w).$$

The dual point configuration consists of the four vertices and four face-centers of the 3-simplex. Fix $a, b, c, d \in \mathbb{C}$ and let

$$\Phi = \left(\frac{x}{y+z+w}\right)^a \left(\frac{y}{x+z+w}\right)^b \left(\frac{z}{x+y+w}\right)^c \left(\frac{w}{x+y+z}\right)^d.$$

The one-form $\omega = d\log \Phi$ belongs to a 4-dimensional component of $\mathcal{R}^2(\mathcal{A})$. If none of a, b, c, d are zero, then $H^1(A, \omega) = 0$ and $H^2(A, \omega) \cong \mathbb{C}$. A nontrivial 2-cocycle for ω is given by

$$\psi = b \cdot \partial(\omega_{167}) + c \cdot \partial(\omega_{257}) + d \cdot \partial(\omega_{347}).$$

One sees that ω is equal to the product

$$[\mathrm{d}x\,\mathrm{d}y\,\mathrm{d}z\,\mathrm{d}w]\begin{bmatrix} \frac{1}{x} & \frac{-1}{x+z+w} & \frac{-1}{x+y+w} & \frac{-1}{x+y+z} \\ \frac{-1}{y+z+w} & \frac{1}{y} & \frac{-1}{x+y+w} & \frac{-1}{x+y+z} \\ \frac{-1}{y+z+w} & \frac{-1}{x+z+w} & \frac{1}{z} & \frac{-1}{x+y+z} \\ \frac{-1}{y+z+w} & \frac{-1}{x+z+w} & \frac{-1}{x+y+w} & \frac{1}{w} \end{bmatrix}\begin{bmatrix} a \\ b \\ c \\ d \end{bmatrix}.$$

Computing the kernel of the matrix, we see that $[x:y:z:w] \in V(\omega)$ if and only if the vector $(x(y+z+w), y(x+z+w), z(x+y+w), w(x+y+z))$ is proportional to (a, b, c, d), giving the equations

$$\frac{x(y+z+w)}{w(x+y+z)} = \frac{a}{d}, \quad \frac{y(x+z+w)}{w(x+y+z)} = \frac{b}{d}, \quad \frac{z(x+y+w)}{w(x+y+z)} = \frac{c}{d}.$$

$V(\omega)$ has a component of codimension two, given by

$$x+y+z+w=0, \quad \frac{a}{x}+\frac{b}{y}+\frac{c}{z}+\frac{d}{w}=0.$$

For general a, b, c, d, the remaining components of $V(\omega)$ consist of four additional points in $\mathbb{P}^3$. It follows from Corollary 2.3 that the cocycle ψ vanishes on the isolated points of $V(\omega)$. This can also be verified here by direct computation.

Example 2.8. The Hessian arrangement consists of the 12 lines through the inflection points of a nonsingular cubic in $\mathbb{P}^2$. It is the only known arrangement of rank greater than two that supports a global component of $\mathcal{R}^1(\mathcal{A})$ of dimension greater than two. That is, there is an element $\omega \in A^1$ which has poles along every hyperplane of $\mathcal{A}$, and satisfies $\dim H^1(A^{\cdot}, \omega) > 1$.

Any nonsingular cubic is equivalent to

$$x^3 + y^3 + z^3 - 3txyz = 0 \tag{2.2}$$

up to projective transformation, for some $t \in \mathbb{C}$. These cubics have the same inflection points. Then, up to projective transformation, the Hessian arrangement $\mathcal{A}$ is defined by

$$\begin{aligned} Q = {} & xyz(x+y+z)(x+y+\zeta z)(x+y+\zeta^2 z)(x+\zeta y+z)(x+\zeta y+\zeta z) \\ & \cdot (x+\zeta y+\zeta^2 z)(x+\zeta^2 y+z)(x+\zeta^2 y+\zeta z)(x+\zeta^2 y+\zeta^2 z), \end{aligned}$$

where $\zeta = e^{\frac{2\pi i}{3}}$. The 12 lines of $\mathcal{A}$ are the irreducible components of the four singular cubics in the family 2.2, corresponding to $t = \infty, 1, \zeta$, and ζ^2. See [19, Example 6.30].

Numbering the hyperplanes in order, these singular cubics are given by

$$\begin{aligned} P_0 &= \alpha_0\alpha_1\alpha_2 &&= xyz, \\ P_1 &= \alpha_3\alpha_8\alpha_{10} &&= x^3+y^3+z^3-3xyz, \\ P_2 &= \alpha_4\alpha_6\alpha_{11} &&= x^3+y^3+z^3-3\zeta xyz, \text{ and} \\ P_3 &= \alpha_5\alpha_7\alpha_9 &&= x^3+y^3+z^3-3\zeta^2 xyz. \end{aligned}$$

For $(a_1,a_2,a_3)\in\mathbb{C}^3$ let $\omega=\omega_{a_1a_2a_3}=\mathrm{d}\log(\Phi)$, where

$$\Phi=\Phi_{a_1a_2a_3}=\left(\frac{P_1}{P_0}\right)^{a_1}\left(\frac{P_2}{P_0}\right)^{a_2}\left(\frac{P_3}{P_0}\right)^{a_3}.$$

Let $D\subseteq A^1$ be the space of all such forms. Then $H^1(A^\cdot,\omega)\cong D/\mathbb{C}\omega$ has dimension two.

As in the previous examples, we write

$$\omega=\left[\mathrm{d}x,\mathrm{d}y,\mathrm{d}z\right]M\begin{bmatrix}a_1\\a_2\\a_3\end{bmatrix},$$

where M is a 3×3 matrix of rational functions, the Jacobian of

$$(\log(P_1/P_0),\log(P_2/P_0),\log(P_3/P_0)).$$

Then $\omega(x)=0$ for $x\in U$ if and only if $a=(a_1,a_2,a_3)$ lies in the kernel of $M(x)$.

The matrix M has rank 1. In fact one finds that $M=vw^T$ where

$$v=\begin{bmatrix}\frac{2x^3-y^3-z^3}{x}\\ \frac{x^3-2y^3+z^3}{y}\\ \frac{x^3+y^3-2z^3}{z}\end{bmatrix}\quad\text{and}\quad w=\begin{bmatrix}\frac{1}{P_1}\\ \frac{1}{P_2}\\ \frac{1}{P_3}\end{bmatrix}.$$

The critical equation becomes $vw^Ta=0$, which is satisfied if and only if $w^Ta=0$ or all components of v vanish. The latter occurs at the points given by $x^3=y^3=z^3$, which are the common inflection points of the cubics (2.2). In particular those points do not lie in the complement U. Thus $\mathrm{crit}(\Phi)$ is defined by the single equation

$$\frac{a_1}{P_1}+\frac{a_2}{P_2}+\frac{a_3}{P_3}=0. \tag{2.3}$$

Then crit(Φ) is empty or has codimension one and degree six. Working in the torus $xyz \neq 0$, set

$$T = \frac{x^3 + y^3 + z^3}{xyz}.$$

Then equation (2.3) is equivalent to

$$\frac{a_1}{T-3} + \frac{a_2}{T-3\zeta} + \frac{a_3}{T-3\zeta^2} = 0, \tag{2.4}$$

which becomes a quadratic $AT^2 + BT + C = 0$ in T.

Then, if $a = (a_1, a_2, a_3)$ is generic, crit(Φ) has two irreducible components. Each component is the intersection with U of a level set $T = 3t$ of T. These are nonsingular fibers in the pencil (2.2), meeting each other and $\mathcal{A}$ at their nine common inflection points. So crit$(\Phi) \subseteq U$ has two connected components. Every pair of nonsingular fibers appears as crit(Φ) for some a. The discriminant $B^2 - 4AC$ defines a hypersurface in (a_1, a_2, a_3)-space for which the corresponding critical locus crit(Φ) has a single nonreduced component, and every nonsingular fiber can appear.

For some values of a, crit(Φ) is empty or has only one reduced component. This is easiest to see by clearing fractions in (2.3), to obtain

$$a_1 P_2 P_3 + a_2 P_1 P_3 + a_3 P_1 P_2 = 0. \tag{2.5}$$

When one component of the variety defined by (2.5) is $P_i = 0$ or $P_0 = xyz = 0$, then crit(Φ) has one reduced component. This occurs if a is a generic point on $a_i = 0$ or $a_1 + a_2 + a_3 = 0$. If $a_i = 0$ for some i and $a_1 + a_2 + a_3 = 0$, or if $a_i \neq 0$ for only one i, then (2.5) becomes $P_i P_j = 0$, and crit$(\Phi) = \emptyset$. If a is a cyclic permutation of $(0, \zeta, 1)$ or equals $(1, \zeta, \zeta^2)$, up to scalar multiple, then (2.5) becomes $P_i^2 = 0$, and crit$(\Phi) = \emptyset$.

Each cocycle of $\omega = \omega_{a_1 a_2 a_3}$ has the form $\psi = \omega_{b_1 b_2 b_3}$ for some b_1, b_2, b_3. So we see that $V(\psi)$ and $V(\omega)$ can have a component in common, but if ω and ψ are not proportional, then $V(\omega) - V(\psi)$ is nonempty and has codimension one.

§3. Decomposable cocycles

We will now assume ψ is a cocycle for $\omega \in A^1$ and ψ is a product of logarithmic one-forms. Then ω is a factor in a vanishing product of $p+1$ one-forms in $A^{\cdot}$. To carry out our geometric analysis it is necessary to work initially over the integers, although eventually the results extend to $\mathbb{C}$-linear combinations of the original integral weights. For that reason we state the result in terms of subspaces of A^1.

We will be dealing with rational functions parametrizing affine or projective algebraic varieties. For that reason we formulate and prove our results algebraically. For any quasi-projective variety X, write $\mathbb{C}[X]$ for the ring of regular functions on X, and $\mathbb{C}(X)$ for the field of rational functions on X. If X is a subvariety of projective space, then elements of $\mathbb{C}[X]$ are represented by homogeneous polynomials and elements of $\mathbb{C}(X)$ by homogeneous rational functions of degree zero. If $\varphi\colon R \to S$ is a homomorphism of $\mathbb{C}$-algebras, denote by $\Omega_{S|R}$ the S-module of Kähler differentials of S over R. Write $\Omega[X] = \Omega_{\mathbb{C}[X]|\mathbb{C}}$ and $\Omega(X) = \Omega_{\mathbb{C}(X)|\mathbb{C}}$. Elements of $\Omega[X]$ (resp. $\Omega(X)$) are polynomial (resp. rational) one-forms on X.

3.1. Singular subspaces of A^1

Let D be a subspace of A^1. We call D a *singular subspace* if the multiplication map $\bigwedge^q(D) \to A^q$ is the zero map, where $q = \dim D$. The *rank* of D is the largest p such that $\bigwedge^p(D) \to A^p$ is not trivial. We say D is *rational* if D has a basis $\{\omega_{\xi_1}, \dots, \omega_{\xi_q}\}$, with $\xi_i = (\xi_{i0}, \dots, \xi_{in}) \in \mathbb{Z}^{n+1}$ for $1 \le i \le q$. Then $\Phi_{\xi_i} = \prod_{j=1}^n f_j^{\xi_{ij}} = \prod_{j=0}^n \alpha_j^{\xi_{ij}}$ is a single-valued rational function on $\mathbb{P}^\ell$, regular on U. (Recall $\sum_{j=0}^n \xi_{ij} = 0$.)

We apply the following general result. It is an easy consequence of the implicit function theorem, but we give an algebraic proof that holds over any algebraically-closed field of characteristic zero. See [13] and [9] for the relevant background on Kähler differentials.

Proposition 3.1. *Suppose $F_1, \dots, F_q$ are rational functions on $\mathbb{C}^\ell$, and*

$$F = (F_1, \dots, F_q)\colon \mathbb{C}^\ell \rightarrowtail \mathbb{C}^q.$$

Then the image of F has dimension less than k if and only if

$$\mathrm{d}F_{i_1} \wedge \cdots \wedge \mathrm{d}F_{i_k} = 0$$

for all $1 \le i_1 < \cdots < i_k \le q$.

Proof. The image of F is a quasi-affine variety, whose function field is isomorphic to $\mathbb{C}(F_1, \dots, F_q)$. Then the dimension p of $\mathrm{im}(F)$ is equal to the transcendence degree of $\mathbb{C}(F_1, \dots, F_q)$ over $\mathbb{C}$. Without loss of generality, suppose $\{F_1, \dots, F_p\}$ is a transcendence base for $\mathbb{C}(F_1, \dots, F_q)$ over $\mathbb{C}$. Then the set $\{\mathrm{d}F_1, \dots, \mathrm{d}F_p\}$ forms a basis for $\Omega(\mathbb{C}(F_1, \dots, F_q))$, a vector space over $\mathbb{C}(F_1, \dots, F_q)$, see [9, Theorem 16.14]. Then

$$\mathrm{d}F_1 \wedge \cdots \wedge \mathrm{d}F_p \neq 0.$$

If $\{F_{i_1},\dots,F_{i_k}\}\subseteq\{F_1,\dots,F_q\}$ with $k>p$, then $\{\mathrm{d}F_{i_1},\dots,\mathrm{d}F_{i_k}\}$ is linearly dependent over $\mathbb{C}(F_1,\dots,F_q)$, hence

$$\mathrm{d}F_{i_1}\wedge\cdots\wedge\mathrm{d}F_{i_k}=0.$$

Q.E.D.

If $\Lambda=(\xi_1,\dots,\xi_q)$ with $\xi_i\in\mathbb{Z}^{n+1}$ satisfying $\sum_{j=0}^n\xi_{ij}=0$, set

$$\Phi_\Lambda:=(\Phi_{\xi_1},\dots,\Phi_{\xi_q})\colon\mathbb{P}^\ell\rightarrowtail\mathbb{C}^q.$$

Proposition 3.1 applies as follows.

Corollary 3.2. *Suppose D is a rational singular subspace of A^1, and Φ_Λ is the rational mapping associated to an ordered integral basis Λ of D. Then $\dim\Phi_\Lambda(U)$ is equal to the rank of D. In particular, $\Phi_\Lambda(U)$ has positive codimension in $\mathbb{C}^q$.*

Let $[z_0:\cdots:z_q]$ be homogeneous coordinates on $\mathbb{P}^q$, and identify $\mathbb{C}^q$ with the $\mathbb{P}^q-\{z_0=0\}$ as above. Let $\overline{Y}$ be the Zariski closure of $Y=\Phi_\Lambda(U)$ in $\mathbb{P}^q$. In homogeneous coordinates, Φ_Λ is given by

$$\Phi_\Lambda=[1:\Phi_{\xi_1}:\cdots:\Phi_{\xi_q}]\colon\mathbb{P}^\ell\rightarrowtail\mathbb{P}^q.$$

Clearing fractions, we have

$$\Phi_\Lambda=[\Phi_{\nu_0}:\Phi_{\nu_1}:\cdots:\Phi_{\nu_q}]$$

where the master functions Φ_{ν_i} are homogeneous polynomials of the same degree d. Moreover we may assume the Φ_{ν_i} have no common factors. Equivalently, the weights $\nu_i=(\nu_{i0},\dots,\nu_{in})$ are non-negative integer vectors with $\sum_{i=0}^n\nu_{ij}=d$ for $0\le i\le q$, whose supports have empty intersection. Then

$$\Phi_{\xi_i}=\frac{\Phi_{\nu_i}}{\Phi_{\nu_0}},\quad \xi_i=\nu_i-\nu_0,\quad\text{and}\quad \omega_{\xi_i}=\omega_{\nu_i}-\omega_{\nu_0}.$$

Definition 3.3. Λ is *essential* if every hyperplane $H\in\mathcal{A}$ appears as a component of $V(\Phi_{\nu_i})$ for some i, $0\le i\le q$.

Henceforth we will tacitly assume Λ is essential. We have $Y\subseteq\overline{Y}\cap(\mathbb{C}^*)^q$ in any case. If Λ is not essential, the inclusion is proper, and may be proper otherwise—see Example 3.16.

The defining ideal $I=I_\Lambda$ of $\overline{Y}$ is generated by homogeneous polynomials $P(z_0,\dots,z_q)$ for which $P(\Phi_{\nu_0},\dots,\Phi_{\nu_q})$ vanishes identically on U, or, equivalently, $P(1,\Phi_{\xi_1},\dots,\Phi_{\xi_q})=0$. We will sometimes refer to I_Λ

as the *syzygy ideal* of Λ. The mapping $z_i \mapsto \Phi_{\nu_i}$ induces an isomorphism of rings

$$\mathbb{C}[\overline{Y}] = \mathbb{C}[z_0, \ldots, z_q]/I \longrightarrow \mathbb{C}[\Phi_{\nu_0}, \ldots, \Phi_{\nu_q}].$$

In particular $\overline{Y}$ is irreducible. Identifying $\mathbb{C}[\overline{Y}]$ with $\mathbb{C}[\Phi_{\nu_0}, \ldots, \Phi_{\nu_q}]$, the dominant rational mapping $\Phi_\Lambda\colon \mathbb{P}^\ell \rightarrowtail \overline{Y}$ corresponds to the field extension

$$\mathbb{C}(\Phi_{\xi_1}, \ldots, \Phi_{\xi_q}) \subseteq \mathbb{C}(x_1, \ldots, x_\ell).$$

The affine ring $\mathbb{C}[Y]$ is isomorphic to a localization of the ring of Laurent polynomials $\mathbb{C}[\Phi_{\nu_0}^{\pm 1}, \ldots, \Phi_{\nu_q}^{\pm 1}]$.

If $\omega \in D - \{0\}$, we write $\omega = \omega_a = \sum_{i=0}^q a_i \omega_{\nu_i} = \sum_{i=1}^n a_i \omega_{\xi_i}$ with $a = (a_0, \ldots a_q) \in \mathbb{C}^{q+1} - \{0\}$, satisfying $\sum_{i=0}^q a_i = 0$. Note that any p-fold wedge product $\psi = \omega_{\xi_{i_1}} \wedge \cdots \wedge \omega_{\xi_{i_p}}$ is a cocycle for ω.

The one-form $\sum_{i=0}^q a_i \mathrm{d}\log(z_i) \in \Omega(\mathbb{C}^{q+1})$ is $\mathbb{C}^*$-invariant, and contracts trivially along the Euler vector field, so it descends to a well-defined rational one-form on $\mathbb{P}^q$. This form restricts to a one-form in $\Omega(\overline{Y})$ which we denote by τ_a. Since $Y \subseteq (\mathbb{C}^*)^q$, τ_a is regular on Y. Note that $\sum_{i=0}^q a_i \mathrm{d}\log(z_i) = \mathrm{d}\log \mu_a$, where $\mu_a = \prod_{i=0}^q z_i^{a_i}$ is a master function for the arrangement of coordinate hyperplanes in $\mathbb{P}^q$.

We show that the zeros of τ_a pull back to zeros of ω_a.

Lemma 3.4. *Let $x \in U$ and $y = \Phi_\Lambda(x) \in Y$. Then $\omega_a(x) = 0$ if and only if $\tau_a(y) \in \ker(\Phi_\Lambda^*)_y$.*

Proof. We have

$$\omega_a = \sum_{i=0}^q a_i \mathrm{d}\log \Phi_{\nu_i} = \mathrm{d}\log \prod_{i=0}^q \Phi_{\nu_i}^{a_i}, = \Phi_\Lambda^*(\mathrm{d}\log \prod_{i=0}^q z_i^{a_i}) = \Phi_\Lambda^*(\tau_a).$$

The result follows upon localization at y. Q.E.D.

Note that

$$\Phi_\Lambda^*(\tau_a) = \sum_{i=0}^q \sum_{j=0}^\ell \frac{a_i}{\Phi_{\nu_i}} \frac{\partial \Phi_{\nu_i}}{\partial x_j} \mathrm{d}x_j.$$

Then $\tau_a(y) \in \ker(\Phi_\Lambda^*)_y$ if and only if $\begin{bmatrix} \frac{a_0}{y_0} & \cdots & \frac{a_q}{y_q} \end{bmatrix}$ lies in the left null space of the Jacobian of Φ_Λ.

Let $\mathrm{Sing}(\Phi_\Lambda)$ denote the singular locus of Φ_Λ, and $\mathrm{Sing}(\overline{Y})$ the singular locus of $\overline{Y}$. Let $S_\Lambda = \mathrm{Sing}(\Phi_\Lambda) \cup \Phi_\Lambda^{-1}(\mathrm{Sing}(\overline{Y})) \subseteq U$.

Theorem 3.5. *$V(\omega_a)$ contains $\Phi_\Lambda^{-1}(V(\tau_a))$, and $V(\omega_a) - \Phi^{-1}(V(\tau_a))$ is a subset of S_Λ.*

Proof. The first statement is immediate from Lemma 3.4. If Φ_Λ is nonsingular at $x \in U$ then the Jacobian of Φ_Λ attains its maximal rank $p = \dim(Y)$ at x. If in addition $\overline{Y}$ is nonsingular at $y = \Phi_\Lambda(x)$, then $\dim(\Omega(Y)_y) = p$. Then $\ker(\Phi_\Lambda^*)_y = 0$. The second statement then follows from Lemma 3.4. Q.E.D.

Corollary 3.6. *Suppose D is a rational singular subspace of A^1, with integral basis Λ. Let p be the rank of D. If* $\mathrm{d}\log(\Phi) \in D$ *then* $\mathrm{crit}(\Phi) \subseteq S_\Lambda$ *or* $\mathrm{codim}(\mathrm{crit}(\Phi)) \leq p$.

Proof. Write $\omega = \mathrm{d}\log(\Phi) = \sum_{i=1}^q a_i\omega_{\xi_i}$. The hypothesis implies $\dim Y = p$, so $V(\tau_a)$ is empty or has codimension at most p in Y. In the first case $V(\omega) \subseteq S_\Lambda$ by the preceding theorem. Otherwise, $V(\omega) \supseteq \Phi_\Lambda^{-1}(V(\tau_a))$ has codimension at most p. Q.E.D.

Corollary 3.7. *Suppose D is a rational singular subspace of A^1, with integral basis Λ. If* $\mathrm{d}\log(\Phi) \in D$, *then* $\mathrm{crit}(\Phi) - S_\Lambda$ *is a union of fibers of Φ_Λ.*

The fibers of Φ_Λ are intersections of level sets of the rational master functions Φ_{ξ_i}, for $1 \leq i \leq q$.

We have not used the assumption that $\{\xi_1, \ldots, \xi_q\}$ is linearly independent, i.e., that the dimension of D is strictly greater than p. This hypothesis rules out a trivial case.

Proposition 3.8. *Suppose $a \neq 0$. Then τ_a is not identically zero on Y.*

Proof. If τ_a is zero on Y, then

$$\Phi_\Lambda^*(\tau_a) = \sum_{i=0}^q a_i \mathrm{d}\log(\Phi_{\nu_i}) = \sum_{i=0}^q a_i\omega_{\nu_i} = \sum_{i=1}^q a_i\omega_{\xi_i}$$

is zero on U. This contradicts the assumption that $\{\omega_{\xi_1}, \ldots, \omega_{\xi_q}\}$ is a basis for D. Q.E.D.

A singular subspace of rank 1 is called an *isotropic* subspace of A^1.

Corollary 3.9. *Suppose Λ is an integral basis of an isotropic subspace of A^1. Then*

(i) *if* $\omega = \mathrm{d}\log(\Phi) \in D$ *then* $\mathrm{Sing}(\Phi_\Lambda) \subseteq \mathrm{crit}(\Phi)$.

(ii) *if* $0 \neq \omega = \mathrm{d}\log(\Phi) \in D$, *then the components of* $\mathrm{crit}(\Phi) - S_\Lambda$ *are disjoint hypersurfaces in U.*

Proof. Since $\dim(Y) = 1$, the Jacobian of Φ_Λ vanishes identically at points of $\mathrm{Sing}(\Phi_\Lambda)$. Then $\mathrm{Sing}(\Phi_\Lambda) \subseteq V(\omega) = \mathrm{crit}(\Phi)$ by Lemma 3.4.

If $a \neq 0$, then τ_a doesn't vanish identically on Y by Proposition 3.8. Then $V(\tau_a)$ is zero-dimensional. Assertion (ii) follows from Theorem 3.5. Q.E.D.

Corollary 3.9(i) can be used to locate singular fibers in Čeva pencils [10, Def. 4.5]—see Example 3.17.

3.2. Zeros of τ_a

We apply the method of Lagrange multipliers to find the zeros of τ_a. The argument applies even if $\overline{Y}$ is singular. Fix a set of homogeneous generators $\{P_1, \dots, P_r\}$ of the defining ideal $I = I_\Lambda \subseteq S$ of $\overline{Y}$. Write ∂_j for $\frac{\partial}{\partial z_j}$. Let

$$J_\Lambda = \left[\partial_j P_i\right]$$

be the Jacobian of $(P_1, \dots, P_r)$. The rank of J_Λ at a nonsingular point $y \in \overline{Y}$ is equal to $q - p$, the codimension of $\overline{Y}$.

Lemma 3.10. *Let $y \in Y$. Then $y \in V(\tau_a)$ if and only if $\left[\frac{a_0}{y_0} \cdots \frac{a_q}{y_q}\right]$ is an element of the row space of $J_\Lambda(y)$.*

Proof. The one-form $d \log \mu_a = \sum_{i=0}^{q} a_i \mathrm{d} \log(z_i) \in \Omega(\mathbb{P}^q)$ restricts to $\tau_a \in \Omega(Y)$. There is an exact sequence of $\mathbb{C}[\overline{Y}]$-modules

$$I/I^2 \xrightarrow{d} \mathbb{C}[\overline{Y}] \otimes_{\mathbb{C}[\mathbb{P}^q]} \Omega[\mathbb{P}^q] \to \Omega[\overline{Y}] \to 0,$$

where d is given by right multiplication by J_Λ [9, Sec. 16.1]. Localization at the maximal ideal corresponding to y preserves exactness of this sequence, so $\tau_a(y) \in \Omega(Y)_y$ vanishes if and only if $\tau_a(y)$ is in the image of d. Q.E.D.

For each $i \geq 0$, let $\mathrm{Fitt}_i(a, I)$ be the variety in $\mathbb{P}^q$ defined by the $(q+1-i) \times (q+1-i)$ minors of the $(r+1) \times (q+1)$ matrix

$$\begin{bmatrix} a_0/z_0 & \cdots & a_q/z_q \\ \partial_0 P_1 & \cdots & \partial_q P_1 \\ \vdots & \ddots & \vdots \\ \partial_0 P_r & \cdots & \partial_q P_r \end{bmatrix}. \tag{3.1}$$

The ideal $\mathrm{Fitt}_i(a, I)$ is independent of the choice of generating set for I. Similarly, let $\mathrm{Fitt}_i(J_\Lambda)$ denote the variety in $\mathbb{P}^q$ defined by the $(q+1-i) \times (q+1-i)$ minors of J_Λ. Let $\overline{Y}_{\mathrm{reg}} = \overline{Y} - \operatorname{Sing} \overline{Y}$. Then $\overline{Y}_{\mathrm{reg}} = \overline{Y} \cap \mathrm{Fitt}_p(J_\Lambda) - \mathrm{Fitt}_{p+1}(J_\Lambda)$, where $p = \dim \overline{Y}$.

If Y is smooth, then $V(\tau_a)$ is a Fitting variety. More generally:

Corollary 3.11. $V(\tau_a) \cap \overline{Y}_{\mathrm{reg}} = \mathrm{Fitt}_p(a, I) \cap \overline{Y}_{\mathrm{reg}}$.

Proof. At any point of $\overline{Y}_{\text{reg}}$ the rank of J_Λ is equal to $q - p = \operatorname{codim} \overline{Y}$, so the rank of matrix (3.1) is at least $q-p$. Then $\begin{bmatrix} \frac{a_0}{y_0} & \cdots & \frac{a_q}{y_q} \end{bmatrix}$ is in the row space of $J(y)$ if and only if (3.1) has rank $q-p$, if and only if all $(q-p+1) \times (q-p+1)$ minors vanish. Q.E.D.

Remark 3.12. In fact, the intersections $\overline{Y} \cap \operatorname{Fitt}_i(J_\Lambda)$, $p \le i \le q$ determine a stratification of $\overline{Y}$ by locally-closed subvarieties, and $V(\tau_a)$ coincides with $\operatorname{Fitt}_i(a, I)$ on the stratum $\operatorname{Fitt}_i(J_\Lambda) - \operatorname{Fitt}_{i+1}(J_\Lambda)$, by the same argument.

In view of Proposition 2.1 we study the zeros of cocycles for τ_a. Since $\dim Y = p$, every $\psi \in \Omega^p(Y)$ is a cocycle for τ_a. For $0 \le i \le q$, set $\tau_i = \frac{\mathrm{d}y_i}{y_i}$, so that $\tau_a = \sum_{i=0}^q a_i\tau_i = \sum_{i=1}^q a_i(\tau_i - \tau_0) = \sum_{i=1}^q \mathrm{d}\log(y_i/y_0)$.

Proposition 3.13. *The intersection*

$$\bigcap \{V(\xi) \mid \xi \in \Omega^p_{\mathbb{C}}(Y), \tau_a \wedge \xi = 0\}$$

is contained in $\operatorname{Sing}(\overline{Y})$.

Proof. For $I = \{i_1 < \cdots < i_p\}_< \subseteq \{1, \ldots, q\}$, consider the p-form

$$\xi_I = (\tau_{i_1} - \tau_0) \wedge \cdots \wedge (\tau_{i_p} - \tau_0).$$

Then $\tau_a \wedge \xi_I = 0$. But $\dim \overline{Y} = p$, so at each point of $\overline{Y}_{\text{reg}}$, ξ_I must be nonzero for some I. Q.E.D.

3.3. The case $q = p + 1$

Suppose D is a singular subspace of rank $\dim(D) - 1$, with integral basis $\Lambda = \{\omega_{\xi_1}, \ldots, \omega_{\xi_q}\}$. Then $\overline{Y}$ is defined by a single homogeneous polynomial $P(z_0, \ldots, z_q)$. This hypothesis holds for all the examples we know, with one exception: the Hessian arrangement, which supports a rational singular subspace of dimension three and rank one—see Examples 2.8 and 4.11.

Consider the rational mapping

$$\rho = [z_0\partial_0 P : \cdots : z_q\partial_q P] \colon \mathbb{P}^q \rightarrowtail \mathbb{P}^q. \tag{3.2}$$

This map has poles along $\operatorname{Sing}(\overline{Y})$ and $\operatorname{Sing}(\overline{Y} \cap \mathbb{C}^I)$ where $\mathbb{C}^I$ is the coordinate subspace $z_i = 0$, $i \in I$. It is regular on $\overline{Y}_{\text{reg}} \cap (\mathbb{C}^*)^q$. By Euler's formula, $\sum_{j=0}^q z_j\partial_j P = \deg(P)P$, so the image of $\overline{Y}$ under ρ is contained in the hyperplane $\sum_{j=0}^q z_j = 0$.

Proposition 3.14. *Suppose $\overline{Y}$ is the hypersurface given by $P = 0$, and ρ is as given in* (3.2). *Then*

(i) *$V(\tau_a) \neq \emptyset$ if and only if $a \in \rho(Y)$.*
(ii) *If $a \in \rho(Y)$ then $V(\tau_a) = \rho^{-1}(a)$.*
(iii) *For generic $a \in \rho(Y)$, $V(\tau_a)$ is nonempty and every component has codimension equal to $\dim \rho(\overline{Y})$.*

Proof. The first two assertions follow from Lemma 3.10, and the third follows from the second. Q.E.D.

Corollary 3.15. *Suppose*
(i) *Φ_Λ is nonsingular on U,*
(ii) *$\overline{Y}$ is nonsingular and intersects the coordinate hyperplanes transversely,*
(iii) *$\dim \rho(\overline{Y}) = \dim \overline{Y}$, and*
(iv) *$\Phi_\Lambda(U) = \overline{Y} \cap (\mathbb{C}^*)^q$.*

If $d\log(\Phi) = \sum_{i=1}^q a_i \omega_{\xi_i}$ and $a_i \neq 0$ for all i, then $\mathrm{crit}(\Phi)$ is nonempty and every component has codimension $q-1$.

Proof. We have observed that $\rho(\overline{Y})$ is contained in the hyperplane Δ defined by $\sum_{i=0}^q z_i = 0$. The second hypothesis ensures that $\mathrm{Sing}(Y \cap \mathbb{C}^I)$ is empty for every $I \subseteq \{0, \dots, q\}$. Then ρ is regular on $\overline{Y}$. By the third condition, $\dim \rho(\overline{Y}) = q - 1 = \dim \Delta$. Since Δ is irreducible, we conclude $\rho(\overline{Y}) = \Delta$. Since $Y = \overline{Y} \cap (\mathbb{C}^*)^q$ by (iii), $\Delta \cap (\mathbb{C}^*)^q = \rho(Y)$. The result then follows from Proposition 3.14 and Theorem 3.5, since (i) and (ii) imply $S_\Lambda = \emptyset$ and the fibers of Φ_Λ all have codimension $q - 1$. Q.E.D.

The hypotheses in Corollary 3.15 are satisfied in many examples. The third condition is automatic in case $q = 2$. The last two conditions hold in every example we know where (i) and (ii) hold.

3.4. Examples

Example 3.16 (Example 2.6, continued). Let D be the rational singular subspace of $A^1 \cong \mathbb{C}^5$ with basis $\Lambda = \{\omega_{010} - \omega_{100}, \omega_{001} - \omega_{100}\}$. We have

$$\Phi_\Lambda = [\Phi_{100} : \Phi_{010} : \Phi_{001}] = [x(y-z) : y(x-z) : z(x-y)].$$

Φ_Λ is nonsingular on U, and $\overline{Y} = Y \cap (\mathbb{C}^*)^5$. The components of Φ_Λ satisfy the homogeneous relation $\Phi_{100} - \Phi_{010} + \Phi_{001} = 0$, so $\overline{Y}$ is the line $z_0 - z_1 + z_2 = 0$ in $\mathbb{P}^2$. Corollary 3.15 implies $\mathrm{crit}(\Phi_{abc})$ is nonempty if and only if $a + b + c = 0$ and a, b and c are nonzero. In this case,

$$\mathrm{crit}(\Phi_{abc}) = (\rho \circ \Phi_\Lambda)^{-1}([a : b : c]) = \Phi_\Lambda^{-1}([a : -b : c]),$$

that is, $\mathrm{crit}(\Phi_{abc})$ is given by

$$[x(y-z):y(x-z):z(x-y)]=[a:-b:c], \text{ or, equivalently}$$
$$[x(y-z):y(z-x)]=[a:b]$$

as we found earlier. It has codimension one in $\mathbb{P}^2$. In this example the map Φ_Λ has connected generic fiber, hence $\mathrm{crit}(\Phi_{abc})$ is connected. If a,b, or c is zero, and $a+b+c=0$, then $\mathrm{crit}(\Phi_{abc})$ is empty.

The basis Λ above has special properties that resulted in the linear syzygy of master functions: we will revisit this in the next section. By way of comparison, consider the basis $\Lambda'=\{\omega_{120}-\omega_{012},\omega_{300}-\omega_{012}\}$. Then

$$\begin{aligned}\Phi_{\Lambda'}&=[\Phi_{010}\Phi_{001}^2:\Phi_{100}\Phi_{010}^2:\Phi_{100}^3]\\&=[yz^2(x-y)^2(x-z):xy^2(x-z)^2(y-z):x^3(y-z)^3].\end{aligned}$$

A *Macaulay 2* calculation [12] shows that $\Phi_{\Lambda'}$ is nonsingular on U. Using the identity $\Phi_{100}-\Phi_{010}+\Phi_{001}=0$, one finds that the Zariski closure $\overline{Y'}$ of $Y'=\Phi_{\Lambda'}(U)$ is defined by

$$z_1^3-z_0^2z_2-4z_0z_1z_2-2z_1^2z_2+z_1z_2^2=0.$$

Then $\overline{Y'}$ is an irreducible cubic with a node at $[z_0:z_1:z_2]=[-2:1:-1]$. This is a point of Y'. It is also in $\widetilde{\Phi}_{\Lambda'}(E)$, where $\widetilde{\Phi}_{\Lambda'}$ is the lift of $\Phi_{\Lambda'}$ to the blow-up of $\mathbb{P}^2$ at the four base points, and E is the exceptional divisor over $[0:1:0]$.

The image of Y' under ρ misses the three points $[0:-1:1]$, $[-2:1:1]$, and $[-2:3:-1]$, corresponding to the three one-forms $\omega_{100}-\omega_{010}$, $\omega_{100}-\omega_{001}$, and $\omega_{010}-\omega_{001}$ in D that have empty zero locus. In particular $Y'\neq\overline{Y'}\cap(\mathbb{C}^*)^5$.

Example 3.17. Let $\mathcal{A}$ be the arrangement with defining equation $Q=Q_0Q_1Q_2$ where

$$\begin{aligned}Q_0&=(x+z)(2x-y-z)(2x+y-z)\\Q_1&=(x-z)(2x+y+z)(2x-y+z), \quad \text{and}\\Q_2&=(y+z)(y-z)z\end{aligned}$$

The image of $\Phi_\Lambda=[Q_0:Q_1:Q_2]\colon\mathbb{P}^2\rightarrowtail\mathbb{P}^2$ is the line $z_0-z_1+2z_2=0$. The fibers are the cubics passing through nine points, three on each of three concurrent lines—$\mathcal{A}$ is a specialization of the Pappus arrangement. One of these cubics is $x(4x^2-y^2-3z^2)=0$. Although $\overline{Y}$ is smooth, Φ_Λ is singular at two points of U, given by $x=y^2+3z^2=0$. These two points lie in $\mathrm{crit}(Q_0^aQ_1^bQ_2^c)$ for every a,b,c with $a+b+c=0$.

Similarly, if $\mathcal{A}$ the subarrangement of the Hessian arrangement (2.8) defined by $P_1P_2P_3 = 0$, then every critical set $\mathrm{crit}(P_1^a P_2^b P_3^c)$, $a+b+c=0$, contains the three points $[1:0:0]$, $[0:1:0]$, $[0:0:1]$ of U where the fourth special fiber $xyz = 0$ is singular. The map $\Phi_\Lambda = [P_1 : P_2 : P_3]$ is singular at these points, although $\overline{Y}$ is smooth, given by $\zeta P_1 + \zeta^2 P_2 + P_3 = 0$. (See also Example 4.11.)

Remark 3.18. In fact, Corollary 3.9 can be used to detect Čeva pencils [10] (see 4.4). For instance the master function

$$\Phi = \frac{x(y-z)}{y(x-z)}$$

has critical points $[0:1:0]$ and $[1:1:0]$, the singular points of the third completely decomposable fiber in Example 2.6.

Example 3.19 (Example 2.7, continued). In this example, the one-form ω has no decomposable 2-cocycle. Indeed there are no singular subspaces of A^1 of rank $p-1$ or $p=2$. For $p=1$ this holds because $\mathcal{A}$ is 2-generic. For $p=2$ one can verify the statement computationally using the approach of [15]; in [2] we give a combinatorial argument based on Theorem 5.4. Setting

$$\Psi = [\frac{x}{y+z+w} : \frac{y}{x+z+w} : \frac{z}{x+y+w} : \frac{w}{x+y+z}] : \mathbb{P}^3 \rightarrowtail \mathbb{P}^3,$$

we have $\omega = \mathrm{d}\log(\Phi) = \Psi^*(\tau)$ where

$$\tau = a\,\mathrm{d}\log(y_0) + b\,\mathrm{d}\log(y_1) + c\,\mathrm{d}\log(y_2) + d\,\mathrm{d}\log(y_3).$$

The map Ψ is dominant. The one-dimensional critical locus of Φ is a fiber of a different map

$$[x(y+z+w) : y(x+z+w) : z(x+y+w) : w(x+y+z)].$$

§4. Linear dependence among master functions

In this section we consider a singular subspace D with an integral basis Λ such that Y_Λ is linear, i.e., the syzygy ideal I_Λ is generated by homogeneous linear forms. We saw this phenomenon in Example 3.16. We start with a trivial example that will be useful for what follows.

4.1. Example: equations for the critical locus

Suppose $\mathcal{A}$ is an essential arrangement of $n+1$ hyperplanes in $\mathbb{P}^\ell$, with $n > \ell$. Then $D = A^1$ is a singular subspace, with integral basis

$\{\omega_i - \omega_0 \mid 1 \leq i \leq n\}$. The corresponding rational mapping is

$$\Phi = \Phi_\Lambda = [\alpha_0 : \cdots : \alpha_n] \colon \mathbb{P}^\ell \rightarrowtail \mathbb{P}^n,$$

and $\overline{Y} = \overline{Y}_\Lambda$ is a linear subvariety of $\mathbb{P}^n$. The reader will recognize that this is the usual identification of a labelled vector configuration, $(\alpha_0, \ldots, \alpha_n)$, with a point in the Grassmannian of ℓ-planes in $\mathbb{P}^n$. Let us denote $\overline{Y}_\Lambda$ by $L_\mathcal{A}$. (This is an abuse of notation; $\overline{Y}_\Lambda$ depends on the choice of defining forms α_i.)

Most of the results of the previous section are vacuous in this situation, but Lemma 3.10 tells us something:

Theorem 4.1. *Let $B = \left[b_{ij}\right]$ be an $(\ell+1) \times (n+1)$ matrix such that $L_\mathcal{A}$ is the kernel of B. Then, for any $\lambda \in \mathbb{C}^n$, the critical locus of Φ_λ is defined by the $\binom{n+1}{\ell+1}$ equations*

$$\sum_{i \in I} \sigma(i, I) b_{I - \{i\}} \frac{\lambda_i}{\alpha_i(x)} = 0$$

where I ranges over the subsets of $\{0, \ldots, n\}$ of size $\ell+1$, the coefficient b_J is given by $b_J = \det\left[b_{ij} \mid j \in J\right]$, and $\sigma(i, I) = \pm 1$, depending on the position of i in I.

Proof. The linear forms defined by the rows of B generate the syzygy ideal I_Λ. Then the Jacobian J_Λ is equal to B. With the observation that $S_\Lambda = \emptyset$, setting $a = \lambda$ in Lemma 3.10 and applying Theorem 3.5 yields the claim. Q.E.D.

The columns of the matrix B above define a realization of the matroid dual to the matroid of $\mathcal{A}$. In Example 2.6,

$$B = \begin{bmatrix} 0 & -1 & 1 & 0 & 0 & 1 \\ -1 & 0 & 1 & 0 & 1 & 0 \\ -1 & 1 & 0 & 1 & 0 & 0 \end{bmatrix}.$$

(The matroid of $\mathcal{A}$ is self-dual.) Theorem 4.1 says that $\operatorname{crit}(\Phi_{abc})$ is defined by the 4×4 minors of

$$\begin{bmatrix} \frac{a}{x} & \frac{b}{y} & \frac{c}{z} & \frac{c}{x-y} & \frac{b}{x-z} & \frac{a}{y-z} \\ 0 & -1 & 1 & 0 & 0 & 1 \\ -1 & 0 & 1 & 0 & 1 & 0 \\ -1 & 1 & 0 & 1 & 0 & 0 \end{bmatrix}.$$

These 15 equations reduce to the single equation found in Example 2.6.

4.2. Linear hypersurfaces

Suppose $\overline{Y}$ is a linear and $p = \text{rank}(D) = q - 1$, i.e., $\overline{Y}$ is a linear hyperplane in $\mathbb{P}^q$. Let $P(z) = \sum_{j=0}^q b_j z_j$ be a generator for the syzygy ideal. It is no loss to assume $b_j \neq 0$ for all j, or equivalently, $\overline{Y}$ is not contained in any coordinate hyperplane. Otherwise some proper subset of $\{\Phi_{\nu_0}, \ldots, \Phi_{\nu_q}\}$ is linearly dependent.

Proposition 4.2. *Suppose $\overline{Y}$ is a hyperplane not contained in any coordinate hyperplane in $\mathbb{P}^q$. Let $\lambda = \sum_{i=0}^q a_i \nu_i$. Then for generic a satisfying $\sum_{i=0}^q a_i = 0$, $\text{crit}(\Phi_\lambda) - S_\Lambda$ is nonempty and every component of $\text{crit}(\Phi_\lambda) - S_\Lambda$ has codimension equal to the rank of D.*

Proof. The syzygy ideal I_Λ is generated by a linear polynomial $P(z) = \sum_{j=0}^q b_j z_j$, and $b_j \neq 0$ for $0 \leq j \leq q$ by hypothesis. The map $\rho = [b_0 z_0 : \cdots : b_q z_q]$ of (3.2) is an automorphism of $\mathbb{P}^q$ since all of the b_j are nonzero. Consequently, ρ maps the hyperplane $\overline{Y}$ isomorphically to the hyperplane Δ defined by $\sum_{i=0}^q z_i = 0$. The result then follows from Proposition 3.14. Q.E.D.

4.3. The general case

Suppose the singular subspace $D \subseteq A^1$ has an integral basis Λ for which $\overline{Y} = \overline{Y}_\Lambda$ is a linear variety in $\mathbb{P}^q$. Choose a linear isomorphism $\varphi : \mathbb{P}^p \to \overline{Y}$, given by a $(q+1) \times (p+1)$ matrix $B = \left[b_{ij}\right]$. Assume D is not contained in any coordinate hyperplane. Then the intersections of $\overline{Y}$ with the coordinate hyperplanes in $\mathbb{P}^q$ determine an essential arrangement $\mathcal{B}$ of $q+1$ not necessarily distinct hyperplanes in $\mathbb{P}^p$, with defining forms $\beta_i(x) = \sum_{i=0}^p b_{ij} x_j$, for $0 \leq i \leq q$. By construction, the subspace $L_\mathcal{B}$ of $\mathbb{P}^q$, as described in Section 4.1, is equal to $\overline{Y}$.

Theorem 4.3. *Suppose $\overline{Y}_\Lambda = L_\mathcal{B}$ is a linear subspace not contained in any coordinate hyperplane in $\mathbb{P}^q$, and $Y = \overline{Y} \cap (\mathbb{C}^*)^q$. Let $\lambda = \sum_{i=0}^q a_i \nu_i$ with $\sum_{i=0}^q a_i = 0$. Let $\Psi_a = \prod_{i=0}^q \beta_i^{a_i}$ be the master function on the complement of the arrangement $\mathcal{B}$ in $\mathbb{P}^p$ corresponding to a. Then*

(i) $\text{crit}(\Phi_\lambda) - S_\Lambda \neq \emptyset$ *if and only if* $\text{crit}(\Psi_a) \neq \emptyset$,
(ii) $\text{crit}(\Phi_\lambda) - S_\Lambda = \Phi_\Lambda^{-1}(\rho(\text{crit}(\Psi_a)))$, *and*
(iii) $\text{codim}(\text{crit}(\Phi_\lambda) - S_\Lambda) \leq \text{codim}\,\text{crit}(\Psi_a)$.

Proof. Just as in Section 4.1, the one-form τ_a on $\overline{Y}$ pulls back to $d\log(\Psi_a)$ under the isomorphism φ, and the assertions follow from Proposition 3.14. Q.E.D.

Example 4.4 (Example 2.6, continued). We saw in Example 3.16 that the variety $\overline{Y}$ is given by $z_0 - z_1 + z_2 = 0$ in $\mathbb{P}^2$, and $Y = \overline{Y} \cap (\mathbb{C}^*)^2$.

We can take $\varphi\colon \mathbb{P}^1 \xrightarrow{\cong} \overline{Y}$ to be given by the matrix

$$B = \begin{bmatrix} 1 & 0 \\ 1 & 1 \\ 0 & 1 \end{bmatrix}.$$

Then associated arrangement $\mathcal{B}$ consists of three points $[1:0]$, $[1:1]$, $[0:1]$ in $\mathbb{P}^1$. The complement of $\mathcal{B}$ has Euler characteristic -1, so a generic $\mathcal{B}$-master function Ψ_{abc} has a single nondegenerate critical point. A computation shows that this holds if a, b, and c are nonzero. We reach the same conclusion as before, that $\mathrm{crit}(\Phi_{abc})$ has codimension one.

4.4. Multinets and codimension-one critical sets

Next we use the main result of [10] to give a complete description of $\mathrm{crit}(\Phi)$ for any $\omega = d\log(\Phi) \in \mathcal{R}^1(\mathcal{A})$. As we observed earlier, if $\omega \in \mathcal{R}^1(\mathcal{A})$, then ω has a nontrivial decomposable 1-cocycle ψ. The statement that $\omega \wedge \psi = 0$ means, for each $x \in U$, $\{\omega(x), \psi(x)\}$ is linearly dependent. Then there are functions a and b on U such that $a(x)\omega(x) + b(x)\psi(x) = 0$ for all $x \in U$. This implies $V(\omega)$ contains $V(b) - V(a)$, which is a hypersurface unless it is empty. It was this observation that led to the current research.

According to [14], the maximal isotropic subspaces D of A^1 of dimension at least two are the components of $\mathcal{R}^1(\mathcal{A})$, and they intersect trivially. By [10, Theorem 3.11], such a component has an integral basis $\Lambda = (\omega_{\xi_1}, \ldots, \omega_{\xi_q})$, with the property that the corresponding polynomial master functions $\Phi_{\nu_0}, \ldots, \Phi_{\nu_q}$ are all collinear in the space of degree d polynomials. Then $\overline{Y} = \overline{Y}_\Lambda$ is a line in $\mathbb{P}^q$. The homogenized basis $\{\omega_{\nu_0}, \ldots, \omega_{\nu_q}\}$ corresponds to the characteristic vectors ν_i of the blocks in a multinet structure on a subarrangement of $\mathcal{A}$, as defined below.

For $X \subseteq \mathbb{P}^\ell$, write $\mathcal{A}_X = \{H \in \mathcal{A} \mid X \subseteq H\}$. A *rank-two flat* of $\mathcal{A}$ is a subspace X of the form $H \cap K$ for some $H, K \in \mathcal{A}, H \neq K$. If $\mathcal{P}$ is a partition of $\mathcal{A}$, the *base locus* of $\mathcal{P}$ is the set of rank-two flats of $\mathcal{A}$ obtained by intersecting hyperplanes from different blocks of $\mathcal{P}$.

Definition 4.5. A $(q+1, d)$-multinet on $\mathcal{A}$ is a pair $(\mathcal{P}, m)$ where $\mathcal{P}$ is a partition $\{\mathcal{A}_0, \ldots, \mathcal{A}_q\}$ of $\mathcal{A}$ into $q+1$ blocks, with base locus $\mathcal{X}$, and $m\colon \mathcal{A} \to \mathbb{Z}_{>0}$ is a multiplicity function, satisfying

(i) $\sum_{H \in \mathcal{A}_i} m(H) = d$ for every i.
(ii) For each $X \in \mathcal{X}$, $\sum_{H \in \mathcal{A}_i \cap \mathcal{A}_X} m(H) = n_X$ for some integer n_X, independent of i.
(iii) For each i, $\bigcup \mathcal{A}_i - \bigcup \mathcal{X}$ is connected.

The third condition says that $\mathcal{P}$ cannot be refined to a (q', d)-multinet with the same multiplicity function, with $q' > q+1$. Given a multinet on $\mathcal{A}$, let $\nu_i = \sum_{H_i \in \mathcal{A}_i} m(H_i)e_i$, for $0 \le i \le q$, and $\xi_i = \nu_i - \nu_0$ for $1 \le i \le q$. We call $\nu_0, \dots, \nu_q$ the *characteristic vectors* of $(\mathcal{P}, m)$. We have the following results from [10].

Theorem 4.6 ([10, Corollary 3.12]). *Suppose D is a maximal isotropic subspace of A^1 of dimension $q \ge 2$. Then there is a subarrangement $\mathcal{A}'$ of $\mathcal{A}$ and a $(q+1, d)$-multinet on $\mathcal{A}'$ whose characteristic vectors $\nu_0, \dots, \nu_q$ yield an integral basis $\omega_{\xi_1}, \dots, \omega_{\xi_q}$ of D.*

Theorem 4.7 ([10, Theorem 3.11]). *Suppose $\nu_0, \dots, \nu_q$ are the characteristic vectors of a $(q+1, d)$-multinet structure on $\mathcal{A}$. Then each of the master functions Φ_{ν_i}, $i \ge 2$, is a linear combination of Φ_{ν_0} and Φ_{ν_1}. Moreover every fiber of the mapping $[\Phi_{\nu_0} : \Phi_{\nu_1}] \colon \mathbb{P}^\ell \rightarrowtail \mathbb{P}^1$ is connected.*

Given this result, the analysis of critical sets proceeds exactly as in the Example 4.4. First, we need a lemma about isolated critical points

Lemma 4.8. *Suppose $\mathcal{A}$ is an affine arrangement of n hyperplanes in $\mathbb{C}^\ell$, and W is a nonempty Zariski-open subset of $\mathbb{C}^\ell$. Then there is a nonempty Zariski-open subset L of $\mathbb{C}^n$ such that $W \cap \operatorname{crit}(\Phi_\lambda)$ consists of $|\chi(U)|$ points, for each $\lambda \in L$.*

Proof. By [20, Theorem 1.1], there is a nonempty Zariski-open subset L' of $\mathbb{C}^n$ for which $\operatorname{crit}(\Phi_\lambda)$ is isolated and consists of $|\chi(U)|$ points, for $\lambda \in L'$. Let

$$\Sigma = \{(\lambda, v) \in \mathbb{C}^n \times U \colon \omega_\lambda(v) = 0\},$$

an n-dimensional smooth complex variety by [20, Prop. 4.1]. Let π_i for $i = 1, 2$ denote its projections onto $\mathbb{C}^n$ and U, respectively. Then $\pi_2^{-1}(U \cap W)$ and $\pi_1^{-1}(L')$ are each nonempty Zariski-open subsets of Σ, as is their intersection Z. Then $\pi_1(Z)$ is a finite union of locally-closed subsets of $\mathbb{C}^n$—see [13, Exercise 3.19]. Since Z is dense in Σ, $\pi_1(Z)$ is dense in $\mathbb{C}^n$. Hence $\pi_1(Z)$ contains a Zariski-open subset L, which has the required property. Q.E.D.

Theorem 4.9. *Suppose $\omega = \mathrm{d}\log(\Phi) \in \mathcal{R}^1(\mathcal{A})$, and D is the maximal isotropic subspace of A^1 containing ω. Let Λ be the integral basis of D arising from the associated multinet. Then*

(i) *For every $\omega \in D$, $\operatorname{Sing}(\Phi_\Lambda) \subseteq \operatorname{crit}(\Phi)$, and,*
(ii) *For generic $\omega \in D$, $\operatorname{crit}(\Phi) - \operatorname{Sing}(\Phi_\Lambda)$ is a union of $\dim(D) - 1$ connected smooth hypersurfaces of the same degree.*

Proof. Write $q = \dim(D)$ and let $\nu_0, \dots, \nu_q$ be the characteristic vectors of the $(q+1, d)$-multinet corresponding to D. Write $\omega =$

$\sum_{i=0}^{q} a_i \omega_{\nu_i}$. By the preceding theorem, for each $2 \le k \le q$, there is a linear relation $\Phi_{\nu_k} = b_k \Phi_{\nu_0} + c_k \Phi_{\nu_1}$. Then $\overline{Y}$ is a line in $\mathbb{P}^q$. The first assertion follows from Corollary 3.9.

We can choose the isomorphism $\varphi\colon \mathbb{P}^1 \xrightarrow{\cong} \overline{Y} \subset \mathbb{P}^q$ to be given by the matrix

$$B = \begin{bmatrix} 1 & 0 \\ 0 & 1 \\ b_2 & c_2 \\ \vdots & \vdots \\ b_q & c_q \end{bmatrix}.$$

The corresponding arrangement $\mathcal{B}$ consists of $q+1$ distinct points in $\mathbb{P}^1$. The Euler characteristic of the complement of $\mathcal{B}$ is $1-q$. Then for generic a the $\mathcal{B}$-master function Ψ_a has $q-1$ isolated, nondegenerate critical points in $\overline{Y} \cap (\mathbb{C}^*)^q$. In fact, since Y is dense in $\overline{Y} \cap (\mathbb{C}^*)^q$, we apply Lemma 4.8 to see that, for generic a, Ψ_a has $q-1$ critical points in Y. Then Corollary 3.7 implies $\operatorname{crit}(\Phi)$ is the union of $q-1$ fibers of Φ_Λ.

The projection $\mathbb{P}^q \rightarrowtail \mathbb{P}^1$ along $z_0 = z_1 = 0$ restricts to an isomorphism on $\overline{Y}$. Then the last statement of Theorem 4.7 implies the fibers of Φ_Λ are connected. These fibers are given by $[\Phi_{\nu_0} : \Phi_{\nu_1}] = [a_0 : a_1]$. Since the Φ_{ν_i} have degree d, $\operatorname{crit}(\Phi_\Lambda)$ is a union of $q-1$ connected hypersurfaces of degree d. The generic fiber of Φ_Λ is smooth by Bertini's Theorem [13, Corollary III.10.9]. Q.E.D.

Corollary 4.10. *For generic* $\omega = \mathrm{d}\log(\Phi) \in \mathcal{R}^1(\mathcal{A})$, *the number of connected components of* $\operatorname{crit}(\Phi_\lambda)$ *is equal to the dimension of* $H^1(A^{\cdot}, \omega)$.

Example 3.17 shows that $\operatorname{crit}(\Phi)$ need not be smooth or irreducible for all $\omega = \mathrm{d}\log(\Phi) \in \mathcal{R}^1(\mathcal{A})$.

Example 4.11. By [22, 31], the maximum number of blocks in a multinet is equal to four. The only known example with four blocks is the multinet on the Hessian arrangement corresponding to the Hesse pencil, Example 2.8. The factors of the polynomial master functions $P_0 = xyz, P_1, P_2, P_3$ define the blocks of a multinet on $\mathcal{A}$ with all multiplicities equal to one. These master functions satisfy two linear syzygies:

$$P_2 = 3(1-\zeta)P_0 + P_1$$
$$P_3 = 3(1-\zeta^2)P_0 + P_1.$$

Then the variety $\overline{Y}_\Lambda$ corresponding to the basis

$$\Lambda = \{\omega_{100}, \omega_{010}, \omega_{001}\}$$

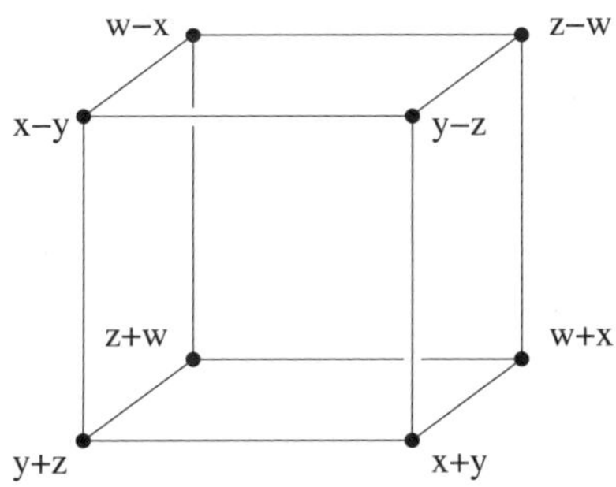

Fig. 2. A rank-four matroid with a linear syzygy of master functions

is the line given by $z_2 = 3(1-\zeta)z_0 + z_1,\ z_3 = 3(1-\zeta^2)z_0 + z_1$ in $\mathbb{P}^3$, which meets the coordinate hyperplanes in the four points corresponding to the singular fibers. The corresponding arrangement $\mathcal{B}$ consists of four points in general position in $\mathbb{P}^1$, given by the rows of the matrix

$$\begin{bmatrix} 1 & 0 \\ 0 & 1 \\ 3(1-\zeta) & 1 \\ 3(1-\zeta^2) & 1 \end{bmatrix}.$$

The complement of $\mathcal{B}$ has Euler characteristic -2, hence a generic $\mathcal{B}$-master function has two isolated critical points. Then, for generic $a = (a_1, a_2, a_3)$, the critical locus of the $\mathcal{A}$-master function $\Phi = \Phi_{a_1 a_2 a_3}$ has two components and codimension one, as found by direct calculation in Example 2.8. This example shows that Theorem 4.8(ii) and Corollary 4.9 may not hold under the weaker hypothesis that $a_i \neq 0$ for all i.

Here is a rank-four example, that has appeared in different form in the lecture of A. Libgober in this volume.

Example 4.12. Let $\mathcal{A}$ be the arrangement with defining polynomial

$$Q = (x+y)(x-y)(y+z)(y-z)(z+w)(z-w)(w+x)(w-x),$$

with the hyperplanes numbered according to the order of factors in Q. Then $\mathcal{A}$ is a 2-generic subarrangement of the Coxeter arrangement of type D_4. Up to lattice-isotopy, the dual projective point configuration consists of the eight vertices of a cube—see Figure 2. Let D be the subspace of A^1 with basis

$$\Lambda = \{(\omega_0+\omega_1)-(\omega_6+\omega_7),\ (\omega_2+\omega_3)-(\omega_6+\omega_7), (\omega_4+\omega_5)-(\omega_6+\omega_7)\}.$$

Then D is a rational singular subspace of rank two; if $\omega \in D$, then $H^1(A^{\cdot}, \omega) = 0$ and $H^2(A^{\cdot}, \omega) \cong D/\mathbb{C}\omega$ has dimension 2. $\overline{Y}_\Lambda$ is the linear hyperplane in $\mathbb{P}^3$ defined by $z_0 + z_1 + z_2 + z_3 = 0$, reflecting the linear syzygy of polynomial master functions

$$(x+y)(x-y) + (y+z)(y-z) + (z+w)(z-w) + (w+x)(w-x) = 0.$$

From Proposition 4.2 we see that

$$\Phi = \left(\frac{x^2 - y^2}{w^2 - x^2}\right)^{a_1} \left(\frac{y^2 - z^2}{w^2 - x^2}\right)^{a_2} \left(\frac{z^2 - w^2}{w^2 - x^2}\right)^{a_3}$$

has nonempty critical set of codimension two in $\mathbb{P}^3$, for generic (a_1, a_2, a_3).

§5. The rank condition

We are left with the problem of finding rational singular subspaces of A^1. The theory of multinets gives a method to find such subspaces of rank one. In this section we give a combinatorial condition for a set Λ of linearly independent integral one-forms to span a singular subspace of A^1 of arbitrary rank, using tropical implicitization and nested sets.

5.1. Tropicalization

The *tropicalization* of a projective variety V in $\mathbb{P}^q$ is a polyhedral fan $\mathrm{trop}(V)$ in tropical projective space $\mathbb{TP}^q = \mathbb{R}^{q+1}/\mathbb{R}(1, \ldots, 1)$, associated to a homogeneous defining ideal I of V. If V is a hypersurface with defining equation $f = 0$, then $\mathrm{trop}(V)$ is the image in $\mathbb{TP}^q$ of the union of the cones of codimension at least one in the normal fan of the Newton polytope of f. In general, $\mathrm{trop}(V)$ is the image of the union of the cones of codimension at least one in the Gröbner fan of I. The set $\mathrm{trop}(V)$ arises geometrically from the lowest-degree terms in Puiseux expansions of curves lying in V. See [7] and the references therein for background on tropical varieties. See [21] for matroid terminology.

We will need several results from tropical geometry. The first is a theorem of Bieri and Groves [3].

Theorem 5.1. *The maximal cones in* $\mathrm{trop}(V)$ *have dimension equal to* $\dim(V)$.

If V is an ℓ-dimensional linear subvariety of $\mathbb{P}^n$, given as the column space of a matrix R, then the tropicalization $\mathrm{trop}(V)$ depends only on the dependence matroid $\mathfrak{G}$ on the rows of R. In our setting the rows of R give the defining forms of a hyperplane arrangement $\mathcal{A}$. The *matroid polytope* $\Delta(\mathfrak{G})$ of $\mathfrak{G}$ is the convex hull of the set

$$\Big\{\sum_{i\in B} e_i \mid B \text{ is a basis of } \mathfrak{G}\Big\}.$$

The tropicalization trop(V), called the *Bergman fan* of $\mathfrak{G}$, is the image in $\mathbb{TP}^n$ of the union of the cones of codimension at least one in the normal fan of $\Delta(\mathfrak{G})$. We denote it by $B(\mathfrak{G})$.

In [11] the Bergman fan is described in terms of *nested set cones.* Let $\mathcal{G}$ be the set of proper connected (i.e., irreducible) flats of $\mathfrak{G}$. These are the flats corresponding to the dense edges of the projective arrangement $\mathcal{A}$. A collection $\mathcal{S} = \{X_1, \dots, X_p\}$ of subsets of $\mathcal{G}$ is a *nested set* if, for every set $\mathcal{T}$ of pairwise incomparable elements of $\mathcal{S}$, the join $\bigvee \mathcal{T}$ is not an element of $\mathcal{G}$. The nested sets form a simplicial complex $\Delta = \Lambda(\mathfrak{G})$, the *nested set complex*, which is pure of dimension $r = \ell - 1$. It is the coarsest of a family of nested set complexes, obtained by replacing $\mathcal{G}$ with larger "building sets." All of these complexes are subdivided by the order complex of the poset of nonempty flats of $\mathfrak{G}$.

If $S \subseteq \mathcal{A}$, set $e_S = \sum_{H_i \in S} e_i$. The *nested set fan* $N(\mathfrak{G})$ is the image in $\mathbb{TP}^n$ of the union of the cones generated by

$$\{e_S \mid S \in \mathcal{S}\}$$

for $\mathcal{S} \in \Delta(\mathfrak{G})$. From [11] we have the following result.

Theorem 5.2. *The nested set fan $N(\mathfrak{G})$ subdivides the Bergman fan $\mathcal{B}(\mathfrak{G})$.*

5.2. Singular subspaces

Let $\mathcal{A}$ be an arrangement in $\mathbb{P}^\ell$ with homogeneous defining linear forms $\alpha_0, \dots, \alpha_n$. Let $\mathfrak{G}$ be the underlying matroid of $\mathcal{A}$, the dependence matroid on $\{\alpha_0, \dots, \alpha_n\}$. Suppose D is a rational subspace of $A^1(\mathcal{A})$, with integral basis $\Lambda = \{\omega_{\xi_1}, \dots, \omega_{\xi_q}\}$. We identify Λ with the $q \times (n+1)$ matrix of integers $[\xi_{ij}]$, and recall that $\sum_{j=0}^n \xi_{ij} = 0$ for $1 \le i \le q$. Let $\overline{Y}_\Lambda$ be the Zariski closure of the image of the associated rational map $\Phi_\Lambda = [1 : \Phi_{\xi_1} : \cdots : \Phi_{\xi_q}] \colon \mathbb{P}^\ell \rightarrowtail \mathbb{P}^q$.

The main observation is that Φ_Λ can be factored as a linear map followed by a monomial map. Assume $\mathcal{A}$ is essential, and let

$$\alpha = \big[\alpha_0 : \cdots : \alpha_n\big] : \mathbb{P}^\ell \to \mathbb{P}^n.$$

Let $\mu = \mu_\Lambda \colon \mathbb{P}^n \rightarrowtail \mathbb{P}^q$ be given by

$$\mu([t_0 : \cdots : t_n]) = [1 : t^{\xi_1} : \cdots : t^{\xi_q}],$$

where we use the usual vector notation for monomials: $t^{(i_0,\ldots,i_n)} = t_0^{i_0} \cdots t_n^{i_n}$. Then the following diagram commutes.

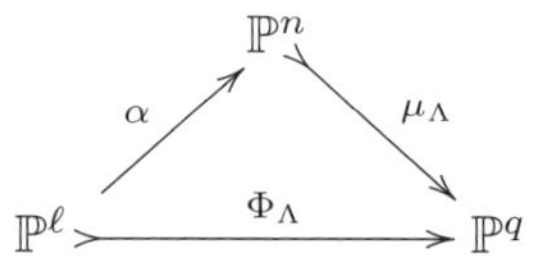

In this situation the diagram tropicalizes faithfully, in the following sense.

Theorem 5.3 ([7, Theorem 3.1]). *The tropicalization* $\operatorname{trop}(\overline{Y}_\Lambda)$ *is equal to the image of the Bergman fan* $B(\mathfrak{G})$ *under the linear map*

$$\mathbb{TP}^n \to \mathbb{TP}^q$$

with matrix Λ.

We obtain the following characterization. Write $\Lambda = [\Lambda_0| \quad \cdots \quad |\Lambda_n]$ with $\Lambda_j \in \mathbb{Z}^q$ for each j. For $S = \{H_{j_1}, \ldots, H_{j_k}\} \subseteq \mathcal{A}$, let $\Lambda_S = \sum_{r=1}^k \Lambda_{j_r}$.

Theorem 5.4. *The subspace D is singular if and only if the rank of the matrix*

$$\Lambda_{\mathcal{S}} = [\Lambda_{S_1}| \quad \cdots \quad |\Lambda_{S_{\ell-1}}]$$

is less than q, for each maximal nested set $\mathcal{S} \in N(\mathfrak{G})$. In this case the rank of D is the maximal rank of $\Lambda_{\mathcal{S}}$ for $\mathcal{S} \in N(\mathfrak{G})$.

Proof. The subspace D is singular if and only if $\dim \overline{Y}_\Lambda < q$. By Theorem 5.1, this occurs if and only if $\dim \operatorname{trop}(\overline{Y}_\Lambda) < q$. The cones of $\operatorname{trop}(\overline{Y}_\Lambda)$ are images of the cones of $B(\mathfrak{G})$ under Λ, by Theorem 5.3. The linear hulls of the cones in $B(\mathfrak{G})$ are the images in $\mathbb{TP}^n$ of the linear spans of the sets $\{e_S \mid S \in \mathcal{S}\}$, for $\mathcal{S} \in N(\mathfrak{G})$, by Theorem 5.2. Since $\Lambda(e_S) = \Lambda_S$, the result follows. The last statement holds because the rank of D is equal to $\dim \overline{Y}_\Lambda$. Q.E.D.

Example 5.5. Consider the arrangement of rank four with defining polynomial

$$Q = xyz(x+y+z)w(x+y+w),$$

with hyperplanes ordered according to the given factorization of Q. The dual point configuration consists of the six vertices of a triangular prism in $\mathbb{P}^3$.

For generic (a,b,c), the master function $\Phi = x^a y^{-a} z^b (x+y+z)^{-b} w^c (x+y+w)^{-c}$ has critical set of codimension two, and $H^2(A,\omega) \cong$

$\mathbb{C}$ for $\omega = d\log(\Phi)$. Based on our other examples, one might suspect that the subspace D with basis $\Lambda = \{\omega_0 - \omega_1,\ \omega_2 - \omega_3,\ \omega_4 - \omega_5\}$ is singular. Among the nested sets of $\mathcal{A}$ is the set $\mathcal{S} = \{0, 02, 024\}$, and

$$\Lambda_{\mathcal{S}} = \begin{bmatrix} 1 & 1 & 1 \\ 0 & 1 & 1 \\ 0 & 0 & 1 \end{bmatrix}$$

does not have rank two. Then by Theorem 5.4, D is not singular. In fact, $\psi = a \cdot \partial(\omega_0\omega_1\omega_5) + b \cdot \partial(\omega_2\omega_3\omega_5)$ is the unique 2-cocycle for ω. ψ is trivial if a or b is zero, and our argument shows that ψ is not decomposable if a and b are both nonzero.

In the forthcoming paper [2], Theorem 5.4 is used to derive combinatorial conditions for p-generic arrangements to support singular subspaces of rank p. Using that approach one can show by combinatorial means that there are no singular subspaces of A^1 of rank two in Example 5.5.

Theorem 5.4 also has the following corollary.

Corollary 5.6. *If $\mathfrak{G}_1$ and $\mathfrak{G}_2$ are loop-free matroids on the ground set $\{1, \ldots, n\}$ and $B(\mathfrak{G}_1) = B(\mathfrak{G}_2)$, then $\mathfrak{G}_1 = \mathfrak{G}_2$.*

Proof. Let $\mathfrak{G}$ be a loop-free matroid on $\{1, \ldots, n\}$, with Orlik–Solomon algebra $A^{\cdot} = A^{\cdot}(\mathfrak{G})$. Let $e_1, \ldots, e_n \in A^1$ denote the canonical generators. Then $S = \{i_1, \ldots, i_k\} \subseteq \{1, \ldots, n\}$ is dependent in $\mathfrak{G}$ if and only if $e_{i_1} \wedge \cdots \wedge e_{i_k} = 0$ in A^k. (This statement holds even if $\mathfrak{G}$ has multiple points.) Equivalently, S is dependent if and only if the coordinate subspace $D \subseteq A^1$ spanned by $\{e_{i_1}, \ldots, e_{i_k}\}$ is singular. By Theorem 5.3, D is singular if and only if the image of the Bergman fan $B(\mathfrak{G}) \subseteq \mathbb{TP}^n$ under the projection $\mathbb{TP}^n \to \mathbb{TP}^S \cong \mathbb{TP}^{k-1}$ has dimension less than $k-1$. Thus $B(\mathfrak{G})$ determines $\mathfrak{G}$. Q.E.D.

Acknowledgements. This work was begun during the semester-long program on Hyperplane Arrangements at MSRI in Fall, 2004. The authors are grateful to the institute for support. We thank Bernd Sturmfels for pointing out the relevance of Theorem 5.3. We also thank an anonymous referee for pointing out the phenomena illustrated in Example 3.17, leading us to correct an earlier version of Theorem 3.5.

References

[1] V. Arnol'd, The cohomology ring of the group of dyed braids, Mat. Zametki, **5** (1969), 227–231, MR0242196.

[2] C. Bibby, M. Falk and I. Williams, Decomposable cocycles for p-generic arrangements, in preparation.

[3] R. Bieri and J. R. J. Groves, The geometry of the set of characters induced by valuations, J. Reine Angew. Math., **347** (1984), 168–195, MR733052.

[4] E. Brieskorn, Sur les groupes de tresses, In: Séminaire Bourbaki, 1971/72, **317** Lecture Notes in Math., Springer-Verlag, 1973, pp. 21–44, MR0422674.

[5] D. Cohen, G. Denham, M. Falk and A. Varchenko, Critical points and resonance of hyperplane arrangements, Canad. J. Math., **63** (2011), 1038–1057.

[6] D. Cohen and A. Varchenko, Resonant local systems and representations of $\mathfrak{sl}_2$, Geom. Dedicata, **101** (2003), 217–233, MR2017904.

[7] A. Dickenstein, E.-M. Feichtner and B. Sturmfels, Tropical discriminants, J. Amer. Math. Soc., **20** (2007), 1111–1133, MR2328718.

[8] A. Dimca, Singularities and Topology of Hypersurfaces, Universitext, Springer-Verlag, 1992, MR1194180.

[9] D. Eisenbud, Commutative Algebra with a View toward Algebraic Geometry, Grad. Texts in Math., **150**, Springer-Verlag, 1995, MR1322960.

[10] M. Falk and S. Yuzvinsky, Multinets, resonance varieties, and pencils of plane curves, Compos. Math., **143** (2007), 1069–1088, MR2339840.

[11] E.-M. Feichtner and B. Sturmfels, Matroid polytopes, nested sets, and Bergman fans, Port. Math., **62** (2005), 437–468, MR2191630.

[12] D. Grayson and M. Stillman, Macaulay 2: a software system for algebraic geometry and commutative algebra, available at www.math.uiuc.edu/Macaulay2.

[13] R. Hartshorne, Algebraic Geometry, Grad. Texts in Math., **52**, Springer-Verlag, 1977, MR0463157.

[14] A. Libgober and S. Yuzvinsky, Cohomology of the Orlik–Solomon algebras and local systems, Compos. Math., **121** (2000), 337–361, MR1761630.

[15] P. Lima-Filho and H. Schenck, Efficient computation of resonance varieties via Grassmannians, J. Pure Appl. Algebra, **213** (2009), 1606–1611, MR2517996.

[16] E. Mukhin and A. Varchenko, Critical points of master functions and flag varieties, Commun. Contemp. Math., **6** (2004), 111–163, MR2048778.

[17] E. Mukhin and A. Varchenko, Norm of a Bethe vector and the Hessian of the master function, Compos. Math., **141** (2005), 1012–1028, MR2148192.

[18] E. Mukhin and A. Varchenko, Miura opers and critical points of master functions, Cent. Eur. J. Math., **3** (2005), 155–182 (electronic), MR2129922.

[19] P. Orlik and H. Terao, Arrangements of Hyperplanes, Grundlehren Math. Wiss., **300**, Springer-Verlag, 1992, MR1217488.

[20] P. Orlik and H. Terao, The number of critical points of a product of powers of linear functions, Invent. Math., **120** (1995), 1–14, MR1323980.

[21] J. Oxley, Matroid Theory, Oxford Univ. Press, New York, 1992, MR1207587.

[22] J. Pereira and S. Yuzvinsky, Completely reducible hypersurfaces in a pencil, Adv. Math., **219** (2008), 672–688, MR2435653.

[23] N. Reshetikhin and A. Varchenko, Quasiclassical asymptotics of solutions to the KZ equations, In: Geometry, Topology, and Physics for Raoul Bott, Conf. Proc. Lecture Notes Geom. Topology, **IV**, International Press, Cambridge, MA, 1995, pp. 293–322, MR1358621.

[24] V. Schechtman and A. Varchenko, Arrangements of hyperplanes and Lie algebra homology, Invent. Math., **106** (1991), 139–194, MR1123378.

[25] I. Scherbak and A. Varchenko, Critical points of functions, $\mathfrak{sl}_2$ representations, and Fuchsian differential equations with only univalued solutions, Mosc. Math. J., **3** (2003), 621–645, MR2025276.

[26] R. Silvotti, On a conjecture of Varchenko, Invent. Math., **126** (1996), 235–248, MR1411130.

[27] A. Varchenko, Critical points of the product of powers of linear functions and families of bases of singular vectors, Compos. Math., **97** (1995), 385–401, MR1353281.

[28] A. Varchenko, Special functions, KZ type equations, and representation theory, CBMS Regional Conf. Ser. in Math., **98**, Amer. Math. Soc. Providence, RI, 2003, MR2016311.

[29] A. Varchenko, Quantum Integrable Model of an Arrangement of Hyperplanes, SIGMA, **7** (2011), 032, 55 pp. http://dx.doi.org/10.3842/SIGMA.2011.032.

[30] S. Yuzvinsky, Cohomology of the Brieskorn-Orlik-Solomon algebras, Comm. Algebra, **23** (1995), 5339–5354, MR1363606.

[31] S. Yuzvinsky, A new bound on the number of special fibers in a pencil of curves. Proc. Amer. Math. Soc., **137** (2008), 1641–1648, MR2470822.

Daniel C. Cohen
Department of Mathematics, Louisiana State University
Baton Rouge, LA 70803, USA
E-mail address: `cohen@math.lsu.edu`

Graham Denham
Department of Mathematics, University of Western Ontario
London, ON N6A 5B7, Canada
E-mail address: `gdenham@uwo.ca`

Michael Falk
Department of Mathematics and Statistics, Northern Arizona University,
Flagstaff, AZ 86011, USA
E-mail address: `michael.falk@nau.edu`

Alexander Varchenko
Department of Mathematics, University of North Carolina at Chapel Hill
Chapel Hill, NC 27599, USA
E-mail address: `anv@email.unc.edu`

Advanced Studies in Pure Mathematics 62, 2012
Arrangements of Hyperplanes—Sapporo 2009
pp. 109–136

Middle convolution for completely integrable systems with logarithmic singularities along hyperplane arrangements

Yoshishige Haraoka

Abstract.

The middle convolution for completely integrable systems with logarithmic singularities along hyperplane arrangements is defined as a natural generalization of the middle convolution for Fuchsian ordinary differential equations. Additivity of the generalized middle convolution is proved. It is observed that the singular locus may increase by the generalized middle convolution. Examples concerning with hypergeometric series in several variables are given.

§1. Introduction

The middle convolution, which is introduced by Katz [7], is an operation for local systems on a punctured complex line, and plays a fundamental role in the theory of rigid local systems. By Riemann–Hilbert correspondence, it induces an operation for Fuchsian ordinary differential equations, which we also call the middle convolution. In this paper we extend the latter one to the operation for completely integrable systems in several variables with logarithmic singularities along hyperplane arrangements. The extended middle convolution possesses similar properties as the original one, such as additivity, and, on the other hand, gives several different features from the original one. In particular, by combining with the prolongation-restriction process, the extended middle convolution may change the index of rigidity, which is invariant under the original one.

Received March 24, 2010.
Revised October 29, 2010.
2010 *Mathematics Subject Classification.* 58A17, 33C65.
Key words and phrases. Rigid local system, integrable system.
Supported by the JSPS grant-in-aid for scientific research B, No.21340038.

According to Dettweiler–Reiter [1], we give a definition of the middle convolution for a Fuchsian system

$$\frac{dU}{dx} = \left(\sum_{i=1}^{r} \frac{A_i}{x-a_i}\right) U, \tag{1}$$

where $a_1, a_2, \ldots, a_r \in \mathbb{C}$ are distinct points, and $A_1, A_2, \ldots, A_r$ are constant $n \times n$-matrices. Take $\lambda \in \mathbb{C}$. Define $rn \times rn$-matrices $G_1, G_2, \ldots, G_r$ by

$$G_i = \sum_{j=1}^{r} E_{ij} \otimes (A_j + \delta_{ij}\lambda) \quad (1 \le i \le r),$$

where E_{ij} denotes the $r \times r$-matrix with the only non-zero entry 1 at (i,j)-th position $(1 \le i, j \le r)$. The operation which sends the system (1) to the system

$$\frac{d\hat{U}}{dx} = \left(\sum_{i=1}^{r} \frac{G_i}{x-a_i}\right) \hat{U} \tag{2}$$

is called the *convolution* with parameter λ, and is denoted by c_λ.

Let $\mathcal{K}$ and $\mathcal{L}$ be the subspaces of $\mathbb{C}^{rn}$ defined by

$$\mathcal{K} = \left\{ \begin{pmatrix} v_1 \\ \vdots \\ v_r \end{pmatrix} ; v_i \in \mathrm{Ker} A_i \ (1 \le i \le r) \right\},$$

$$\mathcal{L} = \mathrm{Ker}(G_1 + G_2 + \cdots + G_r).$$

It is easy to see that $\mathcal{K}$ and $\mathcal{L}$ are invariant subspaces of $\mathbb{C}^{rn}$ for $(G_1, G_2, \ldots, G_r)$. Then $(G_1, G_2, \ldots, G_r)$ induces the action $(\bar{G}_1, \bar{G}_2, \ldots, \bar{G}_r)$ on the quotient space $\mathbb{C}^{rn}/(\mathcal{K}+\mathcal{L})$. The operation which sends (1) to the system

$$\frac{d\bar{U}}{dx} = \left(\sum_{i=1}^{r} \frac{\bar{G}_i}{x-a_i}\right) \bar{U} \tag{3}$$

is called the *middle convolution* with parameter λ, and is denoted by mc_λ.

It is shown that the middle convolution keeps the index of rigidity invariant, and if (1) is irreducible then the middle convolution system (3) is also irreducible. Moreover, if the residue matrices of (1) satisfies the conditions

$$\bigcap_{j \ne i} \mathrm{Ker} A_j \cap \mathrm{Ker}(A_i - c) = 0 \quad (\forall i, \forall c \in \mathbb{C})$$

and

$$\sum_{j\neq i}\mathrm{Im}A_j+\mathrm{Im}(A_i-c)=\mathbb{C}^n \quad (\forall i,\ \forall c\in\mathbb{C}),$$

we have the additivity

$$mc_0=\mathrm{id},\quad mc_{\lambda_2}\circ mc_{\lambda_1}=mc_{\lambda_1+\lambda_2}.$$

It is also shown that the middle convolution keeps the deformation equation invariant ([5]).

One of the main result of the theory of rigid local systems is the following.

Theorem 1.1. ([1, Theorem A.14]) *Any irreducible rigid Fuchsian system can be obtained from a Fuchsian equation of rank 1 by a finite iteration of middle convolutions and additions.*

It is shown in [2] that the middle convolution is analytically realized in the following way. For a solution $U(x)$ of (1), define rn-column vector $V(x)$ by

$$V(x)=\begin{pmatrix}\frac{U(x)}{x-a_1}\\ \frac{U(x)}{x-a_2}\\ \vdots\\ \frac{U(x)}{x-a_r}\end{pmatrix}.$$

Let $\hat{U}(x)$ be a Riemann–Liouville transform with exponent λ of $V(x)$:

$$\hat{U}(x)=\int_\Delta V(t)(t-x)^\lambda\,dt,$$

where Δ is an appropriate 1-cycle. Then $\hat{U}(x)$ makes a solution of the convolution system (2). Solutions of the middle convolution system (3) can be obtained from $\hat{U}(x)$ by the linear transformation induced by the projection $\mathbb{C}^{rn}\to\mathbb{C}^{rn}/(\mathcal{K}+\mathcal{L})$.

We shall use this analytic realization to extend the middle convolution for completely integrable systems in several variables. This will be done in Section 2. In Section 3 we show that the additivity of the middle convolution also holds for the extended case. In Section 4 we give two examples concerning with hypergeometric series in several variables.

In our previous work [4] we defined the middle convolution for completely integrable systems of KZ type. The results in the present paper is a generalization to completely integrable systems with any arrangements of hyperplanes as logarithmic singularities.

§2. Middle convolution

We consider an arrangement $\mathcal{A}$ of hyperplanes in $\mathbb{C}^l$. We fix a coordinate $(x_1, x_2, \ldots, x_l)$ of $\mathbb{C}^l$. For each hyperplane $H \in \mathcal{A}$, let f_H be a defining linear polynomial for H. For each $H \in \mathcal{A}$, we take a constant $n \times n$-matrix A_H.

We consider the Pfaffian system

$$dU = \Omega U, \tag{4}$$

where Ω is a 1-form given by

$$\Omega = \sum_{H \in \mathcal{A}} A_H d \log f_H. \tag{5}$$

We assume that the system (4) is completely integrable:

$$\Omega \wedge \Omega = O.$$

This condition will be stated in terms of A_H in Theorem 2.1 below.

For each i $(1 \leq i \leq l)$, we set

$$\mathcal{A}_{x_i} = \{H \in \mathcal{A};\ (f_H)_{x_i} \neq 0\}.$$

Then the system (4) can be written as

$$\frac{\partial U}{\partial x_i} = \left(\sum_{H \in \mathcal{A}_{x_i}} A_H \frac{(f_H)_{x_i}}{f_H} \right) U \tag{6}$$

for $1 \leq i \leq l$. In this paper we call the system (6) the x_i-equation.

Take any i $(1 \leq i \leq l)$, and set $x_i = x$. For each $H \in \mathcal{A}_x$ we define a_H by

$$f_H = (f_H)_x (x - a_H). \tag{7}$$

Then the x-equation can be written as

$$\frac{\partial U}{\partial x} = \left(\sum_{H \in \mathcal{A}_x} \frac{A_H}{x - a_H} \right) U. \tag{8}$$

Let j $(1 \leq j \leq l)$ be another index, and set $x_j = y$. For each $H \in \mathcal{A}_y$, we define b_H by

$$f_H = (f_H)_y (y - b_H).$$

For $H, H' \in \mathcal{A}_x$ such that $(a_H - a_{H'})_y \neq 0$, we define $c_{HH'}$ by

$$a_H - a_{H'} = (a_H - a_{H'})_y (y - c_{HH'}).$$

Then obviously

$$c_{HH'} = c_{H'H}$$

holds. For $H \in \mathcal{A}_x$ we set

$$\mathcal{C}_{H,y} = \{H' \in \mathcal{A}_x \setminus \{H\}\,;\, (a_H - a_{H'})_y \neq 0\}.$$

Theorem 2.1. *The integrability condition $\Omega \wedge \Omega = O$ for the system (4) holds if and only if, for any pair (i, j) $(1 \leq i, j \leq l)$ of distinct indices, by setting $x_i = x$ and $x_j = y$, the following conditions hold:*

$$\left[A_H, \sum_{\substack{H'' \in \mathcal{C}_{H,y} \\ c_{HH''} = b_{H'}}} A_{H''} + A_{H'} \right] = O \tag{9}$$

for $H \in \mathcal{A}_x \cap \mathcal{A}_y$ and $H' \in \mathcal{A}_x^c \cap \mathcal{A}_y$,

$$\left[A_H, \sum_{\substack{H'' \in \mathcal{C}_{H,y} \\ c_{HH''} = c_{HH'}}} A_{H''} \right] = O \tag{10}$$

for $H \in \mathcal{A}_x \cap \mathcal{A}_y$ and $H' \in \mathcal{C}_{H,y}$ such that $c_{HH'} \neq b_K$ for any $K \in \mathcal{A}_x^c \cap \mathcal{A}_y$,

$$\left[A_H, \sum_{\substack{H'' \in \mathcal{A}_x \cap \mathcal{A}_y \\ c_{HH''} = b_{H'}}} A_{H''} + A_{H'} \right] = O \tag{11}$$

for $H \in \mathcal{A}_x \cap \mathcal{A}_y^c$ and $H' \in \mathcal{A}_x^c \cap \mathcal{A}_y$, and

$$\left[A_H, \sum_{\substack{H'' \in \mathcal{A}_x \cap \mathcal{A}_y \\ c_{HH''} = c_{HH'}}} A_{H''} \right] = O \tag{12}$$

for $H \in \mathcal{A}_x \cap \mathcal{A}_y^c$ and $H' \in \mathcal{A}_x \cap \mathcal{A}_y$ such that $c_{HH'} \neq b_K$ for any $K \in \mathcal{A}_x^c \cap \mathcal{A}_y$.

Proof. We regard the coefficients of the 2-form $\Omega \wedge \Omega$ as rational functions in x. The principal part of the Laurent series expansion at each pole can be regarded as a rational function in y, and then we get the left hand sides of (9) to (12) as the residue matrices of the poles in y, from which the assertion follows. Q.E.D.

Let $\mathcal{F}$ be the local system on $\mathbb{C} \setminus \{a_H\,;\, H \in \mathcal{A}_x\}$ of local solutions of (8). Let λ be a complex number. For each $H \in \mathcal{A}_x$, we define $V_H(x_1, x_2, \ldots, x_l)$ by

$$V_H(x_1, x_2, \ldots, x_l) = \int_\Delta \frac{U(x_1, \ldots, x_{i-1}, t, x_{i+1}, \ldots, x_l)}{t - a_H}(t-x)^\lambda \, dt, \tag{13}$$

where $\Delta \in H_1(\mathbb{C} \setminus \{a_H\,;\, H \in \mathcal{A}_x\}, \mathcal{F})$ and $x = x_i$. We set

$$V(x_1, x_2, \ldots, x_l) = (V_H(x_1, x_2, \ldots, x_l))_{H \in \mathcal{A}_x}. \tag{14}$$

Proposition 2.1. *The function $V = (V_H)_{H \in \mathcal{A}_x}$ defined by (13) satisfies the following systems of differential equations:*
For $x = x_i$, the x-equation for V is given by

$$\frac{\partial V_H}{\partial x} = \frac{1}{x - a_H} \sum_{H' \in \mathcal{A}_x} (A_{H'} + \delta_{HH'}\lambda) V_{H'}. \tag{15}$$

For other x_j, we set $x_j = y$. Then the y-equation for V is given by

$$\begin{aligned} \frac{\partial V_H}{\partial y} &= \frac{1}{y - b_H} \sum_{H' \in \mathcal{A}_x} (A_{H'} + \delta_{HH'}\lambda) V_{H'} \\ &\quad + \sum_{H' \in \mathcal{C}_{H,y}} \frac{A_{H'}}{y - c_{HH'}}(V_H - V_{H'}) + \sum_{H' \in \mathcal{A}_x^c \cap \mathcal{A}_y} \frac{A_{H'}}{y - b_{H'}} V_H \end{aligned} \tag{16}$$

for $H \in \mathcal{A}_x \cap \mathcal{A}_y$, and

$$\frac{\partial V_H}{\partial y} = \sum_{H' \in \mathcal{A}_x \cap \mathcal{A}_y} \frac{A_{H'}}{y - c_{HH'}}(V_H - V_{H'}) + \sum_{H' \in \mathcal{A}_x^c \cap \mathcal{A}_y} \frac{A_{H'}}{y - b_{H'}} V_H \tag{17}$$

for $H \in \mathcal{A}_x \cap \mathcal{A}_y^c$.

Proof. As explained in Introduction, the x-equation for V is nothing but the convolution of the x-equation for U, and then is given by (15).

To obtain the y-equation with $y = x_j$, we regard x_k $(k \neq i, j)$ as a parameter, and use the notation $U(x_1, x_2, \ldots, x_l) = U(x_i, x_j) = U(x, y)$. The partial derivative of V_H with respect to y is

$$\frac{\partial V_H}{\partial y} = \int_\Delta \frac{\partial U}{\partial y}(t, y) \frac{(t-x)^\lambda}{t - a_H}\, dt + \int_\Delta U(t, y)(t-x)^\lambda \frac{\partial}{\partial y}\left(\frac{1}{t - a_H}\right) dt.$$

Note that

$$\frac{\partial U}{\partial y}(t, y) = \left(\sum_{H' \in \mathcal{A}_y} \frac{A_{H'}}{y - b_{H'}}\right) U(t, y).$$

If $H \notin \mathcal{A}_y$, a_H is independent of y, and hence we have

$$\begin{aligned}
\frac{\partial V_H}{\partial y} &= \sum_{H' \in \mathcal{A}_y} A_{H'} \int_\Delta \frac{U(t, y)}{y - b_{H'}} \cdot \frac{(t-x)^\lambda}{t - a_H}\, dt \\
&= \sum_{H' \in \mathcal{A}_x \cap \mathcal{A}_y} A_{H'} \int_\Delta U(t, y)(t-x)^\lambda \frac{(f_{H'})_y}{(f_{H'})_x (t - a_{H'})} \cdot \frac{1}{t - a_H}\, dt \\
&\quad + \sum_{H' \in \mathcal{A}_x^c \cap \mathcal{A}_y} \frac{A_{H'}}{y - b_{H'}} \int_\Delta \frac{U(t, y)}{t - a_H}(t-x)^\lambda\, dt \\
&= \sum_{H' \in \mathcal{A}_x \cap \mathcal{A}_y} \frac{(f_{H'})_y}{(f_{H'})_x} \cdot \frac{A_{H'}}{a_H - a_{H'}}(V_H - V_{H'}) + \sum_{H' \in \mathcal{A}_x^c \cap \mathcal{A}_y} \frac{A_{H'}}{y - b_{H'}} V_H.
\end{aligned}$$

Since $(a_H)_y = 0$, we have

$$\frac{(f_{H'})_y}{(f_{H'})_x} \cdot \frac{1}{a_H - a_{H'}} = \frac{1}{y - c_{HH'}}$$

for $H' \in \mathcal{A}_x \cap \mathcal{A}_y$, and hence we get (17).

If $H \in \mathcal{A}_y$, there appear the two terms

$$\int_\Delta A_H \frac{U(t, y)}{y - b_H} \cdot \frac{(t-x)^\lambda}{t - a_H}\, dt = \int_\Delta A_H \frac{(f_H)_y}{(f_H)_x} \cdot \frac{U(t, y)}{(t - a_H)^2}(t-x)^\lambda\, dt$$

and

$$-\int_\Delta \frac{(f_H)_x}{(f_H)_y} \cdot \frac{U(t, y)}{(y - b_H)^2}(t-x)^\lambda\, dt = -\int_\Delta \frac{(f_H)_y}{(f_H)_x} \cdot \frac{U(t, y)}{(x - a_H)^2}(t-x)^\lambda\, dt$$

in $\partial V_H/\partial y$. Since Δ is a twisted 1-cycle, we have

$$\begin{aligned}
\int_\Delta \frac{U(t,y)}{(t-a_H)^2}(t-x)^\lambda\,dt &= \int_\Delta \frac{\partial}{\partial t}\left(-\frac{1}{t-a_H}\right)U(t,y)(t-x)^\lambda\,dt \\
&= \int_\Delta \frac{1}{t-a_H}\frac{\partial}{\partial t}\left(U(t,y)(t-x)^\lambda\right)\,dt \\
&= \int_\Delta \frac{1}{t-a_H}\Bigg(\sum_{H'\in\mathcal{A}_x}\frac{A_{H'}}{t-a_{H'}}U(t,y)(t-x)^\lambda \\
&\qquad\qquad + \lambda U(t,y)(t-x)^{\lambda-1}\Bigg)\,dt \\
&= \int_\Delta \frac{A_H}{(t-a_H)^2}U(t,y)(t-x)^\lambda\,dt \\
&\qquad + \sum_{H'\in\mathcal{A}_x\setminus\{H\}}\frac{A_{H'}}{a_H-a_{H'}}(V_H-V_{H'}) - \frac{\partial V_H}{\partial x},
\end{aligned}$$

and then we get

$$\begin{aligned}
&\int_\Delta A_H\frac{(f_H)_y}{(f_H)_x}\cdot\frac{U(t,y)}{(t-a_H)^2}(t-x)^\lambda\,dt - \int_\Delta \frac{(f_H)_y}{(f_H)_x}\cdot\frac{U(t,y)}{(x-a_H)^2}(t-x)^\lambda\,dt \\
&= \frac{(f_H)_y}{(f_H)_x}\left\{\frac{\partial V_H}{\partial x} - \sum_{H'\in\mathcal{A}_x\setminus\{H\}}\frac{A_{H'}}{a_H-a_{H'}}(V_H-V_{H'})\right\}.
\end{aligned}$$

Thus we have, for $H\in\mathcal{A}_x\cap\mathcal{A}_y$,

$$\begin{aligned}
\frac{\partial V_H}{\partial y} &= \frac{(f_H)_y}{(f_H)_x}\frac{\partial V_H}{\partial x} - \frac{(f_H)_y}{(f_H)_x}\sum_{H'\in\mathcal{A}_x\setminus\{H\}}\frac{A_{H'}}{a_H-a_{H'}}(V_H-V_{H'}) \\
&\quad + \sum_{H'\in\mathcal{A}_x\cap\mathcal{A}_y\setminus\{H\}}\frac{(f_{H'})_y}{(f_{H'})_x}\frac{A_{H'}}{a_H-a_{H'}}(V_H-V_{H'}) \\
&\quad + \sum_{H'\in\mathcal{A}_x^c\cap\mathcal{A}_y}\frac{A_{H'}}{y-b_{H'}}V_H \\
&= \frac{(f_H)_y}{(f_H)_x}\frac{1}{x-a_H}\sum_{H'\in\mathcal{A}_x}(A_{H'}+\delta_{HH'}\lambda)V_{H'} \\
&\quad + \sum_{H'\in\mathcal{A}_x\cap\mathcal{A}_y\setminus\{H\}}\left(\frac{(f_{H'})_y}{(f_{H'})_x}-\frac{(f_H)_y}{(f_H)_x}\right)\frac{A_{H'}}{a_H-a_{H'}}(V_H-V_{H'}) \\
&\quad - \frac{(f_H)_y}{(f_H)_x}\sum_{H'\in\mathcal{A}_x\cap\mathcal{A}_y^c}\frac{A_{H'}}{a_H-a_{H'}}(V_H-V_{H'}) + \sum_{H'\in\mathcal{A}_x^c\cap\mathcal{A}_y}\frac{A_{H'}}{y-b_{H'}}V_H.
\end{aligned}$$

By the definitions of a_H, b_H and $c_{HH'}$, we have

$$\frac{(f_H)_y}{(f_H)_x}\frac{1}{x-a_H}=\frac{1}{y-b_H}$$

for $H \in \mathcal{A}_x \cap \mathcal{A}_y$,

$$\left(\frac{(f_{H'})_y}{(f_{H'})_x}-\frac{(f_H)_y}{(f_H)_x}\right)\frac{1}{a_H-a_{H'}}=\frac{1}{y-c_{HH'}}$$

for $H \in \mathcal{A}_x \cap \mathcal{A}_y$ and $H' \in \mathcal{C}_{H,y}$,

$$\frac{(f_{H'})_y}{(f_{H'})_x}-\frac{(f_H)_y}{(f_H)_x}=0$$

for $H \in \mathcal{A}_x \cap \mathcal{A}_y$ and $H' \notin \mathcal{C}_{H,y}$, and

$$-\frac{(f_H)_y}{(f_H)_x}\cdot\frac{1}{a_H-a_{H'}}=\frac{1}{y-c_{HH'}}$$

for $H \in \mathcal{A}_x \cap \mathcal{A}_y$ and $H' \in \mathcal{A}_x \cap \mathcal{A}_y^c$. Thus we obtain (16). Q.E.D.

We see that the systems for $V=(V_H)_{H\in\mathcal{A}_x}$ given in Proposition 2.1 make a Pfaffian system

$$dV=\hat{\Omega}V \tag{18}$$

with 1-form $\hat{\Omega}$ of the form

$$\hat{\Omega}=\sum_{\hat{H}\in\hat{\mathcal{A}}} G_{\hat{H}}\, d\log f_{\hat{H}},$$

where $\hat{\mathcal{A}}$ is the arrangement of hyperplanes obtained from $\mathcal{A}$ by adding the hyperplanes of the form $\{y-c_{HH'}=0\}$.

Definition 2.1. We call the operation which sends the Pfaffian system (4) to the system (18) the *convolution* in x_i-direction with parameter λ, and denote it by $c_\lambda^{x_i}$.

Remark 2.1. If there is a pair $H, H' \in \mathcal{A}_x$ such that $H' \in \mathcal{C}_{H,y}$ and $c_{HH'} \neq b_K$ for any $K \in \mathcal{A}_x^c \cap \mathcal{A}_y$, we have $\{y-c_{HH'}=0\} \in \hat{\mathcal{A}} \setminus \mathcal{A}$, and hence $\hat{\mathcal{A}} \supsetneq \mathcal{A}$. Thus the singular locus may increase by the convolution, which is a different feature from the case of one variable.

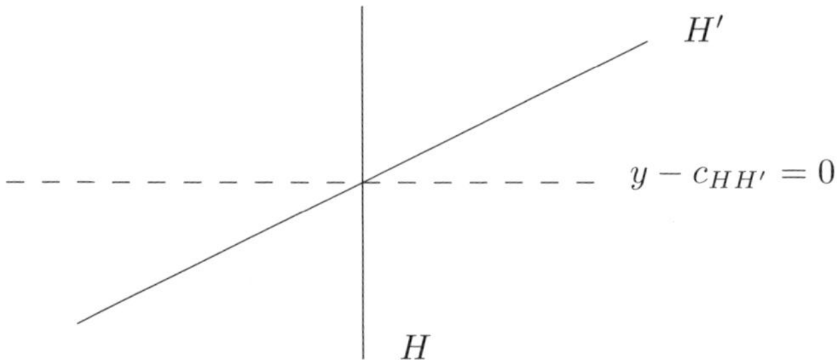

Fig. 1.

Let $\mathcal{K}$ and $\mathcal{L}$ be the subspaces of $(\mathbb{C}^n)^{\mathcal{A}_x} \simeq \mathbb{C}^{n\#\mathcal{A}_x}$ given by

$$\mathcal{K} = \{v = (v_H)_{H\in\mathcal{A}_x}\,;\, v_H \in \mathrm{Ker} A_H\},$$
$$\mathcal{L} = \{v = (v_H)_{H\in\mathcal{A}_x}\,;\, \sum_{H'\in\mathcal{A}_x} A_{H'}v_{H'} + \lambda v_H = 0\ (\forall H \in \mathcal{A}_x)\}.$$

It is easy to see that

$$\mathcal{L} = \{v = (v_0)\,;\, \left(\sum_{H\in\mathcal{A}_x} A_H\right) v_0 + \lambda v_0 = 0\}$$

if $\lambda \neq 0$, and

$$\mathcal{L} = \{v = (v_H)_{H\in\mathcal{A}_x}\,;\, \sum_{H\in\mathcal{A}_x} A_H v_H = 0\}$$

if $\lambda = 0$. As mentioned in Introduction, $\mathcal{K}$ and $\mathcal{L}$ are invariant subspaces for the x-equation of the convolution system (18). For $y = x_j$ $(j \neq i)$, we write the y-equation (16) and (17) of the system (18) as

$$\frac{\partial V}{\partial y} = G(y)V.$$

Proposition 2.2. *For any x_j $(j \neq i)$, the subspaces $\mathcal{K}$ and $\mathcal{L}$ are invariant for the x_j-equation of the convolution system (18). Namely, by setting $x_j = y$,*

$$G(y)\mathcal{K} \subset \mathcal{K}, \quad G(y)\mathcal{L} \subset \mathcal{L}$$

hold for any value of y.

Proof. Take any $v = (v_H)_{H\in\mathcal{A}_x} \in \mathcal{K}$, and set

$$G(y)v = w = (w_H)_{H\in\mathcal{A}_x}.$$

For $H \in \mathcal{A}_x \cap \mathcal{A}_y$, noting that $v_{H'} \in \mathrm{Ker} A_{H'}$, we have

$$\begin{aligned} w_H &= \frac{1}{y - b_H} \sum_{H' \in \mathcal{A}_x} (A_{H'} + \delta_{HH'}\lambda) v_{H'} \\ &\quad + \sum_{H' \in \mathcal{C}_{H,y}} \frac{A_{H'}}{y - c_{HH'}} (v_H - v_{H'}) + \sum_{H' \in \mathcal{A}_x^c \cap \mathcal{A}_y} \frac{A_{H'}}{y - b_{H'}} v_H \\ &= \left(\frac{\lambda}{y - b_H} + \sum_{H' \in \mathcal{C}_{H,y}} \frac{A_{H'}}{y - c_{HH'}} + \sum_{H' \in \mathcal{A}_x^c \cap \mathcal{A}_y} \frac{A_{H'}}{y - b_{H'}} \right) v_H. \end{aligned}$$

By using the integrability conditions (9) and (10), we get

$$\begin{aligned} A_H w_H &= \left(\frac{\lambda}{y - b_H} + \sum_{H' \in \mathcal{C}_{H,y}} \frac{A_{H'}}{y - c_{HH'}} + \sum_{H' \in \mathcal{A}_x^c \cap \mathcal{A}_y} \frac{A_{H'}}{y - b_{H'}} \right) A_H v_H \\ &= 0, \end{aligned}$$

and hence $w_H \in \mathrm{Ker} A_H$. Similarly, by the help of the integrability conditions (11) and (12), we get $w_H \in \mathrm{Ker} A_H$ for $H \in \mathcal{A}_x \cap \mathcal{A}_y^c$. Thus we obtain $w \in \mathcal{K}$.

Next we assume $\lambda \neq 0$, and take any $v = (v_H)_{H \in \mathcal{A}_x} \in \mathcal{L}$. Then we have $v_H = v_0$ $(\forall H)$ and $\left(\sum_{H \in \mathcal{A}_x} A_H\right) v_0 + \lambda v_0 = 0$. Set

$$G(y)v = w = (w_H)_{H \in \mathcal{A}_x}.$$

Then for $H \in \mathcal{A}_x \cap \mathcal{A}_y$, we have

$$\begin{aligned} w_H &= \frac{1}{y - b_H} \sum_{H' \in \mathcal{A}_x} (A_{H'} + \delta_{HH'}\lambda) v_0 \\ &\quad + \sum_{H' \in \mathcal{C}_{H,y}} \frac{A_{H'}}{y - c_{HH'}} (v_0 - v_0) + \sum_{H' \in \mathcal{A}_x^c \cap \mathcal{A}_y} \frac{A_{H'}}{y - b_{H'}} v_0 \\ &= \left(\sum_{H' \in \mathcal{A}_x^c \cap \mathcal{A}_y} \frac{A_{H'}}{y - b_{H'}} \right) v_0. \end{aligned}$$

Similarly, for $H \in \mathcal{A}_x \cap \mathcal{A}_y^c$ we have

$$\begin{aligned} w_H &= \sum_{H' \in \mathcal{A}_x \cap \mathcal{A}_y} \frac{A_{H'}}{y - c_{HH'}} (v_0 - v_0) + \sum_{H' \in \mathcal{A}_x^c \cap \mathcal{A}_y} \frac{A_{H'}}{y - b_{H'}} v_0 \\ &= \left(\sum_{H' \in \mathcal{A}_x^c \cap \mathcal{A}_y} \frac{A_{H'}}{y - b_{H'}} \right) v_0. \end{aligned}$$

Thus, for every $H \in \mathcal{A}_x$, w_H becomes the same vector, which we denote by w_0.

Take any $H' \in \mathcal{A}_x^c \cap \mathcal{A}_y$. Let p be an intersection point of H' and a hyperplane in $\mathcal{A}_x$. We denote by $\mathcal{B}_p$ the set of hyperplanes in $\mathcal{A}_x$ which pass through p. Suppose $H \in \mathcal{B}_p$. If $H \in \mathcal{A}_x \cap \mathcal{A}_y$, we have

$$\{H'' \in \mathcal{C}_{H,y}\,;\, c_{HH''} = b_{H'}\} = \mathcal{B}_p \setminus \{H\}.$$

If $H \in \mathcal{A}_x \cap \mathcal{A}_y^c$, we have

$$\{H'' \in \mathcal{A}_x \cap \mathcal{A}_y\,;\, c_{HH''} = b_{H'}\} = \mathcal{B}_p \setminus \{H\}.$$

Thus the integrability conditions (9) and (11) can be written in the same form as

$$\left[A_H, \sum_{H'' \in \mathcal{B}_p \setminus \{H\}} A_{H''} + A_{H'}\right] = O$$

for every $H \in \mathcal{B}_p$, and hence we get

$$\left[A_H, \sum_{H'' \in \mathcal{B}_p} A_{H''} + A_{H'}\right] = O.$$

Summing up these relations for all $H \in \mathcal{B}_p$, we get

$$\left[\sum_{H \in \mathcal{B}_p} A_H, \sum_{H'' \in \mathcal{B}_p} A_{H''} + A_{H'}\right] = O,$$

from which we obtain

$$\left[\sum_{H \in \mathcal{B}_p} A_H, A_{H'}\right] = O.$$

Since $\bigcup_p \mathcal{B}_p = \mathcal{A}_x$, we have

$$O = \sum_p \left[\sum_{H \in \mathcal{B}_p} A_H, A_{H'}\right] = \left[\sum_{H \in \mathcal{A}_x} A_H, A_{H'}\right]$$

for every $H' \in \mathcal{A}_x^c \cap \mathcal{A}_y$. Using this relation, we get

$$\begin{aligned}\left(\sum_{H\in\mathcal{A}_x} A_H + \lambda\right) w_0 &= \left(\sum_{H\in\mathcal{A}_x} A_H + \lambda\right)\left(\sum_{H'\in\mathcal{A}_x^c\cap\mathcal{A}_y} \frac{A_{H'}}{y-b_{H'}}\right) v_0 \\ &= \left(\sum_{H'\in\mathcal{A}_x^c\cap\mathcal{A}_y} \frac{A_{H'}}{y-b_{H'}}\right)\left(\sum_{H\in\mathcal{A}_x} A_H + \lambda\right) v_0 \\ &= 0,\end{aligned}$$

which shows $w \in \mathcal{L}$.

Finally we consider the case $\lambda = 0$. Take any $v = (v_H)_{H\in\mathcal{A}_x} \in \mathcal{L}$. Then we have

$$\sum_{H\in\mathcal{A}_x} A_H v_H = 0.$$

We set

$$G(y)v = w = (w_H)_{H\in\mathcal{A}_x}.$$

Then we have, for $H \in \mathcal{A}_x \cap \mathcal{A}_y$,

$$\begin{aligned}w_H &= \frac{1}{y-b_H}\sum_{H'\in\mathcal{A}_x} A_{H'}v_{H'} + \sum_{H'\in\mathcal{C}_{H,y}} \frac{A_{H'}}{y-c_{HH'}}(v_H - v_{H'}) \\ &\quad + \sum_{H'\in\mathcal{A}_x^c\cap\mathcal{A}_y} \frac{A_{H'}}{y-b_{H'}} v_{H'} \\ &= \sum_{H'\in\mathcal{C}_{H,y}} \frac{A_{H'}}{y-c_{HH'}}(v_H - v_{H'}) + \sum_{H'\in\mathcal{A}_x^c\cap\mathcal{A}_y} \frac{A_{H'}}{y-b_{H'}} v_{H'},\end{aligned}$$

and for $H \in \mathcal{A}_x \cap \mathcal{A}_y^c$,

$$w_H = \sum_{H'\in\mathcal{A}\cap\mathcal{A}_y} \frac{A_{H'}}{y-c_{HH'}}(v_H - v_{H'}) + \sum_{H'\in\mathcal{A}_x^c\cap\mathcal{A}_y} \frac{A_{H'}}{y-b_{H'}} v_{H'}.$$

Take any $H' \in \mathcal{A}_x^c \cap \mathcal{A}_y$. We calculate the residue of $\sum_{H\in\mathcal{A}_x} A_H w_H$ at the pole $y = b_{H'}$. By the help of the integrability conditions (9) and

(11), the residue becomes

$$
\begin{aligned}
&\sum_{H\in\mathcal{A}_x\cap\mathcal{A}_y} A_H \left\{ \sum_{\substack{H''\in\mathcal{C}_{H,y}\\ c_{HH''}=b_{H'}}} A_{H''}(v_H - v_{H''}) + A_{H'}v_H \right\} \\
&+ \sum_{H\in\mathcal{A}_x\cap\mathcal{A}_y^c} A_H \left\{ \sum_{\substack{H''\in\mathcal{A}_x\cap\mathcal{A}_y\\ c_{HH''}=b_{H'}}} A_{H''}(v_H - v_{H''}) + A_{H'}v_H \right\} \\
=& \sum_{H\in\mathcal{A}_x\cap\mathcal{A}_y} \left(\sum_{\substack{H''\in\mathcal{C}_{H,y}\\ c_{HH''}=b_{H'}}} A_{H''} + A_{H'} \right) A_H v_H \\
&- \sum_{H\in\mathcal{A}_x\cap\mathcal{A}_y} A_H \sum_{\substack{H''\in\mathcal{C}_{H,y}\\ c_{HH''}=b_{H'}}} A_{H''}v_{H''} \\
&+ \sum_{H\in\mathcal{A}_x\cap\mathcal{A}_y^c} \left(\sum_{\substack{H''\in\mathcal{A}_x\cap\mathcal{A}_y\\ c_{HH''}=b_{H'}}} A_{H''} + A_{H'} \right) A_H v_H \\
&- \sum_{H\in\mathcal{A}_x\cap\mathcal{A}_y^c} A_H \sum_{\substack{H''\in\mathcal{A}_x\cap\mathcal{A}_y\\ c_{HH''}=b_{H'}}} A_{H''}v_{H''} \\
=& \sum_{H\in\mathcal{A}_x\cap\mathcal{A}_y} \left(\sum_{\substack{H''\in\mathcal{C}_{H,y}\\ c_{HH''}=b_{H'}}} A_{H''} \right) A_H v_H \\
&- \sum_{H\in\mathcal{A}_x\cap\mathcal{A}_y} A_H \sum_{\substack{H''\in\mathcal{C}_{H,y}\\ c_{HH''}=b_{H'}}} A_{H''}v_{H''} \\
&+ \sum_{H\in\mathcal{A}_x\cap\mathcal{A}_y^c} \left(\sum_{\substack{H''\in\mathcal{A}_x\cap\mathcal{A}_y\\ c_{HH''}=b_{H'}}} A_{H''} \right) A_H v_H \\
&- \sum_{H\in\mathcal{A}_x\cap\mathcal{A}_y^c} A_H \sum_{\substack{H''\in\mathcal{A}_x\cap\mathcal{A}_y\\ c_{HH''}=b_{H'}}} A_{H''}v_{H''}.
\end{aligned}
$$

In the last hand side, the coefficient of $A_H v_H$ for $H \in \mathcal{A}_x \cap \mathcal{A}_y$ is

$$\sum_{\substack{H'' \in \mathcal{C}_{H,y} \\ c_{HH''}=b_{H'}}} A_{H''} - \sum_{\substack{H'' \in \mathcal{C}_{H,y} \cap \mathcal{A}_y \setminus \{H\} \\ c_{HH''}=b_{H'}}} A_{H''} - \sum_{\substack{H'' \in \mathcal{A}_x \cap \mathcal{A}_y^c \\ c_{HH''}=b_{H'}}} A_{H''} = O,$$

and that for $H \in \mathcal{A}_x \cap \mathcal{A}_y^c$ is

$$\sum_{\substack{H'' \in \mathcal{A}_x \cap \mathcal{A}_y \\ c_{HH''}=b_{H'}}} A_{H''} - \sum_{\substack{H'' \in \mathcal{A}_x \cap \mathcal{A}_y \\ c_{HH''}=b_{H'}}} A_{H''} = O.$$

Thus the residue at $y = b_{H'}$ is O. In a similar way, we see that the residue of $\sum_{H \subset \mathcal{A}_x} A_H w_H$ at $y = c_{HH'}$ which does not coincide with any b_K $(K \in \mathcal{A}_x^c \cap \mathcal{A}_y)$ is also O. Hence we have

$$\sum_{H \in \mathcal{A}_x} A_H w_H = O,$$

which shows $w = (w_H)_{H \in \mathcal{A}_x} \in \mathcal{L}$. Q.E.D.

Thanks to Proposition 2.2, we can derive from the convolution system (18) a Pfaffian system

$$d\tilde{V} = \tilde{\Omega}\tilde{V} \tag{19}$$

on the quotient space $(\mathbb{C}^n)^{\mathcal{A}_x}/(\mathcal{K}+\mathcal{L})$. The matrix 1-form $\tilde{\Omega}$ is induced from $\hat{\Omega}$ as the action on $(\mathbb{C}^n)^{\mathcal{A}_x}/(\mathcal{K}+\mathcal{L})$.

Definition 2.2. We call the operation which sends the Pfaffian system (4) to the system (19) the *middle convolution* in x_i-direction with parameter λ, and denote it by $mc_\lambda^{x_i}$.

Theorem 2.2. *Let $\mathcal{A}$ be an arrangement of hyperplanes in $\mathbb{C}^l$, and, for each $H \in \mathcal{A}$, A_H be an $n \times n$-constant matrix. Define the 1-form Ω by (5). Assume that the Pfaffian system (4) is completely integrable.*

Then for any i $(1 \le i \le l)$ and $\lambda \in \mathbb{C}$, the middle convolution system (19) of (4) in x_i-direction with parameter λ is completely integrable, and admits an integral representation of solutions of the form

$$\begin{aligned} &\tilde{V}(x_1, x_2, \ldots, x_l) \\ &= Q\left(\int_\Delta \frac{U(x_1,\ldots,x_{i-1},t,x_{i+1},\ldots,x_l)}{t-a_H}(t-x_i)^\lambda\,dt\right)_{H \in \mathcal{A}_{x_i}}, \end{aligned} \tag{20}$$

where $U(x_1, x_2, \ldots, x_l)$ *is a solution of (4),* Q *a constant linear transformation, and* Δ *a twisted 1-cycle. If the* x_i*-equation of (4) is irreducible, the middle convolution system (18) is also irreducible.*

Remark 2.2. The linear transformation Q is given as follows. We set $N = n \cdot \#\mathcal{A}_{x_i}$ and $M = \dim(\mathcal{K}+\mathcal{L})$. Let P be an $N \times N$ non-singular matrix whose first M columns make a basis of $\mathcal{K}+\mathcal{L}$. Then we can take

$$Q = (O_{N-M,M}, I_{N-M})P^{-1}.$$

Proof. By the construction, solutions $V = (V_H)_{H \in \mathcal{A}_{x_i}}$ of the convolution system (18) are given by the integral (13). This implies that the convolution system (18) is completely integrable. The complete integrability of the middle convolution system (19) follows from Proposition 2.2.

Let P be as in the above Remark. Then, thanks to Proposition 2.2, the Pfaffian system of differential equations for $P^{-1}V$ becomes of block upper triangular form, and the solutions $\tilde{V}$ of the middle convolution system are related with V by

$$P^{-1}V = \begin{pmatrix} * \\ \tilde{V} \end{pmatrix}.$$

Thus we have

$$\tilde{V} = (O_{N-M,M}, I_{N-M})P^{-1}V = QV,$$

from which the expression (20) follows.

Since the x_i-equation of the middle convolution system (19) is just the middle convolution of the x_i-equation of (4) as ordinary differential equations, the irreducibility follows from the result by Katz [7, Theorem 2.9.8]. Q.E.D.

To illustrate the results in this section, we give an example. Let $A_1, A_2, \ldots, A_5$ be constant $m \times m$-matrices. We consider the Pfaffian system

$$dU = \left[A_1 \frac{dx}{x} + A_2 \frac{dy}{y} + A_3 \frac{dx}{x-1} + A_4 \frac{dy}{y-1} + A_5 \frac{d(x-y)}{x-y} \right] U, \tag{21}$$

which is called of KZ- type. We assume that the system (21) is completely integrable. The corresponding arrangement is $\mathcal{A} = \{H_1, H_2, \ldots, H_5\}$ with

$$H_1 = \{x = 0\}, \ H_2 = \{y = 0\}, \ H_3 = \{x = 1\},$$
$$H_4 = \{y = 1\}, \ H_5 = \{x = y\}.$$

Then we have

$$\mathcal{A}_x = \{H_1, H_3, H_5\}, \quad \mathcal{A}_y = \{H_2, H_4, H_5\}.$$

When we consider the (middle) convolution in x-direction, by the definition we have

$$a_{H_1} = 0,\ a_{H_3} = 1,\ b_{H_2} = 0,\ b_{H_4} = 1,$$
$$c_{H_1 H_5} = 0,\ c_{H_3 H_5} = 1.$$

The x-equation of the (middle) convolution of (21) in x-direction is nothing but the (middle) convolution of the x-equation

$$\frac{\partial U}{\partial x} = \left(\frac{A_1}{x} + \frac{A_3}{x-1} + \frac{A_5}{x-y}\right) U$$

of (21) as a system of ordinary differential equations, which is explained in the Introduction. To write down the y-equation (16) and (17) of the convolution in this case, we note that the unknown vector is given by $V = {}^t(V_{H_1}, V_{H_3}, V_{H_5})$, and that

$$\mathcal{A}_x \cap \mathcal{A}_y = \{H_5\}, \quad \mathcal{C}_{H_5,y} = \emptyset.$$

Then the y-equation of the convolution of (21) in x-direction with parameter λ is given by

$$\begin{aligned}
\frac{\partial V_{H_1}}{\partial y} &= \frac{A_5}{y}(V_{H_1} - V_{H_5}) + \frac{A_2}{y} V_{H_1} + \frac{A_4}{y-1} V_{H_1}, \\
\frac{\partial V_{H_3}}{\partial y} &= \frac{A_5}{y-1}(V_{H_3} - V_{H_5}) + \frac{A_2}{y} V_{H_3} + \frac{A_4}{y-1} V_{H_3}, \\
\frac{\partial V_{H_5}}{\partial y} &= \frac{1}{y-x}(A_1 V_{H_1} + A_3 V_{H_3} + (A_5 + \lambda) V_{H_5}) + \frac{A_2}{y} V_{H_5} + \frac{A_4}{y-1} V_{H_5}.
\end{aligned}$$

An explicit example will be given in the last section.

§3. Additivity of the middle convolution

We regard $\mathbb{C}^n$ as $\{A_H\,;\, H \in \mathcal{A}\}$-module, and denote it by $\mathcal{V}$. When we regard $\mathbb{C}^n$ as $\{A_H\,;\, H \in \mathcal{A}_x\}$-module, we use $\mathcal{V}_x$ instead of $\mathcal{V}$. Then the convolution and the middle convolution can be understood as operations for $\mathcal{V}$ or $\mathcal{V}_x$.

We assume

$$\bigcap_{H' \in \mathcal{A}_x \setminus \{H\}} \mathrm{Ker} A_{H'} \cap \mathrm{Ker}(A_H - c) = 0 \quad (\forall H \in \mathcal{A}_x, \forall c \in \mathbb{C}) \tag{22}$$

and

$$\sum_{H' \in \mathcal{A}_x \setminus \{H\}} \mathrm{Im} A_{H'} + \mathrm{Im}(A_H - c) = \mathbb{C}^n \quad (\forall H \in \mathcal{A}_x, \forall c \in \mathbb{C}). \tag{23}$$

Then, as is mentioned in Introduction, we have the additivity

$$mc_0(\mathcal{V}_x) \simeq \mathcal{V}_x, \quad mc_{\lambda_2} \circ mc_{\lambda_1}(\mathcal{V}_x) \simeq mc_{\lambda_1+\lambda_2}(\mathcal{V}_x). \tag{24}$$

We shall show that the additivity also holds for the middle convolution in x_i-direction.

Theorem 3.1. *Let i be an index $(1 \le i \le l)$, and assume the conditions (22) and (23) with $x = x_i$.*

Then we have

$$mc_0^{x_i}(\mathcal{V}) \simeq \mathcal{V}, \quad mc_{\lambda_2}^{x_i} \circ mc_{\lambda_1}^{x_i}(\mathcal{V}) \simeq mc_{\lambda_1+\lambda_2}^{x_i}(\mathcal{V}). \tag{25}$$

Proof. Set $x_i = x$. First we note that $mc_\lambda^x(\mathcal{V}_x) = mc_\lambda(\mathcal{V}_x)$. Then, thanks to (24), there exist the isomorphisms (25) as vector spaces. It remains to show that these isomorphisms are compatible with the action of A_H $(H \in \mathcal{A})$.

As is shown in [1], the isomorphism $mc_0(\mathcal{V}_x) \simeq \mathcal{V}_x$ is induced from the homomorphism

$$\begin{array}{rccc} \phi_1 : & c_0(\mathcal{V}_x) & \to & \mathcal{V}_x \\ & (v_H)_{H \in \mathcal{A}_x} & \mapsto & \displaystyle\sum_{H \in \mathcal{A}_x} A_H v_H. \end{array}$$

Take any $j \neq i$, and set $x_j = y$. Let us write the y-equation of $\mathcal{V}$ as

$$\frac{\partial U}{\partial y} = g(y)U,$$

and that of $c_0^x(\mathcal{V})$ as

$$\frac{\partial V}{\partial y} = G(y)V.$$

In order to show the compatibility of $mc_0^x(\mathcal{V}) \simeq \mathcal{V}$, it is enough to check

$$\phi_1 \circ G(y) = g(y) \circ \phi_1 \tag{26}$$

for any value of y.

Take any $v = (v_H)_{H\in\mathcal{A}_x} \in (\mathbb{C}^n)^{\mathcal{A}_x}$, and set $G(y)v = (w_H)_{H\in\mathcal{A}_x}$. Every w_H can be calculated by using (16) with $\lambda = 0$ and (17), and then by the help of Theorem 2.1 we have

$$
\begin{aligned}
&\phi_1 \circ G(y)v \\
&= \sum_{H\in\mathcal{A}_x} A_H w_H \\
&= \sum_{H\in\mathcal{A}_x\cap\mathcal{A}_y} A_H \Bigg(\frac{1}{y-b_H} \sum_{H'\in\mathcal{A}_x} A_{H'} v_{H'} \\
&\qquad\qquad + \sum_{H'\in\mathcal{C}_{H,y}} \frac{A_{H'}}{y-c_{HH'}}(v_H - v_{H'}) + \sum_{H'\in\mathcal{A}_x^c\cap\mathcal{A}_y} \frac{A_{H'}}{y-b_{H'}} v_H \Bigg) \\
&\quad + \sum_{H\in\mathcal{A}_x\cap\mathcal{A}_y^c} A_H \Bigg(\sum_{H'\in\mathcal{A}_x\cap\mathcal{A}_y} \frac{A_{H'}}{y-c_{HH'}}(v_H - v_{H'}) \\
&\qquad\qquad + \sum_{H'\in\mathcal{A}_x^c\cap\mathcal{A}_y} \frac{A_{H'}}{y-b_{H'}} v_H \Bigg) \\
&= \sum_{H\in\mathcal{A}_x\cap\mathcal{A}_y} \frac{A_H}{y-b_H} \phi_1(v) \\
&\quad + \sum_{H\in\mathcal{A}_x\cap\mathcal{A}_y} \Bigg(\sum_{H'\in\mathcal{C}_{H,y}} \frac{A_{H'}}{y-c_{HH'}} A_H v_H + \sum_{H'\in\mathcal{A}_x^c\cap\mathcal{A}_y} \frac{A_{H'}}{y-b_{H'}} A_H v_H \\
&\qquad\qquad - A_H \sum_{H'\in\mathcal{C}_{H,y}} \frac{A_{H'}}{y-c_{HH'}} v_{H'} \Bigg) \\
&\quad + \sum_{H\in\mathcal{A}_x\cap\mathcal{A}_y^c} \Bigg(\sum_{H'\in\mathcal{A}_x\cap\mathcal{A}_y} \frac{A_{H'}}{y-c_{HH'}} A_H v_H + \sum_{H'\in\mathcal{A}_x^c\cap\mathcal{A}_y} \frac{A_{H'}}{y-b_{H'}} A_H v_H \\
&\qquad\qquad - A_H \sum_{H'\in\mathcal{A}_x\cap\mathcal{A}_y} \frac{A_{H'}}{y-c_{HH'}} v_{H'} \Bigg) \\
&= \sum_{H\in\mathcal{A}_x\cap\mathcal{A}_y} \frac{A_H}{y-b_H} \phi_1(v) \\
&\quad + \sum_{H'\in\mathcal{A}_x^c\cap\mathcal{A}_y} \frac{A_{H'}}{y-b_{H'}} \Bigg(\sum_{H\in\mathcal{A}_x\cap\mathcal{A}_y} A_H v_H + \sum_{H\in\mathcal{A}_x\cap\mathcal{A}_y^c} A_H v_H \Bigg)
\end{aligned}
$$

$$
\begin{aligned}
&= \sum_{H\in\mathcal{A}_x\cap\mathcal{A}_y} \frac{A_H}{y-b_H}\phi_1(v) + \sum_{H\in\mathcal{A}_x^c\cap\mathcal{A}_y} \frac{A_H}{y-b_H}\phi_1(v)\\
&= \sum_{H\in\mathcal{A}_y} \frac{A_H}{y-b_H}\phi_1(v)\\
&= g(y)\phi_1(v).
\end{aligned}
$$

This shows (26).

The isomorphism $mc_{\lambda_2}\circ mc_{\lambda_1}(\mathcal{V}_x)\simeq mc_{\lambda_1+\lambda_2}(\mathcal{V}_x)$ is induced from the homomorphism

$$
\begin{array}{rccc}
\phi_2: & c_{\lambda_2}\circ c_{\lambda_1}(\mathcal{V}_x) & \to & c_{\lambda_1+\lambda_2}(\mathcal{V}_x)\\
 & (\hat{v}_H)_{H\in\mathcal{A}_x} & \mapsto & \displaystyle\sum_{H\in\mathcal{A}_x} G_H\hat{v}_H,
\end{array}
$$

where G_H $(H\in\mathcal{A}_x)$ are the residue matrices of the x-equation

$$
\frac{\partial V}{\partial x} = \left(\sum_{H\in\mathcal{A}_x}\frac{G_H}{x-a_H}\right)V
$$

of $c^x_{\lambda_1}(\mathcal{V})$. We set the y-equations of $c^x_{\lambda_1}(\mathcal{V})$, $c^x_{\lambda_1+\lambda_2}(\mathcal{V})$ and $c^x_{\lambda_2}\circ c^x_{\lambda_1}(\mathcal{V})$ as

$$
\frac{\partial V}{\partial y} = \left(\sum_{H\in\tilde{\mathcal{A}}_y}\frac{G_H}{y-b_H}\right)V,
$$

$$
\frac{\partial V}{\partial y} = \left(\sum_{H\in\tilde{\mathcal{A}}_y}\frac{G'_H}{y-b_H}\right)V,
$$

$$
\frac{\partial \hat{V}}{\partial y} = \left(\sum_{H\in\tilde{\mathcal{A}}_y}\frac{\hat{G}_H}{y-b_H}\right)\hat{V},
$$

respectively, where $\tilde{\mathcal{A}}$ is the arrangement of hyperplanes corresponding to the Pfaffian system (18) of $c^x_{\lambda_1}(\mathcal{V})$. Then, to prove the compatibility of $mc^{x_i}_{\lambda_2}\circ mc^{x_i}_{\lambda_1}(\mathcal{V})\simeq mc^{x_i}_{\lambda_1+\lambda_2}(\mathcal{V})$, it is enough to check

$$
\phi_2\circ\left(\sum_{H\in\tilde{\mathcal{A}}_y}\frac{\hat{G}_H}{y-b_H}\right) = \left(\sum_{H\in\tilde{\mathcal{A}}_y}\frac{G'_H}{y-b_H}\right)\circ\phi_2. \tag{27}
$$

Note that the y-equation of $c^x_{\lambda_1}(\mathcal{V})$ (resp. $c^x_{\lambda_1+\lambda_2}(\mathcal{V})$) can be obtained by replacing λ by λ_1 (resp. $\lambda_1+\lambda_2$) and $\mathcal{A}_y$ by $\tilde{\mathcal{A}}_y$ in (16) and (17).

Then in particular we have

$$G_H = G'_H$$

for any $H \in \mathcal{A}_x^c \cap \tilde{\mathcal{A}}_y$. Similarly, the y-equation of $c_{\lambda_2}^x \circ c_{\lambda_1}^x(\mathcal{V})$ can be obtained by replacing λ by λ_2, $\mathcal{A}_y$ by $\tilde{\mathcal{A}}_y$ and A_H by G_H in (16) and (17).

Take any $\hat{v} = (\hat{v}_H)_{H\in\mathcal{A}_x}$, and set

$$\left(\sum_{H\in\tilde{\mathcal{A}}_y} \frac{\hat{G}_H}{y-b_H}\right)\hat{v} = (\hat{w}_H)_{H\in\mathcal{A}_x}.$$

Then, for $H \in \mathcal{A}_x \cap \tilde{\mathcal{A}}_y$, we have

$$\begin{aligned}\hat{w}_H = {} & \frac{1}{y-b_H}\sum_{H'\in\mathcal{A}_x}(G_{H'} + \delta_{HH'}\lambda_2)\hat{v}_{H'} \\ & + \sum_{H'\in\mathcal{C}_{H,y}} \frac{G_{H'}}{y-c_{HH'}}(\hat{v}_H - \hat{v}_{H'}) + \sum_{H'\in\mathcal{A}_x^c\cap\tilde{\mathcal{A}}_y} \frac{G_{H'}}{y-b_{H'}}\hat{v}_H,\end{aligned}$$

and for $H \in \mathcal{A}_x \cap \tilde{\mathcal{A}}_y^c$ we have

$$\hat{w}_H = \sum_{H'\in\mathcal{A}_x\cap\tilde{\mathcal{A}}_y} \frac{G_{H'}}{y-c_{HH'}}(\hat{v}_H - \hat{v}_{H'}) + \sum_{H'\in\mathcal{A}_x^c\cap\tilde{\mathcal{A}}_y} \frac{G_{H'}}{y-b_{H'}}\hat{v}_H.$$

By using them, we get

$$\begin{aligned}\phi_2 \circ \left(\sum_{H\in\tilde{\mathcal{A}}_y} \frac{\hat{G}_H}{y-b_H}\right)(\hat{v}) &= \sum_{H\in\mathcal{A}_x} G_H \hat{w}_H \\ &= \sum_{H\in\mathcal{A}_x} \frac{G_H}{y-b_H}\phi_2(\hat{v}) + \lambda_2 \sum_{H\in\mathcal{A}_x\cap\tilde{\mathcal{A}}_y} \frac{G_H}{y-b_H}\hat{v}_H.\end{aligned}$$

By (15), we see that

$$G_H = \sum_{H'\in\mathcal{A}_x} E_{HH'} \otimes (A_{H'} + \delta_{HH'}\lambda_1)$$

for $H \in \mathcal{A}_x$. Then, if we set $\hat{v}_H = (v_{HK})_{K\in\mathcal{A}_x}$, we have

$$G_H\hat{v}_H = E_H \otimes \left(\sum_{K\in\mathcal{A}_x} A_K v_{HK} + \lambda_1 v_{HH}\right),$$

where E_H denotes the unit vector in $\mathbb{C}^{\mathcal{A}_x}$ with the only non-zero entry 1 in the H-th position. Moreover we have

$$G_H\phi_2(\hat{v}_H) = E_H \otimes (A_H + \lambda_1)\left(\sum_{K\in\mathcal{A}_x} A_K v_{HK} + \lambda_1 v_{HH}\right).$$

Thus, for $H \in \mathcal{A}_x \cap \tilde{\mathcal{A}}_y$, we have

$$\begin{aligned} G_H\phi_2(\hat{v}) + \lambda_2 G_H \hat{v}_H &= E_H \otimes (A_H + \lambda_1 + \lambda_2)\left(\sum_{K\in\mathcal{A}_x} A_K v_{HK} + \lambda_1 v_{HH}\right) \\ &= G'_H\phi_2(\hat{v}). \end{aligned}$$

As noticed before, for $H \in \mathcal{A}_x^c \cap \tilde{\mathcal{A}}_y$ we have $G_H = G'_H$. Hence we get

$$\phi_2 \circ \left(\sum_{H\in\tilde{\mathcal{A}}_y} \frac{\hat{G}_H}{y - b_H}\right)(\hat{v}) = \left(\sum_{H\in\tilde{\mathcal{A}}_y} \frac{G'_H}{y - b_H}\right) \circ \phi_2(\hat{v}),$$

which proves (27). Q.E.D.

It follows from Theorem 3.1 that the middle convolution in x_i-direction is invertible as a homomorphism of $\{A_H\,;\, H \in \mathcal{A}\}$-modules. Then combining with Theorem 1.1, we get the following assertion.

Theorem 3.2. *Assume that, after a linear transformation of coordinates of $\mathbb{C}^l$ if necessary, for some index i $(1 \le i \le l)$, the x_i-equation of the Pfaffian system (4) is irreducible and rigid as an ordinary differential equation. Then the system (4) can be obtained from a Pfaffian system of rank 1 by a finite iteration of middle convolutions in x_i-direction and additions.*

§4. Examples

Lauricella's hypergeometric series F_D is expressed by the integral

$$F_D(x_1, x_2, \ldots, x_n) = C\int_0^1 t^{a_0}(t-1)^{a_1}\prod_{j=1}^{n}(t - x_j)^{b_j}\,dt$$

with some constant C. A Pfaffian system in n variables for F_D can be obtained by the middle convolution in the following way.

We consider the function

$$u(x_1, x_2, \dots, x_n) = {x_1}^{a_0}(x_1 - 1)^{a_1} \prod_{j=2}^{n} (x_1 - x_j)^{b_j},$$

where $a_0, a_1, b_2, \dots, b_n \in \mathbb{C}$. This function satisfies the Pfaffian system

$$du = \left(a_0 d \log x_1 + a_1 d \log(x_1 - 1) + \sum_{j=2}^{n} b_j d \log(x_1 - x_j) \right) u \tag{28}$$

of rank 1. The singular locus of this system is given by the arrangement $\mathcal{A} = \{H_0, H_1, \dots, H_n\}$, where

$$H_0 = \{x_1 = 0\}, \ H_1 = \{x_1 = 1\}, \ H_j = \{x_1 = x_j\} \quad (2 \le j \le n).$$

The middle convolution of the system (28) in x_1-direction with parameter b_1 makes a Pfaffian system

$$dV = \Omega V, \tag{29}$$

which is a Pfaffian system for F_D.

The singular locus of (29) is obtained as follows. Take any j ($2 \le j \le n$). For each H_k ($0 \le k \le n$, $k \ne j$), we have

$$\mathcal{C}_{H_k, x_j} = \{H_j\}$$

and

$$c_{H_k H_j} = \begin{cases} 0 & (k = 0), \\ 1 & (k = 1), \\ x_k & (2 \le k \le n, \ k \ne j). \end{cases}$$

Then the singular points of the x_j-equation of (29) is given by

$$x_j = x_1, \ x_j = 0, \ x_j = 1, \ x_j = x_k \quad (2 \le k \le n, \ k \ne j).$$

Hence the singular locus of the system (29) is given by the arrangement $\tilde{\mathcal{A}} = \mathcal{A} \cup \{H_{jk} \,;\, 2 \le j \le n, \, 0 \le k \le n, \, k \ne j\}$, where

$$H_{j0} = \{x_j = 0\}, \ H_{j1} = \{x_j = 1\}, \ H_{jk} = \{x_j = x_k\} \quad (2 \le k \le n, \ k \ne j).$$

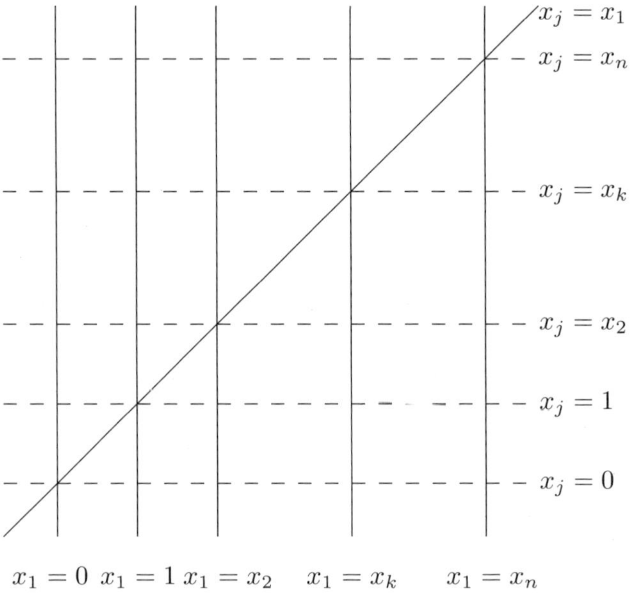

Fig. 2. Singular locus in x_1x_j-plane

We have another example related to Appell's hypergeometric series F_4. We quote the results from our previous work [4].

We consider the KZ-type system (21) of rank 4, where $A_1, A_2, \ldots, A_5$ are given by

$$A_1 = \begin{pmatrix} 0 & 1 & 0 & 0 \\ 0 & 1-\gamma & 0 & 0 \\ 0 & \epsilon & 0 & 1 \\ 0 & 0 & 0 & 1-\gamma \end{pmatrix}, \quad A_2 = \begin{pmatrix} 0 & 0 & 1 & 0 \\ 0 & 0 & \epsilon & 1 \\ 0 & 0 & 1-\gamma & 0 \\ 0 & 0 & 0 & 1-\gamma \end{pmatrix},$$

$$A_3 = \begin{pmatrix} 0 & 0 & 0 & 0 \\ -\alpha\beta & -\gamma' & \epsilon & 0 \\ 0 & 0 & 0 & 0 \\ 0 & 0 & -(\alpha+\epsilon)(\beta+\epsilon) & -\gamma' \end{pmatrix},$$

$$A_4 = \begin{pmatrix} 0 & 0 & 0 & 0 \\ 0 & 0 & 0 & 0 \\ -\alpha\beta & \epsilon & -\gamma' & 0 \\ 0 & -(\alpha+\epsilon)(\beta+\epsilon) & 0 & -\gamma' \end{pmatrix},$$

$$A_5 = \begin{pmatrix} 0 & 0 & 0 & 0 \\ 0 & \epsilon & -\epsilon & 0 \\ 0 & -\epsilon & \epsilon & 0 \\ 0 & 0 & 0 & 0 \end{pmatrix},$$

$\alpha, \beta, \gamma, \gamma' \in \mathbb{C}$, and

$$\epsilon := \gamma + \gamma' - \alpha - \beta - 1.$$

This system is obtained by Kato [6] as a transform of a Pfaffian system for Appell's F_4. The restriction of the system to the singular locus $x = y$ becomes a system of ordinary differential equations of rank 3, which is explicitly given by

$$\frac{du}{dx} = \left(\frac{C_1}{x} + \frac{C_2}{x-1} \right) u \tag{30}$$

with

$$C_1 = \begin{pmatrix} 0 & 2 & 0 \\ 0 & \gamma' - \alpha - \beta & 1 \\ 0 & 0 & 2(1-\gamma) \end{pmatrix},$$
$$C_2 = \begin{pmatrix} 0 & 0 & 0 \\ -\alpha\beta & \gamma - \alpha - \beta - 1 & 0 \\ 0 & -2(\epsilon + \alpha)(\epsilon + \beta) & -2\gamma' \end{pmatrix}.$$

We see that the index of rigidity of the restriction is 0.

The middle convolution of (21) in x-direction with parameter $\lambda \neq 0, \alpha, \beta$ gives the system

$$dW = \left[B_1 \frac{dx}{x} + B_2 \frac{dy}{y} + B_3 \frac{dx}{x-1} + B_4 \frac{dy}{y-1} + B_5 \frac{d(x-y)}{x-y} \right] W \tag{31}$$

with the same singular locus, where

$$B_1 = \begin{pmatrix} \lambda+1-\gamma & 0 & -1 & \epsilon & \epsilon \\ 0 & \lambda+1-\gamma & 0 & -1 & 0 \\ 0 & 0 & 0 & 0 & 0 \\ 0 & 0 & 0 & 0 & 0 \\ 0 & 0 & 0 & 0 & 0 \end{pmatrix},$$

$$B_2 = \begin{pmatrix} \epsilon & (\alpha+\epsilon)(\beta+\epsilon) & 0 & 0 & -\epsilon \\ 0 & 1-\gamma & 0 & 0 & 0 \\ 0 & 0 & 0 & \alpha\beta+\gamma'\epsilon & 0 \\ 0 & 0 & 0 & 1-\gamma & 0 \\ \gamma-1+\epsilon & (\alpha+\epsilon)(\beta+\epsilon) & 0 & 0 & 1-\gamma-\epsilon \end{pmatrix},$$

$$B_3 = \begin{pmatrix} 0 & 0 & 0 & 0 & 0 \\ 0 & 0 & 0 & 0 & 0 \\ \alpha\beta+(1-\gamma)\gamma' & (1-\gamma)\gamma'\epsilon & \lambda-\gamma' & 0 & \gamma'\epsilon \\ \epsilon & (\alpha+\epsilon)(\beta+\epsilon)+(1-\gamma)\gamma' & 0 & \lambda-\gamma' & -\epsilon \\ 0 & 0 & 0 & 0 & 0 \end{pmatrix},$$

$$B_4 = \begin{pmatrix} 0 & 0 & 0 & 0 & 0 \\ -1 & -\gamma' & 0 & 0 & 0 \\ 0 & 0 & 0 & -\gamma'\epsilon & -\gamma'\epsilon \\ 0 & 0 & -1 & -\gamma'+\epsilon & \epsilon \\ 0 & 0 & 1 & -\epsilon & -\gamma'-\epsilon \end{pmatrix},$$

$$B_5 = \begin{pmatrix} 0 & 0 & 0 & 0 & 0 \\ 0 & 0 & 0 & 0 & 0 \\ 0 & 0 & 0 & 0 & 0 \\ 0 & 0 & 0 & 0 & 0 \\ 1-\gamma-\epsilon & -(\alpha+\epsilon)(\beta+\epsilon) & -1 & \epsilon & \lambda+2\epsilon \end{pmatrix}.$$

As a restriction of (31) to the singular locus $x = y$, we get a system of ordinary differential equations of the rank 4. The explicit form is given by

$$\frac{dv}{dx} = \left(\frac{D_1}{x} + \frac{D_2}{x-1}\right) v, \tag{32}$$

where

$$D_1 = \begin{pmatrix} \lambda + 1 - c + \epsilon & \frac{\epsilon}{1-\gamma-\epsilon} & -\frac{1}{1-\gamma-\epsilon} & 0 \\ 0 & 1-\gamma & 0 & 0 \\ 0 & \alpha\beta + \gamma'\epsilon & 0 & 0 \\ 0 & -(\alpha+1-\gamma) & 0 & \lambda + 2(1-\gamma) \end{pmatrix},$$

$$D_2 = \begin{pmatrix} -\gamma' - \epsilon & -\frac{\epsilon}{1-\gamma-\epsilon} & \frac{1}{1-\gamma-\epsilon} & 0 \\ -(\lambda+2\epsilon)\epsilon & \lambda + 2(\epsilon - \gamma') - \frac{(1-\gamma)\epsilon}{1-\gamma-\epsilon} & -\frac{1-\gamma-2\epsilon}{1-\gamma-\epsilon} & -\frac{(\beta+1-\gamma)(1-\gamma)}{1-\gamma-\epsilon} \\ -(\lambda+2\epsilon)(\alpha\beta + (1-\gamma)\gamma') & -\frac{\epsilon(\alpha\beta+\gamma'(2-2\gamma-\epsilon))}{1-\gamma-\epsilon} & \lambda + \frac{\alpha\beta+\gamma'\epsilon}{1-\gamma-\epsilon} & \frac{(\beta+1-\gamma)(\alpha\beta+\gamma'\epsilon)}{1-\gamma-\epsilon} \\ (\lambda+2\epsilon)(\alpha+1-\gamma) & \frac{(\alpha+1-\gamma)\epsilon}{1-\gamma-\epsilon} & -\frac{\alpha+1-\gamma}{1-\gamma-\epsilon} & \frac{(\alpha+1-\gamma)(\beta+1-\gamma)}{1-\gamma-\epsilon} \end{pmatrix}.$$

We observe that the index of rigidity of the last system is -2 unless $\lambda = 2\gamma - 2$ or $\lambda = 2\gamma'$, in which cases we have the index of rigidity 0. This is remarkable, because the index of rigidity is invariant under the middle convolution for ordinary differential equations.

The system (30) appears in the conformal field theory [3], and the system (32) appears in the theory of Heckman–Opdam hypergeometric systems [8]. These systems may also be related to special solutions of difference Painlevé equations. Thus, we expect that the combination of prolongation-restriction process and the middle convolution will be a good tool which relate several interesting equations in various areas of mathematics and physics.

References

[1] M. Dettweiler and S. Reiter, An algorithm of Katz and its application to the inverse Galois problem, J. Symbolic Comput., **30** (2000), 761–798.

[2] M. Dettweiler and S. Reiter, Middle convolution of Fuchsian systems and the construction of rigid differential systems, J. Algebra, **318** (2007), 1–24.

[3] V. S. Dotsenko and V. A. Fateev, Conformal algebra and multipoint correlation function in 2D statistical models, Nuclear Phys. B, **240** (1984), 312–348.

[4] Y. Haraoka, Prolongability of ordinary differential equations, preprint.

[5] Y. Haraoka and G. Filipuk, Middle convolution and deformation for Fuchsian systems, J. London Math. Soc., **76** (2007), 438–450.

[6] M. Kato, Connection formulas for Appell's system F_4 and some applications, Funkcial. Ekvac., **38** (1995), 243–266.
[7] N. M. Katz, Rigid Local Systems, Princeton Univ. Press, Princeton, NJ, 1996.
[8] T. Oshima and N. Shimeno, Heckman–Opdam hypergeometric functions and their specializations, RIMS Kôkyuroku Bessatsu, **B20** (2010), 129–162.

Yoshishige Haraoka
Department of Mathematics, Kumamoto University,
Kumamoto 860-8555, Japan
E-mail address: haraoka@kumamoto-u.ac.jp

Advanced Studies in Pure Mathematics 62, 2012
Arrangements of Hyperplanes—Sapporo 2009
pp. 137–155

On a problem of arrangements related to the hypergeometric integrals of confluent type

Hironobu Kimura

Abstract.

The general hypergeometric integrals on the complex matrix space are considered. We study the twisted algebraic de Rham cohomology group associated with the integrand of the general hypergeometric integrals. After reviewing some fundamental facts, we discuss the exterior product structure of the cohomology group.

§1. Introduction

Let $f_0, f_1, \dots, f_n$ be polynomial functions of degree 1 on $\mathbb{C}^r$ and let $H_j = \ker f_j$ be the hyperplane in $\mathbb{C}^r$ defined as zeros of $f_j = 0$. Then $\mathcal{A} = \{H_0, \dots, H_n\}$ defines an arrangement in $\mathbb{C}^r$. Many authors are interested in the topology of the complement of these hyperplanes. One of the important topics related to this topological problem is the theory of hypergeometric functions on the Grassmannian manifold initiated by K. Aomoto [1] and I.M. Gelfand [3]. It is given by the integral

$$F = \int_C f_0^{\alpha_0} \cdots f_n^{\alpha_n} du_1 \wedge \cdots \wedge du_r,$$

where α_j are complex constants and C is some r-dimensional chain. The integral gives a function on the space of coefficients of the polynomials f_j.

The typical example is the Gauss hypergeometric function

$${}_2F_1(a, b, c; x) = \sum_{m=0}^{\infty} \frac{(a)_m (b)_m}{(c)_m m!} x^m = A \int_0^1 u^{a-1}(1-u)^{c-a-1}(1-xu)^{-b} du,$$

Received May 13, 2010.
Revised January 17, 2011.
2010 *Mathematics Subject Classification.* Primary 33C70; Secondary 44A12.
Key words and phrases. Hypergeometric, de Rham cohomology, Radon transform.

where the polynnomials are $f_0 = 1, f_1 = u, f_2 = 1 - u, f_3 = 1 - xu$ and the chain C is an open segment $(0,1) \subset \mathbb{R} \subset \mathbb{C}$ and A is a constant.

Let $\mathcal{L}$ be the local system on $X = \mathbb{C}^r \setminus \cup_{j=0}^n H_j$ defined by the function $f_0^{-\alpha_0} \cdots f_n^{-\alpha_n}$. Then to investigate the hypergeometric function it is important to study and compute the cohomology group $H^*(X, \mathcal{L})$ and the homology group $H_*(X, \mathcal{L}^\vee)$. Note that elements of the cohomology group $H^*(X, \mathcal{L})$ are represented by rational differential forms by virtue of algebraic de Rham theory due to Deligne and Grothendieck.

This paper concerns an analogous problem of arragements of hyperplanes in the complex affine space $\mathbb{C}^r$ related to the hypergeometric integrals of confluent type. We focus on the computation of the twisted algebraic de Rham cohomology group.

§2. General hypergeometric integrals

2.1. Regular elements of $\mathrm{GL}_N(\mathbb{C})$

To define the general hypergeometric integrals, we use a class of maximal abelian subgroups of $\mathrm{GL}_N(\mathbb{C})$ which are obtained as centralizers of regular elements. We start with explaining about regular elements.

Definition 1. *An element $a \in \mathrm{GL}_N(\mathbb{C})$ is a regular element when the orbit $O(a) = \{gag^{-1} \mid g \in \mathrm{GL}_N(\mathbb{C})\}$ under the adjoint action is of maximum dimension.*

One can see that $a \in \mathrm{GL}_N(\mathbb{C})$ is regular if any two Jordan cells of the Jordan normal form of a have different eigenvalues and in this case $\dim_{\mathbb{C}} O(a) = \dim_{\mathbb{C}} \mathrm{GL}_N(\mathbb{C}) - N$. Then the centralizer $Z(a)$ of a is a maximal abelian subgroup of $\mathrm{GL}_N(\mathbb{C})$ of dimension N. The size of Jordan cells gives a partition of N.

Example 2. *If $a \in \mathrm{GL}_4(\mathbb{C})$ is a regular element, then a is similar to the following Jordan normal form. Partitions of 4 mean possible* *structure of Jordan normal forms.*

$$\begin{pmatrix} a_1 & & & \\ & a_2 & & \\ & & a_3 & \\ & & & a_4 \end{pmatrix} \longleftrightarrow (1,1,1,1)$$

$$\begin{pmatrix} a_1 & 1 & & \\ & a_1 & & \\ & & a_3 & \\ & & & a_4 \end{pmatrix} \longleftrightarrow (2,1,1)$$

$$\begin{pmatrix} a_1 & 1 & & \\ & a_1 & & \\ & & a_3 & 1 \\ & & & a_3 \end{pmatrix} \longleftrightarrow (2,2)$$

$$\begin{pmatrix} a_1 & 1 & & \\ & a_1 & 1 & \\ & & a_1 & \\ & & & a_4 \end{pmatrix} \longleftrightarrow (3,1)$$

$$\begin{pmatrix} a_1 & 1 & & \\ & a_1 & 1 & \\ & & a_1 & 1 \\ & & & a_1 \end{pmatrix} \longleftrightarrow (4)$$

where $a_i \neq a_j$ $(i \neq j)$.

2.2. Maximal abelian groups

We shall describe the centralizer more explicitly. Assume that a regular element a is of Jordan normal form and that the Jordan cells of a have the size $n_1, \dots, n_\ell$ and arrayed in non increasing order: $n_1 \geq n_2 \geq \cdots \geq n_\ell > 0$. Hence the partition $\lambda = (n_1, \dots, n_\ell)$ of N is associated with a. The centralizer $Z(a)$ depends on the partition λ but not on the Jordan normal form. So we denote the centralizer by H_λ.

Lemma 3. *The centralizer* H_λ *is written as*

$$H_\lambda = J(n_1) \times \cdots \times J(n_\ell) \subset \mathrm{GL}_N(\mathbb{C}),$$

where

$$J(n) = \left\{ h = \begin{pmatrix} h_0 & h_1 & \dots & h_{n-1} \\ & \ddots & \ddots & \vdots \\ & & \ddots & h_1 \\ & & & h_0 \end{pmatrix}; h_0 \neq 0 \right\}.$$

An element $h \in H_\lambda$ is denoted as

$$h = (h^{(1)}, \dots, h^{(\ell)}), \quad h^{(k)} = h_0^{(k)} I + h_1^{(k)} \Lambda + \dots + h_{n-1}^{(k)} \Lambda^{n-1} \in J(n_k),$$

where $\Lambda = (\delta_{i+1,j}) \in \mathrm{Mat}_n(\mathbb{C})$ is the shift matrix.

Remark 4. *We have the isomorphism $J(n) \to \left(\mathbb{C}[T]/(T^n)\right)^\times$ by the obvious correspondence*

$$h_0 I + h_1 \Lambda + \dots + h_{n-1} \Lambda^{n-1} \mapsto h_0 + h_1 T + \dots + h_{n-1} T^{n-1} \quad (\mathrm{mod}\, T^n).$$

This identification motivates the definition of generalized Veronese map in Subsection 4.2.

Lemma 5. *There is an isomorphism $J(n) \simeq \mathbb{C}^\times \times \mathbb{C}^{n-1}$ given by*

$$h \mapsto (h_0, \theta_1(h), \dots, \theta_{n-1}(h)),$$

where $\theta_m(h)$ $(m = 0, 1, \dots)$ is defined by

$$\log h = \log(h_0 I + h_1 \Lambda + \dots + h_{n-1} \Lambda^{n-1}) = \sum_{m=0}^{n-1} \theta_m(h) \Lambda^m, \quad \theta_0(h) = \log h_0.$$

It is easily seen that $\theta_m(h), m \geq 1$, are rational functions of $h_0, \dots, h_{n-1}$, with a pole of order m along $h_0 = 0$, given explicitly by

$$\theta_m(h) = \sum (-1)^{k_1 + \dots + k_m - 1} \frac{(k_1 + \dots + k_m - 1)!}{k_1! \cdots k_m!} \left(\frac{h_1}{h_0}\right)^{k_1} \cdots \left(\frac{h_m}{h_0}\right)^{k_m},$$

where the sum is taken over the indices $(k_1, \dots, k_m)$ such that $k_1 + 2k_2 + \dots + mk_m = m$. For example, $\theta_1(h), \theta_2(h), \theta_3(h)$ are written as

$$(1) \qquad \begin{aligned} \theta_1(h) &= \frac{h_1}{h_0}, \\ \theta_2(h) &= \frac{h_2}{h_0} - \frac{1}{2}\left(\frac{h_1}{h_0}\right)^2, \\ \theta_3(h) &= \frac{h_3}{h_0} - \left(\frac{h_1}{h_0}\right)\left(\frac{h_2}{h_0}\right) + \frac{1}{3}\left(\frac{h_1}{h_0}\right)^3. \end{aligned}$$

2.3. Definition of the integrals

By virtue of Lemma 5, the characters of the universal covering group $\tilde{H}_\lambda$ of H_λ are described as follows. First, the character $\chi_n : \tilde{J}(n) \to \mathbb{C}^\times$ is given by

$$\chi_n(h; \alpha) = \exp\left(\alpha_0 \theta_0(h) + \dots + \alpha_{n-1} \theta_{n-1}(h)\right),$$

$$\alpha = (\alpha_0, \ldots, \alpha_{n-1}) \in \mathbb{C}^n.$$

Since H_λ is a product of $J(n_k)$'s, the character $\chi : \tilde{H}_\lambda \to \mathbb{C}^\times$ is written as

$$\chi(h; \alpha) = \prod_{k=1}^{\ell} \chi_{n_k}(h^{(k)}; \alpha^{(k)}) = \prod_{k=1}^{\ell} \exp\left(\sum_{m=0}^{n_k-1} \alpha_m^{(k)} \theta_m(h^{(k)}) \right),$$

$$h = (h^{(1)}, \ldots, h^{(\ell)})$$

with $\alpha = (\alpha^{(1)}, \ldots, \alpha^{(\ell)}), \alpha^{(k)} = (\alpha_0^{(k)}, \ldots, \alpha_{n_k-1}^{(k)}) \in \mathbb{C}^{n_k}$.

We assume that the character χ satisfies the condition

$$\text{(2)} \qquad \sum_{k=1}^{\ell} \alpha_0^{(k)} = -r - 1, \quad \alpha_{n_k-1}^{(k)} \neq 0 \ (n_k \geq 2), \ \notin \mathbb{Z} \ (n_k - 1)$$

and let us consider its Radon transform. Let $u = (u_1, \ldots, u_r)$ be the coordinates of $\mathbb{C}^r$, the variables of integration. Consider N polynomials $f_j^{(k)} = z_{0j}^{(k)} + u_1 z_{1j}^{(k)} + \cdots + u_r z_{rj}^{(k)}$ of degree 1 in u which will be substituted into the character χ. The coefficients of $f_j^{(k)}$ give a column vector $z_j^{(k)} = (z_{0j}^{(k)}, \ldots, z_{rj}^{(k)})^t \in \mathbb{C}^{r+1}$. Hence the polynomials $f_j^{(k)}$ are given by specifying a matrix $z = (z^{(1)}, \ldots, z^{(\ell)}) \in \mathrm{Mat}_{r+1,N}(\mathbb{C})$ with $z^{(k)} = (z_0^{(k)}, \ldots, z_{n_k-1}^{(k)}) \in \mathrm{Mat}_{r+1,n_k}(\mathbb{C})$. Note that the set of polynomials $f_j^{(k)}$ is written as

$$(f_0^{(1)}, \ldots, f_{n_1-1}^{(1)}, \ldots, f_0^{(\ell)}, \ldots, f_{n_\ell-1}^{(\ell)}) = \vec{u} z, \quad \vec{u} = (1, u_1, \ldots, u_r).$$

Let $Z_{r,\lambda} \subset \mathrm{Mat}_{r+1,N}(\mathbb{C})$ be the Zariski open subset consisting of $z = (z^{(1)}, \ldots, z^{(\ell)})$ satisfying the following condition.

Condition(*): For any $(m_1, \ldots, m_\ell) \in \mathbb{Z}_{\geq 0}^{\ell}$ such that $0 \leq m_k \leq n_k$ $(k = 1, \ldots, \ell)$, $m_1 + \cdots + m_\ell = r + 1$, we have

$$\det(z_0^{(1)}, \ldots, z_{m_1-1}^{(1)}, \ldots, z_0^{(\ell)}, \ldots, z_{m_\ell-1}^{(\ell)}) \neq 0.$$

Definition 6. *For a character of $\chi(\cdot\,; \alpha)$ of $\tilde{H}_\lambda$ satisfying (2) and $z \in Z_{r,\lambda}$, the general hypergeometric integral (GHGI) is*

$$\text{(3)} \qquad I(z, \alpha, c) = \int_c \chi(\vec{u} z; \alpha) du, \quad du = du_1 \wedge \cdots \wedge du_r,$$

where c is a cycle of some homology group defined by using $\chi(\vec{u} z; \alpha)$.

Remark 7. *From the explicit form of* θ_m*, the integrand* $\chi(\vec{u}z;\alpha)$ *of the integral* $I(z,\alpha;c)$ *is a multivalued holomorphic function of* $u \in \mathbb{C}^r$ *with the branch locus on the union of hyperplanes* $\cup_{k=1}^{\ell} H^{(k)}$*, where* $H^{(k)} = \{u \in \mathbb{C}^r \mid \vec{u} \cdot z_0^{(k)} = 0\}$.

Remark 8. *For an appropriate chosen cycle* c*, GHGI satisfies a holonomic system of differential equations on* $Z_{r,\lambda}$.

2.4. Classical HGF

We explain how the Gauss hypergeometric function and the associated functions of confluent type are related to the general hypergeometric integral.

Gauss hypergeometric function is a holomorphic solution of

$$x(1-x)y'' + \{c-(a+b+1)x\}y' - aby = 0 \tag{4}$$

at $x = 0$ with the condition $y(0) = 1$. Confluent type functions of ${}_2F_1$ we know are solutions respectively of the following 2nd order linear differential equations on $\mathbb{P}^1$:

(Kummer) $\quad xy'' + (c-x)y' - ay = 0,$

(Bessel) $\quad x^2y'' + xy' + (x^2 - a^2)y = 0,$

(Hermite) $\quad y'' - xy' + ay = 0,$

(Airy) $\quad y'' - xy = 0.$

The solutions of these equations have the integral representations:

(Kummer) $\quad {}_1F_1(a,c;x) = A\int_0^1 e^{xu}u^{a-1}(1-u)^{c-a-1}du,$

(Bessel) $\quad J_a(x) = A\int_C e^{x(u-1/u)}u^{-a-1}du,$

(Hermite) $\quad H_a(x) = A\int_C e^{xu-\frac{1}{2}u^2}u^{-a-1}du,$

(Airy) $\quad \mathrm{Ai}(x) = A\int_C e^{xu-\frac{1}{3}u^3}du,$

where C in each integral is an appropriate chain (see [6]). We would like to explain these functions are all understood as GHGI for various partitions λ of 4. The correspondence between the functions (or differential equations) and the partitions of 4 are

Gauss	Kummer	Bessel	Hermite	Airy
$\updownarrow$	$\updownarrow$	$\updownarrow$	$\updownarrow$	$\updownarrow$
$(1,1,1,1)$	$(2,1,1)$	$(2,2)$	$(3,1)$	(4)

We discuss the correspondence in the cases of Gauss and Airy.

2.4.1. *Gauss case* This is one of the simplest cases of general hypergeometric function considered by K. Aomoto and I. M. Gelfand. For the partition $\lambda = (1,1,1,1)$ we associate the maximal abelian subgroup $H = H_{(1,1,1,1)}$ consisting of diagonal matrices $h = \mathrm{diag}(h_0, \dots, h_3)$. Then the character $\chi : \tilde{H} \to \mathbb{C}^\times$ has the form

$$\chi(h;\alpha) = h_0^{\alpha_0} h_1^{\alpha_1} h_2^{\alpha_2} h_3^{\alpha_3}. \tag{5}$$

Substitute into χ the following linear functions:

$$h_0(u) = 1, \quad h_1(u) = u, \quad h_2(u) = 1-u, \quad h_3(u) = 1-xu. \tag{6}$$

Define the parameters α in χ as $\alpha = (b-c, a-1, c-a-1, -b)$, then we have

$$_2F_1(a,b,c;x) = \frac{\Gamma(c)}{\Gamma(a)\Gamma(c-a)} \int_0^1 \chi(h(u);\alpha)du.$$

As is well known, the right-hand side gives an analytic continuation of $_2F_1$, which is defined in the unit disk $|x| < 1$, along any path in $\mathbb{C} \setminus \{0,1\}$. Note that $x = 0, 1, \infty$ is the set of singular points of the Gauss hypergeometric equation (4).

Let us explain the meaning of the choice of polynomials $h_j(u)$ given in (6) and how far the Gauss hypergeometric integral is generalized by considering GHGI. As explained in Subsection 2.3, the polynomials (6) are determined by the matrix of the form

$$\begin{pmatrix} 1 & 0 & 1 & 1 \\ 0 & 1 & -1 & -x \end{pmatrix}. \tag{7}$$

Note that the point $x = 0, 1$ are excluded as values of x, which are singular points of (4). On the other hand, for GHGI (3), the polynomials f_j to be substituted into the character χ are given by a 2×4 matrix $z \in Z (= Z_{1,(1,1,1,1)})$. Let X be the set of matrices of the form (7) with the condition $x \neq 0, 1$. Then the meaning of the subset $X \subset Z$ is as follows.

Define the left and right action of $\mathrm{GL}_2(\mathbb{C})$ and H on Z by

$$\mathrm{GL}_2(\mathbb{C}) \times Z \times H \ni (g, z, h) \mapsto gzh \in Z. \tag{8}$$

Proposition 9. *For any $z \in Z$, we can find $g \in \mathrm{GL}_2(\mathbb{C})$ and $h \in H$ such that*

$$gzh = \begin{pmatrix} 1 & 0 & 1 & 1 \\ 0 & 1 & -1 & -x \end{pmatrix}, \tag{9}$$

and $x \in \mathbb{C} \setminus \{0, 1\}$ is uniquely determined by z:

$$x = \frac{[1,2][0,3]}{[0,2][1,3]}, \qquad [i,j] := \det(z_i, z_j). \tag{10}$$

This propsition implies that each orbit $O(z)$ in Z by the above action has the unique representative of the form (7). Thus we see that the quotient space $\mathrm{GL}_2(\mathbb{C})\backslash Z/H$ has a realization by the global slice $X \subset Z$. On the level of integrals we can show that GHGI in this setting with appropriately chosen cycle c is related to the Euler integral representation of Gauss hypergeometric function by

$$I(z, \alpha, c) = (\det g)\chi(h; \alpha)^{-1} \int_0^1 u^{a-1}(1-u)^{c-a-1}(1-xu)^{-b} du,$$

where z and x are related by (9) and (10). Thus, not only the Gauss hypergeometric function is obtained by restricting the general hypergeometric integral on X, but also it is essentially the same as GHGI on $Z_{1,(1,1,1,1)}$.

2.4.2. *Airy case* We shall explain that Airy's integral $\mathrm{Ai}(x) = \int_c e^{xu-u^3/3} du$ is essentially the same as GHGI for the partition $\lambda = (4)$. Recall that the maximal abelian subgroup of $\mathrm{GL}_4(\mathbb{C})$ in this case is

$$H_{(4)} = \left\{ h = \begin{pmatrix} h_0 & h_1 & h_2 & h_3 \\ & h_0 & h_1 & h_2 \\ & & h_0 & h_1 \\ & & & h_0 \end{pmatrix} \right\} \subset \mathrm{GL}_4(\mathbb{C}).$$

Let $\chi : \tilde{H}_{(4)} \to \mathbb{C}^\times$ be a character of the universal covering group of $H_{(4)}$. It is written as

$$\chi(h; \alpha) = h_0^{\alpha_0} \exp(\alpha_1\theta_1(h) + \alpha_2\theta_2(h) + \alpha_3\theta_3(h))$$

for some $\alpha = (\alpha_0, \alpha_1, \alpha_2, \alpha_3) \in \mathbb{C}^4$, where $\theta_1, \theta_2, \theta_3$ are those given in (1). To realize $\mathrm{Ai}(x)$ as GHGI, take the character χ with the parameter

$$\alpha = (-2, 0, 0, -1)$$

and substitute into χ the polynomial of u:

$$h_0(u) = 1,\ h_1(u) = u,\ h_2(u) = 0,\ h_3(u) = -xu. \tag{11}$$

And then consider the integral

$$\int \chi(h(u);\alpha)du.$$

That this integral coincides with $\mathrm{Ai}(x)$ is easily seen using the expression of $\theta_3(h)$ given in (1):

$$\chi(h(u);\alpha) = h_0(u)^{-2}\exp(-\theta_3(h(t))) = \exp\left(xu - \frac{1}{3}u^3\right).$$

As in Subsection 2.3, the polynomials $h_j(u)$ in (11) are determined by the matrix of the form

$$\begin{pmatrix} 1 & 0 & 0 & 0 \\ 0 & 1 & 0 & -x \end{pmatrix}. \tag{12}$$

When we consider the Airy integral, there is no restriction on x and can take arbitrary complex number. This is related with the fact that Airy's differential equation has no singular point in $\mathbb{C}$. So in the matrix (12), x is arbitrary complex number. In order to understand why the particular matrix (12) is chosen to obtain the Airy integral from GHGI of type $\lambda = (4)$, we proceed in a similar way as in Gauss hypergeometric case. Let $Z_{(4)} = \{z = (z_0, \dots, z_3) \in \mathrm{Mat}_{2,4}(\mathbb{C}); \det(z_0.z_1) \neq 0\}$ and let $X_{(4)}$ be the subset of $Z_{(4)}$ consisting of the matrices of the form (12) with $x \in \mathbb{C}$. Define the left and right action of $\mathrm{GL}_2(\mathbb{C})$ and $H_{(4)}$ on $Z_{(4)}$by

$$\mathrm{GL}_2(\mathbb{C}) \times Z_{(4)} \times H_{(4)} \ni (g, z, h) \mapsto gzh \in Z_{(4)}. \tag{13}$$

Proposition 10. *For any $z \in Z_{(4)}$, we can find $g \in \mathrm{GL}_2(\mathbb{C})$ and $h \in H_{(4)}$ such that*

$$gzh = \begin{pmatrix} 1 & 0 & 0 & 0 \\ 0 & 1 & 0 & -x \end{pmatrix}, \tag{14}$$

and $x \in \mathbb{C}$ is uniquely determined by z:

$$x = \frac{1}{[0,1]^2}\{[0,1][2,1] + [0,2]^2 - [0,1][0,3]\}, \qquad [i,j] := \det(z_i, z_j). \tag{15}$$

This proposition implies that the quotient space $\mathrm{GL}_2(\mathbb{C})\backslash Z_{(4)}/H_{(4)}$ has a realization as the global slice $X_{(4)} \subset Z_{(4)}$:

$$\mathrm{GL}_2(\mathbb{C})\backslash Z_{(4)}/H_{(4)} \simeq X_{(4)} = \left\{ \begin{pmatrix} 1 & 0 & 0 & 0 \\ 0 & 1 & 0 & -x \end{pmatrix} ; x \in \mathbb{C} \right\}.$$

On the level of integrals we can show that GHGI in this setting with appropriately chosen cycle c is related to the Airy integral by

$$I(z, \alpha, c) = (\det g)\chi(h; \alpha)^{-1} \int_C e^{xu - \frac{1}{3}u^3} du,$$

where z and x are related by (14) and (15). This explains that GHGI for the partition (4) on $Z_{(4)}$ is essentially the same as the classical Airy integral.

Remark 11. *The reason for choosing the parameter α in χ as*

$$\alpha = (-2, 0, 0, -1)$$

is explained by considering the group of symmetry for the generalized Airy function which is given by the analogue of Weyl group $N_{\mathrm{GL}_4(\mathbb{C})}(H_{(4)})/H_{(4)}$. See [9].

§3. Twisted de Rham cohomology

This section concerns the explicit computation of the twisted de Rham cohomology group associated with GHGI (3).

Let $\chi = \chi(\vec{u}z; \alpha)$ with $z \in Z_{r,\lambda}$ and $\alpha \in \mathbb{C}^N$ satisfying (2) and let $\mathcal{A} = \{H^{(1)}, \dots, H^{(\ell)}\}$ be the arrangement of hyperplanes in $\mathbb{C}^r$, where $H^{(k)} = \{u \in \mathbb{C}^r; \vec{u} \cdot z_0^{(k)} = 0\} \subset \mathbb{C}^r$. Put $N(\mathcal{A}) = \bigcup_{k=1}^{\ell} H^{(k)}$. Then χ is a multivalued holomorphic function on $\mathbb{C}^r \setminus N(\mathcal{A})$ whose logarithmic exterior derivative $d\log\chi$ has poles on $H^{(1)}, \dots, H^{(\ell)}$ of order $n_1, \dots, n_\ell$, respectivley.

Let $\Omega^p(*\mathcal{A})$ be the set of rational p-forms having poles at most on $N(\mathcal{A})$. Define the twisted differentiation $\nabla : \Omega^p(*\mathcal{A}) \to \Omega^{p+1}(*\mathcal{A})$ by

$$\nabla(\eta) = (\chi^{-1} \cdot d \cdot \chi)(\eta) = d\eta + \Big(d\log\chi(\vec{u}z; \alpha)\Big) \wedge \eta.$$

Since $d\chi$ has poles only on $N(\mathcal{A})$, ∇ really sends $\Omega^p(*\mathcal{A})$ to $\Omega^{p+1}(*\mathcal{A})$. Note also $\nabla \circ \nabla = \chi^{-1} \cdot d^2 \cdot \chi = 0$. It follows that we have the twisted algebraic de Rham complex

$$C_{z,\alpha}(*\mathcal{A}) : \ \Omega^0(*\mathcal{A}) \xrightarrow{\nabla} \Omega^1(*\mathcal{A}) \xrightarrow{\nabla} \cdots \xrightarrow{\nabla} \Omega^r(*\mathcal{A}) \to 0.$$

We sometimes write $C_{z,\alpha}$ instead of $C_{z,\alpha}(*\mathcal{A})$.

Definition 12. *The twisted algebraic de Rham cohomology is*

$$H^p(C_{z,\alpha}(*\mathcal{A})) := \frac{\mathrm{Ker}\,\{\nabla : \Omega^p(*\mathcal{A}) \to \Omega^{p+1}(*\mathcal{A})\}}{\mathrm{Im}\,\{\nabla : \Omega^{p-1}(*\mathcal{A}) \to \Omega^p(*\mathcal{A})\}} \quad (p \geq 0).$$

We know the following cases about the computation of the cohomology groups which will be described explicitly below.

(1) $r = 1$, i.e. the case where GHGI is defined by 1-dimensional integral.
(2) r is general and $\lambda = (1, \ldots, 1)$. This is the case of Aomoto–Gelfand hypergeometric integral.
(3) r is general and $\lambda = (N)$. The case of generalized Airy integral.
(4) r is general and $\lambda = (q + 1, 1, \ldots, 1)$.

3.1. 1-dimensional case

In the case the general hypergeometric integral is given by 1-dimensional integral, we have the following result.

Proposition 13. *Let* $\lambda = (n_1, \ldots, n_\ell)$ *be a partition of* N. *For any* $z \in Z_{1,\lambda}$, *we have*

(1) $H^p(C_{z,\alpha}) = 0$ *for* $p \neq 1$,
(2) $\dim_{\mathbb{C}} H^1(C_{z,\alpha}) = N - 2$,

As a basis of $H^1(C_{z,\alpha})$ *we can take the following* 1*-forms. In the case* $n_1 = 1$,

$$d\theta_0(\vec{u}z^{(k)}) - d\theta_0(\vec{u}z^{(k+1)}), \quad (2 \leq k \leq N - 1);$$

in the case $n_1 \geq 2$,

(16)
$$d\theta_1(\vec{u}z^{(1)}), \ldots, d\theta_{n_1-2}(\vec{u}z^{(1)}), d\theta_0(\vec{u}z^{(k)}), \ldots, d\theta_{n_k-1}(\vec{u}z^{(k)}), \ (2 \leq k \leq \ell).$$

3.2. Generalized Airy case

The generalized Airy integral is the GHGI for the partition $\lambda = (N)$ of N. The generalized Airy integral is introduced by Gelfand, Retahk and Serganova in [4]. In this case the space of coefficients of polynomial of degree 1 is

$$Z_{r,(N)} = \{z = (z_0, \ldots, z_{N-1}) \in \mathrm{Mat}_{r+1,N}(\mathbb{C}) \mid \det(z_0, \ldots, z_r) \neq 0\}.$$

We assume that

$$(17) \qquad z_0 = \begin{pmatrix} 1 \\ 0 \\ \vdots \\ 0 \end{pmatrix}, \quad \alpha_{N-1} \neq 0.$$

Then the integrand of the generalized Airy integral has the form

$$(18) \qquad \chi(\vec{u}z; \alpha) = e^{f(u)}, \quad f(u) = \sum_{m=1}^{N-1} \alpha_m \theta_m(\vec{u}z).$$

Note that by the assumption (17), f is a polynomial of $u = (u_1, \dots, u_r)$ of degree $N-1$.

Lemma 14. *For any $z \in Z_{r,(N)}$ satisfying (17), f is a polynomial having only isolated critical points with the Milnor number*

$$\mu(f) = \binom{N-2}{r}.$$

This fact can be seen as follows. Put $z = (z', z'')$ with $z' = (z_0, \dots, z_r)$. Since $\det z' \neq 0$ by the definition of $Z_{r,(N)}$, we can make the change of variables $\vec{u} = (1, v_1, \dots, v_r)z'$. If we define the weights of v_k as $wt(v_k) = k$, then the highest weight part of f is $\alpha_{N-1}\theta_{N-1}(\vec{v}(I_{r+1}, 0))$ which is weighted homogeneous of weight $N-1$ having isolated critical point at 0. Then Lemma 14 follows from this fact.

Proposition 15. [7] *For the r-dimensional generalized Airy integral, we assume the condition (17) . Then for any $z \in Z_{r,(N)}$, we have*

(1) $H^p(C_{z,\alpha}) = 0$ *for* $p \neq r$,
(2) $\dim_{\mathbb{C}} H^r(C_{z,\alpha}) = \binom{N-2}{r}$.

To state the result on the choice of a basis of top cohomology group $H^r(C_{z,\alpha})$, we prepare some notations. Let $\mathcal{Y}(r, l)$ be the set of Young diagrams contained in $r \times l$ box, namely $Y \in \mathcal{Y}(r, l)$ if the length is $\ell(Y) \leq r$ and any "parts of Y" is not greater than l. The number of boxes in the diagram Y is denoted by $|Y|$. For any $Y \in \mathcal{Y}(r, l)$, let $s_Y(v)$ be the Schur polynomial of $v_1, \dots, v_r$ associated with Y which is a symmetric polynomial of degree $|Y|$. Since $s_Y(v)$ is a symmetric polynomial, we can write it as a polynomial of the elementary symmetric functions $e_1(v), \dots, e_r(v)$ of v. It is denoted as $S_Y(u)$:

$$s_Y(v) = S_Y(e(v)).$$

Proposition 16. [7] *For the r-dimensional generalized Airy integral, the following r-form gives a basis of $H^r(C_{z,\alpha})$:*

$$S_Y(u)du_1 \wedge \cdots \wedge du_r, \quad Y \in \mathcal{Y}(r, N-r-2). \tag{19}$$

Remark 17. (1) *In the case $r = 1$, namely the integral is one dimensional, the basis (19) above is written as*

$$du, udu, \ldots, u^{N-3}du.$$

(2) *Another choice of a basis is that given in Proposition 13:*

$$d(\theta_1(\vec{u}z)), \ldots, d(\theta_{N-2}(\vec{u}z)).$$

It is an analogue of flat basis of the Jacobi ring of singularity of A_{N-2} type [10].

3.3. $\lambda = (q+1, 1, \ldots, 1)$ case

In the case where the partition of N has the form $\lambda = (q+1, 1, \ldots, 1)$, we can give an explicit form of a basis of the cohomology group. Recall that the space of coefficients of polynomials of degree 1 in the definition of GHGI is

$$Z_{r,\lambda} = \{z = (z_0, \ldots, z_{N-1}) \in \mathrm{Mat}_{r+1,N}(\mathbb{C}) \mid \text{the condition } (*)\},$$

where the condition (*) is that, for any $0 \le k \le q+1$ and $q < j_k < j_{k+1} < \cdots < j_r \le N-1$, there holds

$$\det(z_0, \ldots, z_{k-1}, z_{j_k}, \ldots, z_{j_r}) \neq 0.$$

In this case the integrand of the GHGI has the form $\chi(\vec{u}z; \alpha) = e^{g(u,z)} \prod_{j=q+1}^{N-1} f_j^{\alpha_j}$, where

$$g(u, z) = \sum_{k=0}^{q} \alpha_k \theta_k(f_0, f_1, \ldots, f_q), \quad f_j = \vec{u}z_j, \quad (0 \le j < N).$$

We assume that

$$z_0 = \begin{pmatrix} 1 \\ 0 \\ \vdots \\ 0 \end{pmatrix} \tag{20}$$

and

$$\alpha_0 + \sum_{j>q} \alpha_j = -r-1, \ \alpha_q \neq 0, \ \ \alpha_j \notin \mathbb{Z}, \ (j > q). \tag{21}$$

As a consequence of the assumptions (20) and $\alpha_q \neq 0$, we see that $g(u, z)$ is a polynomial of u of degree q. Put $H_j := \ker f_j$ $(j = q+1, \dots, N-1)$ and $\mathcal{A} = \{H_{q+1}, \dots, H_{N-1}\}$. Moreover from the assumption (21), the integrand $\chi(\vec{u}z; \alpha)$ of GHGI is a multivalued holomorphic function on $\mathbb{C}^r$ with the branch locus $N(\mathcal{A}) = \cup_{j=q+1}^{N-1} H_j$.

Proposition 18. [8] *In the case* $\lambda = (q+1, 1, \dots, 1)$, *we assume for the GHGI the conditions (20) and (21). Then we have*

(1) $H^p(C_{z,\alpha}) = 0$ *for* $p \neq r$,

(2) $\dim_{\mathbb{C}} H^r(C_{z,\alpha}) = \binom{N-2}{r}$.

Let us give a basis for the top cohomology group $H^r(C_{z,\alpha})$. To this end, we use the same notation as in the generalized Airy case for $\mathcal{Y}(r, l)$, $s_Y(v)$, $S_Y(u)$. For $0 \leq k \leq r$, take a Yang diagram $Y \in \mathcal{Y}(k, q-1-k)$ and $J = (j_{k+1}, \dots, j_r)$ satisfying

$$q + 1 \leq j_{k+1} < \cdots < j_r < N$$

and define the logarithmic r-form

$$\omega_{Y,J} := S_Y(f_1, \dots, f_k) df_1 \wedge \cdots \wedge df_k \wedge \frac{df_{j_{k+1}}}{f_{j_{k+1}}} \wedge \cdots \wedge \frac{df_{j_r}}{f_{j_r}} \in \Omega^r(\mathcal{A}).$$

Proposition 19. [8] *Under the assumption (20) and (21), we can take as a basis of* $H^r(C_{z,\alpha})$ *the following set of* r*-forms:*

$$\bigcup_{k=0}^{r} \left\{ \omega_{Y,J} \;\middle|\; \begin{array}{l} Y \in \mathcal{Y}(k, q-1-k), \\ J = (j_1, \dots, j_{r-k}) \text{ such that } q+1 \leq j_1 < \cdots < j_{r-k} \leq N \end{array} \right\}.$$

§4. Exterior power structure

In the case the partition λ of N is general, we compute the cohomology group for particular points z of the parameter space $Z_{r,\lambda}$which will be called the Veronese points. Combining this fact with the holonomicity of the general hypergeometric system of type λ, we can show the purity of the cohomology groups and give the rank of the top cohomology group at any point $z \in Z_{r,\lambda}$. Recall that, in the case of generalized Airy integral, we gave in Proposition 16 a basis of the r-th cohomology group expressed in terms of Schur functions:

$$S_Y(u) du_1 \wedge \cdots \wedge du_r, \qquad Y \in \mathcal{Y}(r, N-r-2).$$

This basis is related to the exterior product structure of the cohomology group at the Veronese points.

4.1. Veronese map for $\lambda = (1, \dots, 1)$

We first recall the Veronese embedding $\mathbb{P}^1 \hookrightarrow \mathbb{P}^r$ in order to motivate the construction of generalized Veronese map given in Subsection 4.2. Consider the map

$$\psi : \mathbb{C}^2 \ni \begin{pmatrix} v_0 \\ v_1 \end{pmatrix} \mapsto \begin{pmatrix} v_0^r \\ v_0^{r-1} v_1 \\ \vdots \\ v_1^r \end{pmatrix} \in \mathbb{C}^{r+1}. \tag{22}$$

It induces the Veronese embedding $\bar{\psi} : \mathbb{P}^1 \ni [v_0, v_1] \mapsto [v_0^r, v_0^{r-1} v_1, \dots, v_1^r] \in \mathbb{P}^r$.

Using the map ψ, we define, in the case $\lambda = (1, \dots, 1)$, the map $\Psi_{(1,\dots,1)} : \mathrm{Mat}_{2,N}(\mathbb{C}) \to \mathrm{Mat}_{r+1,N}(\mathbb{C})$ by $(z_0, z_1, \dots, z_{N-1}) \mapsto (\psi(z_0), \psi(z_1), \dots, \psi(z_{N-1}))$ or by

$$\begin{pmatrix} z_{00} & \cdots & z_{0,N-1} \\ z_{10} & \cdots & z_{1,N-1} \end{pmatrix} \mapsto \begin{pmatrix} (z_{00})^r & \cdots & (z_{0,N-1})^r \\ (z_{00})^{r-1} z_{10} & \cdots & (z_{0,N-1})^{r-1} z_{1,N-1} \\ \vdots & & \vdots \\ (z_{10})^r & \cdots & (z_{1,N-1})^r \end{pmatrix}.$$

We see that it induces the map $\Psi_{(1,\dots,1)} : Z_{1,(1,\dots,1)} \to Z_{r,(1,\dots,1)}$ which we call the Veronese map. This fact is seen as follows. Note that $z = (z_0, \dots, z_{N-1}) \in Z_{1,(1,\dots,1)}$ if and only if $\det(z_i, z_j) \neq 0$ for any $i \neq j$. Take any indices $0 \leq i_0 < i_1 < \cdots < i_r \leq N-1$. Then we see that

$$\det(\psi(z_{i_0}), \psi(z_{i_1}), \dots, \psi(z_{i_r})) = \prod_{k<l} \det(z_{i_k}, z_{i_l}) \neq 0.$$

We will define in the next subsection the map $\Psi_\lambda : Z_{1,\lambda} \to Z_{r,\lambda}$ for any partition λ of N which coincides with $\Psi_{(1,\dots,1)}$ in the case $\lambda = (1, \dots, 1)$. To this end we restate the map ψ as follows. Let V be a vector space of $\dim_{\mathbb{C}} V = 2$ and let $S^r V$ be the r-th symmetric tensor product. Then we have $\dim_{\mathbb{C}} S^r V = r + 1$. Define the map

$$\psi : V \ni v \mapsto \overbrace{v \otimes \cdots \otimes v}^{r \text{ times}} \in S^r V.$$

Let e_0, e_1 be a basis of V, and let $\mathbf{e}_0, \dots, \mathbf{e}_r$ be a basis of $S^r V$ defined by

$$\mathbf{e}_k = \sum_{i_1 + i_2 + \cdots + i_r = k} e_{i_1} \otimes \cdots \otimes e_{i_r}.$$

Then from

$$(v_0e_0 + v_1e_1)^{\otimes r} = \sum_{k=0}^{r} v_0^{r-k} v_1^k \mathbf{e}_k,$$

we see that the map ψ, expressed in terms of the basis e_0, e_1 of V and $\mathbf{e}_0, \ldots, \mathbf{e}_r$ of S^rV, is the same as the map (22).

4.2. Veronese map for $\lambda = (n_1, \ldots, n_\ell)$

Let us generalize the Veronese map $\Psi_{(1,\ldots,1)}$ to the case where the partition λ of N is general. To do this we at first generalize the map ψ given by (22). Let V be a vector space of $\dim_{\mathbb{C}} V = 2$ and let $R_n = \mathbb{C}[T]/(T^n)$ be the quotient of the polynomial ring $\mathbb{C}[T]$ by the ideal (T^n) generated by the n-th power T^n. Consider $V_n := V \otimes R_n$, which is an R_n-module, and its r-th symmetric tensor product S^rV_n as an R_n-module. Define the map

$$\psi_n : V_n \ni v \mapsto \overbrace{v \otimes \cdots \otimes v}^{r} \in S^rV_n. \tag{23}$$

Let us express ψ_n using the basis of V_n as a vector space over $\mathbb{C}$:

$$e_i \otimes T^j \qquad (i = 0, 1,\ 0 \le j < n)$$

and that of S^rV_n:

$$\mathbf{e}_i \otimes T^j \qquad (0 \le i \le r,\ 0 \le j < n).$$

We identify an element $v = \sum_{i,j} v_{ij} e_i \otimes T^j \in V_n$ with the matrix

$$\begin{pmatrix} v_{00} & v_{01} & \cdots & v_{0,n-1} \\ v_{10} & v_{11} & \cdots & v_{1,n-1} \end{pmatrix} \in \mathrm{Mat}_{2,n}(\mathbb{C})$$

and an element $w = \sum_{i,j} w_{ij} \mathbf{e}_i \otimes T^j \in S^rV_n$ with

$$\begin{pmatrix} w_{00} & w_{01} & \cdots & w_{0,n-1} \\ \vdots & \vdots & & \vdots \\ w_{r0} & w_{r1} & \cdots & w_{r,n-1} \end{pmatrix} \in \mathrm{Mat}_{r+1,n}(\mathbb{C}).$$

The map $\psi_n : V_n \to S^rV_n$ defined by (23) induces a map $\psi_n : \mathrm{Mat}_{2,n}(\mathbb{C}) \to \mathrm{Mat}_{r+1,n}(\mathbb{C})$.

Example 20. *Let $r = 2$. Then the map $\psi_3 : \mathrm{Mat}_{2,3}(\mathbb{C}) \to \mathrm{Mat}_{3,3}(\mathbb{C})$ is given by*

$$v \mapsto \begin{pmatrix} v_{00}^2 & 2v_{00}v_{01} & 2v_{00}v_{02} + v_{01}^2 \\ v_{00}v_{01} & v_{00}v_{11} + v_{01}v_{10} & v_{00}v_{12} + v_{02}v_{10} + v_{01}v_{11} \\ v_{01}^2 & 2v_{01}v_{11} & 2v_{10}v_{12} + v_{11}^2 \end{pmatrix}.$$

Definition 21. *For a partition $\lambda = (n_1, \dots, n_\ell)$ of N, define Ψ_λ : $\mathrm{Mat}_{2,N}(\mathbb{C}) \to \mathrm{Mat}_{r+1,N}(\mathbb{C})$ by*

$$z = (z^{(1)}, \dots, z^{(\ell)}) \mapsto (\psi_{n_1}(z^{(1)}), \dots, \psi_{n_\ell}(z^{(\ell)})).$$

We call this map the **Veronese map of type** λ.

Lemma 22. $\Psi_\lambda(Z_{1,\lambda}) \subset Z_{r,\lambda}$.

This fact can be shown as follows. For $z \in Z_{1,\lambda}$, put $w = \Psi_\lambda(z)$ and write it as $w = (w^{(1)}, \dots, w^{(\ell)}), w^{(k)} = (w_0^{(k)}, \dots, w_{n_k-1}^{(k)}) \in \mathrm{Mat}_{r+1,n_k}(\mathbb{C})$. Then for any $\mu = (m_1, \dots, m_\ell)$ such that $0 \le m_k \le n_k$ and $m_1 + \cdots + m_\ell = r + 1$, we have

$$\begin{aligned}\det(w_0^{(1)}, \dots, w_{m_1-1}^{(1)}, \dots, w_0^{(\ell)}, \dots, w_{m_\ell-1}^{(\ell)})\\ = C_\mu \prod_k \det(z_0^{(k)}, z_1^{(k)})^{m_k(m_k-1)/2} \prod_{k<l} \det(z_0^{(k)}, z_0^{(l)})^{m_k m_l}.\end{aligned}$$

with some nonzero constant C_μ. The map $\Psi_\lambda : Z_{1,\lambda} \to Z_{r,\lambda}$ is also called the Veronese map of type λ. The set $\Psi_\lambda(Z_{1,\lambda})$ and its point are called respectively the **Veronese image** and a **Veronese point**.

4.3. Exterior power structure of the cohomology group

The following theorem asserts that the cohomology groups for the r-dimensional GHGI of type λ at Veronese points can be constructed from that for 1 dimensional GHGI of type λ.

Theorem 23. *Let $z \in Z_{1,\lambda}$ be such that $z_0^{(1)} = (1,0)^t$ and let*

$$\tilde{z} = \Psi(z) \in Z_{r,N}, \quad \tilde{\alpha} = \alpha + (-r+1, 0, \dots, 0). \tag{24}$$

Then we have

(1) $H^p(C_{\tilde{z},\tilde{\alpha}}) = 0, \quad (p \neq r)$,

(2) *the top cohomology group has the exterior product structure:*

$$H^r(C_{\tilde{z},\tilde{\alpha}}) \simeq \bigwedge^r H^1(C_{z,\alpha}), \tag{25}$$

(3) $\dim_{\mathbb{C}} H^r(C_{\tilde{z},\tilde{\alpha}}) = \binom{N-2}{r}$.

Example 24. *[Generalized Airy case $\lambda = (N)$]*
Let $z \in Z_{1,(N)}$ and $\tilde{z} = \Psi_{(N)}(z)$. Theorem 23 says the following.

- *If we take a basis of* $H^1(C_{z,\alpha})$ *as*

$$\varphi_i = u^i du, \quad (0 \le i \le N-3),$$

then the isomorphism (25) gives the correspondence

$$\varphi_{i_1} \Box \cdots \Box \varphi_{i_r} \mapsto S_Y(v)dv.$$

Here $\Box$ *denotes the exterior product in* $\bigwedge^r H^1(C_{z,\alpha})$*,*

$$i_1 > i_2 > \cdots > i_r \ge 0, \quad Y = (i_1-r+1, i_2-r+2, \dots, i_r) \in \mathcal{Y}(r, N-r-2)$$

and

$$v = (v_1, \dots, v_r), \quad dv = dv_1 \wedge \cdots \wedge dv_r.$$

- *If one take a basis of* $H^1(C_{z,\alpha})$ *as*

$$\varphi_i = d(\theta_i(\vec{u}z)), \quad (1 \le i \le N-2),$$

then the isomorphism (25) gives the correspondence

$$\varphi_{i_1} \Box \cdots \Box \varphi_{i_r} \mapsto d(\theta_{i_1}(\vec{v}\tilde{z})) \wedge \cdots \wedge d(\theta_{i_r}(\vec{v}\tilde{z})).$$

Example 25. *[*$\lambda = (n_1, \dots, n_\ell)$ *case]*
Let $z \in Z_{1,\lambda}$*. Then as a basis of* $H^1(C_{z,\alpha})$ *we can take, for* $n_1 \ge 2$ *for example,*

$$d(\theta_1(\vec{u}z^{(1)})), \dots, d(\theta_{n_1-2}(\vec{u}z^{(1)}))$$
$$d(\theta_0(\vec{u}z^{(k)})), d(\theta_1(\vec{u}z^{(k)})), \dots, d(\theta_{n_k-1}(\vec{u}z^{(k)})), \quad (2 \le k \le \ell).$$

Put $\tilde{z} = \Psi_\lambda(z)$*. Then as a basis of* $H^r(C_{\tilde{z},\tilde{\alpha}})$ *we can take the* r*-forms obtained by choosing* r *forms* $d(\theta_j(\vec{v}\tilde{z}^{(k)}))$ *and taking the exterior product of them.*

Combining Theorem 23 and the holonomicity of the general hypergeometric system, we can know the purity and the dimension of the cohomology group for any $z \in Z_{r,\lambda}$.

Corollary 26. *Let* $\lambda = (n_1, \dots, n_\ell)$ *be a partition of* N *and let* $\alpha = (\alpha^{(1)}, \dots, \alpha^{(\ell)})$, $\alpha^{(k)} = (\alpha_0^{(k)}, \dots, \alpha_{n_k-1}^{(k)}) \in \mathbb{C}^{n_k}$ *satisfy the condition (2). Then for any point* z *of* $Z_{r,\lambda}$ *satisfying* $z_0^{(1)} = (1, 0, \dots, 0)^t$*, we have*
(1) $H^p(C_{z,\alpha}) = 0$ *for* $p \ne r$*,*
(2) $\dim_{\mathbb{C}} H^r(C_{z,\alpha}) = \binom{N-2}{r}$.

Question: In the case $\lambda = (n_1, \dots, n_\ell)$, we consider the r-forms constructed in Example 25 at any point $z \in Z_{r,\lambda}$. Do these r-forms give a basis for $H^r(C_{z,\alpha})$?

References

[1] K. Aomoto, On the structure of integrals of power product of linear functions, Sci. Papers College Gen. Ed. Univ. Tokyo, **27** (1977), 49–61.

[2] K. Aomoto, M. Kita, P. Orlik and H. Terao, Twisted de Rham cohomology groups of logarithmic forms, Adv. Math., **128** (1997), 119–152.

[3] I. M. Gelfand, General theory of hypergeometric functions, Dokl. Akad. Nauk. SSSR, **288** (1986), 14–18; English translation: Soviet Math. Dokl., **33** (1986), 9–13.

[4] I. M. Gelfand, V. S. Retahk and V. V. Serganova, Generalized Airy functions, Schubert cells, and Jordan groups, Dokl. Akad. Nauk. SSSR, **298** (1988), 17–21;
English translation: Soviet Math. Dokl., **37** (1988), 8–12.

[5] I. Basalaeva, H. Kimura and T. Nakazuka, Cohomological intersection numbers for the generalized Airy functions at Veronese points, Funkcial. Ekvac., **48** (2005), 161–181.

[6] T. Inui, Special Functions, Iwanami Shoten, 1962.

[7] H. Kimura, On rational de Rham cohomology associated with the generalized Airy functions, Annali della Scuola Norm. Sup. di Pisa, **24** (1997), 351–366.

[8] H. Kimura, On the twisted de Rham cohomology group of the general hypergeometric integral of type $(q+1, 1^{N-q})$, J. Math. Sci. Univ. Tokyo, **12** (2005), 165–189.

[9] H. Kimura and T. Koitabashi, Normalizer of maximal abelian subgroups of $GL(n)$ and the general hypergeometric functions, Kumamoto J. Math., **9** (1996), 13–43.

[10] H. Kimura and M. Taneda, Analogue of flat basis and cohomological intersection numbers for general hypergeometric functions, J. Math. Sci. Univ. Tokyo, **6** (1999), 415–436.

Hironobu Kimura
Graduate School of Science and Technology, Kumamoto Univerisity,
2-39-1, Kurokami, Kumamoto, 860-8555, Japan
E-mail address: hiro@sci.kumamoto-u.ac.jp

Advanced Studies in Pure Mathematics 62, 2012
Arrangements of Hyperplanes—Sapporo 2009
pp. 157–174

Hyperplane arrangements, local system homology and iterated integrals

Toshitake Kohno

Abstract.

We review some aspects of the homology of a local system on the complement of a hyperplane arrangement. We describe a relationship between linear representations of the braid groups due to R. Lawrence, D. Krammer and S. Bigelow and the holonomy representations of the KZ connection. We explain a method to describe such representations by the iterated integrals of logarithmic 1-forms.

§1. Introduction

Let $\mathcal{A}$ be an arrangement of hyperplanes in the complex vector space $\mathbf{C}^n$. We denote by $M(\mathcal{A})$ the complement of the union of hyperplanes in $\mathcal{A}$. The purpose of this article is to review developments concerning the homology of a local system on the complement $M(\mathcal{A})$ and its applications.

In the case of a complexification of a real arrangement M. Salvetti [17] constructed a complex which is homotopy equivalent to the complement $M(\mathcal{A})$. First, we briefly review the Salvetti complex and explain how this construction is used to compute the homology of locally finite chains of $M(\mathcal{A})$ with local system coefficients. If a local system $\mathcal{L}$ is generic, then we have a vanishing of homology $H_k(M(\mathcal{A}), \mathcal{L}) \cong 0$ unless $k = n$. We describe such vanishing theorem and a relation to the homology of locally finite chains.

As a typical example we deal with the case of a discriminantal arrangement. This arrangement has an important application to the flat

Received June 16, 2010.
Revised January 14, 2011.
2010 *Mathematics Subject Classification.* 20F36, 52C35, 33C60.
Key words and phrases. Hyperplane arrangement, Salvetti complex, local system, KZ connection, conformal field theory, braid group, configuration space, iterated integral.

connection due to V. G. Knizhnik and A. B. Zamolodchikov [10], which we shall call the KZ connection. It was shown by V. Schechtman and A. Varchenko [18] that the horizontal sections of the KZ connection are expressed by means of hypergeometric integrals over the complement of a discriminantal arrangement. Based on this result we describe a relationship between linear representations of braid groups developed by R. Lawrence, D. Krammer and S. Bigelow and the holonomy representations of the KZ connection. We also investigate a description of the space of conformal blocks in conformal field theory on the Riemann sphere in terms of homology of a local system over the complement of a discriminantal arrangement.

One of the ways to express the holonomy representation of the KZ connection is to use the iterated integrals of logarithmic 1-forms. In the case of the braid arrangement such iterated integral is a prototype of the Kontsevich integral for knots developed in [14]. We review basic facts about such iterated integrals of logarithmic 1-forms in a general situation of the complement of a hyperplane arrangement. We give a criterion so that iterated integrals of logarithmic 1-forms depend only on the homotopy class of loops.

The paper is organized in the following way. In Section 1 we recall basic facts about the homology of a local system on the complement of a hyperplane arrangement. In Section 2 we deal with discriminantal arrangements and describe a relation to the KZ connection. Section 3 is devoted to the iterated integrals of logarithmic 1-forms and holonomy representations of fundamental groups.

§2. Homology of local systems

Let $\mathcal{A} = \{H_1, \cdots, H_\ell\}$ be an arrangement of affine hyperplanes in the complex vector space $\mathbf{C}^n$. We consider the complement

$$M(\mathcal{A}) = \mathbf{C}^n \setminus \bigcup_{H \in \mathcal{A}} H.$$

Let us assume that the hyperplanes $H_1, \cdots, H_\ell$ are defined over $\mathbf{R}$. In this case $H \in \mathcal{A}$ is regarded as a complexification of the real hyperplane $H_{\mathbf{R}}$ in $V_{\mathbf{R}} = \mathbf{R}^n$. The complement $V_{\mathbf{R}} \setminus \bigcup_{H \in \mathcal{A}} H_{\mathbf{R}}$ consists of finitely many connected components called chambers.

The above real hyperplane arrangement $\{H_{\mathbf{R}}\}_{H \in \mathcal{A}}$ determines a natural stratification S of $\mathbf{R}^n$, whose stratum is called a facet. For facets E and F we shall say that $E > F$ if and only if $\overline{E} \supset F$ holds. For an increasing sequence of facets $F_{j_0} < \cdots < F_{j_p}$ we take a point v_{j_k} in each facet F_{j_k}, $0 \le k \le p$, and consider the simplex spanned by the vertices v_{j_k}, $0 \leq k \leq p$. This simplex defined for $F_{j_0} < \cdots < F_{j_p}$ is denoted by

$$\sigma(F_{j_0} < \cdots < F_{j_p}).$$

For a facet F the dual cell is defined by

$$D(F) = \bigcup \sigma(F^i < F^{i-1} < \cdots < F^0)$$

where the union is for all the increasing sequences of facets $F^i < F^{i-1} < \cdots < F^0$ with $F^i = F$ and $\operatorname{codim} F^j = j$.

Let $\pi : M(\mathcal{A}) \longrightarrow \mathbf{R}^n$ be the projection corresponding to the real part. A facet decomposition of the complexified complement $M(\mathcal{A})$ is given by

$$\bigcup_F \pi^{-1}(F).$$

The associated dual complex is called the Salvetti complex $S(\mathcal{A})$, which is an n dimensional CW complex. It was shown by M. Salvetti [17] that the inclusion

$$S(\mathcal{A}) \longrightarrow M(\mathcal{A})$$

is a homotopy equivalence.

Let $\mathcal{L}$ be a complex rank one local system over $M(\mathcal{A})$ associated with a representation of the fundamental group

$$\rho : \pi_1(M(\mathcal{A}), x_0) \longrightarrow \mathbf{C}^*.$$

We shall investigate the homology of $M(\mathcal{A})$ with coefficients in the local system $\mathcal{L}$. For our purpose the homology of locally finite chains $H_*^{lf}(M(\mathcal{A}), \mathcal{L})$ also plays an important role. It was shown by Z. Chen [6] that the complex associated with the facet decomposition $\bigcup_F \pi^{-1}(F)$ of $M(\mathcal{A})$ can be used to compute the homology of locally finite chains $H_*^{lf}(M(\mathcal{A}), \mathcal{L})$.

We briefly summarize basic properties of the above homology groups. For a complex arrangement $\mathcal{A}$ choose a smooth compactification $i : M(\mathcal{A}) \longrightarrow X$ with normal crossing divisors. We shall say that the local system $\mathcal{L}$ is generic if and only if there is an isomorphism

$$i_*\mathcal{L} \cong i_!\mathcal{L}$$

where i_* is the direct image and $i_!$ is the extension by 0. This means that the monodromy of $\mathcal{L}$ along any divisor at infinity is not equal to 1. The following theorem was shown in [12].

Theorem 2.1. *If the local system $\mathcal{L}$ is generic in the above sense, then there is an isomorphism*

$$H_*(M(\mathcal{A}), \mathcal{L}) \cong H_*^{lf}(M(\mathcal{A}), \mathcal{L}).$$

We have $H_k(M(\mathcal{A}), \mathcal{L}) = 0$ for any $k \neq n$.

Proof. In general we have isomorphisms

$$H^*(X, i_*\mathcal{L}) \cong H^*(M(\mathcal{A}), \mathcal{L}), \quad H^*(X, i_!\mathcal{L}) \cong H_c^*(M(\mathcal{A}), \mathcal{L})$$

where H_c denotes cohomology with compact supports.

There are Poincaré duality isomorphisms:

$$H_k^{lf}(M(\mathcal{A}), \mathcal{L}) \cong H^{2n-k}(M(\mathcal{A}), \mathcal{L})$$
$$H_k(M(\mathcal{A}), \mathcal{L}) \cong H_c^{2n-k}(M(\mathcal{A}), \mathcal{L}).$$

By the hypothesis $i_*\mathcal{L} \cong i_!\mathcal{L}$ we obtain an isomorphism

$$H_k^{lf}(M(\mathcal{A}), \mathcal{L}) \cong H_k(M(\mathcal{A}), \mathcal{L}).$$

It follows from the above Poincaré duality isomorpshims and the fact that $M(\mathcal{A})$ has a homotopy type of a CW complex of dimension n we have

$$H_k^{lf}(M(\mathcal{A}), \mathcal{L}) \cong 0, \quad k < n$$
$$H_k(M(\mathcal{A}), \mathcal{L}) \cong 0, \quad k > n.$$

Therefore we obtain $H_k(M(\mathcal{A}), \mathcal{L}) = 0$ for any $k \neq n$. Q.E.D.

In the case of a complexified real arrangement, the $\mathbf{C}$-vector space $H_n^{lf}(M(\mathcal{A}), \mathcal{L})$ is spanned by bounded chambers.

§3. KZ connections and hypergeometric integrals

Let $\mathfrak{g}$ be a complex semi-simple Lie algebra and $\{I_\mu\}$ be an orthonormal basis of $\mathfrak{g}$ with respect to the Cartan–Killing form. We set $\Omega = \sum_\mu I_\mu \otimes I_\mu$. Let $r_i : \mathfrak{g} \to \mathrm{End}(V_i)$, $1 \leq i \leq n$, be representations of the Lie algebra $\mathfrak{g}$. We denote by Ω_{ij} the action of Ω on the i-th and j-th components of the tensor product $V_1 \otimes \cdots \otimes V_n$. By using the fact that the Casimir element $c = \sum_\mu I_\mu \cdot I_\mu$ lies in the center of the universal enveloping algebra $U\mathfrak{g}$, it can be shown that the relations:

$$[\Omega_{ik}, \Omega_{ij} + \Omega_{jk}] = 0, \quad (i, j, k \text{ distinct}),$$
$$[\Omega_{ij}, \Omega_{k\ell}], \quad (i, j, k, \ell \text{ distinct})$$

hold. We define the Knizhnik–Zamolodchikov (KZ) connection as the 1-form

$$\omega = \frac{1}{\kappa} \sum_{1 \leq i < j \leq n} \Omega_{ij} d \log(z_i - z_j)$$

with values in $\mathrm{End}(V_1 \otimes \cdots \otimes V_n)$ for a non-zero complex parameter κ.

We set $\omega_{ij} = d \log(z_i - z_j)$, $1 \leq i, j \leq n$. It follows from the above quadratic relations among Ω_{ij} together with Arnold's relation

$$\omega_{ij} \wedge \omega_{jk} + \omega_{jk} \wedge \omega_{k\ell} + \omega_{k\ell} \wedge \omega_{ij} = 0$$

that $\omega \wedge \omega = 0$ holds. This implies that ω defines a flat connection for a trivial vector bundle over the configuration space

$$X_n = \{(z_1, \cdots, z_n) \in \mathbf{C}^n \ ; \ z_i \neq z_j \text{ if } i \neq j\}$$

with fiber $V_1 \otimes \cdots \otimes V_n$.

The fundamental group of the configuration space X_n is the pure braid group on n strings denoted by P_n. As the holonomy of the connection ω we have a one-parameter family of linear representations of the pure braid group

$$\theta_\kappa : P_n \to \mathrm{GL}(V_1 \otimes \cdots \otimes V_n).$$

The symmetric group S_n acts on X_n by permutations of coordinates. We denote the quotient space X_n/S_n by Y_n. The fundamental group of Y_n is the braid group on n strings denoted by B_n. In the case $V_1 = \cdots = V_n = V$, the symmetric group S_n acts diagonally on the trivial vector bundle over X_n with fiber $V^{\otimes n}$ and the connection ω is invariant by this action. Thus we have a one-parameter family of linear representations of the braid group

$$\theta_\kappa : B_n \to \mathrm{GL}(V^{\otimes n}).$$

Following V. Schechtman and A. Varchenko [18], we shall express the horizontal sections of the KZ connection ω in terms of hypergeometric integrals. Let

$$\pi : X_{m+n} \longrightarrow X_n$$

be the projection map defined by

$$(z_1, \cdots, z_n, t_1, \cdots, t_m) \mapsto (z_1, \cdots, z_n).$$

We denote by $X_{n,m}$ a fiber of π, which is the complement of a discriminantal arrangement.

In this article we deal with the case $\mathfrak{g} = sl_2(\mathbf{C})$. As a complex vector space the Lie algebra $sl_2(\mathbf{C})$ has a basis H, E and F satisfying the relations:

$$[H,E]=2E, \quad [H,F]=-2F, \quad [E,F]=H.$$

For a complex number λ we denote by M_λ the Verma module of $sl_2(\mathbf{C})$ with highest weight λ. Namely, there is a non-zero vector $v \in M_\lambda$ called the highest weight vector satisfying

$$Hv = \lambda v, \ Ev = 0$$

and M_λ is spanned by $F^j v$, $j \geq 0$.

We shall consider the tensor product $M_{\lambda_1} \otimes \cdots \otimes M_{\lambda_n}$ of the Verma modules of $sl_2(\mathbf{C})$. We set $\lambda = \lambda_1 + \cdots + \lambda_n$. For a non-negative integer m we define the space of weight vectors with weight $\lambda - 2m$ by

$$W[\lambda - 2m] = \{x \in M_{\lambda_1} \otimes \cdots M_{\lambda_n} \ ; \ Hx = (\lambda - 2m)x\}$$

and consider the space of coinvariants defined by

$$N[\lambda - 2m] = W[\lambda - 2m]/F \cdot W[\lambda - 2m + 2].$$

The KZ connection ω commutes with the diagonal action of $\mathfrak{g}$ on $V_{\lambda_1} \otimes \cdots \otimes V_{\lambda_n}$, hence it acts on the space of coinvariants $N[\lambda - 2m]$.

For parameters κ and λ we consider the multi-valued function

$$\Phi = \prod_{1 \leq i < j \leq n} (z_i - z_j)^{\frac{\lambda_i \lambda_j}{\kappa}} \prod_{1 \leq i \leq m, 1 \leq \ell \leq n} (t_i - z_\ell)^{-\frac{\lambda_\ell}{\kappa}} \prod_{1 \leq i < j \leq m} (t_i - t_j)^{\frac{2}{\kappa}}$$

defined over X_{n+m}. Let $\mathcal{L}$ denote the local system associated to the multi-valued function Φ. The symmetric group S_m acts on $X_{n,m}$ by permutations of the coordinates $(t_1, \cdots, t_m)$. The function Φ is invariant by the action of S_m. We put $Y_{n,m} = X_{n,m}/S_m$. The local system $\mathcal{L}$ over $X_{n,m}$ defines a local system on $Y_{n,m}$, which we denote by the same letter $\mathcal{L}$. The dual local system is denoted by $\mathcal{L}^*$.

Let us choose the parameters κ and $\lambda_1, \cdots, \lambda_m$ so that the associated local system $\mathcal{L}$ is generic in the sense of the previous section. Under this condition it follows from the vanishing theorem in the previous section that we have

$$H_k(X_{n,m}, \mathcal{L}^*) \cong 0, \quad k \neq m$$

and there is an isomorphism

$$H_m(X_{n,m}, \mathcal{L}^*) \cong H_m^{lf}(X_{n,m}, \mathcal{L}^*).$$

The twisted de Rham complex $(\Omega^*(X_{n,m}), \nabla)$ is a complex with differential $\nabla : \Omega^j(X_{n,m}) \to \Omega^{j+1}(X_{n,m})$ defined by

$$\nabla\omega = d\omega + d\log\Phi \wedge \omega,$$

for $\omega \in \Omega^j(X_{n,m})$. There is a pairing between the homology of the local system $\mathcal{L}^*$ and the cohomology of the twisted de Rham complex

$$H_m(X_{n,m}, \mathcal{L}^*) \times H^m(\Omega^*(X_{n,m}), \nabla) \to \mathbf{C}$$

defined by

$$(c, \omega) \mapsto \int_c \Phi\omega.$$

Such integrals are called hypergeomotric integrals. We refer the reader to [16] for a detailed treatment of hypergeometric integrals in the more general situation of hyperplane arrangements.

We define a map

$$\rho : W[\lambda - 2m] \to \Omega^m(X_{n,m})$$

given by

$$\rho(w) = R_w(t, z) dt_1 \wedge \cdots \wedge dt_m$$

using the rational function $R_w(t, z)$ for $w \in W[\lambda - 2m]$ defined in the following manner. Let $v_k \in M_{\lambda_k}$, $1 \le k \le n$, be the highest weight vectors and we set $v = v_1 \otimes \cdots \otimes v_n$. For an n-tuple of non-negative integers $J = (j_1, \cdots, j_n)$ we set

$$F^J v = F^{j_1} v_1 \otimes \cdots \otimes F^{j_n} v_n.$$

The weight space $W[\lambda - 2m]$ has a basis $F^J v$ for each J with $j_1 + \cdots + j_n = m$. For the sequence of integers $(i_1, \cdots, i_m) = (\underbrace{1, \cdots, 1}_{j_1}, \cdots, \underbrace{n, \cdots, n}_{j_n})$ we set

$$\eta_J(z, t) = \frac{1}{(t_1 - z_{i_1}) \cdots (t_m - z_{i_m})}.$$

We define the rational function $R_J(z, t)$ by

$$R_J(z, t) = \frac{1}{j_1! \cdots j_n!} \sum_{\sigma \in S_m} \eta_J(z_1, \cdots, z_n; t_{\sigma(1)}, \cdots, t_{\sigma(m)}).$$

It turns out that ρ induces a map to the cohomology of the twisted de Rham complex

$$N[\lambda - 2m] \longrightarrow H^m(\Omega^*(X_{n,m}), \nabla).$$

By this construction we obtain a map

$$\phi : H_m(Y_{n,m}, \mathcal{L}^*) \longrightarrow N[\lambda - 2m]^*$$

defined by

$$\langle \phi(c), w \rangle = \int_c \Phi\rho(w).$$

A lot of works have been done on the expression of the solutions of the KZ equation by means of hypergeometric type integrals (see [7] and [18]). According to the formulation due to V. Schechtman and A. Varchenko [18] the integral

$$\int_c \Phi\rho(w)$$

is a horizontal section of the KZ connection with values in $N[\lambda - 2m]$.

Theorem 3.1. *Let us suppose that the local system $\mathcal{L}$ is generic. Then the map ϕ gives an isomophism*

$$H_m(Y_{n,m}, \mathcal{L}^*) \cong N[\lambda - 2m]^*.$$

Moreover, the above isomorphism is equivariant with respect to the action of the pure braid group P_n.

As a consequence we obtain that the following two representations of the pure braid groups are equivalent:

(1) Action of P_n on the twisted homology $H_m(Y_{n,m}, \mathcal{L}^*)$,
(2) Holonomy representation of the KZ equation with values in $N[\lambda - 2m]^*$.

In the case $m = 1$, the above representation of the pure braid group P_n is equivalent to the Gassner representation (see [3]). Let us consider the case $\lambda_1 = \cdots = \lambda_n$ and $m = 2$. We have a two-parameter family of linear representations of the braid group B_n. Here the parameters correspond to λ and κ. The representations of the braid group B_n on $H_2(Y_{n,2}, \mathcal{L}^*)$ is exactly the one investigated by R. Lawrence, D. Krammer and S. Bigelow, which we shall call the LKB representation. The following theorem relates the LKB representation and the holonomy representation of the KZ connection.

Theorem 3.2. *In the case $\lambda_1 = \cdots = \lambda_n$ and $m = 2$, the LKB representation*

$$B_n \to Aut\ H_2(Y_{n,2}, \mathcal{L}^*)$$

is equivalent to the action of B_n obtained as the holonomy of the KZ connection

$$B_n \to Aut\ N[\lambda - 4]^*.$$

We take distinct $n+1$ points $p_1, \cdots, p_{n+1} \in \mathbf{C}P^1$ with $p_{n+1} = \infty$ and we associate to these points the highest weights $\lambda_1, \cdots, \lambda_n, \lambda_{n+1}$. Then the space of coinvariants

$$(M_{\lambda_1} \otimes \cdots \otimes M_{\lambda_n} \otimes M^*_{\lambda_{n+1}})/\mathfrak{g}$$

is identifined with

$$N[\lambda_{n+1}] = N[\lambda - 2m]$$

with

$$m = \frac{1}{2}(\lambda_1 + \cdots + \lambda_n - \lambda_{n+1}).$$

Let us briefly discuss a relation between the space of conformal blocks in conformal field theory on the Riemann sphere and the space of coinvariants $N[\lambda - 2m]$.

First, we recall the definition of the space of conformal blocks. We refer the reader to [13] for an introductory treatment of this subject. We put $\kappa = K + 2$ and assume that K is a positive integer. We suppose that the highest weights $\lambda_1, \cdots, \lambda_{n+1}$ associated with the points $p_1, \cdots, p_{n+1} \in \mathbf{C}P^1$ with $p_{n+1} = \infty$ are non-negative integers and satisfy $0 \leq \lambda_1, \cdots, \lambda_{n+1} \leq K$. We denote by $V_{\lambda_1}, \cdots, V_{\lambda_{n+1}}$ the irreducible representations of $\mathfrak{g} = sl_2(\mathbf{C})$ with highest weights $\lambda_1, \cdots, \lambda_{n+1}$.

Let $\widehat{\mathfrak{g}} = \mathfrak{g} \otimes \mathbf{C}((\xi)) \oplus \mathbf{C}c$ be the affine Lie algebra, which is the central extension of the loop algebra $\mathfrak{g} \otimes \mathbf{C}((\xi))$. Here $\mathbf{C}((\xi))$ denotes the ring of Laurent series. Let $\mathcal{H}_{\lambda_j}$ be the integrable highest weight module of $\widehat{\mathfrak{g}}$ with highest weight λ_j.

We denote by $\mathcal{M}_p$ the set of meromorphic functions on $\mathbf{C}P^1$ with poles at most at $p_1, \cdots, p_{n+1}$. Then $\mathfrak{g} \otimes \mathcal{M}_p$ has a structure of a Lie algebra and acts diagonally on the tensor product $\mathcal{H}_{\lambda_1} \otimes \cdots \otimes \mathcal{H}_{\lambda_n} \otimes \mathcal{H}^*_{\lambda_{n+1}}$ by means of the Laurent expansions of a meromorphic function at the points $p_1, \cdots, p_{n+1} \in \mathbf{C}P^1$.

The space of conformal blocks is defined as the space of coinvariants

$$\mathcal{H}(p, \lambda) = (\mathcal{H}_{\lambda_1} \otimes \cdots \otimes \mathcal{H}_{\lambda_n} \otimes \mathcal{H}^*_{\lambda_{n+1}})/(\mathfrak{g} \otimes \mathcal{M}_p).$$

It turns out that there is a surjective map

$$(V_{\lambda_1} \otimes \cdots \otimes V_{\lambda_n} \otimes V^*_{\lambda_{n+1}})/\mathfrak{g} \longrightarrow \mathcal{H}(p, \lambda)$$

and the kernel is described by some algebraic equations coming from the definition of the space of conformal blocks. It was shown by B. Feigin, V. Schechtman and A. Varchenko [8] that these algebraic equations correspond to exact forms in the twisted de Rham cohomology $\Omega^*(Y_{n,m})$ by the map $\rho : W[\lambda - 2m] \to \Omega^m(Y_{n,m})$ and ρ induces a map

$$\mathcal{H}(p, \lambda) \to H^m(\Omega^*(Y_{n,m}), \nabla).$$

Therefore we obtain a map

$$\phi : H_m(Y_{n,m}, \mathcal{L}^*) \to \mathcal{H}(p, \lambda)^*$$

defined by

$$\langle \phi(c), w \rangle = \int_c \Phi \rho(w).$$

It was shown by A. Varchenko [20] that the above map ϕ is surjective.

It is a basic result in conformal field theory that the spaces of conformal blocks form a vector bundle over the configuration space X_n equipped with the flat KZ connection. Therefore, the pure braid group P_n acts on the space of conformal blocks $\mathcal{H}(p, \lambda)^*$ by means of the holonomy of this connection. The map ϕ is equivariant with respect to the action of the pure braid group P_n.

The map $\phi : H_m(Y_{n,m}, \mathcal{L}^*) \to \mathcal{H}(p, \lambda)^*$ is not in general injective. Our local system $\mathcal{L}$ is no longer generic and there is a subtle point about the structure of the homology $H_m(Y_{n,m}, \mathcal{L}^*)$.

Let us consider the natural map

$$\alpha : H_m(Y_{n,m}, \mathcal{L}^*) \to H_m^{lf}(Y_{n,m}, \mathcal{L}^*)$$

and put $\mathrm{Im}(\alpha) = H_m^{lf}(Y_{n,m}, \mathcal{L}^*)_{reg}$. We call $H_m^{lf}(Y_{n,m}, \mathcal{L}^*)_{reg}$ the space of regularizable cycles.

We describe the relation between the space of conformal blocks and the space of regularizable cycles in the case $n = 2$. This corresponds to the case of 3 points on the Riemann sphere. We fix K a positive integer and suppose that the highest weights $\lambda_1, \lambda_2, \lambda_3$ are non-negative integers satisfying $0 \leq \lambda_1, \lambda_2, \lambda_3 \leq K$.

It follows from a result due to R. Silvotti [19] that ϕ induces an isomorphism

$$H_m^{lf}(Y_{2,m}, \mathcal{L}^*)_{reg} \cong \mathcal{H}(p_1, p_2, p_3; \lambda_1, \lambda_2, \lambda_3)^*.$$

The above homology group $H_m^{lf}(Y_{2,m}, \mathcal{L}^*)_{reg}$ is isomorphic to $\mathbf{C}$ if the quantum Clebsch–Gordan condition

$$\begin{aligned} &|\lambda_1 - \lambda_2| \leq \lambda_3 \leq \lambda_1 + \lambda_2 \\ &\lambda_1 + \lambda_2 + \lambda_3 \in 2\mathbf{Z} \\ &\lambda_1 + \lambda_2 + \lambda_3 \leq 2K \end{aligned}$$

is satisfied and is isomorphic to 0 otherwise.

§4. Iterated integrals of logarithmic forms

In this section we give a description of the holonomy representation of the fundamental group of the complement $M(\mathcal{A})$ of a hyperplane arrangement. We refer the reader to [13] for a more detailed account of this subject.

First, we briefly recall basic definitions for iterated integrals. Let M be a smooth manifold and $\omega_1, \cdots, \omega_k$ be differential forms on M. We fix a base point $x_0 \in M$ and we denote by ΩM the loop space of M based at x_0. Namely, ΩM is the space of piecewise smooth maps $\gamma : I \to M$ such that $\gamma(0) = \gamma(1) = x_0$. Let Δ_k denote the Euclidean simplex defined by

$$\Delta_k = \{(t_1, \cdots, t_k) \in \mathbf{R}^k \ ; \ 0 \leq t_1 \leq \cdots \leq t_k \leq 1\}.$$

There is an evaluation map

$$\varphi : \Delta_k \times \Omega M \to \underbrace{M \times \cdots \times M}_{k}$$

defined by $\varphi(t_1, \cdots, t_k; \gamma) = (\gamma(t_1), \cdots, \gamma(t_k))$. Let $\pi_i : \underbrace{M \times \cdots \times M}_{k} \to M$ be the projection on the i-th factor and put

$$\omega_1 \times \cdots \times \omega_k = \pi_1^* \omega_1 \wedge \cdots \wedge \pi_k^* \omega_k.$$

We define the iterated integral $\int \omega_1 \cdots \omega_k$ by the expression

$$\int_{\Delta_k} \varphi^*(\omega_1 \times \cdots \times \omega_k)$$

which is the integration along the fiber with respect to the projection $p : \Delta_k \times \Omega M \to \Omega M$.

The iterated integral $\int \omega_1 \cdots \omega_k$ is considered to be a differential form on the loop space ΩM with degree $p_1 + \cdots + p_k - k$, where p_j is the degree of the differential form ω_j.

As a differential form on the loop space $d \int \omega_1 \cdots \omega_k$ is expressed as

$$\sum_{j=1}^{k} (-1)^{\nu_{j-1}+1} \int \omega_1 \cdots \omega_{j-1} d\omega_j \ \omega_{j+1} \cdots \omega_k$$

$$+ \sum_{j=1}^{k-1} (-1)^{\nu_j+1} \int \omega_1 \cdots \omega_{j-1} (\omega_j \wedge \omega_{j+1}) \omega_{j+2} \cdots \omega_k$$

where $\nu_j = p_1 + \cdots + p_j - j$.

Let $\mathcal{A} = \{H_1, \cdots, H_\ell\}$ be a hyperplane arrangement in $\mathbf{C}^n$ and we consider the complement

$$M(\mathcal{A}) = \mathbf{C}^n \setminus \bigcup_{H \in \mathcal{A}} H.$$

Let f_j be a linear form defining H_j and we set

$$\omega_j = d \log f_j, \quad 1 \leq j \leq \ell.$$

We denote by $A = \bigoplus_{q \geq 0} A^q$ the Orlik–Solomon algebra of the hyperplane arrangement $\mathcal{A}$, which is isomorphic to the algebra over $\mathbf{C}$ generated by 1 and the logarithmic forms ω_j, $1 \leq j \leq \ell$. It is known from work of E. Brieskorn [4] that there is an isomorphism

$$A \cong H^*(M(\mathcal{A}); \mathbf{C}).$$

For more details about the Orlik–Solomon algebra we refer the reader to Orlik–Terao [15].

We define the reduced complex $\overline{A}$ by shifting the degrees by one as

$$\overline{A}^q = \begin{cases} 0, & q < 0 \\ A^{q+1}, & q \geq 0. \end{cases}$$

The reduced bar complex $\overline{B}^*(A)$ is the tensor algebra

$$\overline{B}^*(A) = \bigoplus_{k \geq 0} \left(\bigotimes^k \overline{A} \right)$$

generated by $\overline{A}$. Then $\overline{B}^*(A)$ has a natural structure of a graded algebra and we introduce the coboundary operator by

$$\begin{aligned} &d(\varphi_1 \otimes \cdots \otimes \varphi_k) \\ &= \sum_{j=1}^{k-1} (-1)^{\nu_j + 1} \varphi_1 \otimes \cdots \otimes (\varphi_j \wedge \varphi_{j+1}) \otimes \cdots \otimes \varphi_k \end{aligned}$$

where $\varphi_j \in \overline{A}^{q_j}$ and we wet $\nu_j = q_1 + \cdots + q_j$. We define the iterated integral map $\mathcal{I}$ from the reduced bar complex to the space of differential forms on the loop space by

$$\mathcal{I}(\varphi_1 \otimes \cdots \otimes \varphi_k) = \int \varphi_1 \cdots \varphi_k.$$

We define the filtration on the reduced bar complex by

$$\mathcal{F}^{-k}(\overline{B}^*(A)) = \bigoplus_{\ell \leq k} (\bigotimes^{\ell} \overline{A}).$$

Let J be the augmentation ideal of the group algebra of the fundamental group of $M = M(\mathcal{A})$. There is a pairing map

$$H^0(\overline{B}^*(A)) \times \mathbf{Z}\pi_1(M, \mathbf{x}_0) \longrightarrow \mathbf{C}$$

given by iterated integrals of logarithmic 1-forms. There is an induced filtration $\mathcal{F}^{-k}$ on $H^0(\overline{B}^*(A))$. By the basic theorem on de Rham fundamental groupd due to K. T. Chen [5] and the fact that the inclusion map from the Orlik–Solomon algebra to the de Rham complex of M induces an isomorphism on cohomology we obtain the following theorem.

Theorem 4.1. [13] *For the reduced bar complex of the Orlik–Solomon algebra the iterated integral map gives an isomorphism*

$$\mathcal{F}^{-k} H^0(\overline{B}^*(A)) \cong \mathrm{Hom}(\mathbf{Z}\pi_1(M, \mathbf{x}_0)/J^{k+1}, \mathbf{C}).$$

Let $\mathcal{L}(X_1, \cdots, X_\ell)$ be the free Lie algebra whose generators are in one to one correspondence with the hyperplanes in $\mathcal{A}$. We define the holonomy Lie algebra by

$$\mathfrak{h}(M) = \mathcal{L}(X_1, \cdots, X_\ell)/\mathfrak{a}$$

where $\mathfrak{a}$ is the ideal generate by

$$[X_{j_p}, X_{j_1} + \cdots + X_{j_k}], \quad 1 \leq p < k$$

for the maximal family of hyperplanes $\{H_{j_1}, \cdots, H_{j_k}\}$ such that

$$\mathrm{codim}_{\mathbf{C}}(H_{j_1} \cap \cdots \cap H_{j_k}) = 2.$$

We put

$$\omega = \sum_{j=1}^{m} \omega_j X_j.$$

Then we have the universal holonomy map

$$\Theta_0 : \pi_1(M, \mathbf{x}_0) \longrightarrow \mathbf{C}\langle\langle X_1, \cdots, X_m \rangle\rangle/\mathfrak{a}$$

defined by

$$\Theta_0(\gamma) = 1 + \sum_{k=1}^{\infty} \int_\gamma \underbrace{\omega \cdots \omega}_{k}.$$

This induces an isomorphism

$$\mathbf{C}\widehat{\pi}_1(M, \mathbf{x}_0) \cong \mathbf{C}\langle\langle X_1, \cdots, X_m\rangle\rangle/\mathfrak{a}.$$

By taking the primitive part, we have an isomorphism between the nilpotent completion of the fundamental group and the completed holonomy Lie algebra over $\mathbf{C}$.

Let

$$\pi_1(M) = \Gamma_1 \supset \Gamma_2 \supset \cdots \supset \Gamma_k \supset \cdots$$

be the lower central series defined by

$$\Gamma_{k+1} = [\Gamma_1, \Gamma_k], \quad k \geq 1.$$

Then there is an isomorphism of graded Lie algebras

$$\bigoplus_{k\geq 1}[\Gamma_k/\Gamma_{k+1}] \otimes \mathbf{C} \cong \mathfrak{h}(M).$$

Because of the integrability of the connection

$$\omega = \sum_{j=1}^{m} \omega_j X_j$$

we have the following linear representations of the fundamental group.

Proposition 4.1. *For any representation of the holonomy Lie algebra $r : \mathfrak{h}(M) \to \mathrm{End}(V)$ there is a linear representation of the fundamental group*

$$\pi_1(M, \mathbf{x}_0) \longrightarrow \mathrm{GL}(V)$$

obtained by substituting the representation r to the universal holonomy homomorphism.

We denote by $C^\infty(\widetilde{M})$ the space of smooth functions on the universal covering of M. We have a map

$$\mathcal{F}^{-k}H^0(\overline{B}^*(A)) \longrightarrow C^\infty(\widetilde{M}).$$

by the iterated integral for paths wth fixed starting point and we denote its image by $F_k(M)$. There is an increasing filtration of functions

$$\mathbf{C} = F_0(M) \subset F_1(M) \subset \cdots \subset F_k(M) \subset \cdots$$

called hyperlogarithms and we have

$$dF_{k+1}(M) \subset F_k(M) \otimes A^1, \quad k = 0, 1, 2, \cdots.$$

We set $H_+(M) = \bigoplus_{p>0} H_p(M;\mathbf{C})$ and equip the tensor algebra $TH_+(M)$ with a structure of a graded algebra such that each non-zero element of $H_p(M;\mathbf{C})$ has degree $p-1$. Let $\{\varphi_j\}$ be a basis of the Orlik–Solomon algebra and $\{Z_j\}$ be its dual basis. We introduce a differential $\delta : TH_+(M)_p \longrightarrow TH_+(M)_{p-1}$ as the dual of the cup product homomorphism. More explicitly, we define the differential δ by

$$\delta Z_k = -\sum_k (-1)^{p_i} c_{ij}^k [Z_i, Z_j]$$

when we have the eqality

$$\varphi_i \wedge \varphi_j = \sum_k c_{ij}^k \varphi_k$$

holds in the Orlik–Solomon algebra. The reduced bar complex $\overline{B}^*(A)$ is considered to be the dual complex of $TH_+(M)$ with the differential δ.

Theorem 4.2. [13] *Let $\mathcal{A} = \{H_1, \cdots, H_\ell\}$ be a hyperplane arrangement. Then linear combinations of iterated integrals of the logarithmic forms ω_j, $1 \leq j \leq \ell$, on the complement of the hyperplane arrangement*

$$\sum_{j_1 \cdots j_k} a_{j_1 \cdots j_k} \int_\gamma \omega_{j_1} \cdots \omega_{j_k}, \quad a_{j_1 \cdots j_k} \in \mathbf{C}$$

depends only on the homotopy class of a loop γ if and only if the correspondence $X_{j_1} \cdots X_{j_k} \mapsto a_{j_1 \cdots j_k}$ defines a linear map $U\mathfrak{h}(M) \otimes \mathbf{C} \longrightarrow \mathbf{C}$.

Proof. We consider the above iterated integral as a function on the loop space ΩM. It will be enough to show that the condition

$$\sum_{j_1 \cdots j_k} a_{j_1 \cdots j_k}\, d \int_\gamma \omega_{j_1} \cdots \omega_{j_k} = 0$$

holds if and only if the correspondence $X_{j_1} \cdots X_{j_k} \mapsto a_{j_1 \cdots j_k}$ defines a linear map $U\mathfrak{h}(M) \otimes \mathbf{C} \longrightarrow \mathbf{C}$. The above condition is equivalent to the equality

$$\sum_{j_1 \cdots j_k} a_{j_1 \cdots j_k}\, d(\omega_{j_1} \otimes \cdots \otimes \omega_{j_k}) = 0$$

in the reduced bar complex $\overline{B}^*(A)$. By means of the duality between $\overline{B}^*(A)$ and $TH_+(M)$ this equality is equivalent to the condition

$$\sum_{j_1 \cdots j_k} \langle a_{j_1 \cdots j_k} \omega_{j_1} \otimes \cdots \otimes \omega_{j_k}, \partial Z \rangle = 0$$

for any $Z \in TH_+(M)_1$. By writing down explicitly the above condition we obtain the desired statement. Q.E.D.

Finally, we describe a correspondence between unipotent representations of the fundamental group of the complement of a hyperplane arrangement and nilpotent connections. The following fact was first obtained by K. Aomoto [1] (see also the work due to R. Hain [9]).

Theorem 4.3. *Let*

$$\rho : \pi_1(M, \mathbf{x}_0) \longrightarrow \mathrm{GL}(V)$$

be a unipotent representation of the fundamental group of the complement of a hyperplane arrangement. Then there exists an integrable connection

$$\omega = \sum_{j=1}^{\ell} A_j \omega_j, \quad A_j \in \mathrm{End}(V)$$

such that each A_j is nilpotent and the monodromy representation of ω coincides with ρ.

Proof. For a unipotent representation $\rho : \pi_1(M, \mathbf{x}_0) \longrightarrow \mathrm{GL}(V)$ there exists a sufficiently large integer k such that ρ induces a homomorphism

$$\widetilde{\rho} : \mathbf{C}\pi_1(M, \mathbf{x}_0)/J^{k+1} \longrightarrow \mathrm{End}(V).$$

On the other hand the universal holonomy homomorphism of the connection $\sum_{j=1}^{\ell} \omega_j X_j$ induces an isomorphism

$$\theta : \mathbf{C}\pi_1(M, \mathbf{x}_0)/J^{k+1} \cong \mathbf{C}\langle\langle X_1, \cdots, X_\ell \rangle\rangle/(\mathfrak{a} + \widehat{J}^{k+1})$$

where $\widehat{J}$ denotes the completed augmentation ideal. We define a homomorphism

$$\alpha : \mathbf{C}\langle\langle X_1, \cdots, X_\ell \rangle\rangle/(\mathfrak{a} + \widehat{J}^{k+1}) \longrightarrow \mathrm{End}(V)$$

by $\alpha = \widetilde{\rho} \circ \theta^{-1}$ and put $A_j = \alpha(X_j)$, $1 \leq j \leq \ell$. Here $A_j^{k+1} = 0$ holds and A_j is a nilpotent endomorphism. Since $\widetilde{\rho} = \alpha \circ \theta$ the representation ρ is obtained from the universal holonomy homomorphism by the substitution $X_j = A_j$. Q.E.D.

References

[1] K. Aomoto, Fonctions hyperlogarithmiques et groupes de monodromie unipotents, J. Fac. Sci. Univ. Tokyo, **25** (1978), 149–156.

[2] S. Bigelow, The Lawrence–Krammer representation, In: Topology and Geometry of Manifolds, Athens, GA, 2001, Proc. Sympos. Pure Math., **71**, Amer. Math. Soc., Providence, RI, 2003, pp. 51–68.

[3] J. Birman, Braids, Links, and Mapping Class Groups, Ann. of Math. Stud., **82**, Princeton Univ. Press, 1974.

[4] E. Brieskorn, Sur les groupes de tresses [d'après V. I. Arnol'd], In: Séminaire Bourbaki 1971/72 Exposé 401, Lecture Notes in Math., **317**, Springer-Verlag, 1973, pp. 21–44.

[5] K. -T. Chen, Iterated path integrals, Bull. of Amer. Math. Soc., **83** (1977), 831–879.

[6] Z. Chen, Stratifications of complementary spaces of hyperplane arrangements and homology groups with local coefficients, Thesis, the University of Tokyo, 2005.

[7] E. Date, M Jimbo, A. Matsuo and T. Miwa, Hypergeomotric type integrals and the $sl(2, \mathbf{C})$ Knizhnik-Zamolodchikov equations, Intern. J. Modern Phys., **B4** (1990), 1049–1057.

[8] B. Feigin, V. Schechtman and A. Varchenko, On algebraic equations satisfied by hyergeometric correlators in WZW models. I, Comm. Math. Phys., **163** (1994), 173–184.

[9] R. M. Hain, On a generalization of Hilbert's 21st problem, Ann. Sci. Ecole Norm. Sup., **19** (1986), 609–627.

[10] V. G. Knizhnik and A. B. Zamolodchikov, Current algebra and Wess–Zumino models in two dimensions, Nuclear Phys. B, **247** (1984), 83–103.

[11] T. Kohno, Homology of a local system on the complement of hyperplanes, Proc. Japan Acad. Ser. A, **62** (1986), 144–147.

[12] T. Kohno, Bar complex of the Orlik–Solomon algebra, Topology Appl., **118** (2002), 147–157.

[13] T. Kohno, Conformal Field Theory and Topology, Transl. Math. Monogr., **210**, Amer. Math. Soc., 2002.

[14] M. Kontsevich, Vassiliev's knot invariants, Adv. Soviet Math., **16** (1993), 137–150 .

[15] P. Orlik and H. Terao, Arrangements of Hyperplanes, Springer-Verlag, 1992.

[16] P. Orlik and H. Terao, Arrangements and Hypergeometric Integrals, MSJ Memoirs, **9**, Math. Soc. Japan, 2001; 2nd ed., 2007.

[17] M. Salvetti, Topology of the complement of real hyperplanes in C^N. Invent. Math., **88** (1987), 603–618.

[18] V. Schechtman and A. Varchenko, Hypergeometric solutions of the Knizhnik–Zamolodchikov equation, Lett. in Math. Phys., **20** (1990), 93–102.

[19] R. Silvotti, Local systems on the complement of hyperplanes and fusion rules in conformal field theory, Internat. Math. Res. Notices, **1994**, no. 3, 111 ff., approx. 17 pp. (electronic).

[20] A. Varchenko, Multidimensional Hypergeometric Functions and Representation Theory of Lie Algebras and Quantum Groups, Adv. Ser. Math. Phys., **21**, World Scientific, 1995.

Toshitake Kohno
IPMU, Graduate School of Mathematical Sciences, the University of Tokyo, Tokyo, 153-8914, Japan
E-mail address: kohno@ms.u-tokyo.ac.jp

Advanced Studies in Pure Mathematics 62, 2012
Arrangements of Hyperplanes—Sapporo 2009
pp. 175–187

On combinatorial invariance of the cohomology of the Milnor fiber of arrangements and the Catalan equation over function fields

Anatoly Libgober

Abstract.

We discuss combinatorial invariance of the betti numbers of the Milnor fiber for arrangements of lines with points of multiplicity two and three and describe a link between this problem and enumeration of solutions of the Catalan equation over function field in the case when its coefficients are products of linear forms and the equation defines an elliptic curve.

§1. Introduction

An interesting problem in the theory of arrangements of hyperplanes over the field of complex numbers is to determine which topological invariants of the complement depend only on the combinatorics of the arrangements i.e. the poset formed by intersections of the hyperplanes of the arrangement. A theorem, going back to Arnold, Orlik and Solomon (cf. [17]), asserts that the cohomology of the complement is a combinatorial invariant. On the other hand, it is known that the homotopy type (cf. [18]) and even the fundamental group (cf. [3]) cannot be determined from the combinatorics of the arrangement i.e. there exist pairs of arrangements with the same combinatorics but different such invariants.

Given an arrangement $\mathcal{A}$ of hyperplanes in $\mathbb{P}^{n+1}$ in which the equations of hyperplanes are $L_i = 0$ where L_i are linear forms, one can consider the affine hypersurface $\Pi L_i = 1$ which is called the Milnor fiber of arrangement. It is known that for the Milnor fiber $M_{\mathcal{A}}$ the homology

Received November 2, 2010.
Revised March 20, 2011.
2010 *Mathematics Subject Classification.* 14H30, 32S22, 52B30, 11F23.
Key words and phrases. Arrangements, combinatorial invariance, Milnor fiber, Catalan equation.
The author was partially supported by NSF grant.

group $H_1(M_{\mathcal{A}}, \mathbb{Z})$ depends only on the fundamental group of the complement to $\mathcal{A}$ and more precisely it is the abelianization of the subgroup of $\pi_1(\mathbb{P}^2 - \mathcal{A})$ which the kernel of the surjection $lk : \pi_1(\mathbb{P}^2 - \mathcal{A}) \to \mathbb{Z}_{\mathrm{Card}\mathcal{A}}$ given by the total linking number with $\mathcal{A}$ cf. ([14]). One can ask about the combinatorial invariance of the cohomology of $M_{\mathcal{A}}$ and in particular $rkH^1(M_{\mathcal{A}}, \mathbb{Q})$ called (the first) Milnor number of the arrangement (for arrangements of lines, this is equivalent to the combinatorial invariance ranks of remaining cohomology groups). To this end we show the following:

Theorem 1.1. *If two arrangements of lines $\mathcal{A}_1$ and $\mathcal{A}_2$ with multiple points of multiplicity three and two only are combinatorially equivalent and if the monodromy on the Milnor fiber of $\mathcal{A}_1$ has an eigenvalue different from one then so does the monodromy on Milnor fiber the arrangement $\mathcal{A}_2$.*

This theorem will be derived from the following more precise statement relating the cohomology of Milnor fiber and other data of arrangement. We say that an arrangement $\mathcal{A}$ is composed of a pencil L of plane curves (cf.[15], [8]) if $L = \{\lambda F + \mu G | degF = degG\}$ contains three curves $\lambda F + \mu G$ with equations which are products of linear forms such that $\mathcal{A}$ is the zero set of the forms defining these curves. A pencil is reduced if equations of the above reducible curved are reduced. (see Section 2 for other definitions needed for this theorem).

Theorem 1.2. *Let $\mathcal{A}$ be an arrangement of lines with double and triple points only. If the monodromy acting on one-dimensional cohomology of the Milnor number of $\mathcal{A}$ has an eigenvalue different from one then the arrangement $\mathcal{A}$ is composed of a reduced pencil. The characteristic variety of the arrangement $\mathcal{A}$ has a component containing the identity $Char\pi_1(\mathbb{P}^2 - \mathcal{A})$. Vice versa, if $\mathcal{A}$ is composed of a reduced pencil then the monodromy has eigenvalue $exp(\frac{2\pi i}{3})$.*

The argument for the proof of above results relies on the relation between the Alexander polynomials of plane curves and Mordell–Weil groups considered recently in [4] and study of the Catalan equation over function field. Catalan equation, i.e. $X^p - Y^q = 1$ or more generally

$$A_1 X^p + A_2 Y^q + A_3 Z^r = 0 \tag{1}$$

has a long history and its properties of course depend on the ground field. Over function field, i.e. the case when $A_i \in \mathbb{C}(\mathcal{X})$ where $\mathcal{X}$ is a projective manifold the equation (1) was considered by J. Silverman. A finiteness results was shown in [20] in the case the equation (1) defines over the function field a curve of genus greater than 1. Elementary treatment of Fermat equation (i.e. $A_i \in \mathbb{C}, p = q = r$) is given in [10]. Finding solutions in the case of genus zero is trivial. Present paper considers the case $p = q = r = 3$ and A_i are product of linear forms in 3 variables. The answer to question about enumeration of solutions depends on the geometry of the curve $A_1 \cdot A_2 \cdot A_3 = 0$ in $\mathbb{P}^2$ and we describe the cases when the number of solutions is infinite (cf. Theorems 3.1 and 3.2).

The paper is organized as follows. In the next section we recall terminology from the theory of characteristic varieties of arrangements and related background information. In Section 3 we prove results on existence and enumeration of solutions to Catalan equation. Proof of theorems are presented in the last section. The paper ends with remark on properties of Mordell Weil groups which would imply combinatorial invariance of the cohomology of Milnor fiber. The author wants to thank S. Kaliman and L. Katzarkov for hospitality during visit to University of Miami where this work was written as well as S. Yuzvinsky for a useful comments. Finally, I am grateful to A. Dimca for pointing out Example 4.1 which helped to correct error in original statement of Theorem 1.2 and to J. I. Cogolludo-Agustin for numerous examples of Catalan equations which helped to correct Theorem 3.1.

§2. Preliminaries

As was already mentioned, Milnor fiber of an arrangement $\mathcal{A}$ of hyperplanes in $\mathbb{P}^{n+1}$ is affine hypersurface $M_{\mathcal{A}}$ in $\mathbb{C}^{n+2}$ given by equation $\Pi L_i = 1$ where L_i are linear forms defining the hyperplanes of the arrangement. The standard map $\mathbb{C}^{n+2} - \mathcal{O} \to \mathbb{P}^{n+1}$ ($\mathcal{O}$ is the origin) restricted on $M_{\mathcal{A}}$ is a cyclic cover

$$M_{\mathcal{A}} \to \mathbb{P}^{n+1} - \mathcal{A} \tag{2}$$

of degree equal to the number of hyperplanes $r = Card\mathcal{A}$ in $\mathcal{A}$. This cyclic cover corresponds to the homomorphism $\pi_1(\mathbb{P}^{n+1} - \mathcal{A}) \to \mathbb{Z}_r$ onto the cyclic group of order $r = \mathrm{Card}\mathcal{A}$ given by the total modulo r linking number with the union of hyperplanes in $\mathcal{A}$. We are interested in $dim H_1(M_{\mathcal{A}}, \mathbb{C})$ and also in the eigenspaces of the generator of the group $\mathbb{Z}_r$ of deck transformations acting on $H_1(M_{\mathcal{A}}, \mathbb{C})$. By Lefschetz theorem, applied to a generic plane H in $\mathbb{P}^{n+1}$, this group is isomorphic to the corresponding group of line arrangement formed by intersections of H with the hyperplanes of $\mathcal{A}$. So from now on we shall assume that $n = 1$. Since the main results in this note deal with the case of arrangements of lines with points of multiplicity 2 and 3 from now on we shall restrict ourselves to the case of such multiplicities as well.

By [12], [13] (cf. also [14]) the rank of the cyclic cover (2) can be found in terms the Alexander polynomial of $\mathbb{P}^2 - \mathcal{A}$. More precisely, the Alexander polynomial of $\mathbb{P}^2 - \mathcal{A}$ is equal to

$$(3) \qquad (t-1)^{\mathrm{Card}\mathcal{A}-2}(t^2+t+1)^s$$

for $s = dimH^1(\mathbb{P}^2, \mathcal{J}_S(\frac{2\mathrm{Card}\mathcal{A}}{3} - 3))$ where $\mathcal{J}_S \subset \mathcal{O}_{\mathbb{P}^2}$ is the ideal sheaf having stalk different from $\mathcal{O}_P$ iff P is a triple point of the arrangement and equal to maximal ideal at those P. Moreover, if r is divisible by 3, then

$$(4) \qquad \mathrm{rk}H_1(M_{\mathcal{A}})_\omega = \mathrm{rk}H_1(M_{\mathcal{A}})_{\omega^2} = s, \quad \mathrm{rk}H_1(M_{\mathcal{A}})_1 = r-2$$

where $H_1(M_{\mathcal{A}})_\zeta$ is the eigenspace of the deck transformation of the covering (2) with eigenvalue ζ and $\omega = exp(\frac{2\pi\sqrt{-1}}{3})$. If r is not divisible by 3 then the dimensions of eigenspaces with eigenvalue ω, ω^2 are equal to zero. Hence from now on we shall assume that r is divisible by 3.

Together with unbranched cover $M_{\mathcal{A}}$ of degree r one can consider a non-singular model of *3-fold* branched cyclic covering $\bar{M}_{\mathcal{A}}$ of $\mathbb{P}^2$ which is a compactification of the quotient of $M_{\mathcal{A}}$ by the action of the subgroup $\mathbb{Z}_{\frac{r}{3}} \subset \mathbb{Z}_r$. The integer $\mathrm{rk}H_1(\bar{M}_{\mathcal{A}}, \mathbb{C})$ is birational invariant and hence depends only on $M_{\mathcal{A}}$. Using results of [12], one has:

$$(5) \qquad \frac{1}{2}\mathrm{rk}H_1(\bar{M}_{\mathcal{A}}) = \mathrm{rk}H_1(\bar{M}_{\mathcal{A}})_\omega = \mathrm{rk}H_1(\bar{M}_{\mathcal{A}})_{\omega^2} = s.$$

Recall the following definition from [4] adapting it to the case of arrangements (rather than arbitrary plane curves).

Definition 2.1. A quasi-toric relation of orbifold type $(3,3,3)$ is the identity

$$F_1 f^3 + F_2 g^3 + F_3 h^3 = 0$$

where $F_i, i = 1,2,3$ is either a constant or a reduced curve such that $F_i = 0$ are without common components and $F_1F_2F_3$ has the arrangement $\mathcal{A}$ as its zero set.

(A'priori, of course for a given arrangement there are several possibilities for F_i's, including $F_i = 1$).

Let E be elliptic curve over $\mathbb{C}$ with j-invariant zero. As a model for E we shall use the 3-fold ramified covering of $\mathbb{C}$ branched at $0,1$ (and infinity) which is a closure of the affine variety in $\mathbb{C}^4$ given by the equations:

$$(6) \qquad c^3 = ab, a = s, b = 1-s$$

(the cyclic covering map is given by $(a,b,c,s) \to s$). As a model of $\bar{M}_{\mathcal{A}}$ for calculation of $H^1(\bar{M}_{\mathcal{A}})$ we shall use the complete intersection in $\mathbb{C}^5$ (with coordinates (x,y,A,B,C)) given by:

$$A = \frac{F_1(x,y)}{F_3(x,y)}, B = \frac{F_2(x,y)}{F_3(x,y)}, C^3 = AB \tag{7}$$

(here F_i are dehomogenized forms used in Definition 2.1). The surface (7) is the threefold cyclic cover of $\mathbb{C}^2$ branched over the curve $F_1F_2F_3 = 0$ corresponding to the homomorphism of the fundamental group sending the classes of loops corresponding to the curves $F_i = 0$ to the same element of $\mathbb{Z}_3$.

There is a correspondence between the $\mathbb{Z}_3$-equivariant maps ϕ : $\bar{M}_{\mathcal{A}} \to E$ and quasi-toric relations given as follows. Note that such a map ϕ takes a fixed point of $\mathbb{Z}_3$ to one of three ramification points of $E \to \mathbb{P}^1$. Denoting the polynomials whose zero sets are the remaining components of preimages of ramification points as f, g, h respectively, one can write the morphism ϕ as a compactification of the restriction on affine portion of $\bar{M}_{\mathcal{A}}$ of $\Phi : \mathbb{C}^5 \to \mathbb{C}^4$ given by:

$$a = A(\frac{f}{h})^3, b = B(\frac{g}{h})^3, c = C\frac{fg}{h^2}, s = A(\frac{f}{h})^3. \tag{8}$$

Let us define the equivalence between quasi-toric relations 2.1 involving F_i, f, g, h and F'_i, f', g', h' saying that two such relations are equivalent if for some $\lambda, \lambda_f, \lambda_g, \lambda_h \in \mathbb{C}$, one has $F_i = \lambda F'_i$, $\frac{\lambda_f}{\lambda_h} \in \mu_3$, $\frac{\lambda_g}{\lambda_h} \in \mu_3$ and $\lambda_f \lambda_g = \lambda_h^2$.

With this we have the following:

Proposition 2.2. *There is one to one correspondence between quasi-toric relations up to above equivalence and non constant equivariant maps* $\bar{M}_{\mathcal{A}} \to E$.

Proof. Indeed it follows from formulas (6), (7) that (f,g,h) appearing in (8) satisfy the quasi-toric relation. Vice versa, if f, g, h satisfy a quasi-toric relation, the formulas (8) yield a map which when restricted on affine portion of $\bar{M}_{\mathcal{A}}$ provides a dominant map $\phi : \bar{M}_{\mathcal{A}} \to E$. Q.E.D.

As in [4] we shall view the maps $\bar{M}_{\mathcal{A}} \to E$ as $\mathbb{C}(\bar{M}_{\mathcal{A}})$-points of the elliptic curve E. As such they form a group with the quotient by the subgroup of constant maps (the Mordell–Weil group) being finitely generated (Mordell–Weil theorem cf. [11]). The following is a special case of a result shown in [4].

Theorem 2.3. *The rank of the Mordell–Weil group $MW(\mathcal{A})$ of $\mathbb{C}(\bar{M}_{\mathcal{A}})$-points of E is equal to $2s$ where s is the degree of the factor (t^2+t+1) in (3).*

Finally we shall need few results on characteristic varieties of arrangements. Recall (cf. [14]) that one can view them as the components of affine subvariety V of the torus $Char\pi_1(\mathbb{P}^2-\mathcal{A})$ consisting of characters χ such that

$$dimH_1(\mathbb{P}^2-\mathcal{A},\chi)>0. \tag{9}$$

These affine subvarieties are finite unions of translated subgroups of $Char\pi_1(\mathbb{P}^2-\mathcal{A})$ by the points of finite order in $\mathrm{Char}\pi_1(\mathbb{P}^2-\mathcal{A})$ (cf. [1],[16]).

The following results are contained in [14] and [15] respectively (cf. also [5]):

Theorem 2.4. *1) The dimension of the ω-eigenspace of the monodromy acting on $H^1(M_{\mathcal{A}})$ of the Milnor number $M_{\mathcal{A}}$ of an arrangement $\mathcal{A}$ is equal to $dimH^1(\mathbb{P}^2-\mathcal{A},L_\omega)$ where L_ω is the local system $\pi_1(\mathbb{P}^2-\mathcal{A})\to\mathbb{C}^*$ sending each loop, having the linking number equal to one with a hyperplane in $\mathcal{A}$ and zero with all the others, to the root of unity ω.*

2)The dimension of $H^1(\mathbb{P}^2-\mathcal{A},L_\chi)$ is not smaller than $d-1$ where d is the dimension of the component of characteristic variety to which it belongs.

More specifically, 2) follows from semi-continuity of ranks of cohomology of local systems (as in [15]) and the fact that generic local system in a component of positive dimension is a twist of the pull back of a local system on $\mathbb{P}^1$ minus a collection of points. The latter does not change generically the rank of H^1 (cf. [1], Prop. 1.7) and relation in 2) is obvious for the local systems on complements to a finite union of point in $\mathbb{P}^1$.

§3. Catalan equation $A_1f^3+A_2g^3+A_3h^3=0$ with coefficients being products of linear forms.

The purpose of this section is to prove the following:

Theorem 3.1. *Let $A_i=\Pi_{j\le r_i}L_{i,j}, i=1,2,3,$ be products of linear forms $L_{i,j}\in\mathbb{C}[x_1,x_2,x_3]$ (r_i are non negative integers). Assume that $L_{i,j}$ is not a multiple of $L_{i',j'}$ for any two pairs $(i,j),(i',j')$ and that lines $L_{i,j}=0$ form an arrangement of lines with points of multiplicity 2 and 3 only.*

1) The equation

$$A_1 f^3 + A_2 g^3 + A_3 h^3 = 0 \tag{10}$$

has solution only if $A_1A_2A_3 = A'_1A'_2A'_3$ where A'_1, A'_2, A'_3 are linear dependent over $\mathbb{C}$. In this case the solutions of (10) with linearly dependent A_i, are the pullbacks of solutions of Catalan equation over $\mathbb{C}(t)$.

2) In the case A_i are linearly dependent over $\mathbb{C}$ the set of solutions to (10) is infinite.

Proof. First we show that if the equation (10) with $A_1 = A_2 = 1$ has a solution then there exist factorization $A_3 = A'_1A'_2A'_3$ such that at least two factors are non-constant and the Catalan equation (10) with these factors as coefficients has a solution. Assume that $A_1 = A_2 = 1$, select f, g, h which form a solution with smallest possible degree of h and such that f, g, h have no common factors. Then $-A_3h^3 = \Pi(f + \omega_i g)$ $(\omega_i^3 = -1)$. Let $\tilde{h}$ be an irreducible factor of h. There is only one factor $f + \omega_i g$ which is divisible by $\tilde{h}$, since if there are two, then both f and g are divisible by $\tilde{h}$ contradicting the assumption of absence of common factors for f, g, h. If each irreducible factor of h divides single factor $f + \omega_i g$ of $f^3 + g^3$, then $f + g = ah'^3A'_3$, $f + \omega_2 g = b(h'')^3A'_1$, $f + \omega_3 g = c(h''')^3A'_2$. So $3degf = (dega + degb + degc) + 3(degh' + degh'' + 3degh''') + degA_3$ i.e. $a = b = c$ are constants. Hence $A'_1h'^3, A'_2h''^3, A'_3h'''^3$ belong to a pencil i.e. h', h'', h''' are solutions to the Catalan equation with coefficients A'_1, A'_2, A'_3. If A'_1, A'_2 are constant then the degree of h''' is strictly smaller than the degree of h. Hence such descent should lead either to Catalan equation with solutions having at least two non-constant coefficients or the Catalan equation $f^3 + g^3 + A_3 = 0$ would have a solution. In the latter case, by unique factorization, each $f + \omega_i g$ is a product of factors of A_3 and the Catalan equation with coefficients $A''_i = f + \omega_i g$ has solution in $\mathbb{C}$ i.e. satisfies conclusion of 1) since A''_i are in the pencil generated by f, g.

Assume now that factors A_1, A_2 in (10) are not constant. Let us consider the restriction of relation (10) on one of the lines $L_{i,j}$, say $L = L_{1,j}$ which equation is a factor of A_1. Denoting the form obtained by restriction on L of a form M as ${}^LM$ we can write the resulting relation as

$${}^LA_2({}^Lg)^3 + {}^LA_3({}^Lh)^3 = 0. \tag{11}$$

Let ${}^LA_i = \Pi l_{i,j}^{a_{i,j}}$. If one has $a_{i,j} > 2$, then the zero of $l_{i,j}$ on L is the point which belongs to more than three lines $L_{i,j} = 0$ forming the zero set of $A_1A_2A_3 = 0$ contradicting the assumption that the multiplicities do not exceed 3. Hence the exponent of a factor of ${}^LA_i$ is either $a_{i,j} = 1$ or $a_{i,j} = 2$. Due to unique factorization in $\mathbb{C}[t]$ we obtain that none of roots of ${}^LA_2$ and ${}^LA_3$ is a root of g or h. Moreover ${}^LA_2$ and ${}^LA_3$ have the same sets of zeros on L with the same multiplicities and $\frac{{}^Lg}{{}^Lh} \in \mathbb{C}$. Hence any intersection point of L and A_i also belongs to $A_{i'}$ for $i = 2,3$ and $i' = 2,3$. This implies that the same is true for any point in $A_1 = 0$, that $degA_2 = degA_3$ and hence the conditions of Noether $AF + BG$ theorem (cf. [9]) are satisfied (the multiplicities of A_i are equal to one). Hence

$$A_1 = \lambda_2 A_2 + \lambda_3 A_3 \quad \text{where} \quad \deg\lambda_i = 0. \tag{12}$$

Next consider rational map $\mathbb{P}^2 \to \mathbb{P}^1$ given by $P \to (A_2(P), A_3(P))$. Its fibers consist of points P such that

$$\frac{A_2}{A_3} = t \in \mathbb{C}. \tag{13}$$

The solution (10) for A_i is also solution to $(\lambda_2 t + \lambda_3)f^3 + tg^3 + h^3 = 0$. Since Fermat equation over the function field of the curve (13) has only constant solutions (cf. [10]) it follows that f, g, h depend only on t. The same argument shows that if $a_{i,j} = 2$ then f, g, h are pullbacks as well. Part 2) is hence proven. Q.E.D.

Next we shall consider solutions to equation (10) over $\mathbb{C}(t)$.

Theorem 3.2. *The solutions to (10) in case $A_i \in \mathbb{C}(t)$ form a group isomorphic to $Hom(Jac(C), E)$ where C is the 3-fold branched cover of $\mathbb{P}^1$ ramified at the set $\mathcal{S}$ which is the union of zeros of A_i $(1 \le i \le 3)$, $Jac(C)$ is its Jacobian and E is the elliptic curve with an automorphism of order 3. In particular if $\mathcal{S}$ contains only three elements then the group of solutions is isomorphic to $End(E) = \mathbb{Z}^2$.*

Proof. Non-constant solutions to (10) correspond to non-constant sections of the elliptic surface (7). This elliptic surface is iso-trivial and is trivialized over the 3-fold cyclic cover C of $\mathbb{P}^1$, containing the t-line as an open set, and totally ramified at the union of zeros of A_i and possibly the infinity. Indeed the holonomy around each zero is trivial on such 3-fold cover. For trivial elliptic surface $C \times E$ over C non constant sections correspond to the maps $C \to E$ by Torelli theorem which correspond to the maps of the Jacobian of C to E. The last claim in the statement 2) and statement 3) follows since the Jacobian of threefold cover of $\mathbb{P}^1$ branched at three points is E. Q.E.D.

Example 3.3. 1) Let $\omega = exp(\frac{2\pi\sqrt{-1}}{3})$ be a primitive root of unity of degree 3. A non constant solution to Catalan equation:

$$(t-\omega)^2(t-\omega^2)^2 f^3 = (1+\omega^2)(t-1)^2(t-\omega^2)^2 g^3 + (-\omega^2)(t-\omega)^2(t-1)^2 h^3$$

is

$$f = t-1, g = t-\omega, h = t-\omega^2.$$

Indeed, after substitution we get $(1+\omega^2 = -\omega)$:

$$\begin{aligned}&(t^3-1)^2[(t-1)-(1+\omega^2)(t-\omega)+\omega^2(t-\omega^2)]\\ &=(t^3-1)[(t-1)+\omega(t-\omega)+\omega^2(t-\omega^2)] = 0.\end{aligned}$$

2) Suppose that $F_3 = F_1 + F_2$ i.e. the arrangement is composed of a pencil. This yields the solution $f = g = h = 1$ to Catalan equation. Elliptic curve $c^3 = s^2 - s$ which is isomorphic to $a = s, b = 1-s, c^3 = ab$ is equivalent to $y^2 = x^3 + 1$ where $s - \frac{1}{2} = y, x = 4^{\frac{1}{3}}c$. Duplication formulas from [21] p 59, are in this case:

$$x_1 := \frac{x^4 - 8x}{4x^3+4}, \qquad y_1 = -y - \frac{3x^2(\frac{x^4-8x}{4x^3+4} - x)}{2y}$$

(here (x_1, y_1) denotes $2(x,y)$ in the sense of addition on elliptic curve). From this we infer that if we take $F_3 = 1$ (which is always the case after replacement $F_1 \to \frac{F_1}{F_3}, F_2 \to \frac{F_2}{F_3}$) start with map $a = F_1, b = F_2$ then we get for doubling:

$$a := \frac{-F_1(F_1-2)^3}{(2F_1-1)^3}, \qquad b = \frac{(F_1-1)(F_1+1)^3}{(2F_1-1)^3}$$

i.e.

$$\frac{f}{h} = \frac{-F_2-1}{2F_1-1}, \qquad \frac{g}{h} = \frac{F_1+1}{2F_1-1}$$

are solutions to Catalan equation (indeed, one easily sees that $a+b=1$).

§4. On Combinatorial invariance

Proof. (Of theorems 1.1 and 1.2) Let $\mathcal{A}_1$ and $\mathcal{A}_2$ be two arrangements as in the statement of the theorem. Recall that the characteristic polynomial of the monodromy of Milnor fiber for each arrangement is $(t-1)^{r(\mathcal{A}_i)-2}(t^2+t+1)^{s(\mathcal{A}_i)}$ where $r(\mathcal{A}_i) = Card(\mathcal{A}_i)$ and $s(\mathcal{A}_i)$ is the superabundance of curves of degree $\frac{2d-9}{3}$ passing through the triple points of the arrangement $\mathcal{A}_i$. Hence we need to show that $s(\mathcal{A}_1) > 0$ implies $s(\mathcal{A}_2) > 0$ as well.

For the rest recall the results from [4]. Let C be a possibly reducible plane curve for which $\omega = exp(\frac{2\pi\sqrt{-1}}{3})$ is a root of the Alexander polynomial. Let $\Pi_{i \in R} F_i = 0$ where R is the set of irreducible components of C be the reduced equation of C. Then the equation of C admits a quasi-toric relation corresponding to ordinary triple point (cf. [4]) i.e. there exist an equality (cf. (2.1)):

$$\Pi_{i \in R_1} F_i f^3 + \Pi_{i \in R_2} F_i g^3 + \Pi_{i \in R_3} F_i h^3 = 0 \tag{14}$$

with distinct F_i and $R = \bigcup_{i=1,2,3} R_i, R_i \cap R_j = \emptyset, i \neq j$ is a partition of the set of components R.

Since $s(\mathcal{A}_1) > 0$, there exist non trivial quasi-toric decomposition corresponding to the curve which is a union of lines in $\mathcal{A}_1$. This means that equations of lines of arrangement can be split into 3 groups which can be used to construct Catalan equation which has solutions. By theorem 3.1, $\mathcal{A}_1$ is composed of reduced pencil.

On the other hand, to each such pencil corresponds the 2-dimensional component of characteristic variety of $\mathcal{A}_1$ containing the identity of $\mathrm{Char}(\pi_1(\mathbb{P}^2 - \mathcal{A}_1))$ (it consists of the pullbacks of local systems from the complement in $\mathbb{P}^1$ to a triple of points). Each component of characteristic variety of $\mathcal{A}_1$ containing the identity of $\mathrm{Char}(\pi_1(\mathbb{P}^2 - \mathcal{A}_1))$ yields a component of the resonance variety of $\mathcal{A}_1$ (cf. [14], [15]). Recall that the resonance variety is the collection of $a \in H^1(\mathbb{P}^2 - \mathcal{A}_1)$ such that the Aomoto complex

$$H^0(\mathbb{P}^2 - \mathcal{A}_1) \xrightarrow{\wedge a} H^1(\mathbb{P}^2 - \mathcal{A}_1) \xrightarrow{\wedge a} H^2(\mathbb{P}^2 - \mathcal{A}_1) \tag{15}$$

has non-zero cohomology at the middle term. Since such cohomology is determined by the cohomology algebra $H^*(\mathbb{P}^2 - \mathcal{A}_1)$ we obtain that the resonance variety of $\mathcal{A}_2$ has the same number of components as $\mathcal{A}_1$. To each component of the resonance variety of $\mathcal{A}_2$ corresponds pencil which is reduced since the corresponding component of resonance variety of $\mathcal{A}_1$ comes from reduced pencil. For a map of $\mathbb{P}^2 - \mathcal{A}_2$ onto the complement in $\mathbb{P}^1$ to a triple of points $\mathcal{P}$, the pull back of the local system on the latter, given by assigning to a standard generator of $\pi_1(\mathbb{P}^1 - \mathcal{P})$ the value $\omega_3 = exp(\frac{2\pi i}{3})$, is the local system on $\mathbb{P}^2 - \mathcal{A}_2$ which cohomology is isomorphic to the eigenspace with eigenvalue ω_3 on $H^1(M_{\mathcal{A}_2})$. Hence the claim follows.

Q.E.D.

Remark 4.1. *The cases when $s > 1$ correspond to arrangements which are composed of several pencils. One such example is the following.*

Let C be an elliptic curve and $\mathcal{A}$ an arrangement of lines in plane dual to the plane containing C consisting of lines dual to nine inflection points of C. Then this arrangement has no double points, 12 triple points and the superabundance of curves of degree 3 containing these 12 points is $s = 2$ (cf. [6],[2]). Indeed, the cohomology sequence corresponding to

$$0 \to \mathcal{J}_S(3) \to \mathcal{O}_{\mathbb{P}^2}(3) \to \mathcal{O}_S \to 0 \tag{16}$$

is the following:

$$0 \to 0 \to \mathbb{C}^{10} \to \mathbb{C}^{12} \to H^1(\mathbb{P}^2, \mathcal{J}_S(3)) \to 0. \tag{17}$$

This arrangement is composed of 4 pencils (cf. [14] Section 3.3 Example 3). For example one can use as explicit equations for this arrangement union of three cubics

$$x^3 - y^3, x^3 - z^3, y^3 - z^3$$

or

$$[(y-z)(x-z)(x-y)], [(y-\omega^2 z)(z-\omega^2 y)(x-\omega z)], [(y-\omega z)(x-\omega^2 z)(x-\omega y)]$$

$(\omega^3 = 1)$ *etc.*

Remark 4.2. *The combinatorial invariance of $s(\mathcal{A})$ would follows from the following refinement of the above argument describing the relation between $s(\mathcal{A})$ and the number $l(\mathcal{A})$ of pencils of which $\mathcal{A}$ is composed. This is the number of essential components of the resonance variety of $\mathcal{A}$. The number of Catalan equations corresponding to an arrangement $\mathcal{A}$ which have solutions by 3.1 is also $l(\mathcal{A})$. To each such pencil correspond collection the map $\bar{M}_{\mathcal{A}} \to E$ (cf. [4]) of 3-fold cyclic covers of $\mathbb{P}^2$ and $\mathbb{P}^1$ respectively defined up to automorphism of E i.e. six elements of the Mordell Weil group $MW(\mathcal{A})$ differ by an automorphism of E. Recall that $MW(\mathcal{A})$ is endowed with the canonical height pairing (cf. [19]). Identification of the number of elliptic pencils on $\bar{M}_{\mathcal{A}}$ with the number of elements of minimal height in $MW(\mathcal{A})$ and showing that the number of elements of minimal height in $MW(\mathcal{A}) = \mathbb{Z}[\omega_3]^{s(\mathcal{A})}$ is determined by $s(\mathcal{A})$ would imply that the number of pencils which composed $\mathcal{A}$ is determined by $s(\mathcal{A})$ and vice versa. Since, as follows from the above argument, the number of pencils composing the arrangement is a combinatorial invariant this would imply the combinatorial invariance of $s(\mathcal{A})$ and hence of the Milnor number.*

Added in proofs. Since this paper was submitted a new work addressing the problem studies here did appear: A. Dimca, Monodromy of triple point line arrangements, arXiv:1107.2214.

References

[1] D. Arapura, Geometry of cohomology support loci for local systems I, J. Algebraic Geom., **6** (1997), 563–597.

[2] N. Budur, A. Dimca and M. Saito, First Milnor cohomology of hyperplane arrangements, arXiv:0905.1284.

[3] E. Artal Bartolo, J. Carmona Ruber, J. I. Cogolludo Agustin and M. A. Marco Buzunariz, Invariants of combinatorial line arrangements and Rybnikov's example, In: Singularity Theory and Its Applications, Adv. Stud. Pure Math., **43**, Math. Soc. Japan, Tokyo, 2006, pp. 1–34.

[4] J. I. Cogolludo Agustin and A. Libgober, Mordell–Weil groups of elliptic threefolds and the Alexander module of plane curves, arXiv:1008.2018.

[5] A. Dimca, Pencils of plane curves and characteristic varieties, arXiv:math/0606442.

[6] A. Dimca and G. Lehrer, Hodge-Deligne equivariant polynomials and monodromy of hyperplane arrangements, arXiv:1006.3462.

[7] M. Falk, Arrangements and cohomology, Ann. Comb., **1** (1997), 135–157.

[8] M. Falk and S. Yuzvinsky, Multinets, resonance varieties, and pencils of plane curves, Compos. Math., **143** (2007), 1069–1088.

[9] W. Fulton, Adjoints and Max Noether's Fundamentalsatz, In: Algebra, Arithmetic and Geometry with Applications, West Lafayette, IN, 2000, Springer-Verlag, 2004, pp. 301–313.

[10] N. Greenleaf, On Fermat's equation in $\mathcal{C}(t)$, Amer. Math. Monthly, **76** (1969), 808–809.

[11] S. Lang and A. Neron, Rational points of abelian varieties over function fields, Amer. J. Math., **81** (1959), 95–118.

[12] A. Libgober, Alexander polynomial of plane algebraic curves and cyclic multiple planes, Duke Math. J., **49** (1982), 833–851.

[13] A. Libgober, Alexander invariants of plane algebraic curves, In: Singularities, Part 2, Arcata, Calif., 1981, Proc. Sympos. Pure Math., **40**, Amer. Math. Soc., Providence, RI, 1983, pp. 135–143.

[14] A. Libgober Characteristic varieties of algebraic curves, In: Applications of Algebraic Geometry to Coding Theory, Physics and Computation, Eilat, 2001, NATO Sci. Ser. II Math. Phys. Chem., **36**, Kluwer Acad. Publ., Dordrecht, 2001, pp. 215–254.

[15] A. Libgober and S. Yuzvinsky, Cohomology of the Orlik–Solomon algebras and local systems, Compositio Math., **121** (2000), 337–361.

[16] A. Libgober, Non vanishing loci of Hodge numbers of local systems, Manuscripta Math., **128** (2009), 1–31.

[17] P. Orlik and L. Solomon, Combinatorics and topology of complements of hyperplanes, Invent. Math., **56** (1980), 167–189.

[18] G. Rybnikov, On the fundamental group of the complement of a complex hyperplane arrangement, Tech. Rep., 94–13, DIMACS, Rutgers Univ., Piscataway, NJ, 1994, preprint, arXiv:math/9805056.

[19] J. P. Serre, Lectures on the Mordell–Weil Theorem, Aspects Math., Friedr. Vieweg and Sohn, 1989.

[20] J. Silverman, The Catalan equation over function fields, Trans. Amer. Math. Soc., **273** (1982), 201–205.

[21] J. Silverman, The Arithmetic of Elliptic Curves, Grad. Texts in Math., **106**, Springer-Verlag, 1986.

Anatoly Libgober
Department of Mathematics, University of Illinois,
Chicago, IL 60607, USA
and
Department of Mathematics, University of Miami,
Coral Gables, FL, USA
E-mail address: `libgober@math.uic.edu`

Advanced Studies in Pure Mathematics 62, 2012
Arrangements of Hyperplanes—Sapporo 2009
pp. 189–231

The KZ system via polydifferentials

Eduard Looijenga

Abstract.

We show that the KZ system has a topological interpretation in the sense that it may be understood as a variation of complex mixed Hodge structure whose successive pure weight quotients are polarized. This in a sense completes and elucidates work of Schechtman–Varchenko done in the early 1990's. A central ingredient is a new realization of the irreducible highest weight representations of a Lie algebra of Kac–Moody type, namely on an algebra of rational polydifferentials on a countable product of Riemann spheres. We also obtain the kind of properties that in the $\mathfrak{sl}(2)$ case are due to Ramadas and are then known to imply the unitarity of the WZW system in genus zero.

§ Introduction

We construct an identification of a given KZ-space with a space constructed out of cohomology with supports of a rank one local system. The latter space is topologically defined, but depends on the choice of n distinct points on the affine line. By letting these points move, we get over the parameter space of such n-tuples a (Gauß–Manin) connection on the trivial bundle with fiber the KZ-space and we show that this connection gets identified with the KZ-connection. In a sense this completes earlier work of Schechtman–Varchenko, who constructed a flat map from the KZ-system to a Gauß–Manin system for the ordinary cohomology (with no supports). Their map factors through the map constructed here, but they were not able to say much about how nontrivial that map might be in any given case (a priori it could be the zero map), whereas ours is always an isomorphism. We should perhaps emphasize that the

---

Received May 3, 2010.
Revised January 11, 2011.
2010 *Mathematics Subject Classification.* Primary 32G34; Secondary 14D07.
Key words and phrases. KZ system, polydifferentials, highest weight module.

topological interpretation tells us a great deal about the monodromy of the KZ-system that may be hard to obtain otherwise: the fundamental group of the base is a colored braid group and we can see it act on the cohomology of a rank one local system over a space that is naturally attached to n distinct points on the complex line. This space is like a partial Deligne–Mumford compactification of the moduli space of genus zero curves that are endowed with $1+n+m$ punctures for some m of which the first $1+n$ ones are fixed and the others freely move. We do not work this out here, but we feel that a closer examination of this representation is not only feasible, but also desirable.

Our approach differs from earlier work in an other respect by putting at the center a bialgebra of rational polydifferentials on an iterated self-product of a Riemann sphere (these are like ordinary differential forms, except that this algebra, rather than obeying a Koszul rule, is plainly commutative). We show that if the number of factors is countably infinite, then such an algebra receives all the irreducible highest weight representations of all Lie algebras of Kac–Moody type, a fact that may have an interest in its own right. This makes for a smooth passage from the KZ system to the Gauß–Manin system and yields precious additional information. Thus the somewhat elusive KZ system is annexed by algebraic geometry, which in the present setting means that the powerful tools of mixed Hodge theory become available to ferret out any finer structure. The strength of this approach manifests itself also in the algebraic domain, witness the new vanishing property Theorem 2.14 and the remarkable invariance property Theorem 2.15.

We expect this set-up to be particularly useful in establishing whether the WZW subsystem of the KZ-system has a flat unitary structure, as is conjectured by physicists and for which the evidence so far is limited to the case $\mathfrak{sl}(2)$ (due to Ramadas [7], see also our version [6]). We characterize the WZW system in terms of the a simple vanishing property Corollary 4.2 that generalizes the one found by Ramadas. (Indeed, as is shown in [6] this result together with the vanishing theorem and the invariance property mentioned above suffice to obtain the unitarity for the $\mathfrak{sl}(2)$-case.)

While this paper may help to resolve one mystery, it creates another and that is the lack of a conceptual explanation for the appearance of polydifferentials. This is not the first place where these do occur in this context: they also do in a paper of Stoyanovski–Feigin [9] and in one by Beilinson–Drinfeld [1] (both are mainly concerned with the WZW system) and our feeling is that especially the first of these might be linked to the present paper.

Although this article owes much to earlier work done in this area, the chosen approach almost demands it to be self-contained. For this reason we have given a complete proof that the vector bundle map between the vector bundles underlying the KZ system to the Gauß–Manin system is flat, despite the fact that the pattern of this proof is quite similar to (and indeed, inspired by) the one of Schechtman–Varchenko [8].

Here is brief review of the separate sections. The first section is preparatory in nature: we introduce here the basic algebra of polydifferentials that is at the center of this paper and derive some of its properties. This is used in Section 2 to show that this algebra contains the highest weight representations of a Lie algebra defined by a generalized Cartan matrix. We also make the transition of its weight spaces to spaces of twisted logarithmic forms. In Section 3 we use the Casimir operator in order to define a Gauß–Manin connection and we show its compatibility with the KZ connection: the main result in this direction, Theorem 3.6, as well as the version that yields a genuine topological interpretation of the KZ system, Theorem 3.7, in terms of cohomology with supports are stated here. As the underlying computations are somewhat delicate, we decided to write these out in full. We state their core as an operator formula, but in order not to interrupt the flow of arguments, we relegated the proof to an appendix. Our motivation was a desire to better understand the WZW-systems in genus zero as subsystems of the KZ-systems and Section 4 testifies to that. We characterize this system in terms of our polydifferentials and Conjecture 4.3 links them to a square integrability property. If true, then this should realize a WZW-system as a flat subbundle of polarized variation of Hodge structure which is of pure bidegree $(m, 0)$ for some m (so that this part is not really varying after all!) and would give the flat unitary structure on this system that physicists since long conjecture to exist.

Conventions. For any finite sequence $I = (i_N, i_{N-1}, \dots, i_1)$ taken from an (index) set $\mathcal{I}$, we denote by $|I| := N$ its length, by $\{I\} := \{i_N, \dots, i_1\} \subset \mathcal{I}$ the set of its terms and by $I^* := (i_1, i_2, \dots, i_N)$ the opposite sequence. If $(f_i)_{i\in\mathcal{I}}$ is an indexed collection of elements of an algebra with unit, then f_I denotes the noncommutative monomial $f_{i_N} \cdots f_{i_1}$ (understood to be 1 if $I = \emptyset$). Similarly, if $(f_i)_{i\in\mathcal{I}}$ is an indexed collection of elements of a Lie algebra, then $[f_I]$ stands for the iterated Lie bracket $\operatorname{ad}(f_{i_N}) \operatorname{ad}(f_{k_{i-1}}) \cdots \operatorname{ad}(f_{i_2})(f_{i_1})$ (read f_i when $I = (i)$ and zero when $I = \emptyset$). But if for a subset $X \subset \mathcal{I}$, $(c_x)_{x\in X}$ is a finite collection elements of an abelian group (usually complex numbers or elements of a vector space), then c_X is sometimes used as an abbreviation for $\sum_{x\in X} c_x$ by c_X. If t_i is a coordinate on a Riemann surface C_i ($i \in \mathcal{I}$), so that

$(t_i)_{i\in\mathcal{I}}$ is a set of coordinates on the product $\prod_{i\in\mathcal{I}} C_i$, then any finite subset $X\subset\mathcal{I}$ defines a locus where all t_x with $x\in X$ are equal to each other; the resulting coordinate on that locus is the denoted by t_X (the remaining coordinates are $(t_i)_{i\in\mathcal{I}-X}$). Strictly speaking these conventions clash, but in practice confusion is unlikely.

The permutation group of a set X is denoted by $\mathfrak{S}(X)$.

§1. Some operators in an algebra of polydifferentials

We introduce an algebra of logarithmic polydifferentials which looks like a shuffle algebra and identify in it certain operators of interest.

Polydifferentials

We consider polydifferentials on $C^{\mathcal{I}}$, where C is a nonsingular complex curve and $\mathcal{I}$ is a finite set. A polydifferential of degree d looks like a d-form on $C^{\mathcal{I}}$, but should not be confused with it as differentials coming from different factors commute rather than anti-commute.

The definition is as follows. Let Ω_C stand for the $\mathcal{O}_C$-module of differentials and denote by $\Omega^\bullet_C$ the sheaf of graded $\mathcal{O}_C$-algebras $\mathcal{O}_C\oplus\Omega_C$. For every $X\subset\mathcal{I}$, we have a natural, $\mathfrak{S}(X)$-equivariant, identification of $\mathcal{O}_{C^X}$-modules

$$\Omega_C^{\boxtimes X}\cong\Omega^{|X|}_{C^X}\otimes \mathrm{or}(X)$$

where $\Omega_C^{\boxtimes X}$ denotes the exterior product $\otimes_{x\in X}\pi_x^*\Omega^1_C$ and $\mathrm{or}(X)$ stands for the orientation module $\wedge^{|X|}\mathbb{Z}^X$. We define the *sheaf of polydifferentials* on $C^{\mathcal{I}}$ as the sheaf of graded $\mathcal{O}_{C^{\mathcal{I}}}$-algebras $(\Omega^\bullet_C)^{\boxtimes\mathcal{I}}$. So its degree N part is $\bigoplus_{X\subset\mathcal{I},|X|=N}\pi_X^*\Omega_C^{\boxtimes X}$, where $\pi_X: C^{\mathcal{I}}\to C^X$ is the evident projection. A total order on $\mathcal{I}$ identifies the graded sheaf of polydifferentials with the graded sheaf of holomorphic differential forms on $C^{\mathcal{I}}$, but there is no natural analogue of the exterior derivative defined on $(\Omega^\bullet_C)^{\boxtimes\mathcal{I}}$. If $\omega\in(\Omega^\bullet_C)^{\boxtimes\mathcal{I}}$, we will write ω^X for its component in $\pi_X^*\Omega_C^{\boxtimes X}$, where $\pi_X: C^{\mathcal{I}}\to C^X$ is the evident projection. So $\omega=\sum_{X\subset\mathcal{I}}\omega^X$. We also write $\omega_X\in\pi_X^*(\Omega^\bullet_C)^{\boxtimes X}$ for the sum $\sum_{Y\subset X}\omega^Y$.

We say that a meromorphic polydifferential on C is *logarithmic* if locally it can be written as a product $(df_1/f_1)\dots(df_N/f_N)$ of logarithmic differentials , with $f_1,\dots,f_N\in\mathcal{O}_{C^{\mathcal{I}}}$.

Residue operators that involve a single factor (and a priori only those) have a meaning for polydifferentials. Recall for any $p\in C$ is defined a residue operator $\mathrm{Res}_{(t=p)}$ as a linear form on the space meromorphic differentials on C at p_i. This extends to polydifferentials: $\mathrm{Res}_{(t_i=p)}$ maps the meromorphic polydifferentials on $\mathcal{C}^{\mathcal{I}}$ to meromorphic polydifferentials on $\mathcal{C}^{\mathcal{I}-\{i\}}$. But a residue along any other type of hypersurface must be treated with care, as it requires an ordering of the index set. For instance, if (i, j) is an ordered pair, then the residue along the diagonal $(t_i = t_j)$ where the ith and jth component are equal is to be understood as viewing t_i as the residue variable and t_j as a parameter, in other words, we take the residue in the t_i-direction and do this at the point with t_i-coordinate equal to t_j. The result is a polydifferential on the diagonal divisor $(t_i = t_j)$. In order to emphasize the asymmetric roles of i and j, we shall denote this by $\mathrm{Res}_{(t_i \to t_j)}$. This is clearly not the same thing as $\mathrm{Res}_{(t_j \to t_i)}$.

A shuffle algebra of polydifferentials

We now take C to be the projective line (which we will denote by $\mathbf{P}$), we pick a point $\infty \in \mathbf{P}$ and denote by $\mathbf{A}$ the affine line $\mathbf{P} - \{\infty\}$. It is helpful to fix an affine coordinate on $\mathbf{A}$ (which depending on the context is denoted t or z), although this is inessential (in practice it will only enter through differentials of the form $(t(p_1) - t(p_2))^{-1}dt$, where $p_1, p_1 \in \mathbf{A}$ are distinct, which is indeed independent of t).

We consider relative polydifferentials on $(\mathbf{P}^{\mathcal{I}})_{\mathbf{A}}$, i.e., on the projection $\mathbf{P}^{\mathcal{I}} \times \mathbf{A} \to \mathbf{A}$. The coordinate of the base $\mathbf{A}$ is denoted z and the affine coordinate of the ith factor $\mathbf{P}$ in the product by t_i.

We first consider the graded $\mathbb{C}$-vector space $\mathcal{B}^\bullet_{\mathcal{I}}$ of the relative polydifferentials $(\mathbf{P}^{\mathcal{I}})_{\mathbf{A}}$ generated by the expressions of the form

$$\zeta_I : z \mapsto \zeta_I(z) := \frac{dt_{i_N} dt_{i_{N-1}} \cdots dt_{i_1}}{(t_{i_N} - t_{i_{N-1}}) \cdots (t_{i_2} - t_{i_1})(t_{i_1} - z)},$$

where $I = (i_N, i_{N-1}, \dots, i_1)$ runs over the finite sequences in $\mathcal{I}$ (the degree of such a polydifferential is N). We stipulate that for $I = \emptyset$ we get 1: $\zeta_\emptyset = 1$. Notice that $\zeta_I = 0$ unless the sequence I is without repetition. In this paper $\mathcal{I}$ shall be fixed (although we will later assume $\mathcal{I}$ to be countably infinite) and so we often write $\mathcal{B}^\bullet$ instead of $\mathcal{B}^\bullet_{\mathcal{I}}$.

We also put

$$\omega_I = \frac{dt_{i_N} dt_{i_{N-1}} \cdots dt_{i_2}}{(t_{i_N} - t_{i_{N-1}}) \cdots (t_{i_2} - t_{i_1})},$$

agreeing that when $I = (i)$, $\omega_{(i)} = \omega_i = 1$ and that $\omega_\emptyset = 0$. So when $j \in \mathcal{I} - \{I\}$, then $\zeta_I(t_j) = \omega_{Ij}$ and when I is nonempty and ends with i, then $\omega_I \zeta_{iJ} = \zeta_{IJ}$. It is clear that if $X \subset \mathcal{I}$, then $\mathcal{B}_{\mathcal{I}}$ is closed under 'taking the X-component'.

For a section a of $\mathbf{P}_{\mathbf{A}}$, the residue along a is a map from the meromorphic differentials on $\mathbf{P}_{\mathbf{A}}$ to $\mathbb{C}(\mathbf{A})$. Given an $x \in \mathcal{I}$, then this extends this in an obvious manner as a map of degree -1 from to the meromorphic relative polydifferentials on $(\mathbf{P}^{\mathcal{I}})_{\mathbf{A}}$ to itself (strictly speaking it is the composite of a residue and the pull-back along the projection that suppresses the xth factor). For certain values of a, this preserves $\mathcal{B}$:

Lemma-definition 1.1. *For $x \in \mathcal{I}$, we put*

$$E_x := -\operatorname{Res}_{(t_x=\infty)}, \quad E'_x := \operatorname{Res}_{(t_x=z)}\,.$$

Then for a sequence I in $\mathcal{I}$ without repetition, we have $E_x(\zeta_I) = 0$ resp. $E'_x(\zeta_I) = 0$ unless I is of the form xJ resp. Jx, in which case we get ζ_J.

If $x, y \in \mathcal{I}$ are distinct, then $\operatorname{Res}_{(t_x \to t_y)} \zeta_I$ is zero unless I is of the form $I''xyI'$ or $I''yxI'$, in which case we get $\zeta_{I''yI'}$ resp. $-\zeta_{I''yI'}$.

The proof is straightforward. This helps us (among other things) to give an intrinsic characterization of $\mathcal{B}^\bullet$. Given a subset $X \subset \mathcal{I}$, denote by D_X the reduced effective divisor on $\mathbf{P}^X \times \mathbf{P}$ that is the sum of the loci $(t_i = \infty)$, $(t_i = z)$ and the diagonals $(t_i = t_j)$ ($i, j \in X$ distinct).

Corollary 1.2. *The polydifferentials ζ_I are linearly independent over $\mathbb{C}[z]$ and (hence) constitute a $\mathbb{C}$-basis of $\mathcal{B}^\bullet$. Moreover, we have a decomposition $\mathcal{B}^N = \oplus_{X \subset \mathcal{I}, |X|=N} \mathcal{B}^N_X$ and when $X \subset \mathcal{I}$ is finite, then the natural map of $\mathbb{C}[z]$-modules*

$$\mathbb{C}[z] \otimes_{\mathbb{C}} \mathcal{B}^{|X|}_X \to H^0(\mathbf{P}^X \times \mathbf{A}, \Omega^{|X|}_{\mathbf{P}^X \times \mathbf{A}/\mathbf{A}}(\log D_X)) \otimes \operatorname{or}(X)$$

is an isomorphism that maps $\mathcal{B}^{|X|}$ onto the space of sections that vanish at $z = \infty$ and restricts for any $z \in \mathbf{A}$ to an isomorphism

$$\mathcal{B}^{|X|}_X \to H^0(\mathbf{P}^X, \Omega^{|X|}_{\mathbf{P}^X}(\log D_X(z))) \otimes \operatorname{or}(X).$$

Proof. The first assertion follows from the observation that for a sequence $I = (i_N, \dots, i_1)$, the iterated residue $E'_{i_N} \cdots E'_{i_1}$ takes on ζ_J the value $\delta_{I,J}$ when $|J| \leq |I|$.

It is clear that $\mathcal{B}^\bullet_{\mathcal{I}} = \oplus_{X \subset \mathcal{I}} \mathcal{B}^{|X|}_X$ and it is easily checked that $\mathcal{B}^N_X$ is a subspace of $H^0(\mathbf{P}^X \times \mathbf{P}, (\Omega^{\boxtimes X}_{\mathbf{P}})_{\mathbf{P}}(D_X - (z = \infty)))$. To see that this inclusion is an equality, we observe that if an element ω of the last space is annihilated by all the iterated residue operators of the type above, then ω must be independent of z. As ω is zero for $z = \infty$, it follows that $\omega = 0$. The last assertion is proved similarly. Q.E.D.

Here is another interesting consequence of Lemma 1.1. Given a sequence $I = (i_N, \dots, i_1)$ in $\mathcal{I}$ without repetition of length $N \geq 2$, then put

$$\operatorname{Res}_I := \operatorname{Res}_{(t_{i_2} \to t_{i_1})} \cdots \operatorname{Res}_{(t_{i_N} \to t_{i_{N-1}})}\,.$$

We regard this operator as taking values in the polydifferentials on the diagonal locus defined by $t_{i_1} = t_{i_2} = \cdots = t_{i_N}$. We recall that $t_{\{I\}}$ denotes the restriction of any t_{i_k} to this locus.

Lemma 1.3. *Let $I = (i_N, \dots, i_1)$ be a sequence without repetition in $\mathcal{I}$. Then the following identities hold for $a \in \{z, \infty\}$ in $\mathcal{B}$:*

$$[[[\cdots[E_{i_N}, E_{i_{N-1}}], \cdots], E_{i_2}], E_{i_1}] = -\operatorname{Res}_{(t_{\{I\}}=\infty)} \operatorname{Res}_I,$$
$$[[[\cdots[E'_{i_N}, E'_{i_{N-1}}], \cdots], E'_{i_2}], E'_{i_1}] = \operatorname{Res}_{(t_{\{I\}}=z)} \operatorname{Res}_I.$$

Proof. The last clause of Lemma 1.1 shows that if $\operatorname{Res}_I \zeta_K \neq 0$, then some permutation $\sigma(I)$ of I must appear as an uninterrupted subsequence of K: $K = K''JK'$, and have the following property: i_N is adjacent to i_{N-1} and removing i_N makes i_{N-1} adjacent to i_{N-2} and so on. In other words, there is a proper subsequence $I' < I$ such that if I'' is the residual subsequence (obtained by removing the terms in I'), then $\sigma(I) = (I')^* I''$; let us call this a *back-forward* permutation of I and denote by $\operatorname{back}(\sigma)$ the length of I'. If we also require that $\operatorname{Res}_{(t_{\{I\}}=z)} \operatorname{Res}_I(\zeta_K) \neq 0$, then $K' = \emptyset$ and its value is then by Lemma 1.1 equal to $(-1)^{\operatorname{back}(\sigma)}\zeta_{K''}$. If on the other hand $-\operatorname{Res}_{(t_{\{I\}}=\infty)} \operatorname{Res}_I(\zeta_K)$ is nonzero, then $K'' = \emptyset$ and its value is then by Lemma 1.1 equal to $(-1)^{\operatorname{back}(\sigma)}\zeta_{K'}$.

Expanding the iterated bracket of residue operators in a straightforward fashion yields

$$[[[\cdots[E'_{i_N}, E'_{i_{N-1}}], \cdots], E'_{i_2}], E'_{i_1}] = \\ = \sum_\sigma (-1)^{\operatorname{back}(\sigma)} E'_{\sigma(i_N)} E'_{\sigma(i_{N-1})} \cdots E'_{\sigma(i_1)},$$

where the sum is over all back-forward permutations σ of I. Its value on ζ_K is nonzero only if K is of the form $K''\sigma(I)$ and is then equal to $\zeta_{K''}$. If we replace E'_{i_k} by E_{i_k} and use the antisymmetry of the bracket: then we find expansion above is also equal to

$$[[[\cdots[E_{i_N}, E_{i_{N-1}}], \cdots], E_{i_2}], E_{i_1}] = \\ = \sum_\sigma (-1)^{\operatorname{back}(\sigma)} E_{\sigma(i_1)} E_{\sigma(i_2)} \cdots E_{(t_{\sigma(i_N)})}.$$

Its value on is ζ_K is nonzero only if K is of the form $\sigma(I)K'$ and is then equal to $(-1)^{\operatorname{back}(\sigma)}\zeta_{K'}$. Q.E.D.

A subset $X \subset \mathcal{I}$ induces via pull-back an injective map from rational polydifferentials on C^X to rational polydifferentials on $C^{\mathcal{I}}$. We now allow $\mathcal{I}$ to be infinite, but countable and define the space of polydifferentials on $C^{\mathcal{I}}$ as the injective limit of the polydifferentials on C^X, where X runs over the finite subsets of $\mathcal{I}$. We denote by $\hat{\mathcal{B}}^m$ the space of (possibly infinite) sums of these relative polydifferentials of degree m:

$$\hat{\mathcal{B}}^m = \prod_{X\subset\mathcal{I},|X|=m} \mathcal{B}^m_X$$

and put $\hat{\mathcal{B}}^\bullet := \oplus_{m=0}^\infty \hat{\mathcal{B}}^m$, but often omit the subscript $\mathcal{I}$ as that has been fixed.

If I and J are sequences such that their juxtaposition IJ is without repetition, then we denote by $I \star J$ the collection of *shuffles* of I and J, i.e., the set of sequences that are permutations of IJ and in which I and J appear as subsequences. This *shuffle product* is associative: we have that $(I \star J) \star K = I \star (J \star K)$ if IJK is without repetition.

Lemma 1.4 (Shuffle rules). *The graded vector space $\hat{\mathcal{B}}$ is closed under product and is as a $\mathbb{C}$-algebra isomorphic to a (completed) shuffle algebra: if I and J are finite sequences in $\mathcal{I}$, then $\zeta_I\zeta_J = \sum_{K\in I\star J}\zeta_K$ (we get zero unless IJ is without repetition). Together with the coproduct*

$$\delta : \hat{\mathcal{B}} \to \hat{\mathcal{B}} \otimes \hat{\mathcal{B}}, \quad \delta(\zeta_I) = \sum_{I=I''I'} \zeta_{I''} \otimes \zeta_{I'},$$

this makes $\hat{\mathcal{B}}$ a commutative bialgebra over $\mathbb{C}$ (with the projection on the degree zero summand as counit).

More generally, if I and J are finite nonempty sequences in $\mathcal{I}$, $k \in \{J\}$ and J is written $J_{<k}J_{\geq k}$, then $\omega_{Ik}\zeta_J = \sum_{K\in I\star J_{<k}} \zeta_{KJ_{\geq k}}$.

Proof. An induction argument shows that it is enough to verify these statements insofar they do not regard the Hopf property when I is a singleton. That case easily follows from repeated use from the simple identity

$$\frac{1}{v-u} - \frac{1}{w-u} = \frac{w-v}{(v-u)(w-u)}.$$

The proof that δ is a coproduct that is compatible with the shuffle product is straightforward. Q.E.D.

So $\mathcal{B}$ can be identified with a shuffle algebra over $\mathbb{C}[t]/(t^2)$ on a set of generators indexed by $\mathcal{I}$. This makes us wonder whether there is a relation with iterated integrals. From Corollary 1.2 and Lemma 1.4 we deduce:

Corollary 1.5. *If $\mathcal{I}$ is a countably infinite set, then $\hat{\mathcal{B}}^{\mathfrak{S}(\mathcal{I})}$ is a polynomial algebra with primitive generator $\zeta := \sum_{i\in\mathcal{I}}\zeta_i$ and we have*

$$\frac{\zeta^N}{N!} = \sum_{\{X\subset\mathcal{I},|X|=N\}} \prod_{x\in X} \frac{dt_x}{t_x - z}.$$

Two-parameter identities

The following two lemmas assert identities in an algebra of meromorphic polydifferentials on **P** depending on two complex variables. They will be needed later (beginning with the construction of the Gauß–Manin connection in Lemma 3.9), but it is convenient to state and prove them now.

Recall that for a sequence I, I^* denotes the opposite sequence.

Lemma 1.6. *Let I be a nonempty sequence in $\mathcal{I}$. If $b(I)$ denotes its first element, then*

$$\frac{z-w}{t_{b(I)}-w}\zeta_I(z) = \sum_{I=I_2I_1} (-1)^{|I_2|}\zeta_{I_1}(z)\cdot\zeta_{I_2^*}(w),$$

or equivalently, the operator $\sum_J(-1)^{|J|}\zeta_J(w)E_J$ (with the sum taken over all finite sequences J in $\mathcal{I}$) sends ζ_I to $\frac{z-w}{t_{b(I)}-w}\zeta_I$.

More generally, if i appears in I so that we can write $I = I''iI'$, then

$$\frac{z-w}{t_i-w}\zeta_I(z) = \zeta_I(z) - \sum_{I'=I_2I_1} (-1)^{|I_2|}\zeta_{I_1}(z)\cdot\omega_{I''i}\zeta_{I_2^*i}(w),$$

and if $j\in\mathcal{I}$, $j\neq i$, then

$$\frac{z-t_j}{t_i-t_j}\zeta_I(z) = \zeta_I(z) - \sum_{I'=I_2I_1} (-1)^{|I_2|}\zeta_{I_1}(z)\cdot\omega_{I''i}\omega_{I_2^*ij}.$$

Proof. The third identity is not really different from the second (take $t_j = w$) and the second follows from the first applied to iI' (and multiply it with $\omega_{I''i}$). So we concentrate on the first identity and prove it with induction on $|I|$. For $I=\emptyset$ there is nothing to show and for $I=(i)$, the lemma states that

$$\frac{z-w}{t_i-w}\zeta_i(z) = \zeta_i(z) - \zeta_i(w),$$

which is a simple consequence of the identity $\frac{z-w}{t_i-w} = 1 - \frac{t_i-z}{t_i-w}$. Now assume I has length >1 and write $I=iI'$ and $I'=jJ'$. So $\zeta_I(z) = \zeta_{ij}\zeta_{jJ}(z) = \frac{dt_i}{t_i-t_j}\zeta_{I'}(z)$. Since

$$\frac{z-w}{(t_i-w)(t_i-t_j)} = \frac{z-w}{t_j-w}\Big(\frac{1}{t_i-t_j} - \frac{1}{t_i-w}\Big),$$

we have

$$\frac{z-w}{t_i-w}\zeta_I(z) = \frac{z-w}{(t_i-w)(t_i-t_j)}dt_i\zeta_{I'}(z)$$
$$= \frac{z-w}{t_j-w}\Big(\frac{dt_i}{t_i-t_j} - \frac{dt_i}{t_i-w}\Big)\zeta_{I'}(z)$$
$$= \frac{z-w}{t_j-w}\zeta_I(z) - \frac{z-w}{t_j-w}\zeta_{I'}(z)\cdot\zeta_i(w),$$

which after invoking the induction hypothesis becomes

$$\zeta_I(z) - \sum_{J'=J_2J_1}(-1)^{|J_2|}\zeta_{J_1}(z)\cdot\omega_{J_2^*j}\zeta_{ij}(w)$$
$$-\Big(\zeta_I'(z) - \sum_{J'=J_2J_1}(-1)^{|J_2|}\zeta_{J_1}(z)\cdot\zeta_{J_2^*j}(w)\Big)\zeta_i(w)$$
$$= \zeta_I(z) + \sum_{J'=J_2J_1}(-1)^{|J_2|}\zeta_{J_1}(z)\Big(-\omega_{J_2^*j}\zeta_{ij}(w) + \zeta_{J_2^*j}(w)\zeta_i(w)\Big)$$
$$-\zeta_{I'}(z)\cdot\zeta_i(w).$$

According to the shuffle rules 1.4, we have $-\omega_{J_2^*j}\zeta_{ij}(w)+\zeta_{J_2^*j}(w)\zeta_i(w) = \zeta_{J_2^*ji}(w) = \zeta_{(jJ_2)^*i}(w)$, so that we get

$$\zeta_I(z) - \sum_{J'=J_2J_1}(-1)^{|jJ_2|}\zeta_{J_1}(z)\zeta_{(jJ_2)^*i}(w) - \zeta_{I'}(z)\cdot\zeta_i(w),$$

which indeed equals $\zeta_I(z) - \sum_{I'=I_2I_1}(-1)^{|I_2|}\zeta_{I_1}(z)\zeta_{I_2^*i}(w)$. Q.E.D.

Remark 1.7. *Notice that the expression $\omega_{I_2^*i}\zeta_{I''i}(w)$ in the right hand side of the preceding lemma is, according to Lemma 1.4, equal to the sum $\sum_L \zeta_{Li}(w)$, where L runs over the shuffles of I_2^* and I''.*

The following lemma generalizes the preceding one.

Lemma 1.8. *If $I := I''iI'$ and $J := J''jJ'$ are sequences as above, then*

$$\frac{z-w}{t_i-t_j}\zeta_I(z)\cdot\zeta_J(w) = \zeta_I(z)\cdot\zeta_J(w)$$
$$-\sum_{I'=I_2I_1}(-1)^{|I_2|}\zeta_{I_1}(z)\cdot\omega_{I_2^*i}\omega_{I''ij}\zeta_J(w)$$
$$-\sum_{J'=J_2J_1}(-1)^{|J_2|}\omega_{J_2^*j}\omega_{J''ji}\zeta_I(z)\cdot\zeta_{J_1}(w).$$

In particular, it is a linear combination of terms of the form $\zeta_K(z)\cdot\zeta_L(w)$.

Proof. The second identity of Lemma 1.6 gives after multiplication by $\zeta_J(w)$

$$\frac{z-t_j}{t_i-t_j}\zeta_I(z)\cdot\zeta_J(w)=\zeta_I(z)\cdot\zeta_J(w)-\sum_{I'=I_2I_1}(-1)^{|I_2|}\zeta_{I_1}(z)\cdot\omega_{I_2^*i}\cdot\omega_{I''ij}\zeta_J(w).$$

Likewise we find

$$\frac{w-t_i}{t_j-t_i}\zeta_I(z)\cdot\zeta_J(w)=\zeta_I(z)\cdot\zeta_J(w)-\sum_{J'=J_2J_1}(-1)^{|J_2|}\omega_{J_2^*j}\omega_{J''ji}\zeta_I(z)\cdot\zeta_{J_1}(w).$$

The lemma then follows by adding these two identities and using that

$$\frac{z-t_j}{t_i-t_j}+\frac{w-t_i}{t_j-t_i}=\frac{z-w}{t_i-t_j}+1.$$

Q.E.D.

The Φ-operators

In the space of rational relative polydifferentials on $(\mathbf{P}^{\mathcal{I}})_{\mathbf{A}}$, we regard dt_i ($i\in I$) not just as an element, but also as the multiplication operator in this space. Its adjoint acts in the ith tensor factor only and sends dt_i to 1 and 1 to zero, hence is the contraction operator $\iota_{\partial/\partial t_i}$.

Let now be given complex numbers $(p_i)_{i\in\mathcal{I}}$ and $(c_{i,j})_{i,j\in\mathcal{I},i\neq j}$. We define for $i\in\mathcal{I}$ an operator Φ_i in the space of relative meromorphic polydifferentials by

$$\Phi_i:=\frac{p_i dt_i}{t_i-z}-\sum_{j\neq i}c_{i,j}\frac{dt_i dt_j}{t_i-t_j}\iota_{\partial/\partial t_j}.$$

So for a finite subset $X\subset\mathcal{I}$, we have

$$\Phi_i(dt_X)=\Big(\frac{p_i}{t_i-z}-\sum_{x\in X}\frac{c_{i,x}}{t_i-t_x}\Big)dt_i dt_X,$$

where it is understood that the right hand side is zero when $i\in X$.

Lemma 1.9. *This operator preserves $\mathcal{B}$ (hence also the completion $\hat{\mathcal{B}}$), for*

$$\Phi_x(\zeta_I)=\sum_{I=I''I'}(p_x-c_{x,I'})\zeta_{I''xI'}.$$

(Observe that the right hand side vanishes if $x\in I$ and that the term indexed by $(I'',I')=(I,\emptyset)$ reduces to $p_x\zeta_{Ix}$.) Furthermore, for any $o\in\mathcal{I}$,

$$[\Phi_x,\omega_{Io}]=\sum_{I=I''I'}-c_{x,I'}\omega_{I''xI'o}.$$

Proof. We have $\Phi_x(\zeta_I) = p_x\zeta_x\zeta_I - \sum_{i\in\{I\}} c_{x,i}\omega_{x,i}\zeta_I$, and the latter is by Lemma 1.4 equal to

$$p_x \sum_{I=I''I'} \zeta_{I''xI'} - \sum_{I=I''I'}\sum_{i\in\{I'\}} c_{x,i}\zeta_{I''xI'} = \sum_{I=I''I'} (p_x - c_{x,I'})\zeta_{I''xI'}.$$

The second identity follows from this. Q.E.D.

Products and powers

Notice that

$$[\Phi_i, dt_X] = \sum_{x\in X} \frac{-c_{i,x}dt_i}{t_i - t_x} dt_X.$$

We also observe that for a sequence $I = (i_N, \dots, i_1)$ in $\mathcal{I}$

$$\begin{aligned}\Phi_I(dt_X) &:= \Phi_{i_N}\Phi_{i_{N-1}}\cdots\Phi_{i_1}(dt_X)\\ &= \prod_{I=I''iI'} dt_i\Big(\frac{p_i}{t_i - z} - \sum_{x\in X}\frac{c_{i,x}}{t_i - t_x} - \sum_{j\in I'}\frac{c_{i,j}}{t_i - t_j}\Big).dt_X\\ &= \prod_{I=I''iI'} \frac{dt_i}{t_i - z}\Big(p_i - \sum_{x\in X} c_{i,x}\frac{t_i - z}{t_i - t_x} - \sum_{j\in\{I'\}} c_{i,j}\frac{t_i - z}{t_i - t_j}\Big).dt_X.\end{aligned}$$

Lemma 1.10. *In* $\mathbb{C}[a,b]$ *we have*

$$\sum_{\sigma\in\mathfrak{S}_N}\prod_{k=1}^{N}\Big(b + a\sum_{l<k}\frac{t_{\sigma(k)} - z}{t_{\sigma(k)} - t_{\sigma(l)}}\Big) = N!b(b+\tfrac{1}{2}a)\cdots(b+\tfrac{1}{2}(N-1)a).$$

Proof. The lemma is evidently true for $N = 1$. We continue with induction on N. Let us write c_N for $N!b(b+a)\cdots(b+(N-1)a)$. Then the left hand side is equal to

$$\begin{aligned}&\sum_{k=1}^{N}\sum_{\substack{\sigma\in\mathfrak{S}_N\\ \sigma(N)=k}}\prod_{k=1}^{N}\Big(b + a\sum_{l<k}\frac{t_{\sigma(k)} - z}{t_{\sigma(k)} - t_{\sigma(l)}}\Big) =\\ &\quad= \sum_{k=1}^{N}\Big(b + a\sum_{l\neq k}\frac{t_k - z}{t_k - t_l}\Big)\sum_{\substack{\sigma\in\mathfrak{S}_N\\ \sigma(N)=k}}\prod_{k=1}^{N-1}\Big(b + a\sum_{l<k}\frac{t_{\sigma(k)} - z}{t_{\sigma(k)} - t_{\sigma(l)}}\Big) =\\ &= \sum_{k=1}^{N}\Big(b + a\sum_{l\neq k}\frac{t_k - z}{t_k - t_l}\Big)c_{N-1} = \Big(Nb + a\sum_{l<k}\big(\frac{t_k - z}{t_k - t_l} + \frac{t_l - z}{t_l - t_k}\big)\Big)c_{N-1} =\\ &\qquad = (Nb + a\sum_{l<k} 1)c_{N-1} = Nc_{N-1}\big(b + \tfrac{1}{2}(N-1)a\big) = c_N.\end{aligned}$$

Q.E.D.

Of special interest is the case when b is a positive integer m and $a = -2$. Then we find that the right hand side of Lemma 1.10 is zero for $N > m$ and equals $\frac{N!m!}{(m-N)!}$ otherwise. The lemma above and the discussion preceding it show:

Corollary 1.11. *Let $\mathcal{J} \subset \mathcal{I}$ be such that for all $i, j \in \mathcal{J}$ we have $c_{i,j} = 2$ and $p_i = m$ for a fixed nonnegative integer m. If $\Phi := \sum_{i\in\mathcal{J}} \Phi_i$, then*

$$\frac{\Phi^N}{N!}(1) = m(m-1)\cdots(m+1-N) \sum_{\{K\subset\mathcal{J}:|K|=N\}} \prod_{k\in\{K\}} \frac{dt_k}{t_k - z},$$

where K runs over all the N-element subsets of $\mathcal{I}'$. In particular, this expression vanishes for $N = m+1$.

Commutators

A straightforward check shows:

Lemma 1.12. *If $x \neq y$, then $[\Phi_x, \Phi_y] = -c_{x,y}\omega_{x,y}\Phi_y + c_{y,x}\omega_{y,x}\Phi_x$.*

We use this to prove:

Lemma 1.13. *Let I be a nonempty sequence in $\mathcal{I}$ without repetition and let $o \in \mathcal{I}$ not occur in I. Assume that for some complex numbers $a, c, \check{c}$ we have $c_{i,j} = a$ for all $i \neq j$ in $\{I\}$ and $c_{i,o} = c$, $c_{o,i} = \check{c}$ for all $i \in \{I\}$. If $\ell(I)$ denotes the last element of I, then we have the operator identity*

$$[\Phi_{Io}] = \prod_{\ell(I)\neq k\in\{I\}} \Big(\frac{c}{t_o - t_k} + \sum_{l\in\{I_{>k}\}} \frac{a}{t_l - t_k}\Big)\cdot \frac{cdt_I\Phi_o + \check{c}\sum_{i\in\{I\}} dt_{I_i}\Phi_i}{t_o - t_{\ell(I)}},$$

where I_i is obtained from I by omitting i.

Proof. Lemma 1.12 gives the asserted identity for $I = (x)$: we have

$$[\Phi_x, \Phi_o] = \frac{1}{t_o - t_x}\big(cdt_x\Phi_o + \check{c}dt_o\Phi_i\big).$$

This verifies the lemma when I is a singleton. We proceed with induction on the length of I. So if we write $I = xJ$, then

$$[\Phi_{Io}] = [\Phi_x, [\Phi_{Jo}]] =$$

$$\prod_{\ell(I)\neq k\in J} \Big(\frac{c}{t_o - t_k} + \sum_{l\in\{J_{>k}\}} \frac{a}{t_l - t_k}\Big)\cdot \frac{c[\Phi_x, dt_J\Phi_o] + \check{c}\sum_{j\in J}[\Phi_x, dt_{J_j o}\Phi_h]}{t_o - t_{\ell(I)}}.$$

So the induction step amounts to showing that the numerator of the last factor, $c[\Phi_x, dt_J\Phi_o] + \check{c}\sum_{j\in J}[\Phi_x, dt_{J_j o}\Phi_h]$, is equal to

$$\Big(\frac{c}{t_o - t_x} + \sum_{l\in J}\frac{a}{t_l - t_x}\Big)\Big(cdt_I\Phi_o + \check{c}\sum_{i\in\{J\}} dt_{I_i o}\Phi_i + \check{c}dt_{Jo}\Phi_x\Big).$$

To this end we expand the brackets in the left hand side using the already verified case $N = 1$:

$$\begin{aligned}[][\Phi_x, dt_J\Phi_o] &= [\Phi_x, dt_J]\Phi_o + dt_J[\Phi_x, \Phi_o] = \\ &= \sum_{l\in\{J\}} a\frac{dt_x}{t_l - t_x}dt_J\Phi_o + \frac{dt_J}{t_o - t_x}(cdt_x\Phi_o + \check{c}dt_o\Phi_x) \\ &= \Big(\frac{c}{t_o - t_x} + \sum_{l\in\{J\}}\frac{a}{t_l - t_x}\Big)dt_I\Phi_o + \check{c}\frac{dt_{Jo}}{t_o - t_x}\Phi_x\end{aligned}$$

and for $i \in J$,

$$\begin{aligned}[][\Phi_x, dt_{J_i o}\Phi_i] &= [\Phi_x, dt_{J_i o}]\Phi_i + dt_{J_i o}[\Phi_x, \Phi_i] = \\ &= \sum_{l\in\{J_i\}} a\frac{dt_{xJ_i o}}{t_l - t_x}\Phi_i + c\frac{dt_{xJ_i o}}{t_o - t_x}\Phi_i + a\frac{dt_{J_i o}}{t_i - t_x}(dt_x\Phi_i + dt_i\Phi_x) = \\ &= \Big(\frac{c}{t_o - t_x} + \sum_{l\in\{J\}}\frac{a}{t_l - t_x}\Big)dt_{I_i o}\Phi_i + a\frac{dt_{Jo}}{t_i - t_x}\Phi_x.\end{aligned}$$

If we substitute these identities in the linear combination $c[\Phi_x, dt_J\Phi_o] + \check{c}\sum_{j\in J}[\Phi_x, dt_{J_j o}\Phi_h]$ we get the desired expression. Q.E.D.

For any $\mathcal{J} \subset \mathcal{I}$, we put $\tau_{\mathcal{J}} := \sum_{x\in\mathcal{J}} dt_x$, so that

$$\frac{\tau_{\mathcal{J}}^N}{N!} = \sum_{X\subset\mathcal{J}, |X|=N} dt_X \quad \text{(read 1 if } N = 0).$$

Corollary 1.14. *Let $\mathcal{J} \subset \mathcal{I}$, $o \in \mathcal{I} - \mathcal{J}$, $c, \check{c} \in \mathbb{C}$ be such that for all $x, y \in \mathcal{J}$, $c_{x,o} = c$, $c_{o,x} = \check{c}$ and $c_{x,y} = 2$. If $\Phi := \sum_{x\in\mathcal{J}}\Phi_x$, then*

$$\frac{(\operatorname{ad}\Phi)^N}{N!}(\Phi_o) = (-c-1)\cdots(-c-(N-1))\Big(c\frac{\tau_{\mathcal{J}}^N}{N!}\Phi_o + \check{c}dt_o\frac{\tau_{\mathcal{J}}^{N-1}}{(N-1)!}\Phi\Big).$$

In particular, the left hand side is zero for $N = -c + 1$.

Proof. First observe that $(\operatorname{ad}\Phi)^N(\Phi_o) = \sum_I[\Phi_{Io}]$, where I runs over the sequences in $\mathcal{J}$ of length N without repetition. Now apply Lemma's 1.10 and 1.13 to each $\mathfrak{S}_N$-orbit of this index set. Q.E.D.

Lemma 1.1 yields:

Lemma 1.15. *Given $x, y \in \mathcal{I}$, then $[E_x, \Phi_y](\zeta_I) = 0$ unless $x = y \notin I$, in which case it multiplies ζ_I by the scalar $p_x - c_{x,I}$.*

Proof. First we notice that $\Phi_y E_x(\zeta_I)$ vanishes unless I is of the form xJ and $x \notin J$: we then get $\sum_{J=J''J'}(p_y - c_{y,J'})\zeta_{J''xJ'}$.

On the other hand $E_x\Phi_y(\zeta_I)$ vanishes unless $y \notin I$ and either I has the form xJ or $x = y$. In the first case, we get $\sum_{J=J''J'}(p_y - c_{y,J'})\zeta_{J''yJ'} = \Phi_y E_x\zeta_I$ and so ζ_I is killed by $[E_x, \Phi_y]$. In the second case, we get $(p_x - c_{x,I})\zeta_I$ and we note that then also $\Phi_y E_x(\zeta_I) = 0$. Q.E.D.

§2. Highest weight representations in spaces of polydifferentials

In this section we show among other things that the highest weight representation of Lie algebras of Kac–Moody type are naturally realized in an algebra of logarithmic polydifferentials.

Kac–Moody Lie algebras

Let $(c_{k,l})_{k,l=1}^r$ be a generalized Cartan matrix, i.e., $c_{k,k} = 2$, and for $k \neq l$, $c_{k,l}$ is a nonpositive integer which is zero if and only if $c_{l,k}$ is zero. Attached to this matrix is the Lie algebra defined by the following presentation: it has generators $\tilde{e}_1, \dots, \tilde{e}_r, \tilde{f}_1, \dots, \tilde{f}_r$ subject to the relations $[\tilde{e}_k, \tilde{f}_l] = 0$ for $k \neq l$ and if we put $\check{\alpha}_k := [\tilde{e}_k, \tilde{f}_k]$, then

$$[\check{\alpha}_k, \tilde{e}_l] = c_{k,l}\tilde{e}_l, \quad [\check{\alpha}_k, \tilde{f}_l] = -c_{k,l}\tilde{f}_l, \quad [\check{\alpha}_k, \check{\alpha}_l] = 0.$$

We define the Lie algebra $\mathfrak{g}$ as a quotient of this Lie algebra by also imposing the *Serre relations* by setting for $k \neq l$, $\mathrm{ad}(\tilde{e}_k)^{1-c_{k,l}}\tilde{e}_l$ and $\mathrm{ad}(\tilde{f}_k)^{1-c_{k,l}}\tilde{f}_l$ equal to zero. We denote the linear span of the $\check{\alpha}_k$'s by $\mathfrak{h}$. (We obtain the (Kac–Moody) Lie algebra as defined in [4] as the quotient of $\mathfrak{g}$ by the maximal ideal of that has zero intersection with $\mathfrak{h}$, but as is shown in *op. cit.*, we have equality in case the generalized Cartan matrix is symmetrizable, a condition that is always fulfilled in the cases of interest.) We denote by $\tilde{\mathfrak{g}}$ the intermediate Lie algebra defined by imposing the latter half of these relations only: so we let $\mathrm{ad}(\tilde{f}_k)^{1-c_{k,l}}\tilde{f}_l = 0$. The images of $\tilde{e}_k$, $\tilde{f}_k$ and $\check{\alpha}_k$ in $\tilde{\mathfrak{g}}$ are denoted by the same symbol (so that is a slight abuse of notation), but in $\mathfrak{g}$ the first two lose their tilde's. The linear span of the $\check{\alpha}_k$'s, which we shall denote by $\mathfrak{h}$, will be regarded as a subalgebra of both $\tilde{\mathfrak{g}}$ and $\mathfrak{g}$. It is a Cartan subalgebra of either. Notice that the simple root $\alpha_l \in \mathfrak{h}^*$, characterized by $[h, e_l] = \alpha_l(h)e_l$, takes on $\check{\alpha}_k$ the value $c_{k,l}$. We denote by $\tilde{\mathfrak{g}}_+ \subset \tilde{\mathfrak{g}}$

the subalgebra generated by the $\tilde{e}_k$'s and by $\tilde{\mathfrak{g}}_- \subset \tilde{\mathfrak{g}}$ the subalgebra generated by the $\tilde{f}_k$'s.

Let $\tilde{V}$ be a representation of $\tilde{g}$ on which $\mathfrak{h}$ acts semisimply (and hence is graded by weights). The *primitive* part $\tilde{V}^{\rm prim} \subset \tilde{V}$ is by definition the set of vectors killed by $\tilde{\mathfrak{g}}_+$ (the biggest subspace on which $\tilde{\mathfrak{g}}_+$ acts trivially), whereas the *coprimitive* part of V is the quotient $\tilde{V}_{\rm coprim} := \tilde{V}/\tilde{\mathfrak{g}}_- V$ of $\tilde{V}$ (the smallest on which the $\tilde{\mathfrak{g}}_-$ acts trivially). Notice that both inherit a semisimple $\mathfrak{h}$-action. The following lemma collects a few simple, but useful facts about the representation $\tilde{V}$.

Lemma 2.1. *The $\tilde{\mathfrak{g}}$-submodule of $\tilde{V}$ generated by its primitive part $\tilde{V}^{\rm prim}$ is in fact a $\mathfrak{g}$-submodule. If $\mathfrak{g}$ is finite dimensional and $v \in \tilde{V}^{\rm prim}$ is a primitive vector that is killed by a large power of $\tilde{f}_k$ ($k = 1, \ldots, r$), then this submodule is a finite dimensional representation of $\mathfrak{g}$, which is irreducible in case v is a weight vector of $\mathfrak{h}$ (this weight is then necessarily dominant).*

Proof. By the PBW-theorem, the $\tilde{\mathfrak{g}}$-submodule of $\tilde{V}^{\rm prim}$ is also the $\tilde{\mathfrak{g}}_-$-submodule generated by $\tilde{V}^{\rm prim}$. In $\tilde{g}$, the Serre element $\mathrm{ad}(\tilde{e}_k)^{1-c_{k,l}}\tilde{e}_l$ ($k \neq l$) commutes with every $\tilde{f}_k$ (see [4], §3.3). Since it kills $\tilde{V}^{\rm prim}$, it must be zero on the $\tilde{\mathfrak{g}}_-$-submodule generated by $\tilde{V}^{\rm prim}$. This proves the first assertion.

The second assertion follows from Lemma 3.4 of [4]. The last assertion is then clear. Q.E.D.

In what follows $\lambda \in \mathfrak{h}^*$ is a dominant weight: for $k = 1, \ldots, r$, $\lambda(\check{\alpha}_k)$ is a nonnegative real number.

Representations in $\hat{\mathcal{B}}$

In what follows we suppose our index set $\mathcal{I}$ endowed with a surjection $\pi : \mathcal{I} \to \{1, \ldots, r\}$ such that each fiber $\mathcal{I}_k := \pi^{-1}(k)$ is countably infinite. We shall often write $\bar{i}$ for $\pi(i)$ and do likewise for the π-image of a sequence in $\mathcal{I}$. We will write $\mathfrak{S}_\bullet$ for $\mathfrak{S}(\mathcal{I}_1) \times \cdots \times \mathfrak{S}(\mathcal{I}_r)$.

If S is any sequence in $\{1, \ldots, r\}$, then we put

$$\zeta(S) := \sum_{\bar{I}=S} \zeta_I,$$

where the sum is over all sequences in $\mathcal{I}$ that map under π to S (for $S = \emptyset$, read 1). The right hand side is an element of $\hat{\mathcal{B}}$ that is invariant under the group $\mathfrak{S}_\bullet$. In fact, these elements give a basis of $\hat{\mathcal{B}}^{\mathfrak{S}_\bullet}$.

We take $c_{i,j} = c_{\bar{i},\bar{j}}$ and $p_i := \lambda(\check{\alpha}_{\bar{i}})$. For this choice of coefficients, we put $\tilde{f}_k := \sum_{i\in\mathcal{I}_k} \Phi_i$. We then have

$$\tilde{f}_k\zeta(S) = \sum_{S=S''S'} (\lambda(\check{\alpha}_k) - c_{k,S'})\zeta(S''kS').$$

Lemma 1.15 suggests to put

$$\tilde{e}_k\zeta(S) := \begin{cases} \zeta(S') & \text{if } S = kS', \\ 0 & \text{otherwise} \end{cases}$$

(which makes $\tilde{e}_k$ independent of λ). It is then clear that $\tilde{e}_k$ and $\tilde{f}_l$ commute when $k \neq l$ and that $[\tilde{e}_k, \tilde{f}_k]$ multiplies $\zeta(S)$ by the scalar $\lambda(\alpha_k) - c_{k,S}$. Lemma 1.1 suggest that we have an interpretation $\tilde{e}_k$ as a sum of residues along divisors at infinity. There is a problem however since $\sum_{i\in\mathcal{I}_k} E_i$ does not make sense as a map defined on $\ddot{\mathcal{B}}$. Indeed, whereas $\tilde{f}_k$ makes sense on $\hat{\mathcal{B}}$, there is no obvious way to define $\tilde{e}_k$ on that space. Yet Lemma 1.15 implies we may define it on $\hat{\mathcal{B}}^{\mathfrak{S}}{}_\bullet$ as follows:

Lemma 2.2. *Let $\zeta \in \hat{\mathcal{B}}^{\mathfrak{S}}{}_\bullet$. Then for every $i \in \mathcal{I}_k$, we have*

$$(\tilde{e}_k(\zeta))_{\mathcal{I}-\{i\}} = E_i(\zeta).$$

In particular, $\tilde{e}_k(\zeta) = 0$ if and only if ζ is regular along every hyperplane at infinity $(t_i = \infty)$ with $i \in \mathcal{I}_k$.

If we combine this with 1.3 and the antisymmetry of the Lie bracket, we obtain a way to express any Lie monomial in the $\tilde{e}_k$'s as an iterated residue:

Corollary 2.3. *Suppose we are in the situation of Lemma 2.2. Then for any sequence $I = (i_N, \dots, i_1)$ of length $N \geq 2$ in $\mathcal{I}$ without repetition we have for $\zeta \in \hat{\mathcal{B}}^{\mathfrak{S}}{}_\bullet$*

$$([\cdots[(\tilde{e}_{\bar{i}_N}, \tilde{e}_{\bar{i}_{N-1}}], \tilde{e}_{\bar{i}_{N-1}}], \cdots, \tilde{e}_{\bar{i}_1}])(\zeta)_{\mathcal{I}-\{I\}} = -\operatorname{Res}_{(t_{\{I\}}=\infty)} \operatorname{Res}_I(\zeta).$$

Proposition 2.4. *The operators $\tilde{e}_k, \tilde{f}_k$, $k = 1, \dots, r$, define a representation of $\tilde{\mathfrak{g}}$ on $\hat{\mathcal{B}}^{\mathfrak{S}}{}_\bullet$ which satisfies for $k \neq l$ and $N \geq 1$ the identity*

$$\frac{(\operatorname{ad} \tilde{f}_k)^N}{N!}\tilde{f}_l = (-c_{k,l}-1)\cdots(-c_{k,l}-(N-1))\Big(c_{k,l}\frac{\tau_k^N}{N!}\tilde{f}_l + c_{l,k}\frac{\tau_l\tau_k^{N-1}}{(N-1)!}\tilde{f}_k\Big),$$

so that indeed the Serre relations $\operatorname{ad}(\tilde{f}_k)^{1-c_{k,l}}\tilde{f}_l = 0$ $(k \neq l)$ are satisfied.

Proof. Put $\check{\alpha}_k := [\tilde{e}_k, \tilde{f}_k]$. We have seen that this operator is semisimple with integral eigenvalues ($\zeta(S)$ is an eigenvector with eigenvalue $\lambda(\check{\alpha}_k) - \sum_{i=1}^N c_{k,s_i}$). So $\zeta(S) \in \hat{\mathcal{B}}^{\mathfrak{S}_\bullet}$ is an eigenvector of $\mathfrak{h}$ of weight $\lambda - \sum_{i=1}^N \alpha_{s_i}$.

The operator $\tilde{f}_l$ changes the weight by $-\alpha_l$. Likewise, $\tilde{e}_l$ changes the weight by α_l and hence all the non-Serre relations are satisfied. Corollary 1.14 shows that the displayed relations also hold. Q.E.D.

When we regard $\hat{\mathcal{B}}^{\mathfrak{S}_\bullet}$ as a $\tilde{\mathfrak{g}}$-module we shall denote it by $\tilde{V}(\lambda)$ and write 1_λ for its generator 1. The $\mathfrak{h}$-grading will be indicated by a subscript, so that in the above proof, $\zeta(S) \in \tilde{V}(\lambda)_{\lambda - \sum_{i=1}^N \alpha_{s_i}}$. In particular, $\tilde{V}(\lambda)$ is a highest weight module of $\tilde{\mathfrak{g}}$ with highest weight λ. Notice that with $\check{\alpha}_k$ as above, we can now write

$$\tilde{f}_k \zeta(S) = \sum_{S=S''S'} \check{\alpha}_k(\zeta(S))\zeta(S''kS').$$

Denote by $V(\lambda)$ the $\tilde{\mathfrak{g}}_-$-submodule of $\tilde{V}(\lambda)$ generated by 1_λ. It follows from the preceding proposition that $V(\lambda)$ is then also invariant under $\tilde{\mathfrak{g}}_+$, hence is a $\tilde{\mathfrak{g}}$-module.

Remark 2.5. *Lemma 2.2 characterizes the primitive part $\tilde{V}(\lambda)^{\mathrm{prim}}$ of $\tilde{V}(\lambda)$ as the subspace of polydifferentials that are regular at the hyperplanes $t_i = \infty$. This is independent of λ (and we may even define $\hat{\mathcal{B}}^{\mathrm{prim}}$ as the space of such polydifferentials, although we did not define an action of $\tilde{\mathfrak{g}}$ on $\hat{\mathcal{B}}$).*

Clearly the generator $1_\lambda \in \tilde{V}(\lambda)$ is primitive.

Theorem 2.6. *The Lie algebra $\tilde{\mathfrak{g}}$ leaves invariant the $\tilde{\mathfrak{g}}_-$-submodule of $\tilde{V}(\lambda)$ generated by the primitive subspace $\tilde{V}(\lambda)^{\mathrm{prim}}$ and acts on that space through $\mathfrak{g}$. In particular, the subrepresentation $V(\lambda)$ generated by 1_λ is a highest weight representation of $\mathfrak{g}$. If λ is an integral weight, then this representation is integrable in the sense that each of the e_k and f_k acts on it in a locally nilpotent fashion. In case the given Cartan matrix is that of a finite dimensional Lie algebra, then $V(\lambda)$ is finite dimensional and irreducible.*

Proof. This is a direct application of Lemma 2.1, where for the last half we invoke Corollary 1.11. Q.E.D.

Notice that we do not claim that $\tilde{V}(\lambda)$ is a representation of $\mathfrak{g}$. Indeed, it is not true in general that $\mathrm{ad}(\tilde{e}_k)^{1-c_{k,l}}\tilde{e}_l$ vanishes on that space.

Example 2.7. *Assume that $r = 1$. Then $\tilde{\mathfrak{g}} = \mathfrak{g} \cong \mathfrak{sl}(2)$ and by Corollary 1.5, $\hat{\mathcal{B}}^{\mathfrak{S}}$ is the polynomial algebra on the generator ζ defined by $\zeta(z) = \sum_{i\in\mathcal{I}}(t_i - z)^{-1}dt_i$. One verifies that the operator e is simply derivation: $e(\zeta^N) = N\zeta^{N-1}$. Let us identify the weight λ that turns $\hat{\mathcal{B}}^{\mathfrak{S}} = \mathbb{C}[\zeta]$ into the $\mathfrak{sl}(2)$ representation $\tilde{V}(\lambda)$ with its value on the unique simple coroot. Then another straightforward computation shows that f then sends ζ^N to $(\lambda - 2N)\zeta^{N+1}$.*

The polar divisor of any $\zeta \in \hat{\mathcal{B}}^{\mathfrak{S}_\bullet}$ is in general much smaller than that of an arbitrary member of $\hat{\mathcal{B}}$. For instance, if $i, j \in \mathcal{I}$ are distinct, but such that $\bar{i} = \bar{j}$, then ζ has no poles along the diagonal hyperplane $t_i = t_j$. To see this observe that $\tilde{\zeta} := (t_i - t_j)\zeta$ has no pole along $t_i = t_j$ and that since ζ is $\mathfrak{S}_\bullet$-invariant, interchanging t_i and t_j turns $\tilde{\zeta}$ into $-\tilde{\zeta}$. So $\tilde{\zeta}$ is divisible by $(t_i - t_j)$ and hence ζ has no pole along $t_i = t_j$.

We can do even better on $V(\lambda)$:

Lemma 2.8. *Let I be a finite sequence in $\mathcal{I}$. If Res_I is nonzero on $V(\lambda)$, then for every initial part J of I, $\alpha_{\{\bar{J}\}}$ is a root.*

Proof. Let $I = (i_N, \dots, i_1)$ and let $\zeta \in V(\lambda)$ be homogeneous, of degree m, say. If $N > m$, then $\mathrm{Res}_I(\zeta) = 0$ and so there is nothing to prove. We proceed with downward induction on N.

If for some $i \in \mathcal{I}$, $\mathrm{Res}_{Ii}\,\zeta \neq 0$, then we may apply our induction hypothesis and conclude that for every initial part J of Ii, $\alpha_{\{\bar{J}\}}$ is a root. It remains to deal with the case when $\mathrm{Res}_{Ii}\,\zeta$ vanishes for all i. Since $\mathrm{Res}_{Ii}\,\zeta = \mathrm{Res}_{(t_{i_1}\to t_i)}\mathrm{Res}_I\,\zeta$, this means that the poles of $\mathrm{Res}_I(\zeta)$ that involve the coordinate $t_{\{I\}}$ can only occur for $t_{\{I\}} = z$ and $t_{\{I\}} = \infty$ and so $-\mathrm{Res}_{(t_{\{I\}}=\infty)}\mathrm{Res}_I\,\zeta = \mathrm{Res}_{(t_{\{I\}}=z)}\mathrm{Res}_I\,\zeta$ by the residue theorem. Recall that $\mathrm{Res}_{(t_{\{I\}}=\infty)}\mathrm{Res}_I\,\zeta = (-e_{\bar{I}}\zeta)_{\mathcal{I}-\{I\}}$ and so if this is nonzero, then for every initial part J of I, $e_{\bar{J}} \neq 0$ and hence $\alpha_{\{\bar{J}\}}$ is a root. Otherwise, $\mathrm{Res}_I\,\zeta$ is regular for a generic value of the coordinates t_j, $j \in \mathcal{I}$. But a Riemann sphere has no nonzero holomorphic differential and since $\mathrm{Res}_I\,\zeta$ involves $dt_{\{I\}}$, this must imply that $\mathrm{Res}_I\,\zeta$ is identically zero. Q.E.D.

We recall that if $\mathfrak{g}$ is simple and finite dimensional, then there is a unique *highest root* $\tilde{\alpha}$ relative to the root basis $(\alpha_1, \dots, \alpha_r)$. It is the unique long root that also dominant and also characterized by the property that for no $k = 1, \dots, r$, $\tilde{\alpha} + \alpha_k$ is a root.

Corollary 2.9. *Suppose that $\mathfrak{g}$ is simple and finite dimensional. If I is a finite sequence in $\mathcal{I}$ such that $\alpha_{\{\bar{I}\}}$ is the highest root, then for every $\zeta \in V(\lambda)$, the poles of $\mathrm{Res}_I(\zeta)$ that involve the coordinate $t_{\{I\}}$*

can only occur for $t_{\{I\}} = z$ *and* $t_{\{I\}} = \infty$ *and we have* $(e_{\tilde{\alpha}}\zeta)_{\mathcal{I}-\{I\}} = \mathrm{Res}_{(t_{\{I\}}=z)} \mathrm{Res}_I(\zeta)$.

The automorphism group $\mathrm{Aut}(\mathbf{P})$ of $\mathbf{P}$ acts on $\mathbf{P}^{\mathcal{I}}$. In fact, it acts on the projection $\mathbf{P}^{\mathcal{I}} \times \mathbf{P} \to \mathbf{P}$. This action preserves the divisors of the form $(t_i = t_j)$, and $(t_i = z)$, but not the divisors $(t_i = \infty)$. The stabilizer of ∞ in $\mathrm{Aut}(\mathbf{P})$ does have that property however and indeed, it leaves every ζ_I invariant and hence acts as the identity on $\mathcal{B}$. We should not expect this action to happen for all of $\mathrm{Aut}(\mathbf{P})$. Indeed, if $\sigma = \left(\begin{smallmatrix} a & b \\ c & d \end{smallmatrix}\right) \in \mathrm{SL}(2,\mathbb{C})$, then a straightforward computation shows that

$$\sigma^*\zeta_I(z) = \frac{cz+d}{ct_{b(I)}+d}\zeta_I(z).$$

According to Lemma 1.6 the left hand side equals (at least for $c \neq 0$)

$$\sum_{I=I''I'} (-1)^{|I''|}\zeta_{I''*}(-d/c).\zeta_{I'}(z).$$

So on $\hat{\mathcal{B}}^{\mathfrak{S}_\bullet}$ we find

$$\sigma^*\zeta(S) = \sum_{S=S''S'} (-1)^{|S''|}\zeta(S''^*)(-d/c).\zeta(S') =$$
$$= \sum_T (-1)^{|T|}\zeta(T)(-d/c).\tilde{e}_T(\zeta(S)),$$

where the sum is over all finite sequences T in $\{1,\ldots,r\}$ (but we get only a nonzero contribution from T if it appears as the initial part of S). To sum up, on $\hat{\mathcal{B}}^{\mathfrak{S}_\bullet}$ we have that

$$\sigma^* = \sum_T (-1)^{|T|}\zeta(T)(-d/c).\tilde{e}_T.$$

On the primitive part of $\tilde{V}(\lambda)^{\mathrm{prim}}$ this reduces to the term corresponding to $T = \emptyset$, which is just $\zeta(S)$. So we find:

Corollary 2.10. *The primitive part of* $\hat{\mathcal{B}}^{\mathfrak{S}_\bullet}$ *is left pointwise fixed under the action of the automorphism group of* $\mathbf{P}$.

In the next section we shall also need to know the infinitesimal (right) action of $\left(\begin{smallmatrix} 0 & 0 \\ 1 & 0 \end{smallmatrix}\right) \in \mathfrak{sl}(2,\mathbb{C})$. on $\hat{\mathcal{B}}^{\mathfrak{S}_\bullet}$. A similar argument shows that

$$\begin{pmatrix} 0 & 0 \\ 1 & 0 \end{pmatrix}^* = \sum_T (-1)^{|T|}\omega(T).\tilde{e}_T,$$

where $\omega(T) = \sum_{\bar{I}=T} \omega_I$.

Tensor products

We generalize the above to the case of a tensor product of highest weight representations. We fix an n-tuple of dominant weights, $\boldsymbol{\lambda} = (\lambda_1, \dots, \lambda_n)$ and instead of working with the base $\mathbf{A}$, we use the base $\mathbf{A}^n$: rather than a single variable z we have n variables $\mathbf{z} = (z_1, \dots, z_n)$. For an n-tuple $\mathbf{I} = (I^1, \dots, I^n)$ of sequences in $\mathcal{I}$ we consider the relative polydifferential

$$\zeta_{\mathbf{I}} : \mathbf{z} = (z_1, \dots, z_n) \mapsto \zeta_{\mathbf{I}}(\mathbf{z}) = \zeta_{I^1}(z_1) \cdot \zeta_{I^2}(z_2) \cdot \cdots \cdot \zeta_{I^n}(z_n).$$

It clear that this polydifferential vanishes unless the sequence $I^1 \cdots I^n$ obtained by juxtaposition is without repetition. We denote by $\mathcal{B}_n$ be the graded vector space spanned by these polydifferentials, by $\hat{\mathcal{B}}_n^d$ the completion of $\mathcal{B}_n^d$ of (form) degree d which allows for infinite sums of these polydifferentials and put $\hat{\mathcal{B}}_n := \oplus_d \hat{\mathcal{B}}_n^d$. It may be worthwhile to observe that $\zeta_{\mathbf{I}}$ is invariant under the stabilizer of ∞ in the automorphism group $\mathbf{P}$, in other words, under the automorphism group of $\mathbf{A}$.

Given an n-tuple $\mathbf{S} = (S^1, \dots, S^n)$ of sequences in $\{1, \dots, r\}$, we observe that

$$\zeta(\mathbf{S})(\mathbf{z}) := \sum_{\bar{\mathbf{I}}=\mathbf{S}} \zeta_{\mathbf{I}}(\mathbf{z}) = \zeta(S_1)(z_1) \cdots \zeta(S_n)(z_n),$$

where the sum is over all n-tuples of sequences $\mathbf{I} = (I^1, \dots, I^n)$ in $\mathcal{I}$ whose juxtaposition is without repetition and map under π to $\mathbf{S}$. These elements form a $\mathbb{C}$-basis of $\hat{\mathcal{B}}_n^{\mathfrak{S}\bullet}$ and so the above factorization defines an isomorphism

$$\hat{\mathcal{B}}_n^{\mathfrak{S}\bullet} \cong \hat{\mathcal{B}}^{\mathfrak{S}\bullet} \otimes_{\mathbb{C}} \cdots \otimes_{\mathbb{C}} \hat{\mathcal{B}}^{\mathfrak{S}\bullet}.$$

Remark 2.11. *Assume that $r = 1$, so that $\mathfrak{g} = \mathfrak{sl}(2)$. It then follows from Corollary 1.5 that $\hat{\mathcal{B}}_n^{\mathfrak{S}}$ is a polynomial algebra on the n generators $\sum_{i \in \mathcal{I}} (t_i - z_\nu)^{-1} dt_i$. This yields the free $\mathbb{C}$-basis*

$$\zeta(1^{k_1}, \dots, 1^{k_n}) = \sum_{\{(X_\bullet), |X_\bullet| = k_\bullet\}} \prod_{\nu=1}^{n} \prod_{x \in X_\nu} \frac{dt_x}{t_x - z_\nu},$$

where the sum is over n-tuples of disjoint subsets $(X_1, \dots, X_n)$ of $\mathcal{I}$ with $|X_\nu| = k_\nu$.

The action of $\tilde{f}_k$ operating on the νth factor with dominant weight $\lambda^{(\nu)}$ is denoted $\tilde{f}_k^{(\nu)}$. The sum $\sum_{\nu=1}^n \tilde{f}_k^{(\nu)}$ acts as $\tilde{f}_k$ in the tensor representation and hence is simply denoted $\tilde{f}_k$. Notice however that we can define $\tilde{f}_k$ without reference to the tensor decomposition above as

$$\tilde{f}_k = \sum_{i\in\mathcal{I}_k} dt_i\Big(\sum_{\nu=1}^{n}\frac{\lambda_\nu(\check{\alpha}_k)}{t_i - z_\nu} - \sum_{j\in\mathcal{I},j\neq i}\frac{c_{k,\bar{j}}}{t_i-t_j}\iota_{\partial/\partial t_j}\Big).$$

For the tensor action of $\tilde{e}_k$ (as $\sum_{\nu=1}^n \tilde{e}_k^{(\nu)}$) the situation is even better, for Lemma 2.2 remains valid in this multivariable setting:

Lemma 2.12. *If $i\in\mathcal{I}_k$ and $\zeta\in\hat{\mathcal{B}}_n^{\mathfrak{S}\bullet}$, then*

$$\tilde{e}_k(\zeta)_{\mathcal{I}-\{i\}} = E_i(\zeta).$$

Proof. It is enough to verify this in case $\zeta=\zeta(\mathbf{S})$, where $\mathbf{S}=(S^1,\dots,S^n)$ is an n-tuple of sequences in $\{1,\dots,r\}$. If $\mathbf{I}=(I^1,\dots,I^n)$ is an n-tuple of sequences in $\mathcal{I}$ whose concatenation is without repetition and which maps under π to $\mathbf{S}$, then the value of E_i on $\zeta_{\mathbf{I}}$ is zero unless i is the first term of some I^ν: $I^\nu = iI'^\nu$, in which case we get $\zeta_{(I^1,\dots,I'^\nu,\dots,I^n)}$. So if we take the sum over such $\mathbf{I}$ we find the $(\mathcal{I}-\{i\})$-component of $\tilde{e}_k(\zeta(\mathbf{S}))$. Q.E.D.

This completely describes $\hat{\mathcal{B}}_n^{\mathfrak{S}\bullet}$ as a tensor product of representations of $\tilde{\mathfrak{g}}$. We will denote it by $\tilde{V}(\boldsymbol{\lambda})$ and write $1_{\boldsymbol{\lambda}}$ for its generator 1. We obtain the following generalization of Theorem 2.6.

Theorem 2.13. *We have a natural identification*

$$\tilde{V}(\boldsymbol{\lambda}) \cong \tilde{V}(\lambda_1)\otimes_{\mathbb{C}}\cdots\otimes_{\mathbb{C}}\tilde{V}(\lambda_n).$$

The primitive subspace $\tilde{V}(\boldsymbol{\lambda})^{\mathrm{prim}}\subset\tilde{V}(\boldsymbol{\lambda})$ is the subspace consisting of forms that are regular at every hyperplane at infinity ($t_i=\infty$), $i\in\mathcal{I}$, and the $\tilde{\mathfrak{g}}_-$-submodule of $\tilde{V}(\boldsymbol{\lambda})$ it generates is acted on by the Lie algebra $\tilde{\mathfrak{g}}$ via $\mathfrak{g}$ (so it is in fact a $\tilde{\mathfrak{g}}_-$-module). In particular,

$$V(\boldsymbol{\lambda}) := V(\lambda_1)\otimes_{\mathbb{C}}\cdots\otimes_{\mathbb{C}} V(\lambda_n)$$

is the smallest subspace of $\hat{\mathcal{B}}_n^{\mathfrak{S}\bullet}$ that contains $1_{\boldsymbol{\lambda}}$ and is invariant under the operators $f_k^{(\nu)}$ and $e_k^{(\nu)}$; it is the tensor product of n highest weight representations of $\mathfrak{g}$. It is integrable if all the λ_k's are integral.

We next state two important properties of the elements of $V(\boldsymbol{\lambda})$ and $V(\boldsymbol{\lambda})^{\mathrm{prim}}$. First, Lemma 2.8 almost immediately generalizes to:

Theorem 2.14. *If $I=(i_N,\dots,i_1)$ is a sequence in $\mathcal{I}$ such that Res_I is nonzero on $V(\boldsymbol{\lambda})$, then for every initial part J of I, $\alpha_{\{\bar{J}\}}$ is a root.*

In case the last element $\alpha_{\{\bar{I}\}}$ is the highest root $\tilde{\alpha}$, then for every $\zeta \in V(\lambda)$, the poles of $\operatorname{Res}_I(\zeta)$ that involve the coordinate $t_{\{I\}}$ can only occur where $t_{\{I\}}$ takes a value in $\{\infty, z_1, \dots z_n\}$, where we may omit ∞ in case $\zeta \in V(\boldsymbol{\lambda})^{\mathrm{prim}}$. We also have $(e_{\tilde{\alpha}}\zeta)_{\mathcal{I}-\{I\}} = \sum_{\nu=1}^n \operatorname{Res}_{(t_{\{I\}}=z_\nu)} \operatorname{Res}_I(\zeta)$.

We also have the obvious extension of Corollary 2.10.

Theorem 2.15. *The primitive part of $\hat{\mathcal{B}}_n^{\mathfrak{S}\bullet}$ is left pointwise fixed under the action of the automorphism group of* $\mathbf{P}$.

Proof. We only need to verify this infinitesimally: that any primitive element of $\hat{\mathcal{B}}_n^{\mathfrak{S}\bullet}$ is killed by the Lie algebra of $\operatorname{Aut}(\mathbf{P})$. This is clearly so for all the standard generators of this Lie algebra except $\binom{0\,0}{1\,0} \in \mathfrak{sl}(2,\mathbb{C})$. Its action is however on a given tensor factor given by the expression $\sum_T (-1)^{|T|}\omega(T).\tilde{e}_T$. Hence the same is true on the full tensor product. The last expression clearly vanishes on $\hat{\mathcal{B}}^{\mathfrak{S}\bullet}$. Q.E.D.

Note: *From now on we assume the weights $\lambda_1, \dots, \lambda_n$ to be integral.*

For the KZ-equation we shall have to consider the subspace $V(\boldsymbol{\lambda})^{\mathfrak{g}}$ of $\mathfrak{g}$-invariants in $V(\boldsymbol{\lambda})$. This is just $V(\boldsymbol{\lambda})_0^{\mathrm{prim}}$, the primitive part of $V(\boldsymbol{\lambda})$ of weight zero. (Note that $V(\boldsymbol{\lambda})_0 \neq 0$ implies that $\sum_\nu \lambda_\nu$ is a sum of positive roots.)

We invoke the representation theory of $\mathfrak{sl}(2)$ to deduce:

Proposition 2.16. *The intersection of $\sum_k \tilde{f}_k \tilde{V}(\boldsymbol{\lambda})$ with $V(\boldsymbol{\lambda})_0$ is the subspace $(\sum_k \tilde{f}_k \tilde{V}(\boldsymbol{\lambda}))_0$. In particular, $V(\boldsymbol{\lambda})^{\mathfrak{g}}$ embeds in $\tilde{V}(\boldsymbol{\lambda})_{0,\mathrm{coprim}}$.*

Proof. The grading of $\tilde{V}(\boldsymbol{\lambda})$ by the weights of $\mathfrak{h}$ shows that we have $\sum_k \tilde{f}_k \tilde{V}(\boldsymbol{\lambda}) \cap V(\boldsymbol{\lambda})_0 = \sum_k \tilde{f}_k \tilde{V}(\boldsymbol{\lambda})_{\alpha_k}$. So it suffices to show that $\tilde{f}_k \tilde{V}(\boldsymbol{\lambda})_{\alpha_k} \cap V(\boldsymbol{\lambda}) = \tilde{f}_k V(\boldsymbol{\lambda})_{\alpha_k}$ for every k. This makes it an issue about $\mathfrak{sl}(2)$. With induction it is easily shown that for every $v \in \tilde{V}(\boldsymbol{\lambda})_{\alpha_k}$ with $\tilde{f}_k v \in V(\boldsymbol{\lambda})$ we have

$$v \equiv \frac{(-1)^p}{p!(p+1)!} \tilde{f}_k^p \tilde{e}_k^p(v) \pmod{V(\boldsymbol{\lambda})}.$$

Since $\tilde{e}_k^p(v) = 0$ for p large, the claim follows. Q.E.D.

The passage to differential forms

In what follows we assume $n \geq 2$. We take as our base variety U_n the set of $(z_1, \dots, z_n) \in \mathbf{A}^n$ with $z_1, \dots, z_n$ pairwise distinct and summing up to $0 \in \mathbf{A}$; in other words U_n is the standard arrangement complement of type A_{n-1}. It is better however to refrain from choosing an origin for $\mathbf{A}$ and to think of U_n in modular terms: if we endow

$\mathbf{A}$ with the constant differential dz, then we easily see that U_n may be identified with the moduli space of smooth genus zero curves C endowed with an embedding of $\{\infty; 1, \dots, n\}$ in C and with a differential dz on the complement of the image of ∞ that is invariant under the automorphism group of that complement. The latter is also equivalent to the choice of a nonzero tangent vector of C at the image of ∞, so this makes U_n a (trivial) $\mathbb{C}^\times$-bundle over $\mathcal{M}_{0,1+n}$.

We now assume that $V(\boldsymbol{\lambda})_0 \neq 0$, so that $\sum_\nu \lambda_\nu$ is a sum of simple roots: $\sum_\nu \lambda_\nu = \sum_{k=1}^r m_k \alpha_k$ with $m_k \in \mathbb{Z}_{\geq 0}$. We put $\mathbf{m} := (m_1, \dots, m_k)$ and $m := \sum_k m_k$, so that $V(\boldsymbol{\lambda})_0$ lies in the homogeneous summand of multi-degree $\mathbf{m}$. Let $M \subset \mathcal{I}$ be a finite subset so that $M_k := M \cap \mathcal{I}_k$ has cardinality m_k. Our use of the symbol M implies that this decomposition is understood. We denote by $\mathrm{or}_M := \wedge^m \mathbb{Z}^M$ the sign representation of $\mathfrak{S}(M)$ and we put $\mathfrak{S}(M_\bullet) := \mathfrak{S}_\bullet \cap \mathfrak{S}(M) = \mathfrak{S}(M_1) \times \cdots \times \mathfrak{S}(M_r)$.

We denote by $\mathcal{B}_{n,M}$ the corresponding graded algebra of relative polydifferentials on $\mathbf{P}^M_{U_n} = (\mathbf{P}^{M_1} \times \cdots \times \mathbf{P}^{M_r} \times U_n)/U_n$. We regard this as a subalgebra of $\mathcal{B}_n$ via pull-back. It is multigraded by r-tuples of nonnegative integers and has $\mathbf{m}$ as highest multi-degree. Consider the homogeneous elements of $\mathcal{B}_{n,M}^{\mathfrak{S}(M_\bullet)}$ defined by

$$\xi_k := \sum_{i \in M_k} \Big(\sum_{\nu=1}^n \frac{\lambda_\nu(\check{\alpha}_k)}{t_i - z_\nu} - \sum_{j \in M - \{i\}} \frac{c_{k,\bar{j}}}{t_i - t_j} \Big) dt_i.$$

Lemma 2.17. *The map which assigns to an element of $\tilde{V}(\boldsymbol{\lambda})$ the sum of its X-components, where X runs over the subsets of M, maps onto $\mathcal{B}_{n,M}^{\mathfrak{S}(M_\bullet)}$ and so identifies the latter with $\tilde{V}_M(\boldsymbol{\lambda})$. It is an isomorphism in nonnegative weights: $\tilde{V}(\boldsymbol{\lambda})_{\geq 0} \cong \tilde{V}_M(\boldsymbol{\lambda})$. The transferred action of $\tilde{e}_k$ to $\tilde{V}_M(\boldsymbol{\lambda})$ is the obvious one (and given by residues as in Lemma 2.12) and the same is true for $\tilde{f}_k$ on the summands of multi-degree strictly lower than $\mathbf{m}$.*

Moreover, if $a_1, \dots, a_k$ are nonzero complex numbers and $m_k \geq 1$ for all k, then the M-component of $\tilde{V}(\boldsymbol{\lambda})_{0,\mathrm{coprim}}$, $\tilde{V}(\boldsymbol{\lambda})^M_{0,\mathrm{coprim}}$, gets identified with that of $\tilde{V}_M(\boldsymbol{\lambda})/((\sum_k a_k \xi_k)\tilde{V}_M(\boldsymbol{\lambda}))$. So this embeds $V(\boldsymbol{\lambda})^{\mathfrak{g}}$ in $\tilde{V}(\boldsymbol{\lambda})^M_{0,\mathrm{coprim}}$.

Proof. All but the last of these assertions follow from the fact that every $\alpha \in \mathcal{B}_{n,M}^{\mathfrak{S}_\bullet}$ of degree d is uniquely written as $\sum_{X \in \mathcal{I}, |X| = d} \pi_X^* \alpha_X$, where α_X is a rational polydifferential on $\mathbf{P}^X_{\mathbf{A}^n}$. To prove the last one, let us first observe that if $\zeta = \prod_{j \in M} dt_j$, then for $k = 1, \dots, r$, then every element of $\mathcal{B}_{n,M}$ of degree $m-1$ is a linear combination of the forms $\iota_{\partial/\partial t_i}\zeta$ with $i \in M$ with rational functions as coefficients. Now if $k \in \{1, \dots, r\}$ and $i \in M_k$, then

$$(\sum_{k=l}^{r} a_l \xi_l)\iota_{\partial/\partial t_i}\zeta = a_k\Big(\sum_{\nu=1}^{n} \frac{\lambda_\nu(\check{\alpha}_k)}{t_i - z_\nu} - \sum_{j\in M-\{i\}} \frac{c_{\bar{i},\bar{j}}}{t_i - t_j}\Big)\zeta$$

is also the M-component of $\tilde{f}_k(\iota_{\partial/\partial t_i}\zeta)$. The last statement now follows easily. Q.E.D.

The moduli space of injections of the disjoint union of $\{1, \dots, n\} \sqcup M$ in $\mathbf{A}$ given up to translations is also the moduli space of triples

$$(C, z \sqcup t : \{\infty, 1, \dots, n\} \sqcup M \hookrightarrow C, dz),$$

where C is a complete smooth curve of genus zero, $z \sqcup t$ is an embedding and dz is a nonzero differential on C invariant under $\mathrm{Aut}(C, z(\infty))$. We denote it by $U_{n,M}$. Ignoring the embedding of M defines an evident morphism $U_{n,M} \to U_n$. We make this morphism proper by means of a relative Deligne–Mumford–Knudsen compactification $U^+_{n,M} \to U_n$. Here $U^+_{n,M}$ is a $\mathbb{C}^\times$-bundle over a moduli space of stable pointed genus zero curves: it parameterizes triples $(C, z \sqcup t, dz)$ as before, where we now allow the pair $(C, z \sqcup t)$ to be a stable pointed curve, but insist that if we ignore the embedding of M and contract the irreducible components of C as to make it stable, the result is an $(1+n)$-pointed curve that is *smooth.* In particular, we get a retraction of C onto a distinguished component C_o such that its composite with z is injective. This component C_o must then contain all but at most one of the images of z and it is on this component that we assume dz is defined. This exhibits the desired morphism $p^+ : U^+_{n,M} \to U_n$. It is proper, indeed. In fact, if we extend the definition of the reduced effective divisors D_X in an obvious manner as (relative) divisors on $\mathbf{P}^M_{U_n}$, or rather $\mathbf{P}^M_{\mathbf{P}^n}$:

$$\begin{aligned} D^f_{n,M} &:= \sum_{i\in M}\sum_{\nu=1}^{n}(t_i = z_\nu) + \sum_{\{i\neq j\}\subset M}(t_i = t_j), \\ D_{n,M} &:= D^f_{n,M} + \sum_{i\in M}(t_i = \infty), \end{aligned}$$

(the superscript in $D^f_{n,M}$ stands for finite), then $U^+_{n,M}$ is obtained from $\mathbf{P}^M \times U_n$ by a blowing up process that is minimal for the property of turning $D_{n,M}$ into a normal crossing divisor $\Delta_{n,M} = U^+_{n,M} - U_{n,M}$. The generic point of an irreducible component of this divisor parameterizes one point unions of two smooth rational curves with the disjoint union of $\{\infty, 1, \dots, n\}$ and M embedded in its smooth part such that besides obeying the usual stability condition (every connected component of the smooth part contains at least two of these points) we have that one of the two connected components meets $\{\infty, 1, \dots, n\}$ either in a singleton or not at all. We denote this irreducible component of the boundary divisor accordingly as $\Delta_\infty(X)$, $\Delta_\nu(X)$ or $\Delta(X)$, where $\nu = 1, \dots, n$ and $X \subset M$ is nonempty and has at least two distinct elements in the last case. In terms of the configuration space of maps $X \to \mathbf{P}$ this corresponds to a confluence of the members X (where in the first case resp. second case the confluence is towards ∞ resp. z_ν). We denote by $\Delta^f_{n,M}$ the 'finite' part of $\Delta_{n,M}$, that is, the sum of the divisors $\Delta_\nu(X)$ and $\Delta(X)$. This is indeed the full preimage of $D^f_{n,M}$ under the blowup. It follows from Corollary 1.2 that we have a natural identification $\mathbb{C}[U_n]\otimes \mathcal{B}_{n,M} \cong \bigoplus_{X\subset M} H^0(\mathbf{P}^X_{U_n}, \Omega^{|X|}_{\mathbf{P}^X_{U_n}}(\log D_{n,X})) \otimes \mathrm{or}(X)$. This gives rise to an isomorphism of $\mathcal{O}_{U_n}$-modules:

$$\mathcal{O}_{U_n} \otimes \mathcal{B}^m_{n,M} \cong (p^+_* \Omega^m_{U^+_{n,M}/U_n}(\log \Delta_{n,M})) \otimes \mathrm{or}(M).$$

If we combine this with Corollary 1.2 and the Lemmas 2.12 and 2.17, we find:

Proposition 2.18. *We have a natural identification of $\tilde{V}(\boldsymbol{\lambda})_0$ with the subspace of $H^0(U_{n,M}, \Omega^m_{U_{n,M}/U_n}(\log \Delta_{M,n})) \otimes_{\mathfrak{S}(M_\bullet)} \mathrm{or}(M)$ of relative logarithmic m-forms that vanish along the hyperplanes $z_\nu = \infty$, $\nu = 1, \dots, n$. This restricts to an isomorphism of $\tilde{V}(\boldsymbol{\lambda})^{\mathrm{prim}}_0$ with the corresponding subspace of $H^0(U_{n,M}, \Omega^m_{U_{n,M}/U_n}(\log \Delta^f_{M,n})) \otimes_{\mathfrak{S}(M_\bullet)} \mathrm{or}(M)$ and yields trivializations of bundles over U_n:*

$$\mathcal{O}_{U_n} \otimes_{\mathbb{C}} \tilde{V}(\boldsymbol{\lambda})_0 \cong (p^+_* \Omega^m_{U^+_{n,M}/U_n}(\log \Delta_{M,n})) \otimes_{\mathfrak{S}(M_\bullet)} \mathrm{or}(M),$$
$$\mathcal{O}_{U_n} \otimes_{\mathbb{C}} \tilde{V}(\boldsymbol{\lambda})^{\mathrm{prim}}_0 \cong (p^+_* \Omega^m_{U^+_{n,M}/U_n}(\log \Delta^f_{M,n})) \otimes_{\mathfrak{S}(M_\bullet)} \mathrm{or}(M).$$

In particular, $\mathcal{O}_{U_n} \otimes_{\mathbb{C}} \tilde{V}(\boldsymbol{\lambda})^{\mathfrak{g}}$ embeds in the last module.

§3. Identification of the KZ connection

We continue with the situation of the previous section. So we have the n-tuple of integral dominant weights $\boldsymbol{\lambda} = (\lambda_1, \dots, \lambda_n)$ and regard $V(\boldsymbol{\lambda}) = V(\lambda_1) \otimes \cdots \otimes \tilde{V}(\lambda_n)$ as a representation of $\mathfrak{g}$. We let $\mathbf{m} = (m_1, \dots, m_r)$ and the finite subset $M \subset \mathcal{I}$ be as in Section 2.

The KZ-connection

The Knizhnik–Zamolodchikov connection requires the choice of a *Casimir element*, that is, a symmetric tensor $C \in \mathfrak{g} \otimes \mathfrak{g}$ that is invariant under $\mathfrak{g}$ (acting adjointly on each factor).

We denote by q^C the quadratic form on $\mathfrak{g}^*$ attached to C: $q^C(a) = \frac{1}{2}C(a,a)$ (note the factor $\frac{1}{2}$). We do not need to assume that C is nondegenerate, but we do suppose that $q^C(\alpha_k) \neq 0$ for every simple root α_k. This implies that the generalized Cartan matrix is symmetrizable and that C is nondegenerate on $(\mathfrak{g}/\mathfrak{h})^* \subset \mathfrak{g}^*$.

We use this tensor C to obtain a slightly different presentation of $\tilde{\mathfrak{g}}$: if α_k is the positive root attached to e_k (so $[\check\alpha_l, f_k] = -\alpha_k(\check\alpha_l)f_k$), then we replace f_k by $\hat f_k := q^C(\alpha_k)f_k = \frac{1}{2}C(\alpha_k,\alpha_k)f_k$. We retain e_k and so $\hat h_k := [e_k, \hat f_k] = \frac{1}{2}C(\alpha_k,\alpha_k)\check\alpha_k$ has now the property that $\lambda(\hat h_k) = C(\lambda, \alpha_k)$ for every $\lambda \in \mathfrak{h}^*$. Then $\tilde{\mathfrak{g}}$ is presented in terms of the symmetric matrix $(C(\alpha_k,\alpha_l))_{k,l}$:

$$[\hat h_k, e_l] = C(\alpha_k,\alpha_l)e_l, \quad [\hat h_k, \hat f_l] = -C(\alpha_k,\alpha_l)\hat f_l, \quad [\hat h_k, \hat h_l] = 0.$$

We still need to impose the Serre relations (which involve the possibly nonsymmetric Cartan matrix), but these are just the ones that make C nondegenerate on the subspace $(\mathfrak{g}/\mathfrak{h})^* \subset \mathfrak{g}^*$. In this setup $\lambda(\check\alpha_k)$ becomes $C(\lambda,\alpha_k)$, $c_{k,l}$ becomes $C(\alpha_k,\alpha_l)$, and Φ_k is replaced by

$$\hat\Phi_k := \sum_{i\in\mathcal{I}_k} dt_i\Big(\sum_\nu \frac{C(\lambda_\nu,\alpha_k)}{t_i - z_\nu} - \sum_{j\neq i} C(\alpha_k,\alpha_{\bar j})\frac{dt_j}{t_i - t_j}\iota_{\partial/\partial t_j}\Big).$$

For $1 \le \nu < \mu \le n$, let $C_{\nu,\mu}$ be the endomorphism of $V(\boldsymbol{\lambda})$ obtained by letting C act trough the tensor factors indexed by ν and μ. This operator commutes with with the diagonal action of $\mathfrak{g}$ and hence preserves the $\mathfrak{g}$-isotypical summands. Then the corresponding KZ connection ∇^C_{KZ} on $\mathcal{O}_{U_n} \otimes_{\mathbb{C}} V(\boldsymbol{\lambda})$ is defined by the $\mathrm{End}(V(\boldsymbol{\lambda}))$-valued differential

$$A^C_{KZ} := \sum_{1\le\nu<\mu\le n} C_{\nu,\mu}\frac{d(z_\nu - z_\mu)}{z_\nu - z_\mu}.$$

This is a connection with logarithmic singularities at infinity. It is easily shown to be flat so that we get a local system $\mathbb{KZ}^C(\boldsymbol{\lambda}) \subset \mathcal{O}_{U_n} \otimes_{\mathbb{C}} V(\boldsymbol{\lambda})$.

Remark 3.1 (Comparison with the SV-map). *This is essentially the situation considered by Schechtman–Varchenko in* [8] *from the outset. Our space* $\tilde V_M(\boldsymbol{\lambda})$ *is basically the one they construct for the case of a symmetrizable Cartan matrix. They identify the action of the operators* $\hat f_1, \dots, \hat f_r$*, but there are no operators* $\tilde e_1, \dots, \tilde e_r$ *acting. So the* *coprimitive quotients can (and do) appear there, but primitive subspaces cannot.*

A local system of rank one

Central in the subsequent discussion will be the following differential associated to C (a formal expression, for this is an infinite sum):

$$\eta^C_{\boldsymbol{\lambda}} := \sum_{\nu=1}^{n}\sum_{i\in\mathcal{I}} C(\alpha_{\bar i},\lambda_\nu)\frac{d(t_i-z_\nu)}{t_i-z_\nu} \\ -\tfrac{1}{2}\sum_{i,j\in\mathcal{I},i\neq j} C(\alpha_{\bar i},\alpha_{\bar j})\frac{d(t_i-t_j)}{t_i-t_j} - \sum_{1\le\nu<\mu\le n} C(\lambda_\nu,\lambda_\mu)\frac{d(z_\nu-z_\mu)}{z_\nu-z_\mu}.$$

The corresponding relative form $\xi^C_{\boldsymbol{\lambda}} := (\eta^C_{\boldsymbol{\lambda}})_{\mathrm{rel}}$ is obtained by ignoring the dz_ν-terms. Consider the finite subsums $\eta^C_{\boldsymbol{\lambda},M}$ and $\xi^C_{\boldsymbol{\lambda},M}$ that involve the factors indexed by M. So

$$\xi^C_{\boldsymbol{\lambda},M} := (\eta^C_{\boldsymbol{\lambda},M})_{\mathrm{rel}} = \sum_{i\in M}\Big(\sum_{\nu=1}^{n}\frac{C(\alpha_{\bar i},\lambda_\nu)}{t_i-z_\nu} - \sum_{j\in M-\{i\}}\frac{C(\alpha_{\bar i},\alpha_{\bar j})}{t_i-t_j}\Big)dt_i \\ = \sum_{k=1}^{r} q^C(\alpha_k)\sum_{i\in M_k}\Big(\sum_{\nu=1}^{n}\frac{\lambda_\nu(\check\alpha_k)}{t_i-z_\nu} - \sum_{j\in M-\{i\}}\frac{c_{k,\bar j}}{t_i-t_j}\Big)dt_i = \sum_{k=1}^{r} q^C(\alpha_k)\xi_k.$$

Since each $q^C(\alpha_k)$ is nonzero, this is an element of $\tilde V(\boldsymbol{\lambda})^M$ of the type that appears in Proposition 2.18.

We can write $\eta^C_{\boldsymbol{\lambda},M}$ as $d\log F^C_{\boldsymbol{\lambda},M}$, where $F^C_{\boldsymbol{\lambda},M}$ is a multivalued function (univalued if the exponents are integral) given by the product

$$\prod_{\substack{i\in M\\ 1\le\nu\le n}}(t_i-z_\nu)^{C(\alpha_{\bar i},\lambda_\nu)}\prod_{i,j\in M,i\neq j}(t_i-t_j)^{-C(\alpha_{\bar i},\alpha_{\bar j})/2}\prod_{\nu<\mu}(z_\nu-z_\mu)^{-C(\lambda_\nu,\lambda_\mu)}.$$

We consider the first order differential operator $d^C := d-\eta^C_{\boldsymbol{\lambda},M}$. So d^C is the ordinary exterior derivative for the multivalued polydifferentials after they get formally multiplied by the inverse of $F^C_{\boldsymbol{\lambda},M}$: $d^C = F^C_{\boldsymbol{\lambda},M}\, d\,(F^C_{\boldsymbol{\lambda},M})^{-1}$. In particular, a d^C-closed form is locally $F^C_{\boldsymbol{\lambda},M}$ times a d-closed form. This amounts to turning the trivial line bundle over $U_{n,M}$ into a local system $\mathbb{L}^C_{\boldsymbol{\lambda},M}$: it is the local system for which $F^C_{\boldsymbol{\lambda},M}$ defines a flat (multivalued) section, in other words, $\mathbb{L}^C_{\boldsymbol{\lambda},M}$ is the structure sheaf of $\mathcal{O}_{U_{n,M}}$ endowed with the connection for which d^C is covariant derivation.

Let $\rho\in\mathfrak{h}^*$ be, as usual in Lie theory, defined by the property that $\rho(\check\alpha_k)=1$ for $k=1,\dots,r$ so that $C(\rho,\alpha_k)=q^C(\alpha_k)$ for every k.

Lemma 3.2. *The differential $\eta^C_{\boldsymbol{\lambda},M}$ has a logarithmic pole along each irreducible component of $\Delta_{n,M}$ and we have*

$$\begin{aligned}
-\operatorname{Res}_{\Delta(X)}\eta^C_{\boldsymbol{\lambda},M} &= q^C(\rho-\alpha_X)-q^C(\rho),\\
-\operatorname{Res}_{\Delta_\infty(X)}\eta^C_{\boldsymbol{\lambda},M} &= q^C(\rho+\alpha_X)-q^C(\rho),\\
-\operatorname{Res}_{\Delta_\nu(X)}\eta^C_{\boldsymbol{\lambda},M} &= q^C((\rho+\lambda_\nu)-\alpha_X)-q^C(\rho+\lambda_\nu),
\end{aligned}$$

where $\alpha_X:=\sum_{x\in X}\alpha_{\bar{x}}$.

Proof. Since $\Delta(X)$ amounts to the confluence of the members of X, its generic point may be described in terms of $\mathbf{P}^M\times U_n$ as the blow up of the diagonal obtained by setting all t_x, $x\in X$, equal to each other (followed by dividing out the action of the translation group of $\mathbf{A}$). The irreducible components of $D_{n,M}$ that pass through this diagonal are defined by $t_x=t_y$, where $\{x,y\}$ runs over the two-element subsets of X. The defining formula for $\eta^C_{\boldsymbol{\lambda},M}$ shows that $\operatorname{Res}_{(t_x=t_y)}\eta^C_{\boldsymbol{\lambda},M}=-\frac{1}{2}C(\alpha_{\bar{x}},\alpha_{\bar{y}})$. We thus find that

$$\begin{aligned}
-\operatorname{Res}_{\Delta(X)}\eta^C_{\boldsymbol{\lambda},M} &= \tfrac{1}{2}\sum_{x\neq y}C(\alpha_{\bar{x}},\alpha_{\bar{y}})\\
&= \tfrac{1}{2}C\Big(\sum_{x\in X}\alpha_{\bar{x}},\sum_{y\in X}\alpha_{\bar{y}}\Big)-\tfrac{1}{2}\sum_{x\in X}C(\alpha_{\bar{x}},\alpha_{\bar{x}})\\
&= \tfrac{1}{2}C(\alpha_X,\alpha_X)-\sum_{x\in X}C(\rho,\alpha_X)=q^C(\alpha_X-\rho)-q^C(\rho).
\end{aligned}$$

In the case $\Delta_\infty(X)$, we also need to include the irreducible components of $D_{n,M}$ defined by $t_x=\infty$, $x\in X$. A straightforward computation shows that $-\operatorname{Res}_{(t_x=\infty)}\eta^C_{\boldsymbol{\lambda},M}=C(\alpha_{\bar{x}},\alpha_{\bar{x}})$ and so we get as additional term $\sum_{x\in X}C(\alpha_{\bar{x}},\alpha_{\bar{x}})=2C(\rho,\alpha_X)$. This yields $\frac{1}{2}C(\alpha_X,\alpha_X)+\sum_{x\in X}C(\rho,\alpha_X)=q^C(\alpha_X+\rho)-q^C(\rho)$, as asserted. Finally, for $\Delta_\nu(X)$, with $\nu=1,\dots,n$, we need to subtract the residues for the divisors $t_x=z_\nu$, that is $\sum_{x\in X}C(\alpha_{\bar{x}},\lambda_\nu)=C(\alpha_X,\lambda_\nu)$ and this gives the last value. Q.E.D.

The associated *Aomoto complex* is the relative De Rham complex of $\mathcal{O}_{U_n}$-modules $(p^+_*\Omega^\bullet_{U^+_{n,M}/U_n}(\log\Delta_{n,M}),d^C)$. Note that since logarithmic forms are d-closed, the relative differential d^C is simply given by the wedge product with $-\xi^C_{\boldsymbol{\lambda},M}$. Proposition 2.18 then tells us that:

Lemma 3.3. *We have a natural isomorphism of $\mathcal{O}_{U_n}$-modules*

$$\mathcal{O}_{U_n}\otimes_{\mathbb{C}}\tilde{V}(\boldsymbol{\lambda})_{0,\mathrm{coprim}}\cong\mathcal{H}^m(p^+_*\Omega^\bullet_{U^+_{n,M}/U_n}(\log\Delta_{n,M}),d^C)\otimes_{\mathfrak{S}(M_\bullet)}\mathrm{or}(M).$$

In particular, $\mathcal{O}_{U_n} \otimes_{\mathbb{C}} V(\boldsymbol{\lambda})^{\mathfrak{g}}$ *embeds in the latter.*

The irreducible components of $\Delta_{n,M}$ along which the local system $\mathbb{L}^C_{\boldsymbol{\lambda},M}$ has trivial monodromy are those on which the residue $\eta^C_{\boldsymbol{\lambda},M}$ is an integer. These play a special role in of a theorem of Esnault, Schechtman and Viehweg [3], or rather our refinement ([5], [6]) which leads to a topological interpretation of the Aomoto cohomology. For this purpose and for later uses, we pause for a moment to discuss the various natural extensions of a rank on local system across a normal crossing divisor.

Extensions across a normal crossing divisor

Let X be a complex manifold of dimension n, D a simple normal crossing divisor on X and $\mathbb{L}$ a rank one local system on $X-D$. If D is irreducible (hence smooth), then there are two basic ways of extending $\mathbb{L}$ to X in the derived category: extension by zero $R^\bullet j_!\mathbb{L}$ (where $j : X-D \subset X$ denotes the inclusion) and and the full direct image $R^\bullet j_*\mathbb{L}$. These are connected by a morphism $R^\bullet j_!\mathbb{L}(=j_!\mathbb{L}) \to R^\bullet j_*\mathbb{L}$, which on global cohomology this gives the map $H^\bullet(X,D;\mathbb{L}) \to H^\bullet(X-D;\mathbb{L})$. (Since $\mathbb{L}$ is not defined on D, a word of explanation is in order: if T is an open regular neighborhood of D in X relative to the Hausdorff topology so that its boundary ∂T lies in $X-D$, then $H^\bullet(X,D;\mathbb{L})$ is understood as $H^m(X-T,\partial T;\mathbb{L})$, or equivalently, as cohomology with supports: $H^\bullet_\Phi(X-D;\mathbb{L})$, where Φ is the collection of closed subsets of $X-D$ that remain closed in $X-D$.) If the monodromy of $\mathbb{L}$ around D is not the identity, then the two extensions coincide. Shifted Verdier duality converts this morphism of extensions into $R^\bullet j_*\mathbb{L}^\vee \leftarrow j_!\mathbb{L}^\vee$.

Suppose D has two irreducible components D' and D'' and we extended $\mathbb{L}$ across the generic point of each of them so that we have an extension over $X-D'\cap D''$. Then there is a natural extension over all of X which is locally along a transversal slice of $D'\cap D''$ like an exterior product of two extensions as above over the complex unit disk. We can obtain it in stages, for instance, by first doing the D'-extension over $X-D''$ and then the D'' extension over X; the opposite order yields the same result. More generally, if D has several irreducible components, then an extension of $\mathbb{L}$ over X in the derived category is specified once we have done so at the generic points of D and its formation is compatible with shifted Verdier duality.

So if we single out a collection of irreducible components of D along which $\mathbb{L}$ has trivial monodromy and denote its union $D^\sharp$, then we have specified an extension of $\mathbb{L}$ over X: at the generic points of $D^\sharp$ we take the full direct image, and at the other generic points of D we extend by zero. We denote that extension $\mathcal{C}^\bullet(\mathbb{L};D^\sharp)$. The cohomology of this extension is $H^\bullet(X-D^\sharp,D-D^\sharp;\mathbb{L})$. Note however, that adding to $D^\sharp$

irreducible components of D at which $\mathbb{L}$ has nontrivial monodromy does no alter $\mathcal{C}^\bullet(\mathbb{L}; D^\sharp)$ as an object in the derived category of constructible sheaves on X and hence will not affect the cohomology. The shifted Verdier dual of $\mathcal{C}^\bullet(\mathbb{L}; D^\sharp)$ is $\mathcal{C}^\bullet(\mathbb{L}^\vee; D^\flat)$, where $D^\flat$ is the union of the irreducible components of D at which $\mathbb{L}$ has trivial monodromy not in $D^\flat$ (but we could equally well take the union of *all* the irreducible components of D not in $D^\sharp$). So if X is of finite type, then we have a perfect pairing

$$H^k(X, \mathcal{C}^\bullet(\mathbb{L}; D^\sharp)) \otimes H_c^{2n-k}(X, \mathcal{C}^\bullet(\mathbb{L}^\vee; D^\flat)) \to \mathbb{C}.$$

An important example is when $\mathbb{L} \subset \mathcal{O}_{X-D}$ is defined by a closed differential η. Given an irreducible component of D, then $\mathbb{L}$ has trivial monodromy at it if and only if η has there an integral residue. We observed in [5] and [6] that the twisted logarithmic De Rham complex $(\Omega_X^\bullet(\log D), d - \eta)$ represents $\mathcal{C}^\bullet(\mathbb{L}^C_{\boldsymbol{\lambda},M}, D^{\geq 0})$, where $D^{\geq 0}$ is the union of the irreducible components of D where η has residue a nonnegative integer.

A Gauß–Manin connection

The next proposition appears in [5] and [6], albeit that it is stated there in an absolute setting.

Proposition 3.4. *Denote by $\Delta^{\geq 0}_{n,M}$ the union of the irreducible components of $\Delta_{n,M}$ where $\eta^C_{\boldsymbol{\lambda},M}$ has residue a nonnative integer. Then we have a natural identification of $\mathcal{O}_{U_n}$-modules*

$$\mathcal{H}^m(p^+_*\Omega^\bullet_{U^+_{n,M}/U_n}(\log \Delta_{n,M}), d^C) \cong \mathcal{O}_{U_n} \otimes R^m p^+_* \mathcal{C}^\bullet(\mathbb{L}^C_{\boldsymbol{\lambda},M}, \Delta^{\geq 0}_{n,M}).$$

Proof. As noted above, the complex $(\Omega^\bullet_{U^+_{n,M}/U_n}(\log \Delta_{n,M}), d^C)$ represents $\mathcal{C}^\bullet(\mathbb{L}^C_{\boldsymbol{\lambda},M}, \Delta^{\geq 0}_{n,M})$. The first lemma of Section 2 of [3] asserts that the direct image $R^q p^+_* \Omega^\bullet_{U^+_{n,M}/U_n}(\log \Delta_{n,M})$ is zero unless $q = 0$. The proposition now follows by taking the mth direct image on U_n. Q.E.D.

Note that the stalk at $\mathbf{z} \in U_n$ of the sheaf that appears in the right hand side of the preceding proposition is equal to the cohomology space $H^m(U^+_{n,M}(\mathbf{z}) - \Delta^{\geq 0}_{n,M}, \Delta_{n,M} - \Delta^{\geq 0}_{n,M}; \mathbb{L}^C_{\boldsymbol{\lambda},M})$. The pair $(U^+_{n,M}, \Delta_{n,M})$ is topologically locally trivial over U_n and so $R^m p^+_* \mathcal{C}^\bullet(\mathbb{L}^C_{\boldsymbol{\lambda},M}, \Delta^{\geq 0}_{n,M})$ is a local system. We conclude that the flat connection d^C on $\mathcal{O}_{U_{M,n}}$ induces one on $\mathcal{O}_{U_n} \otimes R^m p^+_* \mathcal{C}^\bullet(\mathbb{L}^C_{\boldsymbol{\lambda},M}, \Delta^{\geq 0}_{n,M})$ and (via Proposition 3.4) one on the $\mathcal{O}_{U_n}$-module $\mathcal{H}^m(p^+_*\Omega^\bullet_{U^+_{n,M}/U_n}(\log \Delta_{n,M}), d^C)$. We will refer to this connection as the *Gauß–Manin connection* and denote it by ∇_{GM}.

Corollary 3.5. *We have a natural identification*

$$\mathcal{O}_{U_n} \otimes_{\mathbb{C}} \tilde{V}(\boldsymbol{\lambda})_{0,\mathrm{coprim}} \cong \mathcal{O}_{U_n} \otimes R^m p_*^+ \mathcal{C}^\bullet(\mathbb{L}^C_{\boldsymbol{\lambda},M}, \Delta^{\geq 0}_{n,M}) \otimes_{\mathfrak{S}(M_\bullet)} \mathrm{or}(M)$$

as $\mathcal{O}_{U_n}$-modules. (So this makes $\mathcal{O}_{U_n} \otimes V(\boldsymbol{\lambda})^{\mathfrak{g}}$ a trivial subbundle of the right hand side.)

The KZ-connection is a Gauß–Manin connection

Here is the main result:

Theorem 3.6. *The embedding of $\mathcal{O}_{U_n} \otimes V(\boldsymbol{\lambda})^{\mathfrak{g}}$ endowed with the KZ-connection ∇^C_{KZ} in $\mathcal{H}^m(p_*^+\Omega^\bullet_{U^+_{n,M}/U_n}(\log \Delta_{n,M}), d^C)$ endowed with the GM-connection ∇_{GM} is flat and hence induces an embedding of local systems $\mathbb{KZ}^C(\boldsymbol{\lambda}) \hookrightarrow R^m p_*^+ \mathcal{C}^\bullet(\mathbb{L}^C_{\boldsymbol{\lambda},M}, \Delta^{\geq 0}_{n,M}) \otimes_{\mathfrak{S}(M_\bullet)} \mathrm{or}(M)$.*

Before we begin the proof, we show how a bootstrap procedure turns this theorem into a more precise result.

Theorem 3.7. *Let $\Delta^{>0}_{n,M}$ denote the union of irreducible components of $\Delta_{n,M}$ along which the residue of $\eta^C_{\lambda,M}$ is a positive integer. Then $\mathbb{KZ}^C(\boldsymbol{\lambda})$ can be canonically identified with the isotypical part for the sign character of $\mathfrak{S}(M_\bullet)$ of the image of*

$$R^m p_*^+ \mathcal{C}^\bullet(\mathbb{L}^C_{\boldsymbol{\lambda},M}, \Delta^{>0}_{n,M}) \to R^m p_*^+ \mathcal{C}^\bullet(\mathbb{L}^C_{\boldsymbol{\lambda},M}, \Delta^{\geq 0}_{n,M}).$$

For clarity we note that the above map is at $\mathbf{z} \in U_n$ the natural map of cohomology spaces with support

$$H^m_{\Phi^{>0}}(U_{n,M}(\mathbf{z}); \mathbb{L}^C_{\boldsymbol{\lambda},M}) \to H^m_{\Phi^{\geq 0}}(U_{n,M}(\mathbf{z}); \mathbb{L}^C_{\boldsymbol{\lambda},M}),$$

where $\Phi^{>0}$ resp. $\Phi^{\geq 0}$ is the family of closed subsets of $U_{n,M}(\mathbf{z})$ which remain closed in $U^+_{n,M}(\mathbf{z}) - \Delta^{>0}_{n,M}$ resp. $U^+_{n,M}(\mathbf{z}) - \Delta^{\geq 0}_{n,M}$.

Proof. Theorem 3.6 yields an embedding

$$\mathbb{KZ}^C(\boldsymbol{\lambda}) \hookrightarrow R^m p_*^+ \mathcal{C}^\bullet(\mathbb{L}^C_{\boldsymbol{\lambda},M}, \Delta^{\geq 0}_{n,M}) \otimes_{\mathfrak{S}(M_\bullet)} \mathrm{or}(M).$$

Next we note that the local system dual to $\mathbb{KZ}^C(\boldsymbol{\lambda})$ is $\mathbb{KZ}^{-C}(\boldsymbol{\lambda}')$, where the prime $'$ is the *canonical involution* of $\mathfrak{h}^*$, given as $-w_o$, where w_o is the Weyl group element that maps the fundamental chamber to its opposite. This involution preserves the simple roots and (hence) the dominant weights. With this notation, the involution applied to the identity $\lambda_1 + \cdots + \lambda_n = m_1\alpha_1 + \cdots + m_r\alpha_r$ yields $\lambda'_1 + \cdots + \lambda'_n = m_1\alpha'_1 + \cdots + \alpha'_r$ and so the data that served us for $\mathbb{KZ}^C(\boldsymbol{\lambda})$ are for $\mathbb{KZ}^{-C}(\boldsymbol{\lambda}')$ given by $-C$ and the composite π' of $\pi : M \to \{1, \ldots, r\}$

with the involution (also denoted by $'$) of $\{1,\dots,r\}$ that is given by $\alpha'_k = \alpha_{k'}$. We have thus defined $\eta^{-C}_{\boldsymbol{\lambda},\pi'}$. The definition then shows that $\eta^{-C}_{\boldsymbol{\lambda}',\pi'} = -\eta^{C}_{\boldsymbol{\lambda},M}$ so that $\mathbb{L}^{-C}_{\boldsymbol{\lambda}',\pi'}$ may be identified with the dual of $\mathbb{L}^{C}_{\boldsymbol{\lambda},M}$. Let us apply Corollary 3.5 to the triple $(-C,\boldsymbol{\lambda}',\pi')$: we get a natural embedding of local systems

$$\mathbb{KZ}^C(\boldsymbol{\lambda})^\vee \hookrightarrow R^m p^+_* \mathcal{C}^\bullet(\mathbb{L}^C_{\boldsymbol{\lambda},M}, \Delta^{\le 0}_{n,M}) \otimes_{\mathfrak{S}(M_\bullet)} \mathrm{or}(M).$$

By dualizing we obtain a surjection

$$R^m p^+_* \mathcal{C}^\bullet(\mathbb{L}^C_{\boldsymbol{\lambda},M}, \Delta^{>0}_{n,M}) \otimes_{\mathfrak{S}(M_\bullet)} \mathrm{or}(M) \twoheadrightarrow \mathbb{KZ}^C(\boldsymbol{\lambda})$$

of local systems. It remains to observe that the composite of the two relevant displays is the natural map. Q.E.D.

We thus get a genuine topological characterization of the KZ-system. For example, if $\eta^C_{\boldsymbol{\lambda},M}$ has no nonzero integral residues, then we find that $\mathbb{KZ}^C(\boldsymbol{\lambda})_{\mathbf{z}}$ can be identified with isotypical subspace of the sign character of the image of $H^m_c(U_{n,M}(\mathbf{z}),\mathbb{L}^C_{\boldsymbol{\lambda},M}) \to H^m(U_{n,M}(\mathbf{z}),\mathbb{L}^C_{\boldsymbol{\lambda},M})$. If in addition C is defined over $\mathbb{R}$, then $\mathbb{L}^C_{\boldsymbol{\lambda},M}$ has flat unitary metric (that gives $F^C_{\lambda,M}$ norm one) and there results hermitian (intersection) form on this image, which is known to be nondegenerate. This puts on $\mathbb{KZ}^C(\boldsymbol{\lambda})$ a (flat) nondegenerate hermitian form.

Remark 3.8. *Theorem 3.6 tells us what the monodromy of the KZ system is like. If we fix a base point* $\mathbf{z} \in U_n$, *then* $\pi_1(U_n,\mathbf{z})$ *is the colored braid group with* n *strands. It acts on domain and range of the linear map* $H^m_{\Phi>0}(U_{n,M}(\mathbf{z});\mathbb{L}^C_{\boldsymbol{\lambda},M}) \to H^m_{\Phi\ge 0}(U_{n,M}(\mathbf{z});\mathbb{L}^C_{\boldsymbol{\lambda},M})$ *in a manner that makes the map equivariant. It should be worthwhile to investigate such representations in their own right and perhaps make contact with the Kohno–Drinfeld approach via the representation theory of quantum groups. We further note that since the KZ system embeds in a variation of complex mixed Hodge structure, it acquires a (flat) weight filtration. It should be interesting to determine that filtration in terms of the KZ data.*

When the Casimir element is defined over $\mathbb{Q}$

In case C is defined over $\mathbb{Q}$ in the sense that q^C takes rational values on the roots, then $R^m\mathcal{C}^\bullet(\mathbb{L}^C_{\boldsymbol{\lambda},M})$ is a eigen subsystem of a finite cyclic group acting on an ordinary variation of mixed Hodge structure. To be precise, let s be the smallest common denominator of these residues. Then the monodromy of $\mathbb{L}^C_{\boldsymbol{\lambda},M}$ is the group $\boldsymbol{\mu}_s$ of sth roots of unity. It determines an unramified $\boldsymbol{\mu}_s$-cover $\hat{U}_{n,M} \to U_{n,M}$, so that the pull-back of $\mathbb{L}^C_{\boldsymbol{\lambda},M}$ becomes trivial. This means that we may now $F^C_{\boldsymbol{\lambda},M}$ regard as a univalued (invertible) holomorphic function on $\hat{U}_{n,M}$. Its normalization over $U^+_{n,M}$, $\hat{U}^+_{n,M} \to U^+_{n,M}$, is a $\boldsymbol{\mu}_s$-cover of $U^+_{n,M}$ that may have singularities, but these are quotient singularities and hence for our purposes of an innocent nature. The function $F^C_{\boldsymbol{\lambda},M}$ is meromorphic on it and the order of $F^C_{\boldsymbol{\lambda},M}$ along an irreducible component of $\hat{\Delta}_{n,M}$ is given by s times the residue of $\xi^C_{\boldsymbol{\lambda},M}$ along its image in $\Delta_{n,M}$ (indeed an integer).

If $\hat{p}^+ : \hat{U}^+_{n,M} \to U_n$ denotes the projection, then let $\hat{\Delta}^{\geq 0}_{n,M}, \hat{\Delta}^{>0}_{n,M}, \dots$ have the obvious meaning. Then for $\mathbf{z} \in U_n$,

$$(R^m_{\hat{\Phi}\geq 0}\hat{p}_*\mathbb{C}_{U^+_{n,M}})_{\mathbf{z}} = H^m(\hat{U}^+_{n,M} - \hat{\Delta}^{\geq 0}_{n,M}, \hat{\Delta}_{n,M} - \hat{\Delta}^{\geq 0}_{n,M}; \mathbb{C})$$

and $R^m_{\hat{\Phi}\geq 0}\hat{p}_*\mathbb{C}_{U^+_{n,M}}$ comes with the structure of a variation of polarized mixed Hodge structure. There is now a finite group $\hat{\mathfrak{S}}(M_\bullet)$ acting on $\hat{U}^+_{n,M}$ which is an extension of $\mathfrak{S}(M_\bullet)$ by the covering group μ_s. It has a character $\hat{\chi}$ that is tautological on μ_s and lifts the sign character. On the level of stalks this yields the identification of $\mathbb{KZ}^C(\boldsymbol{\lambda})_{\mathbf{z}}$ with the $\hat{\chi}$-isotypical space of the image of $R^m_{\hat{\Phi}>0}\hat{p}_*\mathbb{C}_{U^+_{n,M}} \to R^m_{\hat{\Phi}\geq 0}\hat{p}_*\mathbb{C}_{U^+_{n,M}}$.

Proof of Theorem 3.6

We begin the proof by recalling covariant derivation relative to the Gauß–Manin connection. Covariant derivation with respect to z_ν is exhibited on the form level by Lie derivation of a lift of this vector field to $U_{n,M}$. In order to ensure that logarithmicity is preserved we take a lift that depends on the argument:

Lemma 3.9. *Let $\zeta_{\mathbf{I}}(\mathbf{z}) = \zeta_{I^1}(z_1).\zeta_{I^2}(z_2).\cdots.\zeta_{I^n}(z_n)$ be a basis element of $\mathcal{B}_{n,M}$ and let $\tilde{\partial}_\nu := \partial_\nu + \sum_{i\in\{I^\nu\}}\frac{\partial}{\partial t_i}$ (a vector field on $\mathbf{P}^M \times U_n$ that lifts the vector field ∂_ν to U_n). Then the Lie derivative $\mathcal{L}^C_{\tilde{\partial}_\nu} := d^C\iota_{\tilde{\partial}_\nu} + \iota_{\tilde{\partial}_\nu}d^C$ maps $\zeta_{\mathbf{I}}$ to $-\xi^C_{\boldsymbol{\lambda},M}(\tilde{\partial}_\nu).\omega$ and the latter lies in $\mathbb{C}[U_n]\otimes_{\mathbb{C}}\mathcal{B}_{n,M}$. This map is $\mathfrak{S}_\bullet$-equivariant and defines a connection on $\mathcal{B}_{n,M}$ with logarithmic pole whose form A^C_{GM} lies in*

$$\sum_{\nu<\mu}\frac{d(z_\nu - z_\mu)}{z_\nu - z_\mu}\otimes_{\mathbb{C}}\mathrm{End}_{\mathbf{A}^n}(\mathcal{B}_{n,M}).$$

We shall refer to this as the Gauß–Manin connection.

Proof. We first notice that $\zeta_{\mathbf{I}}$ is invariant under the flow generated by $\tilde{\partial}_\nu$ (which adds to the coordinates $(z_\nu, (t_i)_{i\in\{I^\nu\}})$ the same complex number), in other words, $\mathcal{L}_{\tilde{\partial}_\nu}(\zeta_{\mathbf{I}}) = 0$. Hence

$$\mathcal{L}^C_{\tilde\partial_\nu}(\zeta_{\mathbf{I}}) = d^C\iota_{\tilde\partial_\nu}\zeta_{\mathbf{I}} + \iota_{\tilde\partial_\nu}d^C\zeta_{\mathbf{I}} = (d\iota_{\tilde\partial_\nu}\zeta_{\mathbf{I}} - \eta^C_{\boldsymbol{\lambda},M}\iota_{\tilde\partial_\nu}\zeta_{\mathbf{I}}) + \iota_{\tilde\partial_\nu}(d\zeta_{\mathbf{I}} - \eta^C_{\boldsymbol{\lambda},M}\zeta_{\mathbf{I}})$$
$$= \mathcal{L}_{\tilde\partial_\nu}(\zeta_{\mathbf{I}}) - \eta^C_{\boldsymbol{\lambda},M}\iota_{\tilde\partial_\nu}\zeta_{\mathbf{I}} - \iota_{\tilde\partial_\nu}(\eta^C_{\boldsymbol{\lambda},M}\zeta_{\mathbf{I}}) = -\eta^C_{\boldsymbol{\lambda},M}(\tilde\partial_\nu)\zeta_{\mathbf{I}}.$$

For $i,j \in \{I^\nu\}$, the differentials $dt_i - dt_j$ and $dt_i - dz_\nu$ clearly vanish on $\tilde\partial_\nu$ and so

$$-\eta^C_{\boldsymbol{\lambda},M}(\tilde\partial_\nu) = \sum_{x\notin I^\nu}\frac{C(\alpha_{\bar x},\lambda_\nu)}{t_x - z_\nu} - \sum_{i\in\{I^\nu\}}\sum_{\mu\neq\nu}\frac{C(\alpha_{\bar i},\lambda_\mu)}{t_i - z_\mu}$$
$$+ \sum_{i\in\{I^\nu\}}\sum_{x\in\{I^\mu\},\mu\neq\nu}\frac{C(\alpha_{\bar i},\alpha_{\bar x})}{t_i - t_x} + \sum_{\mu\neq\nu}\frac{C(\lambda_\nu,\lambda_\mu)}{z_\nu - z_\mu}.$$

Lemma's 1.6 and 1.8 show that multiplication of $\zeta_{\mathbf{I}}$ by a factor $(t_i - z_\mu)^{-1}$, $(t_x - z_\nu)^{-1}$ or $(t_i - t_x)^{-1}$ ($i \in \{I^\nu\}$, $x \in \{I^\mu\}$, $\mu \neq \nu$) lands in $\frac{1}{z_\nu - z_\mu}\mathcal{B}_n$. Q.E.D.

As is well-known (and easy to prove), C has the form

$$C = C_0 + \sum_\alpha C_\alpha,$$

with $C_0 \in \mathfrak{h}\otimes\mathfrak{h}$ and $C_\alpha \in \mathfrak{g}_\alpha \otimes \mathfrak{g}_{-\alpha}$, where the sum is over all the roots. Here C_0 can be any symmetric tensor invariant under the Weyl group; it then determines C. Since C is symmetric, $C_{-\alpha}$ is the transpose of C_α.

We put

$$C_+ := \sum_{\alpha>0} C_\alpha \in \prod_{\alpha>0}\mathfrak{g}_\alpha\otimes\mathfrak{g}_{-\alpha} \quad\text{and}\quad C_- := \sum_{\alpha<0} C_\alpha \in \prod_{\alpha<0}\mathfrak{g}_\alpha\otimes\mathfrak{g}_{-\alpha},$$

so that $C = C_+ + C_0 + C_-$. It is easy to check that C_0 acts semisimply in the tensor product of highest weight representations. In fact, for $\chi,\chi' \in \mathfrak{h}^*$, C_0 acts on $V(\lambda)_\chi \otimes V(\lambda')_{\chi'}$ as multiplication by $C_0(\chi,\chi')$. For the proof of Theorem 3.6, we also need a better understanding of C_+. The following lemma is essentially Lemma 7.6.3 of Schechtman and Varchenko [8], and so we omit its proof.

Lemma 3.10. *Let $\tilde C_+ : V(\lambda) \to V(\lambda)\otimes\mathfrak{g}$ be the linear map given by*

$$\tilde C_+(f_S 1_\lambda) = \sum_{\emptyset\neq T\leq S} C(\alpha_{\ell(T)}, \lambda - \alpha_{S_{>\ell(T)}}) f_{S-T}1_\lambda \otimes [f_T],$$

where $\ell(T)$ denotes the last term of T and the sum is taken over all nonempty subsequences of T of S and $S_{>\ell(T)}$ is the largest common tail of

S and $S-T$ (which of course may be empty). If V is any representation of $\mathfrak{g}$, then the action of C_+ on $V(\lambda)\otimes V$ satisfies

$$C_+(f_S 1_\lambda \otimes v) = \tilde{C}_+(f_S)(1_\lambda \otimes v).$$

Proof of Theorem 3.6. In view of the shape of the connections, it suffices to verify this in case $n=2$. We begin working out the computation in the proof of Lemma 3.9 in case $n=2$, $\nu=1$ (so that $\mu=1$). We write (z,w) for (z_1,z_2), (λ,μ) for (λ_1,λ_2), (I,J) for (I^1,I^2) and ζ for $\zeta_I(z)\otimes\zeta_J(w)$.

We have for $\tilde{\partial} = \frac{\partial}{\partial z_\nu} + \sum_{i\in\{I\}}\frac{\partial}{\partial t_i}$:

$$-\eta^C_{\boldsymbol{\lambda},M}(\tilde{\partial}) = \sum_{j\in\{J\}}\frac{C(\alpha_{\bar{j}},\lambda)}{t_j-z} - \sum_{i\in\{I\}}\frac{C(\alpha_{\bar{i}},\mu)}{t_i-w} + \sum_{\substack{i\in\{I\}\\ j\in\{J\}}}\frac{C(\alpha_{\bar{i}},\alpha_{\bar{j}})}{t_i-t_j} + \frac{C(\lambda,\mu)}{z-w}$$

and so

$$\begin{aligned}(z-w)\eta^C_{\boldsymbol{\lambda},M}(\tilde{\partial})\zeta = &-\sum_{j\in\{J\}} C(\alpha_{\bar{j}},\lambda)\frac{w-z}{t_j-z}\zeta_I(z)\otimes\zeta_J(w)\\ &-\sum_{i\in\{I\}} C(\alpha_{\bar{i}},\mu)\frac{z-w}{t_i-w}\zeta_I(z)\otimes\zeta_J(w)\\ &+\sum_{i\in\{I\},j\in\{J\}} C(\alpha_{\bar{i}},\alpha_{\bar{j}})\frac{z-w}{t_i-t_j}\zeta_I(z)\otimes\zeta_J(w) - C(\lambda,\mu)\zeta.\end{aligned}$$

We develop these terms with the help of Lemmas 1.6 and 1.8 and get

$$\begin{aligned}&\sum_{\substack{I=I''iI'\\ J=J''jJ'}} C(\alpha_{\bar{i}},\alpha_{\bar{j}})\cdot\Big(\zeta - \sum_{I''=I_2I_1}(-1)^{|I_2|}\omega_{I_1}(z)\otimes\omega_{I_2^*i}\omega_{I''i}\frac{dt_i}{t_i-t_j}\zeta_J(w)\\ &\qquad - \sum_{J'=J_2J_1}(-1)^{|J_2|}\omega_{J_2^*j}\omega_{J''j}\frac{dt_j}{t_j-t_i}\zeta_I(z)\otimes\zeta_{J_1}(w)\Big)\\ &-\sum_{I=I''iI'} C(\alpha_{\bar{i}},\mu)\Big(\zeta - \sum_{I'=I_2I_1}(-1)^{|I_2|}\zeta_{I_1}(z)\otimes\omega_{I_2^*i}\zeta_{I''i}(w)\zeta_J(w)\Big)\\ &-\sum_{J=J''jJ'} C(\alpha_{\bar{j}},\lambda)\Big(\zeta - \sum_{J'=J_2J_1}(-1)^{|J_2|}\omega_{J_2^*j}\omega_{J''j}(z)\zeta_I(z)\otimes\zeta_{J_1}(w)\Big)+\\ &\qquad\qquad + C(\lambda,\mu),\end{aligned}$$

which after collecting terms becomes

$$C(\lambda - \sum_{i\in\{I\}} \alpha_{\bar{i}}, \mu - \sum_{j\in\{J\}} \alpha_{\bar{j}})\zeta$$

$$+ \sum_{I=I''iI_2I_1} (-1)^{|I_2|}\zeta_{I_1}(z)\otimes\omega_{I_2^*i}\omega_{I''i}\Big(\sum_{j\in\{J\}} \frac{C(\alpha_{\bar{i}},\alpha_{\bar{j}})}{t_i - t_j} - \frac{C(\alpha_{\bar{i}},\mu)}{t_i - w}\Big) dt_i\zeta_J(w)$$

$$+ \sum_{J=J''jJ_2J_1} (-1)^{|J_2|}\omega_{J_2^*j}\omega_{J''j}\cdot\Big(\sum_{i\in N} \frac{C(\alpha_{\bar{i}},\alpha_{\bar{j}})}{t_j - t_i} - \frac{C(\lambda,\alpha_{\bar{j}})}{t_j - z}\Big) dt_j\zeta_I(z)\otimes\zeta_{J_1}(w).$$

Since

$$\Big(\sum_{j\in\{J\}} \frac{C(\alpha_{\bar{i}},\alpha_{\bar{j}})}{t_i - t_j} - \frac{C(\alpha_{\bar{i}},\mu)}{t_i - w}\Big) dt_i\zeta_J(w) = -\tfrac{1}{2}C(\alpha_{\bar{i}},\alpha_{\bar{i}})\Phi_i(\zeta_J(w)),$$

we may also write the previous expression as $\Gamma_0(\zeta) + \Gamma_+(\zeta_I(z))\zeta_J(w) + \Gamma_-(\zeta_J(w))\zeta_I(z)$ with $\Gamma_0(\zeta) = C_0(\lambda - \sum_{i\in I}\alpha_{\bar{i}}, \mu - \sum_{j\in J}\alpha_{\bar{j}})\zeta$ and where $\Gamma_+(\zeta_I(z))$ is the operator from $\mathcal{B}$ to $\mathcal{B}\otimes\mathcal{B}.1_\mu$ defined by

$$\Gamma_+(\zeta_I(z)) = \sum_{I=I''iI_2I_1} (-1)^{|I_2|}\tfrac{1}{2}C(\alpha_{\bar{i}},\alpha_{\bar{i}})\zeta_{I_1}(z)\otimes\omega_{I''i}\omega_{I_2^*i}\Phi_i,$$

and Γ_- is its transpose. It follows from Corollary 5.2 of the appendix that Γ_+ defines a linear map $\Gamma : V(\lambda) \to V(\lambda)\otimes_{\mathbb{C}}\mathfrak{g}$ that coincides with the map $\tilde{C}_+$ that appears in Lemma 3.10. Since Γ_- resp. C_- is the transpose of Γ_+ resp. C_+, we conclude that identity we were after indeed holds: $A^C_{GM} = C_{1,2}\otimes d(z-w)/(z-w) = A^C_{KZ}$. Q.E.D.

§4. The WZW-system

Our discussion of the case when C is defined over $\mathbb{Q}$ covers one of particular interest, namely the one for which is defined the WZW-subsystem of a given level, where it is assumed that $\mathfrak{g}$ is simple and finite dimensional. We recall its definition. Let $\tilde{\alpha}\in\mathfrak{h}^*$ be the highest root relative to the root basis $(\alpha_1,\dots,\alpha_r)$ and let $\tilde{\alpha}^\vee$ be the associated coroot. We fix a generator e of the (one dimensional) root space $\mathfrak{g}_{\tilde{\alpha}}$ and define an $\mathcal{O}_{U_n}$-linear endomorphism $\mathcal{E}$ of $\mathcal{O}_{U_n}\otimes V(\boldsymbol{\lambda})$ by

$$\mathcal{E}(\mathbf{z}) = \sum_{\nu=1}^{n} 1\otimes\cdots\otimes 1\otimes(z_\nu e)\otimes 1\otimes\cdots\otimes 1,$$

where $z_\nu e$ is acting on $V(\lambda_\nu)$. So if for all ν, $z_\nu \neq 0$ (a property we can arrange for by doing a translation in $\mathbf{A}$), and we let $\mathfrak{g}$ act on $V(\lambda_\nu)$ by modifying the given action in terms of the scalar z_ν: $e(z_\nu)_k := z_\nu e_k$ resp. $f(z_\nu)_k := z_\nu^{-1} f_k$, then $\mathcal{E}(\mathbf{z})$ acts on $V(\boldsymbol{\lambda})$ as e.

Lemma 4.1. *Let I be a sequence in $\mathcal{I}$ representing the highest root. Then for a suitable choice of e, we have that for every $\zeta \in \mathcal{O}_{U_n} \otimes V(\boldsymbol{\lambda})$, $(\mathcal{E}(\zeta))_{\mathcal{I}-\{I\}} = \operatorname{Res}_{t_I=\infty} t_I \operatorname{Res}_I$.*

Proof. Let $\zeta \in V(\boldsymbol{\lambda})$ be of the form $\zeta(\mathbf{z}) = \zeta_1(z_1)\cdots\zeta_n(z_n)$ with $\zeta_\nu \in V(\lambda_\nu)$. Then $e(\zeta_\nu(z_\nu))_{\mathcal{I}-\{I\}} = \operatorname{Res}_{(t_I=z_\nu)} \operatorname{Res}_I \zeta_\nu(z_\nu)$ by Corollary 2.9 and so $z_\nu e(\zeta_\nu(z_\nu))_{\mathcal{I}-\{I\}} = \operatorname{Res}_{(t_I=z_\nu)} t_I \operatorname{Res}_I \zeta_\nu(z_\nu)$. It follows that $\mathcal{E}(\zeta(\mathbf{z}))_{\mathcal{I}-\{I\}} = \sum_{\nu=1}^n \operatorname{Res}_{(t_I=z_\nu)} t_I \operatorname{Res}_I \zeta(\mathbf{z}) = -\operatorname{Res}_{(t_I=\infty)} t_I \operatorname{Res}_I \zeta(\mathbf{z})$. Q.E.D.

Let ℓ be a fixed positive integer. We say that a representation V of $\mathfrak{g}$ *is of level* $\leq \ell$ if $e^{1+\ell}$ is the zero endomorphism in V. For $V = V(\lambda)$, with λ dominant integral, this means that $\lambda(\tilde{\alpha}^\vee) \leq \ell$. In what follows we assume that our $V(\lambda_1), \dots, V(\lambda_n)$ are all of level $\leq \ell$. According to Corollary 2.9 this amounts to the condition that for any $(1+\ell)$-tuple of sequences $(I_0, \dots, I_\ell)$ in $\mathcal{I}$ representing the highest root, we have

$$\operatorname{Res}_{(t_{I_0}=z_\nu)} \cdots \operatorname{Res}_{(t_{I_\ell}=z_\nu)} \operatorname{Res}_{I_0} \operatorname{Res}_{I_1} \cdots \operatorname{Res}_{I_\ell} \zeta = 0, \quad (\nu = 1, \dots, n).$$

The *Verlinde space of level* ℓ is defined in a setting which involves a punctured compact Riemann surface as its 'continuous input' so that over the moduli space of such punctured Riemann surfaces these spaces make up a vector bundle, the so-called *WZW-bundle of level* ℓ. (When this bundle is pulled back to a certain $\mathbb{C}^\times$-bundle over that moduli space, it acquires a natural flat connection.) In case of the Riemann sphere, the sheaf of sections of this bundle (or of its dual, depending on convention) may be obtained as a subbundle $\mathcal{W}_\ell(\boldsymbol{\lambda})$ of $\mathcal{O}_{U_n} \otimes V(\boldsymbol{\lambda})^{\mathfrak{g}}$:

$$\mathcal{W}_\ell(\boldsymbol{\lambda}) := \ker(\mathcal{E}^{1+\ell} | \mathcal{O}_{U_n} \otimes V(\boldsymbol{\lambda})^{\mathfrak{g}}).$$

So its fiber over $\mathbf{z}$ yields the vectors in $V(\boldsymbol{\lambda})^{\mathfrak{g}}$ that generate a representation of level $\leq \ell$ relative to the modified $\mathfrak{g}$-representations on the factors.

We recall that the length of the highest root $\tilde{\alpha}$ is one less than the *Coxeter number* h of $\mathfrak{g}$.

Corollary 4.2. *An element of $\zeta \in \mathcal{O}_{U_n} \otimes V(\boldsymbol{\lambda})^{\mathfrak{g}}$ lies in $\mathcal{W}_\ell(\boldsymbol{\lambda})$ if and only if for any $(1+\ell)$-tuple of sequences $(I_0, \dots, I_\ell)$ in $\mathcal{I}$ with each member of length $h-1$, $\operatorname{Res}_{I_0} \operatorname{Res}_{I_1} \cdots \operatorname{Res}_{I_\ell} \zeta$ vanishes on the diagonal locus defined by $\cup_k \{I_k\} \subset \mathcal{I}$.*

Proof. If I is a finite sequence in $\mathcal{I}$ such that Res_I is nonzero on $V(\lambda_\nu)$, then I has length $\leq h-1$ and in case of equality, I represents the highest root $\tilde{\alpha}$. Lemma 4.1 tells us that $\zeta \in \mathcal{O}_{U_n} \otimes V(\boldsymbol{\lambda})^{\mathfrak{g}}$ lies in

$\mathcal{W}_\ell(\boldsymbol{\lambda})$ if and only if for any $(1+\ell)$-tuple of sequences $(I_1,\dots,I_{1+\ell})$ in $\mathcal{I}$ representing the highest root, then

$$\mathrm{Res}_{(t_{\{I_0\}}=\infty)}\cdots\mathrm{Res}_{(t_{\{I_\ell\}}=\infty)}\big(t_{\{I_0\}}t_{\{I_1\}}\cdots t_{\{I_\ell\}}\mathrm{Res}_{I_0}\mathrm{Res}_{I_1}\cdots\mathrm{Res}_{I_\ell}\,\zeta\big)=0.$$

So if we put $\omega := \mathrm{Res}_{I_0}\,\mathrm{Res}_{I_1}\cdots\mathrm{Res}_{I_\ell}\,\zeta$ (a polydifferential on the diagonal with coordinates $t_{\{I_0\}},\dots,t_{\{I_\ell\}},\{t_j\}_{j\in M-\{I\}}$), then it remains to see that the property is equivalent to the vanishing of ω at $t_{\{I_0\}}=\cdots=t_{\{I_\ell\}}$. Since ζ is $\mathrm{Aut}(\mathbf{P})$-invariant, so is ω, and hence it suffices to prove that the latter vanishes on $t_{\{I_0\}}=\cdots=t_{\{I_\ell\}}=\infty$.

Recall that by Theorem 2.14 the polar loci of ω involving the coordinates $t_{\{I_0\}},\dots,t_{\{I_\ell\}}$, are of the type $(t_{\{I_k\}}=z_\nu)$ only. Now let us put for $k=0,\dots,\ell$, $u_k := t^{-1}_{\{I_k\}}$. So ω is regular in the generic point defined by $u_1=\cdots=u_{1+\ell}=0$. The vanishing condition amounts to:

$$\mathrm{Res}_{(u_0=0)}\cdots\mathrm{Res}_{(u_\ell=0)}\frac{\omega}{u_0u_1\cdots u_\ell}=0,$$

This is equivalent to: ω vanishes on the locus $u_0=\cdots=u_\ell=0$.
Q.E.D.

In case $\mathfrak{g}=\mathfrak{sl}(2)$, we have $r=1$ and the highest root is the unique simple root (so that $h-1=1$). Corollary 4.2 then says that $\zeta\in\mathcal{O}_{U_n}\otimes V(\boldsymbol{\lambda})^{\mathfrak{g}}$ lies in $\mathcal{W}_\ell(\boldsymbol{\lambda})$ if and only if ζ vanishes on any diagonal defined by an $(1+\ell)$-element subset of $\mathcal{I}$. This is due to Ramadas [7], who proved this in an entirely different manner. The proof given here is closer in spirit to ours in [6].

Beilinson and Feigin have shown that $\mathcal{W}_\ell(\boldsymbol{\lambda})$ is locally free and flat for the KZ connection $\nabla^{C_\ell}_{KZ}$, where C_ℓ is characterized by the fact that $q^{C_\ell}(\tilde\alpha)=(\check h+\ell)^{-1}$, where $\check h:=1+\rho(\tilde\alpha^\vee)$ is known as the *dual Coxeter number* of $\mathfrak{g}$ (strictly speaking, they prove the dual statement). In particular, C_ℓ is defined over $\mathbb{Q}$. A long standing conjecture in the physicists's community is the existence of a flat unitary metric on this bundle.

Conjecture 4.3. *The subbundle $\mathcal{W}_\ell(\boldsymbol{\lambda})$ maps to the square integrable forms, or what amounts to the same, lands in sign isotypical part of the direct image of the relatively dualizing sheaf, i.e., in $\hat p^+_*\omega_{\hat U^+_{n,M}/U_n}$). In particular, it is of pure bidegree $(m,0)$ and the WZW system has a flat unitary structure.*

For $\mathfrak{g}=\mathfrak{sl}(2)$ this has been proved by Ramadas [7], who derives it from the above vanishing property on codimension ℓ diagonals (see also [6]).

§5. Appendix: an operator formula

We take up the situation of Section 1, but will assume that $c_{i,j} = c_{j,i}$. Recall that if I is a sequence, then I^* denotes that sequence in reverse order.

Lemma 5.1. *Assume that* $c_{i,j} = c_{j,i}$ *and let* $\Gamma : \mathcal{B} \to \mathcal{B} \otimes_{\mathbb{C}} \operatorname{End}(\mathcal{B})$ *be the linear map defined by*

$$\Gamma(\zeta_I) := \sum_{I=LiKJ} (-1)^{|K|} \zeta_J \otimes \omega_{Li}\omega_{K^*i}\Phi_i$$

(so $\Gamma(1) = 0$ *and* $\Gamma(\zeta_i) = 1 \otimes \Phi_i$*). Then for any finite sequence* I *in* $\mathcal{I}$ *and* $x \in \mathcal{I}$*, we have*

$$\Gamma(\Phi_x(\zeta_I)) = (\Phi_x \otimes 1 + 1 \otimes \operatorname{ad}\Phi_x)\Gamma(\zeta_I) + (p_x - c_{x,I})\zeta_I \otimes \Phi_x.$$

Corollary 5.2. *In this situation we have*

$$\Gamma(\Phi_I(1)) = \sum_{\emptyset \neq K \leq I} (p_{\ell(K)} - c_{\ell(K), I_{>\ell(K)}})\Phi_{I-K}(1) \otimes [\Phi_K],$$

where the sum is over all nonempty subsequences of I*, where* $\ell(K)$ *denotes the last term of* K *and* $I_{>\ell(K)}$ *is the largest common tail of* I *and* $I - K$*.*

Proof. We have that $\Phi_I(1)$ is a linear combination of the $\zeta_{\sigma(I)}$, where σ runs over the permutations of I. So in the preceding lemma we may replace ζ_I by $\Phi_I(1)$. Then the claimed identity follows with induction. Q.E.D.

Proof of Lemma 5.1. The proof is a mixture of algebra and bookkeeping and not entirely straightforward.

We derived in Lemma 1.9 the identity

$$\Phi_x(\zeta_I) = \sum_{I=I''I'} (p_x - c_{x,I'})\zeta_{I''xI'}.$$

So each term that appears in $\Gamma(\Phi_x\zeta_I)$ corresponds to a way of writing $I''xI'$ as $\tilde{L}i\tilde{K}\tilde{J}$. We can also express this differently by writing I as $LiKJ$, and then insert x in resp. J, K, L, or write $I = LKJ$ and take $LxKJ$ (this is when $i = x$). We thus get

$$\Gamma(\Phi_x(\zeta_I)) =$$

$$\text{(I)} \qquad \sum_{I=LiKJ, J=J''J'} (-1)^{|K|}(p_x - c_{x,J'})\zeta_{J''xJ'} \otimes \omega_{Li}\omega_{K^*i}\Phi_i +$$

(II) $$\sum_{I=LiKJ,K=K''K'} -(-1)^{|K|}(p_x - c_{x,K'J})\zeta_J \otimes \omega_{Li}\omega_{(K')^*x(K'')^*i}\Phi_i +$$

(III)
$$\sum_{I=LiKJ,L=L''L'} (-1)^{|K|}(p_x - c_{x,L'iKJ})\zeta_J \otimes \omega_{L''xL'i}\omega_{K^*i}\Phi_i +$$

(IV) $$\sum_{I=LKJ} (-1)^{|K|}(p_x - c_{x,KJ})\zeta_J \otimes \omega_{Lx}\omega_{K^*x}\Phi_x.$$

Denoting the subsums appearing above by their roman tags, then we observe that (I) $= (\Phi_x \otimes 1)\Gamma(\zeta_I)$. We rework (IV) by writing it first as

$$\text{(IV)} = \sum_{I=MJ} \zeta_J \otimes \Big(\sum_{M=LK} (-1)^{|K|}(p_x - c_{x,J} - c_{x,K})\omega_{Lx}\omega_{K^*x}\Big)\Phi_x$$

and then continue with the inner sum. The expression $\omega_{Lx}\omega_{K^*x}$ can be written as shuffle product $\sum_{S\in L\star K^*}\omega_{Sx}$. In case S is empty, this reduces to just ω_x, but otherwise such a shuffle appears twice: if S ends with i and M is written $M''iM'$, then either $K = M'$ or $K = iM'$. These terms appear with coefficients $(-1)^{|M'|}(p_x - c_{x,J} - c_{x,M'})$ resp. $(-1)^{|M'|+1}(p_x - c_{x,J} - c_{x,M'} - c_{x,i})$ and so add up to give $(-1)^{|M'|}c_{x,i}$. We conclude that (after substituting L for M'' and K for M'):

$$\text{(IV)} = (p_x - c_{x,I})\zeta_I \otimes \omega_x\Phi_x + \sum_{I=LiKJ} (-1)^{|K|}c_{x,i}\zeta_J \otimes \omega_{Lix}\omega_{K^*ix}\Phi_x.$$

Next we compute

$$(1 \otimes \mathrm{ad}(\Phi_x))\Gamma(\zeta_I) = \sum_{I=LiKJ} (-1)^{|K|}\zeta_J \otimes [\Phi_x, \omega_{Li}\omega_{K^*i}\Phi_i].$$

We work out the expression $[\Phi_x, \omega_{Li}\omega_{K^*i}\Phi_i]$ on the right of the tensor symbol; it is the sum

$$[\Phi_x, \omega_{Li}]\omega_{K^*i}\Phi_i + \omega_{Li}[\Phi_x, \omega_{K^*i}]\Phi_i + \omega_{Li}\omega_{K^*i}[\Phi_x, \Phi_i]$$

and for these terms we have according to Lemma 1.9,

$$[\Phi_x, \omega_{Li}]\omega_{K^*i}\Phi_i = \sum_{L=L''L'} -c_{x,L'}\omega_{L''xL'i}\omega_{K^*i}\Phi_i,$$
$$\omega_{Li}[\Phi_x, \omega_{K^*i}]\Phi_i = \sum_{K=K''K'} -c_{x,K''}\omega_{Li}\omega_{(K')^*x(K'')^*i}\Phi_i.$$

According to Lemma 1.12, $[\Phi_x, \Phi_i] = -c_{x,i}\omega_{(x,i)}\Phi_i + c_{i,x}\omega_{(i,x)}\Phi_x$ and hence we find

$$\omega_{Li}\omega_{K^*i}[\Phi_x, \Phi_i] = -c_{x,i}\omega_{Li}\omega_{K^*i}\omega_{(x,i)}\Phi_i + c_{i,x}\omega_{Li}\omega_{K^*i}\omega_{(i,x)}\Phi_x = \\ - c_{x,i}\omega_{Li}\omega_{K^*i}\omega_{(x,i)}\Phi_i + c_{i,x}\omega_{Lix}\omega_{K^*ix}\Phi_x.$$

It follows that

$$(1 \otimes \mathrm{ad}(\Phi_x))\Gamma(\zeta_I) =$$

$$\text{(V)} \qquad \sum_{I=LiKJ, L=L''L'} -(-1)^{|K|} c_{x,L'}\zeta_J \otimes \omega_{L''xL'i}\omega_{K^*i}\Phi_i +$$

$$\text{(VI)} \qquad \sum_{I=LiKJ, K=K''K'} -(-1)^{|K|} c_{x,K''}\zeta_J \otimes \omega_{Li}\omega_{(K')^*x(K'')^*i}\Phi_i +$$

$$\text{(VII)} \qquad \sum_{I=LiKJ} -(-1)^{|K|} c_{x,i}\zeta_J \otimes \omega_{Li}\omega_{K^*i}\omega_{(x,i)}\Phi_i +$$

$$\text{(VIII)} \qquad \sum_{I=LiKJ} (-1)^{|K|} c_{i,x}\zeta_J \otimes \omega_{Lix}\omega_{K^*ix}\Phi_x.$$

Adhering to our convention of identifying subsums by the corresponding roman tags, we see that

$$\text{(II)}-\text{(VI)}= \sum_{I=LiKJ} -(-1)^{|K|}(p_x - c_{x,KJ})\zeta_J \otimes \omega_{Li} \sum_{K=K''K'} \omega_{(K')^*x(K'')^*i}\Phi_i \\ = \sum_{I=LiKJ} -(-1)^{|K|}(p_x - c_{x,KJ})\zeta_J \otimes \omega_{Li}\omega_{(x,i)}\omega_{K^*i}\Phi_i$$

and similarly

$$\text{(III)}-\text{(V)}-\text{(VII)}= \sum_{I=LiKJ} (-1)^{|K|}(p_x - c_{x,KJ})\zeta_J \otimes \sum_{L=L''L'} \omega_{L''xL'i}\omega_{K^*i}\Phi_i \\ = \sum_{I=LiKJ} (-1)^{|K|}(p_x - c_{x,KJ})\zeta_J \otimes \omega_{Li}\omega_{(x,i)}\omega_{K^*i}\Phi_i,$$

so that (II) + (III) − (V) − (VI) − (VII) = 0.

Since $c_{x,i} = c_{i,x}$, we see that (IV) − (VIII) $= (p_x - c_{x,I})\zeta_I \otimes \Phi_x$. The Lemma follows. Q.E.D.

References

[1] A. Beilinson and V. G. Drinfeld, Affine Kac–Moody algebras and polydifferentials, Internat. Math. Res. Notices, **1** (1994), 1–11.

[2] V. G. Drinfeld, Quasi-Hopf algebras and Knizhnik–Zamolodchikov equations, In: Problems of Modern Quantum Field Theory, Alushta, 1989, Res. Rep. Phys., Springer-Verlag, 1989, pp. 1–13.

[3] H. Esnault, V. Schechtman and E. Viehweg, Cohomology of local sytems on the complement of hyperplanes, Invent. Math., **109** (1992), 557–661, *Erratum*, Invent. Math., **112** (1993), 447.

[4] V. Kac, Infinite dimensional Lie algebras, 3rd ed., Cambridge Univ. Press, Cambridge, 1990.

[5] E. Looijenga, Arrangements, KZ systems and Lie algebra homology, In: Singularity Theory, Liverpool, 1996, London Math. Soc. Lecture Note Ser., **263**, Cambridge Univ. Press, Cambridge, 1999, pp. xvi, 109–130.

[6] E. Looijenga, Unitarity of SL(2)-conformal blocks in genus zero, J. Geom. Phys., **59** (2009), 654–662.

[7] T. R. Ramadas, The "Harder-Narasimhan trace" and unitarity of the KZ/Hitchin connection: genus 0, Ann. of Math. (2), **169** (2009), 1–39.

[8] V. V. Schechtman and A. N. Varchenko, Arrangements of hyperplanes and Lie algebra homology, Invent. Math., **106** (1991), 139–194.

[9] A. V. Stoyanovski and B. L. Feigin, Realization of a modular functor in the space of differentials, and geometric approximation of the manifold of moduli of G-bundles, Funktsional. Anal. i Prilozhen., **28** (1994), 42–65, 95; translation in Funct. Anal. Appl., **28** (1995), 257–275.

Eduard Looijenga
Mathematisch Instituut, Universiteit Utrecht,
PO Box 80.010, 3508 TA Utrecht, Nederland
E-mail address: `E.J.N.Looijenga@uu.nl`

Advanced Studies in Pure Mathematics 62, 2012
Arrangements of Hyperplanes—Sapporo 2009
pp. 233–260

Solutions for some families of Fuchsian differential equations free from accessory parameters in terms of the integral of Euler type

Katsuhisa Mimachi

An ordinary differential equation of regular singular type defined on the Riemann sphere is called the *Fuchsian differential equation free from accessory parameters* or the *rigid Fuchsian differential equation*, if the equation is determined by the set of local data on monodromy, in particular, its spectral type.

About two decades ago, Yokoyama [29] classified such equations into eight types, I, II, III, IV, I*, II*, III*, and IV*, under some conditions from the viewpoint of the differential equation of Okubo type [23] (see also [8]). While the equation of type I is nothing but the generalized hypergeometric equation ${}_{n+1}E_n$ and that of type I* the Jordan–Pochhammer equation, the equations of the other types are new ones. Concerning the latter cases, very little has been understood: a restriction into one variable case of Appel's F_3 satisfies the equation II* of rank 4, the function satisfying the equation II of rank 4 is found in [16], and the functions satisfying the equation III* of rank 5 and of rank 7, the functions satisfying the equation II* of rank 4 and of rank 6, and the functions satisfying the equation II of rank 6 are found in [9].

The purpose of the present paper is to give solutions for the equations of types II, III, IV, II*, III*, and IV* in terms of the integral of Euler type.

In this paper, we frequantly use the symbol

$$e(A) = \exp(2\pi\sqrt{-1}A)$$

for abbreviation.

Received July 24, 2010.
Revised March 8, 2011.
2010 *Mathematics Subject Classification.* 33C70, 34M35.
Key words and phrases. Rigid local system, Yokoyama's classification.

§1. Fuchsian differential equations free from accessory parameters

For a Fuchsian differential equation with regular singular points $\{p_1, \dots, p_N, \infty\}$, let

$$\{\lambda_{j,1}, \lambda_{j,1}+1, \dots, \lambda_{j,1}+m_{j,1}, \lambda_{j,2}, \lambda_{j,2}+1, \dots, \lambda_{j,2}+m_{j,2}, \dots \\ \dots, \lambda_{j,n_j}, \lambda_{j,n_j}+1, \dots, \lambda_{j,n_j}+m_{j,n_j}\}$$

with $\lambda_{j,k}-\lambda_{j,l} \notin \mathbb{Z}$ $(k \neq l)$ be a set of characteristic exponents at p_j for $j = 1, \dots, N$ and ∞. Then $(m_{j,1}, m_{j,2}, \dots, m_{j,n_j})$ is called the *spectral type* at p_j and

$$(m_{1,1}, m_{1,2}, \dots, m_{1,n_1}\,;\, m_{2,1}, m_{2,2}, \dots, m_{2,n_2}\,;\, \cdots \\ \cdots\,;\, m_{N,1}, m_{N,2}, \dots, m_{N,n_N}\,;\, m_{\infty,1}, m_{\infty,2}, \dots, m_{\infty,n_\infty})$$

the spectral type of the equation. For instance, the spectral type of the Gauss hypergeometric equation is, under some genericity condition, $(1,1\,;\,1,1\,;\,1,1)$ and that of the generalized hypergeometric equation ${}_{n+1}E_n$ is, under some genericity condition, $(\underbrace{1,1,\dots,1}_{n+1}\,;\,1,n\,;\,\underbrace{1,1,\dots,1}_{n+1})$, which is also denoted by $(1^{n+1}\,;\,1,n\,;\,1^{n+1})$ or $(1^{n+1}\,;\,1,n\,;\,1^{n+1})$. Actually, the generalized hypergeometric differential equation ${}_{n+1}E_n$ is

$$\left\{\theta_z\Big\{\prod_{1\le i\le n}(\theta_z+\beta_i-1)\Big\} - z\Big\{\prod_{1\le i\le n+1}(\theta_z+\alpha_i)\Big\}\right\}F = 0,$$

where $\theta_z = zd/dz$, and its characteristic exponents are

$$0,\ 1-\beta_1,\ 1-\beta_2,\ \dots,\ 1-\beta_n \quad \text{at} \quad z=0,$$

$$0,\ 1,\ \dots,\ n-1,\ \sum_{1\le i\le n}\beta_i - \sum_{1\le i\le n+1}\alpha_i \quad \text{at} \quad z=1,$$

$$\alpha_1,\ \alpha_2,\ \dots,\ \alpha_{n+1} \quad \text{at} \quad z=\infty.$$

Thus its spectral type is $(1^{n+1}\,;\,1,n\,;\,1^{n+1})$ under the genericity condition

$$\beta_i-\beta_j \notin \mathbb{Z}, \quad 1\le i<j\le n+1, \quad \sum_{i=1}^{n}\beta_i - \sum_{i=1}^{n+1}\alpha_i \notin \mathbb{Z},$$

$$\alpha_i-\alpha_j \notin \mathbb{Z}, \quad 1\le i<j\le n+1,$$

where $\beta_{n+1} = 1$.

Similarly, the spectral type of the Jordan–Pochhamer differential equation

$$\begin{aligned} &Q(z)F^{(n)} - \mu Q'(z)F^{(n-1)} + \frac{\mu(\mu+1)}{2}Q''(z)F^{(n-2)} - \cdots \\ &- R(z)F^{(n-1)} + (\mu+1)R'(z)F^{(n-2)} - \cdots = 0, \end{aligned}$$

where

$$\begin{aligned} Q(z) &= (z-c_1)(z-c_2)\cdots(z-c_n), \\ R(z)/Q(z) &= \sum_{j=1}^{n} \alpha_j/(z-c_j) \end{aligned}$$

is $(1, n-1; 1, n-1; \cdots; 1, n-1)$ under the condition

$$\mu + \alpha_j \notin \mathbb{Z} \quad \text{for} \quad j = 1, \ldots, n, \quad \text{and} \quad \alpha_1 + \cdots + \alpha_n \notin \mathbb{Z},$$

since the characteristic exponents at $z = c_j, j = 1, \ldots, n$ are

$$0, 1, \ldots, n-2, \mu + n - 1 + \alpha_j,$$

and those at $z = \infty$ are

$$-(\mu+1), \ldots, -(\mu+n-1), -(\mu + \alpha_1 + \cdots + \alpha_n).$$

Yokoyama's classification of the Fuchsian differential equations accessory parameter free is asserted as follows [29]:

	rank	# of singularities on $\mathbb{P}^1$	spectral type
I (GHGF)	n	3	$1^n\,;1,n-1\,;1^n$
I* (Pochhammer)	n	$n-1$	$1,n-1;1,n-1;\ldots;1,n-1$
II	$2n$	3	$1^n,n;1^n,n;1,n-1,n$
II*	$2n$	4	$1^n,n;1^{n-1},n+1;1,2n-1;n,n$
III	$2n+1$	3	$1^{n+1},n;1^n,n+1;1,n,n$
III*	$2n+1$	4	$1^n,n+1;1^n,n+1;1,2n;n,n+1$
IV	6	3	$1^2,4;2^3;1^4,2$
IV*	6	4	$1^2,4;1^2,4;1^2,4;2,4$

We consider their solutions.

It is known that the function of the form

$$\int_C \prod_{i=1}^{n} t_i^{\alpha_{i+1}-\beta_i} \prod_{i=1}^{n+1} (t_i - t_{i-1})^{\beta_i-\alpha_i-1}\, dt_1 \cdots dt_n\,, \tag{1.1}$$

where $t_0 = 1,\ t_{n+1} = z$ and C a suitable cycle, satisfies the generalized hypergeometric equation ${}_{n+1}E_n$, which is the equation of type I. See, for instance, [17, 18].

It is also known that the function of the form

$$\int_C \prod_{j=1}^{n+1} (t - c_j)^{\alpha_j-1} dt, \tag{1.2}$$

where $t_{n+1} = z, \alpha_{n+1} = \mu + n$ and C a suitable cycle, satisfies the Jordan–Pochhammer equation, which is the equation of type I*. See, for instance, [28].

In the remaining sections, we give the solutions for all the equations of types II, III, IV, II*, III*, and IV* in terms of the integral of Euler type.

For our purpose, we apply the framework of the twisted homology theory developed by Aomoto in these decades [1, 2]. We refer the reader to [3, 14, 19, 20, 21, 22] for more knowledge about the twisted homology and its application to the integral representation of the solution to differential equations.

§2. The equation of type II

In this section, let $\mathcal{L}_z$ be the locally constant sheaf (the local system) determined by a function

$$u(t) = (t_n - t_0)^{\lambda_{0n}} \prod_{1\le i\le n} t_i^{\lambda_i} \prod_{0\le i\le n} (t_i - t_{i+1})^{\lambda_{i,i+1}} \tag{2.1}$$

on

$$T_z = \mathbb{C}^n \backslash \{ t_n - t_0 = 0 \} \cup \cup_{i=1}^n \{ t_i = 0 \} \cup \cup_{i=0}^n \{ t_i - t_{i+1} = 0 \},$$

where $t_0 = 1$ and $t_{n+1} = z$: the sheaf consisting of the local solutions of $dL = L\omega$ for $\omega = du(t)/u(t)$.

Let $H_n(T_z, \mathcal{L}_z)$ be the n-th homology group with coefficients in $\mathcal{L}_z$, $H_n^{\mathrm{lf}}(T_z, \mathcal{L}_z)$ the n-th locally finite homology group with coefficients in $\mathcal{L}_z$.

After fixing the variable z to be a real number satisfying $0 < z < 1$, the bounded chambers in the real locus $T_{\mathbb{R}}$ of $T = T_z$ are

$$\begin{pmatrix} 0 < t_s < t_{s+1} < \cdots < t_n < z, \\ t_s < t_{s-1} \cdots < t_1 < 1 \end{pmatrix} \tag{2.2}$$

for $1 \le s \le n$ and

$$\begin{pmatrix} z < t_n < 1 \\ 0 < t_s < t_{s+1} < \cdots < t_n, \\ t_s < t_{s-1} \cdots < t_1 < 1 \end{pmatrix} \tag{2.3}$$

for $1 \le s \le n$. Thus it follows from Theorem 5 of [3] that $\dim H_n^{\mathrm{lf}}(T_z, \mathcal{L}_z) = 2n$ under the condition

$$\lambda_{\infty=t_p=\cdots=t_q} = -\sum_{p\le k\le q} \lambda_k - \sum_{p\le k\le q+1} \lambda_{k-1,k} \notin \mathbb{Z}, \quad 1 \le p \le q \le n-1,$$

$$\lambda_{\infty=t_p=\cdots=t_n} = -\lambda_{0n} - \sum_{p\le k\le n} \lambda_k - \sum_{p\le k\le n+1} \lambda_{k-1,k} \notin \mathbb{Z}, \quad 1 \le p \le n.$$

Hereafter we denote the exponent of an irreducible component of the divisor $\widetilde{D} = \pi^{-1}(D)$ by λ_D, where $\pi : \widetilde{(\mathbb{P}^1(\mathbb{C}))^n} \to (\mathbb{P}^1(\mathbb{C}))^n$ is the minimal blow-up along the non-normally crossing loci of D. If λ_D is an integer, the irreducible component or the exponent itself is said to be *resonant*. For instance, $\lambda_{\infty=t_p=\cdots=t_q} \notin \mathbb{Z}$ means that the exponent of $\pi^{-1}\{\infty = t_p = \cdots = t_q\}$ is not resonant.

For each $i = 0, 1$, let γ_i be a simple loop which starts and ends at a base point z for $0 < z < 1$ and surrounds only the singular point i in counterclockwise direction. It corresponds to a generator of $\pi_1(\mathbb{C}\backslash\{0,1\})$, and induces an action of $\pi_1(\mathbb{C}\backslash\{0,1\})$ on the family of the homology group $H_n^{\mathrm{lf}}(T_z, \mathcal{L}_z)$ on $\mathbb{C}\backslash\{0,1\}$.

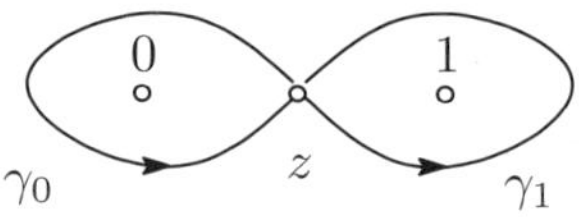

Fig. 1.

Considering the action of γ_0 on the chambers (2.3) shows that the multiplicity of holomorphic solutions around 0, which correspond to the eigenvalue 1, is n; considering the action of γ_0 on the chambers (2.2) shows that the eigenvalue

$$e\Big(\sum_{s \le k \le n} (\lambda_k + \lambda_{k,k+1})\Big) \tag{2.4}$$

for each $1 \le s \le n$ is multiplicity free. Moreover, it is seen that the chamber

$$\begin{pmatrix} 0 < t_s < t_{s+1} < \cdots < t_n < z, \\ 1 < t_1 < \cdots < t_{s-1} < t_s \end{pmatrix}$$

gives the eigenvector for the eigenvalue (2.4).

Similarly, considering the action of γ_1 on the chambers (2.2) shows that the multiplicity of the eigenvalue 1 is $n-1$, considering the action on the chambers (2.3) for $1 \le s \le n-1$ shows that the multiplicity of the eigenvalue

$$e(\lambda_{0n} + \lambda_{n,n+1})$$

is $n-1$, and the action on the chamber (2.3) for $s = n$ shows that the multiplicity of the eigenvalue

$$e(\lambda_{01} + \lambda_{0n} + \sum_{1 \le k \le n} \lambda_{k,k+1})$$

is free.

Next, to know the eigenvalues of the action of $\gamma_\infty = \gamma_0^{-1}\gamma_1^{-1}$ and their multiplicities, we change the variables t_i into t_i^{-1} for $1 \le i \le n$. Then we have

$$u(t) = z^{\lambda_{n,n+1}}(t_0 - t_n)^{\lambda_{0n}} \prod_{1\le i\le n} t_i^{\lambda_i^\infty} \prod_{0\le i\le n} (t_{i+1} - t_i)^{\lambda_{i,i+1}},$$

where

$$\lambda_i^\infty = -\lambda_i - \lambda_{i-1,i} - \lambda_{i,i+1}, \qquad 1 \le i \le n-1,$$

$$\lambda_n^\infty = -\lambda_n - \lambda_{n-1,n} - \lambda_{n,n+1} - \lambda_{0n}$$

with $t_{n+1} = z^{-1}$ and $t_0 = 1$. After fixing the variable z^{-1} to be a real number satisfying $0 < z^{-1} < 1$, the bounded chambers are

$$\begin{pmatrix} 0 < t_s < t_{s+1} < \cdots < t_n < z^{-1}, \\ t_s < t_{s-1} \cdots < t_1 < 1 \end{pmatrix}$$

for $1 \le s \le n$ and

$$\begin{pmatrix} z^{-1} < t_n < 1 \\ 0 < t_s < t_{s+1} < \cdots < t_n, \\ t_s < t_{s-1} \cdots < t_1 < 1 \end{pmatrix}$$

for $1 \le s \le n$. Hence, the multiplicity of the eigenvalue

$$e(-\lambda_{n,n+1})$$

is free and that of

$$e(-\lambda_{0n} - \lambda_{s-1,s} - \sum_{s\le k\le n} (\lambda_k + \lambda_{k,k+1})), \quad 1 \le s \le n$$

is n.

Consequently, the spectral type turns out to be $(1^n, n; n, n-1, 1; 1^n, n)$ under the conditions

$$\sum_{s\le k\le l} (\lambda_k + \lambda_{k,k+1}) \notin \mathbb{Z}, \quad (1 \le s \le l \le n),$$

which is for the separation of the eigenvalues at 0,

$$\lambda_{01} + \lambda_{0n} + \sum_{1\le k\le n} \lambda_{k,k+1} \notin \mathbb{Z},$$
$$\lambda_{0n} + \lambda_{n,n+1} \notin \mathbb{Z},$$
$$\lambda_{01} + \sum_{1\le k\le n-1} \lambda_{k,k+1} \notin \mathbb{Z},$$

which is for the separation of the eigenvalues at 1, and

$$\lambda_{0n} + \sum_{s\le k\le n} (\lambda_k + \lambda_{k-1,k}) \notin \mathbb{Z}, \quad 1 \le s \le n,$$
$$\sum_{s\le k\le l-1} (\lambda_k + \lambda_{k-1,k}) \notin \mathbb{Z}, \quad 1 \le s < l \le n,$$

which is for the separation of the eigenvalues at ∞.

As a result, we have

Theorem 2.1. *The function of the form*

$$\int_C (t_n - t_0)^{\lambda_{0n}} \prod_{1\le i\le n} t_i^{\lambda_i} \prod_{0\le i\le n} (t_i - t_{i+1})^{\lambda_{i,i+1}}\, dt_1 \cdots dt_n,$$

where $t_0 = 1, t_{n+1} = z$ and C a suitable cycle, satisfies the equation of type II, whose rank is $2n$.

§3. The equation of type III

In this section, let $\mathcal{L}_z$ be the locally constant sheaf determined by a function

$$u(t) = (t_n - t_0)^{\lambda_{0n}} \prod_{1\le i\le n-1} t_i^{\lambda_i} \prod_{0\le i\le n} (t_i - t_{i+1})^{\lambda_{i,i+1}} \tag{3.1}$$

on

$$T_z = \mathbb{C}^n \backslash \cup \{ t_n - t_0 = 0 \} \cup \cup_{i=1}^{n-1} \{ t_i = 0 \} \cup \cup_{i=0}^{n} \{ t_i - t_{i+1} = 0 \},$$

where $t_0 = 1$ and $t_{n+1} = z$.

For a fixed variable z such that $0 < z < 1$, the bounded chambers in the real locus $T_{\mathbb{R}}$ of $T = T_z$ are

$$\begin{pmatrix} 0 < t_s < t_{s+1} < \cdots < t_n < z, \\ t_s < t_{s-1} \cdots < t_1 < 1 \end{pmatrix} \tag{3.2}$$

for $1 \leq s \leq n-1$ and

$$(3.3) \qquad \begin{pmatrix} z < t_n < 1 \\ 0 < t_s < t_{s+1} < \cdots < t_n, \\ t_s < t_{s-1} \cdots < t_1 < 1 \end{pmatrix}$$

for $1 \leq s \leq n$ (Remark that the latter one for $s = n$ means $(z < t_n < \cdots < t_1 < 1)$).

Thus $\dim H_n^{\mathrm{lf}}(T_z, \mathcal{L}_z) = 2n - 1$ under the condition

$$\lambda_{\infty = t_p = \cdots = t_q} = -\sum_{p \leq k \leq q} \lambda_k - \sum_{p \leq k \leq q+1} \lambda_{k-1,k} \notin \mathbb{Z}, \quad 1 \leq p \leq q \leq n-1,$$

$$\lambda_{\infty = t_p = \cdots = t_n} = -\lambda_{0n} - \sum_{p \leq k \leq n} \lambda_k - \sum_{p \leq k \leq n+1} \lambda_{k-1,k} \notin \mathbb{Z}, \quad 1 \leq p \leq n.$$

Considering the action of γ_0 on the cycles (3.2) for $1 \leq s \leq n-1$ and (3.3) for $1 \leq s \leq n$ as in Section 2, it is seen that the multiplicity of the eigenvalue

$$e\Big(\sum_{s \leq k \leq n-1} (\lambda_k + \lambda_{k,k+1}) + \lambda_{n,n+1}\Big)$$

for each $1 \leq s \leq n-1$ is free and the multiplicity of the eigenvalue 1 is n.

Similarly, as for the action of γ_1, the multiplicity of the eigenvalue 1 is $n-1$, the multiplicity of the eigenvalue

$$e(\lambda_{0n} + \lambda_{n,n+1})$$

is $n-1$, and the multiplicity of the eigenvalue

$$e\Big(\lambda_{01} + \lambda_{0n} + \sum_{1 \leq k \leq n} \lambda_{k,k+1}\Big)$$

is free.

As for the multiplicities of the eigenvalues of the action of γ_∞, we consider

$$u(t) = z^{\lambda_{n,n+1}} (t_0 - t_n)^{\lambda_{0n}} \prod_{1 \leq i \leq n} t_i^{\lambda_i^\infty} \prod_{0 \leq i \leq n} (t_{i+1} - t_i)^{\lambda_{i,i+1}},$$

where

$$\lambda_i^\infty = -\lambda_i - \lambda_{i-1,i} - \lambda_{i,i+1}, \qquad 1 \leq i \leq n-1,$$

$$\lambda_n^\infty = -\lambda_{0n} - \lambda_{n-1,n} - \lambda_{n,n+1}$$

with $t_{n+1} = z^{-1}$ and $t_0 = 1$. This shows that the multiplicity of the eigenvalue

$$e(-\lambda_{0n} - \sum_{s\le k\le n-1}(\lambda_k + \lambda_{k-1,k}) - \lambda_{n-1,n} - \lambda_{n,n+1})$$

for each $1 \le s \le n$ is free, and the multiplicity of the eigenvalue

$$e(-\lambda_{n,n+1})$$

is $n-1$.

Therefore, the spectral type is $(1^{n-1}, n; 1, n-1, n-1; 1^n, n-1)$ under the condition

$$\sum_{s\le k\le n-1}(\lambda_k + \lambda_{k,k+1}) + \lambda_{n,n+1} \notin \mathbb{Z}, \quad 1 \le s \le n-1,$$

$$\sum_{s\le k\le l-1}(\lambda_k + \lambda_{k,k+1}) \notin \mathbb{Z}, \quad 1 \le s < l \le n-1,$$

which is for the separation of the eigenvalues at 0,

$$\lambda_{01} + \lambda_{0n} + \sum_{1\le k\le n}\lambda_{k,k+1} \notin \mathbb{Z},$$

$$\lambda_{0n} + \lambda_{n,n+1} \notin \mathbb{Z},$$

$$\lambda_{01} + \sum_{1\le k\le n-1}\lambda_{k,k+1} \notin \mathbb{Z},$$

which is for the separation of the eigenvalues at 1, and

$$\lambda_{0,n} + \sum_{s\le k\le n-1}(\lambda_k + \lambda_{k,k+1}) + \lambda_{s-1,s} \notin \mathbb{Z}, \quad 1 \le s \le n,$$

$$\sum_{s\le k\le l-1}(\lambda_k + \lambda_{k-1,k}) \notin \mathbb{Z}, \quad 1 \le s < l \le n,$$

which is for the separation of the eigenvalues at ∞.

Finally we reach the following.

Theorem 3.1. *The function of the form*

$$\int_C (t_n - t_0)^{\lambda_{0n}} \prod_{1\le i\le n-1} t_i^{\lambda_i} \prod_{0\le i\le n}(t_i - t_{i+1})^{\lambda_{i,i+1}}\, dt_1 \cdots dt_n,$$

where $t_0 = 1, t_{n+1} = z$ *and* C *a suitable cycle, satisfies the equation of type III, whose rank is* $2n - 1$.

It is worthwhile to note that the function (3.1) can be obtained from (2.1) by the specialization $\lambda_n = 0$.

§4. The equation of type IV

In this section, let $\mathcal{L}_z$ be the locally constant sheaf determined by a function

$$u(t) = \prod_{i=1,2} t_i^{\lambda_i} \prod_{i=1,3,4} (t_i - t_0)^{\lambda_{0i}} \prod_{i=1,2,3,4} (t_i - t_{i+1})^{\lambda_{i,i+1}} \tag{4.1}$$

on

$$T_z = \mathbb{C}^4 \setminus \cup_{i=1}^{2} \{ t_i - 0 \} \cup \cup_{i=1,3,4} \{ t_i - 1 = 0 \} \cup_{i=1}^{4} \{ t_i - t_{i+1} = 0 \},$$

where $t_0 = 1$ and $t_5 = z$.

For a fixed z such that $0 < z < 1$, the bounded chambers in the real locus $T_{\mathbb{R}}$ of $T = T_z$ are

$$(0 < t_1 < t_2 < t_3 < t_4 < z), \quad \begin{pmatrix} 0 < t_2 < t_3 < t_4 < z, \\ t_2 < t_1 < 1 \end{pmatrix},$$

$$(z < t_4 < t_3 < t_2 < t_1 < 1), \quad \begin{pmatrix} z < t_4 < t_3 < 1, \\ 0 < t_1 < t_2 < t_3 \end{pmatrix},$$

$$\begin{pmatrix} z < t_4 < t_3 < 1, \\ 0 < t_2 < t_3, \\ t_2 < t_1 < 1 \end{pmatrix}, \quad \begin{pmatrix} z < t_4 < 1, \\ 0 < t_1 < t_2 < t_3 < t_4 \end{pmatrix},$$

$$\text{and} \quad \begin{pmatrix} z < t_4 < 1, \\ 0 < t_2 < t_3 < t_4, \\ t_2 < t_1 < 1 \end{pmatrix}.$$

Thus $\dim H_n^{\mathrm{lf}}(T_z, \mathcal{L}_z) = 7$ under the condition

$$\lambda_{\infty=t_p=\cdots=t_q} = -\sum_{p\le k\le q} \lambda_k - \sum_{p\le k\le q+1} \lambda_{k-1,k} \notin \mathbb{Z}, \quad 1\le p\le q\le 2,$$

$$\lambda_{\infty=t_p=\cdots=t_q} = -\sum_{p\le k\le q} \lambda_{0k} - \sum_{p\le k\le q+1} \lambda_{k-1,k} \notin \mathbb{Z}, \quad 3\le p\le q\le 4,$$

$$\lambda_{\infty=t_1=\cdots=t_q} = -\sum_{1\le k\le 2} \lambda_k - \sum_{3\le k\le q} \lambda_{0k} - \sum_{1\le k\le q+1} \lambda_{k-1,k} \notin \mathbb{Z}, \quad 3\le q\le 4,$$

$$\lambda_{\infty=t_2=\cdots=t_q} = -\lambda_2 - \sum_{3\le k\le q} \lambda_{0k} - \sum_{2\le k\le q+1} \lambda_{k-1,k} \notin \mathbb{Z}, \quad 3\le q\le 4.$$

As for the action of γ_0, the multiplicities of the eigenvalues

$$e(\lambda_1+\lambda_2+\sum_{1\le k\le 4}\lambda_{k,k+1}), \quad e(\lambda_2+\sum_{2\le k\le 4}\lambda_{k,k+1})$$

are both multiplicity free, and the multiplicity of the eigenvalue 1 is 5; as for the action of γ_1, the multiplicities of the eigenvalues

$$e(\sum_{k=3,4}(\lambda_{0,k}+\lambda_{k,k+1})), \quad e(\lambda_{04}+\lambda_{45}), \quad 1$$

are all 2, and the multiplicity of the eigenvalue

$$e(\sum_{k=1,3,4}\lambda_{0,k}+\sum_{1\le k\le 4}\lambda_{k,k+1})$$

is free.

As for the action of γ_∞, we consider the function

$$u(t) = z^{\lambda_{45}} \prod_{1\le i\le 4} t_i^{\lambda_i^\infty} \prod_{i=1,3,4} (1-t_i)^{\lambda_{0i}} \prod_{i=1,2,3,4} (t_{i+1}-t_i)^{\lambda_{i,i+1}},$$

where

$$\lambda_i^\infty = \begin{cases} -\lambda_i - \lambda_{i-1,i} - \lambda_{i,i+1}, & i=1,2, \\ -\lambda_{0i} - \lambda_{i-1,i} - \lambda_{i,i+1}, & i=3,4, \end{cases}$$

with $t_5 = z^{-1}$ and $t_0 = 1$. It is seen that the multiplicity of each of the eigenvalues

$$e(-\lambda_{04} - \lambda_{34} - \lambda_{45}) \quad \text{and} \quad e(-\lambda_{45})$$

is 2, and the multiplicity of each of the eigenvalues

$$e(-\lambda_1 - \lambda_2 - \sum_{k=1,3,4} \lambda_{0,k} - \sum_{1\le k\le 4} \lambda_{k,k+1}),$$

$$e(-\lambda_2 - \sum_{k=3,4} \lambda_{0,k} - \sum_{1\le k\le 4} \lambda_{k,k+1}),$$

$$\text{and} \qquad e(-\lambda_{03} - \lambda_{04} - \sum_{2\le k\le 4} \lambda_{k,k+1})$$

is free.

Thus the spectral type is (511; 2221; 22111) under some genericity condition.

At this stage, we impose a resonace condition

$$(4.2)\quad \lambda_{t_1=t_2=t_3=t_4=1} + 3 = \lambda_{01} + \lambda_{12} + \lambda_{23} + \lambda_{03} + \lambda_{34} + \lambda_{04} + 3 = 0.$$

As a result, a subspace which is invariant with respect to the action of the monodromy group emerges; it consists of the regularizable cycles, it has a spectral data (411; 222; 21111), and it gives the solution space of the equation of type IV. We proceed to that point.

First, it is important to note that the resonance (4.2) leads to the nontriviality of the kernel of the map $\iota : H_4(T, \mathcal{L}) \longrightarrow H_4^{\mathrm{lf}}(T, \mathcal{L})$. The dimension of the kernel is one, and thus the dimension of the image is 6. The image $\operatorname{Im} \iota$ is called the space of *regularizable cycles.* (See [19].)

Secondly, the chambers

$$(0 < t_1 < t_2 < t_3 < t_4 < z), \qquad \begin{pmatrix} 0 < t_2 < t_3 < t_4 < z, \\ 1 < t_1 < \infty \end{pmatrix},$$

are regularizable even if the resonance condition (4.2) is imposed, and give the eigenvectors with the eigenvalues

$$e(\lambda_1 + \lambda_2 + \sum_{1\le k\le 4} \lambda_{k,k+1}), \quad e(\lambda_2 + \sum_{2\le k\le 4} \lambda_{k,k+1}),$$

with respect to the action of γ_0. Thus the spectral type at 0 of the space of regularizable cycles must be (114).

On the other hand, the chamber

$$(z < t_4 < t_3 < t_2 < t_1 < 1)$$

cannot be regularizable under the resonance condition (4.2). Thus (222) is the specral type at 1 of the space of regularizable cycles.

Moreover, if we consider

$$u(t) = z^{\lambda_{45}} \prod_{1 \le i \le 4} t_i^{\lambda_i^\infty} \prod_{i=1,3,4} (1-t_i)^{\lambda_{0i}} \prod_{i=1,2,3,4} (t_{i+1}-t_i)^{\lambda_{i,i+1}}$$

for $0 < z^{-1} < 1$, it is seen that the chambers

$$(0 < t_1 < t_2 < t_3 < t_4 < z^{-1}), \quad \begin{pmatrix} 0 < t_2 < t_3 < t_4 < z^{-1}, \\ 1 < t_1 \end{pmatrix},$$

$$\text{and} \quad \begin{pmatrix} 0 < t_3 < t_4 < z^{-1}, \\ 1 < t_1 < t_2 \end{pmatrix},$$

are regularizable, even if the resonance condition (4.2) is imposed. Hence, either the eigenspace with the eigenvalue $e(-\lambda_{04} - \lambda_{34} - \lambda_{45})$ or that with $e(-\lambda_{45})$ becomes one dimensional space. It means that spectral type at ∞ is (11112).

Therefore, the spectral type of the space of regularizable cycles turns out to be (114; 222; 11112).

$$\begin{array}{ll} 1,1,5 & \text{at } 0 \\ 1,2,2,2 & \text{at } 1 \\ 1,1,1,2,2 & \text{at } \infty \end{array} \quad \longrightarrow \quad (\text{IV}) \quad \begin{array}{ll} 1,1,4 & \text{at } 0 \\ 2,2,2 & \text{at } 1 \\ 1,1,1,1,2 & \text{at } \infty. \end{array}$$

Consequently, we have

Theorem 4.1. *The function of the form*

$$\int_C \prod_{i=1,2} t_i^{\lambda_i} \prod_{i=1,3,4} (t_i - t_0)^{\lambda_{0i}} \prod_{i=1,2,3,4} (t_i - t_{i+1})^{\lambda_{i,i+1}} \, dt_1 \cdots dt_4,$$

where

$$\lambda_{01} + \lambda_{12} + \lambda_{23} + \lambda_{03} + \lambda_{34} + \lambda_{04} + 3 = 0, \quad t_0 = 1, \quad t_5 = z,$$

and C a suitable cycle, satisfies the equation of type IV, whose rank is 6.

Remark. *The spectral type of the function* ϕ *which is obtained from the integral in the case III of rank* 5 *multiplied by* $(1-z)^{\lambda_{04}} = (1-t_4)^{\lambda_{04}}$ *is* $(311; 221; 2111)$, *and the characteristic exponent corresponding to the solution around* 1 *whose multiplicity is free is* $\lambda_{01} + \lambda_{12} + \lambda_{23} + \lambda_{03} + \lambda_{34} + \lambda_{04} + 3$. *Euler transfrom of* ϕ *under* (4.2) *is nothing but the integral in Theorem 4.1.*

§5. The equation of type IV*

In this section, let $\mathcal{L}_z$ be the locally constant sheaf determined by a function

$$u(t) = (t_2 - c)^{\lambda_{2c}} \prod_{i=1,2} \left\{ t_i^{\lambda_i} (t_i - t_0)^{\lambda_{0i}} (t_i - t_{i+1})^{\lambda_{i,i+1}} \right\} \tag{5.1}$$

on

$$T_z = \mathbb{C}^2 \backslash \{ t_2 - c = 0 \} \cup \cup_{i=1}^2 (\{ t_i = 0 \} \cup \{ t_i - t_0 = 0 \} \cup \{ t_i - t_{i+1} = 0 \}),$$

where $t_0 = 1$ and $t_3 = z$.

After fixing the variable z and parameter c to be real numbers satisfying $0 < z < c < 1$, the bounded chambers in the real locus $T_{\mathbb{R}}$ of $T = T_z$ are

$$(0 < t_1 < t_2 < z), \qquad (0 < t_2 < z,\ t_2 < t_1 < 1),$$
$$(0 < t_1 < t_2,\ z < t_2 < c), \qquad (z < t_2 < c,\ t_2 < t_1 < 1),$$
$$(0 < t_1 < t_2,\ c < t_2 < 1), \qquad (c < t_2 < t_1 < 1).$$

Thus $\dim H_2^{\mathrm{lf}}(T_z, \mathcal{L}_z) = 6$ under the condition

$$\lambda_{\infty=t_1=t_2} = -\lambda_{2c} - \sum_{1\le k\le 2} (\lambda_k + \lambda_{0k} + \lambda_{k,k+1}) \notin \mathbb{Z},$$

$$\lambda_{\infty=t_1} = -\lambda_1 - \lambda_{01} - \lambda_{12} \notin \mathbb{Z},$$

$$\lambda_{\infty=t_2} = -\lambda_2 - \lambda_{02} - \lambda_{2c} - \lambda_{12} - \lambda_{23} \notin \mathbb{Z}.$$

The eigenvalues of the action of γ_0 are

$$e(\lambda_1 + \lambda_2 + \lambda_{12} + \lambda_{23}), \quad e(\lambda_2 + \lambda_{23}),$$

each with multiplicity free, and 1 with multiplicity 4; the action of γ_c are

$$e(\lambda_{2c} + \lambda_{23})$$

with multiplicity 2, and 1 with multiplicity 4. To know the eigenvalues of the action of γ_1 and γ_∞, fix z and c to be $0 < c < 1 < z < \infty$. Then it turns out that the eigenvalues of the action of γ_1 are

$$e(\lambda_{02} + \lambda_{23}), \quad e(\lambda_{01} + \lambda_{02} + \lambda_{12} + \lambda_{23}),$$

each with multiplicity free, and 1 with multiplicity 4; those of the action of γ_∞ are

$$e(-\sum_{i=1,2} (\lambda_{0i} + \lambda_i + \lambda_{i,i+1}) - \lambda_{2c}), \quad e(-\lambda_{02} - \lambda_2 - \lambda_{23} - \lambda_{2c}),$$

each with multiplicity free, and $e(\lambda_{23})$ with multiplicity 4.

Therefore, the specral type of the space of regularizable cycle is $(411; 42; 411; 411)$ under the conditions:

$$\sum_{k=1,2} (\lambda_k + \lambda_{k,k+1}) \notin \mathbb{Z}, \qquad \lambda_k + \lambda_{k,k+1} \notin \mathbb{Z}, \qquad k = 1, 2,$$

which is for the separation of the eigenvalues at 0,

$$\lambda_{2c} + \lambda_{23} \notin \mathbb{Z},$$

which is for the separation of the eigenvalues at c,

$$\sum_{k=1,2} (\lambda_{0k} + \lambda_{k,k+1}) \notin \mathbb{Z}, \qquad \lambda_{0,k} + \lambda_{k,k+1} \notin \mathbb{Z}, \qquad k = 1, 2,$$

which is for the separation of the eigenvalues at 1, and

$$\lambda_{2c} + \sum_{k=1,2} (\lambda_k + \lambda_{0k} + \lambda_{k,k+1}) \notin \mathbb{Z}, \quad \lambda_{2c} + \lambda_{12} + \sum_{k=1,2} (\lambda_k + \lambda_{0k}) \notin \mathbb{Z},$$

$$\lambda_2 + \lambda_{02} + \lambda_{2c} + \lambda_{12} + \lambda_{23} \notin \mathbb{Z}, \quad \lambda_2 + \lambda_{02} + \lambda_{2c} + \lambda_{12} \notin \mathbb{Z},$$

$$\lambda_1 + \lambda_{01} \notin \mathbb{Z}, \quad \lambda_{23} \notin \mathbb{Z},$$

which is for the separation of the eigenvalues at ∞.

The combination of these facts leads to

Theorem 5.1. *The function of the form*

$$\int_C (t_2 - c)^{\lambda_{2c}} \prod_{i=1,2} \left\{ t_i^{\lambda_i} (t_i - t_0)^{\lambda_{0i}} \, (t_i - t_{i+1})^{\lambda_{i,i+1}} \right\} dt_1 dt_2,$$

where $t_0 = 1, t_3 = z$ and C a suitable cycle, satisfies the equation of type IV^*, whose rank is 6.

§6. The equation of type III*

In this section, let $\mathcal{L}_z$ be the locally constant sheaf determined by a function

$$(6.1)\qquad u(t) = (t_1 - c)^{\lambda_{1c}}(t_n - t_0)^{\lambda_{0n}} \prod_{1\le i\le n} t_i^{\lambda_i} \prod_{1\le i\le n} (t_i - t_{i+1})^{\lambda_{i,i+1}}$$

on

$$T_z = \mathbb{C}^n \backslash \{ t_1 - c = 0 \} \cup \{ t_n - t_0 = 0 \} \cup \cup_{i=1}^n \{\{ t_i = 0 \} \cup \{ t_i - t_{i+1} = 0 \}\},$$

where $t_{n+1} = z$ and $t_0 = 1$. After fixing the variable z and parameter c to be real numbers satisfying $0 < z < c < 1$, the bounded chambers in the real locus $T_{\mathbb{R}}$ of $T = T_z$ are

$$\begin{pmatrix} 0 < t_s < t_{s+1} < \cdots < t_n < z, \\ t_s < t_{s-1} \cdots < t_1 < c \end{pmatrix}$$

for $1 \le s \le n$ and

$$\begin{pmatrix} z < t_n < 1 \\ 0 < t_s < t_{s+1} < \cdots < t_n, \\ t_s < t_{s-1} \cdots < t_1 < c \end{pmatrix}$$

for $1 \le s \le n$ and

$$(z < t_n < \cdots < t_1 < c).$$

Thus $\dim H_n^{\mathrm{lf}}(T_z, \mathcal{L}_z) = 2n + 1$ under the condition

$$\lambda_{\infty=t_1=\cdots=t_q} = -\lambda_{1c} - \sum_{1\le k\le q} (\lambda_k + \lambda_{k,k+1}) \notin \mathbb{Z}, \quad 1 \le q \le n-1,$$

$$\lambda_{\infty=t_1=\cdots=t_n} = -\lambda_{1c} - \lambda_{0n} - \sum_{1\le k\le n} (\lambda_k + \lambda_{k,k+1}) \notin \mathbb{Z},$$

$$\lambda_{\infty=t_p=\cdots=t_q} = -\sum_{p\le k\le q} (\lambda_k + \lambda_{k-1,k}) - \lambda_{q,q+1} \notin \mathbb{Z}, \quad 2 \le p \le q \le n-1,$$

$$\lambda_{\infty=t_p=\cdots=t_n} = -\lambda_{0n} - \sum_{p\le k\le n} (\lambda_k + \lambda_{k-1,k}) - \lambda_{n,n+1} \notin \mathbb{Z}, \quad 2 \le p \le n.$$

The eigenvalues of the action of γ_0 are

$$e(\sum_{s\le k\le n} (\lambda_k + \lambda_{k,k+1})), \quad 1 \le s \le n$$

with multiplicity free and 1 with multiplicity $n+1$; the action of γ_c are

$$e(\lambda_{1c} + \sum_{1\le k\le n} \lambda_{k,k+1})$$

with multiplicity free and 1 with multiplicity $2n$; the eigenvalues of the action of γ_1 are

$$e(\lambda_{0n} + \lambda_{n,n+1})$$

with multiplicity n and 1 with multiplicity $n+1$.

The eigenvalues of the action of γ_∞ is derived from

$$u(t) = c^{\lambda_{1c}} z^{\lambda_{n,n+1}} (c^{-1} - t_1)^{\lambda_{1c}} (t_0 - t_n)^{\lambda_{0n}} \prod_{1\le i\le n} t_i^{\lambda_i^\infty} \prod_{0\le i\le n} (t_{i+1} - t_i)^{\lambda_{i,i+1}},$$

where

$$\lambda_i^\infty = \begin{cases} -\lambda_1 - \lambda_{1c} - \lambda_{12}, & i = 1, \\ -\lambda_i - \lambda_{i-1,i} - \lambda_{i,i+1}, & 2 \le i \le n-1, \\ -\lambda_{0n} - \lambda_n - \lambda_{n-1,n} - \lambda_{n,n+1}, & i = n, \end{cases}$$

with $t_{n+1} = z^{-1}$ and $t_0 = 1$. This shows the eigenvalues of the action of γ_∞ are

$$e(-\lambda_{0n} - \lambda_{1c} - \sum_{1\le k\le n} (\lambda_k + \lambda_{k,k+1})),$$

$$e(-\lambda_{0n} - \sum_{s\le k\le n} (\lambda_k + \lambda_{k-1,k}) - \lambda_{n,n+1}), \quad 2 \le s \le n,$$

each with multiplicity free, and $e(-\lambda_{n,n+1})$ with multiplicity $n+1$.

Thus, the spectral type is $(1^n, n+1; n, n+1; 1, 2n; 1^n, n+1)$ under the conditions

$$\sum_{s\le k\le l-1} (\lambda_k + \lambda_{k,k+1}) \notin \mathbb{Z}, \quad 1 \le s \le l \le n,$$

which is for the separation of the eigenvalues at 0,

$$\lambda_{1c} + \sum_{1\le k\le n} \lambda_{k,k+1} \notin \mathbb{Z},$$

which is for the separation of the eigenvalues at c,

$$\lambda_{0n} + \lambda_{n,n+1} \notin \mathbb{Z},$$

which is for the separation of the eigenvalues at 1, and

$$\lambda_{1c} + \lambda_{0n} + \sum_{1\le k\le n-1} (\lambda_k + \lambda_{k,k+1}) + \lambda_n \notin \mathbb{Z},$$

$$\lambda_{0n} + \sum_{s\le k\le n} (\lambda_k + \lambda_{k-1,k}) \notin \mathbb{Z}, \qquad 2 \le s \le n,$$

$$\lambda_{1c} + \sum_{1\le k\le s-2} (\lambda_k + \lambda_{k,k+1}) + \lambda_{s-1}, \notin \mathbb{Z}, \qquad 2 \le s \le n,$$

which is for the separation of the eigenvalues at ∞.

It means

Theorem 6.1. *The function of the form*

$$\int_C (t_1 - c)^{\lambda_{1c}} (t_n - t_0)^{\lambda_{0n}} \prod_{1\le i\le n} t_i^{\lambda_i} \prod_{1\le i\le n} (t_i - t_{i+1})^{\lambda_{i,i+1}}\, dt_1 \cdots dt_n,$$

where $t_0 = 1, t_{n+1} = z$ and C a suitable cycle, satisfies the equation of type III, whose rank is $2n+1$.*

§7. The equation of type II*

In this section, let $\mathcal{L}_z$ be the locally constant sheaf determined by a function

$$(7.1) \quad u(t) = (t_1 - c)^{\lambda_{1c}} (t_n - t_0)^{\lambda_{0n}} \prod_{1\le i\le n-1} t_i^{\lambda_i} \prod_{1\le i\le n} (t_i - t_{i+1})^{\lambda_{i,i+1}}$$

on

$$T_z = \mathbb{C}^n \backslash \{ t_1 - c = 0 \} \cup \{ t_n - t_0 = 0 \} \cup_{i=1}^{n-1} \{ t_i = 0 \} \cup \cup_{i=1}^{n} \{ t_i - t_{i+1} = 0 \},$$

where $t_{n+1} = z$ and $t_0 = 1$.

After fixing the variable z and parameter c to be real numbers satisfying $0 < z < c < 1$, the bounded chambers in the real locus $T_\mathbb{R}$ of $T = T_z$ are

$$\begin{pmatrix} 0 < t_s < t_{s+1} < \cdots < t_n < z, \\ t_s < t_{s-1} \cdots < t_1 < c \end{pmatrix}$$

for $1 \le s \le n-1$ and

$$\begin{pmatrix} z < t_n < 1 \\ 0 < t_s < t_{s+1} < \cdots < t_n, \\ t_s < t_{s-1} \cdots < t_1 < c \end{pmatrix}$$

for $1 \le s \le n$ and

$$(z < t_n < \cdots < t_1 < c).$$

Thus $\dim H_n^{\mathrm{lf}}(T_z, \mathcal{L}_z) = 2n$ under the condition

$$\lambda_{\infty=t_1=\cdots=t_q} = -\lambda_{1c} - \sum_{1\le k\le q} (\lambda_k + \lambda_{k,k+1}) \notin \mathbb{Z}, \quad 1 \le q \le n-1,$$

$$\lambda_{\infty=t_1=\cdots=t_n} = -\lambda_{1c} - \lambda_{0n} - \sum_{1\le k\le n-1} (\lambda_k + \lambda_{k,k+1}) - \lambda_{n,n+1} \notin \mathbb{Z},$$

$$\lambda_{\infty=t_p=\cdots=t_q} = -\sum_{p\le k\le q} (\lambda_k + \lambda_{k-1,k}) - \lambda_{q,q+1} \notin \mathbb{Z}, \qquad 2 \le p \le q \le n-1,$$

$$\lambda_{\infty=t_p=\cdots=t_n} = -\lambda_{0n} - \sum_{p\le k\le n-1} (\lambda_k + \lambda_{k-1,k}) - \lambda_{n-1,n} - \lambda_{n,n+1} \notin \mathbb{Z}, \qquad 2 \le p \le n.$$

The eigenvalues of the action of γ_0 are

$$e(\lambda_{n,n+1} + \sum_{s\le k\le n-1} (\lambda_k + \lambda_{k,k+1})), \quad 1 \le s \le n-1$$

with multiplicity free and 1 with multiplicity $n+1$; the action of γ_c are

$$e(\lambda_{1c} + \sum_{1\le k\le n} (\lambda_k + \lambda_{k,k+1}))$$

with multiplicity free and 1 with multiplicity $2n-1$; the eigenvalues of the action of γ_1 are

$$e(\lambda_{0n} + \lambda_{n,n+1})$$

with multiplicity n, and 1 with multiplicity n.

The eigenvalues of the action of γ_∞ is derived from

$$u(t) = c^{\lambda_{1c}} z^{\lambda_{n,n+1}} (c^{-1} - t_1)^{\lambda_{1c}} (t_0 - t_n)^{\lambda_{0n}} \prod_{1\le i\le n} t_i^{\lambda_i^\infty} \prod_{0\le i\le n} (t_{i+1} - t_i)^{\lambda_{i,i+1}},$$

where

$$\lambda_i^\infty = \begin{cases} -\lambda_1 - \lambda_{1c} - \lambda_{12}, & i = 1, \\ -\lambda_i - \lambda_{i-1,i} - \lambda_{i,i+1}, & 2 \le i \le n-1, \\ -\lambda_{0n} - \lambda_{n-1,n} - \lambda_{n,n+1}, & i = n, \end{cases}$$

with $t_{n+1} = z^{-1}$ and $t_0 = 1$. This shows the eigenvalues of the action of γ_∞ are

$$e(-\lambda_{0n} - \lambda_{1c} - \sum_{1\le k\le n} (\lambda_k + \lambda_{k,k+1}) - \lambda_{n,n+1}),$$

$$e(-\lambda_{0n} - \sum_{s\le k\le n-1} (\lambda_k + \lambda_{k-1,k}) - \lambda_{n-1,n} - \lambda_{n,n+1}), \quad 2 \le s \le n,$$

each with multiplicity free, and $e(-\lambda_{n,n+1})$ with multiplicity n.

Thus, the spectral type is $(1^{n-1}, n+1; 1, 2n-1; n, n; 1^n, n)$ under the condition

$$\lambda_{n,n+1} + \sum_{s\le k\le n-1} (\lambda_k + \lambda_{k,k+1}) \notin \mathbb{Z}, \quad 1 \le s \le n-1,$$

$$\sum_{s\le k\le l-1} (\lambda_k + \lambda_{k,k+1}) \notin \mathbb{Z}, \quad 1 \le s < l \le n-1,$$

which is for the separation of the eigenvalues at 0,

$$\lambda_{1c} + \sum_{1\le k\le n-1} \lambda_{k,k+1} \notin \mathbb{Z},$$

which is for the separation of the eigenvalues at c,

$$\lambda_{0n} + \lambda_{n,n+1} \notin \mathbb{Z},$$

which is for the separation of the eigenvalues at 1, and

$$\lambda_{1c} + \lambda_{0n} + \sum_{1\le k\le n-1} (\lambda_k + \lambda_{k,k+1}) \notin \mathbb{Z},$$

$$\lambda_{0n} + \sum_{s\le k\le n-1} (\lambda_k + \lambda_{k-1,k}) + \lambda_{n-1,n} \notin \mathbb{Z}, \qquad 2 \le s \le n,$$

$$\lambda_{1c} + \sum_{1\le k\le s-2} (\lambda_k + \lambda_{k,k+1}) + \lambda_{s-1} \notin \mathbb{Z}, \qquad 2 \le s \le n,$$

$$\sum_{s\le k\le l-1} (\lambda_k + \lambda_{k-1,k}) \notin \mathbb{Z}, \qquad 2 \le s < l \le n,$$

which is for the separation of the eigenvalues at ∞.

Finally, we obtain

Theorem 7.1. *The function of the form*

$$\int_C (t_1 - c)^{\lambda_{1c}} (t_n - t_0)^{\lambda_{0n}} \prod_{1\le i\le n-1} t_i^{\lambda_i} \prod_{1\le i\le n} (t_i - t_{i+1})^{\lambda_{i,i+1}}\, dt_1 \cdots dt_n,$$

where $t_0 = 1, t_{n+1} = z$ and C a suitable cycle, satisfies the equation of type II^*, whose rank is $2n$.

§8. Supplements

Apart from the viewpoint of Okubo equations, Oshima recently studies the Fuchsian differential equations free from accessory parameters [24, 25, 26]. He demonstrates that there exist quite many examples of such equations; indeed, the cardinality of them is described in the following tables:

# of irreducible rigid Fuchsian differential systems with 3 singularities on $\mathbb{P}^1$

order	2	3	4	5	6	7	8	9	10	11	12	13	14
#	1	1	3	5	13	20	45	74	142	212	421	588	1004

# of irreducible rigid Fuchsian differential systems

order	2	3	4	5	6	7	8	9	10	11	12	13	14
#	1	2	6	11	28	44	96	157	306	441	857	1117	2032

On the other hand, in addition to Oshima's work, we refer the reader to the works by Katz [13], Simpson [27] Dettweiler and Reiter [5, 6], Kostov [15], Gleizer [7], and Crawley–Boavey and Shaw [4], who study intimately related topics from the viewpoint of rigid local systems (See also [10, 11, 12, 30]).

Finally, for our convenience, we illustrate the diagrams to express the integrands $u(t)$ treated in the present paper. Here $\underset{a}{\circ}\!\!\longrightarrow\!\!\underset{b}{\circ}$ means $(a-b)^{\lambda_{ab}}$ up to a constant factor.

I (Generalized hypergeometric function) :

1^{n+1}	at 0
$1, n$	at 1
1^{n+1}	at ∞

I* (Pochhammmer function):

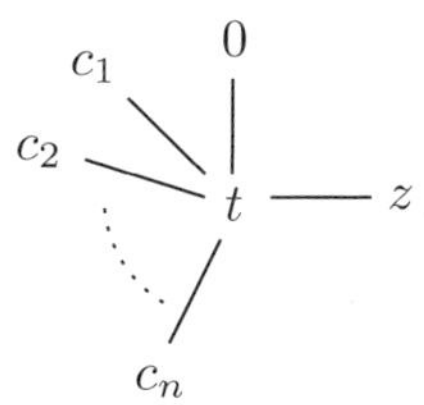

$1, n-1$	at 0
$1, n-1$	at c_1
$\cdots$	$\cdots$
$1, n-1$	at c_n
$1, n-1$	at ∞

II :

$1^n, n$	at 0
$1, n-1, n$	at 1
$1^n, n$	at ∞

II* :

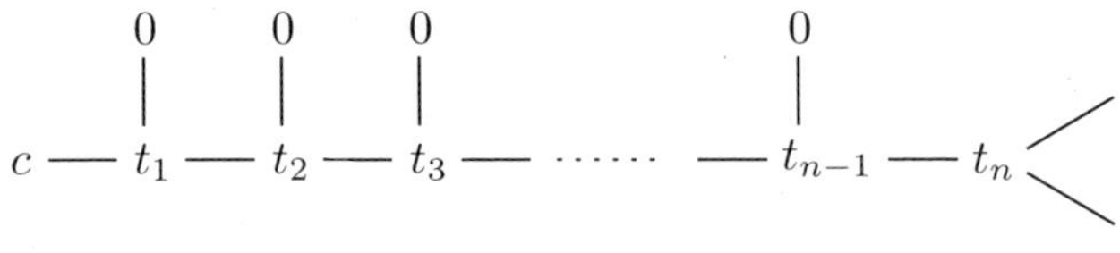

$1^{n-1}, n+1$	at 0
n, n	at 1
$1, 2n-1$	at c
$1^n, n$	at ∞

III :

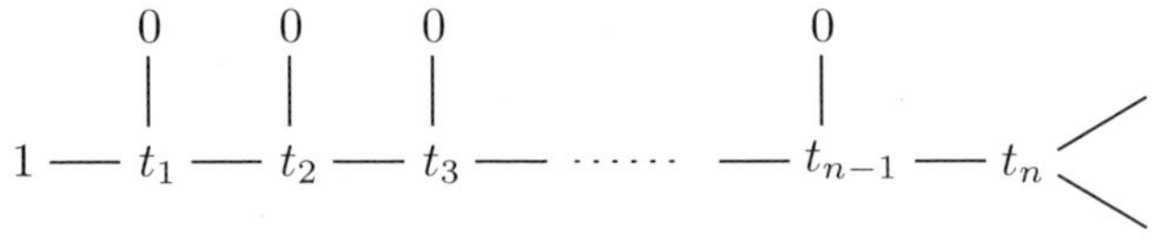

$1^{n-1}, n$	at 0
$1, n-1, n-1$	at 1
$1^n, n-1$	at ∞

III* :

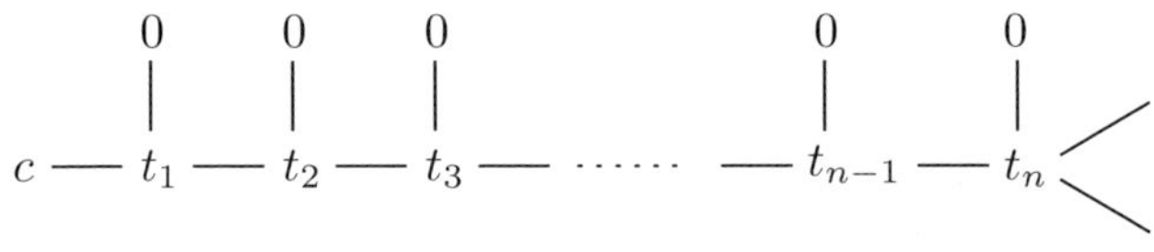

$1^n, n+1$	at 0
$n, n+1$	at 1
$1, 2n$	at c
$1^n, n+1$	at ∞

IV :

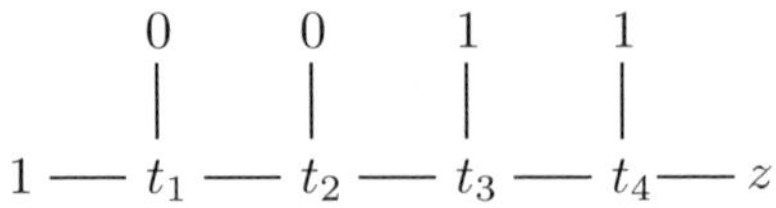

$1^2, 4$	at 0
2^3	at 1
$1^4, 2$	at ∞

(with the resonance around $t_1 = t_2 = t_3 = t_4 = 1$)

IV* :

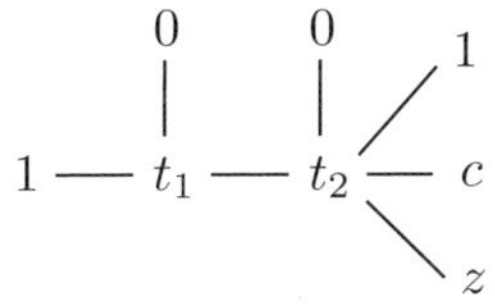

1, 1, 4	at 0
1, 1, 4	at 1
2, 4	at c
1, 1, 4	at ∞

This diagram is very useful in several situations. In fact, when we find our integrands, it played a crucial role.

The integrands $u(t)$ in the case II of rank 4 and that of rank 6, written in subsections 5.5 and 5.10 of [9], are depicted as

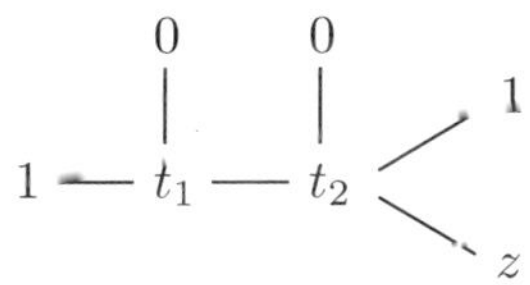

and

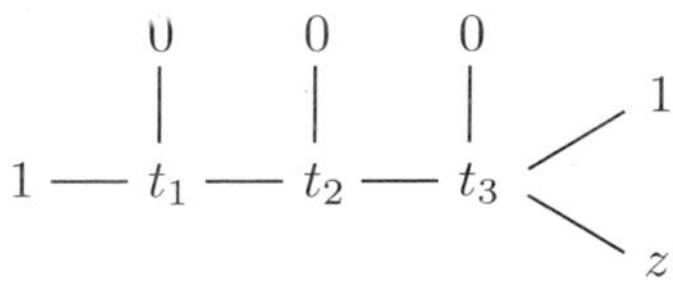

.

It is easy to guess that $u(t)$ for type II of rank $2n$ might be

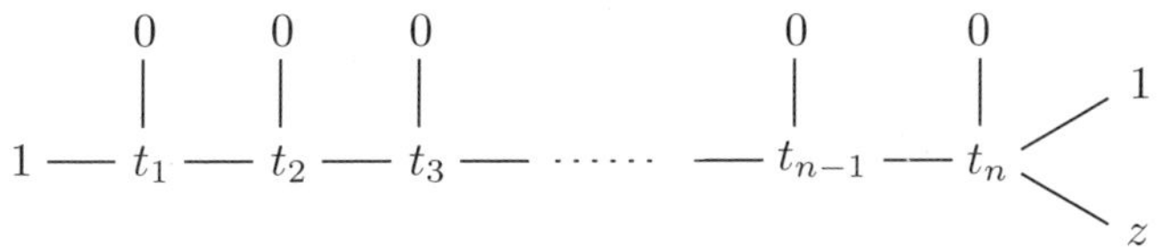

.

The integrands $u(t)$ in the case II* of rank 4 and that of rank 6, written in subsections 5.2 and 5.7 of [9], are depicted as

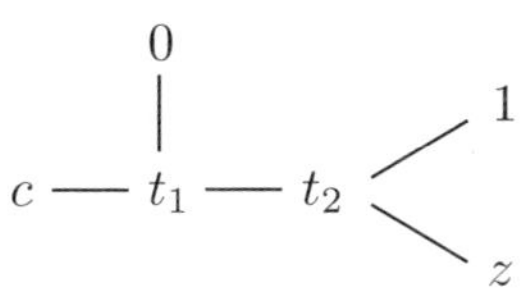

and

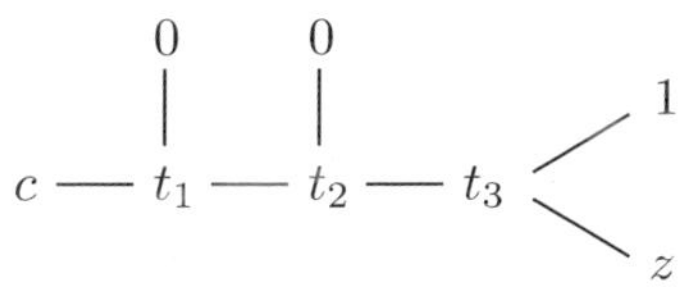

.

It is easy to guess that $u(t)$ for type II* of rank $2n$ might be

Similarly, the integrands $u(t)$ in the case III* of rank 5 and that of rank 7, written in subsections 5.4 and 5.9 of [9],

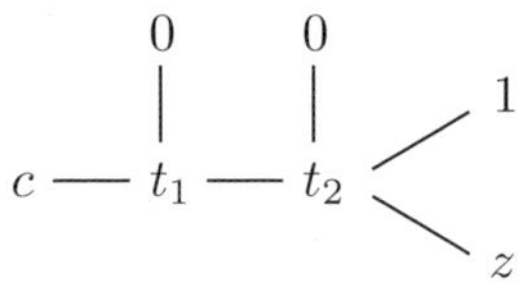

and

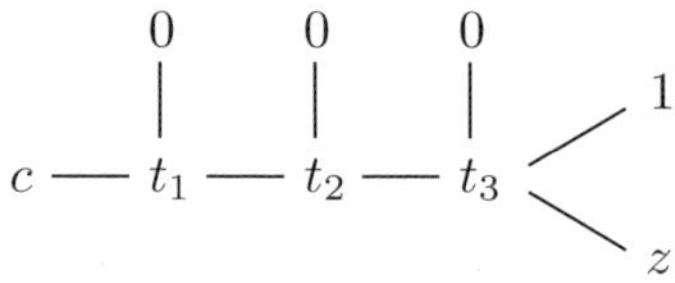

lead to the integrand

for type III* of rank $2n + 1$. Furthermore, considering the fact that the integrand in the case II is obtained from the integrand in the case II* by the specialization $c = 1$, the integrand in the case III of rank $2n + 1$ might be

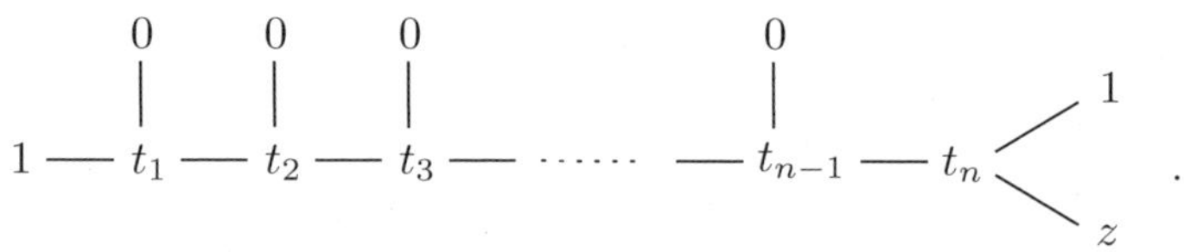

Acknowledgements. The author would like to thank Professor Yoshishige Haraoka for his explaining the rigid local systems, especially their recent work [10] and related works to the middle convolution.

References

[1] K. Aomoto, On vanishing of cohomology attached to certain many valued meromorphic functions, J. Math. Soc. Japan, **27** (1975), 248–255.

[2] K. Aomoto, On the structure of integrals of power product of linear functions, Sci. Papers College Gen. Ed. Univ. Tokyo, **27** (1977), 49–61.

[3] K. Cho, A generalization of Kita and Noumi's vanishing theorems of cohomology groups of local system, Nagoya Math J., **147** (1997), 63–69.

[4] W. Crawley-Boavey and P. Shaw, Multiplicative projective algebras, middle convolution and Deligne–Simpson problem, Adv. Math., **201** (2006), 180–208.

[5] M. Dettweiler and S. Reiter, An algorithm of Katz and its applications to the inverse Galois problems, J. Symbolic Comput., **30** (2000), 761–798.

[6] M. Dettweiler and S. Reiter, Middle convolution of Fuchsian systems and the construction of rigid differential systems, J. Algebra, **318** (2007), 1–24.

[7] O. Gleizer, Explicit solutions of the additive Deligne-Simpson problem and their aplications, Adv Math., **178** (2003), 311–374.

[8] Y. Haraoka, Canonical forms of differential equations free from accessory parameters, SIAM Jour. Math. Anal., **25** (1994), 1203–1226.

[9] Y. Haraoka, Integral representations of solutions of differential equations free from accessory parameters, Adv. Math., **169** (2002), 187–240.

[10] Y. Haraoka and S. Hamaguchi, Topological theory for Selberg type integral associated with rigid Fuchsian systems, to appear in Math. Ann.

[11] Y. Haraoka and K. Mimachi, A connection problem for Simpson's even family of rank four, Funkt. Ekvac., **54** (2011), 495–515.

[12] Y. Haraoka and T. Yokoyama, Construction of rigid local systems and integral representation of their sections, Math. Nachr., **279** (2006), 255–271.

[13] N. Katz, Rigid Local Systems, Ann. of Math. Stud., **139**, Pronceton Univ. Press, 1995.

[14] M. Kita, On hypergeometric functions in several variables 1, New integral representations of Euler type, Japan. J. Math., **18** (1992), 25–74.

[15] V. P. Kostov, The Deligne-Simpson problem—a survey, J. Algebra, **281** (2004), 83–108.

[16] K. Mimachi, An integral representation of the solution of a fourth order Fuchsian differential equation of Okubo type, Funkt. Ekvac., **38** (1995), 411–416.

[17] K. Mimachi, Connection matrices associated with the generalized hypergeometric function ${}_3F_2$, Funkt. Ekvac., **51** (2008), 107–133.

[18] K. Mimachi, Intersection numbers for twisted cycles and the connection problem associated with the generalized hypergeometric function ${}_{n+1}F_n$, Internat. Math. Res. Notices, **2011** (2011), 1757–1781.

[19] K. Mimachi, H. Ochiai and M. Yoshida, Intersection theory for twisted cycles IV—resonant cases, Math. Nachr., **260** (2003), 67–77.

[20] K. Mimachi and M. Yoshida, Intersection numbers of twisted cycles and the correlation functions of the conformal field theory, Commun. Math. Phys., **234** (2003), 339–358.
[21] K. Mimachi and M. Yoshida, Intersection numbers of twisted cycles associated with the Selberg integral and an application to the conformal field theory, Commun. Math. Phys., **250** (2004), 23–45.
[22] K. Mimachi and M. Yoshida, Regularizable cycles associated with a Selberg type integral under some resonance condition, Internat. J. Math., **18** (2007), 395–409.
[23] K. Okubo, On the group of fuchsian equations, Seminar reports of Tokyo Metropolitan Univ., Tokyo, 1987.
[24] T. Oshima, Classification of Fuchsian systems and their connection problem, preprint, arXiv:0811.2916.
[25] T. Oshima, Fractional calculus of Weyl algebra and Fuchsian differential equations, http://akagi.ms.u-tokyo.ac.jp/~oshima/index-j.html.
[26] T. Oshima and N. Shimeno, Heckman-Opdam hypergeometric functions and their specializations, http://akagi.ms.u-tokyo.ac.jp/~oshima/.
[27] C. Simpson, Products of matrices, In: CMS Conference Proc., **12**, Amer. Math. Soc., Providence, RI, 1991, pp. 157–185.
[28] K. Takano and E. Bannai, A global study of Jordan-Pochhammer differential equations, Funk. Ekvac., **19** (1976), 85–99.
[29] T. Yokoyama, On an irreducibility condition for hypergeometric systems, Funkt. Ekvac., **38** (1995), 11–19.
[30] T. Yokoyama, Construction of systems of differential equations of Okubo normal form with rigid monodromy, Math. Nachr., **279** (2006), 327–348.

Katsuhisa Mimachi
Department of Mathematics, Tokyo Institute of Technology,
Oh-okayama, Meguro-ku, Tokyo 152-8551, Japan
E-mail address: `mimachi@math.titech.ac.jp`

Advanced Studies in Pure Mathematics 62, 2012
Arrangements of Hyperplanes—Sapporo 2009
pp. 261–291

Resonance webs of hyperplane arrangements

Jorge Vitório Pereira

Abstract.

Each irreducible component of the first resonance variety of a hyperplane arrangement naturally determines a codimension one foliation on the ambient space. The superposition of these foliations define what we call the resonance web of the arrangement. In this paper we initiate the study of these objects with emphasis on their spaces of abelian relations.

§1. Introduction

Let $\mathsf{A} = \{H_1, \dots, H_r\}$ be an arrangement of $r \geq 1$ hyperplanes in $\mathbb{P}^n$. The complement of A is an affine variety that will be denoted by $M = M(\mathsf{A})$. It is a result of Arnold [2] (for the braid arrangement) and Brieskorn [5] (for and arbitrary hyperplane arrangement) that the cohomology ring of M, $H^\bullet(M, \mathbb{Z})$, is torsion free and generated, as a graded algebra, by the degree one elements determined by the classes of the logarithmic differential forms

$$(2\pi i)^{-1}\left(d\log\frac{h_i}{h_r}\right) \qquad \text{for } i \in \{1, \dots, r-1\}$$

where $h_1, \dots, h_r$ are linear polynomials in $\mathbb{C}[x_0, \dots, x_n]$ defining the hyperplanes in A.

Given $a \in H^1(M) = H^1(M, \mathbb{C})$, consider the complex $(H^\bullet(M), a)$ with arrows given by multiplication by a:

$$0 \longrightarrow H^0(M) \longrightarrow H^1(M) \longrightarrow H^2(M) \longrightarrow \cdots \longrightarrow H^n(M) \longrightarrow 0\,.$$

Received March 31, 2010.
Revised December 24, 2010.
2010 *Mathematics Subject Classification.* 52C35, 53A60.
Key words and phrases. Web geometry, abelian relations, hyperplane arrangements, resonance varieties.

The resonance varieties of M, or A, are defined as

$$\mathsf{R}^i(M) = \mathsf{R}^i(\mathsf{A}) = \{a \in H^1(M), \mid h^i(H^\bullet(M), a) \neq 0\}.$$

The paper [14] provides the following nice description of the first resonance variety $\mathsf{R}^1(M)$. The irreducible components of $\mathsf{R}^1(M)$ are precisely the maximal isotropic subspaces of $H^1(M)$ for the quadratic form

$$\wedge : H^1(M, \mathbb{C}) \otimes H^1(M, \mathbb{C}) \longrightarrow H^2(M, \mathbb{C})$$

having dimension at least two. Moreover, the irreducible components of $\mathsf{R}^1(M)$ of dimension k are in correspondence with pencils of hypersurfaces on $\mathbb{P}^n$ having exactly $k+1$ elements with support contained in the arrangement. In particular, for each irreducible component Σ of the resonance variety there is a unique (singular holomorphic) foliation $\mathcal{F}_\Sigma$ on $\mathbb{P}^n$ defined by the corresponding pencil of hypersurfaces.

The interest of the study of the resonance varieties of a hyperplane arrangement is amplified by its relation with the cohomology jumping loci of rank one local systems on M. The characteristic variety $\mathrm{Char}^i(M)$ of M is the subvariety of $\mathrm{Hom}(\pi_1(M), \mathbb{C}^*)$ defined as

$$\mathrm{Char}^i(M) = \{\rho \in \mathrm{Hom}(\pi_1(M), \mathbb{C}^*) \mid h^i(M, \mathbb{C}_\rho) \neq 0\},$$

where $\mathbb{C}_\rho$ is the rank one local system determined by ρ. The above mentioned relation is given by the following Theorem [1, 11]: the exponential map

$$\begin{aligned} \exp : H^1(M) &\longrightarrow \mathrm{Hom}(\pi_1(M), \mathbb{C}^*) \\ a &\longmapsto \left(\gamma \mapsto \exp\left(2\pi i \int_\gamma a\right)\right) \end{aligned}$$

defines isomorphisms between the germs $(\mathsf{R}^i(M), 0)$ and $(\mathrm{Char}^i(M), \mathbf{1})$, where $\mathbf{1} \in \mathrm{Hom}(\pi_1(M), \mathbb{C}^*)$ is the trivial representation.

The study of the foliation $\mathcal{F}_\Sigma$ for irreducible components $\Sigma \subset \mathsf{R}^1(M)$ led the author and S. Yuzvinsky (see [23]) to bounds for the dimension of Σ. Although there is now (especially after [29]) a reasonably clear picture about each of the irreducible components of $\mathsf{R}^1(M)$, it is not very clear how the totality of them sit inside $H^1(M, \mathbb{C})$. In this paper we propose an approach to produce invariants for arrangements that may turn out to be useful to the study of this question. The underlying idea is fairly simple: instead of looking at the foliations associated to the irreducible components of $\mathsf{R}^1(M)$ one at a time, we should look at all of them at the same time. More precisely, we will associate to an arrangement A what we call its resonance web $\mathcal{W}(\mathsf{A})$—the superposition of all the foliations $\mathcal{F}_\Sigma$ associate to irreducible components of $\mathsf{R}^1(\mathsf{A})$—and will study its space of abelian relations.

Conversely, many relevant examples for web geometry, especially in what concerns the dimension of the space of abelian relations, appear as resonance webs of certain arrangements. The list starts with the Bol exceptional 5-web [4], contains the Spence–Kummer exceptional 9-web [27, 25], and ends with some other exceptional webs presented in [27, 25]. This provides further motivation to pursue the study of resonance webs.

Our main result is Theorem 4.1 which determines the rank of the resonance webs for the braid arrangements which, as we will see in Section 4, correspond to the $\binom{n+3}{4}$-web on $\mathfrak{M}_{0,n+3}$ (the moduli space of $n+3$ distinct ordered points on $\mathbb{P}^1$) defined by the $\binom{n+3}{4}$ natural maps

$$\mathfrak{M}_{0,n+3} \longrightarrow \mathfrak{M}_{0,4} \simeq \mathbb{C} \setminus \{0,1\}\,.$$

We will also draw some general considerations about the abelian relations of resonance webs and use them to study some of the examples of exceptional webs mentioned above. Although we have no major results on the structure of the space of abelian relations of resonance webs for arbitrary arrangements, the blurry picture delineated by these examples is considerably intricate and, we believe, invites further investigation.

Plan of the paper. In Section 2 we define webs, their spaces of abelian relations, and show how to bound the rank of arbitrary codimension one webs. We also review the algebraization results for webs of maximal rank, define exceptional webs, and present Bol's example of exceptional planar 5-web. In Section 3 we define resonance webs and initiate the study of their spaces of abelian relations, more specifically the subspace of polylogarithmic abelian relations generated by collections of iterated integrals of logarithmic 1-forms with poles on the arrangement. The reader will also find in Section 3 a brief presentation of a couple of basic results from Chen's theory of iterated integrals relevant to our study. Section 4 is devoted to the statement and proof of our main result: the determination of the rank of the resonance webs of the braid arrangements. Section 5 studies some of the exceptional planar webs found by Pirio and Robert as resonance webs of line arrangements in $\mathbb{P}^2$.

§2. Web geometry

For us a **germ of codimension one k-web** $\mathcal{W} = \mathcal{F}_1 \boxtimes \cdots \boxtimes \mathcal{F}_k$ on $(\mathbb{C}^n, 0)$ is a collection of k germs of smooth codimension one holomorphic foliations subjected to the condition that any two distinct foliations $\mathcal{F}_i, \mathcal{F}_j$ have distinct tangent spaces at zero.

Usually in the literature a stronger condition is imposed on the tangent spaces at zero. In the terminology of [7] the tangent spaces are usually assumed to be in **strong general position**, meaning that for any $1 \leq m \leq n$ the intersection of tangent spaces at zero of m distinct foliations $\mathcal{F}_i$ have codimension m.

Perhaps the most studied invariant of a germ of codimension one web $\mathcal{W}$ is its **space of abelian relations** $\mathcal{A}(\mathcal{W})$. If we choose integrable 1-forms ω_i inducing the foliations $\mathcal{F}_i$ then $\mathcal{A}(\mathcal{W})$ is equal to

$$\left\{ (\eta_i)_{i=1}^k \in (\Omega^1(\mathbb{C}^n, 0))^k \;\middle|\; \forall i \; d\eta_i = 0 \,,\; \eta_i \wedge \omega_i = 0 \;\text{ and }\; \sum_{i=1}^k \eta_i = 0 \right\}.$$

If $u_i : (\mathbb{C}^n, 0) \to (\mathbb{C}, 0)$ are local submersions defining the foliations $\mathcal{F}_i$ then, after integration, the abelian relations can be read as functional equations of the form $\sum_{i=1}^k g_i(u_i) = 0$ for some germs of holomorphic functions $g_i : (\mathbb{C}, 0) \to (\mathbb{C}, 0)$. Thus we can interpret the abelian relations of $\mathcal{W}$ as functional equations (of a rather special kind) among the first integrals of the foliations defining it.

2.1. Rank of webs

Clearly $\mathcal{A}(\mathcal{W})$ is a vector space and its dimension is commonly called the **rank** of $\mathcal{W}$ and is denoted by $\operatorname{rank}(\mathcal{W})$. We will now explain how one can bound the rank of arbitrary codimension one webs. This is a classical subject in web geometry and has been treated by Bol ($n = 2$) and Chern ($n \geq 3$ for webs in strong general position) in the decade of 1930, and more recently by Cavalier–Lehmann for ordinary webs, see the definition below. Here we will deal with arbitrary codimension one webs. This section is a summary of [21, Section 2.2].

For every $i \in \{1, \ldots, k\}$ let ω_i be a germ of 1-form defining $\mathcal{F}_i$ and satisfying $\omega_i(0) \neq 0$. For any positive integer j define $\mathcal{L}^j(\mathcal{W})$ as the subspace of the $\mathbb{C}$-vector space $\operatorname{Sym}^j(\Omega^1(\mathbb{C}^n, 0))$ generated by the j-th symmetric powers of the exterior forms $\omega_i(0)$ with $i \in \{1, \ldots, k\}$. Set $\ell^j(\mathcal{W}) = \dim \mathcal{L}^j(\mathcal{W})$.

Alternatively, if $u_i : (\mathbb{C}^n, 0) \to (\mathbb{C}, 0)$ are local submersions defining the foliations $\mathcal{F}_i$, and h_i are their linear terms then

$$\ell^j(\mathcal{W}) = \dim\left(\mathbb{C}h_1^j + \cdots + \mathbb{C}h_k^j\right) . \tag{1}$$

Notice that the integer $\ell^j(\mathcal{W})$ is bounded by k and by the dimension of the vector space of homogeneous polynomials of degree j in n variables, i.e.

$$\ell^j(\mathcal{W}) \leq \max\left\{k, \binom{n+j-1}{n-1}\right\} . \tag{2}$$

In the terminology of [7], a germ of k-web $\mathcal{W}$ on $(\mathbb{C}^n, 0)$ is **ordinary** if and only if $\ell^j(\mathcal{W}) = \max\left\{k, \binom{n+j-1}{n-1}\right\}$ for every positive integer j.

A good lower bound is harder to obtain. For webs in strong general position there is a lemma by Castelnuovo [21, Proposition 2.2.2] which says that

$$\ell^j(\mathcal{W}) \geq \min(k, j(n-1)+1). \tag{3}$$

For arbitrary webs, no longer in strong general position, the best possible lower bound is $\ell^j(\mathcal{W}) \geq \min(k, j+1)$.

The argument used to prove the proposition below is borrowed from Trépreau's proof [28] of Chern's bound for the rank of webs in strong general position.

Proposition 2.1. *If $\mathcal{W}$ is an arbitrary k-web on $(\mathbb{C}^n, 0)$ then*

$$\operatorname{rk}(\mathcal{W}) \leq \sum_{j=1}^{\infty} \max(0, k - \ell^j(\mathcal{W}))$$

and the sum involves only finitely many non-zero terms.

Proof. The space of abelian relations of $\mathcal{W}$ admits a natural filtration $\mathcal{A}(\mathcal{W}) = \mathcal{A}^0(\mathcal{W}) \supseteq \mathcal{A}^1(\mathcal{W}) \supseteq \ldots \supseteq \mathcal{A}^j(\mathcal{W}) \supseteq \ldots$, where

$$\mathcal{A}^j(\mathcal{W}) = \ker\left\{\mathcal{A}(\mathcal{W}) \longrightarrow \left(\frac{\Omega^1(\mathbb{C}^n, 0)}{\mathfrak{m}^j \cdot \Omega^1(\mathbb{C}^n, 0)}\right)^k\right\},$$

and $\mathfrak{m}$ is the maximal ideal of $\mathbb{C}\{x_1, \ldots, x_n\}$.

One can easily verify [21, Lemma 2.2.6] that

$$\dim \frac{\mathcal{A}^j(\mathcal{W})}{\mathcal{A}^{j+1}(\mathcal{W})} \leq k - \dim\left(\mathbb{C} \cdot h_1^{j+1} + \cdots + \mathbb{C} \cdot h_k^{j+1}\right) \tag{4}$$

where h_i is as in (1). The bound follows. Moreover, as $\ell^j(\mathcal{W}) \geq \min(k, j+1)$ there are only finitely many non-zero terms at the summation above. Q.E.D.

The proposition above combined with the lower bounds previously discussed allows us to recover the bounds for the rank of germs of codimension one webs available elsewhere.

Corollary 2.1. Let $\mathcal{W}$ be a germ of k-web on $(\mathbb{C}^n, 0)$. The assertions below hold true.

(1) (**Bol's bound**) If $n = 2$ then

$$\operatorname{rank}(\mathcal{W}) \leq \frac{(k-1)(k-2)}{2}.$$

(2) (**Chern's bound**) If $n \geq 3$ and $\mathcal{W}$ is in strong general position then

$$\operatorname{rank}(\mathcal{W}) \leq \sum_{j=1}^{\infty} \max(0, k - j(n-1) - 1).$$

(3) (**Cavalier-Lehmann's bound**) If $n \geq 3$ and $\mathcal{W}$ is an ordinary web then

$$\operatorname{rank}(\mathcal{W}) \leq \sum_{j=1}^{\infty} \max\left(0, k - \binom{n+j-1}{n-1}\right).$$

The number at right-hand side of Chern's bound is Castelnuovo's bound $\pi(n, k)$ for the arithmetic genus of irreducible, non-degenerated curves of degree k on $\mathbb{P}^n$.

2.2. Algebraic and algebraizable webs

An important class of examples of webs is the class of **algebraic webs** which are webs dual to projective curves. If C is a reduced degree k projective curve on $\mathbb{P}^n$ then for every general hyperplane H_0 a germ of codimension one k-web $\mathcal{W}_C$ is naturally defined on $(\check{\mathbb{P}}^n, H_0)$ through projective duality. More precisely $\mathcal{W}_C$ is defined by the germs of submersions $p_i : (\check{\mathbb{P}}^n, H_0) \to C$ characterized by

$$H \cdot C = p_1(H) + p_2(H) + \cdots + p_k(H)$$

for every H sufficiently close to H_0.

If ω_C denotes the sheaf of abelian differentials on C, see for instance [21, Chapter 3, Section 2.2] for the definition, then Abel's addition Theorem says that for every $p_0 \in C$ and every regular 1-form $\omega \in H^0(C, \omega_C)$ the sum

$$\int_{p_0}^{p_1(H)} \omega + \int_{p_0}^{p_2(H)} \omega + \cdots + \int_{p_0}^{p_k(H)} \omega$$

does not depend on H. One can reformulate this statement as

$$\sum_{i=1}^{k} p_i^* \omega = 0 \, .$$

It follows that $(p_1^*\omega, \ldots, p_k^*\omega)$ can be interpreted as an abelian relation of the algebraic web $\mathcal{W}_C$. Consequently there is an injection of $H^0(C, \omega_C)$ into $\mathcal{A}(\mathcal{W}_C)$. There is a converse to Abel's addition Theorem (due to Lie, Poincaré, Darboux, Wirtinger see [21, Chapter 4]) which implies that this injection is indeed an isomorphism.

Since $h^0(C, \omega_C) = (k-1)(k-2)/2$ for any reduced plane curve of degree k, it follows that every algebraic planar web has maximal rank. The same is no longer true in higher dimensions as the (arithmetic) genus of curves is not determined by theirs degrees. The curves giving rise to maximal rank webs are exactly those which have maximal genus in a given degree, the so called Castelnuovo curves.

2.3. Exceptional webs

For k-webs in strong general position on $(\mathbb{C}^n, 0)$, $n \geq 3$, the maximality of rank implies that the web is **algebraizable** (biholomorphic to a web obtained from a projective curve through duality as explained above) when $k \leq n+1$ or $k \geq 2n$. This was proved by Bol for $n = 3$, and for $n > 3$ is a recent result of Trépreau, see [28]. The planar case ($n = 2$) is rather special in what concerns the classification of webs of maximal rank. For $k \leq 4$ it is well-known that planar webs of maximal rank are algebraizable, the proof for $k = 4$ can be traced back to Lie's work on double translation surfaces, and for $k = 3$ is due to Blaschke–Dubourdieu. In sharp contrast, [19] exhibits examples of non-algebraizable planar k-webs of maximal rank for every $k \geq 5$. Further infinite families of examples appear in [20]. Despite recent advances, see for instance [15, 16, 19, 20], the classification of **exceptional** (non-algebraizable and of maximal rank) planar k-webs is wide open. For a short review of these results see [22]. A more leisurely account can be found in [21].

So far the focus was on germs of webs, but we can consider webs globally defined on a complex variety X. For our pourposes it will be sufficient to consider completely decomposable webs, that is webs $\mathcal{W}$ which can be globally presented as the superposition of k pairwise distinct global foliations $\mathcal{F}_1 \boxtimes \cdots \boxtimes \mathcal{F}_k$. For the general definition of global webs see [21, Chapter 1].

Given a global web $\mathcal{W}$, there exists a subvariety $\Lambda = \Lambda(\mathcal{W}) \subset X$ such that for every $x \in X \setminus \Lambda$ the germ of web $\mathcal{W}_x$ obtained by localizing $\mathcal{W}$ at x is a germ of codimension one web and the rank of $\mathcal{W}_x$ is independent of $x \in X \setminus \Lambda$, see [26, Theorem 1.2.2]. More precisely, over $X \setminus \Lambda$ the space of abelian relations of $\mathcal{W}$ is a local system of k-uples of germs of closed 1-forms. Therefore it still makes sense to talk about the rank for global webs.

The first example of exceptional web dates back to the 1936 and was found by Bol, see [4]. It is the global 5-web $\mathcal{B}_5$ on $\mathbb{P}^2$ formed by the superposition of 4 pencils of lines with base points in general position and one pencil of conics through these four base points. We will explain below that this web is naturally associated to an arrangement of lines on $\mathbb{P}^2$.

§3. Resonance webs

Let $\mathsf{A} = \{H_1, \ldots, H_r\}$ be an arrangement of $r \geq 1$ hyperplanes in $\mathbb{P}^n$. Recall from the introduction that $\mathsf{R}^1(\mathsf{A})$, the first resonance variety of A, is the union of the maximal isotropic subspaces of $(H^1(M), \wedge)$ of dimension at least two.

For $i \in \{1, \ldots, r\}$, let $h_i \in \mathbb{C}[x_0, \ldots, x_n]$ be a linear polynomial defining H_i. From now on we will identify $H^\bullet(M)$ with the algebra generated by the logarithmic 1-forms $(d\log h_i - d\log h_r)$ with $i \in \{1, \ldots, r-1\}$.

Before defining our main object of study, let us give a brief idea on how one can associate a pencil of hypersurfaces with $d+1$ completely decomposable fibers to an irreducible component Σ of $\mathsf{R}^1(\mathsf{A})$ of dimension d. Let $\omega_1, \ldots, \omega_d$ be a basis of Σ. As Σ is isotropic there exists non-constant rational functions $h_{ij} \in \mathbb{C}(x_0, \ldots, x_n)$ $(i \neq j)$ such that $\omega_i = h_{ij}\omega_j$. Differentiating this last expression one obtains $0 = dh_{ij} \wedge \omega_j$. Thus the level sets of the functions h_{ij} are tangent to the distribution determined by ω_i for every $i \in \{1, \ldots, d\}$. Stein factorization theorem ensures the existence of a rational map $f : \mathbb{P}^n \dashrightarrow \mathbb{P}^1$ such that $df \wedge dh_{ij} = df \wedge \omega_i = 0$ for every $i, j \in \{1, \ldots, d\}$. Moreover, there exists d linearly independent logarithmic 1-forms on $\mathbb{P}^1$, say $\eta_1, \ldots, \eta_d$, such that $f^*(\eta_i)_\infty \subset \mathsf{A}$, and $f^*(\eta_i) = \omega_i$ for every $i \in \{1, \ldots, d\}$. The maximality of Σ implies that the cardinality of

$$\mathsf{P}_\Sigma = \bigcup_{i\in \underline{d}} (\eta_i)_\infty$$

is equal to $d+1$. Thus the rational map $f = f_\Sigma$ determines a pencil of hypersurfaces with $d+1$ fibers contained in the support of the arrangement. Moreover, the restriction of f to $M = \mathbb{P}^n \setminus \mathsf{A}$ is a regular morphism

$$f_{|M} : M \to C_\Sigma\,,$$

where $C_\Sigma = \mathbb{P}^1 \setminus \mathsf{P}_\Sigma$.

Let $\mathcal{F}_\Sigma$ be the foliation on M (or on $\mathbb{P}^n$) determined by the level sets of f_Σ. We define $\mathcal{W}(\mathsf{A})$, the **resonance web** of A as the global web on M (or on $\mathbb{P}^n$) obtained by the superposition of the foliations $\mathcal{F}_\Sigma$ with Σ ranging over the irreducible components of $\mathsf{R}^1(\mathsf{A})$.

Example 3.1. Consider the arrangement on $\mathbb{P}^2$ defined by the polynomial $\{xyz(x-z)(y-z)(x-y)=0\}$. The points of its complement can be interpreted as isomorphism classes of 5 ordered points on $\mathbb{P}^1$. To wit, the point $(x:y:1)\in\mathbb{P}^2$ satisfying $x-y\neq 0, x\neq 0,1$, and $y\neq 0,1$ naturally correspond to the 5-tuple $(0,1,\infty,x,y)\in(\mathbb{P}^1)^5$. We will denote this arrangement by $\mathsf{A}_{0,5}$ and its complement in $\mathbb{P}^2$ by $\mathfrak{M}_{0,5}$. The resonance variety of $\mathfrak{M}_{0,5}$ has five irreducible components and the associated morphisms are the five forgetting maps $\mathfrak{M}_{0,5} \longrightarrow \mathfrak{M}_{0,4}$, sending isomorphism classes of five ordered points on $\mathbb{P}^1$ to isomorphism classes of four ordered points on $\mathbb{P}^1$. It is a simple exercise to verify that fibers of four of these maps form pencils of lines with base points in general position, and that the fibers of one of them is a pencil of conics through these base points. Thus Bol's exceptional 5-web $\mathcal{B}_5$ is nothing more than $\mathcal{W}(\mathsf{A}_{0,5})$ the resonance web of $\mathsf{A}_{0,5}$.

Due to the important role of Bol's 5-web in the development of web geometry it is natural to enquire about the rank of the resonance webs for arbitrary arrangements. To determine the rank of an arbitrary web is a daunting task. The only general method dates back to Abel (see [26, 25] for a modern account) and involves lengthy algebraic manipulations which lead to linear differential equations of high order that have to be solved. Although implementable, in practice such method is not computationally efficient and cannot deal with k-webs when k is large, say $k>10$.

An alternative approach to compute the rank of certain webs has been devised by Gilles Robert. Loosely speaking it restricts the search

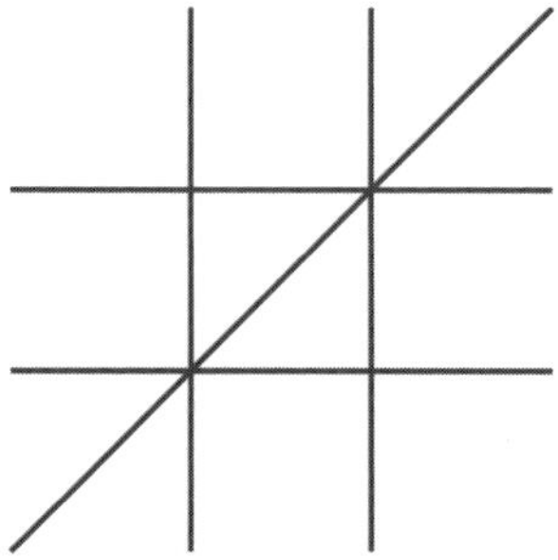

Fig. 1. Affine trace of the arrangement $\mathsf{A}_{0,5}$

of abelian relations to a certain class of differential forms defined from iterated integrals of logarithmic differentials. Here we will explore this approach, adding some topology/combinatorics of arrangements to the picture.

3.1. Logarithmic abelian relations

Consider the morphism

$$\begin{array}{rcl} \Psi_1 : \bigoplus_I H^1(C_\Sigma) & \longrightarrow & H^1(M) \\ (\eta_\Sigma) & \mapsto & \sum f_\Sigma^* \eta_\Sigma \,, \end{array}$$

where Σ ranges over all the irreducible components of $\mathsf{R}^1(\mathsf{A})$. Let $\operatorname{Log}^1 \mathcal{W}(\mathsf{A})$ be its kernel, i.e., $\operatorname{Log}^1 \mathcal{W}(\mathsf{A}) = \ker \Psi_1$.

Proposition 3.1. *The vector space* $\operatorname{Log}^1 \mathcal{W}(A)$ *embeds into the space of abelian relations of* $\mathcal{W}(A)$.

Proof. Let $(\eta_\Sigma) \in \ker \Psi_1$ be a non-zero element. Each η_Σ corresponds to a logarithmic 1-form on $\mathbb{P}^1$. The pull-backs under the corresponding rational map $f_\Sigma^* \eta_\Sigma$ are closed logarithmic 1-forms on $\mathbb{P}^n$, and $f_\Sigma^* \eta_\Sigma$ defines a distribution tangent to the foliation $\mathcal{F}_\Sigma$. Thus $(f_\Sigma^* \eta_\Sigma) \in \mathsf{A}(\mathcal{W}(\mathsf{A}))$ as wanted. Q.E.D.

A natural place to look for further abelian relations is to consider differential forms with logarithmic coefficients, that is, differential forms like

$$\log(z) \frac{dz}{z-1}.$$

A convenient formalism to deal with such objects is Chen's theory of iterated integrals.

3.2. Chen's theory of iterated integrals

In this paragraph M will be an arbitrary connected complex manifold. Given a path $\gamma : [0,1] \to M$ and a collections of 1-forms $\omega_1, \dots, \omega_k$ the iterated integral of $\omega_1 \otimes \cdots \otimes \omega_k$ along γ is defined as

$$\int_\gamma \omega_1 \otimes \cdots \otimes \omega_k = \int_{\Delta_k} p_1^* \omega_1 \wedge \cdots \wedge p_k^* \omega_k$$

where $p_i : M^k \to M$ is the projection on the i-th factor and Δ_k is the image on M^k of the k-simplex in $\mathbb{R}^k$ given by $0 \le x_1 < \cdots \le x_k \le 1$ under the map $\underbrace{\gamma \times \cdots \times \gamma}_{k \text{ times}}$.

Even if the 1-forms ω_i are closed this integral does depend on the path and not just on its homotopy class. It is a result of Chen [8, Theorem 4.1.1] that the elements of $\Omega^1(M)^{\otimes k}$ for which the corresponding iterated integral does not depend on the representative in a given homotopy class are in the intersection of the kernels of the linear maps, $i \in \{1, \dots, k-1\}$,

$$\begin{array}{ccc} \Omega^1(M)^{\otimes k} & \longrightarrow & \Omega^1(M)^{\otimes i-1} \otimes \Omega^2(M) \otimes \Omega^1(M)^{\otimes(k-i-1)} \\ \omega_1 \otimes \cdots \otimes \omega_k & \mapsto & \omega_1 \otimes \cdots \otimes \omega_{i+1} \wedge \omega_{i+2} \otimes \omega_{i+3} \otimes \cdots \otimes \omega_k\,, \end{array}$$

with the kernels of the linear maps, $i \in \{1, \dots, k\}$,

$$\begin{array}{ccc} \Omega^1(M)^{\otimes k} & \longrightarrow & \Omega^1(M)^{\otimes i-1} \otimes \Omega^2(M) \otimes \Omega^1(M)^{\otimes(k-i)} \\ \omega_1 \otimes \cdots \otimes \omega_k & \mapsto & \omega_1 \otimes \cdots \otimes d\omega_{i+1} \otimes \omega_{i+2} \otimes \cdots \otimes \omega_k\,. \end{array}$$

If $B^k(M)$ denotes this intersection then every element of $B^k(M)$ gives rise to a function on the universal covering of M through (iterated) integration. Thus we can interpret the elements of $B^k(M)$ as closed 1-forms on the universal covering of M by considering the differential of this function.

Moreover, Chen also proved that if we consider a vector subspace V of $\Omega^1(M)$ formed by closed 1-forms with no non-zero exact forms then the iterated integrals define an injection of $\oplus_{k\ge1} V^{\otimes k} \cap B^k(M)$ into the space of holomorphic functions on the universal covering of M. In particular, when M is the complement of a hyperplane arrangement, this is the case for $H^1(M)$ seen as a vector subspace of $\Omega^1(M)$. For more about iterated integrals of logarithmic 1-forms on the complement of arrangements see [18].

It is also interesting to observe that $B(M) = \oplus_{k=1}^{\infty} B^k(M)$ admits a natural structure of $\pi_1(M)$-module, with action defined by analytic continuation of the iterated integrals. Notice that the summands $B^k(M)$ are not $\pi_1(M)$-invariant when $k \geq 2$, but the terms of the filtration $F^\bullet : F^k(M) = \bigoplus_{i=1}^k B^i(M)$ are.

3.3. Polylogarithmic abelian relations

It is natural to extend the construction of Section 3.1 to arbitrary iterated integrals of logarithmic 1-forms. For each $i \geq 1$, consider the morphism

$$\begin{aligned} \Psi_i : \bigoplus_I H^1(C_\Sigma)^{\otimes i} &\longrightarrow H^1(M)^{\otimes i} \\ (\eta_\Sigma) &\mapsto \sum f_\Sigma^* \eta_\Sigma \, , \end{aligned}$$

where, as before, Σ ranges over all the irreducible components of $\mathsf{R}^1(\mathsf{A})$. Define $\operatorname{Log}^i \mathcal{W}(\mathsf{A})$ as its kernel, i.e., $\operatorname{Log}^i \mathcal{W}(\mathsf{A}) = \ker \Psi_i$. Define also $\operatorname{Log}^\infty \mathcal{W}(\mathsf{A})$ as the direct sum

$$\operatorname{Log}^\infty \mathcal{W}(\mathsf{A}) = \bigoplus_{i=1}^{\infty} \operatorname{Log}^i \mathcal{W}(\mathsf{A}) \, .$$

Proposition 3.2. *If $\mathcal{W}$ is the localization of the web $\mathcal{W}(A)$ at a generic point of M then the vector space $\operatorname{Log}^\infty \mathcal{W}(A)$ embeds into the space of abelian relations of $\mathcal{W}$. Moreover, the analytic continuation of this embedding gives rise to a local system of abelian relations globally defined on M.*

Proof. Fix $i \in \mathbb{N}$, and let $(\eta_\Sigma) \in \ker \Psi_i$ be a non-zero element. Each η_Σ corresponds to a iterated integral in C_Σ. The pull-backs $f_\Sigma^* \eta_\Sigma$ under the corresponding rational maps are iterated integrals on M, and as such can be interpreted as functions on the universal covering of M. Moreover the (multi-valued) functions on M determined by $f_\Sigma^* \eta_\Sigma$ are (locally) constant along the leaves of the foliation $\mathcal{F}_\Sigma$. As we are considering iterated integrals of logarithmic 1-forms, the properties of iterated integrals recalled on Section 3.2 imply the functions on the universal covering of M coming from the components of $(\eta_\Sigma) \in \ker \Psi_i$ will sum up to zero. Thus the differentials of these functions will define an abelian relation for $\mathcal{W}(\mathsf{A})$. To check injectivity one has just to notice that two different elements in $\oplus_i H^1(C_\Sigma)^{\otimes i}$ will define different functions on the universal covering of C_Σ according to Chen's Theory, see Section 3.2. Q.E.D.

Let $k_d(\mathsf{A})$ be the number of irreducible components of $\mathsf{R}^1(\mathsf{A})$ of dimension d and $k(\mathsf{A})$ be the total number of irreducible components of $\mathsf{R}^1(\mathsf{A})$. Notice that the resonance web of A is a $k(\mathsf{A})$-web.

Corollary 3.1. The following inequalities hold true:

$$\begin{aligned}
\operatorname{rank}(\mathcal{W}(\mathsf{A})) &\geq \dim \operatorname{Log}^\infty(\mathsf{A}) \,,\\
\dim \operatorname{Log}^\infty(\mathsf{A}) &\leq \frac{(k(\mathsf{A})-1)(k(\mathsf{A})-2)}{2} \,,\\
\dim \operatorname{Log}^i(\mathsf{A}) &\geq \sum_d d^i k_d(\mathsf{A}) - \dim B^i(M) \cap H^1(M)^{\otimes i} \,.
\end{aligned}$$

In particular $\operatorname{Log}^\infty(\mathsf{A})$ is a finite dimensional vector space.

Proof. The first inequality follows from Proposition 3.2. The second follows from the first combined with Bol's bound (Corollary 2.1) for the rank of planar webs. To prove the third inequality it suffices to notice that Chen's *integrability conditions* are trivially satisfied by collections of 1-forms on a curve. Thus the morphism Ψ_k factors as in the diagram below

$$\begin{array}{ccc}
 & & N^k(M) \\
 & \nearrow & \downarrow \\
\bigoplus_\Sigma H^1(C_\Sigma)^{\otimes k} & \xrightarrow{\ \Psi_k\ } & H^1(M)^{\otimes k} \\
 & & \downarrow \\
 & & H^2 \otimes {H^1}^{\otimes k-2} \oplus \cdots \oplus {H^1}^{\otimes k-2} \otimes H^2
\end{array}$$

where $H^i = H^i(M)$ and $N^k(M) = B^k(M) \cap H^1(M)^{\otimes k}$. The corollary follows. Q.E.D.

Since M is the complement of a hyperplane arrangement, $H^\bullet(M)$ is generated in degree one. Consequently

$$\dim N^2(M) = h^1(M)^2 - h^2(M) \,.$$

In general we do not know how to control the dimensions of the vector spaces $N^i(M)$ when $i \geq 3$. Nevertheless for fiber type arrangements there is the following Künneth type formula which is a corollary of [6,

Theorem 3.38]: If $\mathcal{A}$ is a fiber type arrangement on $\mathbb{P}^n$ with exponents $\{e_1, \ldots, e_n\}$ then

$$\dim N^i(M) = \sum e_1^{j_1} \cdots e_n^{j_n}$$

where the sum is over all ordered n-uples $0 \leq j_1 \leq \cdots \leq j_n$ with $j_1 + \cdots + j_n = i$.

As will be made clear by the examples in Section 5 the bound for the rank given by Corollary 3.1 is rather crude and does not capture many otherwise easily predicable abelian relations. Nevertheless, we will need not more than these crude bounds to determine the rank of the resonance webs of the braid arrangements.

§4. Resonance webs of the braid arrangements

For $n \geq 2$, let $\mathsf{A}_{0,n+3}$ be the arrangement of hyperplanes on $\mathbb{P}^n$ defined by the vanishing of the polynomial

$$\left(\prod_{i=0}^{n} x_i\right)\left(\prod_{i=1}^{n}(x_i - x_0)\right)\left(\prod_{1\leq i<j\leq n}(x_i - x_j)\right).$$

It is the quotient of the braid arrangement B_{n+2} on $\mathbb{P}^{n+2}$

$$\prod_{0\leq i\leq j\leq n+1}(y_i - y_j)$$

by its center $\{y_0 = y_1 = \ldots = y_{n+1}\}$. The resonance variety of $\mathsf{A}_{0,n+3}$ is isomorphic to the resonance variety of B_{n+2}, and the resonance web of B_{n+2} is a linear pull-back of the resonance web of $\mathsf{A}_{0,n+3}$. Consequently, both webs have isomorphic space (local system) of abelian relations.

The complement of $\mathsf{A}_{0,n+3}$ will be denoted by $\mathfrak{M}_{0,n+3}$ and can be identified with the moduli space of $(n+3)$-uples of pairwise distinct ordered points of $\mathbb{P}^1$. According to [11], the resonance variety of $\mathfrak{M}_{0,n+3}$ has $\binom{n+3}{4}$ irreducible components which are in correspondence with the forgetful maps

$$\mathfrak{M}_{0,n+3} \longrightarrow \mathfrak{M}_{0,4}.$$

Thus the resonance web of $\mathcal{W}(\mathsf{A}_{0,n+3})$ is a $\binom{n+3}{4}$-web on $\mathfrak{M}_{0,n+3} \subset \mathbb{P}^n$. The main result of this paper is the determination of the rank of $\mathcal{W}(\mathsf{A}_{0,n+3})$ given below.

Theorem 4.1. *For every $n \geq 2$ the equality*

$$\operatorname{rank}(\mathcal{W}(\mathsf{A}_{0,n+3})) = 3\binom{n+3}{4} - \binom{n+2}{3} - \binom{n+1}{2} - n$$

holds true.

The remaining of this Section is devoted to the proof of this Theorem. It will be convenient to work in the affine chart $x_0 = 1$.

4.1. Upper bound for the rank

For each ordered 4-uple of ordered integers $1 \leq \alpha < \beta < \gamma < \delta \leq n+3$ consider the map

$$\begin{aligned} \rho_{\alpha\beta\gamma\delta} : \mathfrak{M}_{0,n+3} &\longrightarrow \mathfrak{M}_{0,4} \\ (x_1, \ldots, x_{n+3}) &\longmapsto (x_\alpha, x_\beta, x_\gamma, x_\delta) \end{aligned}$$

where the points in the source and the target represent isomorphism classes

Since $\mathfrak{M}_{0,4} = \mathbb{C} - \{0,1\}$ each of these $\binom{n+3}{4}$ maps define isotropic subspaces of $H^1(\mathfrak{M}_{0,n+3})$, namely $\rho^*_{\alpha\beta\gamma\delta} H^1(\mathfrak{M}_{0,4}) \subset H^1(\mathfrak{M}_{0,n+3})$. It can be verified that these isotropic subspaces are maximal, and that there are no other resonance varieties for $\mathsf{A}_{0,n+3}$, see [11].

Consider now $\mathcal{W} = \mathcal{W}(\mathsf{A}_{0,n+3})$, the germification of the resonance web of $\mathsf{A}_{0,n+3}$ at a generic point of $\mathfrak{M}_{0,n+3} \subset \mathbb{C}^n$. It will be useful to consider the following subwebs:

(1) For each $\alpha \in \underline{n+3}$ let $\mathcal{W}_\alpha$ be the $\binom{n+2}{3}$-subweb defined by the maps ρ_I where I ranges over all the ordered 4-uples containing α. Similarly for distinct α, β and α, β, γ in the above range let $\mathcal{W}_{\alpha,\beta}$ be the $\binom{n+1}{2}$-subweb and $\mathcal{W}_{\alpha,\beta,\gamma}$ be the n-subweb where I ranges over the ordered 4-uples that contains α, β and α, β, γ respectively.
(2) For each $\alpha \in \underline{n+3}$ let $\mathcal{W}^\alpha$ be the pull back of $\mathcal{W}(\mathsf{A}_{0,n+2})$ under the morphism

$$\rho^\alpha : \mathfrak{M}_{0,n+3} \longrightarrow \mathfrak{M}_{0,n+2}$$

that forgets the α-th point.

We combine the two constructions above and define

$$\mathcal{W}^\alpha_{\beta,\gamma} = \mathcal{W}^\alpha \cap \mathcal{W}_{\beta,\gamma}$$

where the intersection of the webs is the web formed by the common foliations of both webs.

Proposition 4.1. *For each $a \in \{1,2,3,4\}$ and every ordered subset I of $\underline{n+3}$ of cardinality $4-a$, $\mathcal{L}^a(\mathcal{W})$ is isomorphic to $\mathcal{L}^a(\mathcal{W}_I)$. Moreover, the dimension of $\mathcal{L}^a(\mathcal{W})$ is given by the formula*

$$\ell^a(\mathcal{W}) = \ell^a(\mathcal{W}_I) = \binom{n+a-1}{n-1}.$$

Proof. The proof will be by a double induction on n and a.

When $n = 2$, $\mathcal{W}_I = \mathcal{W}(\mathsf{A}_{0,5})_I$ is an $(a+1)$-web on $\mathbb{C}^2$. Therefore $\mathcal{L}^a(\mathcal{W})$ contains $\mathcal{L}^a(\mathcal{W}_I)$, and equation (2) implies

$$a + 1 \geq \ell^a(\mathcal{W}) \geq \ell^a(\mathcal{W}_I) = a + 1.$$

The result follows in this case.

When $a = 1$, we can assume that $I = (n+1, n+2, n+3)$ and therefore by normalizing the points of $\mathfrak{M}_{0,n+3}$ in such way that the last three are $0, 1, \infty$ we have that the n foliations defining $\mathcal{W}_I$ are defined by the morphisms $(x_1, \ldots, x_n, 0, 1, \infty) \mapsto (x_i, 0, 1, \infty)$. Clearly $\mathcal{L}^1(\mathcal{W}) = \mathcal{L}^1(\mathcal{W}_I) = \oplus_{i=1}^n \mathbb{C} x_i$.

Suppose now that $a \geq 2$ and $n \geq 3$. Assume $I \subset \{n+1, n+2, n+3\}$, $j \in \{1, \ldots, n\}$, and $(x_{n+1}, x_{n+2}, x_{n+3}) = (0, 1, \infty)$. Consider the linear map defined by the derivation $\frac{\partial}{\partial x_j}$ from $\mathbb{C}_a[x_1, \ldots, x_n] \simeq Sym^a \Omega^1(\mathbb{C}^n, 0)$ to $\mathbb{C}_{a-1}[x_1, \ldots, x_n] \simeq Sym^{a-1}\Omega^1(\mathbb{C}^n, 0)$. It induces the following diagram with exact rows where the unlabeled arrows are the natural inclusions.

$$\begin{array}{ccccccccc}
0 & \longrightarrow & \mathcal{L}^a(\mathcal{W}_I^j) & \longrightarrow & \mathcal{L}^a(\mathcal{W}_I) & \xrightarrow{\frac{\partial}{\partial x_j}} & \mathcal{L}^{a-1}(\mathcal{W}_{I\cup\{j\}}) & \longrightarrow & 0 \\
 & & \downarrow & & \downarrow & & \downarrow & & \\
0 & \longrightarrow & \mathcal{L}^a(\mathcal{W}^j) & \longrightarrow & \mathcal{L}^a(\mathcal{W}) & \xrightarrow{\frac{\partial}{\partial x_j}} & \mathcal{L}^{a-1}(\mathcal{W}_j) & \longrightarrow & 0.
\end{array}$$

Notice that $\mathcal{W}^j$ is isomorphic to $\mathcal{W}(\mathsf{A}_{0,n+2})$ and therefore, by induction hypothesis, the leftmost vertical arrow is an isomorphism. The induction hypothesis also implies that the vector spaces $\mathcal{L}^{a-1}(\mathcal{W}_j)$ and $\mathcal{L}^{a-1}(\mathcal{W}_{I\cup\{j\}})$ are both isomorphic to $\mathcal{L}^{a-1}(\mathcal{W})$. Thus the rightmost arrow is also an isomorphism. It follows that the middle vertical arrow is also an isomorphism and has dimension given by

$$\ell^a(\mathcal{W}) = \binom{(n-1)+a-1}{(n-1)-1} + \binom{n+(a-1)-1}{n-1} = \binom{n+a-1}{n-1}.$$

The proposition follows. Q.E.D.

As a consequence we obtain that for every $n \geq 2$ the webs $\mathcal{W}(\mathsf{A}_{0,n+3})$ are ordinary webs.

Corollary 4.1. For every $n \geq 2$ the inequality

$$\operatorname{rank}(\mathcal{W}(\mathsf{A}_{0,n+3})) \leq 3\binom{n+3}{4} - \binom{n+2}{3} - \binom{n+1}{2} - n$$

holds true.

Proof. It suffices to combine Propositions 2.1 and 4.1. Q.E.D.

4.2. Lower bound for the rank

As there are exactly $\binom{n}{2} + 2n + 1$ hyperplanes in the (projective) arrangement $\mathsf{A}_{0,n+3}$, $h^1(\mathfrak{M}_{0,n+3}) = \binom{n}{2} + 2n$. Notice also that $h^1(\mathfrak{M}_{0,4}) = 2$. Since $k(\mathsf{A}_{0,n+3}) = k_2(\mathsf{A}_{0,n+3})$, Corollary 3.1 implies

$$\dim \mathrm{Log}^1 \mathcal{W}(\mathsf{A}_{0,n+3}) \geq 2\binom{n+3}{4} - h^1(\mathfrak{M}_{0,n+3}) = 2\binom{n+3}{4} - \binom{n}{2} + 2n. \tag{5}$$

Similarly,

$$\dim \mathrm{Log}^2 \mathcal{W}(\mathcal{A}_{0,n+3}) \geq \binom{n+3}{4} \cdot h^1(\mathfrak{M}_{0,4})^2 - \dim N^2(\mathfrak{M}_{0,n+3})\,.$$

It is a result of Arnold [2] that the Poincaré polynomial of $\mathfrak{M}_{0,n+3}$ is $P(t) = (1+2t)(1+3t)\cdots(1+(n+1)t)$. Therefore the dimension of $N^2(\mathfrak{M}_{0,n+3})$ is equal to

$$h^1(\mathfrak{M}_{0,n+3})^2 - h^2(\mathfrak{M}_{0,n+3}) = \left(\binom{n}{2} + 2n\right)^2 - P''(0)/2\,.$$

Consequently

$$\dim AR^2_{log}(\mathcal{A}_{0,n+3}) \geq 4\binom{n+3}{4} - h^1(M_{0,n+3})^2 + h^2(M_{0,n+3})\,. \tag{6}$$

4.3. Proof of Theorem 4.1

It is not hard to prove by induction that summing the right-hand side of the inequalities (5) and (6) one obtains $3\binom{n+3}{4} - \binom{n+2}{3} - \binom{n+1}{2} - n$. Therefore, Proposition 3.2 implies that

$$\begin{aligned} \mathrm{rank}(\mathcal{W}(\mathcal{A}_{0,n+3})) &\geq \dim AR^1_{log}(\mathcal{A}_{0,n+3}) + \dim AR^2_{log}(\mathcal{A}_{0,n+3}) \\ &\geq 3\binom{n+3}{4} - \binom{n+2}{3} - \binom{n+1}{2} - n\,. \end{aligned}$$

Combining this lower bound with the upper bound given in Corollary 4.1 concludes the proof of Theorem 4.1. Q.E.D.

As the webs $\mathcal{W}(\mathsf{A}_{0,n+3})$ attain Cavalier–Lehmann's bound, it is natural to ask if they are algebraizable or, more generally, linearizable. As the leaves of algebraic webs are contained in hypersurfaces, every algebraizable web is linearizable but the converse is not always true. In

[24] it is proved that the webs $\mathcal{W}(\mathsf{A}_{0,n+3})$ are not linearizable. We refer to this work and references therein for more about the linearization of webs.

It is interesting to compare our Theorem 4.1 with Damiano's determination of the rank of the (dimension one) web given by the $(n+3)$ maps [12]

$$\mathfrak{M}_{0,n+3} \longrightarrow \mathfrak{M}_{0,n+2} \,.$$

These webs turn out to attain the corresponding bound for the rank of one-dimensional webs and are also non-linearizable.

§5. Examples

This section is devoted to the study of some exceptional planar webs—first found by Pirio and Robert [25, 27]—which are resonance webs of suitable line arrangements in $\mathbb{P}^2$. We use them to recognize other sources of abelian relations besides iterated integrals with logarithmic forms with poles on the arrangement.

5.1. More polylogarithmic abelian relations

We start with a simple example. Consider A_0 as the arrangement of 8 lines on $\mathbb{P}^2$ with affine trace presented in Figure 2. If we suppose that the triple point is at the origin of $\mathbb{C}^2$ then pencil of lines through it corresponds to an irreducible component of the resonance variety of dimension two. There are other two triple points at the line at infinity and they also correspond to irreducible components of the resonance variety of dimension two. There are no other irreducible components of $\mathsf{R}^1(\mathsf{A}_0)$. The resonance web $\mathcal{W}(\mathsf{A}_0)$ is the 3-web determined by the superposition of the foliations given by the level sets of the functions $x, y, x/y$. It clearly has rank one as

$$d\log x - d\log y - d\log \frac{x}{y} = 0\,,$$

but $\dim \mathrm{Log}^\infty \mathcal{W}(\mathsf{A}_0) = 0$ as one can promptly verify.

Given an arrangement of hyperplanes A on $\mathbb{P}^n$ we define the **resonance closure** of A, denoted by $\overline{\mathsf{A}}$, as the arrangement of hypersurfaces on $\mathbb{P}^n$ characterized by the following property: $H \in \overline{\mathsf{A}}$ if and only $H \in \mathsf{A}$, or there exists two distinct irreducible components Σ_1, Σ_2 of $\mathsf{R}^1(\mathsf{A})$ such that $\dim f_{\Sigma_1}(H) = \dim f_{\Sigma_2}(H) = 0$. In other words either H belongs to the original arrangement A or it is invariant by two distinct foliations of the web $\mathcal{W}(\mathsf{A})$. The complement of $\overline{\mathsf{A}}$ will be denoted by $\underline{M}$.

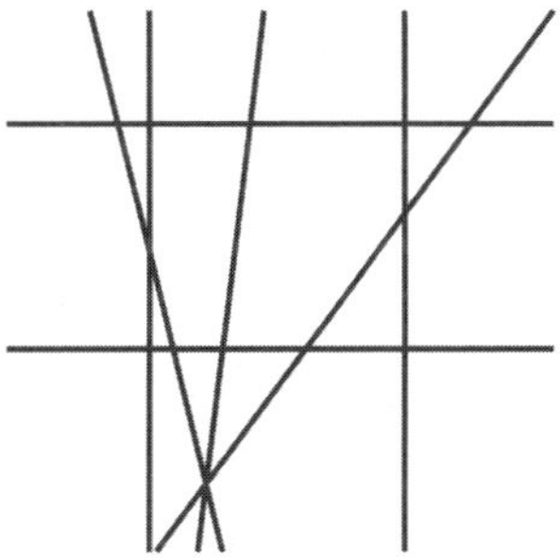

Fig. 2. $1 = \operatorname{rank}(\mathcal{W}(\mathsf{A})) > \dim \operatorname{Log}^{\infty} \mathcal{W}(\mathsf{A}) = 0$

Example 5.1. The resonance closure of the example A_0 is obtained by adding the lines $\{x = 0\}$ and $\{y = 0\}$. A more interesting (family of) examples is obtained by considering the (family of) arrangement(s) K_5 determined by the 10 lines joining 5 points in $\mathbb{P}^2$ in general position. It can be verified that the resonance variety of K_5 has 10 irreducible components: 5 of dimension 3 corresponding to pencils the lines through the points, and 5 of dimension 2 corresponding to the pencils of conics through 4 of the 5 points. If C denotes the conic through the 5 points then $\overline{\mathsf{K}_5} = \mathsf{K}_5 \cup \{C\}$.

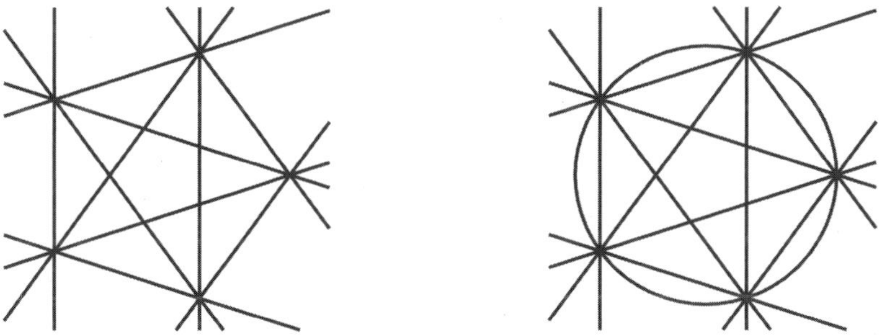

Fig. 3. On the left K_5, on the right $\overline{\mathsf{K}_5}$. The line at infinity is not included in the arrangements above

The relevance of the definition of the closure of an arrangement is put in evidence by the following specialization of [26, Theorem 1.2.2].

Proposition 5.1. *If $\mathcal{W}(\mathsf{A})$ is the resonance web of an arrangement A then every germ of abelian relation of $\mathcal{W}(\mathsf{A})$ extends to the universal covering of $\underline{M}$.*

If Σ is an irreducible component of $\mathsf{R}^1(\mathsf{A})$ we define its closure $\overline{\Sigma}$ as the maximal isotropic subspace of $H^1(\underline{M})$ containing the image of Σ under the natural inclusion $H^1(M) \to H^1(\underline{M})$. The **closure of the resonance variety** of A is the subvariety of $H^1(\underline{M})$ defined as

$$\overline{\mathsf{R}^1(\mathsf{A})} = \bigcup_{\Sigma \subset \mathsf{R}^1(\mathsf{A})} \overline{\Sigma}.$$

Notice that $\overline{\mathsf{R}^1(\mathsf{A})} \subset \mathsf{R}^1(\overline{\mathsf{A}})$ but the equality does not hold in general.

To each irreducible component $\overline{\Sigma} \subset \overline{\mathsf{R}^1(\mathsf{A})}$ there is a set of points of $\mathbb{P}^1$, $\mathsf{P}_{\overline{\Sigma}} \supset \mathsf{P}_\Sigma$ with complement $\underline{C_\Sigma}$ satisfying $(f_\Sigma)^* H^1(\underline{C_\Sigma}) = \overline{\Sigma}$. By analogy with what has been done in Sections 3.1 and 3.3, we can consider

$$\begin{array}{rcl} \overline{\Psi}_i : \bigoplus_{\overline{\Sigma}} H^1(\underline{C_\Sigma})^{\otimes i} & \longrightarrow & H^1(\underline{M})^{\otimes i} \\ (\eta_\Sigma) & \mapsto & \sum f_\Sigma^* \eta_\Sigma \end{array}$$

and define, for every $i \in \mathbb{N}$, $\overline{\mathrm{Log}^i \mathcal{W}(\mathsf{A})} = \ker \overline{\Psi}_i$. We also define $\overline{\mathrm{Log}^\infty \mathcal{W}(\mathsf{A})}$ as the direct sum $\oplus_i \overline{\mathrm{Log}^i \mathcal{W}(\mathsf{A})}$.

Of course, we also have a Proposition 3.2:

Proposition 5.2. *If $\mathcal{W}$ is the localization of the web $\mathcal{W}(\mathsf{A})$ at a generic point of M then the vector space $\overline{\mathrm{Log}^\infty \mathcal{W}(\mathsf{A})}$ embeds into the space of abelian relations of $\mathcal{W}$. Moreover, the analytic continuation of this embedding gives rise to a local system of abelian relations globally defined on $\underline{M}$.*

Of course, we can write lower bounds for $\overline{\mathrm{Log}^i \mathcal{W}(\mathsf{A})}$ analogous to the ones given by Corollary 3.1.

If $\mathsf{A} = \{C_1, \ldots, C_r\}$ is an arrangement of curves on $\mathbb{P}^2$ it is still a simple matter to determine the dimension of the second cohomology group of the complement. If $t_i(\mathsf{A})$ denotes the number of singular points in the support of A through which there exactly i branches, and $\chi(\widetilde{C_i})$ ($i = 1, \ldots, r$) are the Euler characteristic of the normalizations of the curves in the arrangement then [10, Proposition 2.4],

$$(7) \quad h^1(M) = r - 1 \text{ and } h^2(M) = 1 + \sum_{i=2}^{\infty} (i-1) t_i(\mathsf{A}) - \sum_{i=1}^{r} (\chi(\widetilde{C_i}) - 1).$$

We can use these observations (together with some computer algebra) to recover the following (unpublished) joint result of Pirio and Robert.

Theorem 5.1. *The resonance web of* K_5 *is exceptional.*

Proof. Notice that the closures of the irreducible components of $\mathsf{R}^1(\mathsf{K}_5)$ all have dimension 3. As $h^1(\underline{M}) = 10$, we obtain the lower bound

$$\overline{\mathrm{Log}^1}\mathcal{W}(\mathsf{K}_5) \geq 10 \cdot 3 - h^1(\overline{M}) = 30 - 10 = 20\,.$$

To control $\overline{\mathrm{Log}^2}\mathcal{W}(\mathsf{K}_5)$, we need to know the dimensions of $N^2(\underline{M}) = \ker\{H^1(\underline{M})^{\otimes 2} \to H^2(\underline{M})\}$. For an arbitrary arrangement of curves the cohomology algebra of the complement has no reason to be generated in degree 1, but for $\underline{M} = \mathbb{P}^2 \setminus \mathsf{K}_5$, [10, Theorem 2.46] implies it is the case. Thus

$$\overline{\mathrm{Log}^2}\mathcal{W}(\mathsf{K}_5) \geq 10 \cdot 9 - h^1(\overline{M})^2 + h^2(\overline{M}) = 90 - 100 + 25 = 15\,.$$

The inequalities above turn out to be equalities. To determine the dimension of $\overline{\mathrm{Log}^3}\mathcal{W}(\mathsf{K}_5)$ one has *just* to compute the dimension of the kernel of a 3375×243 matrix. A brute-force calculation[1] shows that

$$\dim \overline{\mathrm{Log}^3}\mathcal{W}(\mathsf{K}_5) = 1.$$

Thus $\mathcal{W}(\mathsf{K}_5)$ attains Bol's bound $9 \cdot 8/2 = 36$. To prove it is non-algebraizable it suffices to apply [20, Proposition 2.1]. Q.E.D.

Notice that K_5 is indeed a 2-parameter family of 10-webs as the moduli space of isomorphisms classes of 5 point on $\mathbb{P}^2$ has dimension 2.

5.2. Rational abelian relations

The inclusion of $\overline{\mathrm{Log}^\infty}\mathcal{W}(\mathsf{A})$ into $\mathcal{A}(\mathcal{W}(\mathsf{A}))$ does not exhaust the space of abelian relations of $\mathcal{W}(\mathsf{A})$ in general, unlike when $\mathsf{A} = \mathsf{A}_{0,5}$ or $\mathsf{A} = \mathsf{K}_5$. The simplest example is when A is an arrangement of 9 lines with 3 aligned threefold intersection points and all the other intersections are ordinary. In this case the resonance variety has only three local components, each of them having dimension two. Supposing that these three points are $(0:1:0), (1:0:0)$, and $(1:1:0)$ then the corresponding foliations are defined by the 1-forms dx, dy, and $dx+dy$. As they satisfy $(dx) + (dy) - (dx + dy) = 0$, it is clear that $\mathcal{W}(\mathsf{A})$ has rank one but $\dim \overline{\mathrm{Log}^\infty}\mathcal{W}(\mathsf{A}) = 0$.

---

[1]Maple script available at `www.impa.br/~jvp/artigos.html`.

Define $\mathrm{Rat}\mathcal{W}(\mathsf{A})$ as the kernel of the linear map

$$\begin{aligned} \overline{\Upsilon} : \bigoplus_{\Sigma} \frac{\mathbb{C}(\mathbb{P}^1)}{\mathbb{C}} & \longrightarrow \frac{\mathbb{C}(\mathbb{P}^n)}{\mathbb{C}} \\ (g_\Sigma) & \mapsto \sum f_\Sigma^*(g_\Sigma) , \end{aligned}$$

where $\mathbb{C}(\mathbb{P}^n)$ stands for the field of rational functions on $\mathbb{P}^n$, and the summation Σ run over all the irreducible components of $\mathsf{R}^1(\mathsf{A})$. Coordinate-wise differentiation injects $\ker \Upsilon$ into $\mathcal{A}(\mathcal{W}(\mathsf{A}))$. Notice that its image intersects $\overline{\mathrm{Log}^\infty}\mathcal{W}(\mathsf{A})$ only at zero. Therefore, we have the following lower bound for the rank of $\mathcal{W}(\mathsf{A})$

(8) $$\mathrm{rank}(\mathcal{W}(\mathsf{A})) \geq \dim \overline{\mathrm{Log}^\infty}\mathcal{W}(\mathsf{A}) + \dim \mathrm{Rat}\mathcal{W}(\mathsf{A}) .$$

We do not know how to give general lower bounds for $\dim \mathrm{Rat}\mathcal{W}(\mathsf{A})$. We have only the following simple result.

Lemma 5.1. *Let A be a planar arrangement, and let $\{\ell_1, \ldots, \ell_m\}$ be the lines in its closure. If n_i is the number of local irreducible components $\overline{\Sigma} \subset \overline{R^1}(A)$ containing ℓ_i in its support then*

$$\dim \mathrm{Rat}\mathcal{W}(A) \geq \sum_{i=1}^{n} \frac{(n_i - 1)(n_i - 2)}{2} .$$

Proof. Suppose that ℓ_1 is the line at infinity. The foliations associated to the n_1 components of $\overline{\mathsf{R}^1}(\mathsf{A})$ containing ℓ_1 are defined by $dh_1, \ldots, dh_{n_1}$ where h_i is a linear form. To prove the lemma it suffices to observe that for $p \geq 1$, the kernel of the maps

$$(a_1, \ldots, a_{n_1}) \mapsto \sum_{i=0}^{n_1} a_i (h_i)^p$$

will correspond to rational abelian relations with polar set contained in ℓ_1. Q.E.D.

Of course one can do better by considering rational functions with poles on fibers of f_Σ for non-local components of $\mathsf{R}^1(\mathsf{A})$.

Example 5.2. Let B_{5+n}, $n = 1, 2, 3$, be the arrangements obtained from $\mathsf{A}_{0,5}$ by adding n generic lines, each through a distinct double point, see Figure 4. The resonance webs are the webs $\mathcal{B}_6$, $\mathcal{B}_7$ and $\mathcal{B}_8$ considered by Robert and Pirio, and proved to be exceptional by them.

The rank of $\mathcal{B}_6$ can be computed easily as follows. As it contains $\mathsf{A}_{0,5}$ it has rank at least 6. Adding the pencil of lines through a double

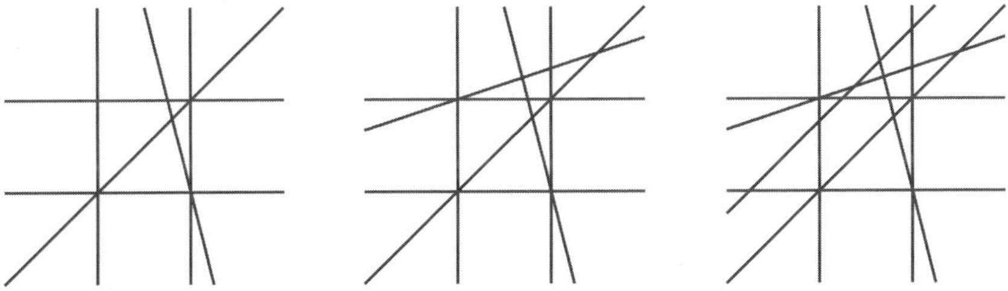

Fig. 4. Arrangements $\mathsf{B}_6, \mathsf{B}_7$, and B_8

point of the support of $\mathsf{A}_{0,5}$ we have that $\overline{\mathsf{B}_6} = \mathsf{B}_6$ contain two lines with three triple points, thus Lemma 5.1 implies $\dim \mathrm{Rat}\, \mathcal{W}(\mathsf{B}_6) \geq 2$. The proof of Theorem 4.1 tell us that $\Psi_1 : \oplus H^1(C_\Sigma) \to H^1(M)$ and $\Psi_2 : \oplus H^1(C_\Sigma)^{\otimes 2} \to H^1(M)^{\otimes 2}$, with $\Sigma \subset \mathsf{R}^1(A_{0,5})$, are both surjective. If h_1, h_2 are the equations of the lines intersecting at the double point under consideration then $d\log h_1 - d\log h_2$ and $(d\log h_1 - d\log h_2)^{\otimes 2}$ belong (respectively) to the image of Ψ_1 and Ψ_2. Thus

$$\dim \mathrm{Log}^\infty \mathcal{W}(\mathsf{B}_6) \geq \dim \mathrm{Log}^\infty \mathcal{W}(\mathsf{A}_{0,5}) + 2\,.$$

Putting all together we deduce that $\operatorname{rank} \mathcal{W}(\mathsf{B}_6) \geq 6+2+2 = 10$. Thus $\mathcal{B}_6 = \mathcal{W}(\mathsf{B}_6)$ is of maximal rank as it attains Bol's bound.

One can deal similarly with B_7 and B_8 but Lemma 5.1 does not suffice. One has to consider also rational first integrals for the foliation associated to the non-local component of $\mathsf{R}^1(\mathsf{B}_6)$ with poles on one (for B_7) or three (for B_8) fibers.

For general arrangements of lines, the inclusion

$$\overline{\mathrm{Log}^\infty \mathcal{W}(\mathsf{A})} \oplus \mathrm{Rat}^\infty \mathcal{W}(\mathsf{A}) \subset \mathcal{A}(\mathcal{W}(\mathsf{A}))$$

is strict. In the next two sections we will consider two examples of line arrangements on $\mathbb{P}^2$ with resonance webs having abelian relations which support this claim.

5.3. Mixed abelian relations

In the same way that we looked for abelian relations among collections of iterated integrals of logarithmic 1-forms we can also look for them in collections of iterated integrals of arbitrary rational 1-forms. In this section we will consider a one-parameter family of arrangements with resonance webs having abelian relations of this form.

Example 5.3. For $\lambda \in \mathbb{C} \setminus \{0,1\}$, let $\mathsf{P} = \mathbb{P}_\lambda$ be the arrangement of 8 lines obtained form $\mathsf{A}_{0,5} = \{xyz(x-y)(x-z)(y-z) = 0\}$ by adding

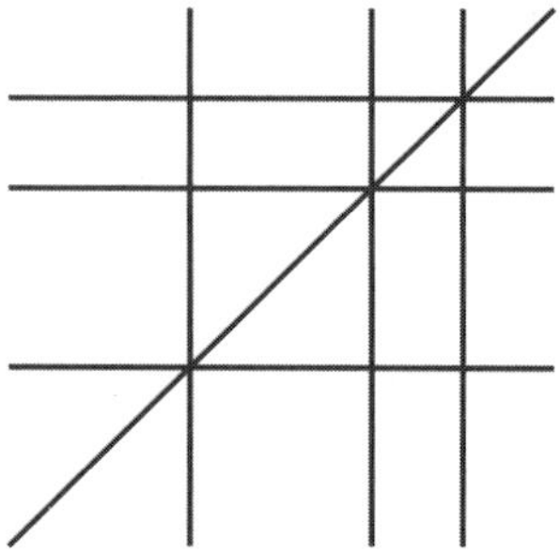

Fig. 5. The arrangement P

the two extra lines $\{(x-\lambda z)(y-\lambda z)=0\}$, see Figure 5. Its resonance variety has 8 irreducible components: two local of dimension three; three local of dimension two; and three non-local of dimension two determined by pencils of conics. It is closed arrangement in the sense that $\mathsf{P}=\overline{\mathsf{P}}$.

The family of webs $\mathcal{W}(\mathsf{P}_\lambda)$ have been studied in [25, 26]. There explicit generators for the corresponding spaces of abelian relations are presented.

From Corollary 3.1 we see that

$$\dim \operatorname{Log}^1 \mathcal{W}(\mathsf{P}) \geq 11 \text{ and } \dim \operatorname{Log}^2 \mathcal{W}(\mathsf{P}) \geq 5 .$$

These inequalities are indeed equalities. Moreover, one can verify that $\dim \operatorname{Rat} \mathcal{W}(\mathsf{P})=4$. Thus we have at least a 20-dimensional subspace of the space of abelian relations.

There is still one extra abelian relation, relation G_{21}^a in Section 3.3 of [25], involving the pull-backs under f_Σ of

$$d\left(\frac{\log(x)}{1-x}\right) = -d\log(x-1) + d\log(x) + \frac{\log(x)}{(1-x)^2}dx .$$

The last summand is neither the differential of a rational function, nor an iterated integral of logarithmic 1-forms. Indeed it can be written as $\frac{dx}{x} \otimes \frac{dx}{(1-x)^2}$, and therefore is an iterated integral of rational 1-forms.

5.4. Twisted logarithmic abelian relations

Of course the class of abelian relations with components being iterated integrals of rational 1-forms encompass all previous classes of abelian relations considered. Note that if we consider all the abelian relations in this class we obtain an unipotent local system over the complement of the closure of the arrangement. We believe that the maximal unipotent local system in $\mathcal{A}(\mathcal{W}(\mathsf{A}))$ is exactly the one generated by the abelian relations given by iterated integrals of rational 1-forms.

Our last example shows that the local system $\mathcal{A}(\mathcal{W}(\mathsf{A}))$ is not in general unipotent.

Example 5.4. Let F be the non-Fano arrangement presented in Figure 6. It is a closed arrangement ($\mathsf{F} = \overline{F}$) and its resonance variety has 9 irreducible components: six of them are local of dimension two, and three of them are determined pencil of conics and also have dimension two. The resonance web $\mathcal{W}(\mathsf{F})$ is the so called Spence–Kummer exceptional 9-web and was studied independently by Pirio [25] and Robert [27]. They proved that $\mathcal{W}(\mathsf{F})$ is an exceptional 9-web. The reference to Spence–Kummer comes from the fact that the foliations of the web are defined, up to a change coordinates, by the rational functions appearing in Spence–Kummer functional equation for the trilogarithm

$$\mathrm{Li}_3(z) = \int \frac{dz}{z} \otimes \frac{dz}{z} \otimes \frac{dz}{1-z} .$$

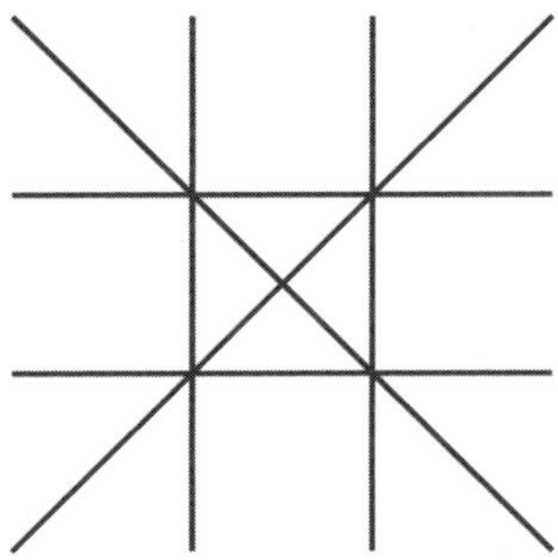

Fig. 6. The non-Fano arrangement

Corollary 3.1 implies

$$\dim \mathrm{Log}^1 \mathcal{W}(\mathsf{F}) \geq 12 \text{ and } \dim \mathrm{Log}^2 \mathcal{W}(\mathsf{F}) \geq 9 ,$$

and Lemma 5.1 implies $\dim \mathrm{Rat}\, \mathcal{W}(\mathsf{F}) \geq 4$. These bounds are indeed equalities. Moreover, brute-force computation yields $\dim \mathrm{Log}^3 \mathcal{W}(\mathsf{F}) = 2$. Thus we have a 27-dimensional subspace of the space of abelian relations.

The missing abelian relation comes from the intersection of the irreducible components of $\mathrm{Char}^1(M)$ determined by the three non-local components. In order to explain this abelian relation we will digress a little.

§

Let A be an arrangement of hypersurfaces on $\mathbb{P}^n$ and M be its complemente. Recall from the introduction the morphism

$$\begin{aligned} \exp : H^1(M) &\longrightarrow \mathrm{Hom}(\pi_1(M), \mathbb{C}^*) \\ a &\longmapsto \left(\gamma \mapsto \exp\left(2\pi i \int_\gamma a\right)\right). \end{aligned}$$

For a 1-form $\omega \in H^1(M)$, let $\varrho_\omega : \pi_1(M) \to \mathbb{C}^*$ be the representation $\exp(\omega)$ and $\mathbb{C}_\omega$ the corresponding rank one local system.

The $\mathbb{C}$-sheaf $\mathbb{C}_\omega$ admits the following resolution

$$0 \longrightarrow \mathbb{C}_\omega \longrightarrow \mathcal{O}_M \xrightarrow{\nabla_\omega} \Omega^1(M) \xrightarrow{\nabla_\omega} \Omega^2(M) \xrightarrow{\nabla_\omega} \cdots \longrightarrow 0$$

where $\Omega^\bullet(M)$ are the sheaves of holomorphic differentials on M and $\nabla_\omega : \Omega^\bullet(M) \to \Omega^{\bullet+1}(M)$ is given by the formula $\nabla_\omega(\alpha) = d\alpha - \omega \wedge \alpha$. Since M is Stein, the sheaves $\Omega^\bullet(M)$ are acyclic and consequently

$$H^i(M, \mathbb{C}_\omega) = \frac{\ker \nabla_\omega : H^0(M, \Omega^i(M)) \to H^0(M, \Omega^{i+1}(M))}{\nabla_\omega(H^0(M, \Omega^{i-1}))}.$$

If $\omega_1 - \omega_2 = d\log F$ for some $F \in H^0(M, \mathcal{O}_M)$ then $\mathbb{C}_{\omega_1} \simeq \mathbb{C}_{\omega_2}$ and the corresponding resolutions relate through the diagram

$$\begin{array}{ccccccccccc} 0 \longrightarrow & \mathbb{C}_{\omega_1} & \longrightarrow & \mathcal{O}_M & \xrightarrow{\nabla_{\omega_1}} & \Omega^1(M) & \xrightarrow{\nabla_{\omega_1}} & \Omega^2(M) & \xrightarrow{\nabla_{\omega_1}} & \cdots & \longrightarrow 0 \\ & & & \downarrow F^{-1} & & \downarrow F^{-1} & & \downarrow F^{-1} & & & \\ 0 \longrightarrow & \mathbb{C}_{\omega_1} & \longrightarrow & \mathcal{O}_M & \xrightarrow{\nabla_{\omega_2}} & \Omega^1(M) & \xrightarrow{\nabla_{\omega_2}} & \Omega^2(M) & \xrightarrow{\nabla_{\omega_2}} & \cdots & \longrightarrow 0 \end{array} \tag{9}$$

where the vertical arrows are multiplication by F^{-1}.

For $\alpha \in \Omega^1(M)$ we have $\nabla_\omega(\alpha) = 0$ if and only if the (multi-valued) 1-form $\exp\left(\int \omega\right) \alpha$ is closed. Moreover, if ω belongs to some irreducible component Σ of $\mathsf{R}^1(\mathsf{A})$ then for every $\alpha \in \Sigma$ we have that the 1-form $\exp\left(\int \omega\right) \alpha$ is closed.

Let Σ be an irreducible component of $\mathsf{R}^1(\mathsf{A})$ and let $\alpha \in \Sigma$ be a logarithmic 1-form with residues in $\mathbb{Q} \setminus \mathbb{Z}_{>0}$. Let $P : N \to \mathbb{P}^n$ be the finite abelian covering determined by $\mathbb{C}_\alpha$. Since the monodromy is finite (residues $\in \mathbb{Q}$), N is a quasi-projective variety.

Lemma 5.2. *If $\beta \in \Sigma$ is not a complex multiple of α then $\exp\left(-\int P^*\alpha\right) P^*\beta$ is a closed rational 1-form on N which is not exact.*

Proof. Let $f = f_\Sigma : \mathbb{P}^n \dashrightarrow \mathbb{P}^1$ be the associate rational map. Recall that for $\alpha, \beta \in \Sigma$ there exist logarithmic 1-forms $\widetilde{\alpha}$, $\widetilde{\beta}$ on $\mathbb{P}^1$ with poles of P_Σ such that $\beta = f_i^* \widetilde{\beta}$.

If $p : C \to \mathbb{P}^1 \setminus \mathsf{P}_\Sigma$ is the finite ramified covering determined by $\exp\left(-\int \widetilde{\alpha}\right)$ then it fits in the commutative diagram

$$\begin{array}{ccc} N & \xrightarrow{P} & M \\ {\scriptstyle \tilde{f}}\downarrow & & \downarrow{\scriptstyle f} \\ C & \xrightarrow{p} & \mathbb{P}^1 \setminus \mathsf{P}_\Sigma . \end{array}$$

Clearly $\exp\left(-\int P^*\alpha\right) P^*\beta$ is equal to $\tilde{f}^* \exp\left(-\int p^*\widetilde{\alpha}\right) \widetilde{\beta}$. Thus if the former is exact, the same holds for the latter.

Suppose $\exp\left(-\int p^*\widetilde{\alpha}\right) \widetilde{\beta} = dg$ for some rational function on C. The rational function $\tilde{h} = \exp\left(\int p^*\widetilde{\alpha}\right) g$ is invariant under the covering transformation, thus is equal to p^*h for some rational function on $\mathbb{P}^1$. It is a simple matter to verify that $\nabla_\alpha(h) = \beta$, and therefore β represents the zero class in $H^1(\mathbb{P}^1 \setminus \mathsf{P}_\Sigma, \mathbb{C}_\alpha)$. But the main result of [13] implies that the complex $(\Omega^\bullet(\mathbb{P}^1 \setminus \mathsf{P}_\Sigma), \nabla_\alpha)$ is quasi-isomorphhic to the complex $(H^\bullet(\mathbb{P}^1 \setminus \mathsf{P}_\sigma, \mathbb{C}), \wedge\alpha)$. Hence the class of β is not zero. Q.E.D.

Proposition 5.3. *Let $\mathcal{A}$ be an arrangement of hyperplanes on $\mathbb{P}^n$ and let $\Sigma_1, \ldots, \Sigma_r$ be irreducible components of the resonance variety $\mathsf{R}^1(\mathcal{A})$. Suppose $\exp(\Sigma_1), \ldots, \exp(\Sigma_r)$ intersect at some $\rho \in \mathrm{Hom}(\pi_1(M), \mathbb{C}^*)$ distinct from the trivial representation. If $\overline{N}$, the projective closure of the finite covering of M determined by ρ, is simply-connected then*

$$\dim \frac{\mathcal{A}(\mathcal{W}(\mathcal{A}))}{\mathrm{Log}^\infty \mathcal{W}(\mathcal{A}) \oplus \mathrm{Rat}\mathcal{W}(\mathcal{A})} \geq -h^1(M, \mathbb{C}_\omega) + \sum_{i=1}^{r} (\dim \Sigma_i - 1) .$$

Proof. Since ω is in the intersection of $\exp(\Sigma_1), \ldots, \exp(\Sigma_r)$ there exist non-zero logarithmic 1-forms $\alpha_i \in \Sigma_i$ $(i = 1, \ldots, r)$ and rational functions $f_{ij} \in \mathcal{O}(M)$ $(i, j = 1, \ldots, r)$ satisfying

$$\alpha_i - \alpha_j = d \log f_{ij} .$$

In particular, $\mathbb{C}_{\alpha_i}$ is isomorphic to $\mathbb{C}_{\alpha_j}$ for every i, j. It is harmless to assume that all the 1-forms α_i have non-integer residues and $\omega = \alpha_1$.

Since $\Sigma_i \subset \ker \nabla_{\alpha_i}$ it follows from (9) that $(f_{ij})^{-1} \cdot \Sigma_j \in \ker \nabla_{\alpha_i}$ for every i, j. Therefore we can define the map

$$\Lambda : \bigoplus_{i=1}^{r} \frac{\Sigma_i}{\mathbb{C} \cdot \alpha_i} \longrightarrow H^1(M, \mathbb{C}_{\alpha_1})$$

$$(\beta_i) \mapsto \beta_1 + \sum_{i=2}^{r} (f_{1i})^{-1} \beta_i \, .$$

Notice that as we explained before the (multi-valued) 1-forms

$$\exp\left(- \int \alpha_1 \right) (f_{1i})^{-1} \beta_i = \exp\left(- \int \alpha_i \right) \beta_i$$

are closed.

If $(\beta_1, \ldots, \beta_r)$ belongs to $\ker \Lambda$ then there exists $g \in H^0(M, \mathcal{O}_M)$ for which

$$\beta_1 + \sum_{i=2}^{r} (f_{1i})^{-1} \beta_i = dg - g\alpha_1$$

or, equivalently, for suitable choices of branches of $\exp\left(\int -\alpha_i\right)$ we have that

$$\sum_{i=1}^{r} \exp\left(\int -\alpha_1 \right) (f_{1i})^{-1} \beta_i = d\left(\exp\left(\int \alpha_1 \right) g \right)$$

If we pull-back this equation to N using the finite covering $P : N \to M$ then all the 1-forms involved are legitime (univalued) closed rational 1-forms. Since $\overline{N}$ is simply-connected, the pull-backs $\widetilde{f_\Sigma}$ of the maps f_Σ have as target rational curves. Thus the 1-forms $\exp\left(\int -\alpha_1\right) (f_{1i})^{-1} \beta_i$ can be uniquely written as the pull-back by $\widetilde{f_\Sigma}$ of the sum of an exact rational differential with an logarithmic 1-form. Discarding the rational component, one obtains an identity as above but with zero right-hand side. Clearly it is an abelian relation. The linear independence of the corresponding abelian relations for $(\beta_1, \ldots, \beta_r)$ varying on a basis of $\ker \Lambda$ follows from Lemma 5.2. Q.E.D.

We do not know if the hypothesis made on the topology of $\overline{N}$ is necessary to prove the proposition above.

§

Back to the analysis of Example 5.4, we have that the exponential of the three non-local components intersect at a representation ρ for which $h^1(M, \mathbb{C}_\rho) = 2$. Moreover, ρ satisfies the hypothesis of Proposition 5.3 as it is non-trivial only along two fibers of the corresponding rational maps

f_Σ, see [11, Example 10.5]. Thus Proposition 5.3 ensures the existence of the *missing* abelian relation of $\mathcal{W}(\mathcal{F})$.

5.5. Final remarks

In the table below we present the dimensions of the subspaces of the space of abelian relations of resonance webs of line arrangements studied in this paper.

	k	Log^1	$\overline{\mathrm{Log}^1}$	Log^2	$\overline{\mathrm{Log}^2}$	Log^3	$\overline{\mathrm{Log}^3}$	Rat	Mixed	Twisted
$\mathsf{A}_{0,5}$	5	5	5	1	1	0	0	0	0	0
B_6	6	6	6	2	2	0	0	2	0	0
B_7	7	7	7	3	3	0	0	5	0	0
B_8	8	8	8	4	4	0	0	9	0	0
P	8	11	11	5	5	0	0	4	1	0
$\mathsf{\Gamma}$	9	12	12	9	9	2	2	4	0	1
K_5	10	16	20	5	15	0	1	0	0	0

Although we have studied resonance webs for hyperplane arrangements one can study resonance webs for arbitrary hypersurfaces arrangements on $\mathbb{P}^n$. Even more generally, if the cohomology algebra $H^\bullet(M)$ is replaced by a finite dimensional algebra of differential forms on a quasi-projective variety then one can still talk about resonance varieties. Its irreducible components are still in correspondence with rational maps to projective curves [3], and consequently one can still define the resonance webs. We are not aware of any exceptional web arising this way that are not pull-backs by rational maps of resonance webs of hyperplane arrangements.

Acknowledgements. This paper would not come to light without the warm reception given to my talk at the 2nd MSJ-SI Arrangements of Hyperplanes. I would like to thank the organizers for the opportunity to participate in this meeting. I am also indebt to Luc Pirio who, through numerous discussions, contributed a lot to this paper. Among other things, he suggested that the resonance webs of $\mathsf{A}_{0,n+3}$ should be looked at, gave a computer-assisted proof of Proposition 4.1 in the case $n = 3$, and brought reference [6] to my knowledge.

References

[1] D. Arapura, Geometry of cohomology support loci for local systems. I, J. Algebraic Geom., **6** (1997), 563–597.

[2] V. I. Arnol'd, The cohomology ring of the group of dyed braids, Mat. Zametki, **5** (1969), 227–231.

[3] I. Bauer, Irrational pencils on non-compact algebraic manifolds, Internat. J. Math., **8** (1997), 441–450.

[4] G. Bol, Über ein bemerkenswertes Fünfgewebe in der Ebene, Abh. Math. Hamb. Univ., **11** (1936), 387–393.

[5] E. Brieskorn, Sur les groupes de tresses [d'après V. I. Arnol'd], In: Séminaire Bourbaki, 24ème année (1971/1972), Exp. No. 401, pp. 21–44; Lecture Notes in Math., **317**.

[6] F. Brown, Multiple zeta values and periods of moduli spaces $\overline{\mathfrak{M}}_{0,n}$, Ann. Sci. Éc. Norm. Supér. (4), **42** (2009), 371–489.

[7] V. Cavalier and D. Lehmann, Ordinary holomorphic webs of codimension one, arXiv:math/0703596v2.

[8] K.-T. Chen, Iterated path integrals, Bull. Amer. Math. Soc., **83** (1977), 831–879.

[9] S.-S. Chern and P. Griffiths, An inequality for the rank of a web and webs of maximum rank, Ann. Scuola Norm. Sup. Pisa Cl. Sci. (4). **5** (1978), 539–557.

[10] J. I. Cogolludo-Agustín, Topological invariants of the complement to arrangements of rational plane curves, Mem. Amer. Math. Soc., **159** (2002), no. 756.

[11] D. Cohen and A. Suciu, Characteristic varieties of arrangements, Math. Proc. Cambridge Philos. Soc., **127** (1999), 33–53.

[12] D. Damiano, Webs and characteristic forms of Grassmann manifolds, Amer. J. Math., **105** (1983), 1325–1345.

[13] H. Esnault, V. Schechtman and E. Viehweg, Cohomology of local systems on the complement of hyperplanes, Invent. Math., **109** (1992), 557–561.

[14] M. Falk and S. Yuzvinsky, Multinets, resonance varieties, and pencils of plane curves, Compos. Math., **143** (2007), 1069–1088.

[15] A. Hénaut, On planar web geometry through abelian relations and connections, Ann. of Math. (2), **159** (2004), 425–445.

[16] A. Hénaut, Planar web geometry through abelian relations and singularities. Inspired by S. S. Chern, Nankai Tracts Math., **11** (2006), 269–295.

[17] A. Hénaut, Formes différentielles abéliennes, bornes de Castelnuovo et géométrie des tissus, Comment. Math. Helv., **79** (2004), 25–57.

[18] T. Kohno, Bar complex of the Orlik–Solomon algebra, In: Arrangements in Boston: A Conference on Hyperplane Arrangements, 1999, Topology Appl., **118**, Elsevier Science B.V., Amsterdam, 2002, pp. 147–157.

[19] D. Marín, J. V. Pereira and L. Pirio, On planar webs with infinitesimal automorphisms. Inspired by S. S. Chern, Nankai Tracts Math., **11** (2006), 351–364.

[20] J. V. Pereira and L. Pirio, The Classification of Exceptional CDQL Webs on Compact Complex Surfaces, International Mathematics Research Notices, 2009, doi:10.1093/imrn/rnp208.

[21] J. V. Pereira and L. Pirio, An invitation to web geometry. From Abel's addition theorem to the algebraization of codimension one webs, In: Publicações Matemáticas do IMPA, 27° Colóquio Brasileiro de Matemática, IMPA, Rio de Janeiro, 2009. xii+245 pp.

[22] J. V. Pereira, Algebraization of codimension one webs, In: Séminaire Bourbaki, Vol. 2006/2007, Astérisque No. 317 (2008), Exp. No. 974, viii, 243–268.

[23] J. V. Pereira and S. Yuzvinsky, Completely reducible hypersurfaces in a pencil, Adv. Math., **219** (2008), 672–688.

[24] L. Pirio, Sur la linearisation des tissus, Enseign. Math. (2), **55** (2009), 285–328.

[25] L. Pirio, Abelian functional equations, planar web geometry and polylogarithms, Selecta Math. (N.S.), **11** (2005), 453–489.

[26] L. Pirio, Équations fonctionnelles abéliennes et géométric des tissus, Thèse de Doctorat de l'Université Paris VI, 2004.

[27] G. Robert, Relations fonctionnelles polylogarithmiques et tissus plans, Prépublication, **146**, Université Bordeaux 1, 2002.

[28] J.-M. Trépreau, Algébrisation des tissus de codimension 1. La généralisation d'un théorème de Bol. Inspired by S. S. Chern, Nankai Tracts Math., **11** (2006), 399–433.

[29] S. Yuzvinsky, A new bound on the number of special fibers in a pencil of curves, Proc. Amer. Math. Soc., **137** (2009), 1641–1648.

Jorge Vitório Pereira
IMPA, Estrada Dona Castorina, 110
22460-320, Rio de Janeiro, RJ, Brazil
E-mail address: jvp@impa.br

Advanced Studies in Pure Mathematics 62, 2012
Arrangements of Hyperplanes—Sapporo 2009
pp. 293–321

Discrete topological methods for subspace arrangements

Francesca Mori and Mario Salvetti*

§1. Introduction

A different (and relatively new) method to deal with the topology of Hyperplane Arrangements (and more generally Subspace Arrangements) is given by a discrete version of Morse Theory, namely the so called Discrete Morse Theory. Practically, this theory originated from the standard theory as a suitable adjustment to spaces usually produced by discrete data, like simplicial complexes and more generally CW-complexes (see [Fo98, Fo02, Ko08]).

In [SaSe07] we considered complement to *real* hyperplane arrangements applying Discrete Morse Theory to a well-known CW-complex with the same homotopy type ([Sa87]). We re-proved the *minimality* of the complement: the complement to a hyperplane arrangement is a *minimal space,* i.e. it has the homotopy type of a CW-complex with as many i-cells as its ith-Betti number ($i \geq 0$). This interesting result was proven independently in [DP03, Ra02] as an existence-like theorem; the explicit structure of the minimal complex was considered before us by [Yo05] and after us by [De08] (see also [DeSe]) .

The construction which uses Discrete Morse Theory is much more precise, even if superabundant in the description of the attaching maps of the cells (new "reduced" descriptions, at least in case of dimension two, were recently found by Yoshinaga himself and, by different method, by the two authors togheter with G. Gaiffi [GMS10]). This combinatorial method allows to produce algebraic complexes which calculate local

Received August 31, 2010.
Revised March 29, 2011.
2010 *Mathematics Subject Classification.* 58K65.
Key words and phrases. Arrangements of subspaces, Discrete Morse Theory.
*Partially supported by M.U.R.S.T. 40%.

systems in general (see formulas in [SaSe07, GaSa09]), which can be considered togheter with the many other well-known constructions which appear in the literature.

In this paper we outline an extension of the use of the combinatorial methods above to a certain class of *subspace arrangements,* which contains interesting spaces like $k(\pi,1)$-spaces for Coxeter groups (see below).

Recall that classical Configuration Spaces in a manifold M are defined as the manifolds $F(M,n)$ of ordered n-tuples of *pairwise different* points in M $(n>0)$. Here we are interested in the case $M=\mathbb{R}^d,\ d>0$. By using coordinates in $(\mathbb{R}^d)^n=\mathbb{R}^{nd}$

$$x_{ij},\ i=1,\ldots,n,\ j=1,\ldots,d,$$

one has

$$F(\mathbb{R}^d,n) \;=\; \mathbb{R}^{nd}\setminus\cup_{i\neq j}\, H_{ij}^{(d)},$$

where $H_{ij}^{(d)}$ is the codimension d-subspace

$$\cap_{k=1,\ldots,d}\ \{x_{ik}=x_{jk}\}.$$

This latter subspace is the intersection of d hyperplanes in $\mathbb{R}^{nd}$, each obtained by the hyperplane $H_{ij}=\{x\in\mathbb{R}^n\ :\ x_i=x_j\}$, considered on the k-th component in $(\mathbb{R}^n)^d=\mathbb{R}^{nd}$, $k=1,\ldots,d$.

We consider here a *Generalized Configuration Space* (for brevity, simply a Configuration Space). Such spaces start from any *Hyperplane Arrangement* $\mathcal{A}$ in $\mathbb{R}^n$. For each $d>0$, one defines a *d-complexification* $\mathcal{A}^{(d)}\subset M^d$ of $\mathcal{A}$, which is given by the collection $\{H^{(d)},\ H\in\mathcal{A}\}$ of the *d-complexified* subspaces. There is an associated *configuration space,* namely the complement to the subspace arrangement

$$\mathcal{M}^{(d)} \;=\; \mathcal{M}(\mathcal{A})^{(d)} := \;(\mathbb{R}^n)^d\setminus\bigcup_{H\in\mathcal{A}} H^{(d)}\;.$$

Notice that for $d=2$ one has the standard complexification of a real hyperplane arrangement. The natural inclusion $\mathcal{M}^{(d)}\hookrightarrow\mathcal{M}^{(d+1)}$ gives rise to a limit space which is contractible. In case of a reflection arrangement relative to a group W, the limit of the orbit space with respect to the action of W is a model of the classifying space of W; see [DS00].

In this paper we use the Discrete Morse Theory for CW-complexes to find an explicit *minimal* CW-complex for the configuration space $\mathcal{M}(\mathcal{A})^{(d)}$, for all $d\geq 1$. That is, we give a CW-complex having as many i-cells as the i-th Betti number B_i of $\mathcal{M}(\mathcal{A})^{(d)}$, $i\geq 0$.

For $d = 1$ the construction is trivial, since $\mathcal{M}^{(1)}$ is a disjoint union of convex sets (the *chambers*). Case $d = 2$ was that considered in [SaSe07]. Even if for $d > 2$ the minimality of configuration spaces can be theoretically deduced by their simply-connectness (see for ex. [Ha02], Prop. 4C.1), our construction is useful in order to produce geometric bases for the cohomology. In fact, we give explicit bases for the homology (and cohomology) of $\mathcal{M}^{(d+1)}$ which we call (d)-*polar bases* (see below). As far as we know, there is no other precise description of a geometric $\mathbb{Z}$-basis in the literature, except for some particular arrangements, even if the $\mathbb{Z}$-module structure derives from the well-known formulas in [GM88]: in fact, the intersection lattice of the d-complexification $\mathcal{A}^{(d)}$ is the same for all $d \geq 1$.

Notice also that the proof of minimality, in case $d > 2$, is straightforward from our construction because of the gaps among the dimensions of the critical cells.

We construct a discrete vector field, depending on a system of *polar coordinates* in $\mathbb{R}^n$, which is generic with respect to the arrangement. We use the coordinates to associate to such system a *total ordering* $\lhd$ (called *polar ordering*) on the set $\Phi := \{F\}$ of all the facets of the stratification of $\mathbb{R}^n$ induced by the arrangement. So, the philosophy of the paper is similar to that used for $d = 2$ in [SaSe07]; however, the extension to the case $d > 2$ is not trivial and technical so all details of the proof will be published elsewhere. For reader convenience, we prefer here to give detailed examples, writing down complete lists of cells of the CW-complexes involved, of the polar ordering introduced, of the pairs in the discrete fields and of the critical cells.

In some details, Section 2 is devoted to recall some notations and results from [DS00], including a description of the complex $\mathbf{S}^{(d)}$.

In Section 3 we introduce one example which will be reconsidered all over the paper: the full reflection arrangement A_3 in $\mathbb{R}^3$.

In Section 4 we briefly give the basic results from Discrete Morse Theory, using the original language from [Fo98],[Fo02], and some reference to the more combinatorial language in [Ko08].

In Section 5 we recall the main constructions of case $d = 2$, considered in [SaSe07].

In Section 7 we introduce the *degree-d discrete polar vector field* and we characterize its critical cells. We obtain that critical dk-cells in $\mathbf{S}^{(d)}$ correspond to chains

$$(C \prec F^k \prec \ldots \prec F^k)$$

(k is the codimension) if d is odd; to

$$(op_{F^k}(C) \prec F^k \prec \ldots \prec F^k)$$

if d is even, where $op_{F^k}(C)$ is the chamber opposite to C with respect to F^k. Here $(C \prec F^k)$ corresponds to a critical cell in $\mathbf{S}^{(1)}$, with respect to polar ordering $\lhd$, so it is characterized by

$$\begin{array}{lll} F \lhd G, & \forall\, G \text{ such that} & F \prec G; \\ H \lhd F, & \forall\, H \text{ such that} & C \prec H \prec F. \end{array}$$

As a final remark, we propose the following conjecture, related in some way to this paper: *torsion-free subspace arrangements are minimal*, that is, when the complement of the arrangement has torsion-free cohomology, then it is a minimal space (see the concluding part 9 for some more details).

§2. CW-structures for the configuration spaces

In this section we use some notations and results from [DS00].

Let $\mathcal{A} = \{H_j\}_{j \in J}$ be a finite arrangement of linear hyperplanes in $M := \mathbb{R}^n$. For brevity, we give all definitions in the case when $\mathcal{A}$ is a central arrangement, remarking that all we say can be generalized to the affine case with little changes.

We introduce a coordinate $x \in M$ and coordinates $(x_1, ..., x_d)$, $x_i \in M$, in M^d, $d > 0$. Each hyperplane is given by a linear equation $H_j = \{x \in M \;:\; a_j \cdot x = 0\}$, $a_j \in M \setminus \{0\}$. For each $d > 0$, one has the *d-complexification* $\mathcal{A}^{(d)} \subset M^d$ of $\mathcal{A}$, given by the collection of linear codimension-d *subspaces*

$$(H_j)^{(d)} := \{(x_1, \ldots, x_d) \;:\; a_j \cdot x_k = 0,\; k = 1, \ldots, d\}$$

(when $d = 2$ one has the standard complexification $\mathcal{A}_{\mathbb{C}} \subset \mathbb{C}^n$).

The *generalized configuration space* associated to $\mathcal{A}$ is the complement to the subspace arrangement

$$\mathcal{M}^{(d)} = \mathcal{M}(\mathcal{A})^{(d)} := M^d \setminus \bigcup_{H \in \mathcal{A}} H^{(d)}\,.$$

It is convenient to introduce the *intersection lattice* $\mathcal{L} := \mathcal{L}(\mathcal{A})$ of $\mathcal{A}$, whose elements are all the subspaces of M of the form

$$L = H_{j_1} \cap \cdots \cap H_{j_k}, \quad H_{j_l} \in \mathcal{A}.$$

The partial ordering in $\mathcal{L}$ is given by

$$L \prec L' \text{ iff } L' \subset L,$$

so there is a minimum element, corresponding to the empty intersection, which is the whole space M, and a maximum $L_0 := \bigcap_{H \in A} H$. The *rank* $rk(L)$ of a subspace $L \in \mathcal{L}$ is its codimension; the rank of $\mathcal{L}$ is the rank of L_0, and we also set $rk(\mathcal{A}) := rk(\mathcal{L}(\mathcal{A}))$. The arrangement is called *essential* when $rk(\mathcal{A}) = n = dim(M)$, i.e. when L_0 reduces to a single point.

In the present situation we can consider a finer poset $\Phi := \Phi(\mathcal{A}) := (\{F\}, \prec)$ whose elements are the strata (also called *facets*) of the stratification induced on M by $\mathcal{A}$, where, as usual:

$$F \prec F' \text{ iff } F' \subset cl(F).$$

The atoms (*chambers*) of $\Phi(\mathcal{A})$ are the connected components of $M^{(1)}$.

We have a map $\Phi \to \mathcal{L}$ which associates to a facet F its *support* $|F|$, which is by definition the subspace generated by F (in a different language, this is the standard map between an oriented matroid and its underlying matroid). We define the rank function on Φ via this map:

$$rk(F) := rk(|F|) = codim(F).$$

Given $L \in \mathcal{L}(\mathcal{A})$, we will use the arrangements $\mathcal{A}_L := \{H \in \mathcal{A} : H \prec L\}$, $\mathcal{A}^L := \{L \cap H : H \in \mathcal{A}, H \not\prec L\}$. The former is an arrangement in M of rank equal to $rk(L)$, the latter is an arrangement inside L itself (of rank $rk(\mathcal{A}) - rk(L)$). Let $\Phi_L := \Phi(\mathcal{A}_L)$, $\Phi^L := \Phi(\mathcal{A}^L)$ be the induced stratifications of M, L respectively. There is a map $pr_L : \Phi \to \Phi_L$, taking F' into the unique stratum containing it, and a map $j^L : \Phi^L \to \Phi$ just given by the inclusion.

Fixing a facet F, set also $\Phi_F = \{F' \in \Phi : F' \prec F\}$. It is easy to see that the restriction $\varphi_F := (pr_{|F|})_{|\Phi_F|} : \Phi_F \to \Phi_{|F|}$ is a dimension-preserving bijection of posets.

Let now Φ^d be the product of d copies of Φ, $d \geq 0$, and let

$$\Phi^{(d)} = \{(F_1, \ldots, F_d) \in \Phi^d : F_1 \prec \cdots \prec F_d\}$$

be the set of d-chains in Φ (repetitions in the chain are allowed). Then $\Phi^{(d)}$ corresponds to a stratification of the space M^d as follows (see [DS00]): to each $\mathcal{F} = (F_1, \ldots, F_d)$ in $\Phi^{(d)}$ it corresponds the stratum $\hat{\mathcal{F}}$ in M^d given by

$$\hat{\mathcal{F}} := \{(x_1, \ldots, x_d) \in M^d : x_1 \in F_d , \ x_k \in \varphi_{F_{d-k+2}}(F_{d-k+1}) , \quad k = 2, \ldots, d\}.$$

One has:

Proposition 2.1. *(i) Each $\hat{\mathcal{F}}$ is homeomorphic to an open cell*
(ii) $\bigcup_{\mathcal{F}\in\Phi^{(d)}} \hat{\mathcal{F}} = M^d$
(iii) $\hat{\mathcal{F}} \cap \hat{\mathcal{G}} = \emptyset$ *if* $\mathcal{F} \neq \mathcal{G}$
(iv) $cl(\hat{\mathcal{F}}) \cap \hat{\mathcal{G}} \neq \emptyset$ *iff* $cl(\hat{\mathcal{F}}) \supset \hat{\mathcal{G}}$
(v) $\mathcal{M}^{(d)} = \bigcup_{\{\mathcal{F}\in\Phi^{(d)}\ :\ F_1 \text{ is a chamber of } \Phi\}} \hat{\mathcal{F}}$.

For $\mathcal{F} = (F_1, \ldots, F_d)$, one has

$$codim(\mathcal{F}) := codim(\hat{\mathcal{F}}) = \sum_{i=1}^{d} codim(F_i).$$

The partial ordering on $\Phi^{(d)}$ is given by

$$\mathcal{F} \prec \mathcal{G} \quad \text{iff} \quad \hat{\mathcal{G}} \subset cl(\hat{\mathcal{F}}).$$

This has the following characterization.

Lemma 2.2. *For $\mathcal{F} = (F_1, \ldots, F_d)$, $\mathcal{G} = (G_1, \ldots, G_d) \in \Phi^{(d)}$ one has*

$$\mathcal{F} \prec \mathcal{G} \quad \text{iff } F_d \prec G_d \quad \text{and } pr_{|F_{i+1}|}(F_i) \prec pr_{|F_{i+1}|}(G_i)$$

in the stratification $\Phi_{|F_{i+1}|}$, $i = d-1, \ldots, 1$.

Part (v) of Proposition 2.1 gives us the poset corresponding to the induced stratification of the generalized configuration space $\mathcal{M}^{(d)}$ which is

$$\Phi_0^{(d)} := \{\mathcal{F} = (F_1, \ldots, F_d) \in \Phi^{(d)} : rk(F_1) = 0\}$$

while the union $\bigcup_{H\in A} H^{(d)}$ of the d-complexified subspaces correspond to the poset

$$\Phi_+^{(d)} := \{\mathcal{F} = (F_1, \ldots, F_d) \in \Phi^{(d)} : rk(F_1) > 0\}.$$

Proposition 2.3. *The set*

$$\mathbf{Q}^{(d)} := \bigcup_{\mathcal{F}\in\Phi^{(d)}} e(\mathcal{F}),$$

where $e(\mathcal{F})$ is the dual *cell to the stratum $\mathcal{F}$, is a cellular $n'd$-ball in M^d (a regular cell complex) dual to the stratification, where $n' := rk(L_0)$.*

Remark. *It follows from Lemma 2.2 that if the first element F_1 of $\mathcal{F}$ is a chamber, then also the first element of any $\mathcal{G} \prec \mathcal{F}$ is a chamber.*

Definition 2.4. *We denote by $\mathbf{S}^{(d)}$ the subcomplex of $\mathbf{Q}^{(d+1)}$ whose cells correspond to $\Phi_0^{(d+1)}$:*

$$\mathbf{S}^{(d)} \ := \ \cup_{\mathcal{F} \in \Phi_0^{(d+1)}} e(\mathcal{F})$$

(case $d = 1$ was introduced in [Sa87]; see also [BZ92], [OT92]).

In general, given a chamber C and a facet F in $\Phi^{(d+1)}$ we will use the notation

$$C.F \ := \ \varphi_F^{-1}(pr_{|F|}(C))$$

which is a uniquely defined chamber containing F in its boundary.

One has:

Theorem 1. *(i) $\mathbf{S}^{(d)}$ is a deformation retract of $M^{(d+1)}$.*
(ii) Writing a $d+1$ chain $\mathcal{F} \in \Phi_0^{(d+1)}$ as $\mathcal{F} = (C, \mathcal{F}')$, $\mathcal{F}' \in \Phi^{(d)}$, we have:

$$\partial(e(C, \mathcal{F}')) \ = \ \bigcup_{\mathcal{F}'' \prec \mathcal{F}',\ codim(\mathcal{F}'') \ = \ codim(\mathcal{F}')-1} e(C.F_1, \mathcal{F}'')$$

($\mathcal{F}'' = (F_1, \ldots)$.)
(ii) $dim\ e(C, \mathcal{F}') \ = \ dim\ e(\mathcal{F}') \ = \ codim_{M^d}(\mathcal{F}')$. In particular, $dim(\mathbf{S}^{(d)}) = dim(\mathbf{Q}^{(d)}) = n'd$.

§3. Examples (1)

We outline here a 3-dimensional example. We use it later in order to illustrate the results of next sections.

Example. The full reflection arrangement of type A_3 is an arrangement $\mathcal{A} \in \mathbb{R}^3 = (x_1, x_2, x_3)$ given by the section of the arrangement $\mathcal{B} = \{\{x_1 = x_2\}, \{x_1 = x_3\}, \{x_2 = x_3\}, \{x_1 = x_4\}, \{x_2 = x_4\}, \{x_3 = x_4\}\} \in \mathbb{R}^4$ with the 3-plane $\{(x_1, x_2, x_3, x_4) | x_1 + x_2 + x_3 + x_4 = 1\}$ orthogonal to the line $l : x_1 = x_2 = x_3 = x_4$. Fig. 1 represents the section of $\mathcal{A}$ with the 2-plane $z = a,\ a > 0$, while Fig. 2 represents the section with the 2-plane $z = -a$. In Fig. 2 we indicate with $\overline{F}$ the opposite of a bounded facet F of Fig. 1. The center of this arrangement is a facet of dimension 0, not indicated in our figures and corresponding to the origin of $\mathbb{R}^3$, that we call P.

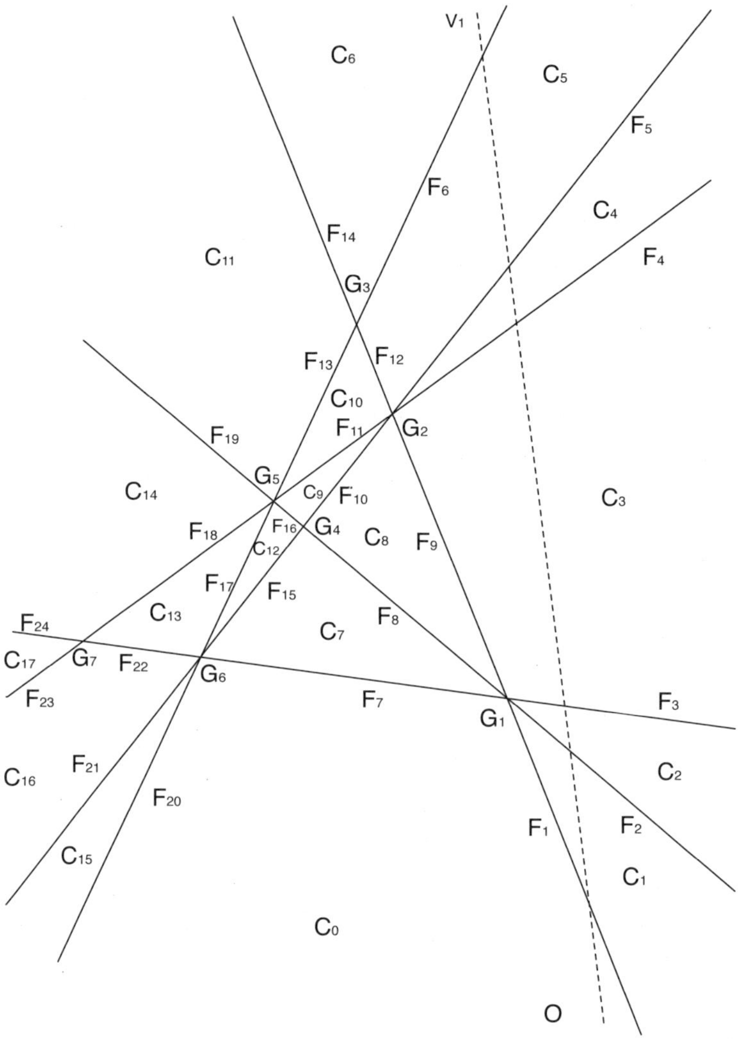

Fig. 1. An upper section of the reflection arrangement A_3

The complex $\mathbf{S}^{(1)}$ is composed by the following cells:

- 24 cells of dimension 0:

$$e(C_i, C_i),\ i = 0, \ldots, 17,$$

more

$$e(\overline{C}_7, \overline{C}_7), e(\overline{C}_8, \overline{C}_8), e(\overline{C}_9, \overline{C}_9), e(\overline{C}_{10}, \overline{C}_{10}), e(\overline{C}_{12}, \overline{C}_{12}), e(\overline{C}_{13}, \overline{C}_{13});$$

- 72 cells of dimension 1: $e(C_0, F_1)$, $e(C_0, F_7)$, $e(C_0, F_{20})$, $e(C_1, F_1)$, $e(C_1, F_2)$, $e(C_1, \overline{F}_{13})$, $e(C_2, F_2)$, $e(C_2, F_3)$, $e(C_2, \overline{F}_{18})$, $e(C_3, F_3)$, $e(C_3, F_4)$, $e(C_3, F_9)$, $e(C_4, F_4)$, $e(C_4, F_5)$, $e(C_4, \overline{F}_{22})$, $e(C_5, F_5)$, $e(C_5, F_6)$, $e(C_5, F_{12})$, $e(C_6, F_6)$, $e(C_6, F_{14})$, $e(C_6, \overline{F}_7)$, $e(C_7, F_7)$, $e(C_7, F_8)$, $e(C_7, F_{15})$, $e(C_8, F_8)$, $e(C_8, F_9)$, $e(C_8, F_{10})$, $e(C_9, F_{10})$, $e(C_9, F_{11})$, $e(C_9, F_{16})$, $e(C_{10}, F_{11})$, $e(C_{10}, F_{12})$, $e(C_{10}, F_{13})$, $e(C_{11}, F_{13})$, $e(C_{11}, F_{14})$, $e(C_{11}, F_{19})$, $e(C_{12}, F_{15})$, $e(C_{12}, F_{16})$, $e(C_{12}, F_{17})$, $e(C_{13}, F_{17})$, $e(C_{13}, F_{18})$, $e(C_{13}, F_{22})$, $e(C_{14}, F_{18})$, $e(C_{14}, F_{19})$, $e(C_{14}, F_{24})$, $e(C_{15}, F_{20})$, $e(C_{15}, F_{21})$, $e(C_{15}, \overline{F}_{12})$, $e(C_{16}, F_{21})$, $e(C_{16}, F_{22})$, $e(C_{16}, F_{23})$, $e(C_{17}, F_{23})$, $e(C_{17}, F_{24})$, $e(C_{17}, \overline{F}_9)$, $e(\overline{C}_7, \overline{F}_{15})$, $e(\overline{C}_7, \overline{F}_7)$, $e(\overline{C}_7, \overline{F}_8)$, $e(\overline{C}_8, \overline{F}_{10})$, $e(\overline{C}_8, \overline{F}_9)$, $e(\overline{C}_8, \overline{F}_8)$, $e(\overline{C}_9, \overline{F}_{16})$, $e(\overline{C}_9, \overline{F}_{10})$, $e(\overline{C}_9, \overline{F}_{11})$, $e(\overline{C}_{10}, \overline{F}_{13})$, $e(\overline{C}_{10}, \overline{F}_{11})$, $e(\overline{C}_{10}, \overline{F}_{12})$, $e(\overline{C}_{12}, \overline{F}_{17})$, $e(\overline{C}_{12}, \overline{F}_{15})$, $e(\overline{C}_{12}, \overline{F}_{16})$, $e(\overline{C}_{13}, \overline{F}_{18})$, $e(\overline{C}_{13}, \overline{F}_{22})$, $e(\overline{C}_{13}, \overline{F}_{17})$;
- 72 cells of dimension 2: $e(C_0, G_1)$, $e(C_0, G_6)$, $e(C_0, \overline{G}_3)$, $e(C_1, G_1)$, $e(C_1, \overline{G}_3)$, $e(C_1, \overline{G}_5)$, $e(C_2, G_1)$, $e(C_2, \overline{G}_5)$, $e(C_2, \overline{G}_7)$, $e(C_3, G_1)$, $e(C_3, G_2)$, $e(C_3, \overline{G}_7)$, $e(C_4, G_2)$, $e(C_4, \overline{G}_7)$, $e(C_4, \overline{G}_6)$, $e(C_5, G_2)$, $e(C_5, G_3)$, $e(C_5, \overline{G}_6)$, $e(C_6, G_3)$, $e(C_6, \overline{G}_6)$, $e(C_6, \overline{G}_1)$, $e(C_7, G_1)$, $e(C_7, G_4)$, $e(C_7, G_6)$, $e(C_8, G_1)$, $e(C_8, G_2)$, $e(C_8, G_4)$, $e(C_9, G_2)$, $e(C_9, G_4)$, $e(C_9, G_5)$, $e(C_{10}, G_2)$, $e(C_{10}, G_3)$, $e(C_{10}, G_5)$, $e(C_{11}, G_3)$, $e(C_{11}, G_5)$, $e(C_{11}, \overline{G}_1)$, $e(C_{12}, G_4)$, $e(C_{12}, G_5)$, $e(C_{12}, G_6)$, $e(C_{13}, G_5)$, $e(C_{13}, G_6)$, $e(C_{13}, G_7)$, $e(C_{14}, G_5)$, $e(C_{14}, G_7)$, $e(C_{14}, \overline{G}_1)$, $e(C_{15}, G_6)$, $e(C_{15}, \overline{G}_2)$, $e(C_{15}, \overline{G}_3)$, $e(C_{16}, G_6)$, $e(C_{16}, G_7)$, $e(C_{16}, \overline{G}_2)$, $e(C_{17}, G_7)$, $e(C_{17}, \overline{G}_1)$, $e(C_{17}, \overline{G}_2)$, $e(\overline{C}_7, \overline{G}_6)$, $e(\overline{C}_7, \overline{G}_1)$, $e(\overline{C}_7, \overline{G}_4)$, $e(\overline{C}_8, \overline{G}_4)$, $e(\overline{C}_8, \overline{G}_1)$, $e(\overline{C}_8, \overline{G}_2)$, $e(\overline{C}_9, \overline{G}_5)$, $e(\overline{C}_9, \overline{G}_4)$, $e(\overline{C}_9, \overline{G}_2)$, $e(\overline{C}_{10}, \overline{G}_5)$, $e(\overline{C}_{10}, \overline{G}_2)$, $e(\overline{C}_{10}, \overline{G}_3)$, $e(\overline{C}_{12}, \overline{G}_6)$, $e(\overline{C}_{12}, \overline{G}_5)$, $e(\overline{C}_{12}, \overline{G}_4)$, $e(\overline{C}_{13}, \overline{G}_7)$, $e(\overline{C}_{13}, \overline{G}_6)$, $e(\overline{C}_{13}, \overline{G}_5)$;

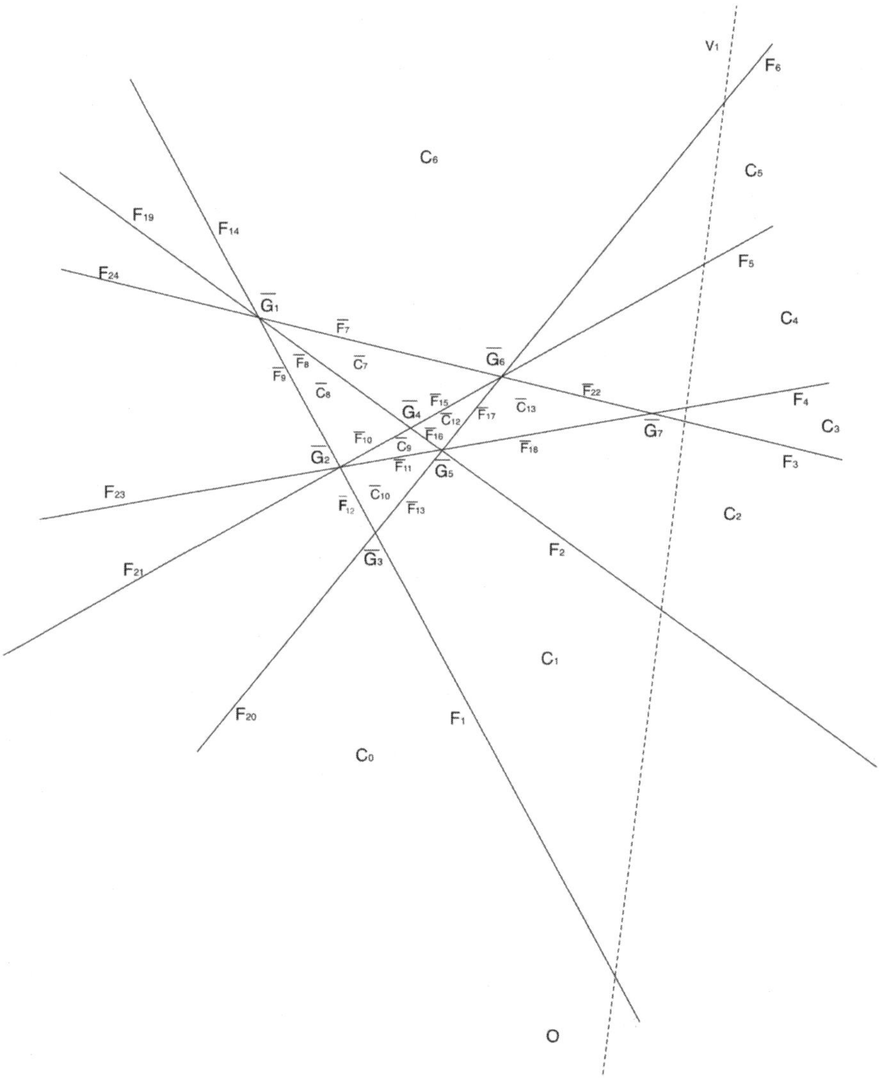

Fig. 2. A lower section of the reflection arrangement A_3

- 24 cells of dimension 3:

$$e(C_i, P), \; i = 0, \ldots, 17,$$

more

$$e(\overline{C}_7, P), e(\overline{C}_8, P), e(\overline{C}_9, P), e(\overline{C}_{10}, P), e(\overline{C}_{12}, P), e(\overline{C}_{13}, P).$$

The complex $\mathbf{S}^{(2)}$ is composed by the following cells:

- 24 cells of dimension 0: The cells of form $e(C_i, C_i, C_i)$ where $C_i \in \Phi(\mathcal{A})$ is a chamber, i.e. a facet of codimension 0.
- 72 cells of dimension 1: The cells of form $e(C_i, C_i, F_j)$ where C_i is a chamber, and $F_j \in \Phi(\mathcal{A})$ is a facet of codimension 1 in the boundary of C_i, i.e. $C_i \prec F_j$.
- 72+72 cells of dimension 2: 72 cells of form $e(C_i, C_i, G_k)$ where C_i is a chamber, and $G_k \subset \Phi(\mathcal{A})$ is a facet of codimension 2 in the boundary of C_i, i.e. $C_i \prec G_k$; 72 cells of form $e(C_i, F_j, F_j)$ where C_i is a chamber, and F_j is a facet of codimension 1 in the boundary of C_i.
- 24+144 cells of dimension 3: 24 cells of form $e(C_i, C_i, P)$ where C_i is a chamber, and $P \in \Phi(\mathcal{A})$ is the center of $\mathcal{A}$, i.e. the only facet of codimension 3 in $\Phi(\mathcal{A})$; 144 cells of form $e(C_i, F_j, G_k)$, where C_i is a chamber, F_j is a facet of codimension 1 in the boundary of C_i, and G_k is a facet of codimension 2 in the boundary of F_j, i.e. $F_j \prec G_k$.
- 72+72 cells of dimension 4: 72 cells of form $e(C_i, F_j, P)$, where C_i is a chamber, F_j is a facet of codimension 1 in the boundary of C_i and P is the center of $\mathcal{A}$; 72 cells of form $e(C_i, G_k, G_k)$, where C_i is a chamber and G_k is a facet of codimension 2 in the boundary of C_i.
- 72 cells of dimension 5: The cells of form $e(C_i, G_k, P)$, where C_i is a chamber, G_k is a facet of codimension 2 in the boundary of C_i and P is the center of $\mathcal{A}$.
- 24 cells of dimension 6: The cells of the form $e(C_i, P, P)$ where C_i is a chamber, and P is the center of $\mathcal{A}$.

The complex $\mathbf{S}^{(3)}$ is composed by the following cells:

- 24 cells of dimension 0: The cells of form $e(C_i, C_i, C_i, C_i)$ where C_i is a chamber.
- 72 cells of dimension 1: The cells of form $e(C_i, C_i, C_i, F_j)$ where C_i is a chamber, and F_j is a facet of codimension 1 in the boundary of C_i.
- 72+72 cells of dimension 2: 72 cells of form $e(C_i, C_i, C_i, G_k)$ where C_i is a chamber, and G_k is a facet of codimension 2 in

the boundary of C_i; 72 cells of form $e(C_i, C_i, F_j, F_j)$ where C_i is a chamber, and F_j is a facet of codimension 1 in the boundary of C_i.

- 24+144+72 cells of dimension 3: 24 cells of form $e(C_i, C_i, C_i, P)$ where C_i is a chamber and P is the center of $\mathcal{A}$; 144 cells of form $e(C_i, C_i, F_j, G_k)$, where C_i is a chamber, F_j is a facet of codimension 1 in the boundary of C_i, and G_k is a facet of codimension 2 in the boundary of F_j; 72 cells of form $e(C_i, F_j, F_j, F_j)$ where C_i is a chamber, and F_j is a facet of codimension 1 in the boundary of C_i.
- 72+72+144 cells of dimension 4: 72 cells of form $e(C_i, C_i, F_j, P)$, where C_i is a chamber, F_j is a facet of codimension 1 in the boundary of C_i and P is the center of $\mathcal{A}$; 72 cells of form $e(C_i, C_i, G_k, G_k)$, where C_i is a chamber and G_k is a facet of codimension 2 in the boundary of C_i; 144 cells of form $e(C_i, F_j, F_j, G_k)$, where C_i is a chamber, F_j is a facet of codimension 1 in the boundary of C_i, and G_k is a facet of codimension 2 in the boundary of F_j.
- 72+72+144 cells of dimension 5: 72 cells of form $e(C_i, C_i, G_k, P)$, where C_i is a chamber, G_k is a facet of codimension 2 in the boundary of C_i and P is the center of $\mathcal{A}$; 72 cells of form $e(C_i, F_j, F_j, P)$, where C_i is a chamber, F_j is a facet of codimension 1 in the boundary of C_i and P is the center of $\mathcal{A}$; 144 cells of form $e(C_i, F_j, G_k, G_k)$, where C_i is a chamber, F_j is a facet of codimension 1 in the boundary of C_i, and G_k is a facet of codimension 2 in the boundary of F_j.
- 24+144+72 cells of dimension 6: 24 cells of form $e(C_i, C_i, P, P)$ where C_i is a chamber, and P is the center of $\mathcal{A}$; 144 cells of form $e(C_i, F_j, G_k, P)$, where C_i is a chamber, F_j is a facet of codimension 1 in the boundary of C_i, and G_k is a facet of codimension 2 in the boundary of F_j and P is the center of $\mathcal{A}$; 72 cells of form $e(C_i, G_k, G_k, G_k)$, where C_i is a chamber, F_j is a facet of codimension 1 in the boundary of C_i.
- 72+72 cells of dimension 7: 72 cells of form $e(C_i, G_k, G_k, P)$, where C_i is a chamber, G_k is a facet of codimension 2 in the boundary of C_i and P is the center of $\mathcal{A}$; 72 cells of form $e(C_i, F_j, P, P)$, where C_i is a chamber, F_j is a facet of codimension 1 in the boundary of C_i and P is the center of $\mathcal{A}$.
- 72 cells of dimension 8: The cells of form $e(C_i, G_k, P, P)$, where C_i is a chamber, G_k is a facet of codimension 2 in the boundary of C_i and P is the center of $\mathcal{A}$.

- 24 cells of dimension 9: The cells of form $e(C_i, P, P, P)$ where C_i is a chamber, and P is the center of $\mathcal{A}$.

§4. Essentials on discrete Morse theory for CW-complexes

We recall here some of the main definitions and results from [Fo98], [Fo02], where Morse theory from a combinatorial viewpoint was first developed. See also [Ko08], where the same theory is re-formulated from a more combinatorial point of view.

Let $\mathcal{C}$ be a finite *regular* CW-complex. Let K be the collection of cells of $\mathcal{C}$, partially ordered by

$$\sigma < \tau \quad \Leftrightarrow \quad \sigma \subset \tau.$$

As usual, denote by K_p the p-skeleton of K.

Definition 4.1. *A discrete Morse function on $\mathcal{C}$ is a function*

$$f : K \longrightarrow \mathbb{R}$$

satisfying for all $\sigma^{(p)} \in K_p$ the following two conditions

$$\begin{array}{lrcl} (i) & \sharp\{\tau^{(p+1)} > \sigma^{(p)} \mid f(\tau^{(p+1)}) \leq f(\sigma^{(p)})\} & \leq & 1 \\ (ii) & \sharp\{v^{(p-1)} < \sigma^{(p)} \mid f(\sigma^{(p)}) \leq f(v^{(p-1)})\} & \leq & 1. \end{array}$$

Actually, one shows that, if f satisfies (i) and (ii) above, then for any given cell of K at least one between (i), (ii) is a strict inequality.
The analog of a critical point of index p in standard Morse theory is here a *critical cell* of dimension p: a p-cell $\sigma^{(p)}$ is critical iff both the cardinalities in i), ii) are zero.

Let $m_p(f)$ denote the number of critical p-cells of f. Then one has

Proposition 4.2. *$\mathcal{C}$ is homotopy equivalent to a CW-complex which has exactly $m_p(f)$ cells of dimension p.*

The *discrete gradient vector field* Γ_f of a Morse function f over K is the set of all pairs of cells for which the exception in Definition 4.1 happens:

$$\Gamma_f = \{(\sigma^{(p)}, \tau^{(p+1)}) |\ \sigma^{(p)} < \tau^{(p+1)},\ f(\tau^{(p+1)}) \leq f(\sigma^{(p)})\}.$$

Since, for any given cell, at most one between (i), (ii) in 4.1 is an equality, it follows that each cell belongs to at most one pair of Γ_f.

A general definition of discrete vector field is the following.

Definition 4.3. *A discrete vector field* Γ *on* $\mathcal{C}$ *is a collection of pairs of cells* $(\sigma^{(p)}, \tau^{(p+1)}) \in \mathcal{C} \times \mathcal{C}$ *such that* $\sigma^{(p)} < \tau^{(p+1)}$ *and such that each cell of* $\mathcal{C}$ *belongs to at most one pair of* Γ.

Remark 4.4. *A discrete vector field over a CW-complex corresponds to a* matching *of the associated poset (see* [Ko08]*).*

For Γ as above, define a Γ-path as a sequence of cells

$$\sigma_0^{(p)}, \tau_0^{(p+1)}, \sigma_1^{(p)}, \tau_1^{(p+1)}, \sigma_2^{(p)}, \cdots, \tau_r^{(p+1)}, \sigma_{r+1}^{(p)} \tag{1}$$

such that for each $i = 0, \cdots, r$ one has $(\sigma_r^{(p)}, \tau_r^{(p+1)}) \in \Gamma$ and $\sigma_i^{(p)} \neq \sigma_{i+1}^{(p)} < \tau_i^{(p+1)}$.
The Γ-path is closed (and non-trivial) iff $\sigma_0^{(p)} = \sigma_{r+1}^{(p)}$, $r \geq 0$. One has:

Theorem 2. *A discrete vector field* Γ *is the gradient vector field of a discrete Morse function on* $\mathcal{C}$ *iff there are no non-trivial closed* Γ*-paths.*

Remark 4.5. *A gradient vector field over a CW complex* $\mathcal{C}$ *corresponds to an* acyclic matching *of the associated poset, i.e. a matching without closed loops. In many cases it is convenient to find directly a gradient vector field, or equivalent an acyclic matching, without passing through an effective discrete Morse function. The critical cells in this case are simply all the cells which do not belong to pairs of the field. The datum of an acyclic matching produces a* collapsing ordering *on the non singular cells of* $\mathcal{C}$*, producing a constructive reduction of* $\mathcal{C}$ *in proposition 4.2 (see* [Ko08]*).*

§5. The standard case of complexified real arrangements

In this part we consider the case of standard complexified arrangements.

The last part of Section 5 is devoted to recall (skipping some details) some of the results in [SaSe07], where we applied the theory to Hyperplane Arrangements. We set in this section $\mathbf{S} := \mathbf{S}^{(1)}$, the case of standard complexification. A k-cell of $\mathbf{S}$ is written as $e(\mathcal{F})$, $\mathcal{F} = (C, F)$, with $codim(F) = k$. We write for brevity also $e(\mathcal{F}) = e(C, F)$. The boundary condition given in Lemma 2.2 specializes here to: $e(D, G)$ is in the boundary of $e(C, F)$ iff

i) $G \prec F$

ii) the chambers C and D are contained in the same chamber of $\mathcal{A}_G$, that is $D = C.G$ in the notations of part 2.

The following constructions are based on a *generic* system of *polar coordinates* in M, which is associated to a *generic* flag of subspaces

$$V_i \;=\; <\mathbf{e}_1, ..., \mathbf{e}_i>, \quad i = 0, ..., n \ (dim(V_i) = i).$$

Consider the pencil of $n-1$-dimensional subspaces of M, $V_{n-1}(\theta)$, $\theta \in \mathbb{R}$, with base V_{n-2} : so $V_{n-1}(0) = V_{n-1}$ and θ grows according to some positive orientation of V/V_{n-2}. By recurrence, let $V_i(\theta, \theta_{i+1}, \ldots, \theta_{n-1})$, $\theta \in \mathbb{R}$, be the pencil of i-dimensional subspaces in $V_{i+1}(\theta_{i+1}, \ldots, \theta_{n-1})$, with base V_{i-1}, $i = 1, \ldots, n-2$. Let also $V_0(\theta, \theta_1, \ldots, \theta_{n-1})$ be the point with distance θ from the origin V_0 inside the line $V_1(\theta_1, \ldots, \theta_{n-1})$.

Each point $P \neq V_0$ is written uniquely as $P = V_0(\theta_0, \theta_1, \ldots, \theta_{n-1})$, so we associate to P the polar coordinates $(\theta_0, \ldots, \theta_{n-1})$.

Definition 5.1. *We say that the above flag is generic with respect to the arrangement $\mathcal{A}$ if it satisfies the following conditions:*

i) the origin V_0 is contained in a chamber C_0 of $\mathcal{A}$;

ii) there exists a $\delta > 0$, $\delta << 1$, such that the set of bounded facets of $\mathcal{A}$ is contained into $\{P \in M : \ 0 < \theta_i(P) < \delta, \ i = 0, \ldots, n-1\}$;

iii) subspaces $V_i(\bar\theta) = V_i(\bar\theta_i, ..., \bar\theta_{n-1})$ which intersect the union of bounded facets are generic with respect to $\mathcal{A}$, in the sense that, for each codim-k subspace $L \in L(\mathcal{A})$,

$$i \geq k \ \Rightarrow \ V_i(\bar\theta) \cap L \cap clos(\tilde B) \neq \emptyset \ \textit{and} \ dim(|V_i(\bar\theta)| \cap L) \;=\; i-k.$$

One proves that a generic system of polar coordinates always exists.

Fix a system of polar coordinates associated to a generic flag.

We consider the subspace $V_i(\bar\theta)$, $\bar\theta = (\bar\theta_i, ..., \bar\theta_{n-1})$. The arrangement $\mathcal{A}$ induces a stratification $\Phi(\bar\theta)$ in $V_i(\bar\theta)$: given a codimension-k facet $F \in \mathcal{S}$, let us denote by

$$F(\bar\theta) := F \cap V_i(\bar\theta), \quad \theta_j \in [0, \delta], \ j = i, ..., n-1.$$

By genericity conditions, if $i \geq k$ then $F(\bar\theta)$ is either empty or it is a codimension $k+n-i$ facet contained in $V_i(\bar\theta)$.

For each such $\bar\theta$ we want to give a *total ordering* on $\Phi(\bar\theta)$.

Definition 5.2. *Given any facet $F(\bar\theta)$ let us denote by*

$$P_{F(\bar\theta)} \in clos(F(\bar\theta))$$

the "minimum" vertex of $clos(F(\bar\theta)) \cap \{\theta_i \geq 0, \ i = 0, \ldots\}$: this is the 0-dimensional facet which has minimum polar coordinates, with respect to the anti-lexicographic ordering of the coordinates (one says that $(\theta_0, \ldots, \theta_{n-1})$ is less than $(\theta'_0, \ldots, \theta'_{n-1})$ iff, being k the last index such that $\theta_h \neq \theta'_h$, one has $\theta_k < \theta'_k$).

Remark that the minimum vertex is well defined by genericity conditions.

Definition 5.3. *We define the* polar ordering *by recurrence on the dimension as follows: given* $F,\ G\ \in \Phi$*, and given* $\bar{\theta} = (\bar{\theta}_i, ..., \bar{\theta}_{n-1})$*,* $0 \le i \le n$*,* $\bar{\theta}_j \in [0, \delta]$ *for* $j \in i, .., n-1$*,* $(\bar{\theta} = \emptyset$ *for* $i = n)$ *such that* $F(\bar{\theta}), G(\bar{\theta}) \neq \emptyset$*, we set*

$$F(\bar{\theta})\ \triangleleft\ G(\bar{\theta})$$

iff one of the following cases occurs:

i) $P_{F(\bar{\theta})} \neq P_{G(\bar{\theta})}$ *and the coordinates of the former point are lower (in the anti-lexicographical ordering) than the coordinates of the latter point.*

ii) $P_{F(\bar{\theta})} = P_{G(\bar{\theta})}$*. Then either*

iia) $dim(F(\bar{\theta})) = 0$ *(so* $P_{F(\bar{\theta})} = F(\bar{\theta})$*) and* $F(\bar{\theta}) \neq G(\bar{\theta})$ *(so* $dim(G(\bar{\theta})) > 0$*)*

or

iib) $dim(F(\bar{\theta})) > 0,\ dim(G(\bar{\theta})) > 0$.

Let $i_0 := max\{j |\ \theta_j(P_{F(\bar{\theta})}) \neq 0\} + 1$*, and let* $i_1 := min\{i_0, i\}$*. If the coordinates of the minimal point are* $P \equiv (\bar{\theta}_0, \ldots, \bar{\theta}_{n-1})$ *then* $\forall \epsilon,\ 0 < \epsilon << \delta$*, it must happen*

$$F(\bar{\theta}_{i_1-1} + \epsilon, \bar{\theta}_{i_1}, ...)\ \triangleleft\ G(\tilde{\theta}_{i_1-1} + \epsilon, \bar{\theta}_{i_1}, ...).$$

The idea in (iib) is to intersect F and G with a lower dimensional subspace contained in $V_i(\bar{\theta})$, very close to the minimum point (see also [DeSe] for a purely combinatorial version of polar ordering).

One shows that polar ordering $\triangleleft$ is a total ordering on the facets of $V_i(\bar{\theta})$, for any given $\bar{\theta} = (\bar{\theta}_i, ..., \bar{\theta}_{n-1})$. In particular (taking $\bar{\theta} = \emptyset$, which corresponds to all $V_n = M$) it gives a total ordering on Φ.

We consider now the regular CW-complex $\mathbf{S} = \mathbf{S}^{(1)}$ and we define a combinatorial *gradient vector field* Γ over $\mathbf{S}$. One can describe Γ as a collection of pairs of cells

$$\Gamma \subset \{(e, f) \in \mathbf{S} \times \mathbf{S} \mid dim(f) = dim(e) + 1,\ e \in \partial(f)\}$$

so that Γ decomposes into its dimension-p components

$$\Gamma = \bigsqcup_{p=1}^{n} \Gamma^p, \qquad \Gamma^p \subset \mathbf{S}_{p-1} \times \mathbf{S}_p$$

($\mathbf{S}_p$ being the p-skeleton of $\mathbf{S}$).

We give the following recursive definition:

Definition 5.4 (Polar Gradient). *We define a combinatorial gradient field* Γ *over* $\mathbf{S}$ *in the following way:*

the $(j+1)$*-th component* Γ^{j+1} *of* Γ, $j = 0, \ldots, n-1$, *is given by all pairs*

$$(e(C, F^j), e(C, F^{j+1})), \quad F^j \prec F^{j+1}$$

(same chamber C*) such that*

1. $F^{j+1} \lhd F^j$
2. $\forall F^{j-1} \prec F^j$ *such that* $C \prec F^{j-1}$ *the pair*

$$(e(C \prec F^{j-1}), e(C \prec F^j)) \notin \Gamma^j.$$

Theorem 3. *One has:*

1. Γ *is a combinatorial vector field on* $\mathbf{S}$ *which is the gradient of a discrete Morse function.*
2. *The pair*

$$(e(C \prec F^j), e(C \prec F^{j+1})), \quad F^j \prec F^{j+1}$$

belongs to Γ *iff the following conditions hold:*
(a) $F^{j+1} \lhd F^j$
(b) $\forall\ F^{j-1}$ *such that* $C \prec F^{j-1} \prec F^j$*, one has* $F^{j-1} \lhd F^j$.

3. *Given* $F^j \in \Phi$*, there exists a chamber* C *such that the cell* $e(C \prec F^j)$ *is the second factor of a pair in* Γ *iff there exists* $F^{j-1} \prec F^j$ *with* $F^j \lhd F^{j-1}$*. More precisely, for each chamber* C *such that there exists* F^{j-1} *with*

$$C \prec F^{j-1} \prec F^j, \quad F^j \lhd F^{j-1} \qquad (*)$$

the pair $(e(C \prec \bar{F}^{j-1}), e(C \prec F^j)) \in \Gamma$, *where* $\bar{F}^{j-1}$ *is the maximum* $(j-1)$*-facet (with respect to polar ordering) satisfying conditions (*).*

4. *The set of* k*-dimensional critical cells is given by*

$$\begin{aligned} &Sing_k(\mathbf{S}) \\ &= \{e(C \prec F^k) \ : \ F^k \cap V_k \neq \emptyset,\ F^j \lhd F^k,\ \forall\ C \prec F^j \precneqq F^k\}. \end{aligned} \tag{2}$$

Equivalently, $F^k \cap V_k$ *is the maximum (in polar ordering) among all facets of* $C \cap V_k$.

In Section 7 we will need some more specific results which were used in the Proof of part (1) of Theorem 3.

Take a Γ-path in $\mathbf{S}$ (according to [Fo02])

$$e(C_1, F_1^k), e(C_1, F_1^{k+1}), \ldots, e(C_m, F_m^k), e(C_m, F_m^{k+1}), e(C_{m+1}, F_{m+1}^k). \tag{3}$$

Here the pair $(e(C_i, F_i^k), e(C_i, F_i^{k+1}))$ is an element of Γ, and $e(C_i, F_i^k)$ is in the boundary of $e(C_{i-1}, F_{i-1}^{k+1})$.

According to a standard result in Discrete Morse Theory ([Fo98]) we have to show that, if the path (3) is closed, (i.e. if $e(C_{m+1}, F_{m+1}^k)$ equals to $e(C_1, F_1^k)$), then it is trivial, i.e. $F_i^k = F_{i+1}^k$, $F_i^{k+1} = F_{i+1}^{k+1}$, and $C_i = C_{i+1}$ $(i = 1, \ldots, m-1)$.

The proof directly follows from the following two claims.

Claim 1. *Given a triple of consecutive cells in (3) of the form:*

$$e(C_i, F_i^{k+1}), e(C_{i+1}, F_{i+1}^k), e(C_{i+1}, F_{i+1}^{k+1}), \tag{4}$$

we have that $F_{i+1}^{k+1} \trianglelefteq F_i^{k+1}$.

Claim 2. *Given a quadruple of consecutive cells in (3) of the form:*

$$e(C_i, F_i^k), e(C_i, F^{k+1}), e(C_{i+1}, F_{i+1}^k), e(C_{i+1}, F^{k+1}), \tag{5}$$

we have $F_i^k \trianglelefteq F_{i+1}^k$.

Remark 5.5. *1) Once a polar ordering is assigned, the set of singular cells is described only in terms of it by*

$Sing_k(\mathbf{S}) := \{e(C \prec F^k) :$

a) $F^k \triangleleft F^{k+1}, \quad \forall F^{k+1}$ *s.t.* $F^k \prec F^{k+1}$

b) $F' \triangleleft F^k, \quad \forall F'$ *s.t.* $C \prec F' \prec F^k \}$.

2) The construction of Theorem 3 gives an explicit additive basis for the homology and for the cohomology in terms of the singular cells in $\mathbf{S}$. *We can call it a* polar basis *(relative to a given system of generic polar coordinates).*

3) The minimality of the associated Morse complex is obtained by the one-to-one correspondence between singular cells and the set of all the chambers of Φ, *and the well-known formula* $\sum B_i = |\{chambers\}|$ *(see [Za75]). Here* B_i *means the ith-Betti number of the complement.*

§6. Examples (2)

Consider now the arrangement $\mathcal{A} \in \mathbb{R}^3$ of Example 1 (see Fig. 1 and 2). The system of polar coordinates here is given by O, V_1 and V_2. Here

V_2 is the 2-plane in $\mathbb{R}^3$ which corresponds to the plane $z = a$, $a > 0$, i.e. the plane of Fig. 1, while V_1 is the line indicated in Fig. 1.

To give the total ordering in $\Phi(\mathcal{A})$ we proceed as follows: the cells intersecting the plane V_2 comes before all the others cells, and they are ordered as in the case of a line arrangement with O, V_1 as polar coordinates system. The center P comes after these cells and before of the other ones. For ordering the other ones we have to move the plane V_2 fixing the line V_1, until meeting the center P, and then we have to pass to the other side and ordering all the cells comparing in Fig. 2 as in the case of a line arrangement with O, V_1 as polar coordinates system. So the polar ordering in A_3 is given by:

$$C_0 \lhd F_1 \lhd C_1 \lhd F_2 \lhd C_2 \lhd F_3 \lhd C_3 \lhd F_4 \lhd C_4 \lhd F_5 \lhd C_5 \lhd F_6 \lhd C_6 \lhd G_1 \lhd F_7 \lhd C_7 \lhd F_8 \lhd C_8 \lhd F_9 \lhd G_2 \lhd F_{10} \lhd C_9 \lhd F_{11} \lhd C_{10} \lhd F_{12} \lhd G_3 \lhd F_{13} \lhd C_{11} \lhd F_{14} \lhd G_4 \lhd F_{15} \lhd C_{12} \lhd F_{16} \lhd G_5 \lhd F_{17} \lhd C_{13} \lhd F_{18} \lhd C_{14} \lhd F_{19} \lhd G_6 \lhd F_{20} \lhd C_{15} \lhd F_{21} \lhd C_{16} \lhd F_{22} \lhd G_7 \lhd F_{23} \lhd C_{17} \lhd F_{24} \lhd P \lhd \overline{G}_7 \lhd \overline{F}_{18} \lhd \overline{C}_{13} \lhd \overline{F}_{22} \lhd \overline{G}_6 \lhd \overline{F}_{17} \lhd \overline{C}_{12} \lhd F_{15} \lhd \overline{C}_7 \lhd \overline{F}_7 \lhd \overline{G}_5 \lhd \overline{F}_{13} \lhd \overline{C}_{10} \lhd \overline{F}_{11} \lhd \overline{C}_9 \lhd \overline{F}_{16} \lhd \overline{G}_4 \lhd \overline{F}_{10} \lhd \overline{C}_8 \lhd \overline{F}_8 \lhd \overline{G}_3 \lhd \overline{F}_{12} \lhd \overline{G}_2 \lhd \overline{F}_9 \lhd \overline{G}_1.$$

Recall that the complex $\mathbf{S}^{(1)}$ has 192 cells. The polar gradient has components: $\Gamma = \Gamma_1 \sqcup \Gamma_2 \sqcup \Gamma_3$. Following Definition 5.4 we can see that:

- Γ_1 is composed by 23 pairs of type $(e(C_i, C_i), e(C_i, F_j))$, with $F_j \lhd C_i$;
- Γ_2 is composed by 43 pairs of type $(e(C_i, F_j), e(C_i, G_k))$, with $G_k \lhd F_j$ and $e(C_i, F_j)$ does not belong to any pair of Γ_1.
- Γ_3 is composed by 18 pairs of type $(e(C_i, G_k), e(C_i, P))$, with $P \lhd G_k$ and $e(C_i, G_k)$ does not belong to any pair of Γ_2.

So the set of critical cells $Sing(\mathbf{S})$ is composed by the cells not appearing in any pair of Γ, i.e. by the following cells:

- 1 critical 0-cell $e(C_0, C_0)$;
- 6 critical 1-cells $e(C_0, F_1), e(C_1, F_2), e(C_2, F_3), e(C_3, F_4), e(C_4, F_5), e(C_5, F_6)$, which correspond to those 1-cells $e(C_i, F_j)$ with $F_j \cap V_1 \neq \emptyset$ and $F_j \cap V_1$ is the maximum (in polar ordering) among all facets of $C \cap V_1$;
- 11 critical 2-cells $e(C_1, G_1)$, $e(C_2, G_1)$, $e(C_3, G_2)$, $e(C_4, G_2)$, $e(C_5, G_3)$, $e(C_7, G_6)$, $e(C_8, G_4)$, $e(C_9, G_5)$, $e(C_{10}, G_5)$, $e(C_{12}, G_6)$, $e(C_{13}, G_7)$, which correspond to those 2-cells $e(C_i, G_k)$ with $G_k \cap V_2 \neq \emptyset$ and $G_k \cap V_2$ is the maximum (in polar ordering) among all facets of $C \cap V_2$.
- 6 critical 3-cells $e(C_7, P), e(C_8, P), e(C_9, P), e(C_{10}, P), e(C_{12}, P), e(C_{13}, P)$, which correspond to those 2-cells $e(C_i, P)$ with P is the maximum (in polar ordering) among all facets of C.

Remark 6.1. *The number of cells is* $1+6+11+6+2*(23+43+18) = 192$.

§7. Discrete methods on configuration spaces

We now consider general configuration spaces as in part 2, and we generalize the theory of part 5. We fix a polar ordering $\lhd$ induced by a generic system of polar coordinates $V_0, V_1, \ldots, V_n$ as in part 5.

Lemma 7.1. *Let $L \in \mathcal{L}(\mathcal{A})$ be a codimension k subspace. Then the system $V_0, \ldots, V_k$ gives a generic system of polar coordinates for the arrangement $\mathcal{A}_L \cap V_k := \{H \cap V_k | \ H \in \mathcal{A}_L\}$ in V_k, so it induces a polar ordering $\lhd_L$ on Φ_L. The system $V_k \cap L, \ldots, V_n \cap L$ gives a generic system of polar coordinates for the arrangement $\mathcal{A}^L$ on L, inducing a polar ordering $\lhd^L$ on Φ^L.*

One has that $\lhd^L$ coincides with the restriction $\lhd_{|\Phi^L}$ of the polar ordering $\lhd$ to $\Phi^L \subset \Phi$.

Definition 7.2. *Given $G \in \Phi$, we set also $\lhd_G := \lhd_{|G|}$.*

Definition 7.3. *Given $G \in \Phi$, and being Φ_G as in Section 2, we define an involution:*

$$\begin{array}{rcl} op_G : \Phi_G & \rightarrow & \Phi_G \\ F & \mapsto & op_G(F) \end{array}$$

where $op_G(F)$ is the unique facet which is symmetric to F with respect to $supp(G)$. In other terms (using the maps φ_G, $pr_{|G|}$ defined in Section 2):

$$op_G(F) \ := \ \varphi_G^{-1}(-(pr_{|G|}(F))).$$

Here we notice that, for a central arrangement, every facet F has a unique opposite $\overline{F}$ with respect to the center.

Definition 7.4. *Define the* opposite *polar ordering $\lhd_G^{op}$ in Φ_G as:*

$$F \lhd_G^{op} F' \Leftrightarrow op_G(F) \lhd_G op_G(F')$$

$F, F' \in \Phi_G$.

Definition 7.5. *For all arrangements $\mathcal{A}$, all polar orderings $\lhd$ on $\mathcal{A}$, and all $d \geq 1$, we define the degree-d discrete field*

$$\Gamma_{(d)} := \Gamma_{(d)}(\mathcal{A}, \lhd)$$

on the complex $\mathbf{S}^{(d)}(\mathcal{A})$. *Assume by recurrence that* $\Gamma_{(d')}(\mathcal{A}^L)$ *has been defined for* $d' < d$, *for any* $\mathcal{A}^L \subset \mathcal{A}$, $L \in \mathcal{L}(\mathcal{A})$, *for the induced polar ordering* $\triangleleft^L$ *(see Lemma 7.1). Then the k-dimensional part* $\Gamma^k_{(d)}(\mathcal{A})$ *is given by the set of pairs of cells in* $\mathbf{S}^{(d)}(\mathcal{A})$

$$(e(\mathcal{F}), e(\mathcal{F}'))$$

where $dim(e(\mathcal{F})) = k - 1$, $dim(e(\mathcal{F}')) = k$, $\mathcal{F}' \prec \mathcal{F}$ *(so* $e(\mathcal{F}) \subset \partial(e(\mathcal{F}'))$*), and the two flags differ only in a single position:*

$$\mathcal{F} = (C, F_1, \ldots, F_{i-1}, F_i^j, F_{i+1}, \ldots, F_d),$$

$$\mathcal{F}' = (C, F_1, \ldots, F_{i-1}, F_i^{j+1}, F_{i+1}, \ldots, F_d)$$

with $F_i^j \prec F_i^{j+1}$ *(*j, $j+1$ *denote codimensions). Moreover, such pairs satisfy the following conditions (6),(7):*

(6) *for* $i < d$, $e(F_i^j, F_{i+1}, \ldots, F_d)$ *is a critical cell in the complex* $\mathbf{S}^{(d-i)}(\mathcal{A}^L)$, *endowed with the discrete* $(d-i)$*-vector field* $\Gamma_{(d-i)}(\mathcal{A}^L)$, *with* $L := |F_i^j|$ *(i.e.,* $e(F_i^j, F_{i+1}, \ldots, F_d) \notin \Gamma_{(d-i)}(\mathcal{A}^L)$, *see Remark (4.5))*

Set $l = d - i$; *then:*

(7) *for even* l

$$F_i^{j+1} \triangleleft_{F_{i+1}} F_i^j \quad \text{and} \quad F_i^j = max_{\triangleleft_{F_{i+1}}}\{F \mid F_{i-1} \prec F \prec F_i^j\};$$

for odd l

$$F_i^{j+1} \triangleleft^{op}_{F_{i+1}} F_i^j \quad \text{and} \quad F_i^j = max_{\triangleleft^{op}_{F_{i+1}}}\{F \mid F_{i-1} \prec F \prec F_i^j\}.$$

For $i = d$ *there is no* F_{i+1} *in the first condition of (7), which is to be considered in this case as defined by using the given polar ordering* $\triangleleft$.

We have $\Gamma_{(d)} = \oplus_{k=1}^{n'd}\Gamma^k_{(d)}$, $n' = rk(\mathcal{A})$.

Definition 7.6. *Let* $L \in \mathcal{L}(\mathcal{A})$ *be a codimension k subspace.*
Set $\Gamma^L_{(d)}$ *as the degree-d discrete field of the arrangement* $\mathcal{A}^L$ *with respect to the polar ordering* $\triangleleft^L = \triangleleft_{|L}$.
Set $\Gamma_{L,(d)}$ *as the degree-d discrete field of the arrangement* $\mathcal{A}_L \cap V^k$ *with respect to the polar ordering* $\triangleleft_L$ *(see Lemma (7.1)).*

The following main theorem describes the minimal complex in terms of the field $\Gamma_{(d)}$, exhibiting its critical cells.

Theorem 4. *One has:*

1. $\Gamma_{(d)}$ *is a discrete vector field;*
2. $\Gamma_{(d)}$ *is a gradient field of a discrete Morse function;*
3. *the critical cells of* $\Gamma_{(d)}$ *are the following ones, depending on the parity of* d*:*

$$e(C, F^k, \dots, F^k) \tag{8}$$

with $e(C, F^k) \in \mathbf{S}^{(1)}$ *critical cell for* $(\Gamma_{(1)}, \lhd)$, *if* d *is odd;*

$$e(op_{F^k}(C), F^k, \dots, F^k) \tag{9}$$

with $e(C, F^k) \in \mathbf{S}^{(1)}$ *critical cell for* $(\Gamma_{(1)}, \lhd)$, *if* d *is even.*

As an immediate consequence of Theorem 4 we have

Theorem 5. 1. *The configuration space* $\mathcal{M}^{(d)}(\mathcal{A})$ *is a minimal space* $(d \geq 1)$.
2. *The cohomology of* $\mathbf{S}^{(d)}$ *(or of* $\mathcal{M}^{(d+1)}$*),* $d \geq 1$, *is concentrated in dimension*

$$id\ ,\ i = 0 \dots n\ .$$

The Betti numbers are given by

$$B_{id}(\mathbf{S}^{(d)}) = B_i(\mathbf{S}^{(1)}).$$

Proof of Theorem 5. Case $d = 1$ is considered in Section 5. For $d > 1$ minimality follows immediately from the gap between the dimensions of the critical cells. Q.E.D.

The Proof of Theorem 4 is very technical and will appear elsewhere.

Remark. *For* $d > 1$ *the configuration space is simply connected, therefore its minimality follows by general means. For* $d = 1$ *there is also a non-trivial Morse complex for the cohomology with local coefficients (see* [SaSe07],[GaSa09]*).*

Remark. *Notice the essential difference between the case when* d *is even and the case when* d *is odd, while the shape of the flags is essentially the same when the parity is the same. This fact reflects similar well known phenomena about the representation of a reflection group* W *onto the cohomology of the corresponding configuration spaces: for odd complexifications one has the regular representation of* W, *while in the even case one obtain an induced representation (see* [Ma96],[Le00]*).*
Q.E.D.

§8. Examples (3)

We consider again now the arrangement $\mathcal{A} \subset \mathbb{R}^3$ of Example 1 (see Fig. 3 and 2).

Case $d = 2$.

Recall that the complex $\mathbf{S}^{(2)}$ has 648 cells.

The polar gradient $\Gamma_{(2)}$ has components: $\Gamma_{(2)} = \bigsqcup_{i=1}^{6} \Gamma_{(2)}^{i}$. By Definition 7.5 we have:

- $\Gamma_{(2)}^1$ is composed by 23 pairs of type $(e(C_i, C_i, C_i), e(C_i, C_i, F_j))$, with $F_j \lhd C_i$, i.e. $(e(C_i, C_i), e(C_i, F_j)) \in \Gamma_{(1)}^1$;
- $\Gamma_{(2)}^2$ is composed by two types of pairs: 43 pairs of type $(e(C_i, C_i, F_j),\ e(C_i,\ C_i, G_k))$, with $G_k \lhd F_j$ and $e(C_i, C_i, F_j)$ does not belong to any pair of $\Gamma_{(2)}^1$, i.e. $(e(C_i, F_j),\ e(C_i, G_k)) \in \Gamma_{(1)}^2$; 6 pairs of type $(e(C_i, C_i, F_j),\ e(C_i, F_j, F_j))$ with $e(C_i, F_j)$ critical for $\Gamma_{(1)}$;
- $\Gamma_{(2)}^3$ is composed by three types of pairs: 60 pairs of type $(e(C_i, F_j, F_j),\ e(C_i, F_j, G_k))$, with $G_k \lhd F_j$; 11 pairs of type $(e(C_i, C_i, G_k), e(C_i, F_j, G_k))$ with $F_j \lhd_{G_k}^{op} C_i$, and $e(C_i, G_k)$ is critical for $\Gamma_{(1)}$; 18 pairs of type $(e(C_i, C_i,\ G_k),\ e(C_i, C_i, P))$, with $P \lhd G_k$ and $e(C_i, C_i, G_k)$ does not belong to any pair of $\Gamma_{(2)}^2$, i.e. $(e(C_i, G_k), e(C_i, P)) \in \Gamma_{(1)}^3$;
- $\Gamma_{(2)}^4$ is composed by three types of pairs: 48 pairs of type $(e(C_i, F_j, G_k), e(C_i, F_j, P))$ with $P \lhd G_k$ and $G_k = max_{\lhd}\{F | C_i \prec F \prec G_k\}$; 25 pairs of type $(e(C_i, F_j, G_k), e(C_i, G_k, G_k))$ with $G_k \lhd_{G_k}^{op} F_j$ and $e(F_j, G_k)$ critical for $\Gamma_{(1)}^{|F_j|}$, i.e. $G_k \lhd P$; 6 pairs of type $(e(C_i, C_i, P), e(C_i, F_j, P))$ with $F_j \lhd_P^{op} C_i$ and $e(C_i, P)$ critical for $\Gamma_{(1)}$, i.e. $C_i \cap V_2 \neq \emptyset$ and bounded;
- $\Gamma_{(2)}^5$ is composed by two types of pairs: 18 pairs of type $(e(C_i, F_j, P), e(C_i, G_k, P))$ with $e(F_j, P)$ critical for $\Gamma_{(1)}(\mathcal{A}^{|F_j|})$, i.e. $F_j \cap V_2 \neq \emptyset$ and bounded, and $G_k \lhd_P^{op} F_j$; 36 pairs of type $(e(C_i, G_k, G_k), e(C_i, G_k, P))$ with $P \lhd G_k$;
- $\Gamma_{(2)}^6$ is composed by 18 pairs of type $(e(C_i, G_k, P), e(C_i, P, P))$, with $P \lhd_P^{op} G_k$, such that $e(C_i, G_k, P)$ is not in a pair of $\Gamma_{(2)}^3$;

So the set of critical cells $Sing(\mathbf{S})$ is given by the cells not appearing in any pair of $\Gamma_{(3)}$, i.e. by the following cells:

- 1 critical 0-cell $e(C_0, C_0, C_0)$;
- 6 critical 2-cells $e(C_1, F_1, F_1)$, $e(C_2, F_2, F_2)$, $e(C_3, F_3, F_3)$, $e(C_4, F_4, F_4)$, $e(C_5, F_5, F_5)$, $e(C_6, F_6, F_6)$;

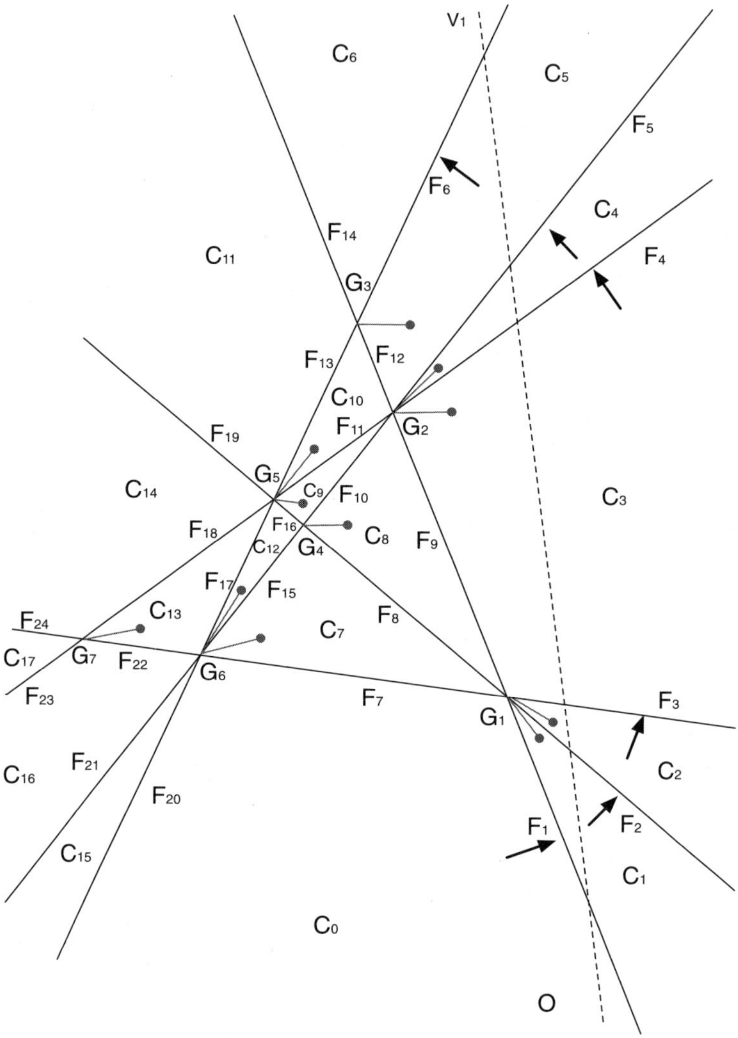

Fig. 3. An upper section of A_3 and critical cells in case d odd

- 11 critical 4-cells $e(C_7, G_1, G_1)$, $e(C_8, G_1, G_1)$, $e(C_9, G_2, G_2)$, $e(C_{10}, G_2, G_2)$, $e(C_{11}, G_3, G_3)$, $e(C_{12}, G_4, G_4)$, $e(C_{13}, G_5, G_5)$, $e(C_{14}, G_5, G_5)$, $e(C_{15}, G_6, G_6)$, $e(C_{16}, G_6, G_6)$, $e(C_{17}, G_7, G_7)$;

- 6 critical 6-cells $e(\overline{C}_7, P, P)$, $e(\overline{C}_8, P, P)$, $e(\overline{C}_9, P, P)$, $e(\overline{C}_{10}, P, P)$, $e(\overline{C}_{12}, P, P)$, $e(\overline{C}_{13}, P, P)$.

Remark 8.1. *As in previous examples we can calculate the number of cells as* $1 + 6 + 11 + 6 + 2 * (23 + 43 + 6 + 60 + 11 + 18 + 48 + 25 + 6 + 18 + 36 + 18) = 648$.

Case $d = 3$.

Recall that the complex $\mathbf{S}^{(3)}$ has 1536 cells.

The polar gradient $\Gamma_{(3)}$ has components: $\Gamma_{(3)} = \bigsqcup_{i=1}^{9} \Gamma^i_{(3)}$. Following Definition 7.5 we can see that:

- $\Gamma^1_{(3)}$ is composed by 23 pairs of type $(e(C_i, C_i, C_i, C_i), e(C_i, C_i, C_i, F_j))$, with $F_j \lhd C_i$, i.e. $(e(C_i, C_i, C_i), e(C_i, C_i, F_j)) \in \Gamma^1_{(2)}$;
- $\Gamma^2_{(3)}$ is composed by two types of pairs: 43 pairs of type $(e(C_i, C_i, C_i, F_j), e(C_i, C_i, C_i, G_k))$, with $G_k \lhd F_j$ and $e(C_i, C_i, C_i, F_j)$ does not belong to any pair of $\Gamma^1_{(3)}$, i.e. $(e(C_i, C_i, F_j), e(C_i, C_i, G_k)) \in \Gamma^2_{(2)}$; 6 pairs of type $(e(C_i, C_i, C_i, F_j), e(C_i, C_i, F_j, F_j))$ with $e(C_i, F_j)$ critical for $\Gamma_{(1)}$, i.e. $(e(C_i, C_i, F_j), e(C_i, F_j, F_j)) \in \Gamma^2_{(2)}$;
- $\Gamma^3_{(3)}$ is composed by four types of pairs: 60 pairs of type $(e(C_i, C_i, F_j, F_j), e(C_i, C_i, F_j, G_k))$, with $G_k \lhd F_j$, i.e. $(e(C_i, F_j, F_j), e(C_i, F_j, G_k)) \in \Gamma^3_{(2)}$; 11 pairs of type $(e(C_i, C_i, C_i, G_k), e(C_i, C_i, F_j, G_k))$ with $F_j \lhd^{op}_{G_k} C_i$, and $e(C_i, G_k)$ is critical for $\Gamma^1_{(1)}$, i.e. $(e(C_i, C_i, G_k), e(C_i, F_j, G_k)) \in \Gamma^3_{(2)}$; 18 pairs of type $(e(C_i, C_i, C_i, G_k), e(C_i, C_i, C_i, P))$, with $P \lhd G_k$ and $e(C_i, C_i, C_i, G_k)$ does not belong to any pair of $\Gamma^2_{(3)}$, i.e. $(e(C_i, C_i, G_k), e(C_i, C_i, P)) \in \Gamma^3_{(2)}$; 6 pairs of type $(e(C_i, C_i, F_j, F_j), e(C_i, F_j, F_j, F_j))$ with $e(C_i, F_j, F_j)$ critical for $\Gamma_{(2)}$;
- $\Gamma^4_{(3)}$ is composed by four types of pairs: 48 pairs of type $(e(C_i, C_i, F_j, G_k), e(C_i, C_i, F_j, P))$ with $P \lhd G_k$ and $G_k = max_{\lhd}\{F | C_i \prec F \prec G_k\}$, i.e. $(e(C_i, F_j, G_k), e(C_i, F_j, P)) \in \Gamma^4_{(2)}$; 25 pairs of type $(e(C_i, C_i, F_j, G_k), e(C_i, C_i, G_k, G_k))$ with $G_k \lhd^{op}_{G_k} F_j$ and $e(F_j, G_k)$ critical for $\Gamma^{|F_j|}_{(1)}$, i.e. $(e(C_i, F_j, G_k), e(C_i, G_k, G_k)) \in \Gamma^4_{(2)}$; 6 pairs of type $(e(C_i, C_i, C_i, P), e(C_i, C_i, F_j, P))$ with $F_j \lhd^{op}_P C_i$ and $e(C_i, P)$ critical for $\Gamma_{(1)}$, i.e. $(e(C_i, C_i, P), e(C_i, F_j, P)) \in \Gamma^4_{(2)}$; 60 pairs of type $(e(C_i, F_j, F_j, F_j), e(C_i, F_j, F_j, G_k))$, with $G_k \lhd F_j$. Notice that in $\Gamma_{(3)}$ there are no pair of type $(e(C_i, C_i, F_j, G_k), e(C_i, F_j, F_j, G_k))$ because no cell of the form $e(C_i, F_j, G_k)$ is critical for $\Gamma_{(2)}$;

- $\Gamma^5_{(3)}$ is composed by five types of pairs: 18 pairs of type $(e(C_i, C_i, F_j, P),\ e(C_i, C_i, G_k, P))$ with $e(F_j, P)$ critical for $\Gamma_{(1)}(\mathcal{A}^{|F_j|})$, and $G_k \lhd^{op}_P F_j$, i.e. $(e(C_i, F_j, P),\ e(C_i, G_k, P)) \in \Gamma^5_{(2)}$; 36 pairs of type $(e(C_i, C_i, G_k, G_k),\ e(C_i, C_i, G_k, P))$ with $P \lhd G_k$, i.e. $(e(C_i, G_k, G_k),\ e(C_i, G_k, P)) \in \Gamma^5_{(2)}$; 11 pairs of type $(e(C_i, C_i, G_k, G_k),\ e(C_i, F_j, G_k, G_k))$ with $F_j \lhd C_i$, and $e(C_i, G_k, G_k)$ is critical for $\Gamma_{(2)}$; 48 pairs of type $(e(C_i, F_j, F_j, G_k),\ e(C_i, F_j, F_j, P))$ with $P \lhd G_k$ and $G_k = max_{\lhd}\{F|F_j \prec F \prec G_k\}$; 36 pairs of type $(e(C_i, F_j, F_j, G_k),\ e(C_i, F_j, G_k, G_k))$ with $e(F_j, G_k)$ critical for $\Gamma_{(1)}(\mathcal{A}^{|F_j|})$, $G_k \lhd^{op}_{G_k} F_j$. Notice that in $\Gamma_{(3)}$ there are no pair of type $(e(C_i, C_i, F_j, P),\ e(C_i, F_j, F_j, P))$ because no cell of the form $e(C_i, F_j, P)$ is critical for $\Gamma_{(2)}$;
- $\Gamma^6_{(3)}$ is composed by four types of pairs: 18 pairs of type $(e(C_i, C_i, G_k, P), e(C_i, C_i, P, P))$, with $P \lhd^{op}_P G_k$, such that $e(C_i, C_i, G_k, P)$ is not in $\Gamma^5_{(3)}$, i.e. $(e(C_i, G_k, P), e(C_i, P, P)) \in \Gamma^6_{(2)}$; 72 pairs of type $(e(C_i, F_j, G_k, G_k),\ e(C_i, F_j, G_k, P))$ with $P \lhd G_k$; 25 pairs of type $(e(C_i, F_j, G_k, G_k), e(C_i, G_k, G_k, G_k))$ with $G_k \lhd F_j$ and $e(F_j, G_k, G_k)$ critical for $\Gamma^{|F_j|}_{(2)}$, i.e. $G_k \lhd P$; 24 pairs of type $(e(C_i, F_j, F_j, P), e(C_i, F_j, G_k, P))$ with $e(F_j, P)$ critical for $\Gamma_{(1)}(\mathcal{A}^{|F_j|})$, i.e. $F_j \cap V_2 \neq \emptyset$ and bounded, and $G_k \lhd^{op}_P F_j$. Notice that in $\Gamma_{(3)}$ there are no pair of type $(e(C_i, C_i, G_k, P),\ e(C_i, F_j, G_k, P))$ because no cell of the form $e(C_i, G_k, P)$ is critical for $\Gamma_{(2)}$;
- $\Gamma^7_{(3)}$ is composed by three types of pairs: 6 pairs of type $(e(C_i, C_i, P, P),\ e(C_i, F_j, P, P))$ with $F_j \lhd C_i$, and $e(C_i, P, P)$ is critical for $\Gamma_{(2)}$; 36 pairs of type $(e(C_i, G_k, G_k, G_k), e(C_i, G_k, G_k, P))$ with $P \lhd G_k$; 48 pairs of type $(e(C_i, F_j, G_k, P),\ e(C_i, F_j, P, P))$, with $P \lhd^{op}_P G_k$, and $G_k = max_{\lhd^{op}_P}\{F|F_j \prec F \prec G_k\}$. Notice that in $\Gamma_{(3)}$ there are no pair of type $(e(C_i, F_j, G_k, P),\ e(C_i, G_k, G_k, P))$ because no cell of the form $e(F_j, G_k, P)$ is critical for $\Gamma^{|F_j|}_{(2)}$;
- $\Gamma^8_{(3)}$ is composed by two types of pairs: 18 pairs of type $(e(C_i, F_j, P, P),\ e(C_i, G_k, P, P))$ with $e(F_j, P, P)$ critical for $\Gamma_{(2)}(\mathcal{A}^{|F_j|})$, i.e. $F_j \cap V_2 = \emptyset$, and $G_k \lhd F_j$; 36 pairs of type $(e(C_i, G_k, G_k, P),\ e(C_i, G_k, P, P))$ with $P \lhd^{op}_P G_k$;
- $\Gamma^9_{(3)}$ is composed by 18 pairs of type $(e(C_i, G_k, P, P), e(C_i, P, P, P))$, with $P \lhd G_k$, such that $e(C_i, G_k, P, P)$ is not in a pair of $\Gamma^8_{(3)}$;

So the set of critical cells $Sing(\mathbf{S})$ is composed by the cells not appearing in any pair of Γ, i.e. by the following cells:

- 1 critical 0-cell $e(C_0, C_0, C_0, C_0)$;
- 6 critical 3-cells $e(C_0, F_1, F_1, F_1), e(C_1, F_2, F_2, F_2), e(C_2, F_3, F_3, F_3)$, $e(C_3, F_4, F_4, F_4)$, $e(C_4, F_5, F_5, F_5), e(C_5, F_6, F_6, F_6)$;
- 11 critical 6-cells $e(C_1, G_1, G_1, G_1)$, $e(C_2, G_1, G_1, G_1)$, $e(C_3, G_2, G_2, G_2)$, $e(C_4, G_2, G_2, G_2)$, $e(C_5, G_3, G_3, G_3)$, $e(C_8, G_4, G_4, G_4)$, $e(C_9, G_5, G_5, G_5)$, $e(C_{10}, G_5, G_5, G_5)$, $e(C_7, G_6, G_6, G_6)$, $e(C_{12}, G_6, G_6, G_6)$, $e(C_{13}, G_7, G_7, G_7)$;
- 6 critical 9-cells $e(C_7, P, P, P), e(C_8, P, P, P), e(C_9, P, P, P), e(C_{10}, P, P, P)$, $e(C_{12}, P, P, P)$, $e(C_{13}, P, P, P)$.

Remark 8.2. *The number of cells is* $1 + 6 + 11 + 6 + 2 * (23 + 43 + 6 + 60 + 11 + 18 + 6 + 48 + 25 + 6 + 60 + 18 + 36 + 11 + 48 + 36 + 18 + 72 + 25 + 24 + 6 + 36 + 48 + 18 + 36 + 18) = 1536$.

§9. Concluding remarks

As we mentioned in the introduction, it would be interesting to analyze the situation for general subspace arrangements. Of course, minimality condition requires that the cohomology is torsion-free. For Hyperplane Arrangements, such property was known much before than the minimality condition.

For general subspace arrangements, it is well known that the complement can have torsion. An explicit example can be found in [J94]. Let $V = \mathbb{R}^{10}$. By using coordinates $(x_1, \ldots, x_{10})$ in V, the arrangement is given by $\mathcal{A} = \{A_i\}_{i=1\cdots6}$, where $A_1, \ldots, A_6$ are the coordinate-subspaces:

$A_1 = \{x_1 = x_2 = x_3 = x_4 = x_5 = 0\}$
$A_2 = \{x_1 = x_2 = x_6 = x_7 = x_8 = 0\}$
$A_3 = \{x_1 = x_3 = x_6 = x_9 = x_{10} = 0\}$
$A_4 = \{x_2 = x_4 = x_7 = x_9 = x_{10} = 0\}$
$A_5 = \{x_3 = x_5 = x_7 = x_8 = x_9 = 0\}$
$A_6 = \{x_4 = x_5 = x_6 = x_8 = x_{10} = 0\}$.

One can verify (for example by using the Goreski–MacPherson formula) that the complement has 2-torsion in dimension 7.

The previous example can be generalized by using the recent theory of *moment-angle complexes*, of the special type $Z(K, (D^2, S^1))$. For every simplicial complex K, such complexes realize the complement of certain coordinate-subspace arrangements ([BP00]).

The non-trivial part of the conjecture stated in the introduction deals with non-torsion and non-simply connected subspace arrangements.

The previous examples in general produce simply-connected subspace arrangements.

References

[BZ92] A. Bjorner and G. Ziegler, Combinatorial stratifications of complex arrangements, J. Amer. Math. Soc., **5** (1992), 105–149.

[BP00] V. M. Bukhshtaber and T. E. Panov, Torus actions, equivariant moment-angle complexes, and configurations of coordinate subspaces, transl. in J. Math. Sci. (N.Y.), **113** (2003), 558–568.

[DS00] C. De Concini and M. Salvetti, Cohomology of Coxeter groups and Artin groups, Math. Res. Lett., **7** (2000), 213–232.

[De08] E. Delucchi, Shelling-Type Orderings of Regular CW-Complexes and Acyclic Martchings of the Salvetti Complex, Int. Math. Res. Notices, **6** (2008), 39 pages.

[DeSe] E. Delucchi and S. Settepanella, Combinatorial polar ordering and recursively orderable arrangements, to appear in Adv. in Appl. Math.

[DP03] A. Dimca and S. Papadima, Hypersurface complements. Milnor fibers and higher homotopy groups of arrangements, Ann. of Math. (2), **158** (2003), 473–507.

[Fo98] R. Forman, Morse theory for cell complexes, Adv. in Math., **134** (1998), 90–145.

[Fo02] R. Forman, A User's guide to discrete Morse theory, Sém. Lothar. Combin., **48** (2002).

[GaSa09] G. Gaiffi and M. Salvetti, The Morse complex of a line arrangement, J. Algebra, **321** (2009), 316–337.

[GMS10] G. Gaiffi, F. Mori and M.Salvetti, to appear.

[GM88] M. Goreski and R. MacPherson, Stratified Morse Theory, Ergeb. Math. Grenzgeb. (3), **14**, Springer-Verlag, 1988.

[Ha02] A. Hatcher, Algebraic Topology, http://www.math.cornell.edu/hatcher/AT/AT.pdf.

[J94] K. Jewell, Complements of sphere and subspace arrangements, Topology Appl., **56** (1994), 199–214.

[Ko08] D. Kozlov, Combinatorial Algebraic Topology, Algorithms Comput. Math., **31**, Springer-Verlag, 2008.

[Le00] G. I. Lehrer, Equivariant cohomology of configurations in R^d, Algebr. Represent. Theory, **3** (2000), 377–384.

[Ma96] O. Mathieu, Hidden Σ_{n+1}-actions, Comm. Math. Phys., **176** (1996), 467–474.

[OT92] P. Orlik and H. Terao, Arrangements of Hyperplanes, Grundlehren Math. Wiss., **300**, Springer-Verlag, 1992.

[Ra02] R. Randell, Morse theory, Milnor fibers and minimality of a complex hyperplane arrangement, Proc. Amer. Math. Soc., **130** (2002), 2737–2743.

[Sa87] M. Salvetti, Topology of the complement of real hyperplanes in $\mathbb{C}^n$, Invent. Math., **88** (1987), 603–618.

[SaSe07] M. Salvetti and S. Settepanella, Combinatorial Morse theory and minimality of hyperplane arrangements, Geom. Topol., **11** (2007), 1733–1765.

[Za75] T. Zaslavsky, Facing Up to Arrangements: Face Count Formulas for Partitions of Space by Hyperplanes, Mem. Amer. Math. Soc., **1** (1975), no.154.

[Yo05] M. Yoshinaga, Hyperplane arrangements and Lefschetz's hyperplane section theorem, Kodai Math. J., **30** (2007), 157–194.

Francesca Mori
Dipartimento di Matematica, Università di Pisa,
"L. Tonelli", Largo B. Pontecorvo 5, 56127, Pisa, Italy
E-mail address: `mori@mail.dm.unipi.it`

Mario Salvetti
Dipartimento di Matematica, Università di Pisa,
"L. Tonelli", Largo B. Pontecorvo 5, 56127, Pisa, Italy
E-mail address: `salvetti@dm.unipi.it`

Advanced Studies in Pure Mathematics 62, 2012
Arrangements of Hyperplanes—Sapporo 2009
pp. 323–358

Hyperplane arrangements: computations and conjectures

Hal Schenck

Abstract.

This paper provides an overview of selected results and open problems in the theory of hyperplane arrangements, with an emphasis on computations and examples. We give an introduction to many of the essential tools used in the area, such as Koszul and Lie algebra methods, homological techniques, and the Bernstein–Gelfand–Gelfand correspondence, all illustrated with concrete calculations. We also explore connections of arrangements to other areas, such as De Concini–Procesi wonderful models, the Feichtner–Yuzvinsky algebra of an atomic lattice, fatpoints and blowups of projective space, and plane curve singularities.

Contents

Received October 29, 2009.
Revised March 1, 2011.
2010 *Mathematics Subject Classification.* Primary 52C35; Secondary 13D02, 16E05, 16S37, 20F14.
Schenck supported by NSF 0707667.

§1. Introduction and algebraic preliminaries

There are a number of wonderful sources available on hyperplane arrangements, most notably Orlik–Terao's landmark 1992 text [58]. In the last decade alone several excellent surveys have appeared: Suciu's paper on aspects of the fundamental group [84], Yuzvinsky's paper on Orlik–Solomon algebras and local systems cohomology [97], and several monographs devoted to connections to areas such as hypergeometric integrals [59], mathematical physics [91], as well as proceedings from conferences at Sapporo [38], Northeastern [16] and Istanbul [30].

The aim of this note is to provide an overview of some recent results and open problems, with a special emphasis on connections to computation. The paper also gives a concrete and example driven introduction for non-specialists, but there is enough breadth here that even experts should find something new. There are few proofs, but rather pointers to original source material. We also explore connections of arrangements to other areas, such as De Concini–Procesi wonderful models, the Feichtner–Yuzvinsky algebra of an atomic lattice, the Orlik–Terao algebra and blowups, and plane curve singularities. All computations in this survey can be performed using Macaulay2 [45], available at: `http://www.math.uiuc.edu/Macaulay2/`, and the arrangements package by Denham and Smith [22].

Let $V = \mathbb{K}^{\ell}$, and let S be the symmetric algebra on V^*: $S = \bigoplus_{i \in \mathbb{Z}} S_i$ is a $\mathbb{Z}$-graded ring, which means that if $s_i \in S_i$ and $s_j \in S_j$, then $s_i \cdot s_j \in S_{i+j}$. A graded S-module M is defined in similar fashion. Of special interest is the case where S_0 is a field $\mathbb{K}$, so that each M_i is a $\mathbb{K}$–vector space. The free S module with generator in degree i is written $S(-i)$, and in general $M(i)_j = M_{i+j}$.

Definition 1. *The Hilbert function* $HF(M,i) = \dim_{\mathbb{K}} M_i$.

Definition 2. *The Hilbert series* $HS(M,i) = \sum_{\mathbb{Z}} \dim_{\mathbb{K}} M_i t^i$.

Example 3. $S = \mathbb{K}[x,y]$, $M = S/\langle x^2, xy\rangle$. Then

i	M_i	$M(-2)_i$
0	1	0
1	x,y	0
2	y^2	1
3	y^3	x,y
4	y^4	y^2
n	y^n	y^{n-2}

The respective Hilbert series are

$$HS(M,i) = \frac{1-2t^2+t^3}{(1-t)^2} \text{ and } HS(M(-2),i) = \frac{t^2(1-2t^2+t^3)}{(1-t)^2}.$$

An induction shows that $HS(S(-i),t) = t^i/(1-t)^\ell$; this makes it easy to compute the Hilbert series of an arbitrary graded module from a free resolution. For $S/\langle x^2, xy\rangle$, a minimal free resolution is

$$0 \longrightarrow S(-3) \xrightarrow{\begin{bmatrix} y \\ -x \end{bmatrix}} S(-2)^2 \xrightarrow{\begin{bmatrix} x^2 & xy \end{bmatrix}} S \longrightarrow S/I \longrightarrow 0.$$

$$\text{The map } [x^2, xy] \text{ sends } \begin{array}{l} e_1 \mapsto x^2 \\ e_2 \mapsto xy, \end{array}$$

so in order to have a map of graded modules, the basis elements of the source must have degree two, explaining the shifts in the free resolution. Taking the alternating sum of the Hilbert series yields

$$HS(M,i) = \frac{t^3-2t^2+1}{(1-t)^2}$$

which agrees with the previous computation.

Example 4. The 2×2 minors of $\begin{bmatrix} x & y & z \\ y & z & w \end{bmatrix}$ define the twisted cubic $I \subseteq S = \mathbb{K}[x,y,z,w]$.

$$0 \longrightarrow S(-3)^2 \xrightarrow{\begin{bmatrix} -z & w \\ y & -z \\ -x & y \end{bmatrix}} S(-2)^3 \xrightarrow{\begin{bmatrix} y^2-xz & yz-xw & z^2-yw \end{bmatrix}} S \longrightarrow S/I$$

The numerical information in a free resolution may be compactly displayed as a *betti table*:

$$b_{ij} = \dim_{\mathbb{K}} \operatorname{Tor}_i^S(M,\mathbb{K})_{i+j}.$$

total	1	3	2
0	1	–	–
1	–	3	2

In particular, the indexing begins at position $(0,0)$ and is read over and down. So for the twisted cubic, $b_{21}(S/I) = \dim_{\mathbb{K}} \operatorname{Tor}_2^S(S/I,\mathbb{K})_3 = 2$.

We now give a quick review of arrangements. Let $\mathcal{A} = \{H_1, \ldots, H_n\}$ be an arrangement of complex hyperplanes in $\mathbb{C}^\ell$. We assume $\mathcal{A}$ is *central* and *essential*: the ℓ_i with $H_i = V(\ell_i)$ are homogeneous, and the common zero locus $V(\ell_1, \ldots, \ell_n) = 0 \in \mathbb{C}^\ell$. The central condition means that $\mathcal{A}$ also defines an arrangement in $\mathbb{P}^{\ell-1}$. The main combinatorial object associated to $\mathcal{A}$ is the intersection lattice $L_\mathcal{A}$, which consists of the intersections of elements of $\mathcal{A}$, ordered by reverse inclusion. $\mathbb{C}^n$ is the lattice element $\hat{0}$ and the rank one elements of $L_\mathcal{A}$ are the hyperplanes.

Definition 5. *The Möbius function* $\mu : L_\mathcal{A} \longrightarrow \mathbb{Z}$ *is defined by*

$$\begin{aligned} \mu(\hat{0}) &= 1 \\ \mu(t) &= -\sum_{s<t} \mu(s), \text{ if } \hat{0} < t. \end{aligned}$$

The Poincaré and characteristic polynomials of $\mathcal{A}$ are defined as

$$\pi(\mathcal{A}, t) = \sum_{x \in L(\mathcal{A})} \mu(x) \cdot (-t)^{\operatorname{rank}(x)}, \text{ and } \chi(\mathcal{A}, t) = t^{\operatorname{rk}(\mathcal{A})} \pi(\mathcal{A}, \frac{-1}{t}).$$

Example 6. The A_3 arrangement is $\bigcup_{1 \le i < j \le 4} V(x_i - x_j) \subseteq \mathbb{C}^4$. Projecting along $(1,1,1,1)$ gives a central arrangement in $\mathbb{C}^3$, hence a configuration of lines in $\mathbb{P}^2$. This configuration corresponds to the figure below, but with the line at infinity (which bounds the figure) omitted.

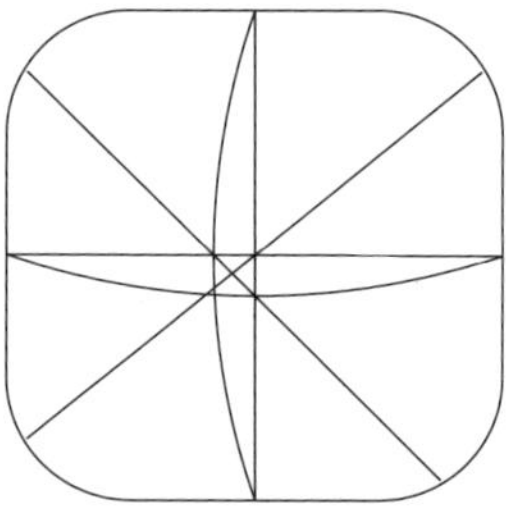

For the 7 rank two elements of $L(A_3)$, the four corresponding to triple points have $\mu = 2$, and the three normal crossings have $\mu = 1$. Thus, $\pi(A_3, t) = 1 + 6t + 11t^2 + 6t^3$. Adding the bounding line gives the non-Fano arrangement NF, with $\pi(NF, t) = 1 + 7t + 15t^2 + 9t^3$.

In [57], Orlik and Solomon showed that the cohomology ring of the complement $M_\mathcal{A} = \mathbb{C}^n \setminus \bigcup_{i=1}^d H_i$ has presentation $H^*(M_\mathcal{A}, \mathbb{Z}) = \bigwedge(\mathbb{Z}^n)/I$, with generators $e_1, \ldots, e_n$ in degree 1 and

$$I = \langle \sum_q (-1)^{q-1} e_{i_1} \cdots \widehat{e_{i_q}} \cdots e_{i_r} \mid \operatorname{codim} H_{i_1} \cap \cdots \cap H_{i_r} < r \rangle.$$

For additional background on arrangements, see [58].

§2. $D(\mathcal{A})$ and freeness

Let $\mathcal{A} = \bigcup_{i=1}^{n} H_i \subseteq V = \mathbb{C}^\ell$ be a central arrangement. For each i, fix $V(l_i) = H_i \in \mathcal{A}$, and define $Q_\mathcal{A} = \prod_{i=1}^{n} l_i \in S = \mathbb{C}[x_1, \ldots, x_\ell]$.

Definition 7. *The module of $\mathcal{A}$-derivations (or Terao module) is the submodule of $Der_\mathbb{C}(S)$ consisting of vector fields tangent to $\mathcal{A}$:*

$$D(\mathcal{A}) = \{\theta \in Der_\mathbb{C}(S) | \theta(l_i) \in \langle l_i \rangle \text{ for all } l_i \text{ with } V(l_i) \in \mathcal{A}\}.$$

An arrangement is free when $D(\mathcal{A})$ is a free S-module. In this case, the degrees of the generators of $D(\mathcal{A})$ are called the exponents of the arrangement. Note that $D(\mathcal{A})$ is always nonzero, since the Euler derivation $\theta_E = \sum_{i=1}^{\ell} x_i \partial/\partial x_i \in D(\mathcal{A})$. It is easy to show that

$$D(\mathcal{A}) \simeq S \cdot \theta_E \oplus syz(J_\mathcal{A}),$$

where $J_\mathcal{A}$ is the Jacobian ideal of $Q_\mathcal{A}$, and syz denotes the module of syzygies on $J_\mathcal{A}$: polynomial relations on the generators of $J_\mathcal{A}$.

Theorem 8 (Saito [72]). *$\mathcal{A}$ is free iff there exist ℓ elements*

$$\theta_i = \sum_{j=1}^{\ell} f_{ij} \frac{\partial}{\partial x_j} \in D(\mathcal{A}),$$

such that $\det([f_{ij}]) = c \cdot Q_\mathcal{A}$, *for some* $c \neq 0$.

Example 9. For Example 6, a computation shows that

$$\begin{aligned} D(A_3) &\simeq S(-1) \oplus S(-2) \oplus S(-3) \\ D(NF) &\simeq S(-1) \oplus S(-3) \oplus S(-3). \end{aligned}$$

Interestingly, the respective Poincaré polynomials factor, as

$$\pi(A_3, t) = (1+t)(1+2t)(1+3t), \text{ and } \pi(NF, t) = (1+t)(1+3t)^2.$$

This suggests the possibility of a connection between the exponents of a free arrangement and the Poincaré polynomial.

A landmark result in arrangements is:

Theorem 10 (Terao [86]). *If* $D(\mathcal{A}) \simeq \bigoplus_{i=1}^{\ell} S(-a_i)$, *then*

$$\pi(\mathcal{A}, t) = \prod (1 + a_i t) = \sum \dim_\mathbb{C} H^i(\mathbb{C}^\ell \setminus \mathcal{A}) t^i.$$

Example 11. [Stanley] For $\mathcal{A}$ below, $\pi(\mathcal{A}, t) = (1+t)(1+3t)^2$.

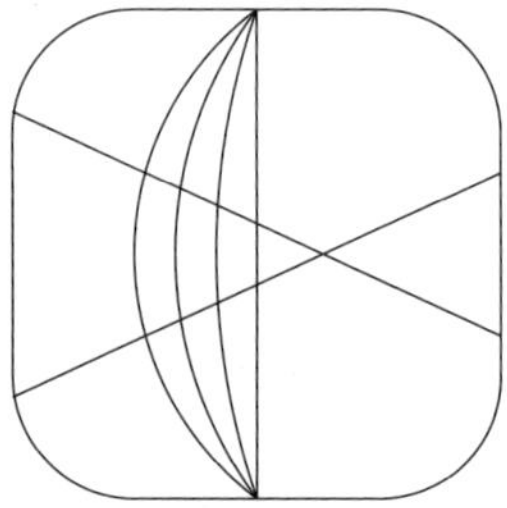

A computation shows that $\mathcal{A}$ is not free, so factorization of $\pi(\mathcal{A}, t)$ is a necessary but not sufficient for freeness of $\mathcal{A}$.

A famous open conjecture in the field of arrangements is:

Conjecture 12 (Terao). *If $char(\mathbb{K}) = 0$, then freeness of $D(\mathcal{A})$ depends only on $L_{\mathcal{A}}$.*

Example 13. [Ziegler's pair [101]] Let $\mathcal{A}$ be an arrangement of 9 lines in $\mathbb{P}^2$, as below.

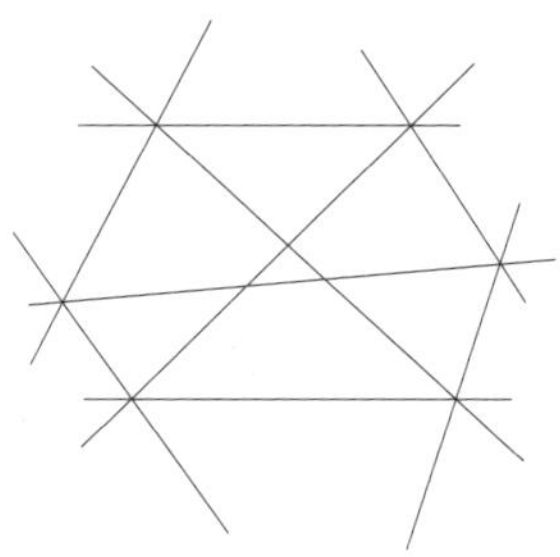

Then $D(\mathcal{A})$ depends on nonlinear geometry: if the six triple points lie on a smooth conic, we compute:

$$0 \longrightarrow S(-7) \oplus S(-8) \longrightarrow S(-5) \oplus S^3(-6) \longrightarrow syz(J_{\mathcal{A}}) \longrightarrow 0\ ,$$

while if six triple points are not on a smooth conic, the resolution is:

$$0 \longrightarrow S^4(-7) \longrightarrow S^6(-6) \longrightarrow syz(J_{\mathcal{A}}) \longrightarrow 0\ .$$

A version of Terao's theorem applies to any arrangement:

Definition 14. *$D^p(\mathcal{A}) \subseteq \Lambda^p(Der_{\mathbb{K}}(S))$ consists of θ such that*

$$\theta(l_i, f_2, \ldots, f_p) \in \langle l_i \rangle, \forall\ V(l_i) \in \mathcal{A}, f_i \in S.$$

Theorem 15 (Solomon–Terao, [82]).

$$\chi(\mathcal{A},t) = (-1)^{\ell} \lim_{x\to 1} \sum_{p\geq 0} HS(D^p(\mathcal{A});x)(t(x-1)-1)^p.$$

Problem 16. Relate the modules $D^p(\mathcal{A})$, for $p \geq 2$, to $L_{\mathcal{A}}$.

A closed subarrangement $\hat{\mathcal{A}} \subseteq A$ is a subarrangement such that $\hat{\mathcal{A}} = \mathcal{A}_X$ for some flat X. The best result relating $D(\mathcal{A})$ to $L_{\mathcal{A}}$ is:

Theorem 17 (Terao, [89]). *If* $\hat{\mathcal{A}} \subset \mathcal{A}$ *is a closed subarrangement, then* $\operatorname{pdim} D(\mathcal{A}) \geq \operatorname{pdim} D(\hat{\mathcal{A}})$.

Problem 18. Find bounds on $\operatorname{pdim} D(\mathcal{A})$ depending on $L_{\mathcal{A}}$.

A particularly interesting class of arrangements are graphic arrangements, which are subarrangements of A_n. Given a simple (no loops or multiple edges) graph G, with ℓ vertices and edge set E, we define $\mathcal{A}_G = \{z_i - z_j = 0 \mid (i,j) \in \mathsf{E} \subseteq \mathbb{C}^{\ell}\}$.

Theorem 19 (Stanley [83]). $\mathcal{A}_G$ *is supersolvable iff* G *is chordal.*

Theorem 20 (Kung–Schenck [52]). *If* $\mathcal{A}_G$ *has an induced* k*-cycle, then* $\operatorname{pdim} D(\mathcal{A}_G) \geq k-3$.

Example 21. The largest induced cycle of G below is a 6-cycle.

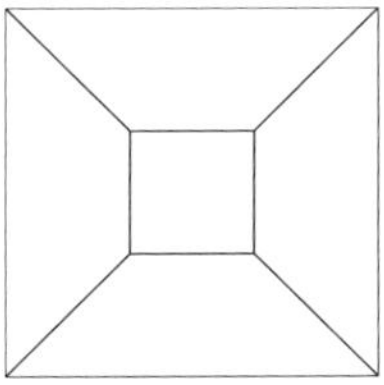

A computation shows $\operatorname{pdim}(D(\mathcal{A})) = 3$.

Example 22. The largest induced cycle of G below is a 4-cycle.

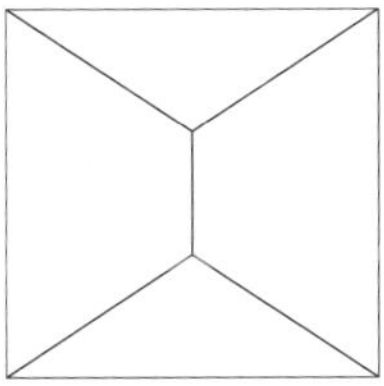

A computation shows $\operatorname{pdim}(D(\mathcal{A})) = 2$.

Problem 23. Find a formula for $\operatorname{pdim} D(\mathcal{A}_G)$.

Definition 24. *A triple $(\mathcal{A}', \mathcal{A}, \mathcal{A}'')$ of arrangements consists of a choice of $H \in \mathcal{A}$, with $\mathcal{A}' = \mathcal{A} \setminus H, \mathcal{A}'' = \mathcal{A}|_H$.*

A main tool for proving freeness is Terao's addition-deletion theorem.

Theorem 25 (Terao [87]). *For a triple, any two of the following imply the third*

(1) $D(\mathcal{A}) \simeq \oplus_{i=1}^{n} S(-b_i)$.
(2) $D(\mathcal{A}') \simeq S(-b_n + 1) \oplus_{i=1}^{n-1} S(-b_i)$.
(3) $D(\mathcal{A}'') \simeq \oplus_{i=1}^{n-1} S/L(-b_i)$.

Example 26. In Example 6, the A_3 arrangement is free with exponents $\{1, 2, 3\}$. Let H be the line at infinity, which meets A_3 in four points. Then $D(\mathcal{A}'')$ is free, with exponents $\{1, 2\}$, so the non-Fano arrangement is free with exponents $\{1, 3, 3\}$, which agrees with our earlier computation. Example 4.59 of [58] gives a free arrangement for which the addition-deletion theorem does not apply.

As a corollary of Theorem 25, Terao showed that supersolvable arrangements are free.

Definition 27. *An element X of a lattice is modular if for all $Y \in L$ and all $Z < Y$, $Z \vee (X \wedge Y) = (Z \vee X) \wedge Y$. A central arrangement $\mathcal{A}$ is supersolvable if there exists a maximal chain $\hat{0} = X_0 < X_1 < \cdots < X_n = \hat{1}$ of modular elements in $L(\mathcal{A})$.*

For line configurations in $\mathbb{P}^2$, the supersolvability condition simply means there is a singular point $p \in \mathcal{A}$ such that every other singularity of $\mathcal{A}$ lies on a line of $\mathcal{A}$ which passes through p. For example, the A_3 arrangement is supersolvable, since any triple point is such a singularity. For arrangements in $\mathbb{P}^2$, there is a beautiful characterization of freeness involving multiarrangements.

Definition 28. *A multiarrangement $(\mathcal{A}, \mathbf{m})$ consists of an arrangement $\mathcal{A}$, along with a multiplicity $m_i \in \mathbb{N}$ for each $H \in \mathcal{A}$.*

$$D(\mathcal{A}, \mathbf{m}) = \{\theta \mid \theta(l_i) \in \langle l_i^{m_i} \rangle\}.$$

Theorem 29. *$\mathcal{A} \subseteq \mathbb{P}^2$ is free if and only if*

(1) $\pi(\mathcal{A}, t) = (1 + t)(1 + at)(1 + bt)$ *and*
(2) $D(\mathcal{A}|_H, \mathbf{m}) \simeq S/L(-a) \oplus S/L(-b)$,

where (2) *holds for all $H = V(L) \in \mathcal{A}$, with $\mathbf{m}(H_i) = \mu_{\mathcal{A}}(H \cap H_i)$.*

The necessity of these conditions was shown by Ziegler in [100], and sufficiency was proved by Yoshinaga in [94]. In [93], Yoshinaga gives a generalization to higher dimensions.

§3. Multiarrangements

The exponents of free multiarrangements are not combinatorial:

Example 30. [Ziegler, [100]] Consider the two multiarrangements in $\mathbb{P}^1$, with underlying arrangements defined by

$$\begin{aligned} \mathcal{A}_1 &= V(x \cdot y \cdot (x+y) \cdot (x-y)) \\ \mathcal{A}_2 &= V(x \cdot y \cdot (x+y) \cdot (x-ay)), \end{aligned}$$

with $a \neq 1$. To compute $D(\mathcal{A}_1, (1,1,3,3))$, we must find all

$$\theta = f_1(x,y)\partial/\partial x + f_2\partial/\partial y$$

such that

$$\begin{aligned} &\theta(x) \in \langle x \rangle, \ \theta(x+y) \in \langle x+y \rangle^3 \\ &\theta(y) \in \langle y \rangle, \ \theta(x-y) \in \langle x-y \rangle^3. \end{aligned}$$

Thus, $D(\mathcal{A}_1, (1,1,3,3))$ is the kernel of the matrix

$$\begin{bmatrix} 1 & 0 & x & 0 & 0 & 0 \\ 0 & 1 & 0 & y & 0 & 0 \\ 1 & 1 & 0 & 0 & (x+y)^3 & 0 \\ 1 & -1 & 0 & 0 & 0 & (x-y)^3 \end{bmatrix}.$$

Computations show that $D(\mathcal{A}_1, (1,1,3,3))$ has exponents $\{3,5\}$, and $D(\mathcal{A}_2, (1,1,3,3))$ has exponents $\{4,4\}$.

There is an analog of Theorem 15 for multiarrangements.

Definition 31. *$D^p(\mathcal{A}, \mathbf{m}) \subseteq \Lambda^p(Der_{\mathbb{K}}(S))$ consists of θ such that*

$$\theta(l_i, f_2, \ldots, f_p) \in \langle l_i \rangle^{m(l_i)}, \forall\ V(l_i) \in \mathcal{A}, f_i \in S.$$

Theorem 32 (Abe–Terao–Wakefield [2]). *Define*

$$\begin{aligned} \Psi(\mathcal{A}, \mathbf{m}, t, x) &= \sum_{p=0}^{\ell} HS(D^p(\mathcal{A}, \mathbf{m}), x)(t(x-1)-1)^p. \\ \chi((\mathcal{A}, \mathbf{m}), t) &= (-1)^{\ell} \lim_{x \to 1} \Psi(\mathcal{A}, \mathbf{m}, t, 1). \end{aligned}$$

If $D^1(\mathcal{A}, \mathbf{m}) \simeq \oplus S(-d_i)$, then $\chi((\mathcal{A}, \mathbf{m}), t) = \prod_{i=1}^{\ell} (1 + d_i t)$.

In [1], Abe–Terao–Wakefield prove an addition-deletion theorem for multiarrangements by introducing *Euler multiplicity* for the restriction. It follows from the Hilbert–Burch theorem that any $(\mathcal{A}, \mathbf{m}) \subseteq \mathbb{P}^1$ is free, which leads to the question of whether there exist other arrangements which are free for any $\mathbf{m}$. In [3], Abe–Terao–Yoshinaga prove that any such arrangement is a product of one- and two-dimensional arrangements. Nevertheless, several natural questions arise:

Problem 33. Characterize the projective dimension of $D(\mathcal{A}, \mathbf{m})$.

Problem 34. Define supersolvability for multiarrangements.

§4. Arrangements of plane curves

For a collection of hypersurfaces

$$\mathcal{C} = \bigcup_i V(f_i) \subseteq \mathbb{P}^n,$$

the module of derivations $D(\mathcal{C})$ is obtained by substituting f_i for l_i in Definition 7. It is not hard to prove that Saito's criterion still applies. Are there other freeness theorems?

Example 35. For the arrangement $\mathcal{C} \subseteq \mathbb{P}^2$ depicted below

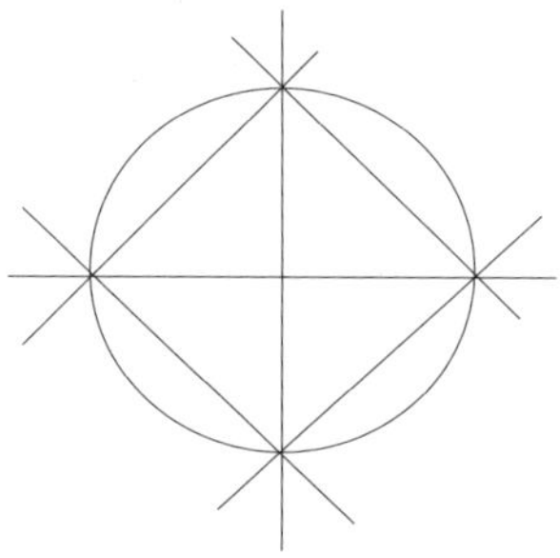

we compute that $D(\mathcal{C}) \simeq S(-1) \oplus S(-2) \oplus S(-5)$.

This example can be explained by an addition-deletion theorem [79], but there is subtle behavior related to singular points. For the remainder of this section, $C = \cup_i V(f_i) \subseteq \mathbb{C}^2$ is reduced plane curve, and if $p \in C$ is a singular point, translate so $p = (0,0)$.

Definition 36. *A plane curve singularity is quasihomogeneous if and only if there exists a holomorphic change of variables so that* $f(x,y) = \sum c_{ij}x^iy^j$ *is weighted homogeneous: there exists* $\alpha, \beta \in \mathbb{Q}$ *such that* $\sum c_{ij}x^{i\cdot\alpha}y^{j\cdot\beta}$ *is homogeneous.*

Definition 37. *The Milnor number at* $(0,0)$ *is*

$$\mu_{(0,0)}(C) = \dim_{\mathbb{C}} \mathbb{C}\{x,y\}/\langle \frac{\partial f}{\partial x}, \frac{\partial f}{\partial y} \rangle.$$

The Tjurina number at $(0,0)$ *is*

$$\tau_{(0,0)}(C) = \dim_{\mathbb{C}} \mathbb{C}\{x,y\}/\langle \frac{\partial f}{\partial x}, \frac{\partial f}{\partial y}, f \rangle.$$

For a projective plane curve $V(Q) \subseteq \mathbb{P}^2$, it is easy to see that the degree of $Jac(Q) = \sum_{p \in sing(V(Q))} \tau_p$.

Example 38. Let $\mathcal{C}$ be as below:

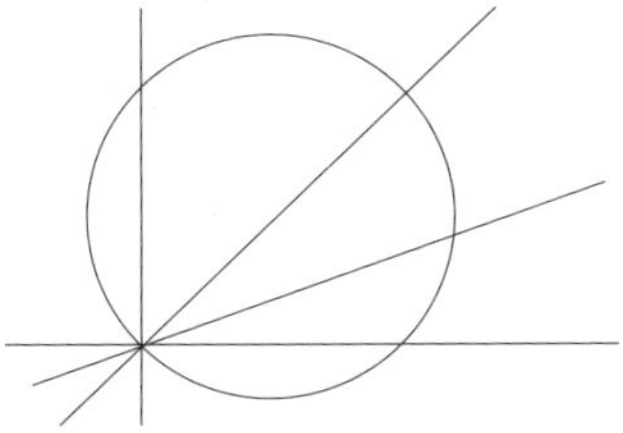

If p is an ordinary singularity with k distinct branches, then $\mu_p(C) = (k-1)^2$, so the sum of the Milnor numbers is 20. However, a computation shows that $\deg(J_\mathcal{C}) = 19$. All singularities are ordinary, but the singularity at the origin is not quasihomogeneous.

Theorem 39 (Saito [71]). *If $C = V(f)$ has an isolated singularity at the origin, then $f \in Jac(f)$ iff f is quasihomogeneous.*

For arrangements of lines and conics such that every singular point is quasihomogeneous, [79] proves an addition/deletion theorem; [78] generalizes the result to curves of higher genus.

Example 40. Let $\mathcal{C}$ be as below:

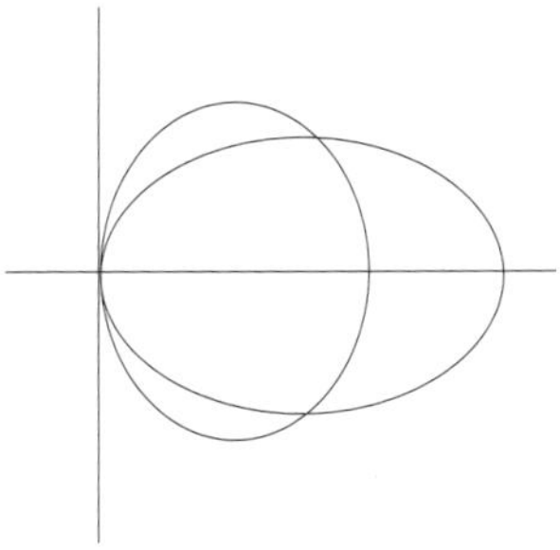

$D(\mathcal{C})$ has exponents $\{1,2,3\}$, which can be shown using the aforementioned addition-deletion theorem. Change $\mathcal{C}$ to $\mathcal{C}'$ via:

$$y = 0 \longrightarrow x - 13y = 0.$$

A computation shows that $D(\mathcal{C}')$ is not free. Thus, for line-conic arrangements, freeness is not combinatorial.

Problem 41. Define supersolvability for hypersurface arrangements.

Problem 42. Give combinatorial bounds on $\operatorname{pdim} D(\mathcal{C})$.

Problem 43. Analyze associated primes and Ext modules of $D(\mathcal{C})$.

§5. The Orlik–Terao algebra and blowups

The Orlik–Terao algebra is a symmetric analog of the Orlik–Solomon algebra. While the Orlik–Solomon algebra records the existence of dependencies among sets of hyperplanes, the Orlik–Terao algebra records the actual dependencies. If $\operatorname{codim} \cap_{j=1}^{m} H_{i_j} < m$, then there exist c_{i_j} with

$$\sum_{j=1}^{m} c_{i_j} \cdot l_{i_j} = 0 \text{ a dependency.}$$

Definition 44. *The Orlik–Terao ideal*

$$I_{\mathcal{A}} = \langle \sum_{j=1}^{m} c_{i_j}(y_{i_1} \cdots \hat{y}_{i_j} \cdots y_{i_m}) \mid \textit{ over all dependencies} \rangle$$

The Orlik–Terao algebra is $C(\mathcal{A}) = \mathbb{K}[x_1, \dots, x_n]/I_{\mathcal{A}}$.

Example 45. $\mathcal{A} = V(x_1 \cdot x_2 \cdot x_3 \cdot (x_1 + x_2 + x_3))$, the only dependency is $l_1 + l_2 + l_3 - l_4 = 0$, so $I_{\mathcal{A}} = \langle y_2 y_3 y_4 + y_1 y_3 y_4 + y_1 y_2 y_4 - y_1 y_2 y_3 \rangle$.

In [60], Orlik and Terao answer a question of Aomoto by considering the quotient AOT of $C(\mathcal{A})$ by $\langle x_1^2, \dots, x_n^2 \rangle$. They prove:

Theorem 46 (Orlik–Terao [60]). $HS(AOT, t) = \pi(\mathcal{A}, t)$.

Theorem 47 (Terao [90]).

$$HS(C(\mathcal{A}), t) = \pi\Big(\mathcal{A}, \frac{t}{1-t}\Big).$$

It is not hard to show that

$$0 \to I_{\mathcal{A}} \to \mathbb{K}[x_1, \dots, x_n] \xrightarrow{\phi} \mathbb{K}\left[\frac{1}{l_1}, \dots, \frac{1}{l_n}\right] \to 0$$

is exact, so $V(I_{\mathcal{A}}) \subseteq \mathbb{P}^{n-1}$ is irreducible and rational. In any situation where weights of dependencies play a role, the Orlik–Terao algebra is the natural candidate to study. One such situation involves 2-formality:

Definition 48. *$\mathcal{A}$ is 2-formal if all dependencies are generated by dependencies among three hyperplanes.*

Theorem 49 (Falk–Randell [37]). *If $\mathcal{A}$ is $K(\pi, 1)$, $\mathcal{A}$ is 2-formal.*

Theorem 50 (Yuzvinsky [95]). *If $\mathcal{A}$ is free, $\mathcal{A}$ is 2-formal.*

One reason that formality is interesting is that it is not a combinatorial invariant: in Example 13, the arrangement for which the six triple points lie on a smooth conic is not 2-formal, and the arrangement for which the points do not lie on a smooth conic is 2-formal.

Theorem 51 ([79]). *$\mathcal{A}$ is 2-formal iff* $\operatorname{codim}(I_{\mathcal{A}})_2 = n - \ell$.

In [7], Brandt and Terao generalized the notion of 2-formality to k-formality: $\mathcal{A}$ is k-formal if all dependencies are generated by dependencies among $k+1$ or fewer hyperplanes. Brandt and Terao prove that every free arrangement is k-formal.

Problem 52. Find an analog of Theorem 51 for k-formality.

Example 53. The configuration of Example 45 consists of four generic lines:

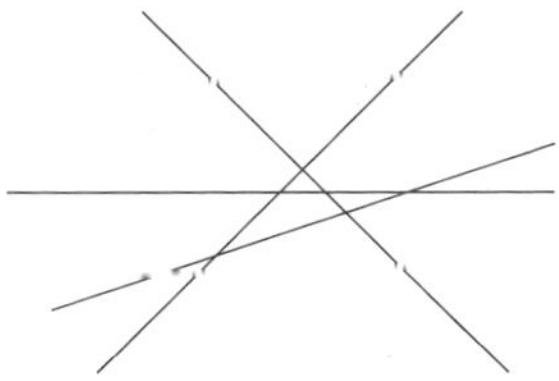

The Orlik–Terao ideal defines a cubic surface in $\mathbb{P}^3$, and a computation shows that $V(I_{\mathcal{A}})$ has four singular points.

This can be interpreted in terms of a rational map. Let $\alpha_i = Q_{\mathcal{A}}/l_i$, and define $\phi_{\mathcal{A}} = [\alpha_1, \ldots, \alpha_n]$.

$$\mathbb{P}^{\ell-1} \xrightarrow{\phi_{\mathcal{A}}} \mathbb{P}^{n-1},$$

Restrict to the case $\mathcal{A} \subseteq \mathbb{P}^2$, and let $X_{\mathcal{A}} \xrightarrow{\pi} \mathbb{P}^2$ denote the blowup of $\mathbb{P}^2$ at the singular points of $\mathcal{A}$, with E_0 denoting the pullback to $X_{\mathcal{A}}$ of the class of a line on $\mathbb{P}^2$, and E_i the exceptional divisors over singular points of $\mathcal{A}$. Let

$$D_{\mathcal{A}} = (n-1)E_0 - \sum_{p_i \in L_2(\mathcal{A})} \mu(p_i)E_i.$$

Utilizing results of Proudfoot–Speyer [67] showing that $C(\mathcal{A})$ is Cohen–Macaulay and the Riemann–Roch theorem, [75] shows that the map $\phi_{\mathcal{A}}$ is determined by the global sections of $D_{\mathcal{A}}$, and that $\phi_{\mathcal{A}}$

(1) is an isomorphism on $\pi^*(\mathbb{P}^2 \setminus \mathcal{A})$
(2) contracts the lines of $\mathcal{A}$ to points
(3) blows up the singularities of $\mathcal{A}$.

Definition 54. *A graded S-module N has Castelnuovo–Mumford regularity d if* $\mathrm{Ext}^j(N,S)_n = 0$ *for all j and all* $n \le -d-j-1$.

In terms of the betti table, the regularity of N is the label of the last non-zero row, so in Example 4, S/I has Castelnuovo–Mumford regularity one. The regularity of $C(\mathcal{A})$ is determined in [75]:

Theorem 55 ([75]). *For* $\mathcal{A} \subseteq \mathbb{P}^{\ell-1}$, $C(\mathcal{A})$ *is* $\ell-1$*-regular.*

To see this, note that since $C(\mathcal{A})$ is Cohen–Macaulay, quotienting $C(\mathcal{A})$ by ℓ generic linear forms yields an Artinian ring whose Hilbert series is the numerator of the Hilbert series of $C(\mathcal{A})$. The regularity of an Artinian module is equal to the length of the module, so the result follows from Theorem 47.

A main motivation for studying $C(\mathcal{A})$ is a surprising connection to nets and resonance varieties, which are the subject of §9. First, the definition of a net:

Definition 56. *Let* $3 \le k \in \mathbb{Z}$. *A k-net in* $\mathbb{P}^2$ *is a partition of the lines of an arrangement* $\mathcal{A}$ *into k subsets* $\mathcal{A}_i$, *together with a choice of points* $Z \subseteq \mathcal{A}$, *such that:*

(1) *for every* $i \ne j$ *and every* $L \in \mathcal{A}_i$, $L' \in \mathcal{A}_j$, $L \cap L' \in Z$.
(2) $\forall\, p \in Z$ *and every* $i \in \{1,\dots,k\}$, *there exists a unique* $L \in A_i$ *with* $Z \in L$.

In [53], Libgober and Yuzvinsky show that nets are related to the first resonance variety $R^1(\mathcal{A})$. The definition of a net forces each subset $\mathcal{A}_i$ to have the same cardinality, and if $m = |\mathcal{A}_i|$, the net is called a (k,m)-net. Using work of [53] and [39], it is shown in [75] that

Theorem 57. *Existence of a* (k,m) *net implies that there is a decomposition* $D_{\mathcal{A}} = A + B$ *with* $h^0(A) = 2$ *and* $h^0(B) = km - \binom{m+1}{2}$.

Definition 58. *A matrix of linear forms is* 1*-generic if it has no zero entry, and cannot be transformed by row and column operations to have a zero entry.*

In [26], Eisenbud shows that if a divisor D on a smooth curve X factors as $D \simeq A+B$, with A having m-sections and B having n-sections, then the ideal of the image of X under the map defined by the global sections of D will contain the 2×2 minors of a 1-generic matrix. Using this result and Theorem 57, it can be shown that $I_{\mathcal{A}}$ contains the ideal $I_2(M)$ of 2×2 minors of a 1-generic $2 \times \left(km - \binom{m+1}{2}\right)$ matrix M. So if $G = S(-1)^{km-\binom{m+1}{2}}$, the Eagon–Northcott complex [27]

$$\cdots \to S_2(S^2)^* \otimes \Lambda^4 G \to (S^2)^* \otimes \Lambda^3 G \to \Lambda^2 G \to \Lambda^2 S^2 \to S/I_2(M) \to 0$$

is a subcomplex of resolution of $S/I_{\mathcal{A}}$. The geometric content of Theorem 57 is that it implies $V(I_{\mathcal{A}})$ lies on a *scroll* [27].

Example 59. For the A_3 arrangement, the set of triple points Z gives a $(3,2)$ net, where the A_i correspond to normal crossing points: $A_1 = 12 \mid 34$, $A_2 = 13 \mid 24$, $A_3 = 14 \mid 23$.

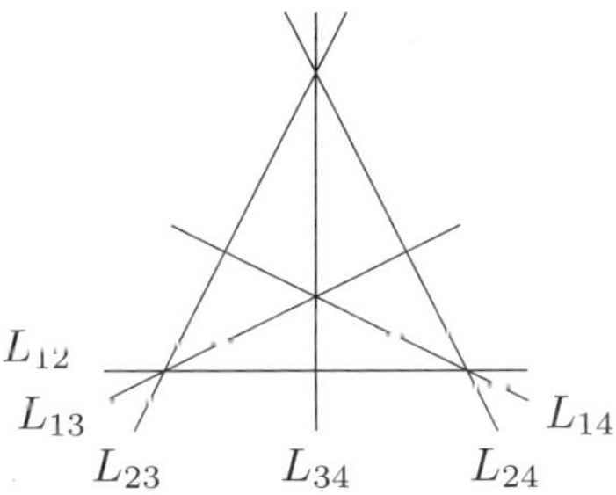

$$\text{Let } A = 2E_0 - \sum_{\{p|\mu(p)=2\}} E_p \text{ and } B = 3E_0 - \sum_{p \in L_2(\mathcal{A})} E_p.$$

So $n - \binom{m+1}{2} = 6-3 = 3$ and I contains the 2×2 minors of a 2×3 matrix, whose resolution appears in Example 4. The graded betti diagram for $\mathbb{C}[x_0,\ldots,x_5]/I_A$ is

total	1	4	5	2
0	1	–	–	–
1	–	4	2	–
2	–	–	3	2

From this, it follows that the free resolution of S/I_A is a mapping cone resolution [27]. The geometric meaning is that $X_{\mathcal{A}}$ is the intersection of a generic quadric hypersurface with the scroll.

Since $D_{\mathcal{A}}$ contracts proper transforms of lines to points, it is not very ample. However, it follows from [75] that $D_{\mathcal{A}} + E_0$ is very ample, and gives a De Concini–Procesi wonderful model (see next section) for the blowup.

Problem 60. Determine the graded betti numbers of $C(\mathcal{A})$.

Problem 61. Relate $R^k(\mathcal{A})$ to the graded betti numbers of $C(\mathcal{A})$.

§6. Compactifications

In [44], Fulton–MacPherson provide a compactification $F(X,n)$ for the configuration space of n marked points on an algebraic variety X. The construction is quite involved, but the combinatorial data is that of A_n. In a related vein, in [18], De Concini–Procesi construct a wonderful model X for a subspace complement $M_{\mathcal{A}} = \mathbb{C}^\ell \setminus \mathcal{A}$: a smooth, compact X such that $X \setminus M_{\mathcal{A}}$ is a normal crossing divisor. Here it is the combinatorics which are complex. A key object in their construction is

$$M_{\mathcal{A}} \longrightarrow \mathbb{C}^\ell \times \prod_{D \in G} \mathbb{P}(\mathbb{C}^\ell/D),$$

where G is a building set. In [41], Feichtner–Kozlov generalize the construction of [18] to a purely lattice-theoretic setting. See [40] for additional background on this section.

Definition 62. *For a lattice L, a building set G is a subset of L, such that for all $x \in L$, $\max\{G_{\leq x}\} = \{x_1, \ldots, x_m\}$ satisfies $[\hat{0}, x] \simeq \prod_{j=1}^m [\hat{0}, x_j]$. A building set contains all irreducible $x \in L$.*

Definition 63. *A subset N of a building set G is nested if for any set of incomparable $\{x_1, \ldots, x_p\} \subseteq N$ with $p \geq 2$, $x_1 \vee x_2 \vee \cdots \vee x_p$ exists in L, but is not in G.*

Nested sets form a simplicial complex $N(G)$, with vertices the elements of G (which are vacuously nested).

Example 64. The minimal building set for A_3 consists of the hyperplanes themselves, the triple intersections in L_2, and the element $\hat{1}$. Since $\hat{1}$ is a member of every face of $N(G)$, the nested set complex $N(G)$ is the cone over

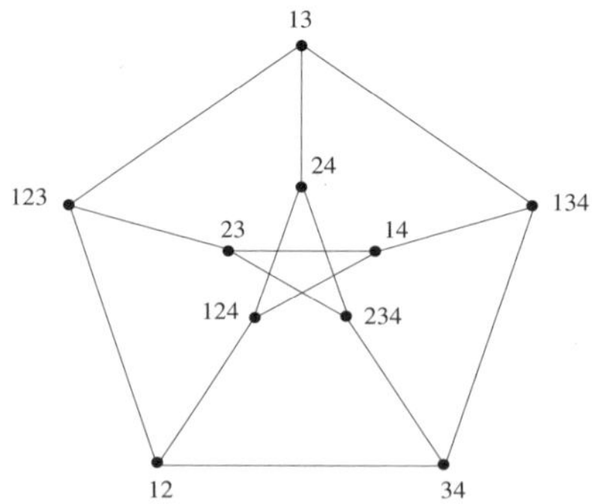

There is an edge $\overline{(12),(123)}$ because there are no incomparable subsets with at least two elements, while $\overline{(12),(34)}$ is an edge because $(12) \vee (34)$ exists in L (it is a normal crossing), but is not in G.

Suppose L is an atomic lattice, and G a building set in L. In [42], Feichtner and Yuzvinsky study a certain algebra associated to the pair L, G:

$$D(L,G) = \mathbb{Z}[x_g | g \in G]/I, \text{ with } x_g \text{ of degree 2.}$$

where I is generated by

$$\prod_{\{g_1,\ldots,g_n\} \notin N(G)\}} x_{g_i} \text{ and } \sum_{g_i \geq H \in L_1} x_{g_i}$$

Theorem 65 (Feichtner–Yuzvinsky [42]). *If $\mathcal{A}$ is a hyperplane arrangement and G a building set containing $\hat{1}$, then*

$$D(L,G) \simeq H^*(Y_{\mathcal{A},G}, \mathbb{Z}),$$

where $Y_{\mathcal{A},G}$ is the wonderful model arising from the building set G.

The importance of this is the relation to the Knudson–Mumford compactification $\overline{M_{0,n}}$ of the moduli space of n marked points on $\mathbb{P}^1$.

Theorem 66 (De Concini–Procesi [19]).

$$\overline{M_{0,n}} \simeq Y_{A_{n-2},G},$$

where G is the minimal building set for A_{n-2}.

A presentation for the cohomology ring of $\overline{M_{0,n}}$ was first described by Keel in [49]; the description which follows from [42] is very economic.

Example 67. By Theorem 65 and [19],

$$H^*(\overline{M_{0,5}}, \mathbb{Z}) \simeq D(L(A_3), G_{min}).$$

The nested set complex for A_3 and G_{min} appears in Example 64, so that $D(L(A_3), G_{min})$ is the quotient of a polynomial ring S with eleven generators by an ideal consisting of 6 linear forms (one form for each hyperplane) and 19 quadrics. To see that there are 19 quadrics, note that the space of quadrics in 11-variables has dimension 45, and $N(G_{min})$ has $15 + 11 = 26$ edges (recall that $\hat{1}$ is not pictured). A computation shows that

$$D(L(A_3), G_{min}) \simeq \mathbb{Z}[x_1, \ldots, x_5]/I,$$

where I consists of all but one quadric of S (and includes all squares of variables). This meshes with the intuitive picture: to obtain a wonderful model, simply blow up the four triple points, so that $\overline{M_{0,5}}$ is the corresponding Del Pezzo surface X_4, which has $\sum h^i(X_4, \mathbb{Z})t^i = 1 + 5t^2 + t^4$, agreeing with the computation.

Problem 68. Analyze $D(L,G)$ for other lattices.

§7. Associated Lie algebra of π_1 and LCS ranks

Let G be a finitely-generated group, with normal subgroups,

$$G = G_1 \geq G_2 \geq G_3 \geq \cdots,$$

defined inductively by $G_k = [G_{k-1}, G]$. We obtain an associated Lie algebra

$$gr(G) \otimes \mathbb{Q} := \bigoplus_{k=1}^{\infty} G_k/G_{k+1} \otimes \mathbb{Q},$$

with Lie bracket induced by the commutator map. Let $\phi_k = \phi_k(G)$ denote the rank of the k-th quotient. Presentations for $\pi_1(M_{\mathcal{A}})$ are given by Randell [68], Salvetti [73], Arvola [5], and Cohen–Suciu [13]. For computations, the braid monodromy presentation of [13] is easiest to implement. For a detailed survey of $\pi_1(M_{\mathcal{A}})$, see Suciu's survey [84]. The fundamental group is quite delicate, and in this section, we investigate properties of $\pi_1(M_{\mathcal{A}})$ via the associated graded Lie algebra

$$\mathfrak{g} = gr(\pi_1(M_{\mathcal{A}})) \otimes \mathbb{Q}.$$

The Lefschetz-type theorem of Hamm–Le [46] implies that taking a generic two dimensional slice gives an isomorphism on π_1. Thus, to study $\pi_1(M_{\mathcal{A}})$, we may assume $\mathcal{A} \subseteq \mathbb{C}^2$ or $\mathbb{P}^2$. As shown by Rybnikov [70], $\pi_1(M_{\mathcal{A}})$ is not determined by $L_{\mathcal{A}}$; whereas the Orlik–Solomon algebra $H^*(M_{\mathcal{A}}, \mathbb{Z})$ is determined by $L_{\mathcal{A}}$.

Example 69. In Example 6, we saw that the Hilbert series for A_3 is $1 + 6t + 11t^2 + 6t^3$. A computation shows that the LCS ranks begin

$$6 \quad 4 \quad 10 \quad 21 \quad 54 \quad \cdots.$$

For higher k, $\phi_k(\pi_1(A_3)) = w_k(2) + w_k(3)$, where w_k is a Witt number. In general, we may encode the LCS ranks via

$$\prod_{k=1}^{\infty} \frac{1}{(1-t^k)^{\phi_k}}.$$

For A_3, this is

$$\frac{1}{(1-t)^6} \frac{1}{(1-t^2)^4} \frac{1}{(1-t^3)^{10}} \frac{1}{(1-t^4)^{21}} \frac{1}{(1-t^5)^{54}} \cdots.$$

Expanding this and writing out the first few terms yields

$$1 + 6t + 25t^2 + 90t^3 + 301t^4 + 966t^5 + 3025t^6 + \cdots.$$

If we multiply this with

$$\pi(A_3, -t) = 1 - 6t + 11t^2 - 6t^3,$$

the result is 1, and is part of a general pattern.

Theorem 70 (Kohno's LCS formula [51]). *For the arrangement A_{n-1} (graphic arrangement of K_n)*

$$\prod_{k=1}^{\infty}(1-t^k)^{\phi_k} = \prod_{i=1}^{n-1}(1-it).$$

This explains the computation of Example 69. We now compute the free resolution of the residue field A/m as an A-module, where $m = \langle E_1 \rangle$. Let

$$b_{ij} = \dim_{\mathbb{Q}} Tor_i^A(\mathbb{Q}, \mathbb{Q})_j.$$

Example 71. For A_3, we compute $b_{ij} = 0$ if $i \neq j$, and

$$\sum_i b_{ii}t^i = 1 + 6t + 25t^2 + 90t^3 + 301t^4 + 966t^5 + 3025t^6 + \cdots.$$

The b_{ii} are the coefficients of the formal power series in Example 69!

Kohno's result was the first of a long line of results on LCS formulas for certain special families of arrangements

(1) Braid arrangements: Kohno [51]
(2) Fiber type arrangements: Falk–Randell [36]
(3) Supersolvable arrangements: Terao [88]
(4) Lower bound for ϕ_k: Falk [33]
(5) Koszul arrangements: Shelton–Yuzvinsky [81]
(6) Hypersolvable arrangements: Jambu–Papadima [48]
(7) Rational $K(\pi, 1)$ arrangements: Papadima–Yuzvinsky [64]
(8) MLS arrangements: Papadima–Suciu [61]
(9) Graphic arrangements: Lima-Filho–Schenck [54]
(10) No such formula in general: Peeva [65]

Let $\mathbb{L}(H_1(M_{\mathcal{A}}, \mathbb{K}))$ denote the free Lie algebra on $H_1(M_{\mathcal{A}}, \mathbb{K})$. Dualizing the cup product gives a map

$$H_2(M_{\mathcal{A}}, \mathbb{Q}) \stackrel{c}{\hookrightarrow} H_1(M_{\mathcal{A}}, \mathbb{Q}) \wedge H_1(M_{\mathcal{A}}, \mathbb{Q}) \longrightarrow \mathbb{L}(H_1(M_{\mathcal{A}}, \mathbb{Q})).$$

Following Chen [10], define the holonomy Lie algebra

$$\mathfrak{h}_{\mathcal{A}} = \mathbb{L}(H_1(M_{\mathcal{A}}, \mathbb{K}))/I_{\mathcal{A}},$$

where $I_{\mathcal{A}}$ is the Lie ideal generated by $\mathrm{Im}(c)$. As noted by Kohno in [50], taking transpose of cup product shows that the image of c is generated by

$$[x_j, \sum_{i=1}^{k} x_i],$$

where x_i is a generator of $\mathbb{L}(H_1(X,\mathbb{K}))$ corresponding to H_i, and the set $\{H_1, \dots, H_k\}$ is a maximal dependent set of codimension two, so corresponds to an element of $L_2(\mathcal{A})$. The upshot is that

$$\prod_{k=1}^{\infty} \frac{1}{(1-t^k)^{\phi_k}} = \sum_{i=0}^{\infty} \dim_{\mathbb{Q}} Tor_i^A(\mathbb{Q},\mathbb{Q})_i t^i.$$

This was first made explicit by Peeva in [65]; the proof runs as follows. First, Brieskorn [8] showed that $M_{\mathcal{A}}$ is formal, in the sense of [85]. Using Sullivan's work and an analysis of the bigrading on Hirsch extensions, Kohno proved

Theorem 72 (Kohno). $\phi_k(\mathfrak{g}) = \phi_k(\mathfrak{h}_{\mathcal{A}})$.

Thus

(1) $\prod_{k=1}^{\infty} \frac{1}{(1-t^k)^{\phi_k}} = HS(U(\mathfrak{h}_{\mathcal{A}}, t))$, which follows from Kohnos work and Poincaré–Birkhoff–Witt.
(2) Shelton–Yuzvinsky show in [81] that $U(\mathfrak{h}_{\mathcal{A}}) = \overline{A}^!$ is the quadratic dual of the quadratic Orlik–Solomon algebra.
(3) Results of Priddy–Löfwall show that the quadratic dual is related to diagonal Yoneda Ext-algebra via

$$\overline{A}^! \cong \bigoplus_i \mathrm{Ext}^i_{\overline{A}}(\mathbb{Q},\mathbb{Q})_i.$$

Results of Peeva [65] and Roos [69] show that in general there does not exist a standard graded algebra R such that $\prod_{k=1}^{\infty}(1-t^k)^{\phi_k} = HS(R,-t)$. For any quotient of a free Lie algebra, can we:

Problem 73. Find spaces for which there is a simple generating function for ϕ_k.

Problem 74. Relate $\mathfrak{h}_{\mathcal{A}}$ to $\bigoplus_{X \in L_2} \mathfrak{h}_{\mathcal{A}_X}$, as in [61].

As Shelton–Yuzvinsky proved in [81], the natural class of arrangements for which an LCS formula holds are arrangements for which A is a Koszul algebra, which we tackle next.

§8. Koszul algebras

Let $T(V)$ denote the tensor algebra on V.

Definition 75. *A quadratic algebra is $T(V)/I$, with $I \subseteq V \otimes V$.*

A quadratic algebra A has a quadratic dual $A^{\perp} = T(V^*)/I^{\perp}$:

$$\langle \alpha \otimes \beta \mid \alpha(a) \cdot \beta(b) = 0 \mid \forall a \otimes b \in I \rangle = I^{\perp} \subseteq V^* \otimes V^*.$$

Definition 76. *A is Koszul if $Tor_i^A(\mathbb{K}, \mathbb{K})_j = 0$, $i \neq j$.*

A quadratic algebra A is Koszul iff the minimal free resolution of the residue field over A has matrices with only linear entries.

Example 77. The Hilbert series of $S = T(\mathbb{K}^n)/\langle x_i \otimes x_j - x_j \otimes x_i \rangle$ is $1/(1-t)^n$, and a computation shows that the minimal free resolution of $\mathbb{K}$ over S is the Koszul complex, so $\dim_{\mathbb{K}} Tor_i^S(\mathbb{K}, \mathbb{K})_i = \binom{n}{i}$. Since

$$I^{\perp} = \langle x_i \otimes x_j + x_j \otimes x_i \rangle,$$

we see that $S^! = E$. The Hilbert series of E is $(1+t)^n = \sum_{i=0}^{n} \binom{n}{i} t^i$. A computation shows that $\dim_{\mathbb{K}} Tor_i^E(\mathbb{K}, \mathbb{K})_i = \binom{n-1+i}{i}$, which are the coefficients in an expansion of $1/(1-t)^n$.

Fröberg [43] proved that if I is a quadratic monomial ideal then S/I is Koszul. By uppersemicontinuity [47], this means S/I is Koszul if I has a quadratic Grobner basis (QGB). See Example 81 below for a Koszul algebra having no QGB. Both S and E are Koszul, and the relation between their Hilbert series is explained by:

Theorem 78. *If A is Koszul, so is $A^!$, and*

$$HS(A, t) \cdot HS(A^!, -t) = 1.$$

Theorem 79 (Björner–Ziegler [6])**.** *The Orlik–Solomon algebra has a QGB iff $\mathcal{A}$ is supersolvable.*

Example 80. A computation shows that the Orlik–Solomon algebra of A_3 has a quadratic Grobner basis, so is Koszul. For the non-Fano arrangement, $\dim_{\mathbb{K}} Tor_3^A(\mathbb{K}, \mathbb{K})_4 = 1$, so A is not Koszul.

Example 81. [Caviglia [9]] Map $R = \mathbb{K}[a_1, \ldots a_9] \xrightarrow{\phi} \mathbb{K}[x, y, z]$ using all cubic monomials of $\mathbb{K}[x, y, z]$ except xyz, and let $I = \ker(\phi)$. Then R/I is Koszul, but has no quadratic Grobner basis.

Problem 82. For Orlik–Solomon algebras, does Koszul imply supersolvable? In the case of graphic arrangements, it does [76].

Problem 83. Find a combinatorial description of $Tor_i^A(\mathbb{K}, \mathbb{K})_j$.

§9. Resonance varieties

Let A be the Orlik–Solomon algebra of $M_{\mathcal{A}}$, with $|\mathcal{A}| = n$. For each $a = \sum a_i e_i \in A_1$, we consider the Aomoto complex (A, a), whose i^{th} term is A_i, and differential is $\wedge a$:

$$(A, a)\colon\ 0 \longrightarrow A_0 \xrightarrow{\ a\ } A_1 \xrightarrow{\ a\ } A_2 \xrightarrow{\ a\ } \cdots \xrightarrow{\ a\ } A_\ell \longrightarrow 0\,.$$

This complex arose in Aomoto's work [4] on hypergeometric functions, as well as in the study of cohomology with local system coefficients [31], [74]. In [96], Yuzvinsky showed that for a generic a, the Aomoto complex is exact; the resonance varieties of $\mathcal{A}$ are the loci of points $a = \sum_{i=1}^{n} a_i e_i \leftrightarrow (a_1 : \cdots : a_n) \in \mathbb{P}^{n-1}$ for which (A, a) fails to be exact, that is:

Definition 84. *For each $k \geq 1$,*

$$R^k(\mathcal{A}) = \{a \in \mathbb{P}^{n-1} \mid H^k(A, a) \neq 0\}.$$

In [34], Falk gave necessary and sufficient conditions for $a \in R^1(\mathcal{A})$.

Definition 85. *A partition Π of $\mathcal{A}$ is neighborly if for all $Y \in L_2(\mathcal{A})$ and π a block of Π,*

$$\mu(Y) \leq |Y \cap \pi| \Longrightarrow Y \subseteq \pi.$$

Falk proved that components of $R^1(\mathcal{A})$ arise from neighborly partitions; he conjectured that $R^1(\mathcal{A})$ is a union of linear components. This was established (essentially simultaneously) by Cohen–Suciu [14] and Libgober–Yuzvinsky [53]. Libgober and Yuzvinsky also showed that $R^1(\mathcal{A})$ is a disjoint union of positive dimensional subspaces in $\mathbb{P}(E_1)$, and Cohen–Orlik [12] show that $R^{k \geq 2}(\mathcal{A})$ is also a subspace arrangement.

On the other hand, as shown by Falk in [35], in positive characteristic, the components of $R^1(\mathcal{A})$ can meet, and need not be linear. The approach of Libgober–Yuzvinsky involves connecting $R^1(\mathcal{A})$ to pencils/nets/webs and there is much recent work in the area, e.g. [39], [64] [99]. Of special interest is the following conjecture relating $R^1(\mathcal{A})$ and the LCS ranks ϕ_k:

Conjecture 86 (Suciu [84]). *If $\phi_4 = \theta_4$, then*

$$\prod_{k \geq 1} (1 - t^k)^{\phi_k} = \prod_{L_i \in R^1(\mathcal{A})} (1 - (dim(L_i)t)),$$

where θ_4 is the fourth Chen rank (Definition 88).

Example 87. Let $\mathcal{A} = V(xy(x-y)z) \subseteq \mathbb{P}^2$, and $E = \Lambda(\mathbb{K}^4)$, with generators $e_1, \ldots, e_4$, so that

$$A = E/\langle \partial(e_1e_2e_3), \partial(e_1e_2e_3e_4)\rangle.$$

Since $\partial(e_1e_2e_3e_4) = e_1 \wedge \partial(e_1e_2e_3) - e_4\partial(e_1e_2e_3)$, the second relation is unnecessary. To compute $R^1(\mathcal{A})$, we need only the first two differentials in the Aomoto complex. Using $e_{13}, e_{14}, e_{23}, e_{24}, e_{34}$ as a basis for A_2, we find that $e_1 \mapsto e_1 \wedge (\sum_{i=1}^{4} a_ie_i) = a_2e_{12} + a_3e_{13} + a_4e_{14}$. Since

$$\partial(e_1e_2e_3) = e_1 \wedge e_2 - e_1 \wedge e_3 + e_2 \wedge e_3,$$

in A, $e_{12} = e_{13} - e_{23}$, so that

$$a_2e_{12} = a_2(e_{13} - e_{23}).$$

This means $e_1 \mapsto (a_2 + a_3)e_{13} + a_4e_{14} - a_2e_{23}$; similar computations for the other e_i show that the Aomoto complex is

$$0 \longrightarrow \mathbb{K}^1 \xrightarrow{\begin{bmatrix} a_1 \\ a_2 \\ a_3 \\ a_4 \end{bmatrix}} \mathbb{K}^4 \xrightarrow{\begin{bmatrix} a_2+a_3 & -a_1 & -a_1 & 0 \\ a_4 & 0 & 0 & -a_1 \\ -a_2 & a_1+a_3 & -a_2 & 0 \\ 0 & a_4 & 0 & -a_2 \\ 0 & 0 & a_4 & -a_3 \end{bmatrix}} \mathbb{K}^5.$$

The rank of the first map is always one, $R^1(A) \subseteq \mathbb{P}^3$ is the locus where the second matrix has kernel of dimension at least two, so the 3×3 minors must vanish. A computation shows this locus is $\langle a_4, a_1 + a_2 + a_3\rangle$.

Letting $a = \sum_{i=1}^{n} a_ie_i$, observe that $a \in R^1(\mathcal{A})$ iff there exists a $b \in E_1$ so that $a \wedge b \in I_2$, so that $R^1(\mathcal{A})$ is the locus of decomposable 2-tensors in I_2. Since I_2 is determined by the intersection lattice $L(\mathcal{A})$ in rank ≤ 2, to study $R^1(\mathcal{A})$, it can be assumed that $\mathcal{A} \subseteq \mathbb{P}^2$.

While the first resonance variety is conjecturally connected (under certain conditions) to the LCS ranks, $R^1(A)$ is *always* connected to the Chen ranks introduced by Chen in [10].

Definition 88. *The Chen ranks of G are the LCS ranks of the maximal metabelian quotient of G:*

$$\theta_k(G) := \phi_k(G/G''),$$

where $G' = [G, G]$.

Conjecture 89 (Suciu [84]). *Let $G = G(\mathcal{A})$ be an arrangement group, and let h_r be the number of components of $R^1(\mathcal{A})$ of dimension r. Then, for $k \gg 0$:*

$$\theta_k(G) = (k-1)\sum_{r\geq 1} h_r \binom{r+k-1}{k}.$$

For the previous example, $R^1(\mathcal{A}) = V(a_4, a_1+a_2+a_3) \simeq \mathbb{P}^1$, so

$$\theta_k(G) = (k-1).$$

To discuss the Chen ranks, we need some background. The Alexander invariant G'/G'' is a module over $\mathbb{Z}[G/G']$. For arrangements, $\mathbb{Z}[G/G'] =$ Laurent polynomials in n-variables. In [55], Massey showed that

$$\sum_{k\geq 0} \theta_{k+2}\, t^k = HS(\text{gr } G'/G'' \otimes \mathbb{Q}, t).$$

It turns out to be easier to work with the linearized Alexander invariant B introduced by Cohen–Suciu in [15]

$$(A_2 \oplus E_3) \otimes S \xrightarrow{\Delta} E_2 \otimes S \longrightarrow B \longrightarrow 0,$$

where Δ is built from the Koszul differential and $(E_2 \to A_2)^t$.

Theorem 90 (Cohen–Suciu [15]).

$$V(ann\ B) = R^1(\mathcal{A}).$$

Theorem 91 (Papadima–Suciu [62]). *For $k \geq 2$,*

$$\sum_{k\geq 2} \theta_k\, t^k = HS(B,t).$$

This shows that the Chen ranks are combinatorially determined, and depend only on $L(\mathcal{A})$ in rank ≤ 2.

Example 92. For the A_3 arrangement depicted in Example 59, write $e_0 = L_{12}, e_1 = L_{13}, e_2 = L_{23}, e_3 = L_{24}, e_4 = L_{14}, e_5 = L_{34}$. With this labelling

$$I_2 = \langle \partial(e_1e_4e_5), \partial(e_0e_1e_2), \partial(e_2e_3e_5), \partial(e_0e_3e_4)\rangle,$$

from which a presentation for B can be computed:

$$S^{14} \to S^4 \to B \to 0.$$

A computation shows that $R^1(A_3)$ is

$$
\begin{gathered}
V(x_1 + x_4 + x_5, x_0, x_2, x_3) \sqcup \\
V(x_2 + x_3 + x_5, x_0, x_1, x_4) \sqcup \\
V(x_0 + x_3 + x_4, x_1, x_2, x_4) \sqcup \\
V(x_0 + x_1 + x_2, x_3, x_4, x_5) \sqcup \\
V(x_0 + x_1 + x_2, x_0 - x_5, x_1 - x_3, x_2 - x_4),
\end{gathered}
$$

and that the Hilbert Series of B is:

$$(4t^2 + 2t^3 - t^4)/(1-t)^2 = 4t^2 + 10t^3 + 15t^4 + 20t^5 + \cdots .$$

On the other hand, the graded betti numbers $Tor_i^E(A_3, \mathbb{K})_j$ are

total	1	4	10	21	45	91
0	1	–	–	–	–	–
1		4	10	15	20	25
2	–	–	–	6	25	66

So the Hilbert series for B encodes the ranks of $Tor_i^E(A_3, \mathbb{K})_{i+1}$. This suggests a connection between $R^1(A)$ and $Tor_i^E(A_3, \mathbb{K})_{i+1}$, which we tackle in the next section.

Besides the connection to resonance varieties, there is a second reason to study $Tor_i^E(A, \mathbb{K})$: the numbers $b_{ij} = \dim_{\mathbb{K}} Tor_i^A(\mathbb{K}, \mathbb{K})_j$ studied in §7 grow very fast, while the numbers $b'_{ij} = \dim_{\mathbb{K}} Tor_i^E(A, \mathbb{K})_j$ grow at a much slower rate.

Example 93. For the non-Fano arrangement of Example 6

total	1	7	23	63	165	387
0	1	–	–	–	–	–
1	–	6	17	27	36	45
2	–	1	6	36	129	342
3	–	–	–	–	–	–

b'_{ij},

total	1	7	35	156	664	2773	*
0	1	7	34	143	560	2108	*
1	–	–	1	13	103	646	*
2	–	–	–	–	1	19	*
3	–	–	–	–	–	–	1

b_{ij}.

The spaces $Tor_i^E(A,\mathbb{K})$ and $Tor_i^A(\mathbb{K},\mathbb{K})$ are related via the change of rings spectral sequence:

$$\mathrm{Tor}_i^A\left(\mathrm{Tor}_j^E(A,\mathbb{K}),\mathbb{K}\right) \Longrightarrow \mathrm{Tor}_{i+j}^E(\mathbb{K},\mathbb{K}).$$

For arrangements, details of this relationship are investigated in [76].

Problem 94. Find a combinatorial description of $Tor_i^E(A,\mathbb{K})_j$.

Problem 95. If A is Koszul, does this provide data on $Tor_i^E(A,\mathbb{K})_j$?

§10. Linear syzygies

It is fairly easy to see that there is a connection between $R^1(A)$ and linear syzygies, that is, to the module $Tor_2^E(A,\mathbb{K})_3$. Since

$$a \wedge b \in I_2 \longrightarrow a \wedge b = \sum c_i f_i,\ c_i \in \mathbb{K}, f_i \in I_2,$$

the relations $a \wedge a \wedge b = 0 = b \wedge a \wedge b$ yield linear syzygies on I_2:

$$\sum ac_i f_i = 0 = \sum bc_i f_i.$$

Example 87, continued. Since $\partial(e_1e_2e_3) = (e_1 - e_2) \wedge (e_2 - e_3)$, both $(e_1 - e_2)$ and $(e_2 - e_3)$ are in $R^1(A)$, as is the line connecting them:

$$s(e_1 - e_2) + t(e_2 - e_3) \subseteq R^1(\mathcal{A}) \subseteq \mathbb{P}(E_1).$$

Parametrically, this may be written

$$(s : t - s : -t : 0) = V(a_4, a_1 + a_2 + a_3),$$

so $s(e_1 - e_2) + t(e_2 - e_3) \wedge \partial(e_1e_2e_3) = 0$ gives a family of linear syzygies on I_2, parameterized by $\mathbb{P}^1$. $\diamond$

To make the connection between linear syzygies and the module B precise, we need the following result:

Theorem 96 (Eisenbud–Popescu–Yuzvinsky [29]). *For an arrangement $\mathcal{A}$, the Aomoto complex is exact, as a sequence of S-modules:*

$$0 \longrightarrow A_0 \otimes S \xrightarrow{\cdot a} A_1 \otimes S \xrightarrow{\cdot a} \cdots \xrightarrow{\cdot a} A_\ell \otimes S \longrightarrow F(A) \longrightarrow 0\,.$$

Theorem 97 (Schenck–Suciu [77]). *The linearized Alexander invariant B is determined by $F(A)$:*

$$B \cong \mathrm{Ext}_S^{\ell-1}(F(A), S).$$

Furthermore, for $k \geq 2$, $\dim_{\mathbb{K}} B_k = \dim_{\mathbb{K}} \mathrm{Tor}_{k-1}^E(A,\mathbb{K})_k$.

Using this, it is possible to prove one direction of Conjecture 89.

Theorem 98 (Schenck–Suciu [77]). *For $k \gg 0$,*

$$\theta_k(G) \geq (k-1) \sum_{L_i \in R^1(\mathcal{A})} \binom{\dim L_i + k - 1}{k}.$$

Problem 99. Prove the remaining direction of Conjecture 89.

What makes all this work is the Bernstein–Gelfand–Gelfand correspondence, which is our final topic.

§11. Bernstein–Gelfand–Gelfand correspondence

Let $S = Sym(V^*)$ and $E = \bigwedge(V)$. The Bernstein–Gelfand–Gelfand correspondence is an isomorphism between derived categories of bounded complexes of coherent sheaves on $\mathbb{P}(V^*)$ and bounded complexes of finitely generated, graded E-modules. Although this sounds exotic, from this it is possible to extract functors

R: finitely generated, graded S-modules $\longrightarrow$ linear free E-complexes.
L: finitely generated, graded E-modules $\longrightarrow$ linear free S-complexes.

The point is that problems can be translated to a (possibly) simpler setting. For example, BGG yields a very fast way to compute sheaf cohomology, using Tate resolutions.

Definition 100. *Let P be a finitely generated, graded E-module. Then $\mathbf{L}(P)$ is the complex*

$$\cdots \longrightarrow S \otimes P_{i+1} \xrightarrow{\cdot a} S \otimes P_i \xrightarrow{\cdot a} S \otimes P_{i-1} \xrightarrow{\cdot a} \cdots,$$

where $a = \sum_{i=1}^{n} x_i \otimes e_i$, so that $1 \otimes p \mapsto \sum x_i \otimes e_i \wedge p$.

Note that elements of V^* have degree $= 1$, and elements of V have degree $= -1$.

Example 101. $P = E = \bigwedge \mathbb{K}^3$. Then we have

$$0 \longrightarrow S \otimes E_0 \longrightarrow S \otimes E_1 \longrightarrow S \otimes E_2 \longrightarrow S \otimes E_3 \longrightarrow 0.$$

Clearly $1 \mapsto \sum_1^3 x_i \otimes e_i$. For d_1

$$\begin{aligned} e_1 &\mapsto -x_2 e_{12} - x_3 e_{13} \\ e_2 &\mapsto x_1 e_{12} - x_3 e_{23} \\ e_3 &\mapsto x_1 e_{13} + x_2 e_{23} \end{aligned}$$

$$d_2 : e_{12} \mapsto x_3 e_{123}, \; e_{13} \mapsto -x_2 e_{123} \; e_{23} \mapsto x_1 e_{123}.$$

Thus, $\mathbf{L}(E)$ is

$$S^1 \xrightarrow{\begin{bmatrix} x_1 \\ x_2 \\ x_3 \end{bmatrix}} S^3 \xrightarrow{\begin{bmatrix} -x_2 & x_1 & 0 \\ -x_3 & 0 & x_1 \\ 0 & -x_3 & x_2 \end{bmatrix}} S^3 \xrightarrow{\begin{bmatrix} x_3 & -x_2 & x_1 \end{bmatrix}} S^1.$$

This is simply the Koszul complex.

If M is a finitely generated, graded S-module, then $\mathbf{R}(M)$ is the complex

$$\cdots \longrightarrow \hat{E} \otimes M_{i-1} \xrightarrow{\cdot a} \hat{E} \otimes M_i \xrightarrow{\cdot a} \hat{E} \otimes M_{i+1} \xrightarrow{\cdot a} \cdots,$$

where $a = \sum_{i=1}^{n} e_i \otimes x_i$, so $1 \otimes m \mapsto \sum e_i \otimes x_i \cdot m$, and $\hat{E}$ is $\mathbb{K}$-dual to E:

$$\hat{E} \simeq E(n) = Hom_{\mathbb{K}}(E, \mathbb{K}).$$

Just as $\mathbf{L}(P) = S \otimes_{\mathbb{K}} P$, $\mathbf{R}(M) = Hom_{\mathbb{K}}(E, M)$.

Example 3, continued. If $M = \mathbb{K}[x_0, x_1]/\langle x_0 x_1, x_0^2 \rangle$, then

$$\begin{aligned} 1 &\mapsto e_0 \otimes x_0 + e_1 \otimes x_1 \\ x_0 &\mapsto e_0 \otimes x_0^2 + e_1 \otimes x_0 x_1 \\ x_1 &\mapsto e_0 \otimes x_0 x_1 + e_1 \otimes x_1^2 \\ x_1^n &\mapsto e_0 \otimes x_0 x_1^n + e_1 \otimes x_1^{n+1}. \end{aligned}$$

Thus, $\mathbf{R}(M)$ is

$$E(2)^1 \xrightarrow{\begin{bmatrix} e_0 \\ e_1 \end{bmatrix}} E(3)^2 \xrightarrow{\begin{bmatrix} 0 & e_1 \end{bmatrix}} E(4)^1 \xrightarrow{\begin{bmatrix} e_1 \end{bmatrix}} E(5)^1 \xrightarrow{\begin{bmatrix} e_1 \end{bmatrix}} \cdots.$$

This complex is exact, except at the second step. The kernel of

$$\begin{bmatrix} 0 & e_1 \end{bmatrix}$$

is generated by $\alpha = [1, 0]$ and $\beta = [0, e_1]$, with relations $im(d_1) = \beta + e_0 \alpha = 0, e_1 \beta = 0$, so that

$$H^1(\mathbf{R}(M)) \simeq E(3)/e_0 \wedge e_1.$$

The betti table for M is:

total	1	2	1
0	1	–	–
1	–	2	1

Note that in this example, M is 1-regular.

Theorem 102 (Eisenbud–Fløystad–Schreyer [28]).

$$H^j(\mathbf{R}(M))_{i+j} = Tor_i^S(M, \mathbb{K})_{i+j}.$$

Corollary 103. *The Castelnuovo–Mumford regularity of M is $\leq d$ iff $H^i(\mathbf{R}(M)) = 0$ for all $i > d$.*

What can be said about higher resonance varieties? In [12], Cohen–Orlik prove that for $k \geq 2$,

$$R^k(\mathcal{A}) = \bigcup L_i \text{ linear.}$$

As observed by Suciu, in general the union need not be disjoint.

Theorem 104 (Eisenbud–Popescu–Yuzvinsky [29]). $R^k(\mathcal{A}) \subseteq R^{k+1}(\mathcal{A})$.

The key point is that

$$H^k(A, a) \neq 0 \text{ iff } Tor^S_{\ell-k}(F(A), S/I(p)) \neq 0.$$

The result follows from interpreting this in terms of Koszul cohomology.

Theorem 105 (Denham–Schenck [21]). *Higher resonance may be interpreted via* Ext*:*

$$R^k(\mathcal{A}) = \bigcup_{k' \leq k} V(\operatorname{ann} \operatorname{Ext}^{\ell-k'}(F(A), S)).$$

Furthermore, the differentials in free resolution of A over E can be analyzed using BGG and the Grothendieck spectral sequence.

For any coherent sheaf $\mathcal{F}$ on $\mathbb{P}^d$, there is a finitely generated, graded *saturated* S-module M whose sheafification is $\mathcal{F}$. If $\mathcal{F}$ has Castelnuovo–Mumford regularity r, then the Tate resolution of $\mathcal{F}$ is obtained by splicing the complex $\mathbf{R}(M_{\geq r})$:

$$0 \longrightarrow \hat{E} \otimes M_r \xrightarrow{d^r} \hat{E} \otimes M_{r+1} \longrightarrow \hat{E} \otimes M_{r+2} \longrightarrow \cdots,$$

with a free resolution $P_\bullet$ for the kernel of d^r:

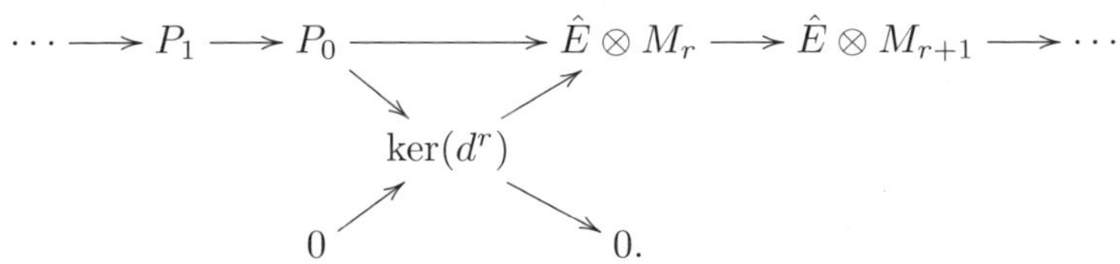

By Corollary 103, $\mathbf{R}(M_{\geq r})$ is exact except at the first step, so this yields an exact complex of free E-modules.

Example 106. Since $M = S$ has regularity zero, we obtain Cartan resolutions in both directions, and the splice map $E \to \widehat{E} = E(d+1)$ is multiplication by $e_0 \wedge e_1 \wedge \cdots \wedge e_d = \ker \begin{bmatrix} e_0, & \cdots & , e_d \end{bmatrix}^t$.

Theorem 107 (Eisenbud–Fløystad–Schreyer [28]). *The ith free module T^i in a Tate resolution for $\mathcal{F}$ satisfies*

$$T^i = \bigoplus_j \widehat{E} \otimes H^j(\mathcal{F}(i-j)).$$

Example 4, continued. The betti table for the twisted cubic shows that S/I has regularity one, which provides us the information needed to compute the Tate resolution. Plugging the resulting numbers into Theorem 107 shows that

i	-3	-2	-1	0	1	2
$h^1(\mathcal{F}(i))$	8	5	2	0	0	0
$h^0(\mathcal{F}(i))$	0	0	0	1	4	7

Does this make sense? Since $\mathcal{F} = \mathcal{O}_X = \mathcal{O}_{\mathbb{P}^1}(3)$,

$$h^1(\mathcal{F}(i)) = h^1(\mathcal{O}_{\mathbb{P}^1}(3i)) = h^0(\mathcal{O}_{\mathbb{P}^1}(-3i-2))$$

and

$$h^0(\mathcal{F}(i)) = h^0(\mathcal{O}_{\mathbb{P}^1}(3i)) = 3i+1, \ i \geq 0,$$

which agrees with our earlier computation. ◇

Problem 108. Investigate the Tate resolution for $D(\mathcal{A})$ and $C(\mathcal{A})$.

Conclusion. In this note we have surveyed a number of open problems in arrangements. The beauty of the area is that these problems are all interconnected. Perhaps the most central objects are the resonance varieties, which are related to both the LCS ranks studied in §7 and §8 using Koszul and Lie algebras, and to the Chen ranks. The results of §5 tie resonance to the Orlik–Terao algebra, and [75] notes that $J_{\mathcal{A}} \subseteq H^0(D_{\mathcal{A}})$, so the Orlik–Terao algebra is also linked to $D(\mathcal{A})$ and freeness. But freeness ties in to multiarrangements, and can be generalized to hypersurface arrangements, the topics of §3 and §4. To complete the circle, recent work of Cohen–Denham–Falk–Varchenko [11] relates freeness to $R^1(\mathcal{A})$. In short, everything is connected!

Acknowledgements. Many thanks are due to the organizers of the Mathematical Society of Japan Seasonal Institute: Takuro Abe, Hiroaki Terao, Masahiko Yoshinaga and Sergey Yuzvinsky organized a wonderful meeting. I also am grateful to my research collaborators: G. Denham, M. Mustata, A. Suciu, H. Terao, S. Tohaneanu, and M. Yoshinaga. As noted in the introduction, the calculations carried out in this survey may be performed using the hyperplane arrangements package of Denham and Smith [22] in Macaulay2 [45]. Code for the individual examples is available at `http://www.math.uiuc.edu/~schenck/`.

References

[1] T. Abe, H. Terao and M. Wakefield, The Euler multiplicity and addition-deletion theorems for multiarrangements, J. Lond. Math. Soc. (2), **77** (2008), 335–348.

[2] T. Abe, H. Terao and M. Wakefield, The characteristic polynomial of a multiarrangement, Adv. Math., **215** (2007), 825–838

[3] T. Abe, H. Terao and M. Yoshinaga, Totally free arrangements of hyperplanes, Proc. Amer. Math. Soc., **137** (2009), 1405–1410.

[4] K. Aomoto, Un théorème du type de Matsushima–Murakami concernant l'integrale des fonctions multiformes, J. Math. Pures Appl. (9), **52** (1973), 1–11.

[5] W. Arvola, The fundamental group of the complement of an arrangement of complex hyperplanes, Topology, **31** (1992), 757–765.

[6] A. Björner and G. Ziegler, Broken circuit complexes: factorizations and generalizations, J. Combin. Theory Ser. B, **51** (1991), 96–126.

[7] K. Brandt and H. Terao, Free Arrangements and Relation Spaces, Discrete Comput. Geom., **12** (1994), 49–63.

[8] E. Brieskorn, Sur les groupes de tresses, In: Séminaire Bourbaki, 1971/72, Lecture Notes in Math., **317**, Springer-Verlag, 1973, pp. 21–44.

[9] G. Caviglia, The pinched veronese is Koszul, J. Algebraic Combin., **30** (2009), 539–548.

[10] K. Chen, Iterated integrals of differential forms and loop space cohomology, Ann. of Math. (2), **97** (1973), 217–246.

[11] D. Cohen, G. Denham, M. Falk and A. Varchenko, Critical points and resonance of hyperplane arrangements, Canad. J. Math., **63** (2011), 1038–1057.

[12] D. Cohen and P. Orlik, Arrangements and local systems, Math. Res. Lett., **7** (2000), 299–316.

[13] D. Cohen and A. Suciu, The braid monodromy of plane algebraic curves and hyperplane arrangements, Comment. Math. Helv., **72** (1997), 285–315.

[14] D. Cohen and A. Suciu, Characteristic varieties of arrangements, Math. Proc. Cambridge Phil. Soc., **127** (1999), 33–53.

[15] D. Cohen and A. Suciu, Alexander invariants of complex hyperplane arrangements, Trans. Amer. Math. Soc., **351** (1999), 4043–4067.

[16] D. Cohen and A. Suciu (eds.), Arrangements in Boston: A Conference on Hyperplane Arrangements, Topology Appl., **118**, Elsevier Sci. B.V., 2002.

[17] W. Decker and D. Eisenbud, Sheaf algorithms using the exterior algebra, In: Computations in Algebraic Geometry using Macaulay 2, Springer-Verlag, 2002.

[18] C. De Concini and C. Procesi, Wonderful models of subspace arrangements, Selecta Math. (N.S.), **1** (1995), 459–494.

[19] C. De Concini and C. Procesi, Hyperplane arrangements and holonomy equations, Selecta Math. (N.S.), **1** (1995), 495–535.

[20] G. Denham, Homological aspects of hyperplane arrangements, In: Arrangements, Local Systems and Singularities, Birkhauser, 2010.

[21] G. Denham and H. Schenck, The double Ext spectral sequence and rank varieties, preprint, 2010.

[22] G. Denham and G. G. Smith, Hyperplane Arrangements: A Macaulay2 Package http://www.math.uiuc.edu/Macaulay2/.

[23] G. Denham and A. Suciu, On the homotopy Lie algebra of an arrangement, Michigan Math. J., **54** (2006), 319–340.

[24] G. Denham and S. Yuzvinsky, Annihilators of Orlik–Solomon relations, Adv. in Appl. Math., **28** (2002), 231–249.

[25] A. Dimca and S. Yuzvinsky, Lectures on Orlik–Solomon algebras, in: Arrangements, Local Systems and Singularities, Birkhauser, 2010.

[26] D. Eisenbud, Linear sections of determinantal varieties, Amer. J. Math., **110** (1988), 541–575.

[27] D. Eisenbud, The Geometry of Syzygies, Springer-Verlag, 2004.

[28] D. Eisenbud, G. Fløystad and F.-O. Schreyer, Sheaf cohomology and free resolutions over exterior algebras, Trans. Amer. Math. Soc., **355** (2003), 4397–4426.

[29] D. Eisenbud, S. Popescu and S. Yuzvinsky, Hyperplane arrangement cohomology and monomials in the exterior algebra, Trans. Amer. Math. Soc., **355** (2003), 4365–4383.

[30] F. El Zein, A. Suciu, M. Tosun, M Uludag and S. Yuzvinsky (eds.), Arrangements, Local Systems and Singularities, Progr. Math., **283**, Birkhauser, 2010.

[31] H. Esnault, V. Schechtman and E. Viehweg, Cohomology of local systems on the complement of hyperplanes, Invent. Math., **109** (1992), 557–561.

[32] M. Falk, The minimal model of the complement of an arrangement of hyperplanes, Trans. Amer. Math. Soc., **309** (1988), 543–556.

[33] M. Falk, The cohomology and fundamental group of a hyperplane complement, In: Singularities, Contemporary Math., **90**, Amer. Math. Soc, Providence, RI, 1989, pp. 55–72.

[34] M. Falk, Arrangements and cohomology, Ann. Combin., **1** (1997), 135–157.

[35] M. Falk, Resonance varieties over fields of positive characteristic, Int. Math. Res. Not. IMRN, **2007**, no. 3.

[36] M. Falk and R. Randell, The lower central series of a fiber-type arrangement, Invent. Math., **82** (1985), 77–88.

[37] M. Falk and R. Randell, On the homotopy theory of arrangements II, In: Arrangements—Tokyo 1998, Adv. Stud. Pure. Math., **27**, Kinokuniya, Tokyo, 2000, pp. 93–125.

[38] M. Falk and H. Terao (eds.), Arrangements—Tokyo 1998, Proceedings of the Workshop on Mathematics Related with Arrangements of Hyperplanes, celebrating the 60th birthday of Peter Orlik, Math. Soc. Japan, Tokyo, 2000.

[39] M. Falk and S. Yuzvinsky, Multinets, resonance varieties, and pencils of plane curves, Compos. Math., **143** (2007), 1069–1088.

[40] E. Feichtner, De Concini–Procesi wonderful arrangement models: a discrete geometer's point of view, In: Combinatorial and Computational Geometry, Math. Sci. Res. Inst. Publ., **52**, Cambridge Univ. Press, Cambridge, 2005, ppp. 333–360.

[41] E. Feichtner and D. Kozlov, Incidence combinatorics of resolutions, Selecta Math., **10** (2004), 37–60.

[42] E. Feichtner and S. Yuzvinsky, Chow rings of toric varieties defined by atomic lattices, Invent. Math., **155** (2004), 515–536.

[43] R. Fröberg, Determination of a class of Poincaré series, Math. Scand., **37** (1975), 29–39.

[44] W. Fulton and R. MacPherson, A compactification of configuration spaces, Ann. of Math. (2), **139** (1994), 183–225.

[45] D. Grayson and M. Stillman, Macaulay2: a software package for commutative algebra http://www.math.uiuc.edu/Macaulay2/.

[46] H. Hamm and D. T. Lê, Un théorème de Zariski du type de Lefschetz, Ann. Sci. École Norm. Sup., **6** (1973), 317–366.

[47] J. Herzog, Finite free resolutions, In: Computational Commutative and Non-Commutative Algebraic Geometry, NATO Sci. Ser. III, Comput. Syst. Sci., **196**, IOS, Amsterdam, 2005, pp. 118–144.

[48] M. Jambu and S. Papadima, A generalization of fiber-type arrangements and a new deformation method, Topology, **37** (1998), 1135–1164.

[49] S. Keel, Intersection theory of moduli space of stable n-pointed curves of genus zero, Trans. Amer. Math. Soc., **330** (1992), 545–574.

[50] T. Kohno, On the holonomy Lie algebra and the nilpotent completion of the fundamental group of the complement of hypersurfaces, Nagoya Math J., **92** (1983), 21–37.

[51] T. Kohno, Série de Poincaré–Koszul associée aux groupes de tresses pures, Invent. Math., **82** (1985), 57–75.

[52] J. Kung and H. Schenck, Derivation modules of orthogonal duals of hyperplane arrangements, J. Algebraic Combin., **24** (2006), 253–262.

[53] A. Libgober and S. Yuzvinsky, Cohomology of the Orlik–Solomon algebras and local systems, Compositio Math., **121** (2000), 337–361.

[54] P. Lima-Filho and H. Schenck, Holonomy Lie algebras and the LCS formula for subarrangements of A_n, Int. Math. Res. Not. IMRN, **2009**, no. 8, 1421–1432.

[55] W. Massey, Completion of link modules, Duke Math. J., **47** (1980), 399–420.

[56] J. Morgan, The algebraic topology on smooth algebraic varieties, Inst. Hautes Études Sci. Publ. Math., **48** (1978), 137–204.

[57] P. Orlik and L. Solomon, Combinatorics and topology of complements of hyperplanes, Invent. Math., **56** (1980), 167–189.

[58] P. Orlik and H. Terao, Arrangements of Hyperplanes, Grundlehren Math. Wiss., **300**, Springer-Verlag, 1992.

[59] P. Orlik and H. Terao, Arrangements and Hypergeometric Integrals, MSJ Memoirs, **9**, Math. Soc. Japan, Tokyo, 2007.

[60] P. Orlik and H. Terao, Commutative algebras for arrangements, Nagoya Math. J., **134** (1994), 65–73.

[61] S. Papadima and A. Suciu, When does the associated graded Lie algebra of an arrangement group decompose?, Comment. Math. Helv., **81** (2006), 859–875.

[62] S. Papadima and A. Suciu, Chen Lie algebras, Int. Math. Res. Not., **2004**, no. 21, 1057–1086.

[63] S. Papadima and A. Suciu, Higher homotopy groups of complements of complex hyperplane arrangements, Adv. Math., **165** (2002), 71–100.

[64] S. Papadima and S. Yuzvinsky, On rational $K[\pi, 1]$ spaces and Koszul algebras, J. Pure Applied Algebra, **144** (1999), 157–167.

[65] I. Peeva, Hyperplane arrangements and linear strands in resolutions, Trans. Amer. Math. Soc., **355** (2003), 609–618.

[66] J. Pereira and S. Yuzvinsky, Completely reducible hypersurfaces in a pencil, Adv. Math., **219** (2008), 672–688.

[67] N. Proudfoot and D. Speyer, A broken circuit ring, Beiträge Algebra Geom., **47** (2006), 161–166.

[68] R. Randell, The fundamental group of the complement of a union of complex hyperplanes, Invent. Math., **69** (1982), 103–108.

[69] J. E. Roos, The homotopy Lie algebra of a complex hyperplane arrangement is not necessarily finitely generated, Experiment. Math., **17** (2008), 129–143.

[70] G. Rybnikov, On the fundamental group of the complement of a complex hyperplane arrangement, DIMACS Tech. Report 94-13 (1994), 33–50.

[71] K. Saito, Quasihomogene isolierte Singularitäten von Hyperflächen, Invent. Math., **14** (1971), 123–142.

[72] K. Saito, Theory of logarithmic differential forms and logarithmic vector fields, J. Fac. Sci. Univ. Tokyo Sect. IA Math., **27** (1980), 265–291.

[73] M. Salvetti, Topology of the complement of real hyperplanes in $\mathbb{C}^N$, Invent. Math., **88** (1987), 603–618.

[74] V. Schechtman, H. Terao and A. Varchenko, Local systems over complements of hyperplanes and the Kac–Kazhdan conditions for singular vectors, J. Pure Appl. Algebra, **100** (1995), 93–102.

[75] H. Schenck, Resonance varieties via blowups of P^2 and scrolls, Int. Math. Res. Not. IMRN, **20** (2011), 4756–4778.

[76] H. Schenck and A. Suciu, Lower central series and free resolutions of hyperplane arrangements, Trans. Amer. Math. Soc., **354** (2002), 3409–3433.

[77] H. Schenck and A. Suciu, Resonance, linear syzygies, Chen groups, and the Bernstein–Gelfand–Gelfand correspondence, Trans. Amer. Math. Soc., **358** (2006), 2269–2289.

[78] H. Schenck, H. Terao, and M. Yoshinaga, Logarithmic forms for quasihomogeneous curve configurations in $\mathbb{P}^2$, in preparation (2009).

[79] H. Schenck and S. Tohaneanu, Freeness of Conic–Line arrangements in $\mathbb{P}^2$, Comment. Math. Helv., **84** (2009), 235–258.

[80] H. Schenck and S. Tohaneanu, The Orlik–Terao algebra and 2-formality, Math. Res. Lett., **16** (2009), 171–182.

[81] B. Shelton and S. Yuzvinsky, Koszul algebras from graphs and hyperplane arrangements, J. London Math. Soc. (2), **56** (1997), 477–490.

[82] L. Solomon and H. Terao, A formula for the characteristic polynomial of an arrangement, Adv. in Math., **64** (1987), 305–325.

[83] R. Stanley, Supersolvable lattices, Algebra Universalis, **2** (1972), 197–217.

[84] A. Suciu, Fundamental groups of line arrangements: Enumerative aspects, In: Advances in Algebraic Geometry Motivated by Physics, Contemporary Math., **276**, Amer. Math. Soc, Providence, RI, 2001, pp. 43–79.

[85] D. Sullivan, Infinitesimal computations in topology, Inst. Hautes Études Sci. Publ. Math., **47** (1977), 269–331.

[86] H. Terao, Generalized exponents of a free arrangement of hyperplanes and Shepard–Todd–Brieskorn formula, Invent. Math., **63** (1981), 159–179.

[87] H. Terao, Arrangements of hyperplanes and their freeness I, J. Fac. Science Univ. Tokyo, **27** (1980), 293–312.

[88] H. Terao, Modular elements of lattices and topological fibration, Adv. Math., **62** (1986), 135–154.

[89] H. Terao, On the homological dimensions of arrangements, Unpublished manuscript, 1990.

[90] H. Terao, Algebras generated by reciprocals of linear forms, J. Algebra, **250** (2002), 549–558.

[91] A. Varchenko, Special Functions, KZ Type Equations, and Representation Theory, CBMS Conference Series, **98**, Amer. Math. Soc., 2003.

[92] M. Wakefield and S. Yuzvinsky, Derivations of an effective divisor on the complex projective line, Trans. Amer. Math. Soc., **359** (2007), 4389–4403.

[93] M. Yoshinaga, Characterization of a free arrangement and conjecture of Edelman and Reiner, Invent. Math., **157** (2004), 449–454.

[94] M. Yoshinaga, On the freeness of 3-arrangements, Bull. London Math. Soc., **37** (2005), 126–134.

[95] S. Yuzvinsky, First two obstructions to the freeness of arrangements, Trans. Amer. Math. Soc., **335** (1993), 231–244.

[96] S. Yuzvinsky, Cohomology of Brieskorn–Orlik–Solomon algebras, Comm. Algebra, **23** (1995), 5339–5354.

[97] S. Yuzvinsky, Orlik–Solomon algebras in algebra and topology, Russian Math. Surveys, **56** (2001), 293–364.

[98] S. Yuzvinsky, Realization of finite Abelian groups by nets in $\mathbb{P}^2$, Compos. Math., **140** (2004), 1614–1624.

[99] S. Yuzvinsky, A new bound on the number of special fibers in a pencil of plane curves, Proc. Amer. Math. Soc., **137** (2009), 1641–1648.

[100] G. Ziegler, Multiarrangements of hyperplanes and their freeness, In: Singularities, Contemp. Math., **90**, Amer. Math. Soc., Providence, 1989, pp. 345–359.

[101] G. Ziegler, Combinatorial construction of logarithmic differential forms, Adv. Math., **76** (1989), 116–154.

Hal Schenck
Mathematics Department, University of Illinois Urbana-Champaign, Urbana IL 61801, USA
E-mail address: `schenck@math.uiuc.edu`

Advanced Studies in Pure Mathematics 62, 2012
Arrangements of Hyperplanes—Sapporo 2009
pp. 359–398

Resonance varieties and Dwyer–Fried invariants

Alexander I. Suciu

Abstract.

The Dwyer–Fried invariants of a finite cell complex X are the subsets $\Omega^i_r(X)$ of the Grassmannian of r-planes in $H^1(X,\mathbb{Q})$ which parametrize the regular $\mathbb{Z}^r$-covers of X having finite Betti numbers up to degree i. In previous work, we showed that each Ω-invariant is contained in the complement of a union of Schubert varieties associated to a certain subspace arrangement in $H^1(X,\mathbb{Q})$. Here, we identify a class of spaces for which this inclusion holds as equality. For such "straight" spaces X, all the data required to compute the Ω-invariants can be extracted from the resonance varieties associated to the cohomology ring $H^*(X,\mathbb{Q})$. In general, though, translated components in the characteristic varieties affect the answer.

Contents

Received July 20, 2010.
Revised December 12, 2010.
2010 *Mathematics Subject Classification.* Primary 20J05, 55N25; Secondary 14F35, 32S22, 55R80, 57M07.
Key words and phrases. Free abelian cover, characteristic variety, resonance variety, tangent cone, Dwyer–Fried set, special Schubert variety, toric complex, Kähler manifold, hyperplane arrangement.
Partially supported by NSA grant H98230-09-1-0021 and NSF grant DMS–1010298.

§1. Introduction

One of the most fruitful ideas to arise from arrangement theory is that of turning the cohomology ring of a space into a family of cochain complexes, parametrized by the cohomology group in degree 1, and extracting certain varieties from these data, as the loci where the cohomology of those cochain complexes jumps. What makes these "resonance" varieties really useful is their close connection with a different kind of jumping loci: the "characteristic" varieties, which record the jumps in homology with coefficients in rank 1 local systems.

In this paper, we use the geometry of the cohomology jump loci to study a classical problem in topology: determining which infinite covers of a space have finite Betti numbers. Restricting our attention to regular, free abelian covers of a fixed rank allows us to state the problem in terms of a suitable filtration on the rational Grassmannian. Under favorable circumstances, the finiteness of the Betti numbers of such covers is exclusively controlled by the incidence varieties to the resonance varieties of our given space.

1.1. Cohomology jump loci and straightness

Let X be a connected CW-complex with finite skeleta. To such a space, we associate two types of jump loci. The first are the *resonance varieties* $\mathcal{R}^i(X)$. These are homogeneous subvarieties of the affine space $H^1(X, \mathbb{C}) = \mathbb{C}^n$, where $n = b_1(X)$, and they are defined in terms of the cohomology algebra $A = H^*(X, \mathbb{C})$, as follows. For each $a \in A^1$, left-multiplication by a defines a cochain complex $(A, \cdot a)$. Then

$$(1) \qquad \mathcal{R}^i(X) = \{a \in H^1(X, \mathbb{C}) \mid H^j(A, \cdot a) \neq 0, \text{ for some } j \leq i\}.$$

The second type of jump loci we consider here are the *characteristic varieties* $\mathcal{W}^i(X)$. These are Zariski closed subsets of the complex algebraic torus $H^1(X, \mathbb{C}^\times)^0 = \mathrm{Hom}(\pi_1(X), \mathbb{C}^\times)^0 = (\mathbb{C}^\times)^n$, defined as follows. Each character $\rho\colon \pi_1(X) \to \mathbb{C}^\times$ gives rise to a rank 1 local system on X, call it $\mathcal{L}_\rho$. Then

$$(2) \qquad \mathcal{W}^i(X) = \{\rho \in H^1(X, \mathbb{C}^\times)^0 \mid H_j(X, \mathcal{L}_\rho) \neq 0, \text{ for some } j \leq i\}.$$

One of our goals in this paper is to isolate a class of spaces for which the resonance and characteristic varieties have a rather simple nature, and are intimately related to each other.

We say that X is *locally k-straight* if, for each $i \leq k$, all components of $\mathcal{W}^i(X)$ passing through the origin 1 are algebraic subtori, and the tangent cone at 1 to $\mathcal{W}^i(X)$ equals $\mathcal{R}^i(X)$. If, moreover, all positive-dimensional components of $\mathcal{W}^i(X)$ contain the origin, we say X is *k-straight.* For locally straight spaces, the resonance varieties $\mathcal{R}^i(X)$ are finite unions of rationally defined linear subspaces.

Examples of straight spaces include Riemann surfaces, tori, and knot complements. Under some further assumptions, the straightness properties behave well with respect to finite direct products and wedges.

A related notion is Sullivan's k-formality. Using the tangent cone formula from [11], it is readily seen that 1-formal spaces are locally 1-straight. In general though, 1-formality does not imply 1-straightness, and the converse does not hold, either.

1.2. Dwyer–Fried invariants

The second goal of this paper is to analyze the homological finiteness properties of all regular, free abelian covers of a given space X, and relate these properties to the resonance varieties of X, under a straightness assumption.

The connected, regular $\mathbb{Z}^r$-covers $\hat{X} \to X$ are parametrized by the Grassmannian of r-planes in the vector space $H^1(X, \mathbb{Q})$. Moving about this variety, and recording when all the Betti numbers $b_1(\hat{X}), \dots, b_i(\hat{X})$ are finite defines subsets

$$\Omega^i_r(X) \subset \operatorname{Gr}_r(H^1(X, \mathbb{Q})), \tag{3}$$

which we call the *Dwyer–Fried invariants* of X. These sets depend only on the homotopy type of X. Consequently, if G is a finitely generated group, the sets $\Omega^i_r(G) := \Omega^i_r(K(G,1))$ are well-defined.

In [13], Dwyer and Fried showed that the support varieties of the Alexander invariants of a finite cell complex X completely determine the Ω-sets of X. In [29] and [34], this foundational result was refined and reinterpreted in terms of the characteristic varieties of X, as follows. Let $\exp\colon H^1(X, \mathbb{C}) \to H^1(X, \mathbb{C}^{\times})$ be the coefficient homomorphism induced by the exponential map $\exp\colon \mathbb{C} \to \mathbb{C}^{\times}$. Then,

$$\Omega^i_r(X) = \big\{P \in \operatorname{Gr}_r(\mathbb{Q}^n) \bigm| \dim_{\mathbb{C}}(\exp(P \otimes \mathbb{C}) \cap \mathcal{W}^i(X)) = 0\big\}. \tag{4}$$

We pursue this study here, by investigating the relationship between the Dwyer–Fried sets and the resonance varieties. Given a homogeneous variety $V \subset \Bbbk^n$, let $\sigma_r(V) \subset \operatorname{Gr}_r(\Bbbk^n)$ be the variety of r-planes incident to V. Our main result reads as follows.

Theorem 1.1. *Let X be a connected CW-complex with finite k-skeleton.*

(1) *Suppose X is locally k-straight. Then, for all $i \leq k$ and $r \geq 1$,*

$$\Omega_r^i(X) \subseteq \mathrm{Gr}_r(H^1(X,\mathbb{Q})) \setminus \sigma_r(\mathcal{R}^i(X,\mathbb{Q})).$$

(2) *Suppose X is k-straight. Then, for all $i \leq k$ and $r \geq 1$,*

$$\Omega_r^i(X) = \mathrm{Gr}_r(H^1(X,\mathbb{Q})) \setminus \sigma_r(\mathcal{R}^i(X,\mathbb{Q})).$$

As a consequence, if X is k-straight, then each set $\Omega_r^i(X)$ with $i \leq k$ is the complement of a finite union of special Schubert varieties in the Grassmannian of r-planes in $\mathbb{Q}^n$, where $n = b_1(X)$. In particular, $\Omega_r^i(X)$ is a Zariski open subset of $\mathrm{Gr}_r(\mathbb{Q}^n)$.

1.3. Applications

We illustrate our techniques with a broad variety of examples, coming from low-dimensional topology, toric topology, algebraic geometry, and the theory of hyperplane arrangements.

One class of spaces for which things work out very well are the toric complexes. Every simplicial complex K on n vertices determines a subcomplex T_K of the n-torus, with fundamental group the right-angled Artin group associated to the 1-skeleton of K. It turns out that all toric complexes are straight (that is, k-straight, for all k). As shown by Papadima and Suciu in [26], the resonance varieties of a toric complex are unions of coordinate subspaces, which can be read off directly from the corresponding simplicial complex. This leads to some explicit formulas for the Ω-sets of toric complexes and right-angled Artin groups.

The characteristic varieties of (quasi-) Kähler manifolds are fairly well understood, due to work of Beauville, Green–Lazarsfeld, Simpson, Campana, and finally Arapura [1]. In particular, if X is a smooth, quasi-projective variety, then all the components of $\mathcal{W}^1(X)$ are torsion-translated subtori. Recent work of Dimca, Papadima, and Suciu [11] sheds light on the first resonance variety of such varieties: if X is also 1-formal (e.g., if X is compact), then all the components of $\mathcal{R}^1(X)$ are rationally defined linear subspaces. It follows that X is locally 1-straight, and $\Omega_r^1(X) \subseteq \sigma_r(\mathcal{R}^1(X,\mathbb{Q}))^{\complement}$. In general, though, the inclusion can be strict. For instance, we prove in Theorem 10.11 the following: if $\mathcal{W}^1(X)$ has a 1-dimensional component not passing through 1, and $\mathcal{R}^1(X)$ has no codimension-1 components, then $\Omega_2^1(X) \neq \sigma_2(\mathcal{R}^1(X,\mathbb{Q}))^{\complement}$.

Hyperplane arrangements have been the main driving force behind the development of the theory of cohomology jump loci, and still provide a rich source of motivational examples for this theory. If $\mathcal{A}$ is an arrangement of hyperplanes in $\mathbb{C}^\ell$, then its complement, $X = \mathbb{C}^\ell \setminus \bigcup_{H\in\mathcal{A}} H$, is a connected, smooth, quasi-projective variety. It turns out that X is also formal, locally straight, but not always straight. The first resonance variety of the arrangement is completely understood, owing to work of Falk [17], Cohen, Libgober, Suciu, Yuzvinsky, and others, with the state of the art being the recent work of Falk, Pereira, and Yuzvinsky [19, 30, 36]. This allows for explicit computations of the Dwyer–Fried invariants of various classes of arrangements. For the deleted B_3 arrangement, though, the computation is much more subtle, due to the presence of a translated component in the first characteristic variety.

1.4. Organization of the paper

The paper is divided in roughly two parts. In the first part (§§2–6), we recall some of the basic theory of cohomology jump loci, and develop the notion of straightness. In the second part (§§7–11), we develop the Dwyer–Fried theory in the straight context, and apply it in a variety of settings.

In §2 we define the Aomoto complex of a space X, and study it in more detail in the case when X admits a minimal cell structure. We use the Aomoto complex in §3 to define the resonance varieties, and establish some of the basic properties of these varieties.

In §4 we define two types of tangent cones to a subvariety of $(\mathbb{C}^\times)^n$, and recall some of their features. In §5, we introduce the characteristic varieties, and review some pertinent facts about their tangent cones. Finally, in §6 we define and study the various notions of straightness, based on the geometry of the jump loci.

We start §7 with a review of the Dwyer–Fried invariants, and the way they relate to the characteristic varieties, after which we prove Theorem 1.1. In §8, we discuss the relevance of formality in this setting.

Finally, we show how our techniques work for three classes of spaces: toric complexes in §9, Kähler and quasi-Kähler manifolds in §10, and complements of hyperplane arrangements in §11. In each case, we explain what is known about the cohomology jump loci of those spaces, and use this knowledge to determine their straightness properties, and to compute some of their Ω-invariants.

§2. The Aomoto complex

We start by recalling the definition of the (universal) Aomoto complex associated to the cohomology ring of a space X. When X admits a minimal cell structure, this cochain complex can be read off from the equivariant cellular chain complex of the universal abelian cover, X^{ab}.

2.1. A cochain complex from the cohomology ring

Let X be a connected CW-complex with finite k-skeleton, for some $k \geq 1$. Consider the cohomology algebra $A = H^*(X, \mathbb{C})$, with product operation given by the cup product of cohomology classes. For each $a \in A^1$, we have $a^2 = 0$, by graded-commutativity of the cup product.

Definition 2.1. The *Aomoto complex* of A (with respect to $a \in A^1$) is the cochain complex of finite-dimensional, complex vector spaces,

$$(A, a)\colon\ A^0 \xrightarrow{a} A^1 \xrightarrow{a} A^2 \xrightarrow{a} \cdots, \tag{5}$$

with differentials given by left-multiplication by a.

Here is an alternative interpretation. Pick a basis $\{e_1, \dots, e_n\}$ for the complex vector space $A^1 = H^1(X, \mathbb{C})$, and let $\{x_1, \dots, x_n\}$ be the Kronecker dual basis for $A_1 = H_1(X, \mathbb{C})$. Identify the symmetric algebra $\mathrm{Sym}(A_1)$ with the polynomial ring $S = \mathbb{C}[x_1, \dots, x_n]$.

Definition 2.2. The *universal Aomoto complex* of A is the cochain complex of free S-modules,

$$\mathbf{A}\colon\ \cdots \longrightarrow A^i \otimes_{\mathbb{C}} S \xrightarrow{d^i} A^{i+1} \otimes_{\mathbb{C}} S \xrightarrow{d^{i+1}} A^{i+2} \otimes_{\mathbb{C}} S \longrightarrow \cdots, \tag{6}$$

where the differentials are defined by $d^i(u \otimes 1) = \sum_{j=1}^n e_j u \otimes x_j$ for $u \in A^i$, and then extended by S-linearity.

The fact that $\mathbf{A}$ is a cochain complex is verified as follows:

$$\begin{aligned} d^{i+1} d^i(u \otimes 1) &= \sum_{k=1}^n \sum_{j=1}^n e_k e_j u \otimes x_j x_k \\ &= \sum_{j<k} (e_k e_j + e_j e_k) u \otimes x_j x_k = 0. \end{aligned}$$

The relationship between the two definitions is given by the following well-known lemma.

Lemma 2.3. *The evaluation of the universal Aomoto complex* $\mathbf{A}$ *at an element* $a \in A^1$ *coincides with the Aomoto complex* (A, a).

Proof. Write $a = \sum_{j=1}^n a_j e_j \in A^1$, and let $\mathrm{ev}_a\colon S \to \mathbb{C}$ be the ring morphism given by $g \mapsto g(a_1, \dots, a_n)$. The resulting cochain complex,

$\mathbf{A}(a) = \mathbf{A} \otimes_S \mathbb{C}$, has differentials $d^i(a) := d^i \otimes \mathrm{id}_{\mathbb{C}}$ given by

$$d^i(a)(u) = \sum_{j=1}^{n} e_j u \otimes \mathrm{ev}_a(x_j) = \sum_{j=1}^{n} e_j u \cdot a_j = a \cdot u. \tag{7}$$

Thus, $\mathbf{A}(a) = (A, a)$. Q.E.D.

2.2. Minimality and the Aomoto complex

As shown by Papadima and Suciu in [28], the universal Aomoto complex of $H^*(X, \mathbb{C})$ is functorially determined by the equivariant chain complex of the universal abelian cover X^{ab}, provided X admits a minimal cell structure.

More precisely, suppose X is a connected, finite-type CW-complex. We say the CW-structure on X is *minimal* if the number of i-cells of X equals the Betti number $b_i(X)$, for every $i \geq 0$. Equivalently, the boundary maps in the cellular chain complex $C_\bullet(X, \mathbb{Z})$ are all zero maps. In particular, the homology groups $H_i(X, \mathbb{Z})$ are all torsion-free.

Theorem 2.4 ([28]). *Let X be a minimal CW-complex. Then the linearization of the cochain complex $C^\bullet(X^{\mathrm{ab}}, \mathbb{C})$ coincides with the universal Aomoto complex of $H^*(X, \mathbb{C})$.*

Let us explain in more detail how this theorem works. Pick an isomorphism $H_1(X, \mathbb{Z}) \cong \mathbb{Z}^n$, and identify $\mathbb{C}[\mathbb{Z}^n]$ with $\Lambda = \mathbb{C}[t_1^{\pm 1}, \dots, t_n^{\pm 1}]$. Next, filter Λ by powers of the maximal ideal $I = (t_1 - 1, \dots, t_n - 1)$, and identify the associated graded ring, $\mathrm{gr}(\Lambda)$, with the polynomial ring $S = \mathbb{C}[x_1, \dots, x_n]$, via the ring map $t_i - 1 \mapsto x_i$.

The minimality hypothesis allows us to identify $C_i(X^{\mathrm{ab}}, \mathbb{C})$ with $\Lambda \otimes_{\mathbb{C}} H_i(X, \mathbb{C})$ and $C^i(X^{\mathrm{ab}}, \mathbb{C})$ with $A^i \otimes_{\mathbb{C}} \Lambda$. Under these identifications, the boundary map $\partial_{i+1}^{\mathrm{ab}} \colon C_{i+1}(X^{\mathrm{ab}}, \mathbb{C}) \to C_i(X^{\mathrm{ab}}, \mathbb{C})$ dualizes to a map $\delta^i \colon A^i \otimes_{\mathbb{C}} \Lambda \to A^{i+1} \otimes_{\mathbb{C}} \Lambda$. Let $\mathrm{gr}(\delta^i) \colon A^i \otimes_{\mathbb{C}} S \to A^{i+1} \otimes_{\mathbb{C}} S$ be the associated graded of δ^i, and let $\mathrm{gr}(\delta^i)^{\mathrm{lin}}$ be its linear part. Theorem 2.4 then provides an identification

$$\mathrm{gr}(\delta^i)^{\mathrm{lin}} = d^i \colon A^i \otimes_{\mathbb{C}} S \to A^{i+1} \otimes_{\mathbb{C}} S, \tag{8}$$

for each $i \geq 0$.

Example 2.5. Let $X = T^n$ be the n-dimensional torus, with the standard product cell structure. Then X is a minimal cell complex, and $C_\bullet(X^{\mathrm{ab}}, \mathbb{C})$ is the Koszul complex $K(t_1 - 1, \dots, t_n - 1)$ over the ring Λ. The cohomology ring $H^*(T^n, \mathbb{C})$ is the exterior algebra E on variables $e_1, \dots, e_n$, and the universal Aomoto complex $E \otimes_{\mathbb{C}} S$ is the Koszul complex $K(x_1, \dots, x_n)$ over the ring S. In this case, Theorem 2.4

simply says that the substitution $t_i - 1 \mapsto x_i$ takes one Koszul complex to the other.

Example 2.6. Let $Y = S^1 \vee S^2$, and identify $\pi_1(Y) = \mathbb{Z}$, with generator t, and $\pi_2(Y) = \mathbb{Z}[t^{\pm 1}]$. Given a polynomial $f \in \mathbb{Z}[t]$, let $\varphi\colon S^2 \to Y$ be a map representing f, and attach a 3-cell to Y along φ to obtain a CW-complex X_f. For instance, if $f(t) = t - 1$, then $X_f \simeq S^1 \times S^2$. More generally, X_f is minimal if and only if $f(1) = 0$, in which case $H_*(X_f, \mathbb{Z}) \cong H_*(S^1 \times S^2, \mathbb{Z})$.

Now identify $\pi_1(X_f) = \mathbb{Z}$ and $\mathbb{C}[\mathbb{Z}]$ with $\Lambda = \mathbb{C}[t^{\pm 1}]$. The chain complex $C_\bullet(X_f^{\mathrm{ab}}, \mathbb{C})$ can then be written as

$$\Lambda \xrightarrow{f(t)} \Lambda \xrightarrow{0} \Lambda \xrightarrow{t-1} \Lambda \,. \tag{9}$$

Finally, identify $\mathrm{gr}(\Lambda)$ with $S = \mathbb{C}[x]$, and set $g(x) = f(1+x)$. Suppose $f(1) = 0$, so that X_f is minimal. Then, the linear term of $g(x)$ is $f'(1)\cdot x$, and so the universal Aomoto complex of $H^*(X_f, \mathbb{C})$ is

$$S \xrightarrow{x} S \xrightarrow{0} S \xrightarrow{f'(1)x} S \,. \tag{10}$$

§3. Resonance varieties

In this section, we review the definition and basic properties of the resonance varieties of a space X, which measure the deviation from exactness of the Aomoto complexes associated to the cohomology ring $H^*(X, \mathbb{C})$.

3.1. Jump loci for the Aomoto–Betti numbers

As usual, let X be a connected CW-complex with finite k-skeleton. Denote by A the cohomology algebra $H^*(X, \mathbb{C})$. Computing the homology of the Aomoto complexes (A, a) for various values of the parameter $a \in A^1$, and recording the resulting Betti numbers, carves out some very interesting subsets of the affine space $A^1 = \mathbb{C}^n$, where $n = b_1(X)$.

Definition 3.1. The *resonance varieties* of X are the sets

$$\mathcal{R}^i_d(X) = \{a \in A^1 \mid \dim_{\mathbb{C}} H^i(A, \cdot a) \geq d\},$$

defined for all integers $0 \leq i \leq k$ and $d > 0$.

The degree-1, depth-1 resonance variety is especially easy to describe: $\mathcal{R}^1_1(X)$ consists of those elements $a \in A^1$ for which there exists an element $b \in A^1$, not proportional to a, and such that $b \cdot a = 0$.

The terminology from Definition 3.1 is justified by the following well-known lemma. For completeness, we include a proof.

Lemma 3.2. *The sets $\mathcal{R}^i_d(X)$ are homogeneous algebraic subvarieties of the affine space $A^1 = H^1(X,\mathbb{C})$.*

Proof. By definition, an element $a \in A^1$ belongs to $\mathcal{R}^i_d(X)$ if and only if $\operatorname{rank}\delta^{i-1}(a)+\operatorname{rank}\delta^i(a) \le c_i - d$, where $c_i = c_i(X)$ is the number of i-cells of X. As a set, then, $\mathcal{R}^i_d(X)$ can be written as the intersection

$$\bigcap_{\substack{r+s=c_i-d+1\\ r,s\geq 0}} \{a \in A^1 \mid \operatorname{rank}\delta^{i-1}(a) \le r-1 \text{ or } \operatorname{rank}\delta^i(a)\le s-1\}.$$

Using this description, we may rewrite $\mathcal{R}^i_d(X)$ as the zero-set of a sum of products of determinantal ideals,

$$\mathcal{R}^i_d(X) = V\Bigg(\sum_{p+q=c_{i+1}+d-1} E_p(\delta^{i-1})\cdot E_q(\delta^i)\Bigg). \tag{11}$$

Clearly, $a \in \mathcal{R}^i_d(X)$ if and only if $\lambda a \subset \mathcal{R}^i_d(X)$, for all $\lambda \in \mathbb{C}^\times$. Thus, $\mathcal{R}^i_d(X)$ is a homogeneous variety. Q.E.D.

Sometimes it will be more convenient to consider the projectivization $\overline{\mathcal{R}}^i_d(X)$, viewed as a subvariety of $\mathbb{P}(A^1) = \mathbb{CP}^{n-1}$.

The resonance varieties $\mathcal{R}^i_d(X)$ are homotopy-type invariants of the space X. The following (folklore) result makes this more precise.

Lemma 3.3. *Suppose $X \simeq X'$. There is then a linear isomorphism $H^1(X',\mathbb{C}) \cong H^1(X,\mathbb{C})$ which restricts to isomorphisms $\mathcal{R}^i_d(X') \cong \mathcal{R}^i_d(X)$, for all $i \le k$ and $d > 0$.*

Proof. Let $f\colon X \to X'$ be a homotopy equivalence. The induced homomorphism in cohomology, $f^*\colon H^*(X',\mathbb{C}) \xrightarrow{\simeq} H^*(X,\mathbb{C})$, defines isomorphisms $(H^*(X',\mathbb{C}),a) \xrightarrow{\simeq} (H^*(X,\mathbb{C}), f^*(a))$ between the respective Aomoto complexes, for all $a \in H^1(X',\mathbb{C})$. Hence, $f^*\colon H^1(X',\mathbb{C}) \xrightarrow{\simeq} H^1(X,\mathbb{C})$ restricts to isomorphisms $\mathcal{R}^i_d(X') \xrightarrow{\simeq} \mathcal{R}^i_d(X)$. Q.E.D.

3.2. Discussion

In each degree $i \ge 0$, the resonance varieties provide a descending filtration,

$$H^1(X,\mathbb{C}) = \mathcal{R}^i_0(X) \supseteq \mathcal{R}^i_1(X) \supseteq \mathcal{R}^i_2(X) \supseteq \cdots. \tag{12}$$

Note that $0 \in \mathcal{R}^i_d(X)$, for $d \le b_i(X)$, but $\mathcal{R}^i_d(X) = \emptyset$, for $d > b_i(X)$. In degree 0, we have $\mathcal{R}^0_1(X) = \{0\}$, and $\mathcal{R}^0_d(X) = \emptyset$, for $d > 1$. In degree 1, the varieties $\mathcal{R}^1_d(X)$ depend only on the group $G = \pi_1(X, x_0)$—in fact, only on the cup-product map $\cup\colon H^1(G,\mathbb{C}) \wedge H^1(G,\mathbb{C}) \to H^2(G,\mathbb{C})$. Accordingly, we will sometimes write $\mathcal{R}^1_d(G)$ for $\mathcal{R}^1_d(X)$.

Example 3.4. Let T^n be the n-dimensional torus. Using the exactness of the Koszul complex from Example 2.5, we see that $\mathcal{R}^i_d(T^n)$ equals $\{0\}$ if $d \leq \binom{n}{i}$, and is empty, otherwise.

Example 3.5. Let S_g be the compact, connected, orientable surface of genus $g > 1$. With suitable identifications $H^1(S_g, \mathbb{C}) = \mathbb{C}^{2g}$ and $H^2(S_g, \mathbb{C}) = \mathbb{C}$, the cup-product map $\cup\colon H^1(S_g, \mathbb{C}) \wedge H^1(S_g, \mathbb{C}) \to H^2(S_g, \mathbb{C})$ is the standard symplectic form. A computation then shows

$$\mathcal{R}^i_d(S_g) = \begin{cases} \mathbb{C}^{2g} & \text{if } i = 1,\ d < 2g-1, \\ \{0\} & \text{if } i = 1,\ d \in \{2g-1, 2g\}, \text{ or } i \in \{0,2\},\ d = 1, \\ \emptyset & \text{otherwise.} \end{cases} \tag{13}$$

One may extend the definition of resonance varieties to arbitrary fields $\Bbbk$, provided $H_1(X, \mathbb{Z})$ is torsion-free, if $\operatorname{char} \Bbbk = 2$. The resulting varieties, $\mathcal{R}^i_d(X, \Bbbk)$, behave well under field extensions: if $\Bbbk \subseteq \mathbb{K}$, then $\mathcal{R}^i_d(X, \Bbbk) = \mathcal{R}^i_d(X, \mathbb{K}) \cap H^1(X, \Bbbk)$. In particular, $\mathcal{R}^i_d(X, \mathbb{Q})$ is just the set of rational points on the integrally defined variety $\mathcal{R}^i_d(X) = \mathcal{R}^i_d(X, \mathbb{C})$.

3.3. Depth one resonance varieties

Most important for us are the depth-1 resonance varieties, $\mathcal{R}^i_1(X)$, and their unions up to a fixed degree, $\mathcal{R}^i(X) = \bigcup_{j=0}^{i} \mathcal{R}^j_1(X)$. The latter varieties can be written as

$$\mathcal{R}^i(X) = \{a \in A^1 \mid H^j(A, \cdot a) \neq 0, \text{ for some } j \leq i\}. \tag{14}$$

These sets provide an ascending filtration of the first cohomology group,

$$\{0\} = \mathcal{R}^0(X) \subseteq \mathcal{R}^1(X) \subseteq \cdots \subseteq \mathcal{R}^k(X) \subseteq H^1(X, \mathbb{C}) = \mathbb{C}^n. \tag{15}$$

For low values of $n = b_1(X)$, the variety $\mathcal{R}^1(X)$ is easy to describe.

Proposition 3.6. *If $n \leq 1$, then $\mathcal{R}^1(X) = \{0\}$. If $n = 2$, then $\mathcal{R}^1(X) = \mathbb{C}^2$ or $\{0\}$, according to whether the cup product vanishes on $H^1(X, \mathbb{C})$ or not.*

For $n \geq 3$, the resonance variety $\mathcal{R}^1(X)$ can be much more complicated; in particular, it may have irreducible components which are not linear subspaces. The following example (a particular case of a more general construction described in [11]) illustrates this phenomenon.

Example 3.7. Let $X = F(T^2, 3)$ be the configuration space of 3 labeled points on the torus. The cohomology ring of X is the exterior algebra on generators $a_1, a_2, a_3, b_1, b_2, b_3$ in degree 1, modulo the ideal

$\langle (a_1 - a_2)(b_1 - b_2), (a_1 - a_3)(b_1 - b_3), (a_2 - a_3)(b_2 - b_3)\rangle$. A calculation reveals that

$$\mathcal{R}^1(X) = \{(a, b) \in \mathbb{C}^6 \mid a_1 + a_2 + a_3 = b_1 + b_2 + b_3 = a_1 b_2 - a_2 b_1 = 0\}.$$

Hence, $\mathcal{R}^1(X)$ is isomorphic to $Q = V(a_1 b_2 - a_2 b_1)$, a smooth quadric hypersurface in $\mathbb{C}^4$. (The projectivization of Q is the image of the Segré embedding of $\mathbb{CP}^1 \times \mathbb{CP}^1$ in $\mathbb{CP}^3$.)

The depth-1 resonance varieties of a product or a wedge of two spaces can be expressed in terms of the resonance varieties of the factors. Start with a product $X = X_1 \times X_2$, where both X_1 and X_2 have finite k-skeleton, and identify $H^1(X, \mathbb{C}) = H^1(X_1, \mathbb{C}) \times H^1(X_2, \mathbb{C})$.

Proposition 3.8 ([29]). *For all $i \leq k$,*

$$\mathcal{R}^i_1(X_1 \times X_2) = \bigcup_{p+q=i} \mathcal{R}^p_1(X_1) \times \mathcal{R}^q_1(X_2).$$

Corollary 3.9. *We have $\mathcal{R}^i(X_1 \times X_2) = \bigcup_{p+q=i} \mathcal{R}^p(X_1) \times \mathcal{R}^q(X_2)$, for all $i \leq k$.*

Example 3.10. For a Riemann surface of genus $g > 1$, formula (13) yields $\mathcal{R}^0(S_g) = \{0\}$ and $\mathcal{R}^i(S_g) = \mathbb{C}^{2g}$, for all $i \geq 1$. For a product of two such surfaces, Corollary 3.9 now gives

$$\mathcal{R}^i(S_g \times S_h) = \begin{cases} \{0\} & \text{if } i = 0, \\ \mathbb{C}^{2g} \times \{0\} \cup \{0\} \times \mathbb{C}^{2h} & \text{if } i = 1, \\ \mathbb{C}^{2(g+h)} & \text{if } i \geq 2. \end{cases} \tag{16}$$

Next, consider a wedge $X = X_1 \vee X_2$, and identify $H^1(X, \mathbb{C}) = H^1(X_1, \mathbb{C}) \times H^1(X_2, \mathbb{C})$.

Proposition 3.11 ([29]). *Suppose X_1 and X_2 have positive first Betti numbers. Then*

$$\mathcal{R}^i_1(X) = \begin{cases} H^1(X, \mathbb{C}) & \text{if } i = 1, \\ \mathcal{R}^i_1(X_1) \times H^1(X_2, \mathbb{C}) \cup H^1(X_1, \mathbb{C}) \times \mathcal{R}^i_1(X_2) & \text{if } 1 < i \leq k. \end{cases}$$

Corollary 3.12. *Let $X = X_1 \vee X_2$, where X_1 and X_2 have positive first Betti numbers. Then $\mathcal{R}^i(X) = H^1(X, \mathbb{C})$, for all $1 \leq i \leq k$.*

§4. Tangent cones to affine varieties

We now discuss two versions of the tangent cone to a subvariety of the complex algebraic torus $(\mathbb{C}^\times)^n$.

4.1. The tangent cone

We start by reviewing a well-known notion in algebraic geometry (see [20, pp. 251–256]). Let $W \subset (\mathbb{C}^\times)^n$ be a Zariski closed subset, defined by an ideal I in the Laurent polynomial ring $\Lambda = \mathbb{C}[t_1^{\pm 1}, \dots, t_n^{\pm 1}]$. Picking a finite generating set for I, and multiplying these generators with suitable monomials if necessary, we see that W may also be defined by the ideal $I \cap R$ in the polynomial ring $R = \mathbb{C}[t_1, \dots, t_n]$.

Now consider the polynomial ring $S = \mathbb{C}[z_1, \dots, z_n]$, and let J be the ideal generated by the polynomials

$$g(z_1, \dots, z_n) = f(z_1 + 1, \dots, z_n + 1), \tag{17}$$

for all $f \in I \cap R$. Finally, let $\operatorname{in}(J)$ be the ideal in S generated by the initial forms of all non-zero elements from J.

Definition 4.1. The *tangent cone* of W at 1 is the algebraic subset $\mathrm{TC}_1(W) \subset \mathbb{C}^n$ defined by the initial ideal $\operatorname{in}(J) \subset S$.

The tangent cone $\mathrm{TC}_1(W)$ is a homogeneous subvariety of $\mathbb{C}^n$, depending only on the analytic germ of W at the identity. In particular, $\mathrm{TC}_1(W) \neq \emptyset$ if and only if $1 \in W$. Moreover, TC_1 commutes with finite unions, but not necessarily with intersections. Explicit equations for the tangent cone to a variety can be found using the Gröbner basis algorithm described in [14, Proposition 15.28].

4.2. The exponential tangent cone

A competing notion of tangent cone was introduced by Dimca, Papadima and Suciu in [11], and further studied in [29] and [34].

Definition 4.2. The *exponential tangent cone* of W at 1 is the homogeneous subvariety $\tau_1(W)$ of $\mathbb{C}^n$, defined by

$$\tau_1(W) = \{z \in \mathbb{C}^n \mid \exp(\lambda z) \in W, \text{ for all } \lambda \in \mathbb{C}\}.$$

The exponential tangent cone $\tau_1(W)$ depends only on the analytic germ of W at the identity. In particular, $\tau_1(W) \neq \emptyset$ if and only if $1 \in W$. Moreover, τ_1 commutes with finite unions, as well as arbitrary intersections. The most important property of this construction is given in the following result from [11] (see [34] for full details).

Theorem 4.3 ([11]). *The exponential tangent cone $\tau_1(W)$ is a finite union of rationally defined linear subspaces of $\mathbb{C}^n$.*

These subspaces can be described explicitly. By the above remarks, we may assume $W = V(f)$, where f is a non-zero Laurent polynomial with $f(1) = 0$. Write $f = \sum_{a \in S} c_a t_1^{a_1} \cdots t_n^{a_n}$, where S is a finite subset of

$\mathbb{Z}^n$, and $c_a \in \mathbb{C} \setminus \{0\}$ for each $a = (a_1, \dots, a_n) \in S$. We say a partition $\mathsf{p} = (\mathsf{p}_1 \mid \cdots \mid \mathsf{p}_q)$ of the support S is *admissible* if $\sum_{a \in \mathsf{p}_i} c_a = 0$, for all $1 \le i \le q$. For each such partition, let $L(\mathsf{p})$ be the rational linear subspace consisting of all points $x \in \mathbb{Q}^n$ for which the dot product $(a - b) \cdot x$ vanishes, for all $a, b \in \mathsf{p}_i$ and all $1 \le i \le q$. Then

$$\tau_1(W) = \bigcup_{\mathsf{p}} L(\mathsf{p}) \otimes \mathbb{C}, \tag{18}$$

where the union is taken over all (maximal) admissible partitions of S.

4.3. Relating the two tangent cones

The next lemma (first noted in [11]) records a general relationship between the two kinds of tangent cones discussed here. For completeness, we include a proof, along the lines of the proof given in [29], though slightly modified to fit the definition of TC_1 given here.

Lemma 4.4 ([11, 29]). *For every Zariski closed subset $W \subset (\mathbb{C}^\times)^n$, we have $\tau_1(W) \subseteq \mathrm{TC}_1(W)$.*

Proof. Without loss of generality, we may assume $1 \in W$. Let $f \in R$ be a non-zero polynomial in $I = I(W)$, let $g \in S$ be the polynomial defined by (17), and let $g_0 = \mathrm{in}(g)$. We then have

$$\begin{aligned} f(e^{\lambda z_1}, \dots, e^{\lambda z_n}) &= f(1 + \lambda z_1 + O(\lambda^2), \dots, 1 + \lambda z_n + O(\lambda^2)) \\ &= g(\lambda z_1 + O(\lambda^2), \dots, \lambda z_n + O(\lambda^2)) \\ &= g_0(\lambda z_1, \dots, \lambda z_n) + \text{higher order terms.} \end{aligned}$$

Now suppose $z \in \tau_1(W)$, that it to say, $f(e^{\lambda z_1}, \dots, e^{\lambda z_n}) = 0$, for all $\lambda \in \mathbb{C}$. The above calculation shows that $g_0(\lambda z) = 0$, for all λ; in particular, $g_0(z) = 0$. We conclude that $z \in \mathrm{TC}_1(W)$. Q.E.D.

If W is an algebraic subtorus of $(\mathbb{C}^\times)^n$, then $\tau_1(W) = \mathrm{TC}_1(W)$, and both coincide with the tangent space at the origin, $T_1(W)$. In general, though, the inclusion from Lemma 4.4 can be strict, even when 1 is a smooth point of W. Here is an example illustrating this phenomenon.

Example 4.5. Let W be the hypersurface in $(\mathbb{C}^\times)^3$ with equation $t_1 + t_2 + t_3 - t_1 t_2 - t_1 t_3 - t_2 t_3 = 0$. Then $\tau_1(W)$ is a union of 3 lines in $\mathbb{C}^3$, given by the equations $z_1 = z_2 + z_3 = 0$, $z_2 = z_1 + z_3 = 0$, and $z_3 = z_1 + z_2 = 0$. On the other hand, $\mathrm{TC}_1(W)$ is a plane in $\mathbb{C}^3$, with equation $z_1 + z_2 + z_3 = 0$. Hence, $\mathrm{TC}_1(W)$ strictly contains $\tau_1(W)$.

§5. Characteristic varieties

In this section, we briefly review the characteristic varieties of a space, and their relation to the resonance varieties.

5.1. Jump loci for twisted homology

Let X be a connected CW-complex with finite k-skeleton, $k \geq 1$. Without loss of generality, we may assume X has a single 0-cell, call it x_0. Let $G = \pi_1(X, x_0)$ be the fundamental group of X, and let $\widehat{G} = \operatorname{Hom}(G, \mathbb{C}^\times)$ be the group of complex characters of G. Clearly, $\widehat{G} = \widehat{G}_{\mathrm{ab}}$, where $G_{\mathrm{ab}} = H_1(X, \mathbb{Z})$ is the abelianization of G. Thus, the universal coefficient theorem allows us to identify

$$\widehat{G} = H^1(X, \mathbb{C}^\times). \tag{19}$$

Each character $\rho\colon G \to \mathbb{C}^\times$ determines a rank 1 local system $\mathcal{L}_\rho$ on our space X. Computing the homology groups of X with coefficients in such local systems leads to a natural filtration of the character group.

Definition 5.1. The *characteristic varieties* of X are the sets

$$\mathcal{V}^i_d(X) = \{\, \rho \in H^1(X, \mathbb{C}^\times) \mid \dim_{\mathbb{C}} H_i(X, \mathcal{L}_\rho) \geq d \,\},$$

defined for all $0 \leq i \leq k$ and all $d > 0$.

Clearly, $1 \in \mathcal{V}^i_d(X)$ if and only if $d \leq b_i(X)$. In degree 0, we have $\mathcal{V}^0_1(X) = \{1\}$ and $\mathcal{V}^0_d(X) = \emptyset$, for $d > 1$. In degree 1, the sets $\mathcal{V}^i_d(X)$ depend only on the group $G = \pi_1(X, x_0)$—in fact, only on its maximal metabelian quotient, G/G''.

The jump loci $\mathcal{V}^i_d(X)$ are Zariski closed subsets of the algebraic group $H^1(X, \mathbb{C}^\times)$; moreover, these varieties are homotopy-type invariants of X. For details and further references, see [34].

5.2. Depth one characteristic varieties

Most important for us are the depth one characteristic varieties, $\mathcal{V}^i_1(X)$, and their unions up to a fixed degree, $\mathcal{V}^i(X) = \bigcup_{j=0}^i \mathcal{V}^j_1(X)$. These varieties yield an ascending filtration of the character group,

$$\{1\} = \mathcal{V}^0(X) \subseteq \mathcal{V}^1(X) \subseteq \cdots \subseteq \mathcal{V}^k(X) \subseteq H^1(X, \mathbb{C}^\times). \tag{20}$$

Let $\widehat{G}^0 = H^1(X, \mathbb{C}^\times)^0$ be the identity component of the character group $\widehat{G}$. Writing $n = b_1(X)$, we may identify $\widehat{G}^0$ with the complex algebraic torus $(\mathbb{C}^\times)^n$. Set $\mathcal{W}^i_d(X) = \mathcal{V}^i_d(X) \cap \widehat{G}^0$, and $\mathcal{W}^i(X) = \bigcup_{j=0}^i \mathcal{W}^j_1(X)$. We then have

$$\mathcal{W}^i(X) = \mathcal{V}^i(X) \cap \widehat{G}^0. \tag{21}$$

Example 5.2. Let L be an n-component link in S^3, with complement X. Using a basis for $H_1(X,\mathbb{Z})=\mathbb{Z}^n$ given by (oriented) meridians, we may identify $H^1(X,\mathbb{C}^\times)=(\mathbb{C}^\times)^n$. Then

$$\mathcal{W}^1(X)=\{z\in(\mathbb{C}^\times)^n \mid \Delta_L(z)=0\}\cup\{1\}, \tag{22}$$

where $\Delta_L=\Delta_L(t_1,\dots,t_n)$ is the Alexander polynomial of the link. For details and references, see [33].

As shown in [29] (see also [34]), the characteristic varieties satisfy product and wedge formulas similar to those satisfied by the resonance varieties.

Proposition 5.3. *Let X_1 and X_2 be connected CW-complexes with finite k skeleton, and fix an integer $1\le i\le k$. Then*

(1) $\mathcal{W}^i(X_1\times X_2)=\bigcup_{p+q=i}\mathcal{W}^p(X_1)\times\mathcal{W}^q(X_2)$.

(2) *Suppose, moreover, that $b_1(X_1)>0$ and $b_1(X_2)>0$. Then* $\mathcal{W}^i(X_1\vee X_2)=H^1(X_1\vee X_2,\mathbb{C}^\times)^0$.

5.3. Characteristic subspace arrangements

As before, let X be a connected CW-complex with finite k-skeleton. Set $n=b_1(X)$, and identify $H^1(X,\mathbb{C})=\mathbb{C}^n$ and $H^1(X,\mathbb{C}^\times)^0=(\mathbb{C}^\times)^n$. Applying Theorem 4.3 to the characteristic varieties $\mathcal{W}^i(X)\subseteq(\mathbb{C}^\times)^n$ leads to the following definition.

Definition 5.4. For each $i\le k$, the *i-th characteristic arrangement* of X, denoted $\mathcal{C}_i(X)$, is the subspace arrangement in $H^1(X,\mathbb{Q})$ whose complexified union is the exponential tangent cone to $\mathcal{W}^i(X)$:

$$\tau_1(\mathcal{W}^i(X))=\bigcup_{L\in\mathcal{C}_i(X)} L\otimes\mathbb{C}.$$

We thus have a sequence $\mathcal{C}_0(X),\mathcal{C}_1(X),\dots,\mathcal{C}_k(X)$ of rational subspace arrangements, all lying in the same affine space $H^1(X,\mathbb{Q})=\mathbb{Q}^n$. As noted in [34, Lemma 6.7], these subspace arrangements depend only on the homotopy type of X.

5.4. Tangent cone and resonance

Of great importance in the theory of cohomology jumping loci is the relationship between characteristic and resonance varieties, based on the tangent cone construction. A foundational result in this direction is the following theorem of Libgober [21].

Theorem 5.5 ([21]). *The tangent cone at 1 to $\mathcal{W}^i_d(X)$ is included in $\mathcal{R}^i_d(X)$, for all $i\le k$ and $d>0$.*

Since tangent cones commute with finite unions, we get:

$$\mathrm{TC}_1(\mathcal{W}^i(X)) \subseteq \mathcal{R}^i(X), \quad \text{for all } i \leq k. \tag{23}$$

For many spaces of interest, inclusion (23) holds as an equality. In general, though, the inclusion is strict.

Example 5.6. Let M be the 3-dimensional Heisenberg nilmanifold. It is easily seen that $\mathcal{W}^1(M) = \{1\}$; hence, $\mathrm{TC}_1(\mathcal{W}^1(M)) = \{0\}$. On the other hand, $\mathcal{R}^1(M) = \mathbb{C}^2$, since the cup product vanishes on $H^1(M, \mathbb{C})$.

§6. Straight spaces

We now delineate a class of spaces for which the characteristic and resonance varieties have a simple nature, and are intimately related to each other via the tangent cone constructions discussed in §4.

6.1. Straightness

As before, let X be a connected CW-complex with finite k-skeleton. For each $i \leq k$, consider the following conditions on the varieties $\mathcal{W}^i(X)$ and $\mathcal{R}^i(X)$:

(a) All components of $\mathcal{W}^i(X)$ passing through the origin are algebraic subtori.
(b) $\mathrm{TC}_1(\mathcal{W}^i(X)) = \mathcal{R}^i(X)$.
(c) All components of $\mathcal{W}^i(X)$ not passing through the origin are 0-dimensional.

Definition 6.1. We say X is *locally k-straight* if conditions (a) and (b) hold, for each $i \leq k$. If these conditions hold for all $k \geq 1$, we say X is a *locally straight space.*

Definition 6.2. We say X is *k-straight* if conditions (a), (b), and (c) hold, for each $i \leq k$. If these conditions hold for all $k \geq 1$, we say X is a *straight space.*

Clearly, the k-straightness property depends only on the homotopy type of a given space. In view of this observation, we may declare a group G to be k-straight if there is a classifying space $K(G, 1)$ which is k-straight; in particular, such a group G must be of type F_k, i.e., have a $K(G, 1)$ with finite k-skeleton. Note that a space is 1-straight if and only if its fundamental group is 1-straight. Similar considerations apply to local straightness.

The straightness conditions are quite stringent, and thus easily violated. Here are a few examples when this happens, all for $k = 1$.

Example 6.3. The closed three-link chain L (the link 6^3_1 from Rolfsen's tables) has Alexander polynomial $\Delta_L = t_1+t_2+t_3-t_1t_2-t_1t_3-t_2t_3$. Let $X = S^3 \setminus L$ be the link complement. By (22), the characteristic variety $W = \mathcal{W}^1(X)$ has equation $\Delta_L = 0$. On the other hand, we know from Example 4.5 that $\tau_1(W) \neq \mathrm{TC}_1(W)$. Thus, W is not an algebraic torus, and so X is not locally 1-straight: condition (a) fails.

Example 6.4. The Heisenberg nilmanifold M from Example 5.6 is not locally 1-straight: condition (a) is met, but not condition (b).

Example 6.5. Let $G = \langle x_1, x_2 \mid x_1^2x_2 = x_2x_1^2\rangle$ be the group from [29, Example 6.4]. Then $\mathcal{W}^1(G) = \{1\} \cup \{(t_1,t_2) \in (\mathbb{C}^\times)^2 \mid t_1 = -1\}$, and $\mathcal{R}^1(G) = \{0\}$. Thus, G is locally 1-straight, but not 1-straight: conditions (a) and (b) are met, but not condition (c).

Nevertheless, as we will see below, there is an abundance of interesting spaces which are, to a degree on another, straight.

6.2. Examples and discussion

To start with, let us consider the case when the first characteristic variety $\mathcal{W}^1(X)$ is as big as it can be.

Lemma 6.6. *Let X be a finite-type CW-complex. Suppose that $\mathcal{W}^1(X) = H^1(X, \mathbb{C}^\times)^0$. Then X is straight.*

Proof. Our assumption forces $\mathcal{W}^i(X) = H^1(X, \mathbb{C}^\times)^0$, for all $i \geq 1$. Straightness conditions (a) and (c) are automatically met, while condition (b) follows from (23). Q.E.D.

Examples of spaces satisfying the hypothesis of the above lemma include the Riemann surfaces S_g with $g \geq 2$.

Lemma 6.7. *Let X be a CW-complex with finite k-skeleton, $k \geq 1$. Suppose that $\mathcal{W}^k(X)$ is finite, and $\mathcal{R}^k(X) = \{0\}$. Then X is k-straight.*

Proof. Since $\mathcal{W}^k(X)$ is finite, conditions (a) and (c) are trivially satisfied. By formula (23), we have $\mathrm{TC}_1(\mathcal{W}^i(X)) \subseteq \mathcal{R}^i(X) = \{0\}$, for all $i \leq k$. Since $1 \in \mathcal{W}^i(X)$, equality must hold, i.e., condition (b) is satisfied. Q.E.D.

Examples of spaces satisfying the hypothesis of the above lemma include the tori T^n; these spaces are, in fact, straight.

Evidently, all spaces X with $b_1(X) = 0$ are straight. Next, we deal with the case $b_1(X) = 1$. The following example shows that not all finite CW-complexes with first Betti number 1 are (locally) straight.

Example 6.8. Let f be a polynomial in $\mathbb{Z}[t]$ with $f(1) = 0$, and let $X_f = (S^1 \vee S^2) \cup_\varphi e^3$ be the corresponding minimal CW-complex constructed in Example 2.6. A calculation with the equivariant chain complex (9) shows that $\mathcal{W}^1(X_f) = \{1\}$ and $\mathcal{W}^2(X_f) = V(f)$; clearly, both these sets are finite subsets of $H^1(X, \mathbb{C}^\times) = \mathbb{C}^\times$.

An analogous calculation with the universal Aomoto cochain complex (10) shows that $\mathcal{R}^1(X_f) = \{0\}$, yet $\mathcal{R}^2(X_f) = \{0\}$ only if $f'(1) \neq 0$, but $\mathcal{R}^2(X_f) = \mathbb{C}$, otherwise. Therefore, X_f is always 1-straight, but

$$X_f \text{ is locally 2-straight} \iff f'(1) \neq 0. \tag{24}$$

Applying a similar reasoning to cell complexes of the form $X_f = (S^1 \vee S^k) \cup_\varphi e^{k+1}$, with attaching maps corresponding to polynomials $f \in \mathbb{Z}[t]$ with $f(1) = f'(1) = 0$, we obtain the following result.

Proposition 6.9. *For each $k \geq 2$, there is a minimal CW-complex which has the integral homology of $S^1 \times S^k$ and which is $(k-1)$-straight, but not locally k-straight.*

Nevertheless, we have the following positive result, guaranteeing straightness in a certain range for spaces with first Betti number 1.

Proposition 6.10. *Let X be a CW-complex with finite k-skeleton. Assume $b_1(X) = 1$. Then,*

(1) *X is 1-straight.*
(2) *If, moreover, $b_i(X) = 0$ for $1 < i \leq k$, then X is k-straight.*

Proof. As noted in Proposition 3.6, the fact that $b_1(X) = 1$ implies $\mathcal{R}^1(X) = \{0\}$. Identify $H^1(X, \mathbb{C}^\times) = \mathbb{C}^\times$. By formula (23), we have $\mathrm{TC}_1(\mathcal{W}^1(X)) = \{0\}$. Hence, $\mathcal{W}^1(X)$ is a proper subvariety of $\mathbb{C}^\times$; consequently, it is a finite set. Part (1) now follows from Lemma 6.7.

For Part (2), note that $\mathcal{R}^i(X) = \{0\}$, for all $i \leq k$. As above, we conclude that $\mathcal{W}^k(X)$ is finite, and hence X is k-straight. Q.E.D.

Corollary 6.11. *Let K be a smoothly embedded d-sphere in S^{d+2}, and let $X = S^{d+2} \setminus K$ be its complement. Then*

(1) *$\mathcal{R}^k(X) = \{0\}$ and $\mathcal{W}^k(X)$ is finite, for all $k \geq 1$.*
(2) *X is straight.*

Proof. The knot complement has the homotopy type of a $(d+1)$-dimensional CW-complex, with the integral homology of S^1. The desired conclusions follow from Proposition 6.10 and its proof. Q.E.D.

For a knot K in S^3, the variety $\mathcal{W}^1(X) \subset \mathbb{C}^\times$ consists of 1, together with all the roots of the Alexander polynomial, Δ_K (these roots are always different from 1).

6.3. Products and wedges

We now look at how the straightness properties behave with respect to (finite) products and wedges of spaces.

Proposition 6.12. *Let X_1 and X_2 be CW-complexes with finite k-skeleta and positive first Betti numbers. Then the space $X = X_1 \vee X_2$ is k-straight.*

Proof. Follows from Proposition 5.3(2) and Lemma 6.6. Q.E.D.

In particular, a wedge of knot complements is straight.

Proposition 6.13. *Let X_1 and X_2 be two CW-complexes.*

(1) *If X_1 and X_2 are locally k-straight, then so is $X_1 \times X_2$.*
(2) *If X_1 and X_2 are 1-straight, then so is $X_1 \times X_2$.*

Proof. Both assertions follow from the product formulas for resonance and characteristic varieties (Proposition 5.3(1) and Corollary 3.9, respectively). Q.E.D.

Proposition 6.14. *Let X_1 and X_2 be CW-complexes with finite k-skeleton. Suppose $\mathcal{R}^k(X_j) = \{0\}$ and $\mathcal{W}^k(X_j)$ is finite, for $j = 1, 2$. Then $X_1 \times X_2$ is k-straight.*

Proof. From Lemma 6.7, we know that both X_1 and X_2 are k-straight. Now, Corollary 3.9 gives that $\mathcal{R}^k(X_1 \times X_2) = \{0\}$, while Proposition 5.3(1) gives that $\mathcal{W}^k(X_1 \times X_2)$ is finite. Hence, again by Lemma 6.7, $X_1 \times X_2$ is k-straight. Q.E.D.

In particular, a product of knot complements is straight. In general, though, a product of straight spaces need not be straight.

Example 6.15. Let K be a knot in S^3, with Alexander polynomial Δ_K not equal to 1 (for instance, the trefoil knot), and let X be its complement. Let $Y = \bigvee^n S^1$ be a wedge of n circles, with $n \geq 2$. By Corollary 6.11 and Proposition 6.12, both X and Y are straight.

By Proposition 6.13(2), the product $X \times Y$ is 1-straight. Nevertheless, $X \times Y$ is not 2-straight. Indeed, the variety $\mathcal{W}^2(X \times Y)$ has irreducible components of the form $\{\rho\} \times (\mathbb{C}^\times)^n$, where ρ runs through the roots of Δ_K. As these components do not pass through the origin, straightness condition (c) fails for $X \times Y$ in degree $i = 2$.

6.4. Rationality properties

Before proceeding, let us give an alternative characterization of straightness. As usual, let X be a connected CW-complex with finite k-skeleton, and let $\exp\colon H^1(X, \mathbb{C}) \to H^1(X, \mathbb{C}^\times)^0$ be the exponential map (or rather, its corestriction to its image).

Theorem 6.16. *A space X as above is locally k-straight if and only if the following equalities hold, for all $i \leq k$:*

$$(\alpha) \qquad \mathcal{W}^i(X) = \Bigg(\bigcup_{L \in \mathcal{C}_i(X)} \exp(L \otimes \mathbb{C})\Bigg) \cup Z_i$$

$$(\beta) \qquad \mathcal{R}^i(X) = \bigcup_{L \in \mathcal{C}_i(X)} L \otimes \mathbb{C},$$

for some algebraic subsets $Z_i \subset H^1(X, \mathbb{C}^\times)^0$ not containing the origin.

The space X is k-straight if and only if, in addition, the sets Z_i are finite.

Proof. Suppose X is locally k-straight, and fix an index $i \leq k$. By hypothesis (a) from Definition 6.2, the variety $\mathcal{W}^i(X)$ admits a decomposition into irreducible components of the form

$$(25) \qquad \mathcal{W}^i(X) = \bigcup_{\alpha \in \Lambda} T_\alpha \cup Z_i,$$

for some algebraic subtori $T_\alpha = \exp(P_\alpha \otimes \mathbb{C})$ and some algebraic sets Z_i with $1 \notin Z_i$. Hence,

$$(26) \qquad \tau_1(\mathcal{W}^i(X)) = \bigcup_{\alpha \in \Lambda} P_\alpha \otimes \mathbb{C}.$$

On the other hand, by the definition of characteristic subspace arrangements, we also have $\tau_1(\mathcal{W}^i(X)) = \bigcup_{L \in \mathcal{C}_i(X)} L \otimes \mathbb{C}$. By uniqueness of decomposition into irreducible components, the arrangement $\{P_\alpha\}_{\alpha \in \Lambda}$ must coincide with $\mathcal{C}_i(X)$. Consequently, equation (25) yields (α).

Equation (25) also implies $\tau_1(\mathcal{W}^i(X)) = \mathrm{TC}_1(\mathcal{W}^i(X))$. Hypothesis (b) then gives $\tau_1(\mathcal{W}^i(X)) = \mathcal{R}^i(X)$, which is precisely condition (β).

Conversely, condition (α) implies (a) and conditions (α) and (β) together imply (b).

Finally, hypothesis (c) is satisfied if and only if the sets $Z_1, \dots, Z_k$ are all finite. Q.E.D.

Corollary 6.17. *Let X be a locally k-straight space. Then, for all $i \leq k$,*

(1) $\tau_1(\mathcal{W}^i(X)) = \mathrm{TC}_1(\mathcal{W}^i(X)) = \mathcal{R}^i(X)$.
(2) $\mathcal{R}^i(X, \mathbb{Q}) = \bigcup_{L \in \mathcal{C}_i(X)} L$.

In particular, the resonance varieties $\mathcal{R}^i(X)$ are unions of rationally defined subspaces.

The next example (adapted from [11]) illustrates how this rationality property may be used to detect non-straightness.

Example 6.18. Consider the group G with generators x_1, x_2, x_3, x_4 and relators $r_1 = [x_1, x_2]$, $r_2 = [x_1, x_4][x_2^{-2}, x_3]$, $r_3 = [x_1^{-1}, x_3][x_2, x_4]$. Direct computation shows that

$$\mathcal{R}^1(G) = \{z \in \mathbb{C}^4 \mid z_1^2 - 2z_2^2 = 0\}.$$

Evidently, this variety splits into two linear subspaces defined over $\mathbb{R}$, but not over $\mathbb{Q}$. Thus, G is not (locally) 1-straight.

§7. The Dwyer–Fried invariants

In this section, we recall the definition of the Dwyer–Fried sets, and the way these sets relate to the characteristic varieties of a space.

7.1. Betti numbers of free abelian covers

As before, let X be a connected CW-complex with finite k-skeleton, and let $G = \pi_1(X, x_0)$. Denote by $n = b_1(X)$ the first Betti number of X. (We may as well assume $n > 0$, otherwise the whole theory is empty of content.) Fix an integer r between 1 and n, and consider the regular covers of X, with group of deck-transformations $\mathbb{Z}^r$.

Each such cover, $X^\nu \to X$, is determined by an epimorphism $\nu\colon G \twoheadrightarrow \mathbb{Z}^r$. The induced homomorphism in cohomology, $\nu^*\colon H^1(\mathbb{Z}^r, \mathbb{Q}) \hookrightarrow H^1(G, \mathbb{Q})$, defines an r-dimensional subspace, $P_\nu = \operatorname{im}(\nu^*)$, in the rational vector space $H^1(G, \mathbb{Q}) = \mathbb{Q}^n$. Conversely, each r-dimensional subspace $P \subset \mathbb{Q}^n$ can be written as $P = P_\nu$, for some epimorphism $\nu\colon G \twoheadrightarrow \mathbb{Z}^r$, and thus defines a regular $\mathbb{Z}^r$-cover of X.

To recap, the regular $\mathbb{Z}^r$-covers of X are parametrized by the Grassmannian of r-planes in $H^1(X, \mathbb{Q})$, via the correspondence

$$\big\{\mathbb{Z}^r\text{-covers } X^\nu \to X\big\} \longleftrightarrow \big\{r\text{-planes } P_\nu := \operatorname{im}(\nu^*) \text{ in } H^1(X, \mathbb{Q})\big\}.$$

Moving about the rational Grassmannian and recording how the Betti numbers of the corresponding covers vary leads to the following definition.

Definition 7.1. The *Dwyer–Fried invariants* of X are the subsets

$$\Omega_r^i(X) = \big\{P_\nu \in \operatorname{Gr}_r(H^1(X, \mathbb{Q})) \bigm| b_j(X^\nu) < \infty \text{ for } j \le i\big\}. \tag{27}$$

Set $n = b_1(X)$. For a fixed r between 1 and n, these sets form a descending filtration of the Grassmannian of r-planes in $H^1(X, \mathbb{Q}) = \mathbb{Q}^n$,

$$\operatorname{Gr}_r(\mathbb{Q}^n) = \Omega_r^0(X) \supseteq \Omega_r^1(X) \supseteq \Omega_r^2(X) \supseteq \cdots . \tag{28}$$

If $r > n$, we adopt the convention that $\mathrm{Gr}_r(\mathbb{Q}^n) = \emptyset$ and define $\Omega^i_r(X) = \emptyset$ in this range.

As noted in [34], the Ω-sets are homotopy-type invariants. More precisely, if $f\colon X \to Y$ is a homotopy equivalence, the induced isomorphism in cohomology, $f^*\colon H^1(Y,\mathbb{Q}) \to H^1(X,\mathbb{Q})$, defines isomorphisms $f^*_r\colon \mathrm{Gr}_r(H^1(Y,\mathbb{Q})) \to \mathrm{Gr}_r(H^1(X,\mathbb{Q}))$, which send each subset $\Omega^i_r(Y)$ bijectively onto $\Omega^i_r(X)$.

Example 7.2. Let T^n be the n-dimensional torus. Since every connected cover of T^n is homotopy equivalent to a k-torus, for some $0 \le k \le n$, we conclude that $\Omega^i_r(T^n) = \mathrm{Gr}_r(\mathbb{Q}^n)$, for all $i \ge 0$ and $r \ge 1$.

7.2. Dwyer–Fried invariants and characteristic varieties

The next theorem reduces the computation of the Ω-sets to a more standard computation in algebraic geometry. The theorem was proved by Dwyer and Fried in [13], using the support loci for the Alexander invariants, and was recast in a slightly more general context by Papadima and Suciu in [29], using the characteristic varieties. We state this result in the form most convenient for our purposes, namely, the one established in [34].

Theorem 7.3 ([13, 29, 34])**.** *For all $i \le k$ and $1 \le r \le n$,*

$$\Omega^i_r(X) = \big\{P \in \mathrm{Gr}_r(\mathbb{Q}^n) \bigm| \#\big(\exp(P \otimes \mathbb{C}) \cap \mathcal{W}^i(X)\big) < \infty\big\}.$$

In other words, an r-plane $P \subset \mathbb{Q}^n$ belongs to $\Omega^i_r(X)$ if and only if the algebraic torus $T = \exp(P \otimes \mathbb{C})$ intersects the characteristic variety $W = \mathcal{W}^i(X)$ only in finitely many points. When this happens, the exponential tangent cone $\tau_1(T \cap W)$ equals $\{0\}$, forcing $P \cap L = \{0\}$, for every subspace $L \subset \mathbb{Q}^n$ in the characteristic subspace arrangement $\mathcal{C}_i(X)$. As in [34, Theorem 7.1], we obtain the following "upper bound" for the Dwyer–Fried invariants of X:

$$\Omega^i_r(X) \subseteq \Big(\bigcup_{L \in \mathcal{C}_i(X)} \{P \in \mathrm{Gr}_r(H^1(X,\mathbb{Q})) \mid P \cap L \ne \{0\}\} \Big)^{\complement}. \tag{29}$$

7.3. The incidence correspondence

The right side of (29) may be reinterpreted in terms of the classical incidence correspondence from algebraic geometry.

Let V be a homogeneous variety in $\Bbbk^n$. Consider the locus of r-planes in $\Bbbk^n$ meeting V,

$$\sigma_r(V) = \big\{P \in \mathrm{Gr}_r(\Bbbk^n) \bigm| P \cap V \ne \{0\}\big\}. \tag{30}$$

This set is a Zariski closed subset of the Grassmannian $\mathrm{Gr}_r(\Bbbk^n)$, called the *variety of incident r-planes* to V.

Particularly manageable is the case when V is a non-zero linear subspace $L \subset \Bbbk^n$. The corresponding incidence variety, $\sigma_r(L)$, is known as the *special Schubert variety* defined by L. If L has codimension d in $\Bbbk^n$, then $\sigma_r(L)$ has codimension $d - r + 1$ in $\mathrm{Gr}_r(\Bbbk^n)$.

Theorem 7.4 ([34]). *Let X be a CW-complex with finite k-skeleton. Then*

$$\Omega^i_r(X) \subseteq \mathrm{Gr}_r(H^1(X, \mathbb{Q})) \setminus \bigcup_{L \in \mathcal{C}_i(X)} \sigma_r(L),$$

for all $i \leq k$ and $r \geq 1$.

In other words, each Dwyer–Fried set $\Omega^i_r(X)$ is contained in the complement to the variety of incident r-planes to the i-th characteristic arrangement of X.

7.4. Straightness and the Dwyer–Fried invariants

Under a (local) straightness assumption, the bound from Theorem 7.4 can be expressed in terms of simpler, purely cohomological data. The next result proves Theorem 1.1, part (1) from the Introduction.

Corollary 7.5. *Suppose X is locally k-straight. Then, for all $i \leq k$ and $r \geq 1$,*

$$\Omega^i_r(X) \subseteq \mathrm{Gr}_r(H^1(X, \mathbb{Q})) \setminus \sigma_r(\mathcal{R}^i(X, \mathbb{Q})).$$

Proof. By Corollary 6.17, we have $\tau_1(\mathcal{W}^i(X)) = \mathcal{R}^i(X)$. Hence, $\bigcup_{L \in \mathcal{C}_i(X)} L = \mathcal{R}^i(X, \mathbb{Q})$, and the desired conclusion follows from Theorem 7.4. Q.E.D.

Under a more stringent straightness assumption, the above inclusion holds as an equality. The next result proves Theorem 1.1, part (2) from the Introduction.

Theorem 7.6. *Suppose X is k-straight. Then, for all $i \leq k$ and $r \geq 1$,*

$$\Omega^i_r(X) = \mathrm{Gr}_r(H^1(X, \mathbb{Q})) \setminus \sigma_r(\mathcal{R}^i(X, \mathbb{Q})).$$

Proof. By Theorem 7.3, we have

$$\Omega^i_r(X) = \big\{P \bigm| \exp(P \otimes \mathbb{C}) \cap \mathcal{W}^i(X) \text{ is finite}\big\}. \tag{31}$$

Since X is k-straight, Theorem 6.16, part (α) yields

$$\Omega^i_r(X) = \big\{P \bigm| P \cap L = \{0\}, \text{ for all } L \in \mathcal{C}_i(X)\big\}. \tag{32}$$

By Theorem 6.16, part (β), the right side of (32) is the complement to $\sigma_r(\mathcal{R}^i(X, \mathbb{Q}))$, and we are done. Q.E.D.

Particularly interesting is the case when all the components of $\mathcal{R}^i(X)$ have the same codimension, say, r. In this situation, $\Omega^i_r(X)$ is the complement of the rational Chow divisor of $\mathcal{R}^i(X,\mathbb{Q})$.

The previous theorem yields a noteworthy qualitative result about the Dwyer–Fried sets of straight spaces, in arbitrary ranks $r \geq 1$.

Corollary 7.7. *Let X be a k-straight space. Then each set $\Omega^i_r(X)$ is the complement of a finite union of special Schubert varieties in the Grassmannian of r-planes in $H^1(X,\mathbb{Q})$. In particular, $\Omega^i_r(X)$ is a Zariski open set in $\operatorname{Gr}_r(H^1(X,\mathbb{Q}))$.*

The straightness hypothesis is crucial for Theorem 7.6 to hold. The next example shows the necessity of condition (c) from Definition 6.2.

Example 7.8. Let G be the group from Example 6.5. Recall we have $\mathcal{W}^1(G) = \{1\} \cup \{t \in (\mathbb{C}^\times)^2 \mid t_1 = -1\}$, but $\mathcal{R}^1(G) = \{0\}$. Thus, $\Omega^1_2(G) = \emptyset$, yet $\sigma_2(\mathcal{R}^1(G,\mathbb{Q}))^{\complement} = \{\mathrm{pt}\}$.

§8. The influence of formality

An important property that bridges the gap between the tangent cone to a characteristic variety and the corresponding resonance variety is that of formality.

8.1. Formality

As before, let X be connected CW-complex with finite 1-skeleton. In [35], Sullivan constructs an algebra $A_{\mathrm{PL}}(X,\mathbb{Q})$ of polynomial differential forms on X with coefficients in $\mathbb{Q}$, and provides it with a natural commutative differential graded algebra (cdga) structure.

Let $H^*(X,\mathbb{Q})$ be the rational cohomology algebra of X, endowed with the zero differential. The space X is said to be *formal* if there is a zig-zag of cdga morphisms connecting $A_{\mathrm{PL}}(X,\mathbb{Q})$ to $H^*(X,\mathbb{Q})$, with each such morphism inducing an isomorphism in cohomology. The space X is merely *k-formal* (for some $k \geq 1$) if each of these morphisms induces an isomorphism in degrees up to k, and a monomorphism in degree $k+1$.

Examples of formal spaces include rational cohomology tori, surfaces, compact connected Lie groups, as well as their classifying spaces. On the other hand, the only nilmanifolds which are formal are tori. Formality is preserved under wedges and products of spaces, and connected sums of manifolds.

The 1-minimality property of a space X depends only on its fundamental group, $G = \pi_1(X, x_0)$. Alternatively, a finitely generated group G is 1-formal if and only if its Malcev Lie algebra admits a quadratic presentation. Examples of 1-formal groups include free groups and free abelian groups of finite rank, surface groups, and groups with first Betti number equal to 0 or 1. The 1-formality property is preserved under free products and direct products.

8.2. The tangent cone formula

The main connection between the formality property and the cohomology jump loci is provided by the following theorem from [11]. For more details and references, we refer to the recent survey [27].

Theorem 8.1 ([11]). *Let X be a 1-formal space. For each $d > 0$, the exponential map* $\exp\colon H^1(X,\mathbb{C}) \to H^1(X,\mathbb{C}^\times)$ *restricts to an isomorphism of analytic germs,* $\exp\colon (\mathcal{R}^1_d(X),0) \xrightarrow{\simeq} (\mathcal{V}^1_d(X),1)$. *Thus, the following "tangent cone formula" holds:*

$$\tau_1(\mathcal{V}^1_d(X)) = \mathrm{TC}_1(\mathcal{V}^1_d(X)) = \mathcal{R}^1_d(X).$$

As a consequence, the irreducible components of $\mathcal{R}^1_d(X)$ are all rationally defined subspaces, while the components of $\mathcal{V}^1_d(X)$ passing through the origin are all rational subtori of the form $\exp(L)$, with L running through the irreducible components of $\mathcal{R}^1_d(X)$. The next corollary is immediate.

Corollary 8.2. *Every* 1*-formal space is locally* 1*-straight.*

In general, though, 1-formal spaces need not be 1-straight, as we shall see in Example 11.8. Conversely, 1-straight spaces need not be 1-formal, as we shall see in Example 10.9.

8.3. Formality and the Dwyer–Fried invariants

The formality of a space has definite implications on the nature of its Ω-invariants, and their relationship to the resonance varieties.

Corollary 8.3. *If X is* 1*-formal, then* $\Omega^1_r(X) \subseteq \sigma_r(\mathcal{R}^1(X,\mathbb{Q}))^{\mathsf{c}}$, *for all $r \geq 1$.*

Proof. Follows from Corollaries 7.5 and 8.2. Q.E.D.

Corollary 8.4. *Let X be a* 1*-formal space. Suppose all positive-dimensional components of* $\mathcal{W}^1(X)$ *pass through* 1*. Then:*

(1) *X is* 1*-straight.*
(2) $\Omega^1_r(X) = \sigma_r(\mathcal{R}^1(X,\mathbb{Q}))^{\mathsf{c}}$, *for all $r \geq 1$.*

Proof. Part (1) follows from Corollary 8.2 and the additional hypothesis. Part (2) now follows from Theorem 7.6 (with $k = 1$). Q.E.D.

The hypothesis on the components of $\mathcal{W}^1(X)$ is really needed in this corollary. Indeed, if X is the presentation 2-complex for the group G from Examples 6.5 and 7.8, then $H^*(X,\mathbb{Q}) \cong H^*(T^2,\mathbb{Q})$, and so X is formal. Yet, as we know, X is not 1-straight, and $\Omega^1_2(X) \neq \sigma_2(\mathcal{R}^1(X,\mathbb{Q}))^{\complement}$. We shall see another instance of this phenomenon in Example 11.8.

8.4. The case of infinite cyclic covers

In the case when $r = 1$, Theorem 7.3 has the following consequence.

Corollary 8.5 ([29]). *Let $\nu\colon \pi_1(X) \twoheadrightarrow \mathbb{Z}$ be an epimorphism, and let $\bar\nu \in H^1(X,\mathbb{Z}) \subset H^1(X,\mathbb{C})$ be the corresponding cohomology class. If the exponential map restricts to an isomorphism of analytic germs, $(\mathcal{R}^i_1(X),0) \cong (\mathcal{V}^i_1(X),1)$, for all $i \leq k$, then*

$$\sum_{i\leq k} b_i(X^\nu) < \infty \Longleftrightarrow \bar\nu \notin \mathcal{R}^k(X).$$

Using Theorem 8.1 together with Corollary 8.5, we obtain the following immediate consequence.

Corollary 8.6. *Suppose G is a 1-formal group. Then*

$$\Omega^1_1(G) = \overline{\mathcal{R}}^1(G,\mathbb{Q})^{\complement}.$$

In other words, if $X^\nu \to X$ is a regular, infinite cyclic cover of a 1-formal space, then $b_1(X^\nu)$ is finite if and only if the corresponding cohomology class, $\bar\nu \in H^1(X,\mathbb{Z})$, is non-resonant.

§9. Toric complexes

In this section, we illustrate our techniques on a class of CW-complexes that are carved out of an n-torus in a manner prescribed by a simplicial complex on n vertices. Such "toric" complexes are minimal, formal, and straight—thus, ideal from our point of view.

9.1. Toric complexes and right-angled Artin groups

Let T^n be the n-torus, endowed with the standard cell decomposition, and with basepoint $*$ the unique 0-cell. For a simplex $\sigma \in [n]$, let $T^\sigma \subset T^n$ be the cellular subcomplex $T^\sigma = \{x \in T^n \mid x_i = * \text{ if } i \notin \sigma\}$.

Definition 9.1. Let K be a simplicial complex on n vertices. The associated *toric complex*, T_K, is the union of all T^σ, with σ running through the simplices of K.

The k-cells of T_K are in one-to-one correspondence with the $(k-1)$-simplices of K. Since the toric complex is a subcomplex of T^n, all boundary maps in $C_\bullet(T_K, \mathbb{Z})$ vanish; thus, T_K is a minimal cell complex. Evidently, $H_k(T_K, \mathbb{Z})$ is isomorphic to $C_{k-1}^{\mathrm{simp}}(K, \mathbb{Z})$, the free abelian group on the $(k-1)$-simplices of K.

Denote by V be the set of 0-cells of K, and by E the set of 1-cells of K. The fundamental group of the toric complex T_K is then the *right-angled Artin group* associated to the graph $\Gamma = K^{(1)}$,

$$G_\Gamma = \langle v \in \mathsf{V} \mid vw = wv \text{ if } \{v, w\} \in \mathsf{E} \rangle. \tag{33}$$

Groups of this sort interpolate between $G_\Gamma = \mathbb{Z}^n$ in case Γ is a complete graph, and $G_\Gamma = F_n$ in case Γ is a discrete graph.

The toric complex construction behaves well with respect to simplicial joins: $T_{K*K'} = T_K \times T_{K'}$. Consequently, $G_{\Gamma*\Gamma'} = G_\Gamma \times G_{\Gamma'}$.

A classifying space for the group G_K is the toric complex T_Δ, where $\Delta = \Delta_K$ is the flag complex of K, i.e., the maximal simplicial complex with 1-skeleton equal to the graph $\Gamma = K^{(1)}$.

Finally, it is known from the work of Notbohm and Ray [24] that all toric complexes are formal spaces. In particular, all right-angled Artin groups are 1-formal, a fact also proved in [25].

9.2. Cohomology jump loci

As noted above, $H_1(T_K, \mathbb{Z}) = \mathbb{Z}^n$, with generators indexed by the vertex set $\mathsf{V} = [n]$. Thus, we may identify $H^1(T_K, \mathbb{C})$ with the vector space $\mathbb{C}^{\mathsf{V}} = \mathbb{C}^n$, and $H^1(T_K, \mathbb{C}^\times)$ with the algebraic torus $(\mathbb{C}^\times)^{\mathsf{V}} = (\mathbb{C}^\times)^n$. For each subset $\mathsf{W} \subseteq \mathsf{V}$, let $\mathbb{C}^{\mathsf{W}}$ be the respective coordinate subspace, and let $(\mathbb{C}^\times)^{\mathsf{W}} = \exp(\mathbb{C}^{\mathsf{W}})$ be the respective algebraic subtorus.

Theorem 9.2 ([26]). *With notation as above,*

$$\mathcal{R}^i_d(T_K) = \bigcup_{\mathsf{W}} \mathbb{C}^{\mathsf{W}} \quad \textit{and} \quad \mathcal{V}^i_d(T_K) = \bigcup_{\mathsf{W}} (\mathbb{C}^\times)^{\mathsf{W}},$$

where, in both cases, the union is taken over all subsets $\mathsf{W} \subset \mathsf{V}$ for which $\sum_{\sigma \in K_{\mathsf{V}\setminus\mathsf{W}}} \dim_{\mathbb{C}} \widetilde{H}_{i-1-|\sigma|}(\mathrm{lk}_{K_{\mathsf{W}}}(\sigma), \mathbb{C}) \geq d$.

In the above, K_{W} denotes the simplicial subcomplex induced by K on W, and $\mathrm{lk}_L(\sigma)$ denotes the link of a simplex σ in a subcomplex $L \subseteq K$.

In homological degree 1, the resonance formula from Theorem 9.2 takes a simpler form, already noted in [25]. Namely, $\mathcal{R}^1(G_\Gamma) = \bigcup_{\mathsf{W}} \mathbb{C}^{\mathsf{W}}$, where the union is taken over all (maximal) subsets $\mathsf{W} \subset \mathsf{V}$ for which the induced graph Γ_{W} is disconnected. In particular, the codimension of the resonance variety $\mathcal{R}^1(G_\Gamma)$ equals the connectivity of the graph Γ.

Corollary 9.3. *All toric complexes T_K are straight spaces.*

Proof. By the above theorem, each resonance variety $\mathcal{R}^i(T_K)$ is the union of a coordinate subspace arrangement, and each characteristic variety $\mathcal{V}^i(T_K)$ is the union of the corresponding arrangement of coordinate subtori. Consequently, all components of $\mathcal{V}^i(T_K)$ are algebraic subtori, and $\mathrm{TC}_1(\mathcal{V}^i(T_K)) = \mathcal{R}^i(T_K)$. Q.E.D.

9.3. Ω-invariants

In their landmark paper [3], Bestvina and Brady studied the geometric finiteness properties of certain subgroups of right-angled Artin groups G_Γ, arising as kernels of "diagonal" homomorphisms $G_\Gamma \twoheadrightarrow \mathbb{Z}$. In [26], Papadima and Suciu computed the homology of such subgroups, and, more generally, the homology of regular $\mathbb{Z}$-covers of toric complexes. In a related vein, Denham [8] investigated the homology of covers of toric complexes T_K corresponding to "coordinate" homomorphisms $\pi_1(T_K) \twoheadrightarrow \mathbb{Z}^r$.

The study of homological finiteness properties of regular, free abelian covers of toric complexes was pursued in [29], where a general formula for the Dwyer–Fried sets of such complexes was given. In our setting, this result may be restated as follows.

Theorem 9.4 ([29]). *Let T_K be a toric complex. Then*

$$\Omega_r^k(T_K) = \sigma_r(\mathcal{R}^k(T_K, \mathbb{Q}))^{\complement},$$

for all $k, r \geq 1$. In particular, $\Omega_1^k(T_K) = \overline{\mathcal{R}}^k(T_K, \mathbb{Q})^{\complement}$.

Proof. Follows from Theorem 7.6 and Corollary 9.3. Q.E.D.

As a consequence, all the Ω-invariants of a toric complex T_K are Zariski open subsets of the Grassmannian $\mathrm{Gr}_r(\mathbb{Q}^n)$.

When combined with Theorem 9.2 and the discussion following it, Theorem 9.4 allows us to compute very explicitly the Dwyer–Fried sets of toric complexes. Let us provide one such computation.

Corollary 9.5. *Let Γ be a finite simple graph, and let κ be the connectivity of Γ. Then $\Omega_r^1(G_\Gamma) = \emptyset$, for all $r \geq \kappa + 1$.*

Proof. Recall that $\operatorname{codim} \mathcal{R}^1(G_\Gamma) = \kappa$. Thus, $\operatorname{codim} \sigma_r(\mathcal{R}^1(G_\Gamma)) = \kappa - r + 1$. The desired conclusion follows from Theorem 9.4. Q.E.D.

In particular, if Γ is disconnected, then $\Omega_r^1(G_\Gamma) = \emptyset$, for all $r \geq 1$.

Example 9.6. Let Γ be a tree on $n \geq 3$ vertices. Label the non-terminal vertices as $v_1, \dots, v_s$, and the terminal vertices as $v_{s+1}, \dots, v_n$.

The cut sets of Γ are all singletons, consisting of the non-terminal vertices. Thus, the resonance variety $\mathcal{R}^1(G_\Gamma, \mathbb{Q})$ is the union of the coordinate hyperplanes $L_j = \{z \in \mathbb{Q}^n \mid z_j = 0\}$, with $1 \leq j \leq s$. Hence,

$$\Omega^1_r(G_\Gamma) = \begin{cases} \mathbb{QP}^{n-1} \setminus \bigcup_{j=1}^s \overline{L}_j & \text{if } r = 1, \\ \emptyset & \text{if } r \geq 2. \end{cases} \tag{34}$$

§10. Kähler and quasi-Kähler manifolds

We now discuss the cohomology jumping loci and the Dwyer–Fried invariants of Kähler and quasi-Kähler manifolds.

10.1. Cohomology ring and formality

Let M be a compact, connected, complex manifold of complex dimension m. Such a manifold is called a *Kähler manifold* if it admits a Hermitian metric h for which the imaginary part $\omega = \Im(h)$ is a closed 2-form. The class of Kähler manifolds, includes smooth, complex projective varieties, such as Riemann surfaces. This class is closed under finite direct products and finite covers.

Hodge theory provides two sets of data on the cohomology ring of M. The first data, known as the Hodge decomposition on $H^i(M, \mathbb{C})$, depend only on the complex structure on M. The second data, known as the Lefschetz isomorphism and the Lefschetz decomposition on $H^i(M, \mathbb{R})$, depend only on the choice of a Kähler class $[\omega] \in H^{1,1}(M, \mathbb{C})$.

These data impose strong conditions on the possible Betti numbers $b_i = b_i(M)$, beyond the symmetry property $b_i = b_{2m-i}$ imposed by Poincaré duality. For example, the odd Betti numbers b_{2i+1} must be even, and increasing in the range $2i + 1 \leq m$, while the even Betti numbers b_{2i} must be increasing in the range $2i \leq m$.

Another constraint on the topology of compact Kähler manifolds was established by Deligne, Griffiths, Morgan, and Sullivan in [7]. For such a manifold M, let d be the exterior derivative, J the complex structure, and $d^c = J^{-1}dJ$. Then the following holds: If η is a form which is closed for both d and d^c, and exact for either d or d^c, then η is exact for dd^c. As a consequence, all compact Kähler manifolds are formal.

A manifold X is said to be a *quasi-Kähler manifold* if there is a compact Kähler manifold $\overline{X}$ and a normal-crossings divisor D such that $X = \overline{X} \setminus D$. The class of quasi-Kähler manifolds includes smooth, irreducible, quasi-projective complex varieties, such as complements of plane algebraic curves.

Each quasi-Kähler manifold X inherits a mixed Hodge structure from its compactification $\overline{X}$. If X is a smooth, quasi-projective variety with vanishing degree 1 weight filtration on $H^1(X,\mathbb{C})$, then X is 1-formal. This happens, for instance, when X admits a non-singular compactification $\overline{X}$ with $b_1(\overline{X}) = 0$, e.g., when X is the complement of a hypersurface in $\mathbb{CP}^m$. In general, though, smooth, quasi-projective varieties need not be 1-formal. For a detailed treatment of the subject, we refer to Morgan [22].

10.2. Characteristic varieties

Foundational results on the structure of the cohomology support loci for local systems on smooth projective varieties, and more generally, on compact Kähler manifolds were obtained by Beauville, Green–Lazarsfeld, Simpson, and Campana. A further, wide-ranging generalization was obtained by Arapura in [1].

Theorem 10.1 ([1]). *Let $X = \overline{X} \setminus D$, where $\overline{X}$ is a compact Kähler manifold and D is a normal-crossings divisor. If either $D = \emptyset$ or $b_1(\overline{X}) = 0$, then each characteristic variety $\mathcal{V}^i_d(X)$ is a finite union of unitary translates of algebraic subtori of $H^1(X,\mathbb{C}^\times)$.*

In other words, for each $i \geq 0$ and $d > 0$, the characteristic variety $\mathcal{V}^i_d(X)$ admits a decomposition into irreducible components of the form $W = \rho \cdot T$, with

- direction subtorus $T = \operatorname{dir}(W)$, a connected algebraic subgroup of the character torus $\widehat{G}^0$, where $G = \pi_1(X)$;
- translation factor $\rho\colon G \to S^1$, a unitary character of G.

In degree 1 and depth 1, the condition that $b_1(\overline{X}) = 0$ if $D \neq \emptyset$ may be lifted. Furthermore, each positive-dimensional component of $\mathcal{V}^1_1(X)$ is of the form $\rho \cdot T$, with ρ a *torsion* character.

In the quasi-projective setting, more can be said. The next theorem summarizes several recent results in this direction: the first two parts are from [10], the third part is from [4] and [2], while the last part is from [9] and [2].

Theorem 10.2 ([9, 10, 4, 2]). *Let X be a smooth, quasi-projective variety. Then:*

(1) *If W and W' are two distinct components of $\mathcal{V}^1(X)$, then either* $\operatorname{dir}(W) = \operatorname{dir}(W')$, *or* $T_1 \operatorname{dir}(W) \cap T_1 \operatorname{dir}(W') = \{0\}$.
(2) *For each pair of distinct components, W and W', the intersection $W \cap W'$ is a finite set of torsion characters.*
(3) *The isolated points in $\mathcal{V}^1(X)$ are also torsion characters.*

(4) *If* $W = \rho T$, *with* $\dim T = 1$ *and* $\rho \neq 1$, *then* T *is not a component of* $\mathcal{V}^1(X)$.

10.3. Resonance varieties

In the presence of 1-formality, the quasi-Kähler condition also imposes stringent conditions on the degree 1 resonance varieties.

Theorem 10.3 ([11]). *Let* X *be a* 1*-formal, quasi-Kähler manifold, and let* $\{L_\alpha\}$ *be the collection of positive-dimensional, irreducible components of* $\mathcal{R}^1_1(X)$. *Then:*

(1) *Each* L_α *is a linear subspace of* $H^1(X,\mathbb{C})$ *of dimension at least* $2\varepsilon(\alpha)+2$, *for some* $\varepsilon(\alpha)\in\{0,1\}$.
(2) *The restriction of the cup-product map* $H^1(X,\mathbb{C})\wedge H^1(X,\mathbb{C})\to H^2(X,\mathbb{C})$ *to* $L_\alpha\wedge L_\alpha$ *has rank* $\varepsilon(\alpha)$.
(3) *If* $\alpha\neq\beta$, *then* $L_\alpha\cap L_\beta=\{0\}$.
(4) $\mathcal{R}^1_d(X)=\{0\}\cup\bigcup_{\alpha: d\le \dim L_\alpha - \varepsilon(\alpha)} L_\alpha$.

Remark 10.4. Suppose X is a smooth quasi-projective variety with $W_1(H^1(X,\mathbb{C}))=0$. Then, as mentioned in §10.1, X is 1-formal. In this situation, each subspace L_α is isotropic, i.e., $\epsilon(\alpha)=0$.

Remark 10.5. Suppose M is a compact Kähler manifold. Then of course M is formal, and Theorem 10.3 again applies. In this situation, each subspace L_α has dimension $2g(\alpha)$, for some $g(\alpha)\ge 2$, and the restriction of the cup-product map to $L_\alpha\wedge L_\alpha$ has rank $\epsilon(\alpha)=1$.

It is now a straightforward exercise to enumerate the possibilities for the first resonance variety of a Kähler manifold M, at least for low values of $n=b_1(M)$:

- If $n=0$ or 2, then $\mathcal{R}^1(M)=\{0\}$.
- If $n=4$, then $\mathcal{R}^1(M)=\{0\}$ or $\mathbb{C}^4$.
- If $n=6$, then $\mathcal{R}^1(M)=\{0\}$, $\mathbb{C}^4$, or $\mathbb{C}^6$.
- If $n=8$, then $\mathcal{R}^1(M)=\{0\}$, $\mathbb{C}^4$, $\mathbb{C}^6$, $\mathbb{C}^8$, or $\mathcal{R}^1(M)$ consists of two transverse, 4-dimensional subspaces.

Using the computations from Examples 3.4 and 3.10, it is readily seen that all these possibilities can be realized by manifolds of the form $M=S_g$ or $M=S_g\times S_h$, with S_g and S_h Riemann surfaces of suitable genera $g,h\ge 0$.

10.4. Straightness

We now discuss in detail the straightness properties of Kähler and quasi-Kähler manifolds, and how these properties relate to 1-formality.

Proposition 10.6. *Let* X *be a* 1*-formal, quasi-Kähler manifold (for instance, a compact Kähler manifold). Then:*

(1) *X is locally 1-straight.*
(2) *X is 1-straight if and only if $\mathcal{W}^1(X)$ contains no positive-dimensional translated subtori.*

Proof. Part (1) follows from Corollary 8.2, using only the 1-formality assumption, while part (2) follows from Theorem 10.1. Q.E.D.

In general, quasi-Kähler manifolds may fail to be locally 1-straight, as illustrated in Example 10.7 below. Perhaps more surprisingly, compact Kähler manifolds may fail to be 1-straight; for a concrete example, we refer to [34].

Example 10.7. Let X be the complex Heisenberg manifold, i.e., the total space of the $\mathbb{C}^\times$-bundle over $\mathbb{C}^\times \times \mathbb{C}^\times$ with Euler number 1. Then X is a smooth, quasi-projective variety with $\mathcal{V}^1(X) = \{1\}$, yet $\mathcal{R}^1(X) = \mathbb{C}^2$. Thus, the tangent cone formula fails in this instance, and so X is neither 1-formal, nor locally 1-straight.

Let $(Y, 0)$ be a quasi-homogeneous isolated surface singularity. Then $X = Y \setminus \{0\}$ is a smooth, quasi-projective variety which supports a "good" $\mathbb{C}^\times$-action, with orbit space a smooth projective curve. Moreover, X deform-retracts onto the singularity link, which is a closed, smooth, orientable 3-manifold.

Proposition 10.8. *Suppose the curve $X/\mathbb{C}^\times$ has genus $g > 1$. Then X is straight, yet X is not 1-formal.*

Proof. According to formula (14) from [12], we have $\mathcal{V}^1(X) = H^1(X, \mathbb{C}^\times)$. Hence, by Lemma 6.6, X is straight. The fact that X is not 1-formal follows from [12, Proposition 7.1]. Q.E.D.

We illustrate this result with a concrete family of examples.

Example 10.9. Let X be the total space of the $\mathbb{C}^\times$-bundle over the Riemann surface of genus g, with Euler number -1. Then X is homotopy equivalent to the Brieskorn manifold $\Sigma(2, 2g+1, 2(2g+1))$. Assume now $g > 1$; then clearly X fits into the setup of Proposition 10.8. Hence, X is straight, but not 1-formal.

10.5. Dwyer–Fried invariants

We conclude this section with a discussion of the Ω-invariants of (quasi-) Kähler manifolds, and the extent to which these invariants are determined by the resonance varieties.

Theorem 10.10. *Let X be a 1-formal, quasi-Kähler manifold (for instance, a compact Kähler manifold). Then:*

(1) $\Omega^1_1(X) = \overline{\mathcal{R}}^1(X,\mathbb{Q})^{\complement}$ *and* $\Omega^1_r(X) \subseteq \sigma_r(\mathcal{R}^1(X,\mathbb{Q}))^{\complement}$, *for* $r \geq 2$.
(2) *If* $\mathcal{W}^1(X)$ *contains no positive-dimensional translated subtori, then* $\Omega^1_r(X) = \sigma_r(\mathcal{R}^1(X,\mathbb{Q}))^{\complement}$, *for all* $r \geq 1$.

Proof. Part (1). Using only the 1-formality assumption, the two statements follow from Corollaries 8.6 and 8.3, respectively.

Part (2). In view of Theorem 10.1, the hypothesis is equivalent to $\mathcal{W}^1(X)$ containing no positive-dimensional component not passing through 1. The conclusion follows from Corollary 8.4. Q.E.D.

In general, though, the inclusion $\Omega^1_r(X) \subseteq \sigma_r(\mathcal{R}^1(X,\mathbb{Q}))^{\complement}$ can be strict, provided $2 \leq r \leq b_1(X)$. The next theorem identifies a fairly broad class of smooth, quasi-projective varieties for which this is the case. An explicit example will be given at the end of Section 11.

Theorem 10.11. *Let* X *be a 1-formal, smooth, quasi-projective variety. Suppose*
(1) $\mathcal{W}^1(X)$ *has a* 1*-dimensional component not passing through* 1*;*
(2) $\mathcal{R}^1(X)$ *has no codimension-*1 *components.*
Then $\Omega^1_2(X)$ *is strictly contained in* $\sigma_2(\mathcal{R}^1(X,\mathbb{Q}))^{\complement}$.

Proof. Set $n = b_1(X)$, and identify $H = H^1(X,\mathbb{Q})$ with $\mathbb{Q}^n$. By assumption (1), the characteristic variety $\mathcal{W}^1(X) \subset (\mathbb{C}^{\times})^n$ has a component of the form $W = \rho \cdot T$, with

- $T = \exp(\ell \otimes \mathbb{C})$, where ℓ is a 1-dimensional subspace in H;
- $\rho = \exp(2\pi i q)$, where q is a vector in $H \setminus \ell$.

We claim that the line ℓ is not contained in the resonance variety $\mathcal{R}^1(X,\mathbb{Q})$. For, if it were, the 1-dimensional algebraic torus T would be contained in $\mathcal{W}^1(X)$. But we know from Theorem 10.2(4) that T is not a component of $\mathcal{W}^1(X)$. Hence, there would exist a component W' with $T \subsetneq W'$. Therefore,

$$\begin{aligned}&\operatorname{dir}(W) = T \neq W' = \operatorname{dir}(W'), \quad \text{and}\\ &T_1(\operatorname{dir}(W)) \cap T_1(\operatorname{dir}(W')) = \ell \otimes \mathbb{C} \neq \{0\}.\end{aligned}$$

This contradicts Theorem 10.2(1), thereby establishing the claim.

In view of the above, and of assumption (2), the resonance variety $\mathcal{R}^1(X,\mathbb{R})$ has codimension at least 2 in $H_{\mathbb{R}} = H^1(X,\mathbb{R})$. Therefore, the set

$$Z = \{P \in \operatorname{Gr}_2(H_{\mathbb{R}}) \mid \ell \subset P \text{ and } P \cap \mathcal{R}^1(X,\mathbb{R}) \neq \{0\}\}$$

is a proper subvariety of $\operatorname{Gr}_2(H_{\mathbb{R}})$. Hence, there is a non-zero vector $r_0 \in \mathbb{R}^n$, and an open cone U containing r_0, such that, for all $r \in U$, the plane P spanned by r and ℓ intersects $\mathcal{R}^1(X,\mathbb{R})$ only at 0.

Let $\pi\colon \mathbb{R}^n \backslash \{0\} \to \mathbb{RP}^{n-1}$ be the projection map. Clearly, $\pi(\mathbb{Z}^n \backslash \{0\})$ is a dense subset of $\mathbb{RP}^{n-1}$. Thus, $\pi(q + \mathbb{Z}^n)$ is also dense, and so intersects $\pi(U)$. Hence, there is a lattice point $\lambda \in \mathbb{Z}^n$ such that $\pi(q+\lambda)$ belongs to $\pi(U)$. The rational vector $q_0 := q + \lambda$ then belongs to U.

Let P_0 be the 2-dimensional subspace of H spanned by ℓ and q_0. By construction, $P_0 \cap \mathcal{R}^1(X, \mathbb{Q}) = \{0\}$, and so $P_0 \in \sigma_2(\mathcal{R}^1(X, \mathbb{Q}))^{\mathsf{c}}$. On the other hand, the algebraic torus $T_0 = \exp(P_0 \otimes \mathbb{C})$ contains both $\exp(\ell \otimes \mathbb{C}) = T$ and $\exp(2\pi i q_0) = \rho$; therefore, $T_0 \supset \rho T$. Consequently, $\dim(T_0 \cap \mathcal{W}^1(X)) > 0$, showing that $P_0 \notin \Omega_2^1(X)$. Q.E.D.

§11. Hyperplane arrangements

We conclude with a class of spaces exhibiting a strong interplay between the resonance varieties and the finiteness properties of free abelian covers. These spaces, obtained by deleting a finite number of hyperplanes from a complex affine space, are minimal, formal, and locally straight, but not always straight.

11.1. Cohomology jump loci

Let $\mathcal{A}$ be an arrangement of hyperplanes in $\mathbb{C}^\ell$. To start with, we will assume that all hyperplanes in $\mathcal{A}$ pass through the origin; non-central arrangements can be handled much the same way, using standard coning and deconing constructions.

Let $X = X(\mathcal{A})$ be the complement of the union of the hyperplanes comprising $\mathcal{A}$. Then X can be viewed as the complement of a normal-crossing divisor in a suitably blown-up $\mathbb{CP}^\ell$. In particular, X has the homotopy type of an ℓ-dimensional CW-complex. Moreover, this CW-complex can be chosen to be minimal (Dimca–Papadima, Randell).

Let $G = G(\mathcal{A})$ be the fundamental group of the complement. Its abelianization, G_{ab}, is the free abelian group of rank $n = |\mathcal{A}|$. Thus, we may identify the character group $\widehat{G}$ with the complex algebraic torus $(\mathbb{C}^\times)^n$. Let $\mathcal{V}_d^i(\mathcal{A}) = \mathcal{V}_d^i(X(\mathcal{A}))$ be the characteristic varieties of the arrangement. By Arapura's work [1], these varieties consist of subtori of $(\mathbb{C}^\times)^n$, possibly translated by unitary characters, together with a finite number of isolated unitary characters.

The cohomology ring $A = H^*(X(\mathcal{A}), \mathbb{Z})$ was computed by Brieskorn in the early 1970s, building on pioneering work of Arnol'd on the cohomology ring of the braid arrangement. It follows from Brieskorn's work that the space X is formal; in particular, the fundamental group of X is 1-formal. In 1980, Orlik and Solomon gave a simple description of the ring A, solely in terms of the intersection lattice $L(\mathcal{A})$, i.e., the poset of all intersections of $\mathcal{A}$, ordered by reverse inclusion.

The resonance varieties $\mathcal{R}^i_d(\mathcal{A}) = \mathcal{R}^i_d(X(\mathcal{A}))$ were first defined and studied by Falk in [17]. Identifying $H^1(X, \mathbb{C})$ with $\mathbb{C}^n$, we may view the resonance varieties of $\mathcal{A}$ as homogeneous subvarieties of $\mathbb{C}^n$. It is known from work of Yuzvinsky that these varieties actually lie in the hyperplane $\{x \in \mathbb{C}^n \mid \sum_{i=1}^n x_i = 0\}$. Moreover, it follows from [15] that resonance "propagates" for the Orlik–Solomon algebra. More precisely, if $i \leq k \leq \ell$, then $\mathcal{R}^i_1(\mathcal{A}) \subseteq \mathcal{R}^k_1(\mathcal{A})$; in particular, $\mathcal{R}^k(\mathcal{A}) = \mathcal{R}^k_1(\mathcal{A})$.

11.2. Straightness

As above, let $X = X(\mathcal{A})$ be an arrangement complement. Using work of Esnault, Schechtman, and Viehweg [16], one can show that the exponential map, $\exp\colon H^1(X, \mathbb{C}) \to H^1(X, \mathbb{C}^\times)$, induces an isomorphism of analytic germs from $(\mathcal{R}^i_d(X), 0)$ to $(\mathcal{V}^i_d(X), 1)$, for all $i \geq 0$ and $d > 0$. We then have

$$\mathrm{TC}_1(\mathcal{V}^i_d(\mathcal{A})) = \mathcal{R}^i_d(\mathcal{A}), \quad \text{for all } d > 0. \tag{35}$$

In particular, all the resonance varieties $\mathcal{R}^i_d(\mathcal{A})$ are finite unions of rationally defined linear subspaces.

The tangent cone formula (35) was first proved in degree $i = 1$ (using different methods) by Cohen and Suciu [6] and Libgober [21], and was generalized to the higher-degree jump loci by Cohen and Orlik [5].

Proposition 11.1. *Let $\mathcal{A}$ be a hyperplane arrangement in $\mathbb{C}^\ell$, with complement X.*

(1) *X is locally straight.*
(2) *X is k-straight if and only if $\mathcal{V}^k(X)$ contains no positive-dimensional translated tori.*

Proof. For part (1), we must verify the two properties from Definition 6.1. Property (a) follows at once from Arapura's theorem 10.1, while property (b) follows from the tangent cone formula (35).

For part (2), we must verify the additional property (c) from Definition 6.2. The conclusion follows again from Theorem 10.1. Q.E.D.

As first noted in [31], there do exist arrangements $\mathcal{A}$ for which $\mathcal{V}^1(\mathcal{A})$ contains positive-dimensional translated components. By Proposition 11.1, such arrangements are not 1-straight. We will come back to this point in Example 11.8.

11.3. Dwyer–Fried invariants

Let us define the Dwyer–Fried invariants of an arrangement $\mathcal{A}$ as $\Omega^i_r(\mathcal{A}) = \Omega^i_r(X(\mathcal{A}))$. Suppose $\mathcal{A}$ consists of n hyperplanes in $\mathbb{C}^\ell$. For each $1 \leq r \leq n$, there is then a filtration

$$\mathrm{Gr}_r(\mathbb{Q}^n) = \Omega^0_r(\mathcal{A}) \supseteq \Omega^1_r(\mathcal{A}) \supseteq \Omega^2_r(\mathcal{A}) \supseteq \cdots \supseteq \Omega^\ell_r(\mathcal{A}). \tag{36}$$

The next result establishes a comparison between the terms of this filtration and the incidence varieties to the resonance varieties of $\mathcal{A}$.

Theorem 11.2. *Let $\mathcal{A}$ be an arrangement of n hyperplanes, and fix an integer $k \geq 1$. Then,*

(1) $\Omega^k_r(\mathcal{A}) \subseteq \mathrm{Gr}_r(\mathbb{Q}^n) \setminus \sigma_r(\mathcal{R}^k(\mathcal{A}, \mathbb{Q}))$, *for all* $r \geq 1$.
(2) $\Omega^1_1(\mathcal{A}) = \mathbb{QP}^{n-1} \setminus \overline{\mathcal{R}}^1(\mathcal{A}, \mathbb{Q})$.
(3) *If $\mathcal{V}^k(\mathcal{A})$ contains no positive-dimensional translated tori, then* $\Omega^k_r(\mathcal{A}) = \mathrm{Gr}_r(\mathbb{Q}^n) \setminus \sigma_r(\mathcal{R}^k(\mathcal{A}, \mathbb{Q}))$, *for all* $r \geq 1$.

Proof. Part (1) follows from Proposition 11.1(1) and Corollary 7.5.

Part (2) follows from the 1-formality of the complement of $\mathcal{A}$ and Corollary 8.6.

Part (3) follows from Proposition 11.1(2) and Theorem 7.6. Q.E.D.

In other words, each Dwyer–Fried invariant $\Omega^k_r(\mathcal{A})$ is included in the complement of a union of special Schubert varieties of the form $\sigma_r(L)$, where L runs through the components of $\mathcal{R}^k(\mathcal{A}, \mathbb{Q})$, with the inclusion being an equality when $\mathcal{V}^k(\mathcal{A}) = \bigcup_L \exp(L \otimes \mathbb{C})$.

Remark 11.3. In [18], Falk gives a decomposition of the resonance variety $\mathcal{R}^1(\mathcal{A})$ into combinatorial pieces and shows that, projectively, each of these pieces is the ruled variety corresponding to an intersection of special Schubert varieties in special position in the Grassmannian of lines in projective space. It would be interesting to see if Falk's description sheds additional light on the Dwyer–Fried invariants $\Omega^1_r(\mathcal{A})$.

11.4. Resonance varieties of line arrangements

For the rest of this section, we will concentrate on the resonance varieties $\mathcal{R}^1(\mathcal{A}) = \mathcal{R}^1(G(\mathcal{A}))$ and their relation to the Dwyer–Fried invariants $\Omega^1_r(\mathcal{A}) = \Omega^1_r(G(\mathcal{A}))$. We start with a brief review of the former.

By the Lefschetz-type theorem of Hamm and Lê, taking a generic two-dimensional section does not change the group of the arrangement. Thus, we may assume $\mathcal{A} = \{\ell_1, \dots, \ell_n\}$ is an affine line arrangement in $\mathbb{C}^2$, for which no two lines are parallel.

The variety $\mathcal{R}^1(\mathcal{A})$ is a union of linear subspaces in $\mathbb{C}^n$. Each subspace has dimension at least 2, and each pair of subspaces meets transversely at 0. The simplest components of $\mathcal{R}^1(\mathcal{A})$ are the *local* components: to an intersection point $v_J = \bigcap_{j \in J} \ell_j$ of multiplicity $|J| \geq 3$, there corresponds a subspace L_J of dimension $|J| - 1$, given by equations of the form $\sum_{j\in J} x_j = 0$, and $x_i = 0$ if $i \notin J$. The remaining components correspond to certain "neighborly partitions" of sub-arrangements of $\mathcal{A}$.

If $|\mathcal{A}| \leq 5$, then all components of $\mathcal{R}^1(\mathcal{A})$ are local. For $|\mathcal{A}| \geq 6$, though, the resonance variety $\mathcal{R}^1(\mathcal{A})$ may have non-local components.

Example 11.4. Let $\mathcal{A}$ be the braid arrangement, with defining polynomial $Q(\mathcal{A}) = z_0z_1z_2(z_0-z_1)(z_0-z_2)(z_1-z_2)$. Take a generic plane section, and label the corresponding lines as $6, 2, 4, 3, 5, 1$. The variety $\mathcal{R}_1(\mathcal{A}) \subset \mathbb{C}^6$ has 4 local components, corresponding to the triple points $124, 135, 236, 456$, and one non-local component, $L_\Pi = \{x_1 + x_2 + x_3 = x_1 - x_6 = x_2 - x_5 = x_3 - x_4 = 0\}$, corresponding to the neighborly partition $\Pi = (16|25|34)$.

For an arbitrary arrangement $\mathcal{A}$, work of Falk, Pereira, and Yuzvinsky [19, 30, 36] shows that any non local component in $\mathcal{R}^1(\mathcal{A})$ has dimension either 2 or 3.

11.5. Ω-invariants of line arrangements

We are in a position now where we can compute explicitly the Dwyer–Fried invariants of some line arrangements. We start with a simple example.

Example 11.5. Let $\mathcal{A}$ be the arrangement with defining polynomial $Q(\mathcal{A}) = z_0z_1z_2(z_1 - z_2)$. The variety $\mathcal{R}^1(\mathcal{A}) \subset \mathbb{C}^4$ has a single component, namely, the 2-plane $L = \{x \mid x_1 + x_2 + x_3 = x_4 = 0\}$, while $\mathcal{V}^1(\mathcal{A}) = \exp(L)$. Using Theorem 11.2, we obtain

$$\Omega^1_r(\mathcal{A}) = \begin{cases} \mathbb{QP}^3 \setminus \overline{L} & \text{if } r = 1, \\ \mathrm{Gr}_2(\mathbb{Q}^4) \setminus \sigma_2(L) & \text{if } r = 2, \\ \emptyset & \text{if } r \geq 3. \end{cases}$$

Here, the Grassmannian $\mathrm{Gr}_2(\mathbb{Q}^4)$ is the hypersurface in $\mathbb{P}(\bigwedge^2 \mathbb{Q}^4)$ with equation $p_{12}p_{34} - p_{13}p_{24} + p_{23}p_{14} = 0$, while the Schubert variety $\sigma_2(L)$ is the 3-fold in $\mathrm{Gr}_2(\mathbb{Q}^4)$ cut out by the hyperplane $p_{12} - p_{13} + p_{23} = 0$.

Proposition 11.6. *Let $\mathcal{A}$ be an arrangement of n lines in $\mathbb{C}^2$, and let m be the maximum multiplicity of its intersection points.*

(1) *If $m = 2$, then $\Omega^1_r(\mathcal{A}) = \mathrm{Gr}_r(\mathbb{Q}^n)$, for all $r \geq 1$.*
(2) *If $m \geq 3$, then $\Omega^1_r(\mathcal{A}) = \emptyset$, for all $r \geq n - m + 2$.*

Proof. If $\mathcal{A}$ has only double points, then $G(\mathcal{A}) = \mathbb{Z}^n$, by a well-known theorem of Zariski. Assertion (1) follows from Example 7.2.

Now suppose $\mathcal{A}$ has an intersection point of multiplicity $m \geq 3$. Then $\mathcal{R}^1(\mathcal{A})$ has a (local) component L of dimension $m - 1$. The corresponding Schubert variety, $\sigma_r(L)$, has codimension $n - m + 2 - r$. Assertion (2) follows from Theorem 11.2(1). Q.E.D.

Proposition 11.7. *Let $\mathcal{A}$ be an arrangement of n lines in $\mathbb{C}^2$. Suppose $\mathcal{A}$ has 1 or 2 lines which contain all the intersection points of multiplicity 3 and higher. Then $\Omega^1_r(\mathcal{A}) = \sigma_r(\mathcal{R}^1(\mathcal{A}, \mathbb{Q}))^{\complement}$, for all $1 \leq r \leq n$.*

Proof. As shown by Nazir and Raza in [23], the characteristic variety $\mathcal{V}^1(\mathcal{A})$ of such an arrangement has no translated components. The conclusion follows from Theorem 11.2(3). Q.E.D.

In general, though, the "resonance upper bound" for the Dwyer–Fried invariants of arrangements is not attained. We illustrate this claim with the smallest possible example.

Example 11.8. Let $\mathcal{A}$ be the deleted B_3 arrangement, with defining polynomial $Q(\mathcal{A}) = z_0z_1(z_0^2 - z_1^2)(z_0^2 - z_2^2)(z_1^2 - z_2^2)$. The jump loci of $\mathcal{A}$ were computed in [32]. Briefly, the resonance variety $\mathcal{R}^1(\mathcal{A}) \subset \mathbb{C}^8$ contains 7 local components, corresponding to 6 triple points and one quadruple point, and 5 non-local components, corresponding to braid sub-arrangements. In particular, $\operatorname{codim} \mathcal{R}^1(\mathcal{A}) = 5$.

In addition to the 12 subtori arising from the subspaces in $\mathcal{R}^1(\mathcal{A})$, the characteristic variety $\mathcal{V}^1(\mathcal{A}) \subset (\mathbb{C}^{\times})^8$ also contains a component of the form $\rho \cdot T$, where T is a 1-dimensional algebraic subtorus, and ρ is a root of unity of order 2.

Let $X = X(\mathcal{A})$ be complement of the arrangement. Then X is formal, yet not 1-straight. Moreover, the hypothesis of Theorem 10.11 are satisfied for X. We conclude that $\Omega^1_2(\mathcal{A})$ is strictly contained in $\sigma_2(\mathcal{R}^1(\mathcal{A}))^{\complement}$.

§12. Acknowledgements

An incipient version of this work was presented at the Mathematical Society of Japan Seasonal Institute on *Arrangements of Hyperplanes*, held at Hokkaido University in August 2009. I wish to thank the organizers for the opportunity to participate in such an interesting meeting, and for their warm hospitality.

A fuller version of this work was presented at the Centro di Ricerca Matematica Ennio De Giorgi in Pisa, in May–June 2010. I wish to thank the organizers of the Intensive Research Period on *Configuration Spaces: Geometry, Combinatorics and Topology* for their friendly hospitality, and for providing an inspiring mathematical environment.

Finally, I am grateful to Stefan Papadima for many illuminating discussions on the topics presented here, and to Graham Denham for help with the proof of Theorem 10.11.

References

[1] D. Arapura, Geometry of cohomology support loci for local systems. I., J. Algebraic Geom., **6** (1997), 563–597, MR1487227.

[2] E. Artal Bartolo, J. I. Cogolludo-Agustín and D. Matei, Characteristic varieties of quasi-projective manifolds and orbifolds, arXiv:1005.4761.

[3] M. Bestvina and N. Brady, Morse theory and finiteness properties of groups, Invent. Math., **129** (1997), 445–470, MR1465330.

[4] N. Budur, Unitary local systems, multiplier ideals, and polynomial periodicity of Hodge numbers, Adv. Math., **221** (2009), 217–250, MR2509325.

[5] D. Cohen and P. Orlik, Arrangements and local systems, Math. Res. Lett., **7** (2000), 299–316, MR1764324.

[6] D. Cohen and A. Suciu, Characteristic varieties of arrangements, Math Proc. Cambridge Phil. Soc., **127** (1999), 33–53, MR1692519.

[7] P. Deligne, P. Griffiths, J. Morgan and D. Sullivan, Real homotopy theory of Kähler manifolds, Invent. Math., **29** (1975), 245–274, MR0382702.

[8] G. Denham, Homology of subgroups of right-angled Artin groups, arXiv:math/0612748.

[9] A. Dimca, On the irreducible components of characteristic varieties, An. Ştiinţ. Univ. "Ovidius" Constanţa Ser. Mat., **15** (2007), 67–74, MR2396745.

[10] A. Dimca, S. Papadima and A. Suciu, Alexander polynomials: Essential variables and multiplicities, Int. Math. Res. Notices, **2008**, no. 3, Art. ID rnm119, 36 pp, MR2416998.

[11] A. Dimca, S. Papadima and A. Suciu, Topology and geometry of cohomology jump loci, Duke Math. J., **148** (2009), 405–457, MR2527322.

[12] A. Dimca, S. Papadima, and A. Suciu, Quasi–Kähler groups, 3-manifold groups, and formality, Math. Z., **268** (2011), 169–186, MR2805428.

[13] W. G. Dwyer and D. Fried, Homology of free abelian covers. I, Bull. London Math. Soc., **19** (1987), 350–352, MR0887774.

[14] D. Eisenbud, Commutative algebra with a view towards algebraic geometry, Grad. Texts in Math., **150**, Springer-Verlag, 1995, MR1322960.

[15] D. Eisenbud, S. Popescu and S. Yuzvinsky, Hyperplane arrangement cohomology and monomials in the exterior algebra, Trans. Amer. Math. Soc., **55** (2003), 4365–4383, MR1986506.

[16] H. Esnault, V. Schechtman and E. Viehweg, Cohomology of local systems of the complement of hyperplanes, Invent. Math., **109** (1992), 557–561, MR1176205.

[17] M. Falk, Arrangements and cohomology, Ann. Combin., **1** (1997), 135–157, MR1629681.

[18] M. Falk, The line geometry of resonance varieties, arXiv:math/0405210.

[19] M. Falk and S. Yuzvinsky, Multinets, resonance varieties, and pencils of plane curves, Compositio Math., **143** (2007), 1069–1088, MR2339840.

[20] J. Harris, Algebraic geometry, Grad. Texts in Math, **133**, Springer-Verlag, 1992, MR1182558.

[21] A. Libgober, First order deformations for rank one local systems with a non-vanishing cohomology, Topology Appl., **118** (2002), 159–168, MR1877722.

[22] J. W. Morgan, The algebraic topology of smooth algebraic varieties, Inst. Hautes Études Sci. Publ. Math., **48** (1978), 137–204, MR0516917.

[23] S. Nazir and Z. Raza, Admissible local systems for a class of line arrangements, Proc. Amer. Math. Soc., **137** (2009), 1307–1313, MR2465653.

[24] D. Notbohm and N. Ray, On Davis–Januszkiewicz homotopy types. I. formality and rationalisation, Algebr. Geom. Topol., **5** (2005), 31–51, MR2135544.

[25] S. Papadima and A. Suciu, Algebraic invariants for right-angled Artin groups, Math. Ann., **334** (2006), 533–555, MR2207874.

[26] S. Papadima and A. Suciu, Toric complexes and Artin kernels, Adv. Math., **220** (2009), 441–477, MR2466422.

[27] S. Papadima and A. Suciu, Geometric and algebraic aspects of 1-formality, Bull. Math. Soc. Sci. Math. Roumanie, **52** (2009), 355–375, MR2554494.

[28] S. Papadima and A. Suciu, The spectral sequence of an equivariant chain complex and homology with local coefficients, Trans. Amer. Math. Soc., **362** (2010), 2685–2721, MR2584616.

[29] S. Papadima and A. Suciu, Bieri–Neumann–Strebel–Renz invariants and homology jumping loci, Proc. London Math. Soc., **100** (2010), 795–834, MR2640291.

[30] J. Pereira and S. Yuzvinsky, Completely reducible hypersurfaces in a pencil, Adv. Math., **219** (2008), 672–688, MR2435653.

[31] A. Suciu, Fundamental groups of line arrangements: Enumerative aspects, In: Advances in Algebraic Geometry Motivated by Physics, Lowell, MA, 2000, Contemp. Math., **276**, Amer. Math. Soc., Providence, RI, 2001, pp. 43–79, MR1837109.

[32] A. Suciu, Translated tori in the characteristic varieties of complex hyperplane arrangements, Topology Appl., **118** (2002), 209–223, MR1877726.

[33] A. Suciu, Fundamental groups, Alexander invariants, and cohomology jumping loci, In: Topology of Algebraic Varieties and Singularities, Contemp. Math., **538**, Amer. Math. Soc., Providence, RI, 2011, pp. 179–223, MR2777821.

[34] A. Suciu, Characteristic varieties and Betti numbers of free abelian covers, arXiv:1111.5803.

[35] D. Sullivan, Infinitesimal computations in topology, Inst. Hautes Études Sci. Publ. Math., **47** (1977), 269–331, MR0646078.

[36] S. Yuzvinsky, A new bound on the number of special fibers in a pencil of curves, Proc. Amer. Math. Soc., **137** (2009), 1641–1648, MR2470822.

Alexander I. Suciu
Department of Mathematics, Northeastern University,
Boston, MA 02115, USA
E-mail address: a.suciu@neu.edu

Advanced Studies in Pure Mathematics 62, 2012
Arrangements of Hyperplanes—Sapporo 2009
pp. 399–415

Application of arrangement theory to unfolding models

Hidehiko Kamiya, Akimichi Takemura and Norihide Tokushige

Abstract.

Arrangement theory plays an essential role in the study of the unfolding model used in many fields. This paper describes how arrangement theory can be usefully employed in solving the problems of counting (i) the number of admissible rankings in an unfolding model and (ii) the number of ranking patterns generated by unfolding models. The paper is mostly expository but also contains some new results such as simple upper and lower bounds for the number of ranking patterns in the unidimensional case.

§1. Introduction

The unfolding model (Coombs [6], De Leeuw [8]) is a model for preference rankings in psychometrics. It is now widely applied not only in psychometrics (De Soete, Feger, and Klauer [10]) but also in other fields such as marketing science (DeSarbo and Hoffman [9]) and voting theory (Clinton, Jackman, and Rivers [5]). The model is also used as a submodel for more complex models, as in item response theory for unfolding (Andrich [1, 2]). Moreover, in the context of Voronoi diagrams, this model can be regarded as a higher-order Voronoi diagram (Okabe, Boots, Sugihara, and Chiu [22]).

The unfolding model describes the ranking process in which judges rank a set of objects in order of preference. In this model, judges and objects are assumed to be represented by points in the Euclidean space

Received March 25, 2010.
Revised August 6, 2010.
2010 *Mathematics Subject Classification.* 32S22, 52C35, 62F07.
Key words and phrases. All-subset arrangement, braid arrangement, chamber, characteristic polynomial, finite field method, hyperplane arrangement, intersection poset, mid-hyperplane arrangement, partition lattice, ranking pattern, unfolding model.

$\mathbb{R}^n$. Suppose a judge $y \in \mathbb{R}^n$ ranks m objects $x_1, \ldots, x_m \in \mathbb{R}^n$. According to the unfolding model, y ranks $x_1, \ldots, x_m$ in descending order of proximity in the usual Euclidean distance. Hence, y likes x_{i_1} best, x_{i_2} second best, and so on, iff $\|y - x_{i_1}\| < \|y - x_{i_2}\| < \cdots < \|y - x_{i_m}\|$. In this case, we will say y gives ranking $(i_1 i_2 \cdots i_m)$.

For a given m-tuple $(x_1, \ldots, x_m)$ of objects, let $\mathrm{RP}^{\mathrm{UF}}(x_1, \ldots, x_m)$ be the set of admissible rankings, i.e., $(i_1 \cdots i_m)$ such that $\|y - x_{i_1}\| < \cdots < \|y - x_{i_m}\|$ for some $y \in \mathbb{R}^n$. We call $\mathrm{RP}^{\mathrm{UF}}(x_1, \ldots, x_m)$ the ranking pattern of the unfolding model with m-tuple $(x_1, \ldots, x_m)$. In the psychometric literature, there has not been much study on the structure of the ranking pattern. In this paper, we investigate the ranking pattern by using the theory of hyperplane arrangements (Orlik and Terao [23]). Specifically, we consider the following two problems:

(i) Find the cardinality of $\mathrm{RP}^{\mathrm{UF}}(x_1, \ldots, x_m)$ for a given generic m-tuple $(x_1, \ldots, x_m)$;
(ii) Find the cardinality of

$$\{\mathrm{RP}^{\mathrm{UF}}(x_1, \ldots, x_m) : (x_1, \ldots, x_m) \text{ is a generic } m\text{-tuple}\}.$$

The first problem asks how many rankings are admissible in one unfolding model, and the second inquires how many ranking patterns are possible by using different unfolding models (that is, by taking different choices of m-tuples of objects). As we will see, these problems can be reduced to those of counting the numbers of chambers of some real arrangements; moreover, the latter problems can be solved by employing general results in the theory of hyperplane arragements (e.g., Zaslavsky's result on the number of chambers of a real arrangement, the finite field method, etc.). In this sense, arrangement theory plays an essential role in the study of the unfolding model.

This paper gives a survey of recent results ([13], [14], [15], [19]) on the problems stated above. It also contains new results on upper and lower bounds for the number of ranking patterns in the unidimensional case $n = 1$. In addition, the problem of counting inequivalent ranking patterns (i.e., those which cannot be obtained from one another by just the relabeling of the objects) when $n = 1$ was not dealt with specifically in [13] but is discussed fully in the present paper.

The organization of the paper is as follows. In Section 2, we define genericness of the unfolding model, and give the answer to problem (i) above, i.e., the number of admissible rankings of the unfolding model with generic objects. Next, in Section 3 we discuss the problem of counting the number of ranking patterns (problem (ii)). In Subsection 3.1, we deal with the unidimensional case, and give the number of ranking patterns in terms of the number of chambers of the mid-hyperplane arrangement. We also provide explicit upper and lower bounds for the number of ranking patterns. In Subsection 3.2, we treat the unfolding model of codimension one, where the restriction by dimension is weakest. In this case, we describe how the number of ranking patterns can be expressed by the number of chambers of an arrangement called the all-subset arrangement.

§2. Number of admissible rankings

In this section, we define genericness of the unfolding model, and discuss the problem of counting the number of admissible rankings generated by the unfolding model with generic objects.

Suppose we are given $x_1, \ldots, x_m \in \mathbb{R}^n$ with $m \geq 3$ and $n \leq m-2$.

In general, for m distinct points $z_1, \ldots, z_m \in \mathbb{R}^\nu$ ($m \geq \nu + 1$), let $\overline{z_i z_j}$ denote the one-simplex connecting two points z_i and z_j ($i < j$). Consider the following condition:

(A) The union of ν distinct one-simplices $\overline{z_{i_k} z_{j_k}}$ ($i_k < j_k$, $k = 1, \ldots, \nu$) contains no loop if and only if the corresponding vectors $z_{i_k} - z_{j_k}$ ($k = 1, \ldots, \nu$) are linearly independent.

We assume $x_1, \ldots, x_m \in \mathbb{R}^n$ ($n \leq m-2$) are generic in the sense that they satisfy the following two conditions:

(A1) The m points $x_1, \ldots, x_m \in \mathbb{R}^n$ satisfy condition (A).
(A2) The m points $(x_1^T, \|x_1\|^2)^T, \ldots, (x_m^T, \|x_m\|^2)^T \in \mathbb{R}^{n+1}$ satisfy condition (A).

Now, according to the unfolding model, judge $y \in \mathbb{R}^n$ prefers x_i to x_j ($i \neq j$) iff $\|y - x_i\| < \|y - x_j\|$. This condition is equivalent to y being on the same side as x_i of the perpendicular bisector

$$\begin{aligned} H_{ij} &:= \{y \in \mathbb{R}^n : \|y - x_i\| = \|y - x_j\|\} \\ &= \{y \in \mathbb{R}^n : (x_i - x_j)^T(y - \frac{x_i + x_j}{2}) = 0\} \end{aligned}$$

of the line segment $\overline{x_i x_j}$ joining x_i and x_j. Let us define a hyperplane arrangement

$$\mathcal{A}_{m,n} = \mathcal{A}_{m,n}(x_1, \ldots, x_m) := \{H_{ij} : 1 \leq i < j \leq m\}$$

in $\mathbb{R}^n$. We call $\mathcal{A}_{m,n}$ the unfolding arrangement.

Then $\mathcal{A}_{m,n}$, like any real hyperplane arrangement, cuts $\mathbb{R}^n$ into chambers, i.e., connected components of the complement $\mathbb{R}^n \setminus \bigcup \mathcal{A}_{m,n}$, where $\bigcup \mathcal{A}_{m,n} := \bigcup_{H \in \mathcal{A}_{m,n}} H$. Moreover, each of these chambers is of the form

$$C_{i_1 \cdots i_m} := \{\|y - x_{i_1}\| < \cdots < \|y - x_{i_m}\|\} \neq \emptyset$$

for some admissible ranking $(i_1 \cdots i_m) \in \mathbb{P}_m$, where $\mathbb{P}_m$ denotes the set of permutations of $[m] := \{1, \ldots, m\}$.

We observe that $y \in \mathbb{R}^n$ gives ranking $(i_1 \cdots i_m) \in \mathbb{P}_m$ if and only if $y \in C_{i_1 \cdots i_m} \neq \emptyset$. Thus there is a one-to-one correspondence between the set of admissible rankings and the set of chambers $\mathbf{Ch}(\mathcal{A}_{m,n})$ of $\mathcal{A}_{m,n}$:

$$(i_1 \cdots i_m) \leftrightarrow C_{i_1 \cdots i_m}$$

for $(i_1 \cdots i_m)$ such that $C_{i_1 \cdots i_m} \neq \emptyset$. This implies that the problem of counting the number of admissible rankings reduces to that of counting the number of chambers of $\mathcal{A}_{m,n}$. The answer to the latter problem is given by the theorem below. Let $\mathcal{S}_k^m$ $(k \in \mathbb{Z})$ be the signless Stirling numbers of the first kind: $t(t+1) \cdots (t+m-1) = \sum_k \mathcal{S}_k^m t^k$.

Theorem 1 (Good and Tideman [11], Kamiya and Takemura [14, 15], Zaslavsky [30]). *Suppose $x_1, \ldots, x_m \in \mathbb{R}^n$ $(n \leq m-2)$ are generic. Then, the number of chambers of $\mathcal{A}_{m,n} = \mathcal{A}_{m,n}(x_1, \ldots, x_m)$ is*

$$|\mathbf{Ch}(\mathcal{A}_{m,n})| = \mathcal{S}_{m-n}^m + \mathcal{S}_{m-n+1}^m + \cdots + \mathcal{S}_m^m.$$

Furthermore, the number of bounded chambers of $\mathcal{A}_{m,n}$ is

$$\mathcal{S}_{m-n}^m - \mathcal{S}_{m-n+1}^m + \mathcal{S}_{m-n+2}^m - \cdots + (-1)^n \mathcal{S}_m^m.$$

The proof of Theorem 1 is based on Zaslavsky's general result on the number of chambers of an arrangement (Zaslavsky [29]) and the following proposition. Denote by Π_m the partition lattice, consisting of partitions of $[m]$ and ordered by refinement. Further, let Π_m^n stand for the rank n truncation of Π_m, i.e., the subposet of Π_m comprising elements of rank ($= m -$ # of blocks) at most n.

Proposition 1 (Kamiya and Takemura [14, 15]). *The intersection poset $L(\mathcal{A}_{m,n})$ of the unfolding arrangement $\mathcal{A}_{m,n}$ is isomorphic to Π_m^n:*

$$L(\mathcal{A}_{m,n}) \cong \Pi_m^n.$$

The isomorphism is given by

$$L(\mathcal{A}_{m,n}) \ni X \mapsto I_X \in \Pi_m^n,$$

where I_X is the partition of $[m]$ into equivalence classes under the equivalence relation $\sim_X$ defined by $i \sim_X j \overset{\text{def}}{\Longleftrightarrow} X \subseteq H_{ij}$ $(H_{ii} := \mathbb{R}^n)$.

Remark 1. *When $n \geq m-1$, and $x_1, \ldots, x_m \in \mathbb{R}^n$ satisfy condition (A1) with the $\nu = n$ in (A) replaced by $m-1$, we can easily see that $|\mathbf{Ch}(\mathcal{A}_{m,n})| = m!$ and that the number of bounded chambers of $\mathcal{A}_{m,n}$ is zero (so the results in Theorem 1 continue to be valid). Therefore, all $m!$ rankings arise as unbounded chambers of $\mathcal{A}_{m,n}$ in this case.*

§3. Number of ranking patterns

In this section, we consider the problem of counting the number of ranking patterns. We treat two extreme cases—the unidimensional unfolding model: $n = 1$ (Subsection 3.1) and the unfolding model of codimension one: $n = m-2$ (Subsection 3.2).

3.1. Unidimensional unfolding models

In this subsection, we look into the problem of counting the number of ranking patterns of unidimensional unfolding models: $n = 1$. A related problem is studied in Stanley [24].

In this case $n = 1$, objects are m points on the real line: $x_1, \ldots, x_m \in \mathbb{R}$. We assume $x_1, \ldots, x_m$ are generic, i.e., the midpoints $x_{ij} := (x_i + x_j)/2$, $1 \leq i < j \leq m$, are all distinct. This condition can be written as

$$(x_1, \ldots, x_m) \in \mathbb{R}^m \setminus \bigcup \mathcal{M}_m,$$

where $\mathcal{M}_m := \mathcal{B}_m \cup \mathcal{N}_m$ is the mid-hyperplane arrangement (Kamiya, Orlik, Takemura, and Terao [13]) with

$$\begin{aligned}
\mathcal{B}_m &:= \{K_{ij} : 1 \leq i < j \leq m\}, \\
K_{ij} &:= \{(x_1, \ldots, x_m) \in \mathbb{R}^m : x_i = x_j\}, \\
\mathcal{N}_m &:= \{H_{ijkl} : (i,j,k,l) \in I_4\}, \\
H_{ijkl} &:= \{(x_1, \ldots, x_m) \in \mathbb{R}^m : x_i + x_j = x_k + x_l\}, \\
I_4 &:= \{(i,j,k,l) : i,j,k,l \text{ are all distinct}, \\
&\qquad\qquad 1 \leq i < j \leq m,\ i < k < l \leq m\}.
\end{aligned}$$

(In this subsection, we write elements of $\mathbb{R}^m$ as row vectors.) Note that $\mathcal{B}_m$ is the braid arrangement. We have $H_{ij} = \{x_{ij}\}$, $1 \leq i < j \leq m$, and $\mathcal{A}_{m,1} = \{\{x_{ij}\} : 1 \leq i < j \leq m\}$.

An m-tuple $\mathbf{x} := (x_1, \ldots, x_m)$ of objects gives the ranking pattern

$$\mathrm{RP}^{\mathrm{UF}}(\mathbf{x}) = \{(i_1 \cdots i_m) \in \mathbb{P}_m : |y - x_{i_1}| < \cdots < |y - x_{i_m}| \text{ for some } y \in \mathbb{R}\}.$$

We want to know

$$r(m) := |\{\mathrm{RP}^{\mathrm{UF}}(\mathbf{x}) : \mathbf{x} \in \mathbb{R}^m \setminus \bigcup \mathcal{M}_m\}|. \tag{1}$$

The braid arrangement $\mathcal{B}_m$ has a chamber $C_0 \in \mathbf{Ch}(\mathcal{B}_m)$ defined by $x_1 < \cdots < x_m$:

$$C_0 := \{(x_1, \ldots, x_m) \in \mathbb{R}^m : x_1 < \cdots < x_m\}.$$

Let us concentrate our attention on C_0. For $\mathbf{x} = (x_1, \ldots, x_m) \in C_0 \setminus \bigcup \mathcal{N}_m$ and $\mathbf{x}' = (x'_1, \ldots, x'_m) \in C_0 \setminus \bigcup \mathcal{N}_m$, we can easily see that $\mathrm{RP}^{\mathrm{UF}}(\mathbf{x}) = \mathrm{RP}^{\mathrm{UF}}(\mathbf{x}')$ if and only if the order of the midpoints on $\mathbb{R}$ is the same for $\mathbf{x}$ and $\mathbf{x}'$ (i.e., $\forall(i,j,k,l) \in I_4 : x_{ij} < x_{kl} \iff x'_{ij} < x'_{kl}$). Noting that $x_{ij} < x_{kl}$ iff $(x_1, \ldots, x_m) \in H^-_{ijkl} := \{(x_1, \ldots, x_m) \in \mathbb{R}^m : x_i + x_j < x_k + x_l\}$, we obtain the following lemma.

Lemma 1 (Kamiya, Orlik, Takemura, and Terao [13]). *For $\mathbf{x}, \mathbf{x}' \in C_0 \setminus \bigcup \mathcal{N}_m$, we have $\mathrm{RP}^{\mathrm{UF}}(\mathbf{x}) = \mathrm{RP}^{\mathrm{UF}}(\mathbf{x}')$ if and only if $\mathbf{x}$ and $\mathbf{x}'$ are in the same chamber of $\mathcal{N}_m$.*

Put

$$r_0(m) := |\{\mathrm{RP}^{\mathrm{UF}}(\mathbf{x}) : \mathbf{x} \in C_0 \setminus \bigcup \mathcal{N}_m\}|,$$

i.e., the number of ranking patterns of unidimensional unfolding models with generic m-tuples such that $x_1 < \cdots < x_m$. Then, by Lemma 1 we have

$$r_0(m) = \frac{|\mathbf{Ch}(\mathcal{M}_m)|}{m!} \tag{2}$$

(Kamiya, Orlik, Takemura, and Terao [13]).

Now consider $r(m)$ in (1). For $\mathbf{x} = (x_1, \ldots, x_m) \in \mathbb{R}^m \setminus \bigcup \mathcal{M}_m$, define $-\mathbf{x} := (-x_1, \ldots, -x_m) \in \mathbb{R}^m \setminus \bigcup \mathcal{M}_m$. Then, clearly we have $\mathrm{RP}^{\mathrm{UF}}(\mathbf{x}) = \mathrm{RP}^{\mathrm{UF}}(-\mathbf{x})$. On the other hand, for $C, C' \in \mathbf{Ch}(\mathcal{M}_m)$ such that $C' \neq \pm C$ ($-C := \{-\mathbf{x} : \mathbf{x} \in C\}$), we can easily see that $\mathrm{RP}^{\mathrm{UF}}(\mathbf{x}) \neq \mathrm{RP}^{\mathrm{UF}}(\mathbf{x}')$ for $\mathbf{x} \in C$ and $\mathbf{x}' \in C'$. These two facts, together with Lemma 1, yield the following theorem.

Theorem 2. *The number of ranking patterns of unidimensional unfolding models with generic m-tuples of objects is*

$$r(m) = \frac{m!}{2} r_0(m) = \frac{|\mathbf{Ch}(\mathcal{M}_m)|}{2}, \quad m \geq 3.$$

Let us define equivalence of ranking patterns by saying that two ranking patterns $\mathrm{RP}^{\mathrm{UF}}(\mathbf{x})$ and $\mathrm{RP}^{\mathrm{UF}}(\mathbf{x}')$ are equivalent iff

$$\mathrm{RP}^{\mathrm{UF}}(\mathbf{x}) = \sigma \mathrm{RP}^{\mathrm{UF}}(\mathbf{x}') \text{ for some } \sigma \in \mathfrak{S}_m, \tag{3}$$

where $\mathfrak{S}_m$ is the symmetric group on m letters, consisting of all bijections: $[m] \to [m]$, and $\sigma \mathrm{RP}^{\mathrm{UF}}(\mathbf{x}') := \{(\sigma(i_1) \cdots \sigma(i_m)) : (i_1 \cdots i_m) \in$

$\mathrm{RP}^{\mathrm{UF}}(\mathbf{x}')\}$. We want to find the number of inequivalent ranking patterns.

Let $r_{\mathrm{IE}}(m)$ be the number of inequivalent ranking patterns of unidimensional unfolding models with generic m-tuples of objects:

$$r_{\mathrm{IE}}(m) := |\{[\mathrm{RP}^{\mathrm{UF}}(\mathbf{x})] : \mathbf{x} \in \mathbb{R}^m \setminus \bigcup \mathcal{M}_m\}|,$$

where $[\cdot]$ stands for the equivalence class under the equivalence relation defined by (3). We will see that $r_{\mathrm{IE}}(m)$ is half of $r_0(m)$ for $m \geq 4$. Suppose we are given $\mathbf{x} = (x_1, \ldots, x_m) \in C_0 \setminus \bigcup \mathcal{N}_m$ with $m \geq 4$. Then $\mathbf{x}' = (x_1', \ldots, x_m') := (-x_m, \ldots, -x_1)$ also lies in $C_0 \setminus \bigcup \mathcal{N}_m$: $\mathbf{x}' \in C_0 \setminus \bigcup \mathcal{N}_m$. Moreover, since $m \geq 4$, four indices $1, 2, m-1, m$ are all distinct and we have $x_{1m} < x_{2,m-1}$ iff $x_{1m}' > x_{2,m-1}'$. This means $\mathrm{RP}^{\mathrm{UF}}(\mathbf{x}) \neq \mathrm{RP}^{\mathrm{UF}}(\mathbf{x}')$ by Lemma 1. However, $[\mathrm{RP}^{\mathrm{UF}}(\mathbf{x})] = [\mathrm{RP}^{\mathrm{UF}}(\mathbf{x}')]$ since $\mathrm{RP}^{\mathrm{UF}}(\mathbf{x}) = \mathrm{RP}^{\mathrm{UF}}(-\mathbf{x})$. Next, it can be seen that any $\mathbf{x}'' \in C_0 \setminus \bigcup \mathcal{N}_m$ such that $\mathrm{RP}^{\mathrm{UF}}(\mathbf{x}'') \neq \mathrm{RP}^{\mathrm{UF}}(\mathbf{x})$ and $[\mathrm{RP}^{\mathrm{UF}}(\mathbf{x}'')] = [\mathrm{RP}^{\mathrm{UF}}(\mathbf{x})]$ satisfies $\mathrm{RP}^{\mathrm{UF}}(\mathbf{x}'') = \mathrm{RP}^{\mathrm{UF}}(\mathbf{x}')$. These arguments lead to the following theorem.

Theorem 3. *The number of inequivalent ranking patterns of unidimensional unfolding models with generic m-tuples of objects is*

$$r_{\mathrm{IE}}(m) = \begin{cases} r_0(3) = \frac{|\mathbf{Ch}(\mathcal{B}_3)|}{3!} = 1 & \text{if } m = 3, \\ \frac{r_0(m)}{2} = \frac{|\mathbf{Ch}(\mathcal{M}_m)|}{2 \cdot m!} & \text{if } m \geq 4. \end{cases}$$

So far, we have expressed the number of ranking patterns in terms of the number of chambers of an arrangement. We can use the finite field method (Athanasiadis [3, 4], Crapo and Rota [7], Kamiya, Takemura, and Terao [16, 17, 18], Stanley [25, Lecture 5]) to calculate specific values of $r_0(m)$, $m \leq 10$:

$$r_0(4) = 2,\ r_0(5) = 12,\ r_0(6) = 168,\ r_0(7) = 4680,$$
$$r_0(8) = 229386,\ r_0(9) = 18330206,\ r_0(10) = 2241662282.$$

The values of $r(m)$ for $m \leq 8$ are given in Kamiya, Orlik, Takemura, and Terao [13] along with the characteristic polynomials $\chi(\mathcal{M}_m, t)$ of $\mathcal{M}_m$, $m \leq 8$. After [13], the second author of the present paper, Takemura [26], improved on Lemma 3.3 of [13] and calculated $\chi(\mathcal{M}_9, t)$ and

$r_0(9)$; later Ishiwata [12] obtained $\chi(\mathcal{M}_{10},t)$ and $r_0(10)$ after an extensive computation. The characteristic polynomials found by them are:

$$\begin{aligned}\chi(\mathcal{M}_9,t) &= t(t-1)(t^7-413t^6+73780t^5\\ &\qquad -7387310t^4+447514669t^3\\ &\qquad -16393719797t^2+336081719070t\\ &\qquad -2972902161600),\\ \chi(\mathcal{M}_{10},t) &= t(t-1)(t^8-674t^7+201481t^6-34896134t^5\\ &\qquad +3830348179t^4-272839984046t^3\\ &\qquad +12315189583899t^2-321989533359786t\\ &\qquad +3732690616086600).\end{aligned}$$

However, for large values of m, the finite field method is not feasible. We will provide simple upper and lower bounds for $r_0(m)$.

Theorem 4. *For all $m \ge 4$, we have*

$$2\left(\frac{3}{4}\right)^{m-4}\{(m-3)!\}^2 \le r_0(m) < \frac{2}{m!}\left\{\frac{em(m-1)^2}{8}\right\}^{m-2}.$$

Proof. First, we derive the upper bound in the theorem.

Define $H_0 := \{(x_1,\dots,x_m)\in\mathbb{R}^m : x_1+\cdots+x_m=0\}$, and consider the essentialization (Stanley [25, p.392]) $\mathcal{M}_m^0 := \{H\cap H_0 : H\in\mathcal{M}_m\}$ of $\mathcal{M}_m$. Since $L(\mathcal{M}_m^0)\cong L(\mathcal{M}_m)$, we may consider the essential, central arrangement $\mathcal{M}_m^0$ in H_0 ($\dim H_0 = m-1$) instead of $\mathcal{M}_m$.

Recall, in general, that h hyperplanes divide $\mathbb{R}^d$ into at most $\sum_{i=0}^d\binom{h}{i}\le (eh/d)^d =: c(h,d)$ chambers (see, e.g., [20, Proposition 6.1.1] and [21, Theorem 3.6.1]). Thus, $\tilde{h}$ linear hyperplanes divide $\mathbb{R}^{\tilde{d}}$ into at most $2c(\tilde{h}-1,\tilde{d}-1)$ chambers.

In our case, $\mathcal{M}_m^0$ is central, so we can take $\tilde{h}=|\mathcal{M}_m|=|\mathcal{B}_m|+|\mathcal{N}_m|=\binom{m}{2}+3\binom{m}{4}\le m(m-1)^2(m-2)/8$ ($m\ge 4$) and $\tilde{d}=m-1$. Hence, we have

$$\begin{aligned}|\mathbf{Ch}(\mathcal{M}_m^0)| &\le 2c(\tilde{h}-1,\tilde{d}-1)\\ &\le 2\times\left\{\frac{e\left(\frac{m(m-1)^2(m-2)}{8}-1\right)}{m-2}\right\}^{m-2}\\ &< 2\times\left\{\frac{em(m-1)^2}{8}\right\}^{m-2}.\end{aligned}$$

This together with (2) and $|\mathbf{Ch}(\mathcal{M}_m)|=|\mathbf{Ch}(\mathcal{M}_m^0)|$ gives the upper bound of $r_0(m)$ in the theorem.

Next, we will obtain the lower bound in the theorem.

Let $\mathbf{x} = (x_1, \ldots, x_m)$, $x_1 < \cdots < x_m$ be fixed. We add one more object $y = x_m + 2t$ $(t > 0)$ to $\mathbf{x}$, and we will count the number of ranking patterns arising from $\mathbf{y}_t = (\mathbf{x}, y)$, $t > 0$. Let $M = \{x_{ij} : 1 \le i < j \le m\}$ be the set of midpoints for $\mathbf{x}$, and $Y_t = \{x_{im} + t : 1 \le i \le m\}$ the set of midpoints of x_i $(1 \le i \le m)$ and y. Then $M \cup Y_t$ is the set of midpoints for $\mathbf{y}_t$. To guarantee all these midpoints are distinct, we require the following. First, by perturbing each x_i without changing the ranking pattern of $\mathbf{x}$, we may assume that $x_1, \ldots, x_m$ are independent over $\mathbb{Q}$. Then we have $|M \cap Y_t| \le 1$ for all $t > 0$. Next, let $T_0 = \{t > 0 : |M \cap Y_t| = 1\}$, $T_1 = (0, \infty) \setminus T_0$, and we only consider $t \in T_1$. Then $M \cup Y_t$ is legal, i.e., all midpoints are distinct.

Now the crucial observation is as follows: $|\{\mathrm{RP}^{\mathrm{UF}}(\mathbf{y}_t) : t \in T_1\}| = 1 + |T_0|$. Moreover, we have $|T_0| = \sum_{i=1}^{m-1} |V_i|$, where $V_i = \{v \in M : x_{im} < v\}$. Using $|V_i| \ge m - 1 - i$ obtained by $V_i \supset \{x_{jm} : i < j < m\}$, we have

$$|\{\mathrm{RP}^{\mathrm{UF}}(\mathbf{y}_t) : t \in T_1\}| = 1 + \sum_{i=1}^{m-1} |V_i| \ge 1 + |V_1| + \frac{(m-3)(m-2)}{2} =: N.$$

Namely, N is a lower bound for the number of ranking patterns arising from $\mathbf{y}_t$, $t \in T_1$.

Applying exactly the same argument to $\mathbf{x}' = (-x_m, \ldots, -x_1)$ instead of $\mathbf{x}$, we see that the number of ranking patterns arising from $(\mathbf{x}', -x_1 + 2t)$, $t > 0$ (or equivalently, $(x_1 - 2t, \mathbf{x})$, $t > 0$) is at least $N' = 1 + |V_1'| + (m-3)(m-2)/2$, where $|V_1'| = |\{u \in M : u < x_{1m}\}| = \binom{m}{2} - |V_1| - 1$. Notice that $N + N' = 1 + \binom{m}{2} + (m-3)(m-2) > (3/2)(m-2)^2$. Therefore, by the averaging argument, we have

$$r_0(m+1) \ge r_0(m) \times \tfrac{1}{2}(N + N') > \tfrac{3}{4}(m-2)^2\, r_0(m).$$

So the induction starting from $r_0(4) = 2$ gives the desired lower bound. Q.E.D.

Let $\ell(m)$ and $u(m)$ be the lower and upper bounds in the theorem, respectively. A computation shows $\{u(m)\}^{1/m}/m^2 \to e^2/8 \approx 0.92$ and $\{\ell(m)\}^{1/m}/m^2 \to 3/(4e^2) \approx 0.1$ as $m \to \infty$. It would be interesting to prove (or disprove) the existence of $\lim\{r_0(m)\}^{1/m}/m^2$.

Strangely enough, $r_0(m) = a(m)$ holds for $4 \le m \le 7$, where

$$a(m) := \frac{(m-2)\{(m-2)^{m-3} - 1\} \cdot (m-4)!}{m-3},$$

but $r_0(8) > a(8)$, $r_0(9) > a(9)$, $r_0(10) > a(10)$. Also, $a(m)$ satisfies $\{a(m)\}^{1/m}/m^2 \to 1/e \approx 0.37$. We mention that $a(m)/\{(m-3)!\} =$

m	$r_0(m)$	$a(m)$	$\ell(m)$	$u(m)$	$f(m)$
4	2	2	2	12	2
5	12	12	6	334	12
6	168	168	41	$18,744$	286
7	$4,680$	$4,680$	486	1.82×10^6	$33,592$
8	$229,386$	$223,920$	$9,113$	2.76×10^8	$23,178,480$
9	$18,330,206$	$16,470,720$	$246,038$	6.06×10^{10}	$108,995,910,720$
10	$2,241,662,282$	$1,725,655,680$	9.05×10^6	1.81×10^{13}	$3,973,186,258,569,120$

Table 1. $r_0(m)$, $a(m)$, $\ell(m)$, $u(m)$, $f(m)$, $4 \leq m \leq 10$.

$(m-2)\{(m-2)^{m-3}-1\}/(m-3)^2$ $(m \geq 4)$ is the number of acyclic-function digraphs on $m-2$ vertices (Walsh [28], OEIS id:A058128).

Thrall [27] gave an upper bound $f(m)$ for $r_0(m)$:

$$f(m) := \frac{\{\frac{m(m-1)}{2}\}! \prod_{i=1}^{m-2} i!}{\prod_{i=1}^{m-1} (2i-1)!}.$$

Here, $f(m)$ is the number of mappings $\{(i,j) : 1 \leq i < j \leq m\} \ni (i,j) \mapsto d(i,j) \in \{1,2,\ldots,m(m-1)/2\}$ satisfying the condition that $d(i,j)$ be increasing in i for each fixed j as well as increasing in j for each fixed i. He obtained this number by considering a problem similar to that of counting the number of standard Young tableaux. Since for $\mathbf{x} = (x_1,\ldots,x_m) \in C_0 \setminus \bigcup \mathcal{N}_m$, the ranks $d_{\mathbf{x}}(i,j)$ of the midpoints $x_{ij} = (x_i+x_j)/2$ from left to right on the real line $\mathbb{R}$ meet this condition, $f(m)$ is an upper bound for $r_0(m)$. We can see our $u(m)$ satisfies $f(m) < u(m)$ for $m \leq 8$, $f(m) > u(m)$ for $m \geq 9$, and $u(m) = o(f(m))$. For m such that $f(m) < u(m)$, we know the exact values $r_0(m)$ anyway, so the upper bound $u(m)$ based on arrangement theory may be said to be better than $f(m)$.

We list the values of $r_0(m)$, $a(m)$, $f(m)$ and approximate values of $\ell(m)$, $u(m)$ for $m = 4,\ldots,10$ in Table 1. (For $\ell(m)$, $m \leq 9$, and $u(m)$, $m \leq 6$, we exhibit $\lceil \ell(m) \rceil$ and $\lfloor u(m) \rfloor$, respectively. For $\ell(10)$, we display $\lceil \ell(m) \times 10^{-4} \rceil \times 10^4$, and similarly using $\lfloor \cdot \rfloor$ for $u(m)$, $m \geq 7$.)

3.2. Unfolding models of codimension one

In this subsection, we deal with the problem of counting the number of ranking patterns of unfolding models of codimension one: $n = m-2$ (i.e., when the restriction by dimension is weakest).

First, let us forget the unfolding model for a while and consider the ranking patterns of braid slices.

We begin by defining the ranking pattern of a braid slice. For

$$H_0 = \{x = (x_1, \ldots, x_m)^T \in \mathbb{R}^m : x_1 + \cdots + x_m = 0\},$$

consider the essential arrangement

$$\mathcal{B}_m^0 := \{H \cap H_0 : H \in \mathcal{B}_m\}$$

in H_0, and write its chambers as

$$B_{i_1 \cdots i_m} := \{x = (x_1, \ldots, x_m)^T \in H_0 : x_{i_1} > \cdots > x_{i_m}\} \in \mathbf{Ch}(\mathcal{B}_m^0)$$

for $(i_1 \cdots i_m) \in \mathbb{P}_m$. Moreover, define a hyperplane

$$K_v := \{x \in H_0 : v^T x = 1\}$$

in H_0 for each $v \in \mathbb{S}^{m-2} := \{x \in H_0 : \|x\| = 1\}$. Now we call the subset

$$\mathrm{RP}(v) := \{(i_1 \cdots i_m) \in \mathbb{P}_m : K_v \cap B_{i_1 \cdots i_m} \neq \emptyset\}, \quad v \in \mathbb{S}^{m-2},$$

of $\mathbb{P}_m$ the ranking pattern of the braid slice by K_v.

Next, let us define genericness of the braid slice as follows. For the all-subset arrangement (Kamiya, Takemura, and Terao [19])

$$\mathcal{A}_m := \{H_I : I \subseteq [m],\ |I| \geq 1\}$$

with $H_I := \{x = (x_1, \ldots, x_m)^T \in \mathbb{R}^m : \sum_{i \in I} x_i = 0\}$, $\emptyset \neq I \subseteq [m]$, consider its restriction to $H_0 = H_{[m]}$:

$$\mathcal{A}_m^0 := \mathcal{A}_m^{H_0} = \{H_I^0 : I \subset [m],\ 1 \leq |I| \leq m-1\},$$
$$H_I^0 := H_I \cap H_0 \quad (1 \leq |I| \leq m-1).$$

Then define

$$\mathcal{V} := (H_0 \setminus \bigcup \mathcal{A}_m^0) \cap \mathbb{S}^{m-2}.$$

We will say $v \in \mathbb{S}^{m-2}$, or the braid slice by K_v, is generic if $v \in \mathcal{V}$.

Now, we will see that the set of ranking patterns $\mathrm{RP}(v)$ for generic v's is in one-to-one correspondence with the set of chambers of $\mathcal{A}_m^0$. Write $\mathcal{V}$ as $\mathcal{V} = \bigsqcup_{D \in \mathbf{D}(\mathcal{A}_m^0)} D$ (disjoint union), where

$$\mathbf{D}(\mathcal{A}_m^0) := \{D = \tilde{D} \cap \mathbb{S}^{m-2} : \tilde{D} \in \mathbf{Ch}(\mathcal{A}_m^0)\},$$

which clearly is in one-to-one correspondence with $\mathbf{Ch}(\mathcal{A}_m^0)$. Then, we can prove (Kamiya, Takemura, and Terao [19]) that there is a bijection from $\mathbf{D}(\mathcal{A}_m^0)$ to $\{\mathrm{RP}(v) : v \in \mathcal{V}\}$ given by

$$\mathbf{D}(\mathcal{A}_m^0) \ni D \mapsto \mathrm{RP}(v),\ v \in D. \tag{4}$$

Hence,

$$\mathrm{RP}_D := \mathrm{RP}(v) \text{ for } v \in D \in \mathbf{D}(\mathcal{A}_m^0)$$

is well-defined, and the mapping $\mathbf{D}(\mathcal{A}_m^0) \to \{\mathrm{RP}_D : D \in \mathbf{D}(\mathcal{A}_m^0)\} = \{\mathrm{RP}(v) : v \in \mathcal{V}\} : D \mapsto \mathrm{RP}_D$ is bijective.

Let us get back to the unfolding model and consider the ranking pattern of the unfolding model of codimension one.

Suppose we are given $x_1, \dots, x_m \in \mathbb{R}^n$ with $n = m - 2 \geq 1$. We assume $x_1, \dots, x_m$ are generic in the sense that they satisfy (A1) and (A2) in Section 2. We call the unfolding model with such $x_1, \dots, x_m \in \mathbb{R}^{m-2}$ the unfolding model of codimension one (for the reason stated below). In addition, we will assume without loss of generality that $x_1, \dots, x_m$ are taken so that $\sum_{i=1}^m x_i = 0$, $\sum_{i=1}^m \|x_i\|^2/m = 1$.

We will see that the ranking pattern of the unfolding model of codimension one with m-tuple $(x_1, \dots, x_m)$:

$$\begin{aligned}\mathrm{RP}^{\mathrm{UF}}(x_1, \dots, x_m) \quad &= \quad \{(i_1 \cdots i_m) \in \mathbb{P}_m : \|y - x_{i_1}\| < \cdots < \|y - x_{i_m}\| \\ &\qquad \text{for some } y \in \mathbb{R}^{m-2}\}\end{aligned} \tag{5}$$

can be expressed as the ranking pattern of a braid slice.

Define

$$\begin{aligned} W \quad &= \quad \mathrm{W}(x_1, \dots, x_m) = (w_1, \dots, w_{m-2}) := \begin{pmatrix} x_1^T \\ \vdots \\ x_m^T \end{pmatrix} \in \mathrm{Mat}_{m \times (m-2)}(\mathbb{R}), \\ u \quad &= \quad \mathrm{u}(x_1, \dots, x_m) := -\frac{1}{2} \begin{pmatrix} \|x_1\|^2 - 1 \\ \vdots \\ \|x_m\|^2 - 1 \end{pmatrix} \in \mathbb{R}^m, \end{aligned}$$

where $\mathrm{Mat}_{m \times (m-2)}(\mathbb{R})$ denotes the set of $m \times (m-2)$ matrices with real entries. For the affine map $\kappa : \mathbb{R}^{m-2} \to \mathbb{R}^m$ defined by $\kappa(y) := Wy + u$, $y \in \mathbb{R}^{m-2}$, consider the image $K := \operatorname{im} \kappa = \{k(y) : y \in \mathbb{R}^{m-2}\}$ of κ. Then we have

$$K = u + \operatorname{col} W \subset H_0,$$

where $\operatorname{col} W$ stands for the column space of W. Using this K, we can easily see that $\mathrm{RP}^{\mathrm{UF}}(x_1, \dots, x_m)$ in (5) can be expressed as

$$\mathrm{RP}^{\mathrm{UF}}(x_1, \dots, x_m) = \{(i_1 \cdots i_m) \in \mathbb{P}_m : K \cap B_{i_1 \cdots i_m} \neq \emptyset\}. \tag{6}$$

We have $\dim K = \dim H_0 - 1$ and $u \notin \operatorname{col} W$ by (A1) and (A2), respectively. That is, K is an affine hyperplane of H_0. For this reason, we called the unfolding model with generic $x_1, \ldots, x_m \in \mathbb{R}^{m-2}$ the unfolding model of codimension one.

Write the affine hyperplane $K \subset H_0$ as

$$K = K_{\tilde{v}} := \{x \in H_0 : \tilde{v}^T x = \|\tilde{v}\|^2\}$$

using the orthogonal projection of $u \in H_0$ on $(\operatorname{col} W)^\perp := \{x \in H_0 : x^T W = 0\}$:

$$\tilde{v} := \tilde{\mathrm{v}}(x_1, \ldots, x_m) = u - \operatorname{proj}_{\operatorname{col} W}(u), \quad u = \mathrm{u}(x_1, \ldots, x_m),$$

where $\operatorname{proj}_{\operatorname{col} W}$ denotes the orthogonal projection on $\operatorname{col} W$. Noting $\tilde{v} \neq 0$, we can represent (6) as

(7)
$$\mathrm{RP}^{\mathrm{UF}}(x_1, \ldots, x_m) = \{(i_1 \cdots i_m) \in \mathbb{P}_m : K_{\mathrm{v}(x_1,\ldots,x_m)} \cap B_{i_1 \cdots i_m} \neq \emptyset\},$$
$$\mathrm{v}(x_1, \ldots, x_m) := \frac{1}{\|\tilde{v}\|}\tilde{v} \in \mathbb{S}^{m-2},$$

in terms of $K_{\mathrm{v}(x_1,\ldots,x_m)} = \{x \in H_0 : \mathrm{v}(x_1, \ldots, x_m)^T x = 1\}$ instead of $K = K_{\tilde{v}}$. The right-hand side of (7) is the ranking pattern of the braid slice by $K_{\mathrm{v}(x_1,\ldots,x_m)}$: $\mathrm{RP}(\mathrm{v}(x_1, \ldots, x_m))$. Besides, it can be seen that $\mathrm{v}(x_1, \ldots, x_m) \in \mathcal{V}$.

Proposition 2 (Kamiya, Takemura, and Terao [19]). *For generic $x_1, \ldots, x_m \in \mathbb{R}^{m-2}$, we have $\mathrm{v}(x_1, \ldots, x_m) \in \mathcal{V}$ and*

$$\mathrm{RP}^{\mathrm{UF}}(x_1, \ldots, x_m) = \mathrm{RP}(\mathrm{v}(x_1, \ldots, x_m)).$$

Proposition 2 and bijection (4) tell us that in order to find the number of ranking patterns of unfolding models of codimension one, we need to study the image of the mapping $\mathrm{v} : \{(x_1, \ldots, x_m) : x_1, \ldots, x_m \in \mathbb{R}^{m-2} \text{ are generic}\} \to \mathcal{V} = \bigsqcup_{D \in \mathbf{D}(\mathcal{A}_m^0)} D$, $(x_1, \ldots, x_m) \mapsto \mathrm{v}(x_1, \ldots, x_m)$. In their main theorem (Theorem 4.1), Kamiya, Takemura, and Terao [19] proved that the image $\operatorname{im} \mathrm{v}$ is given by

$$\operatorname{im} \mathrm{v} = \mathcal{V}_2 \sqcup D_1 \sqcup \cdots \sqcup D_m = \mathcal{V} \setminus ((-D_1) \sqcup \cdots \sqcup (-D_m)), \tag{8}$$

where

$$\mathcal{V}_2 := \{v = (v_1, \ldots, v_m)^T \in \mathcal{V} : v_j > 0 \text{ for at least two } j \in [m] \text{ and } v_k < 0 \text{ for at least two } k \in [m]\}$$

and

$$\begin{aligned} D_i &:= \{v = (v_1, \ldots, v_m)^T \in \mathcal{V} : v_i > 0,\ v_j < 0\ (j \neq i)\} \in \mathbf{D}(\mathcal{A}_m^0), \\ -D_i &:= \{-v : v \in D_i\} \\ &= \{v = (v_1, \ldots, v_m)^T \in \mathcal{V} : v_i < 0,\ v_j > 0\ (j \neq i)\} \in \mathbf{D}(\mathcal{A}_m^0) \end{aligned}$$

for $i \in [m]$.

By Proposition 2 and im v in (8), we obtain the number of ranking patterns of unfolding models of codimension one, which is denoted by

$$q(m) := |\{\mathrm{RP}^{\mathrm{UF}}(x_1, \ldots, x_m) : \text{ generic } x_1, \ldots, x_m \in \mathbb{R}^{m-2}\}|.$$

Theorem 5 (Kamiya, Takemura, and Terao [19]). *The number $q(m)$ of ranking patterns of unfolding models of codimension one is given by*

$$q(m) = |\mathbf{Ch}(\mathcal{A}_m^0)| - m.$$

Kamiya, Takemura, and Terao [19, Lemma 5.3] obtained the characteristic polynomials $\chi(\mathcal{A}_m^0, t)$ of $\mathcal{A}_m^0$ for $m \leq 8$ by the finite field method. Then $q(m)$ can be calculated by $q(m) = (-1)^{m-1}\chi(\mathcal{A}_m^0, -1) - m$:

$$q(3) = 3,\ q(4) = 28,\ q(5) = 365,$$
$$q(6) = 11286,\ q(7) = 1066037,\ q(8) = 347326344$$

([19, Corollary 5.5]).

We end this subsection by looking at the problem of finding the number of inequivalent ranking patterns of unfolding models of codimension one.

In (3), we defined equivalence of ranking patterns of unidimensional unfolding models. We define equivalence of ranking patterns of unfolding models of codimension one in an obvious similar manner. At the moment, we can only give an upper bound for the number $q_{\mathrm{IE}}(m)$ of inequivalent ranking patterns of unfolding models of codimension one:

(9)
$$q_{\mathrm{IE}}(m) \leq \frac{|\mathbf{Ch}(\mathcal{A}_m^0 \cup \mathcal{B}_m^0)|}{m!} - 1 = |\mathbf{D}^{1\cdots m}(\mathcal{A}_m^0)| - 1 = |\mathbf{D}_2^{1\cdots m}(\mathcal{A}_m^0)| + 1$$

for $m \geq 3$ (Kamiya, Takemura, and Terao [19]), where $\mathbf{D}^{1\cdots m}(\mathcal{A}_m^0) := \{D \in \mathbf{D}(\mathcal{A}_m^0) : D \cap B_{1\cdots m} \neq \emptyset\}$ and $\mathbf{D}_2^{1\cdots m}(\mathcal{A}_m^0) := \{D \in \mathbf{D}(\mathcal{A}_m^0) : D \subset \mathcal{V}_2,\ D \cap B_{1\cdots m} \neq \emptyset\} = \mathbf{D}^{1\cdots m}(\mathcal{A}_m^0) \setminus \{D_1, -D_m\}$. It is shown in [19], however, that the upper bound in (9) is actually the exact number for $m \leq 6$. The specific values are

$$q_{\mathrm{IE}}(3) = 1,\ q_{\mathrm{IE}}(4) = 3,\ q_{\mathrm{IE}}(5) = 11,\ q_{\mathrm{IE}}(6) = 55$$

([19, Subsection 6.2]).

Open problem: Does the upper bound in (9) agree with the exact number $q_{\mathrm{IE}}(m)$ for all m?

Acknowledgments. The authors are very grateful to an anonymous referee for his/her valuable comments on an earlier version of this paper. The first author is supported by JSPS KAKENHI (19540131, 22540134). The second author is supported by JST CREST. The last author is supported by JSPS KAKENHI (20340022).

References

[1] D. Andrich, The application of an unfolding model of the PIRT type to the measurement of attitude, Appl. Psychol. Meas., **12** (1988), 33–51.

[2] D. Andrich, A probabilistic IRT model for unfolding preference data, Appl. Psychol. Meas., **13** (1989), 193–216.

[3] C. A. Athanasiadis, Algebraic Combinatorics of Graph Spectra, Subspace Arrangements and Tutte Polynomials, Ph.D. thesis, MIT, 1996.

[4] C. A. Athanasiadis, Characteristic polynomials of subspace arrangements and finite fields, Adv. Math., **122** (1996), 193–233.

[5] J. Clinton, S. Jackman and D. Rivers, The statistical analysis of roll call data, Amer. Polit. Sci. Rev., **98** (2004), 355–370.

[6] C. H. Coombs, Psychological scaling without a unit of measurement, Psychol. Rev., **57** (1950), 145–158.

[7] H. Crapo and G.-C. Rota, On the Foundations of Combinatorial Theory: Combinatorial Geometries, preliminary ed., MIT Press, Cambridge, MA, 1970.

[8] J. De Leeuw, Multidimensional unfolding, In: Encyclopedia of Statistics in Behavioral Science, (eds. B. S. Everitt and D. C. Howell), John Wiley & Sons, Chichester, West Sussex, 2005, pp. 1289–1294.

[9] W. S. DeSarbo and D. L. Hoffman, Constructing MDS joint spaces from binary choice data: A multidimensional unfolding threshold model for marketing research, J. Marketing Res., **24** (1987), 40–54.

[10] G. De Soete, H. Feger and K. C. Klauer, New Developments in Psychological Choice Modeling, North-Holland, Amsterdam, 1989.

[11] I. J. Good and T. N. Tideman, Stirling numbers and a geometric structure from voting theory, J. Combin. Theory A, **23** (1977), 34–45.

[12] M. Ishiwata, Mid-Hyperplane Arrangement and Ranking Patterns, (in Japanese), Master's thesis, Tokyo Metropolitan Univ., 2007.

[13] H. Kamiya, P. Orlik, A. Takemura and H. Terao, Arrangements and ranking patterns, Ann. Combin., **10** (2006), 219–235.

[14] H. Kamiya and A. Takemura, On rankings generated by pairwise linear discriminant analysis of m populations, J. Multivariate Anal., **61** (1997), 1–28.

[15] H. Kamiya and A. Takemura, Characterization of rankings generated by linear discriminant analysis, J. Multivariate Anal., **92** (2005), 343–358.
[16] H. Kamiya, A. Takemura and H. Terao, Periodicity of hyperplane arrangements with integral coefficients modulo positive integers, J. Algebraic Combin., **27** (2008), 317–330.
[17] H. Kamiya, A. Takemura and H. Terao, The characteristic quasi-polynomials of the arrangements of root systems and mid-hyperplane arrangements, In: Arrangements, Local Systems and Singularities, CIMPA Summer School, Galatasaray Univ., Istanbul, 2007, (eds. F. El Zein, A. I. Suciu, M. Tosun, A. M. Uludağ and S. Yuzvinsky), Progr. Math., **283**, Birkhäuser Verlag, Basel, 2009, pp. 177–190.
[18] H. Kamiya, A. Takemura and H. Terao, Periodicity of non-central integral arrangements modulo positive integers, Ann. Combin., **15** (2011), 449–464.
[19] H. Kamiya, A. Takemura and H. Terao, Ranking patterns of unfolding models of codimension one, Adv. Appl. Math., **47** (2011), 379–400.
[20] J. Matoušek, Lectures on Discrete Geometry, Grad. Texts in Math.,**212**, Springer-Verlag, 2002.
[21] J. Matoušek and J. Nešetřil, Invitation to Discrete Mathematics, 2nd ed., Oxford Univ. Press, New York, 2009.
[22] A. Okabe, B. Boots, K. Sugihara and S. N. Chiu, Spatial Tessellations: Concepts and Applications of Voronoi Diagrams, 2nd ed., John Wiley & Sons, Chichester, West Sussex, 2000.
[23] P. Orlik and H. Terao, Arrangements of Hyperplanes, Springer-Verlag, 1992.
[24] R. Stanley, Ordering events in Minkowski space, Adv. Appl. Math., **37** (2006), 514–525.
[25] R. Stanley, An introduction to hyperplane arrangements, In: Geometric Combinatorics, (eds. E. Miller, V. Reiner and B. Sturmfels), IAS/Park City Math. Ser., **13**, Amer. Math. Soc., Providence, RI, 2007, pp. 389–496.
[26] A. Takemura, unpublished results, 2005.
[27] R. M. Thrall, A combinatorial problem, Michigan Math. J., **1** (1952), 81–88.
[28] D. P. Walsh, Notes on acyclic functions and their directed graphs, available at `http://frank.mtsu.edu/~dwalsh/acyclic/acycnote.html`.
[29] T. Zaslavsky, Facing up to Arrangements: Face-Count Formulas for Partitions of Space by Hyperplanes, Mem. Amer. Math. Soc., **1** (1975), issue 1, no. 154, vii+102 pp.
[30] T. Zaslavsky, Perpendicular dissection of space, Discrete Comput. Geom., **27** (2002), 303–351.

Hidehiko Kamiya
Graduate School of Economics, Nagoya University
Nagoya 464-8601, Japan
E-mail address: kamiya@soec.nagoya-u.ac.jp

Akimichi Takemura
Graduate School of Information Science and Technology,
University of Tokyo,
Tokyo 113-0033, Japan
E-mail address: takemura@stat.t.u-tokyo.ac.jp

Norihide Tokushige
College of Education, Ryukyu University
Nishihara, Okinawa 903-0213, Japan
E-mail address: hide@edu.u-ryukyu.ac.jp

Advanced Studies in Pure Mathematics 62, 2012
Arrangements of Hyperplanes—Sapporo 2009
pp. 417–457

On the homology of configuration spaces associated to centers of mass

Dai Tamaki

Abstract.

The aim of this paper is to make sample computations with the Salvetti complex of the "center of mass" arrangement introduced in [CK07] by Cohen and Kamiyama. We compute the homology of the Salvetti complex of these arrangements with coefficients in the sign representation of the symmetric group on $\mathbb{F}_p$ in the case of four particles. We show, when p is an odd prime, the homology is isomorphic to the homology of the configuration space $F(\mathbb{C}, 4)$ of distinct four points in $\mathbb{C}$ with the same coefficients. When $p = 2$, we show the homology is different from the equivariant homology of $F(\mathbb{C}, 4)$, hence we obtain an alternative and more direct proof of a theorem of Cohen and Kamiyama in [CK07].

§1. Introduction

The configuration spaces of distinct points in $\mathbb{C}$

$$F(\mathbb{C}, n) = \{(z_1, \cdots, z_n) \in \mathbb{C}^n \mid z_i \neq z_j \text{ if } i \neq j\}$$

have been playing important roles in homotopy theory. For example, F. Cohen [Coh83] used the weak equivalence

$$\Omega^2\Sigma^2 X \underset{w}{\simeq} \left(\coprod_n F(\mathbb{C}, n) \times_{\Sigma_n} X^n\right)\Big/_{\sim}$$

to construct an unstable splitting map

$$\Sigma^{2n}\Omega^2\Sigma^2 X \longrightarrow \Sigma^{2n} F(\mathbb{C}, n)_+ \wedge_{\Sigma_n} X^{\wedge n}. \tag{1}$$

Received March 31, 2010.
Revised August 3, 2010.
2010 *Mathematics Subject Classification.* Primary 52C35; Secondary 55P35.
Key words and phrases. Salvetti complex, braid arrangement, loop space.

This is a desuspension of the well-known stable splitting

$$\Sigma^\infty\Omega^2\Sigma^2X \underset{w}{\simeq} \Sigma^\infty\left(\bigvee_n F(\mathbb{C},n)_+ \wedge_{\Sigma_n} X^{\wedge n}\right)$$

due to Snaith [Sna74].

These stable and unstable splitting maps can be used to construct important maps in unstable homotopy theory. See [Mah77, Coh83], for example. It is also known that the map (1) cannot be desuspended further [CM82]. There is a chance of desuspending this map, however, if we localize at an appropriate prime. B. Gray observed in [Gra93a, Gra93b] that if we could construct a map

$$\Sigma^2\Omega^2S^3 \longrightarrow \Sigma^2F(\mathbb{C},p)_+ \wedge_{\Sigma_p} S^p \tag{2}$$

after localizing at an odd prime p, we would be able to refine results of Cohen, Moore, and Neisendorfer [CMN79b, CMN79a] and construct higher order EHP sequences.

The difficulty is to construct a localized model of Ω^2S^3 in terms of configuration spaces. We do not know very much about localizations of configuration spaces. As an attempt to construct such a localized model, F. Cohen and Kamiyama introduced a subspace $M_\ell(\mathbb{C},n)$ of $F(\mathbb{C},n)$ in [CK07], for natural numbers n and ℓ with $\ell < n$. It can be defined as the complement in $\mathbb{C}^n$ of the complexification of the real central hyperplane arrangement defined by

$$\mathcal{C}_{n-1}^\ell = \{L_{I,J} \mid I,J \subset \{1,\cdots,n\}, |I| = |J| = \ell, I \neq J\},$$

where

$$L_{I,J} = \left\{(x_1,\cdots,x_n) \in \mathbb{R}^n \;\middle|\; |J|\sum_{i\in I} x_i = |I|\sum_{j\in J} x_j\right\}$$

for $I,J \subset \{1,\cdots,n\}$. For $\ell \geq n$, we define $\mathcal{C}_{n-1}^\ell$ to be the braid arrangement $\mathcal{A}_{n-1}$.

Notice that

$$\mathcal{C}_{n-1}^\ell = \begin{cases} \{L_{I,J} \mid I,J \subset \{1,\cdots,n\}, \\ \qquad |I| = |J| \leq \ell, I\cap J = \emptyset\}, & \ell \leq \frac{n}{2}, \\ \{L_{I,J} \mid I,J \subset \{1,\cdots,n\}, \\ \qquad |I| = |J| \leq n-\ell, I\cap J = \emptyset\}, & \ell > \frac{n}{2} \end{cases}$$

and we have the duality

$$\mathcal{C}_{n-1}^\ell = \mathcal{C}_{n-1}^{n-\ell}$$

for $\ell < n$. Thus we have the following inclusions of arrangements:

$$\mathcal{A}_{n-1} = \mathcal{C}_{n-1}^1 \subset \mathcal{C}_{n-1}^2 \subset \cdots \subset \mathcal{C}_{n-1}^{[\frac{n}{2}]} \supset \cdots \supset \mathcal{C}_{n-1}^{n-2} \supset \mathcal{C}_{n-1}^{n-1} = \mathcal{A}_{n-1}.$$

By taking the complements, we obtain

$$(3) \quad M_\ell(\mathbb{C}, n) \subset M_{\ell-1}(\mathbb{C}, n) \subset \cdots \subset M_2(\mathbb{C}, n) \subset M_1(\mathbb{C}, n) = F(\mathbb{C}, n)$$

if $\ell \leq [\frac{n}{2}]$. When $n > \ell > [\frac{n}{2}]$, we have

$$(4) \quad \begin{aligned} M_\ell(\mathbb{C}, n) &= M_{n-\ell}(\mathbb{C}, n) \subset M_{n-\ell-1}(\mathbb{C}, n) \subset \\ &\cdots \subset M_2(\mathbb{C}, n) \subset M_1(\mathbb{C}, n) = F(\mathbb{C}, n). \end{aligned}$$

We denote these inclusions by $i_{\ell,n} : M_\ell(\mathbb{C}, n) \hookrightarrow F(\mathbb{C}, n)$.

Conjecture 1.1 (Cohen–Kamiyama). *For an odd prime p, the natural inclusion*

$$i_{p,n} : M_p(\mathbb{C}, n) \hookrightarrow F(\mathbb{C}, n)$$

induces an isomorphism

$$(i_{p,n})_* : H_*(S_*(M_p(\mathbb{C}, n)) \otimes_{\Sigma_n} \mathbb{F}_p(\pm 1)) \longrightarrow H_*(S_*(F(\mathbb{C}, n)) \otimes_{\Sigma_n} \mathbb{F}_p(\pm 1))$$

for all n, where $S_*(-)$ is the singular chain complex functor, Σ_n is the symmetric group on n letters, and $\mathbb{F}_p(\pm 1)$ is $\mathbb{F}_p$ regarded as a Σ_n-module via sign representation.

They proved that the above conjecture implies the existence of the desired map (2). They also initiated the analysis of the homology of $M_p(\mathbb{C}, n)$ and proved the statement of the conjecture does not hold when $p = 2$. Notice that $M_2(\mathbb{C}, 2) = F(\mathbb{C}, 2)$ by definition and $M_2(\mathbb{C}, 3) = M_1(\mathbb{C}, 3) = F(\mathbb{C}, 3)$ by the duality. Thus $M_2(\mathbb{C}, 4)$ is the first nontrivial case.

Theorem 1.2. *The class in $H_3(S_*(F(\mathbb{C}, 4)) \otimes_{\Sigma_4} \mathbb{F}_2)$ corresponding to $Q_1^2(x)$ in $H_*(\Omega^2 S^{n+2}; \mathbb{F}_2)$ is not in the image of the map*

$$H_3(S_*(M_2(\mathbb{C}, 4)) \otimes_{\Sigma_4} \mathbb{F}_2) \longrightarrow H_3(S_*(F(\mathbb{C}, 4)) \otimes_{\Sigma_4} \mathbb{F}_2)$$

induced by the natural inclusion. Hence this map is not surjective.

Their method is indirect in the sense that they proved it by contradiction by calculating homology and cohomology operations. In order to find a way to attack the conjecture, a more direct method is desirable.

For a real central hyperplane arrangement $\mathcal{A}$ in general, Salvetti [Sal87] constructed a finite cell complex $\mathrm{Sal}(\mathcal{A})$ embedded in the complement of the complexification of $\mathcal{A}$ as a deformation retract. The aim of this paper is to determine the maps induced on homology groups in the last step of the inclusions (3) and (4) by using the Salvetti complex.

We reprove Theorem 1.2 by analyzing the cellular structure of the Salvetti complex. In fact, we show that $H_3(S_*(M_2(\mathbb{C},4)) \otimes_{\Sigma_4} \mathbb{F}_2) \cong \mathbb{F}_2$ and the map takes the generator to 0.

For odd primes, we obtain the following result.

Theorem 1.3. *For an odd prime p, the inclusion $M_2(\mathbb{C},4) \hookrightarrow F(\mathbb{C},4)$ induces an isomorphism*

$$H_*(S_*(M_2(\mathbb{C},4)) \otimes_{\Sigma_4} \mathbb{F}_p(\pm 1)) \cong H_*(S_*(F(\mathbb{C},4)) \otimes_{\Sigma_4} \mathbb{F}_p(\pm 1)).$$

The paper is organized as follows:

- Basic properties of the Salvetti complex used in this paper are recalled in §2.
- We describe the cell decomposition of the Salvetti complex for the braid arrangement in §3 by using the notations in [Tam].
- The Salvetti complex for the center of mass arrangement is studied in §4, including the computation of the homology of $\mathcal{C}_3^2$. We also include a computation with trivial $\mathbb{F}_p$ coefficients for p an odd prime, following the suggestion by the referee.

Acknowledgments. The calculations in §4 were done while the author was visiting the University of Aberdeen and the National University of Singapore. He really appreciates the hospitality of the members of these institutes, especially, Jelena Grbic, Stephen Theriault, Ran Levi, Jon Berrick, and Jie Wu. The author would also like to thank the organizers of the MSJ-SI conference "Arrangements of Hyperplanes" for invitation, during which the author learned the results in [Sal94] from Professor Salvetti himself. It helped the author to understand and clarify the boundary formula proved in [Tam].

This work is partially supported by Grants-in-Aid for Scientific Research, Ministry of Education, Culture, Sports, Science and Technology, Japan: 17540070.

§2. Salvetti complex and oriented matroid

Let us recall the definition and basic properties of the Salvetti complex used in this paper.

2.1. Salvetti complex for real central arrangements

A hyperplane arrangement $\mathcal{A}$ in a real vector space V defines a stratification of V by

$$\begin{aligned} S_0 &= V - \bigcup_{L\in\mathcal{A}} L \\ S_1 &= \bigcup_{L\in\mathcal{A}} L - \bigcup_{L,L'\in\mathcal{A}} L\cap L' \\ &\vdots \\ S_{|\mathcal{L}|} &= \bigcap_{L\in\mathcal{A}} L. \end{aligned}$$

Each stratum is a disjoint union of convex regions. These connected components are called faces and the faces in the top stratum are called chambers or topes. The set of faces is denoted by $\mathcal{L}(\mathcal{A})$ and the subset of chambers is denoted by $\mathcal{L}^{(0)}(\mathcal{A})$. $\mathcal{L}(\mathcal{A})$ has a structure of poset by

$$F \le G \Longleftrightarrow F \subset \overline{G}$$

and is called the face poset of $\mathcal{A}$.

When $\mathcal{A}$ is central, i.e. hyperplanes in $\mathcal{A}$ are vector subspaces, Salvetti [Sal87] constructed a finite regular cell complex $\mathrm{Sal}(\mathcal{A})$ embedded in $V\otimes\mathbb{C} - \bigcup_{L\in\mathcal{A}} L\otimes\mathbb{C}$ as a deformation retract. Its cellular structure is determined by the face poset $\mathcal{L}(\mathcal{A})$ of the real arrangement $\mathcal{A}$.

A rough idea of the original construction of the Salvetti complex $\mathrm{Sal}(\mathcal{A})$ is as follows: For each face F in $\mathcal{A}$, choose a point $w(F)$ in F. For each pair of a face F and a chamber C with $F \le C$, define a point in $V\otimes\mathbb{C}$ by

$$v(F,C) = w(F)\otimes 1 + (w(C) - w(F))\otimes i.$$

The set of all such points is denoted by

$$\mathrm{sk}_0\,\mathrm{Sal}(\mathcal{A}) = \{v(F,C) \mid F\in\mathcal{L}(\mathcal{A}), C\in\mathcal{L}^{(0)}(\mathcal{A}), F\le C\}.$$

The Salvetti complex $\mathrm{Sal}(\mathcal{A})$ is constructed as a Euclidean simplicial complex embedded in $V\otimes\mathbb{C} - \bigcup_{L\in\mathcal{A}} L\otimes\mathbb{C}$ by forming simplices by choosing vertices from these points in a certain manner. Salvetti defines a structure of a finite cell complex on $\mathrm{Sal}(\mathcal{A})$ by combining several simplices together. Since all we need is this cell decomposition, we recall this cellular structure, instead of describing the rule for simplices. The cell decomposition can be described in terms of the face-chamber pairing, or the matroid product. The definition of the matroid product can be found in [Sal87, Arv91]. An alternative description will be given later in Lemma 2.4.

Definition 2.1. For $F \in \mathcal{L}(\mathcal{A})$ and $C \in \mathcal{L}^{(0)}(\mathcal{A})$ with $F \leq C$, define a subset of $\mathrm{sk}_0(\mathrm{Sal}(\mathcal{A}))$ by

$$\mathcal{D}(F, C) = \{v(G, G \circ C) \mid G \geq F\},$$

where $\circ$ is the matroid product. This set is regarded as a poset by

$$v(G, G \circ C) \leq v(H, H \circ C) \Longleftrightarrow G \leq H.$$

The (geometric realization of the) order complex of $\mathcal{D}(F, C)$ is denoted by $D(F, C)$.

Lemma 2.2. *The complex $D(F, C)$ has the following properties:*

(1) *The inclusion of vertices induces a simplicial embedding*

$$D(F, C) \hookrightarrow \mathrm{Sal}(\mathcal{A}).$$

(2) *$D(F, C)$ is homeomorphic to a disk of dimension* $\operatorname{codim} F$.

(3) *The boundary of $D(F, C)$ is given by*

$$\partial D(F, C) = \bigcup_{G > F} D(G, G \circ C).$$

(4) *The decomposition*

$$\mathrm{Sal}(\mathcal{A}) = \bigcup_{v(F,C) \in \mathrm{sk}_0(\mathrm{Sal}(\mathcal{A} \otimes \mathbb{C}))} (D(F, C) - \partial D(F, C))$$

defines a structure of a finite regular cell complex on $\mathrm{Sal}(\mathcal{A})$.

In order to compute the boundary, therefore, we need to understand the matroid product. The following elementary fact is very useful.

Lemma 2.3. *Given a real central hyperplane arrangement $\mathcal{A} = \{L_1, \cdots, L_n\}$ in a real inner product space $(V, \langle -, - \rangle)$, choose a normal vector a_i for each hyperplane L_i*

$$L_i = \{x \in V \mid \langle a_i, x \rangle = 0\}$$

and define $\mathcal{V}(\mathcal{A}) = \{a_1, \cdots, a_n\}$.

Let $S_1 = \{0, +1, -1\}$ be the poset with $0 < +1, -1$. For $F \in \mathcal{L}(\mathcal{A})$, define

$$\tau_F : \mathcal{V}(\mathcal{A}) \longrightarrow S_1$$

by

$$\tau_F(a) = \mathrm{sign}\langle a, F\rangle,$$

where

$$\mathrm{sign} : \mathbb{R} \longrightarrow S_1$$

is the sign function.

Then we obtain an embedding

$$\tau : \mathcal{L}(\mathcal{A}) \hookrightarrow \mathrm{Map}(\mathcal{V}(\mathcal{A}), S_1).$$

It is possible to give an explicit description of the subset $\tau(\mathcal{L}(\mathcal{A}))$ by using the language of oriented matroid. See the paper [GR89] by Gel′fand and Rybnikov or the book on oriented matroids [BLVS+99] by five authors for more details.

With this identification, the face-chamber pairing in the face lattice $\mathcal{L}(\mathcal{A})$ can be translated into the following product of functions.

Lemma 2.4. *Let E be a set. For $\varphi, \psi \in \mathrm{Map}(E, S_1)$ define $\varphi \circ \psi \in \mathrm{Map}(E, S_1)$ by*

$$(\varphi \circ \psi)(x) = \begin{cases} \varphi(x), & \varphi(x) \neq 0 \\ \psi(x), & \varphi(x) = 0. \end{cases}$$

Then for a real central arrangement $\mathcal{A}$, we have

$$\tau_F \circ \tau_G = \tau_{F \circ G}$$

for $F, G \in \mathcal{L}(\mathcal{A})$.

We may formally complexify the poset $\mathcal{L}(\mathcal{A})$ by using the poset $S_2 = \{0, +1, -1, +i, -i\}$ with ordering $0 < \pm 1 < \pm i$.

Definition 2.5. Define Σ_2-equivariant inclusions

$$i_1, i_2 : S_1 \hookrightarrow S_2$$

by

$$\begin{aligned} i_1(0) &= i_2(0) = 0 \\ i_1(\pm 1) &= \pm 1 \\ i_2(\pm 1) &= \pm i. \end{aligned}$$

Definition 2.6. Let E be a set and L be a subset of $\mathrm{Map}(E, S_1)$. Define a subposet $L \otimes \mathbb{C}$ of $\mathrm{Map}(E, S_2)$ by

$$L \otimes \mathbb{C} = \{(i_1)_*(X) \circ (i_2)_*(Y) \mid X, Y \in L\},$$

where

$$(i_1)_*, (i_2)_* : \mathrm{Map}(E, S_1) \longrightarrow \mathrm{Map}(E, S_2)$$

are the maps induced by i_1 and i_2 and the matroid product $\circ$ in $\mathrm{Map}(E, S_2)$ is defined by

$$(\varphi \circ \psi)(x) = \begin{cases} \varphi(x), & \varphi(x) \not\leq \psi(x) \\ \psi(x), & \varphi(x) \leq \psi(x). \end{cases}$$

Another useful way of describing the Salvetti complex is as follows. See [BZ92], for example.

Proposition 2.7. *Let $\mathcal{A}$ be a real central arrangement in an inner product space V and $\mathcal{V}(\mathcal{A})$ be a set of unit normal vectors of hyperplanes in $\mathcal{A}$. Define a subposet $\mathcal{L}^{(1)}(\mathcal{A})$ of $\mathcal{L}(\mathcal{A}) \otimes \mathbb{C}$ by*

$$\mathcal{L}^{(1)}(\mathcal{A}) = \{X \in \mathcal{L}(\mathcal{A}) \otimes \mathbb{C} \mid X(v) \neq 0 \text{ for all } v \in \mathcal{V}(\mathcal{A})\}.$$

Then $\mathcal{L}^{(1)}(\mathcal{A})$ is isomorphic to the face poset $F(\mathrm{Sal}(A))$ of the Salvetti complex of $\mathcal{A}$ as posets. Thus we have an isomorphism of simplicial complexes

$$B\mathcal{L}^{(1)}(\mathcal{A}) \cong \mathrm{Sd}(\mathrm{Sal}(A)),$$

where $B(-)$ is the classifying space (order complex) functor and $\mathrm{Sd}(-)$ is the barycentric subdivision.

2.2. Salvetti complex for reflection arrangements

When $\mathcal{A}$ is a reflection arrangement, Salvetti analyzed the cellular structure of $\mathrm{Sal}(\mathcal{A})/G(\mathcal{A})$ in [Sal94], where $G(\mathcal{A})$ is the reflection group associated with $\mathcal{A}$. In particular, he described the boundary in the cellular cochain complex. Salvetti's work includes the case of affine reflection groups. Here we only consider central arrangements.

Let $\mathcal{A}$ be a real central reflection arrangement in V and $\mathcal{L}(\mathcal{A})$ be the face poset. We have a cellular decomposition (stratification) of V

$$V = \coprod_{F \in \mathcal{L}(\mathcal{A})} F.$$

The cell complex dual to this cellular decomposition is denoted by $C(\mathcal{A})$, whose face poset is denoted by $F(C(\mathcal{A}))$. One of the ways to construct $C(\mathcal{A})$ is to choose a chamber $C_0 \in \mathcal{L}^{(0)}(\mathcal{A})$ and a point v_0 inside of C_0 and to take the convex hull of the G-orbit of v_0

$$C(\mathcal{A}) = \mathrm{Conv}(gv_0 \mid g \in G(\mathcal{A})).$$

The cellular structure of $C(\mathcal{A})$ is given by that of this convex polytope.

When $e \in F(C(\mathcal{A}))$ is the face dual to $F \in \mathcal{L}(\mathcal{A})$ we denote $e = F^*$ and $F = e^*$.

Definition 2.8. For each face $e \in F(C(\mathcal{A}))$, let $\gamma(e) \in G(\mathcal{A})$ be the unique element of minimal length such that

$$\gamma(e)^{-1}(e^*) \subset \overline{C_0}.$$

Salvetti found the following description of $\mathrm{Sal}(\mathcal{A})/G(\mathcal{A})$.

Theorem 2.9. *The cell complex* $\mathrm{Sal}(\mathcal{A})/G(\mathcal{A})$ *can be identified with the cell complex given by identifying those two cells* e, e' *in* $C(\mathcal{A})$ *which are in the same* $G(\mathcal{A})$*-orbit by using the homeomorphism induced by the element* $\gamma(e')\gamma(e)^{-1}$.

In order to describe the boundary homomorphism in the cellular chain complex, Salvetti defined an orientation on each cell in $C(\mathcal{A})$.

Definition 2.10. Fix an ordering of hyperplanes $H_1, \cdots, H_n$ bounding the chosen chamber C_0. Let v_i be the projection of v_0 onto H_i. Define

$$\begin{aligned} F(C_0) &= \{F \in \mathcal{L}(\mathcal{A}) \mid F \subset \overline{C_0}\}, \\ F^*(C_0) &= \{F^* \in F(C(\mathcal{A})) \mid F \in F(C_0)\}. \end{aligned}$$

For a cell $e \in F^*(C_0)$, define an orientation on e as follows. Let $H_{i_1}, \cdots, H_{i_k}$ be hyperplanes with

$$e^* \subset H_{i_1} \cap \cdots \cap H_{i_k}$$

and $i_1 < \cdots < i_k$. The orientation of e is induced from the ordering $v_0, v_{i_1}, \cdots, v_{i_k}$ under the inclusion

$$\mathrm{Conv}(v_0, v_{i_1}, \cdots, v_{i_k}) \subset \overline{e}.$$

In general, define an orientation on $e \in C(\mathcal{A})$ in such a way that $\gamma(e)^{-1}$ is orientation preserving.

Under the above orientations, the incidence numbers among cells in $C(\mathcal{A})$ are described as follows.

Proposition 2.11. *Let* $F \in F(C_0)$ *and* $G \in \mathcal{L}(\mathcal{A})$ *with*

$$\overline{gG} \supset F$$

for an element $g \in G(\mathcal{A})$ *of the shortest length and* $\dim G = \dim F + 1$. *Then*

$$[F^* : G^*] = (-1)^{\ell(g)}[F^*, (gG)^*].$$

2.3. Maps between Salvetti complexes

The "center of mass" arrangement $\mathcal{C}_{n-1}^{\ell}$ is obtained by adding hyperplanes to the braid arrangement $\mathcal{A}_{n-1}$. And contravariantly we have an inclusion

$$M_\ell(\mathbb{C}, n) \hookrightarrow F(\mathbb{C}, n)$$

of the complements. We also have corresponding maps on the Salvetti complexes.

Lemma 2.12. *Let $\mathcal{A}$ be a real central arrangement in a vector space V and $\mathcal{B} \subset \mathcal{A}$ be a subarrangement. Then the inclusion*

$$i : \mathcal{B} \hookrightarrow \mathcal{A}$$

induces a cellular map

$$i^* : \mathrm{Sal}(\mathcal{A}) \longrightarrow \mathrm{Sal}(\mathcal{B})$$

which makes the following diagram commutative up to homotopy

$$\begin{array}{ccc} \mathrm{Sal}(\mathcal{A}) & \xrightarrow{\quad i^* \quad} & \mathrm{Sal}(\mathcal{B}) \\ \downarrow & & \downarrow \\ V \otimes \mathbb{C} - \bigcup_{H \in \mathcal{A}} H \otimes \mathbb{C} & \longrightarrow & V \otimes \mathbb{C} - \bigcup_{H \in \mathcal{B}} H \otimes \mathbb{C}. \end{array}$$

Proof. The inclusion $i : \mathcal{B} \hookrightarrow \mathcal{A}$ induces a map of face posets

$$i^* : \mathcal{L}(\mathcal{A}) \longrightarrow \mathcal{L}(\mathcal{B})$$

(i.e. it induces a strong map between oriented matroids) which induces a map of posets

$$i^* : \mathcal{L}(\mathcal{A}) \otimes \mathbb{C} \longrightarrow \mathcal{L}(\mathcal{B}) \otimes \mathbb{C}.$$

Since i^* is given by restriction, we obtain

$$i^* : \mathcal{L}^{(1)}(\mathcal{A}) \longrightarrow \mathcal{L}^{(1)}(\mathcal{B})$$

and hence a map

$$i^* : \mathrm{Sal}(\mathcal{A}) \longrightarrow \mathrm{Sal}(\mathcal{B})$$

by Proposition 2.7.

The embeddings of the Salvetti complexes depend on choices of simplicial vertices corresponding to faces in the face posets. We obtain embeddings by choosing $w(F)$ for $F \in \mathcal{L}(\mathcal{A})$ first and then by choosing vertices for $\mathcal{B}$ among $\{w(F)\}$ which make the required diagram commutative. Q.E.D.

We would like to know the behavior of the chain map

$$i_* : C_*(\mathrm{Sal}(\mathcal{A})) \longrightarrow C_*(\mathrm{Sal}(\mathcal{B})).$$

Lemma 2.13. *Let $\mathcal{A}$ be a real central arrangement in V and $\mathcal{B} \subset \mathcal{A}$ be a subarrangement. Then the inclusion*

$$i : \mathcal{B} \hookrightarrow \mathcal{A}$$

induces a surjective chain map

$$i^* : C_*(\mathrm{Sal}(\mathcal{A})) \longrightarrow C_*(\mathrm{Sal}(\mathcal{B})).$$

Proof. Generators of $C_*(\mathrm{Sal}(\mathcal{B}))$ are in one-to-one correspondence with pairs (F, C) of a face F and a chamber C in $\mathcal{L}(\mathcal{B})$. Since

$$i^* : \mathcal{L}(\mathcal{A}) \longrightarrow \mathcal{L}(\mathcal{B})$$

is surjective, it induces a surjective map on the cellular chain complexes of Salvetti complexes. Q.E.D.

§3. The Salvetti complex for the braid arrangement

We need to understand the cellular structure of the Salvetti complex of the braid arrangement in order to compare it with that of the center of mass arrangement.

3.1. The structure of cell complex

The braid arrangement is a typical example of reflection arrangements and the results of §2.2 apply. In particular, the cell structure of the Salvetti complex for the braid arrangement can be described in terms of partitions. The following symbols are introduced in [Tam].

Definition 3.1. A partition of $\{1, \cdots, n\}$ is a surjective map

$$\lambda : \{1, \cdots, n\} \longrightarrow \{1, \cdots, n-r\}$$

for some $0 \leq r < n$. The number r is called the rank of this partition.

The set of partitions of $\{1, \cdots, n\}$ is denoted by Π_n. The subset of rank r partitions is denoted by $\Pi_{n,r}$. Π_n becomes a poset under refinement. Note that rank 0 partitions are nothing but elements of Σ_n.

Definition 3.2. For a partition $\lambda \in \Pi_n$ of rank r and $\sigma \in \Sigma_n$ with $\sigma \geq \lambda$, define a symbol $S(\lambda, \sigma)$ as follows:

(1) For each $1 \leq i \leq n-r$, draw vertically stacked squares S_i of length $|\lambda^{-1}(i)|$.

(2) Order $\lambda^{-1}(i)$ according to σ and label each square in S_i from bottom to top by elements in $\lambda^{-1}(i)$. For example, when $\lambda^{-1}(i) = \{i_1, i_2, i_3, i_4, i_5\}$ and if these numbers appear in $(\sigma(1), \cdots, \sigma(n))$ in the order

$$i_1, i_2, i_3, i_4, i_5$$

then S_i is labeled as

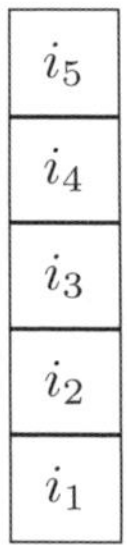

(3) Place $S_1, \cdots, S_{n-r}$ side by side from left to right. $S(\lambda, \sigma)$ is the resulting picture.

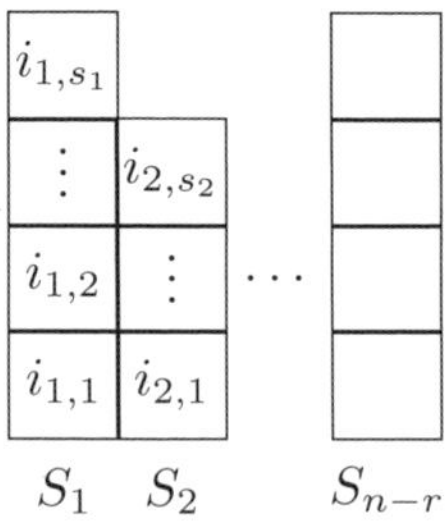

The following observation played an essential role in [Tam].

Lemma 3.3. *There is a bijection between the set of vertices (of the simplicial structure)* $\mathrm{sk}_0\,\mathrm{Sal}(\mathcal{A}_{n-1})$ *and the set of symbols*

$$\{S(\lambda, \sigma) \mid \lambda \in \Pi_n, \sigma \in \Sigma_n, \lambda \leq \sigma\}.$$

Thus we obtain a bijection between the cells of $\mathrm{Sal}(\mathcal{A}_{n-1})$ *and the symbols* $S(\lambda, \sigma)$.

In order to compute the boundary maps of the cellular chain complex of $\mathrm{Sal}(\mathcal{A}_{n-1})$, we need to fix orientations of cells. We follow the orientations defined in §2.2.

We choose the chamber

$$C_0 = \{(x_1, \cdots, x_n) \in \mathbb{R}^n \mid x_1 < x_2 < \cdots < x_n\}$$

and define $v_0 = (1, 2, \cdots, n) \in C_0$. Then

$$C(\mathcal{A}_{n-1}) = \mathrm{Conv}(v_0\sigma \mid \sigma \in \Sigma_n).$$

We have the following refinement of Theorem 2.9.

Proposition 3.4. *We have the following isomorphism of* $\mathbb{F}_p$*-modules*

$$\begin{aligned} C_*(\mathrm{Sal}(\mathcal{A}_{n-1})) \otimes_{\Sigma_n} \mathbb{F}_p(\pm 1) &\cong \mathbb{F}_p \langle [e] \mid [e] \in F(C(\mathcal{A}_{n-1}))/\Sigma_n \rangle \\ &\cong \mathbb{F}_p \langle e \mid e \in F^*(C_0) \rangle . \end{aligned}$$

Proof. Let S be a set with an action of Σ_n. Then we have an isomorphism of $\mathbb{F}_p$-modules

$$\mathbb{Z}\langle S \rangle \otimes_{\Sigma_n} \mathbb{F}_p(\pm 1) \cong \mathbb{Z}\langle S/\Sigma_n \rangle \otimes \mathbb{F}_p.$$

And the result follows by the idenfitications of cells in the proof of Theorem 2.9 in [Sal94]. Q.E.D.

The bounding hyperplanes of C_0 are

$$L_{1,2}, L_{2,3}, \cdots, L_{n,n-1}.$$

This ordering of hyperplanes determines orientations of cells in $C(\mathcal{A}_{n-1})$. Under the correspondence in Lemma 3.3, cells in $C(\mathcal{A}_{n-1})$ correspond to $S(\lambda, 1)$ with $\lambda \in \Pi_n$ and $\lambda \leq 1$. Those cells in $F^*(C_0)$ corresponds to ordered partitions.

Definition 3.5. An order preserving surjective map

$$\lambda : \{1, \cdots, k\} \longrightarrow \{1, \cdots, k-r\}$$

is called an ordered partition of rank r. The set of ordered partitions of $\{1, \cdots, k\}$ of rank r is denoted by $O_{k,r}$.

Corollary 3.6. *Under the identification in Lemma 3.3, we have the following isomorphism of* $\mathbb{F}_p$*-modules*

$$C_s(\mathrm{Sal}(\mathcal{A}_{n-1})) \otimes_{\Sigma_n} \mathbb{F}_p(\pm 1) \cong \mathbb{F}_p\langle D(\lambda, (1|\cdots|n)) \mid \lambda \in O_{n,n-s}\rangle.$$

The following formula for the boundary map follows from Proposition 2.11.

Lemma 3.7. *For $\lambda \in O_{n,n-s}$, we have*

$$\partial(D(\lambda,(1|\cdots|n))) = \sum_{\tau\in O_{n,n-s-1},\lambda<\tau}\sum_{\sigma\in\Sigma(\lambda)} \operatorname{sgn}(\sigma)[D(\lambda,(1|\cdots|n)) : D(\tau,(1|\cdots|n))]D(\tau\sigma,\sigma)$$

in $C_*(\mathrm{Sal}(\mathcal{A}_{n-1}))$, where

$$\Sigma(\lambda) = \{\sigma\in\Sigma_n \mid \lambda\sigma = \lambda\}$$

is the set of permutations preserving the partition λ.

Proof. The set of faces in $\mathrm{Sal}(\mathcal{A}_{n-1})$ contained in $D(\lambda,(1|\cdots|n))$ as a face is given by $D(\tau\sigma,\sigma)$ for $\lambda<\tau$ and $\sigma\in\Sigma(\lambda)$. Thus

$$\begin{aligned} & \partial D(\lambda,(1|\cdots|n)) \\ = & \sum_{\tau\in O_{n,n-s-1}\lambda<\tau}\sum_{\sigma\in\Sigma(\tau)} [D(\lambda,(1|\cdots|n)) : D(\tau\sigma,\sigma)]D(\tau\sigma,\sigma) \\ = & \sum_{\tau\in O_{n,n-s-1}\lambda<\tau} \\ & \sum_{\sigma\in\Sigma(\tau)} \operatorname{sgn}(\sigma)[D(\lambda,(1|\cdots|n)) : D(\tau,(1|\cdots|n))]D(\tau\sigma,\sigma) \end{aligned}$$

by Proposition 2.11. Q.E.D.

Note that the incidence number $[D(\lambda,(1|\cdots|n)) : D(\tau,(1|\cdots|n))]$ can be determined by comparing the "positions of =" in λ and τ.

3.2. The homology of $F(\mathbb{C},4)$

The homology $H_*(S_*(F(\mathbb{C},4))\otimes_{\Sigma_4}\mathbb{F}_p(\pm1))$ is well-known. We need, however, an explicit description in order to compare it with $H_*(C_*(\mathrm{Sal}(\mathcal{C}_3^2))\otimes_{\Sigma_4}\mathbb{F}_p(\pm1))$ in the next section.

Let us compute $H_*(C_*(\mathrm{Sal}(\mathcal{A}_3))\otimes_{\Sigma_4}\mathbb{F}_p(\pm1))$ by using the symbols introduced in the previous section. $\mathrm{Sal}(\mathcal{A}_3)$ has the following cells:

- 0-cells are in one-to-one correspondence with the symbols

$\sigma(1)$	$\sigma(2)$	$\sigma(3)$	$\sigma(4)$

.

- 1-cells are in one-to-one correspondence with the symbols

$$\begin{array}{ccc} \boxed{\sigma(2)} & & \\ \boxed{\sigma(1)} & \boxed{\sigma(3)} & \boxed{\sigma(4)} \end{array}, \quad \begin{array}{ccc} & \boxed{\sigma(3)} & \\ \boxed{\sigma(1)} & \boxed{\sigma(2)} & \boxed{\sigma(4)} \end{array}, \quad \begin{array}{ccc} & & \boxed{\sigma(4)} \\ \boxed{\sigma(1)} & \boxed{\sigma(2)} & \boxed{\sigma(3)} \end{array}.$$

- 2-cells are in one-to-one correspondence with the symbols

$$\begin{array}{cc} & \boxed{\sigma(4)} \\ & \boxed{\sigma(3)} \\ \boxed{\sigma(1)} & \boxed{\sigma(2)} \end{array}, \quad \begin{array}{cc} \boxed{\sigma(2)} & \boxed{\sigma(4)} \\ \boxed{\sigma(1)} & \boxed{\sigma(3)} \end{array}, \quad \begin{array}{cc} \boxed{\sigma(3)} & \\ \boxed{\sigma(2)} & \\ \boxed{\sigma(1)} & \boxed{\sigma(4)} \end{array}.$$

- 3-cells are in one-to-one correspondence with the symbols

$$\begin{array}{c} \boxed{\sigma(4)} \\ \boxed{\sigma(3)} \\ \boxed{\sigma(2)} \\ \boxed{\sigma(1)} \end{array}.$$

Thus the chain complex $C_*(\mathrm{Sal}(\mathcal{A}_3)) \otimes_{\Sigma_4} \mathbb{F}_p(\pm 1)$ has the following basis.

$$C_0(\mathrm{Sal}(\mathcal{A}_3)) \otimes_{\Sigma_4} \mathbb{F}_p(\pm 1) = \left\langle \begin{array}{cccc} \boxed{1} & \boxed{2} & \boxed{3} & \boxed{4} \end{array} \right\rangle$$

$$C_1(\mathrm{Sal}(\mathcal{A}_3)) \otimes_{\Sigma_4} \mathbb{F}_p(\pm 1) = \left\langle \begin{array}{ccc} \boxed{2} & & \\ \boxed{1} & \boxed{3} & \boxed{4} \end{array}, \begin{array}{ccc} & \boxed{3} & \\ \boxed{1} & \boxed{2} & \boxed{4} \end{array}, \begin{array}{ccc} & & \boxed{4} \\ \boxed{1} & \boxed{2} & \boxed{3} \end{array} \right\rangle$$

$$C_2(\mathrm{Sal}(\mathcal{A}_3)) \otimes_{\Sigma_4} \mathbb{F}_p(\pm 1) = \left\langle \begin{array}{cc} & \boxed{4} \\ & \boxed{3} \\ \boxed{1} & \boxed{2} \end{array}, \begin{array}{cc} \boxed{2} & \boxed{4} \\ \boxed{1} & \boxed{3} \end{array}, \begin{array}{cc} \boxed{3} & \\ \boxed{2} & \\ \boxed{1} & \boxed{4} \end{array} \right\rangle$$

$$C_3(\mathrm{Sal}(\mathcal{A}_3)) \otimes_{\Sigma_4} \mathbb{F}_p(\pm 1) = \left\langle \begin{array}{c} \boxed{4} \\ \boxed{3} \\ \boxed{2} \\ \boxed{1} \end{array} \right\rangle.$$

The boundaries can be computed by using Lemma 3.7 as follows. For 0-cells, we obviously have

$$\partial_0 \left(\begin{array}{cccc} \boxed{1} & \boxed{2} & \boxed{3} & \boxed{4} \end{array} \right) = 0.$$

For 1-cells, we have

$$\partial_1\left(\begin{ytableau} 2 \\ 1 & 3 & 4 \end{ytableau}\right) = \begin{ytableau} 1 & 2 & 3 & 4 \end{ytableau} - \begin{ytableau} 2 & 1 & 3 & 4 \end{ytableau} = 2\begin{ytableau} 1 & 2 & 3 & 4 \end{ytableau}.$$

Similarly, we have

$$\partial_1\left(\begin{ytableau} \none & 3 \\ 1 & 2 & 4 \end{ytableau}\right) = 2\begin{ytableau} 1 & 2 & 3 & 4 \end{ytableau}$$

$$\partial_1\left(\begin{ytableau} \none & \none & 4 \\ 1 & 2 & 3 \end{ytableau}\right) = 2\begin{ytableau} 1 & 2 & 3 & 4 \end{ytableau}.$$

For 2-cells, we have

$$\partial_2\left(\begin{ytableau} \none & 4 \\ \none & 3 \\ 1 & 2 \end{ytableau}\right) = 3\begin{ytableau} \none & \none & 4 \\ 1 & 2 & 3 \end{ytableau} - 3\begin{ytableau} \none & 3 \\ 1 & 2 & 4 \end{ytableau}$$

$$\partial_2\left(\begin{ytableau} 2 & 4 \\ 1 & 3 \end{ytableau}\right) = 2\begin{ytableau} \none & \none & 4 \\ 1 & 2 & 3 \end{ytableau} - 2\begin{ytableau} 2 \\ 1 & 3 & 4 \end{ytableau}$$

$$\partial_2\left(\begin{ytableau} 3 \\ 2 \\ 1 & 4 \end{ytableau}\right) = 3\begin{ytableau} \none & 3 \\ 1 & 2 & 4 \end{ytableau} - 3\begin{ytableau} 2 \\ 1 & 3 & 4 \end{ytableau}.$$

Finally, for 3-cells, we have

$$\partial_3\left(\begin{ytableau} 4 \\ 3 \\ 2 \\ 1 \end{ytableau}\right) = 4\begin{ytableau} \none & 4 \\ \none & 3 \\ 1 & 2 \end{ytableau} - 6\begin{ytableau} 2 & 4 \\ 1 & 3 \end{ytableau} + 4\begin{ytableau} 3 \\ 2 \\ 1 & 4 \end{ytableau}.$$

Thus we have the following well-known result.

Proposition 3.8. *When $p > 3$, we have*

$$H_i(C_*(\mathrm{Sal}(\mathcal{A}_3)) \otimes_{\Sigma_4} \mathbb{F}_p(\pm 1)) = 0$$

for all i.

Proof. Since $p \neq 2$,

$$
\begin{aligned}
\operatorname{Im}\partial_1 &= \left\langle 2\,\begin{ytableau} 1 & 2 & 3 & 4 \end{ytableau} \right\rangle = \left\langle \begin{ytableau} 1 & 2 & 3 & 4 \end{ytableau} \right\rangle \\
\operatorname{Ker}\partial_1 &= \left\langle \begin{ytableau} \none & 3 \\ 1 & 2 & 4 \end{ytableau} - \begin{ytableau} \none & \none & 4 \\ 1 & 2 & 3 \end{ytableau},\ \begin{ytableau} \none & 3 \\ 1 & 2 & 4 \end{ytableau} - \begin{ytableau} 2 \\ 1 & 3 & 4 \end{ytableau} \right\rangle.
\end{aligned}
$$

Thus we have

$$H_0(C_*(\mathrm{Sal}(\mathcal{C}_3^2)) \otimes_{\Sigma_4} \mathbb{F}_p(\pm 1)) = 0.$$

Since $p \neq 3$, we have

$$
\begin{aligned}
\operatorname{Im}\partial_2 &= \left\langle \begin{ytableau} \none & 3 \\ 1 & 2 & 4 \end{ytableau} - \begin{ytableau} \none & \none & 4 \\ 1 & 2 & 3 \end{ytableau},\ \begin{ytableau} \none & 3 \\ 1 & 2 & 4 \end{ytableau} - \begin{ytableau} 2 \\ 1 & 3 & 4 \end{ytableau} \right\rangle \\
&= \operatorname{Ker}\partial_1
\end{aligned}
$$

and

$$H_1(C_*(\mathrm{Sal}(\mathcal{C}_3^2)) \otimes_{\Sigma_4} \mathbb{F}_p(\pm 1)) = 0.$$

We also have

$$\operatorname{Ker}\partial_2 = \left\langle \frac{1}{3}\,\begin{ytableau} \none & 4 \\ \none & 3 \\ 1 & 2 \end{ytableau} - \frac{1}{2}\,\begin{ytableau} 2 & 4 \\ 1 & 3 \end{ytableau} + \frac{1}{3}\,\begin{ytableau} 3 \\ 2 \\ 1 & 4 \end{ytableau} \right\rangle = \operatorname{Im}\partial_3$$

and

$$H_2(C_*(\mathrm{Sal}(\mathcal{C}_3^2)) \otimes_{\Sigma_4} \mathbb{F}_p(\pm 1)) = 0.$$

Since $\operatorname{Ker}\partial_3 = 0$,

$$H_3(C_*(\mathrm{Sal}(\mathcal{C}_3^2)) \otimes_{\Sigma_4} \mathbb{F}_p(\pm 1)) = 0.$$

Q.E.D.

Proposition 3.9. *When $p = 3$, we have*

$$
\begin{aligned}
H_0(C_*(\mathrm{Sal}(\mathcal{C}_3^2)) \otimes_{\Sigma_4} \mathbb{F}_3(\pm 1)) &= 0 \\
H_1(C_*(\mathrm{Sal}(\mathcal{C}_3^2)) \otimes_{\Sigma_4} \mathbb{F}_3(\pm 1)) &= \left\langle \left[\begin{ytableau} \none & 3 \\ 1 & 2 & 4 \end{ytableau} - \begin{ytableau} \none & \none & 4 \\ 1 & 2 & 3 \end{ytableau} \right] \right\rangle \cong \mathbb{F}_3 \\
H_2(C_*(\mathrm{Sal}(\mathcal{C}_3^2)) \otimes_{\Sigma_4} \mathbb{F}_3(\pm 1)) &= \left\langle \left[\begin{ytableau} \none & 4 \\ \none & 3 \\ 1 & 2 \end{ytableau} - \begin{ytableau} 3 \\ 2 \\ 1 & 4 \end{ytableau} \right] \right\rangle \cong \mathbb{F}_3 \\
H_3(C_*(\mathrm{Sal}(\mathcal{C}_3^2)) \otimes_{\Sigma_4} \mathbb{F}_3(\pm 1)) &= 0.
\end{aligned}
$$

Proof. The differences are the computation of H_2 and H_3. The result follows from

$$\operatorname{Ker}\partial_2 = \left\langle \begin{array}{|c|c|} \hline \multicolumn{1}{c|}{} & 4 \\ \cline{2-2} \multicolumn{1}{c|}{} & 3 \\ \hline 1 & 2 \\ \hline \end{array} \, , \, \begin{array}{|c|c|} \cline{1-1} 3 & \multicolumn{1}{c}{} \\ \cline{1-1} 2 & \multicolumn{1}{c}{} \\ \hline 1 & 4 \\ \hline \end{array} \right\rangle .$$

The details are omitted. Q.E.D.

Remark 3.10. It is well-known that

$$H_*(\Omega^2 S^{2n+1}; \mathbb{F}_3) \cong \Lambda(Q_1^a(x_{2n-1}) \mid a \geq 0) \otimes \mathbb{F}_3[\beta Q_1^{a+1}(x) \mid a \geq 0].$$

Under the Snaith splitting (and dimension shifts), the generators in $H_1(C_*(\mathrm{Sal}(\mathcal{C}_3^2))\otimes_{\Sigma_4}\mathbb{F}_3(\pm 1))$ and $H_2(C_*(\mathrm{Sal}(\mathcal{C}_3^2))\otimes_{\Sigma_4}\mathbb{F}_3(\pm 1))$ correspond to $x_{2n-1}\beta Q_1(x_{2n-1})$ and $x_{2n-1}Q_1(x_{2n-1})$, respectively, since

$$\begin{aligned} \deg x_{2n-1}\beta Q_1(x_{2n-1}) &= (2n-1) + 3(2n-1) + (p-2) \\ &= 4(2n-1) + 1 \\ \deg x_{2n-1} Q_1(x_{2n-1}) &= (2n-1) + 3(2n-1) + (p-1) \\ &= 4(2n-1) + 2. \end{aligned}$$

The 2-primary case is simpler, since we don't have to worry about the signs. We have

$$\begin{aligned} H_*(C_*(\mathrm{Sal}(\mathcal{A}_3)) \otimes_{\Sigma_4} \mathbb{F}_2(\pm 1)) &= H_*(C_*(\mathrm{Sal}(\mathcal{A}_3)) \otimes_{\Sigma_4} \mathbb{F}_2) \\ &= H_*(C_*(\mathrm{Sal}(\mathcal{A}_3)/\Sigma_4) \otimes \mathbb{F}_2) \\ &= H_*(F(\mathbb{C}, 4)/\Sigma_4; \mathbb{F}_2). \end{aligned}$$

We obtain the following well-known result by elementary calculations. Details are omitted.

Proposition 3.11. *The homology $H_*(F(\mathbb{C},4)/\Sigma_4;\mathbb{F}_2)$ has the following description:*

$$
\begin{aligned}
H_0(F(\mathbb{C},4)/\Sigma_4;\mathbb{F}_2) &= \left\langle \begin{array}{|c|c|c|c|}\hline 1&2&3&4\\\hline\end{array} \right\rangle \\
&\cong \mathbb{F}_2 \\
H_1(F(\mathbb{C},4)/\Sigma_4;\mathbb{F}_2) &= \left\langle \begin{array}{|c|c|c|}\cline{1-1} 2 \\ \hline 1&3&4\\\hline\end{array} = \begin{array}{|c|c|c|}\cline{2-2} \multicolumn{1}{c|}{} & 3 \\ \hline 1&2&4\\\hline\end{array} = \begin{array}{|c|c|c|}\cline{3-3} \multicolumn{2}{c|}{} & 4 \\ \hline 1&2&3\\\hline\end{array} \right\rangle \\
&\cong \mathbb{F}_2 \\
H_2(F(\mathbb{C},4)/\Sigma_4;\mathbb{F}_2) &= \left\langle \begin{array}{|c|c|}\hline 2&4\\\hline 1&3\\\hline\end{array} \right\rangle \\
&\cong \mathbb{F}_2 \\
H_3(F(\mathbb{C},4)/\Sigma_4;\mathbb{F}_2) &= \left\langle \begin{array}{|c|}\hline 4\\\hline 3\\\hline 2\\\hline 1\\\hline\end{array} \right\rangle \\
&\cong \mathbb{F}_2
\end{aligned}
$$

Remark 3.12. Under the stable splitting

$$\Omega^2 S^{n+2} \underset{S}{\simeq} \bigvee_j F(\mathbb{C},j)_+ \wedge_{\Sigma_j} (S^n)^{\wedge j},$$

the elements in the mod 2 homology of $F(\mathbb{C},4)/\Sigma_4$, up to a shift of degree, correspond to elements in $H_*(\Omega^2 S^{n+2};\mathbb{F}_2)$ as follows:

$$
\begin{aligned}
Q_0^2(x) &\longleftrightarrow \begin{array}{|c|c|c|c|}\hline 1&2&3&4\\\hline\end{array} \\
Q_0(x)Q_1(x) &\longleftrightarrow \begin{array}{|c|c|c|}\cline{1-1} 2 \\ \hline 1&3&4\\\hline\end{array} = \begin{array}{|c|c|c|}\cline{2-2} \multicolumn{1}{c|}{} & 3 \\ \hline 1&2&4\\\hline\end{array} = \begin{array}{|c|c|c|}\cline{3-3} \multicolumn{2}{c|}{} & 4 \\ \hline 1&2&3\\\hline\end{array} \\
Q_0Q_1(x) &\longleftrightarrow \begin{array}{|c|c|}\hline 2&4\\\hline 1&3\\\hline\end{array} \\
Q_1^2(x) &\longleftrightarrow \begin{array}{|c|}\hline 4\\\hline 3\\\hline 2\\\hline 1\\\hline\end{array}
\end{aligned}
$$

§4. The center of mass configuration

Let us recall the definition of the center of mass configuration space introduced by F. Cohen and Kamiyama in [CK07].

Definition 4.1. For $I, J \subset \{1, \cdots, n\}$, define

$$L_{I,J} = \left\{ (x_1, \cdots, x_n) \in \mathbb{R}^n \;\middle|\; |J| \sum_{i \in I} x_i = |I| \sum_{j \in J} x_j \right\}.$$

For $\ell < n$, define a real central hyperplane arrangement $\mathcal{C}^{\ell}_{n-1}$ by

$$\mathcal{C}^{\ell}_{n-1} = \{ L_{I,J} \mid I, J \subset \{1, \cdots, n\}, |I| = |J| = p, I \neq J \}.$$

The configuration space of n points with distinct center of mass of ℓ points is defined as the complement of the complexification of $\mathcal{C}^{\ell}_{n-1}$

$$M_{\ell}(\mathbb{C}, n) = \mathbb{C}^n - \bigcup_{L_{I,J} \in \mathcal{C}^{\ell}_{n-1}} L_{I,J} \otimes \mathbb{C}.$$

As we have seen in §1, we have the following inclusions of arrangements

$$\mathcal{A}_{n-1} = \mathcal{C}^1_{n-1} \subset \mathcal{C}^2_{n-1} \subset \cdots \subset \mathcal{C}^{[\frac{n}{2}]}_{n-1} \supset \cdots \supset \mathcal{C}^{n-1}_{n-1} = \mathcal{A}_{n-1}.$$

By Proposition 2.7, we obtain a sequence of maps between the Salvetti complexes

$$\mathrm{Sal}(\mathcal{C}^p_{n-1}) \to \cdots \to \mathrm{Sal}(\mathcal{C}^2_{n-1}) \to \mathrm{Sal}(\mathcal{C}^1_{n-1}) = \mathrm{Sal}(\mathcal{A}_{n-1})$$

when $p \leq [\frac{n}{2}]$. We also have

$$\mathrm{Sal}(\mathcal{C}^p_{n-1}) = \mathrm{Sal}(\mathcal{C}^{n-p}_{n-1}) \to \cdots \to \mathrm{Sal}(\mathcal{C}^2_{n-1}) \to \mathrm{Sal}(\mathcal{C}^1_{n-1}) = \mathrm{Sal}(\mathcal{A}_{n-1})$$

when $p > [\frac{n}{2}]$.

We would like to know if these inclusions induce isomorphisms of homology groups with coefficients in $\mathbb{F}_p(\pm 1)$. Our strategy is to compute the homology of the kernel of the map

$$i^k_n : C_*(\mathrm{Sal}(\mathcal{C}^k_{n-1})) \otimes_{\Sigma_n} \mathbb{F}_p(\pm 1) \longrightarrow C_*(\mathrm{Sal}(\mathcal{C}^{k-1}_{n-1})) \otimes_{\Sigma_n} \mathbb{F}_p(\pm 1),$$

for $k \leq [\frac{n}{2}]$. By Lemma 2.13, these chain maps are surjective.

Corollary 4.2. *For $k \leq [\frac{n}{2}]$, the map*

$$i^k_n : C_*(\mathrm{Sal}(\mathcal{C}^k_{n-1})) \otimes_{\Sigma_n} \mathbb{F}_p(\pm 1) \longrightarrow C_*(\mathrm{Sal}(\mathcal{C}^{k-1}_{n-1})) \otimes_{\Sigma_n} \mathbb{F}_p(\pm 1),$$

is surjective.

In the rest of this article, we consider the first stage, i.e.

$$i_n^2 : C_*(\mathrm{Sal}(\mathcal{C}_{n-1}^2)) \otimes_{\Sigma_n} \mathbb{F}_p(\pm 1) \longrightarrow C_*(\mathrm{Sal}(\mathcal{C}_{n-1}^1)) \otimes_{\Sigma_n} \mathbb{F}_p(\pm 1) = C_*(\mathrm{Sal}(\mathcal{A}_{n-1})) \otimes_{\Sigma_n} \mathbb{F}_p(\pm 1).$$

When $n = 3$,

$$\mathcal{C}_2^2 = \mathcal{C}_2^{3-2} = \mathcal{C}_2^1 = \mathcal{A}_2$$

and there is nothing to compute. The first nontrivial case is

$$i_4^2 : C_*(\mathrm{Sal}(\mathcal{C}_3^2)) \otimes_{\Sigma_4} \mathbb{F}_p(\pm 1) \longrightarrow C_*(\mathrm{Sal}(\mathcal{C}_3^1)) \otimes_{\Sigma_4} \mathbb{F}_p(\pm 1) = C_*(\mathrm{Sal}(\mathcal{A}_3)) \otimes_{\Sigma_4} \mathbb{F}_p(\pm 1).$$

By Corollary 4.2, it suffices to calculate the kernel of i_4^2 in order to compare $H_*(C_*(\mathrm{Sal}(\mathcal{C}_3^2)) \otimes_{\Sigma_4} \mathbb{F}_p(\pm 1))$ and $H_*(C_*(\mathrm{Sal}(\mathcal{A}_3)) \otimes_{\Sigma_4} \mathbb{F}_p(\pm 1))$.

Definition 4.3. We denote

$$K_*^{n,k} = \mathrm{Ker}(i_n^k : C_*(\mathrm{Sal}(\mathcal{C}_{n-1}^k)) \otimes_{\Sigma_n} \mathbb{F}_p(\pm 1) \to C_*(\mathrm{Sal}(\mathcal{C}_{n-1}^{k-1})) \otimes_{\Sigma_n} \mathbb{F}_p(\pm 1)).$$

For simplicity, we also abbreviate

$$C_*^{n,k} = C_*(\mathrm{Sal}(\mathcal{C}_{n-1}^k)) \otimes_{\Sigma_n} \mathbb{F}_p(\pm 1).$$

Thus we have a short exact sequence of chain complexes

$$0 \longrightarrow K_*^{n,k} \longrightarrow C_*^{n,k} \xrightarrow{i_n^k} C_*^{n,k-1} \longrightarrow 0.$$

4.1. The face poset of $\mathcal{C}_3^2$

In order to compute $H_*(\mathrm{Sal}(\mathcal{C}_3^2) \otimes_{\Sigma_4} \mathbb{F}_p(\pm 1))$, the first step is to determine the face poset of $\mathcal{C}_3^2$.

Since $\mathcal{C}_3^2 = \mathcal{A}_3 \cup \{L_{\{1,2\},\{3,4\}}, L_{\{1,3\},\{2,4\}}, L_{\{1,4\},\{2,3\}}\}$, the faces of $\mathcal{C}_3^2$ are given by splitting the faces of $\mathcal{A}_3$ by the hyperplanes

$$L_{\{1,2\},\{3,4\}}, L_{\{1,3\},\{2,4\}}, L_{\{1,4\},\{2,3\}}.$$

In order to understand these cuttings, let us see how the chamber

$$\{(x_1, x_2, x_3, x_4) \in \mathbb{R}^4 \mid x_1 < x_2 < x_3 < x_4\}$$

is cut. Notice that under the action of Σ_4, the cells in $M_2(\mathbb{C}, 4)$ can be represented by cells related to this chamber of $\mathcal{A}_3$.

The only hyperplane among $L_{\{1,2\},\{3,4\}}, L_{\{1,3\},\{2,4\}}, L_{\{1,4\},\{2,3\}}$ that intersects with this chamber is $L_{\{1,4\},\{2,3\}}$ and the chamber is cut into two pieces:

$$\begin{aligned}&\{(x_1,x_2,x_3,x_4)\in\mathbb{R}^4 \mid x_1<x_2<x_3<x_4\}\\ &\quad=\{(x_1,x_2,x_3,x_4)\in\mathbb{R}^4 \mid x_1<x_2<x_3<x_4, x_1+x_4\leq x_2+x_3\}\\ &\quad\cup\{(x_1,x_2,x_3,x_4)\in\mathbb{R}^4 \mid x_1<x_2<x_3<x_4, x_1+x_4\geq x_2+x_3\}.\end{aligned}$$

We denote these chambers by the following symbols:

$$\begin{array}{|c|c|c|c|c|}\hline 1 & & 2 & 3 & 4\\ \hline\end{array}$$
$$=\{(x_1,x_2,x_3,x_4)\in\mathbb{R}^4 \mid x_1<x_2<x_3<x_4, x_1+x_4<x_2+x_3\}$$

$$\begin{array}{|c|c|c|c|c|}\hline 1 & 2 & 3 & & 4\\ \hline\end{array}$$
$$=\{(x_1,x_2,x_3,x_4)\in\mathbb{R}^4 \mid x_1<x_2<x_3<x_4, x_1+x_4>x_2+x_3\}.$$

The faces of these chambers of $\mathcal{C}_3^2$ are also denoted by analogous symbols. The chamber $\begin{array}{|c|c|c|c|c|}\hline 1 & & 2 & 3 & 4\\ \hline\end{array}$ has five 3-dimensional faces, but under the action of Σ_4, we only need the following three faces:

$$\begin{array}{cccc} & & & \boxed{4}\\ \boxed{1} & \boxed{} & \boxed{2} & \boxed{3}\end{array}=\{(x_1,x_2,x_3,x_4)\in\mathbb{R}^4 \mid x_1<x_2<x_3=x_4\},$$

$$\begin{array}{cccc} & & \boxed{3} & \\ \boxed{1} & \boxed{} & \boxed{2} & \boxed{4}\end{array}$$
$$=\{(x_1,x_2,x_3,x_4)\in\mathbb{R}^4 \mid x_1<x_2=x_3<x_4, x_1+x_4<x_2+x_3\},$$

$$\begin{array}{|c|c|c|c|c|}\hline 1 & 2 & & 3 & 4\\ \hline\end{array}$$
$$=\{(x_1,x_2,x_3,x_4)\in\mathbb{R}^4 \mid x_1<x_2<x_3<x_4, x_1+x_4=x_2+x_3\}.$$

Similarly, we need the following three faces for the chamber $\begin{array}{|c|c|c|c|c|}\hline 1 & 2 & 3 & & 4\\ \hline\end{array}$.

$$\begin{array}{cccc} \boxed{2} & & & \\ \boxed{1} & \boxed{3} & \boxed{} & \boxed{4}\end{array}=\{(x_1,x_2,x_3,x_4)\in\mathbb{R}^4 \mid x_1=x_2<x_3<x_4\},$$

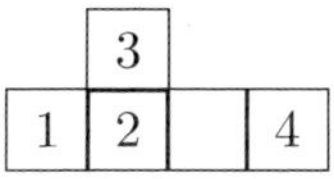

$$= \{(x_1, x_2, x_3, x_4) \in \mathbb{R}^4 \mid x_1 < x_2 = x_3 < x_4, x_1 + x_4 > x_2 + x_3\},$$

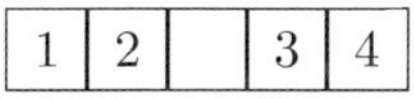

$$= \{(x_1, x_2, x_3, x_4) \in \mathbb{R}^4 \mid x_1 < x_2 < x_3 < x_4, x_1 + x_4 = x_2 + x_3\}.$$

The 2-dimensional faces we need are the following:

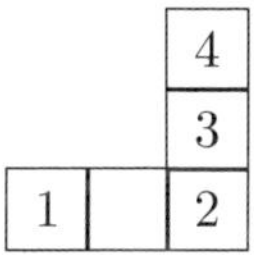

$$= \{(x_1, x_2, x_3, x_4) \in \mathbb{R}^4 \mid x_1 < x_2 = x_3 = x_4\},$$

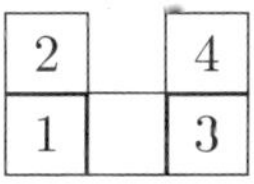

$$= \{(x_1, x_2, x_3, x_4) \in \mathbb{R}^4 \mid x_1 = x_2 < x_3 = x_4, x_1 + x_4 = x_2 + x_3\},$$

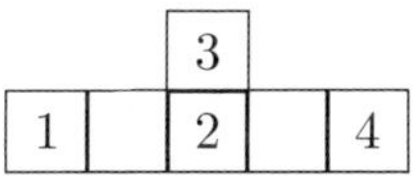

$$= \{(x_1, x_2, x_3, x_4) \in \mathbb{R}^4 \mid x_1 < x_2 = x_3 < x_4, x_1 + x_4 = x_2 + x_3\},$$

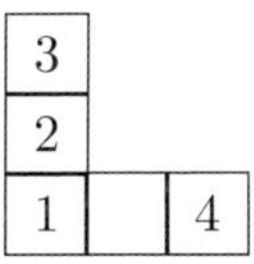

$$= \{(x_1, x_2, x_3, x_4) \in \mathbb{R}^4 \mid x_1 = x_2 = x_3 < x_4\}.$$

All these faces have the following 1-dimensional face in common.

$$= \{(x_1, x_2, x_3, x_4) \in \mathbb{R}^4 \mid x_1 = x_2 = x_3 = x_4\}.$$

Notice that Σ_4 acts on $\mathcal{L}(\mathcal{C}_3^2)$ and the action is compatible with the ordering. We have the following description of the poset $\mathcal{L}(\mathcal{C}_3^2)/\Sigma_4$.

Lemma 4.4. *The poset $\mathcal{L}(\mathcal{C}_3^2)/\Sigma_4$ has the following structure:*

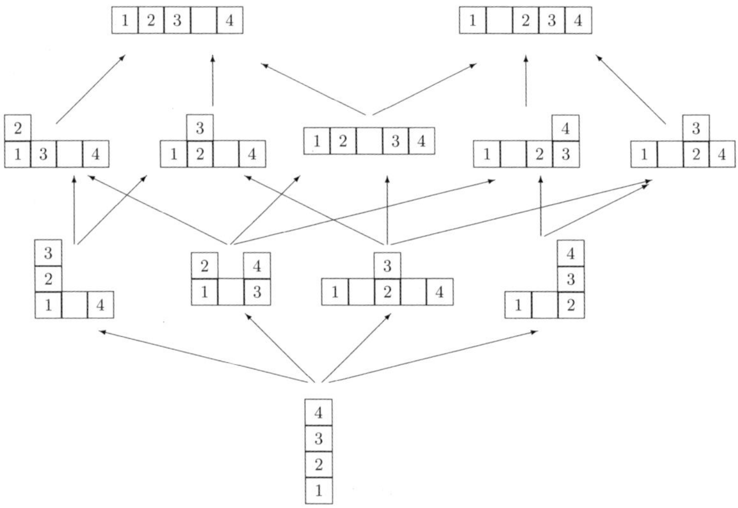

4.2. The Cellular Structure on $\mathrm{Sal}(\mathcal{C}_3^2)/\Sigma_4$

Let us determine the cellular structure of $\mathrm{Sal}(\mathcal{C}_3^2)/\Sigma_4$. The cell decomposition of the Salvetti complex for $\mathcal{C}_3^2$ is compatible with the action of Σ_4 and the quotient $\mathrm{Sal}(\mathcal{C}_3^2)/\Sigma_4$ has the induced cell decomposition.

The cells of the Salvetti complex are labeled by pairs of a face F and a chamber C with $C \geq F$. The cell for the pair (F, C) is denoted by $D(F, C)$ in §2.1. Thus the cells of $\mathrm{Sal}(\mathcal{C}_3^2)/\Sigma_4$ are in one-to-one correspondence with elements in

$$\{([F], [C]) \mid [F] \in \mathcal{L}(\mathcal{C}_3^2)/\Sigma_4, [C] \in \mathcal{L}^{(0)}(\mathcal{C}_3^2)/\Sigma_4, F \leq C\}.$$

In the case of $\mathcal{C}_3^2$, there are only two chambers in $\mathcal{L}^{(0)}(\mathcal{C}_3^2)/\Sigma_4$, and we denote the cells corresponding to the pair $\left([F], \left[\begin{array}{|c|c|c|c|c|}\hline 1&2&3&&4\\\hline\end{array}\right]\right)$ and $\left([F], \left[\begin{array}{|c|c|c|c|c|}\hline 1&&2&3&4\\\hline\end{array}\right]\right)$ by F^+ and F^-, respectively. To be more efficient, we simply denote them by F when F is contained in only one chamber.

More explicitly,

Lemma 4.5. $\mathrm{Sal}(\mathcal{C}_3^2)/\Sigma_4$ *has*

- *two* 0-*cells*

$$\begin{array}{|c|c|c|c|c|}\hline 1&2&3&&4\\\hline\end{array}\,,\ \begin{array}{|c|c|c|c|c|}\hline 1&&2&3&4\\\hline\end{array}\,.$$

- *six* 1-*cells*

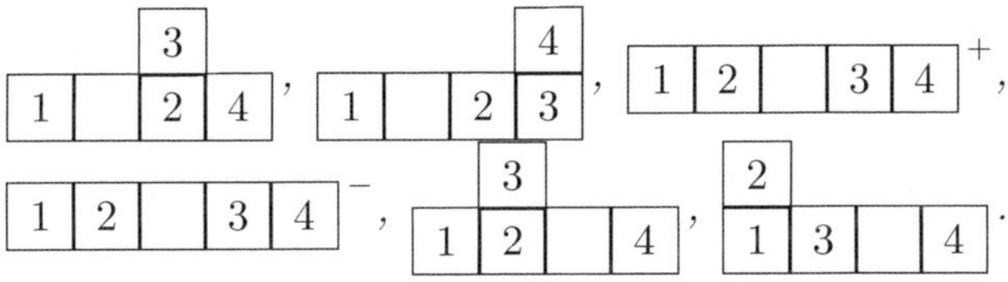

- *six 2-cells*

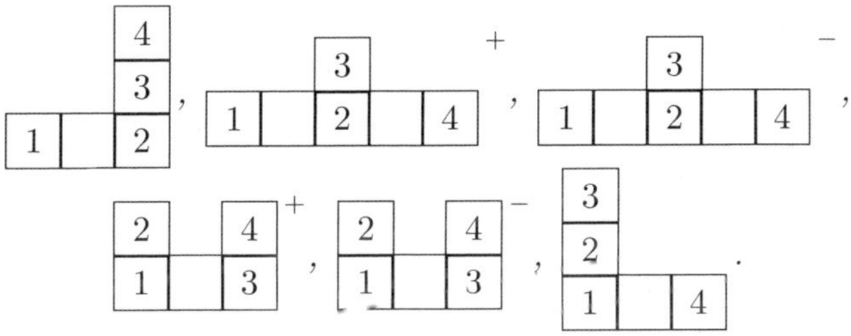

- *two 3-cells*

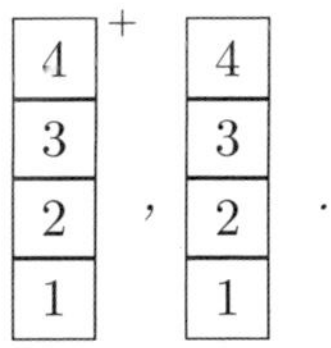

We use above symbols as representatives of cells in $\mathrm{Sal}(\mathcal{C}_3^2)$. We define orientations on these cells and then transfer orientations to other cells in $\mathrm{Sal}(\mathcal{C}_3^2)$ via the action of Σ_4. Those cells which are mapped to cells of the same dimensions in $\mathrm{Sal}(\mathcal{A}_3)$ by the map i_4^2 are oriented in such a way i_4^2 is orientation preserving. Then remaining four types of cells

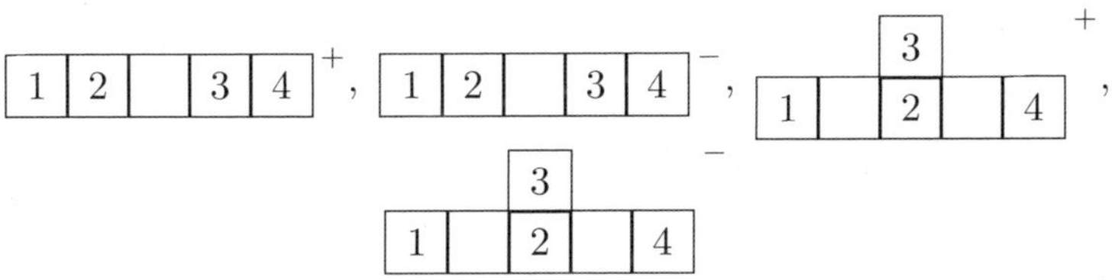

are oriented as follows. As we will see below, the first two 1-cells have 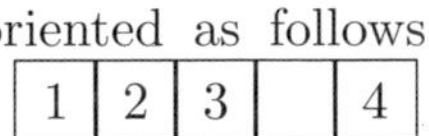 and 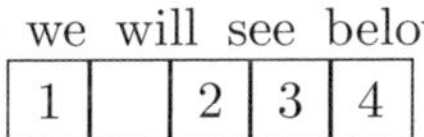 as vertices. We orient these 1-cells from 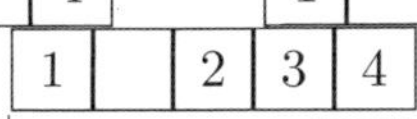 to 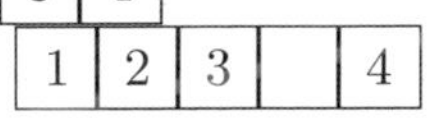. The 2-cells 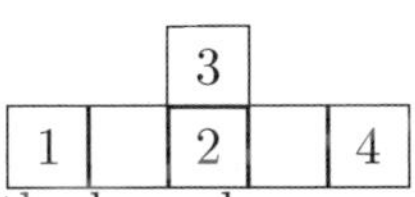 and 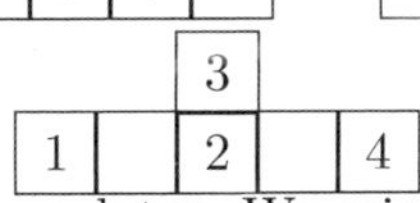 contains 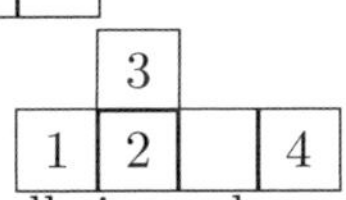 in the boundary, as we will see later. We orient these 2-cells in such a way the incidence number to this 1-cell is positive.

Now we are ready to consider the boundaries. This can be done by using the formula for the boundary in Lemma 2.2 and a formula analogous to the case of the braid arrangement (Lemma 3.7).

The first nontrivial case is the boundaries of 1-cells.

Lemma 4.6. *We have the following formula in* $C_*^{4,2}$.

$$\partial_1\left(\begin{array}{cccc} & & \boxed{3} & \\ \boxed{1} & \boxed{} & \boxed{2} & \boxed{4}\end{array}\right) = 2\,\boxed{1}\boxed{}\boxed{2}\boxed{3}\boxed{4}$$

$$\partial_1\left(\begin{array}{cccc} & & & \boxed{4} \\ \boxed{1} & \boxed{} & \boxed{2} & \boxed{3}\end{array}\right) = 2\,\boxed{1}\boxed{}\boxed{2}\boxed{3}\boxed{4}$$

$$\partial_1\left(\boxed{1}\boxed{2}\boxed{}\boxed{3}\boxed{4}^{\,+}\right) = \boxed{1}\boxed{2}\boxed{3}\boxed{}\boxed{4} - \boxed{1}\boxed{}\boxed{2}\boxed{3}\boxed{4}$$

$$\partial_1\left(\boxed{1}\boxed{2}\boxed{}\boxed{3}\boxed{4}^{\,-}\right) = \boxed{1}\boxed{2}\boxed{3}\boxed{}\boxed{4} - \boxed{1}\boxed{}\boxed{2}\boxed{3}\boxed{4}$$

$$\partial_1\left(\begin{array}{cccc} & \boxed{3} & & \\ \boxed{1} & \boxed{2} & \boxed{} & \boxed{4}\end{array}\right) = 2\,\boxed{1}\boxed{2}\boxed{3}\boxed{}\boxed{4}$$

$$\partial_1\left(\begin{array}{cccc} \boxed{2} & & & \\ \boxed{1} & \boxed{3} & \boxed{} & \boxed{4}\end{array}\right) = 2\,\boxed{1}\boxed{2}\boxed{3}\boxed{}\boxed{4}.$$

Proof. By Lemma 2.2

$$\partial_1\left(\begin{array}{cccc} & & \boxed{3} & \\ \boxed{1} & \boxed{} & \boxed{2} & \boxed{4}\end{array}\right) = \sum_F \varepsilon_F D\left(F, F\circ \boxed{1}\boxed{}\boxed{2}\boxed{3}\boxed{4}\right)$$

where F runs over all faces containing $\begin{array}{cccc} & & \boxed{3} & \\ \boxed{1} & \boxed{} & \boxed{2} & \boxed{4}\end{array}$ in $\mathcal{L}(\mathcal{C}_3^2)$ and

$$\varepsilon_F = \left[\begin{array}{cccc} & & \boxed{3} & \\ \boxed{1} & \boxed{} & \boxed{2} & \boxed{4}\end{array} : D\left(F, F\circ \boxed{1}\boxed{}\boxed{2}\boxed{3}\boxed{4}\right)\right].$$

In this case, we have

$$\begin{aligned}
&\partial_1\left(\begin{array}{cccc} & & \boxed{3} & \\ \boxed{1} & \boxed{} & \boxed{2} & \boxed{4}\end{array}\right)\\
&= \varepsilon_1 D\left(\boxed{1}\boxed{}\boxed{2}\boxed{3}\boxed{4}, \boxed{1}\boxed{}\boxed{2}\boxed{3}\boxed{4}\circ\boxed{1}\boxed{}\boxed{2}\boxed{3}\boxed{4}\right)\\
&\quad +\varepsilon_2 D\left(\boxed{1}\boxed{}\boxed{3}\boxed{2}\boxed{4}, \boxed{1}\boxed{}\boxed{3}\boxed{2}\boxed{4}\circ\boxed{1}\boxed{}\boxed{2}\boxed{3}\boxed{4}\right),
\end{aligned}$$

where $\varepsilon_1, \varepsilon_2$ are appropriate incidence numbers.

In order to compute the boundaries, therefore, we need to understand the matroid product in the face lattice $\mathcal{L}(\mathcal{C}_3^2)$.

As we have recalled in §2, the cellular structure of the Salvetti complex was originally described by face-chamber pairings. For computations, however, it is much more convenient to regard faces as functions from the set of normal vectors to the poset of three elements $S_1 = \{0, +1, -1\}$ and use the matroid product of these functions, as we have seen in Lemma 2.3 and 2.4.

We choose the following set of normal vectors for the arrangement $\mathcal{C}_3^2$:

$$\begin{aligned}
\boldsymbol{a}_1 &= (1,-1,0,0) \\
\boldsymbol{a}_2 &= (0,1,-1,0) \\
\boldsymbol{a}_3 &= (0,0,1,-1) \\
\boldsymbol{a}_4 &= (1,0,-1,0) \\
\boldsymbol{a}_5 &= (1,0,0,-1) \\
\boldsymbol{a}_6 &= (0,1,0,-1) \\
\boldsymbol{a}_7 &= (1,1,-1,-1) \\
\boldsymbol{a}_8 &= (1,-1,1,-1) \\
\boldsymbol{a}_9 &= (1,-1,-1,1).
\end{aligned}$$

Then a face $F \in \mathcal{L}(\mathcal{C}_3^2)$ can be regarded as a function

$$\tau_F : \{\boldsymbol{a}_1, \boldsymbol{a}_2, \cdots, \boldsymbol{a}_9\} \longrightarrow S_1.$$

For simplicity, we denote elements $+1, -1$ in S_1 by $+, -$, respectively, and denote the above function by the symbol

$$(\tau_F(\boldsymbol{a}_1), \tau_F(\boldsymbol{a}_2), \cdots, \tau_F(\boldsymbol{a}_6) \mid \tau_F(\boldsymbol{a}_7), \tau_F(\boldsymbol{a}_8), \tau_F(\boldsymbol{a}_9)).$$

For example, the face $\begin{array}{|c|c|c|c|c|}\hline 1 & \; & 2 & 3 & 4 \\ \hline\end{array}$ corresponds to the symbol

$$(-,-,-,-,-,- \mid -,-,-).$$

With these notations, the matroid product of a face and a chamber is given by replacing 0's in the face by the values at the same position in the chamber. For example,

$$\begin{aligned}
& \begin{array}{|c|c|c|c|c|}\hline 1 & \; & 3 & 2 & 4 \\ \hline\end{array} \circ \begin{array}{|c|c|c|c|c|}\hline 1 & \; & 2 & 3 & 4 \\ \hline\end{array} \\
= \ & (-,+,-,-,-,- \mid -,-,-) \circ (-,-,-,-,-,- \mid -,-,-) \\
= \ & (-,+,-,-,-,- \mid -,-,-) \\
= \ & \begin{array}{|c|c|c|c|c|}\hline 1 & \; & 3 & 2 & 4 \\ \hline\end{array}.
\end{aligned}$$

Thus, by taking the orientations into an account, we have

$$\begin{aligned}\partial_1\left(\begin{array}{|c|c|c|c|}\cline{3-3}\multicolumn{2}{c|}{} & 3 & \multicolumn{1}{c}{}\\ \hline 1 & & 2 & 4\\ \hline\end{array}\right) &= \begin{array}{|c|c|c|c|c|}\hline 1 & & 2 & 3 & 4\\ \hline\end{array} - \begin{array}{|c|c|c|c|c|}\hline 1 & & 3 & 2 & 4\\ \hline\end{array}\\ &= 2\begin{array}{|c|c|c|c|c|}\hline 1 & & 2 & 3 & 4\\ \hline\end{array}.\end{aligned}$$

By analogous calculations, we obtain

$$\begin{aligned}\partial_1\left(\begin{array}{|c|c|c|c|}\cline{4-4}\multicolumn{3}{c|}{} & 4\\ \hline 1 & & 2 & 3\\ \hline\end{array}\right) &= 2\begin{array}{|c|c|c|c|c|}\hline 1 & & 2 & 3 & 4\\ \hline\end{array}\\ \partial_1\left(\begin{array}{|c|c|c|c|}\cline{2-2}\multicolumn{1}{c|}{} & 3 & \multicolumn{2}{c}{}\\ \hline 1 & 2 & & 4\\ \hline\end{array}\right) &= 2\begin{array}{|c|c|c|c|c|}\hline 1 & 2 & 3 & & 4\\ \hline\end{array}\\ \partial_1\left(\begin{array}{|c|c|c|c|}\cline{1-1} 2 & \multicolumn{3}{c}{}\\ \hline 1 & 3 & & 4\\ \hline\end{array}\right) &= 2\begin{array}{|c|c|c|c|c|}\hline 1 & 2 & 3 & & 4\\ \hline\end{array}.\end{aligned}$$

On the other hand, $\begin{array}{|c|c|c|c|c|}\hline 1 & 2 & & 3 & 4\\ \hline\end{array}^{+}$ has

$$D(\begin{array}{|c|c|c|c|c|}\hline 1 & 2 & 3 & & 4\\ \hline\end{array}, \begin{array}{|c|c|c|c|c|}\hline 1 & 2 & 3 & & 4\\ \hline\end{array} \circ \begin{array}{|c|c|c|c|c|}\hline 1 & 2 & 3 & & 4\\ \hline\end{array})$$

and

$$D(\begin{array}{|c|c|c|c|c|}\hline 1 & & 2 & 3 & 4\\ \hline\end{array}, \begin{array}{|c|c|c|c|c|}\hline 1 & & 2 & 3 & 4\\ \hline\end{array} \circ \begin{array}{|c|c|c|c|c|}\hline 1 & 2 & 3 & & 4\\ \hline\end{array})$$

as its boundary. By computing the matroid products, we see these 0-cells are $\begin{array}{|c|c|c|c|c|}\hline 1 & 2 & 3 & & 4\\ \hline\end{array}$ and $\begin{array}{|c|c|c|c|c|}\hline 1 & & 2 & 3 & 4\\ \hline\end{array}$, respectively. By the definition of the orientation, we obtain

$$\begin{aligned}\partial_1\left(\begin{array}{|c|c|c|c|c|}\hline 1 & 2 & & 3 & 4\\ \hline\end{array}^{+}\right) &= \begin{array}{|c|c|c|c|c|}\hline 1 & 2 & 3 & & 4\\ \hline\end{array} - \begin{array}{|c|c|c|c|c|}\hline 1 & & 2 & 3 & 4\\ \hline\end{array}\\ \partial_1\left(\begin{array}{|c|c|c|c|c|}\hline 1 & 2 & & 3 & 4\\ \hline\end{array}^{-}\right) &= \begin{array}{|c|c|c|c|c|}\hline 1 & 2 & 3 & & 4\\ \hline\end{array} - \begin{array}{|c|c|c|c|c|}\hline 1 & & 2 & 3 & 4\\ \hline\end{array}.\end{aligned}$$

Q.E.D.

Remark 4.7. Note that in the above proof, the computation of $\partial_1\left(\begin{array}{|c|c|c|c|}\cline{3-3}\multicolumn{2}{c|}{} & 3 & \multicolumn{1}{c}{}\\ \hline 1 & & 2 & 4\\ \hline\end{array}\right)$ is exactly the same as that of $\partial_1\left(\begin{array}{|c|c|c|}\cline{2-2}\multicolumn{1}{c|}{} & 3 & \multicolumn{1}{c}{}\\ \hline 1 & 2 & 4\\ \hline\end{array}\right)$ in $C_*(\mathrm{Sal}(\mathcal{A}_3)) \otimes_{\Sigma_4} \mathbb{F}_p(\pm 1)$.

In general, boundaries of those cells which do not hav $+$ or $-$ sign on the shoulder can be computed by the same formulas for the corresponding cells in the braid arrangement.

Let us consider 2-cells next.

Lemma 4.8. *We have the following formula in* $C_*^{4,2}$.

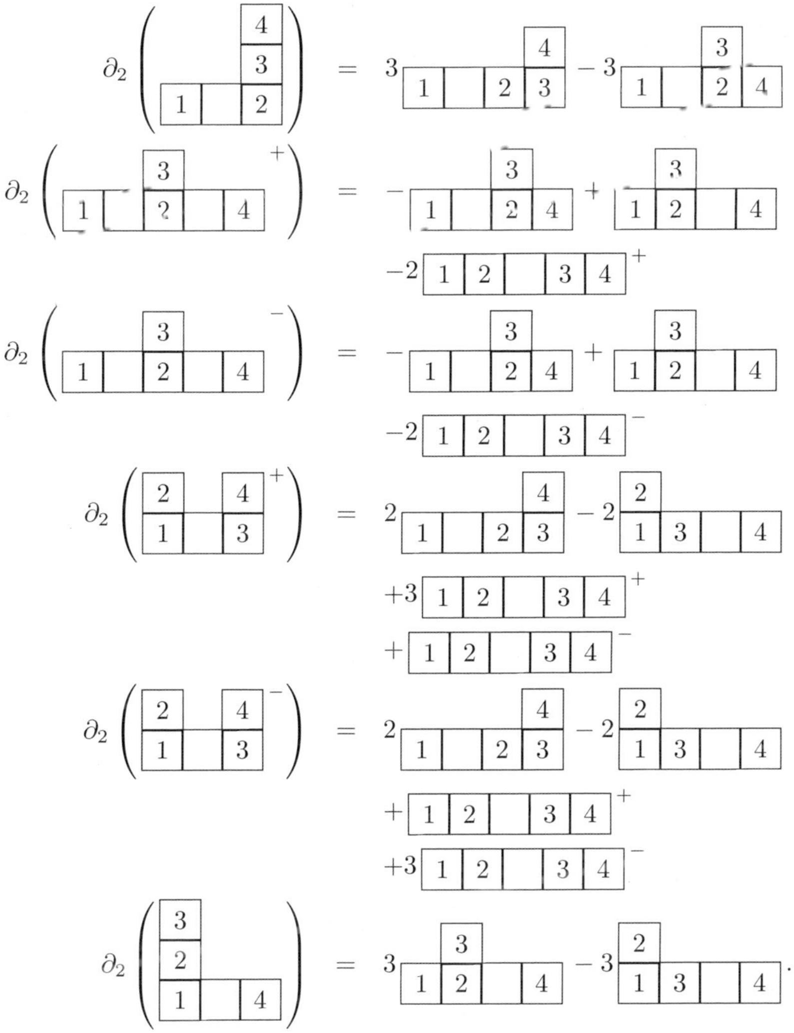

Proof. The computations of $\partial_2\left(\begin{smallmatrix} & & 4\\ & & 3\\ 1 & \square & 2\end{smallmatrix}\right)$ and $\partial_2\left(\begin{smallmatrix}3 & & \\ 2 & & \\ 1 & \square & 4\end{smallmatrix}\right)$ are essentially the same as those of $\partial_2\left(\begin{smallmatrix} & 4\\ & 3\\ 1 & 2\end{smallmatrix}\right)$ and $\partial_2\left(\begin{smallmatrix}3 & \\ 2 & \\ 1 & 4\end{smallmatrix}\right)$ in §3.2 as is noted in Remark 4.7 and are omitted.

There are four faces in $\mathcal{L}(\mathcal{C}_3^2)$ that contain $\begin{smallmatrix} & & 3 & & \\ 1 & \square & 2 & \square & 4\end{smallmatrix}$ as faces. Thus

$$
\begin{aligned}
&\partial_2\left(\begin{smallmatrix} & & 3 & & \\ 1 & \square & 2 & \square & 4\end{smallmatrix}^{+}\right)\\
= \ & \varepsilon_1 D\left(\begin{smallmatrix} & & 3 & \\ 1 & \square & 2 & 4\end{smallmatrix}, \begin{smallmatrix} & & 3 & \\ 1 & \square & 2 & 4\end{smallmatrix} \circ \begin{smallmatrix}1 & 2 & 3 & \square & 4\end{smallmatrix}\right)\\
&+\varepsilon_2 D\left(\begin{smallmatrix} & 3 & & \\ 1 & 2 & \square & 4\end{smallmatrix}, \begin{smallmatrix} & 3 & & \\ 1 & 2 & \square & 4\end{smallmatrix} \circ \begin{smallmatrix}1 & 2 & 3 & \square & 4\end{smallmatrix}\right)\\
&+\varepsilon_3 D\left(\begin{smallmatrix}1 & 2 & \square & 3 & 4\end{smallmatrix}, \begin{smallmatrix}1 & 2 & \square & 3 & 4\end{smallmatrix} \circ \begin{smallmatrix}1 & 2 & 3 & \square & 4\end{smallmatrix}\right)\\
&+\varepsilon_4 D\left(\begin{smallmatrix}1 & 3 & \square & 2 & 4\end{smallmatrix}, \begin{smallmatrix}1 & 3 & \square & 2 & 4\end{smallmatrix} \circ \begin{smallmatrix}1 & 2 & 3 & \square & 4\end{smallmatrix}\right)\\
= \ & \varepsilon_1 D\left(\begin{smallmatrix} & & 3 & \\ 1 & \square & 2 & 4\end{smallmatrix}, \begin{smallmatrix}1 & \square & 2 & 3 & 4\end{smallmatrix}\right)\\
&+\varepsilon_2 D\left(\begin{smallmatrix} & 3 & & \\ 1 & 2 & \square & 4\end{smallmatrix}, \begin{smallmatrix}1 & 2 & 3 & \square & 4\end{smallmatrix}\right)\\
&+(\varepsilon_3+\varepsilon_4) D\left(\begin{smallmatrix}1 & 2 & \square & 3 & 4\end{smallmatrix}, \begin{smallmatrix}1 & 2 & 3 & \square & 4\end{smallmatrix}\right)\\
= \ & \varepsilon_1 \begin{smallmatrix} & & 3 & \\ 1 & \square & 2 & 4\end{smallmatrix} + \varepsilon_2 \begin{smallmatrix} & 3 & & \\ 1 & 2 & \square & 4\end{smallmatrix} + (\varepsilon_3+\varepsilon_4)\begin{smallmatrix}1 & 2 & \square & 3 & 4\end{smallmatrix}^{+},
\end{aligned}
$$

where $\varepsilon_i \in \{\pm 1\}$ $(i = 1, 2, 3, 4)$ are certain signs. By the definition of the orientation of $\begin{smallmatrix} & & 3 & & \\ 1 & \square & 2 & \square & 4\end{smallmatrix}^{+}$, we have $\varepsilon_2 = 1$. Since $\partial_1 \circ \partial_2 = 0$, we see that $\varepsilon_1 = -1$ and $\varepsilon_3 + \varepsilon_4 = 2$.

By the same calculation, we also have

$$\partial_2\left(\begin{array}{@{}c@{}c@{}c@{}c@{}c@{}} & & \boxed{3} & & \\ \boxed{1} & \boxed{} & \boxed{2} & \boxed{} & \boxed{4} \end{array}^{-}\right) =$$

$$- \begin{array}{@{}c@{}c@{}c@{}c@{}} & & \boxed{3} & \\ \boxed{1} & \boxed{} & \boxed{2} & \boxed{4} \end{array} + \begin{array}{@{}c@{}c@{}c@{}c@{}} & \boxed{3} & & \\ \boxed{1} & \boxed{2} & \boxed{} & \boxed{4} \end{array} + 2\,\boxed{1}\boxed{2}\boxed{}\boxed{3}\boxed{4}^{-}.$$

Finally, we have

$$\begin{aligned}
&\partial_2\left(\begin{array}{@{}c@{}c@{}c@{}} \boxed{2} & & \boxed{4} \\ \boxed{1} & \boxed{} & \boxed{3} \end{array}^{+}\right)\\
= &-2D\left(\begin{array}{@{}c@{}c@{}c@{}c@{}} \boxed{2} & & & \\ \boxed{1} & \boxed{3} & \boxed{} & \boxed{4} \end{array}, \begin{array}{@{}c@{}c@{}c@{}c@{}} \boxed{2} & & & \\ \boxed{1} & \boxed{3} & \boxed{} & \boxed{4} \end{array} \circ \boxed{1}\boxed{2}\boxed{3}\boxed{}\boxed{4}\right)\\
&+D\left(\boxed{1}\boxed{2}\boxed{}\boxed{3}\boxed{4}, \boxed{1}\boxed{2}\boxed{}\boxed{3}\boxed{4} \circ \boxed{1}\boxed{2}\boxed{3}\boxed{}\boxed{4}\right)\\
&-D\left(\boxed{2}\boxed{1}\boxed{}\boxed{3}\boxed{4}, \boxed{2}\boxed{1}\boxed{}\boxed{3}\boxed{4} \circ \boxed{1}\boxed{2}\boxed{3}\boxed{}\boxed{4}\right)\\
&-D\left(\boxed{1}\boxed{2}\boxed{}\boxed{4}\boxed{3}, \boxed{1}\boxed{2}\boxed{}\boxed{4}\boxed{3} \circ \boxed{1}\boxed{2}\boxed{3}\boxed{}\boxed{4}\right)\\
&+D\left(\boxed{2}\boxed{1}\boxed{}\boxed{4}\boxed{3}, \boxed{2}\boxed{1}\boxed{}\boxed{4}\boxed{3} \circ \boxed{1}\boxed{2}\boxed{3}\boxed{}\boxed{4}\right)\\
&+2D\left(\begin{array}{@{}c@{}c@{}c@{}c@{}} & & & \boxed{4} \\ \boxed{1} & \boxed{2} & \boxed{} & \boxed{3} \end{array}, \begin{array}{@{}c@{}c@{}c@{}c@{}} & & & \boxed{4} \\ \boxed{1} & \boxed{2} & \boxed{} & \boxed{3} \end{array} \circ \boxed{1}\boxed{2}\boxed{3}\boxed{}\boxed{4}\right)\\
= &-2\begin{array}{@{}c@{}c@{}c@{}c@{}} \boxed{2} & & & \\ \boxed{1} & \boxed{3} & \boxed{} & \boxed{4} \end{array} + \boxed{1}\boxed{2}\boxed{}\boxed{3}\boxed{4}^{+} - \boxed{2}\boxed{1}\boxed{}\boxed{3}\boxed{4}^{+}\\
&-\boxed{1}\boxed{2}\boxed{}\boxed{4}\boxed{3}^{-} + \boxed{2}\boxed{1}\boxed{}\boxed{4}\boxed{3}^{+} + 2\begin{array}{@{}c@{}c@{}c@{}c@{}} & & & \boxed{4} \\ \boxed{1} & \boxed{2} & \boxed{} & \boxed{3} \end{array}\\
= &-2\begin{array}{@{}c@{}c@{}c@{}c@{}} \boxed{2} & & & \\ \boxed{1} & \boxed{3} & \boxed{} & \boxed{4} \end{array} - +2\begin{array}{@{}c@{}c@{}c@{}c@{}} & & & \boxed{4} \\ \boxed{1} & \boxed{2} & \boxed{} & \boxed{3} \end{array}\\
&+3\,\boxed{1}\boxed{2}\boxed{}\boxed{3}\boxed{4}^{+} + \boxed{1}\boxed{2}\boxed{}\boxed{3}\boxed{4}^{-},
\end{aligned}$$

where coefficients (signs) of the first two terms are determined by comparing with the braid arrangement. The coefficients of the last two terms are determined in the same way as in the previous calculation.

The calculation of $\partial_2\left(\begin{array}{ccc}\boxed{2} & & \boxed{4} \\ \boxed{1} & \boxed{} & \boxed{3}\end{array}^{-}\right)$ can be done in the same manner. Q.E.D.

It remains to compute boundaries on 3-cells.

Lemma 4.9. *We have*

$$
\begin{aligned}
\partial_3\left(\begin{array}{c}\boxed{4}\\ \boxed{3}\\ \boxed{2}\\ \boxed{1}\end{array}^{+}\right) = {} & 4\begin{array}{ccc} & & \boxed{4}\\ & & \boxed{3}\\ \boxed{1} & \boxed{} & \boxed{2}\end{array} + 4\begin{array}{ccc}\boxed{3} & & \\ \boxed{2} & & \\ \boxed{1} & \boxed{} & \boxed{4}\end{array} \\
& -5\begin{array}{ccc}\boxed{2} & & \boxed{4} \\ \boxed{1} & \boxed{} & \boxed{3}\end{array}^{+} - \begin{array}{ccc}\boxed{2} & & \boxed{4} \\ \boxed{1} & \boxed{} & \boxed{3}\end{array}^{-} \\
& +8\begin{array}{ccccc} & & \boxed{3} & & \\ \boxed{1} & \boxed{} & \boxed{2} & \boxed{} & \boxed{4}\end{array}^{+} + 4\begin{array}{ccccc} & & \boxed{3} & & \\ \boxed{1} & \boxed{} & \boxed{2} & \boxed{} & \boxed{4}\end{array}^{-}, \\
\partial_3\left(\begin{array}{c}\boxed{4}\\ \boxed{3}\\ \boxed{2}\\ \boxed{1}\end{array}^{-}\right) = {} & 4\begin{array}{ccc} & & \boxed{4}\\ & & \boxed{3}\\ \boxed{1} & \boxed{} & \boxed{2}\end{array} + 4\begin{array}{ccc}\boxed{3} & & \\ \boxed{2} & & \\ \boxed{1} & \boxed{} & \boxed{4}\end{array} \\
& -\begin{array}{ccc}\boxed{2} & & \boxed{4} \\ \boxed{1} & \boxed{} & \boxed{3}\end{array}^{+} - 5\begin{array}{ccc}\boxed{2} & & \boxed{4} \\ \boxed{1} & \boxed{} & \boxed{3}\end{array}^{-} \\
& +4\begin{array}{ccccc} & & \boxed{3} & & \\ \boxed{1} & \boxed{} & \boxed{2} & \boxed{} & \boxed{4}\end{array}^{+} + 8\begin{array}{ccccc} & & \boxed{3} & & \\ \boxed{1} & \boxed{} & \boxed{2} & \boxed{} & \boxed{4}\end{array}^{-}.
\end{aligned}
$$

Proof. Let us consider $\partial_3\left(\begin{array}{c}\boxed{4}\\ \boxed{3}\\ \boxed{2}\\ \boxed{1}\end{array}^{+}\right)$. There are many faces containing $\begin{array}{c}\boxed{4}\\ \boxed{3}\\ \boxed{2}\\ \boxed{1}\end{array}$. For example, there are $\binom{4}{2} = 6$ faces of the shape $\begin{array}{ccc}\boxed{2} & & \boxed{4} \\ \boxed{1} & \boxed{} & \boxed{3}\end{array}$.

Matroid products can be computed, for example,

$$
\begin{aligned}
& \begin{array}{|c|c|c|}\cline{1-1}\cline{3-3} 3 & \multicolumn{1}{c|}{} & 4 \\ \hline 2 & & 1 \\ \hline \end{array} \circ \begin{array}{|c|c|c|c|c|}\hline 1 & 2 & 3 & & 4 \\ \hline \end{array} \\
= \; & (+,0,-,+,0,- \mid 0,0,+) \circ (-,-,-,-,-,- \mid -,-,+) \\
= \; & (+,-,-,+,-,- \mid -,-,+) \\
= \; & \left\{ (x_1,x_2,x_3,x_4) \in \mathbb{R}^4 \;\middle|\; \begin{array}{l} x_1 > x_2,\ x_2 < x_3,\ x_3 < x_4, \\ x_1 > x_3,\ x_1 < x_4,\ x_2 < x_4, \\ x_1 + x_2 < x_3 + x_4, \\ x_1 + x_3 < x_2 + x_4, \\ x_1 + x_4 > x_2 + x_3 \end{array} \right\} \\
= \; & \left\{ (x_1,x_2,x_3,x_4) \in \mathbb{R}^4 \;\middle|\; \begin{array}{l} x_2 < x_3 < x_1 < x_4, \\ x_1 + x_3 < x_2 + x_4 \end{array} \right\} \\
= \; & \begin{array}{|c|c|c|c|c|}\hline 2 & 3 & 1 & & 4 \\ \hline \end{array}.
\end{aligned}
$$

Similarly, we have

$$
\begin{aligned}
\begin{array}{|c|c|c|}\cline{1-1}\cline{3-3} 2 & \multicolumn{1}{c|}{} & 4 \\ \hline 1 & & 3 \\ \hline \end{array} \circ \begin{array}{|c|c|c|c|c|}\hline 1 & 2 & 3 & & 4 \\ \hline \end{array} &= \begin{array}{|c|c|c|c|c|}\hline 1 & 2 & 3 & & 4 \\ \hline \end{array} \\
\begin{array}{|c|c|c|}\cline{1-1}\cline{3-3} 3 & \multicolumn{1}{c|}{} & 4 \\ \hline 1 & & 2 \\ \hline \end{array} \circ \begin{array}{|c|c|c|c|c|}\hline 1 & 2 & 3 & & 4 \\ \hline \end{array} &= \begin{array}{|c|c|c|c|c|}\hline 1 & 3 & 2 & & 4 \\ \hline \end{array} \\
\begin{array}{|c|c|c|}\cline{1-1}\cline{3-3} 4 & \multicolumn{1}{c|}{} & 3 \\ \hline 1 & & 2 \\ \hline \end{array} \circ \begin{array}{|c|c|c|c|c|}\hline 1 & & 2 & 3 & 4 \\ \hline \end{array} &= \begin{array}{|c|c|c|c|c|}\hline 1 & & 4 & 2 & 3 \\ \hline \end{array} \\
\begin{array}{|c|c|c|}\cline{1-1}\cline{3-3} 4 & \multicolumn{1}{c|}{} & 3 \\ \hline 2 & & 1 \\ \hline \end{array} \circ \begin{array}{|c|c|c|c|c|}\hline 1 & & 2 & 3 & 4 \\ \hline \end{array} &= \begin{array}{|c|c|c|c|c|}\hline 2 & 4 & 1 & & 3 \\ \hline \end{array} \\
\begin{array}{|c|c|c|}\cline{1-1}\cline{3-3} 4 & \multicolumn{1}{c|}{} & 2 \\ \hline 3 & & 1 \\ \hline \end{array} \circ \begin{array}{|c|c|c|c|c|}\hline 1 & & 2 & 3 & 4 \\ \hline \end{array} &= \begin{array}{|c|c|c|c|c|}\hline 3 & 4 & 1 & & 2 \\ \hline \end{array}.
\end{aligned}
$$

This implies that, in $C_3(\mathrm{Sal}(\mathcal{C}_3^2)) \otimes_{\Sigma_4} \mathbb{F}_p$, if we write

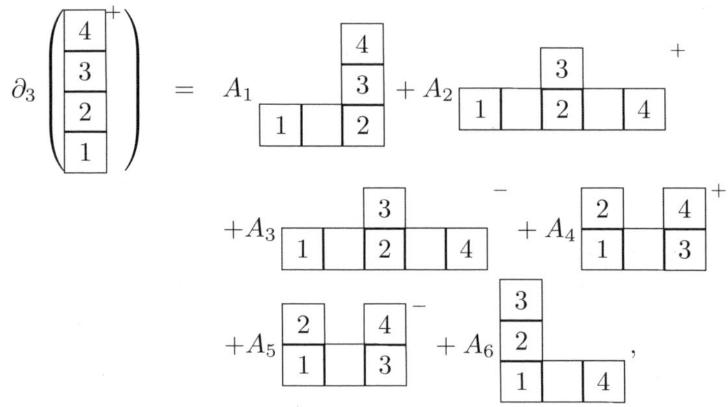

we have

$$A_4 = -5, \ A_5 = -1.$$

By similar calculations of matroid products and by that fact $\partial_2\partial_3 = 0$, we can determine other coefficients and we have

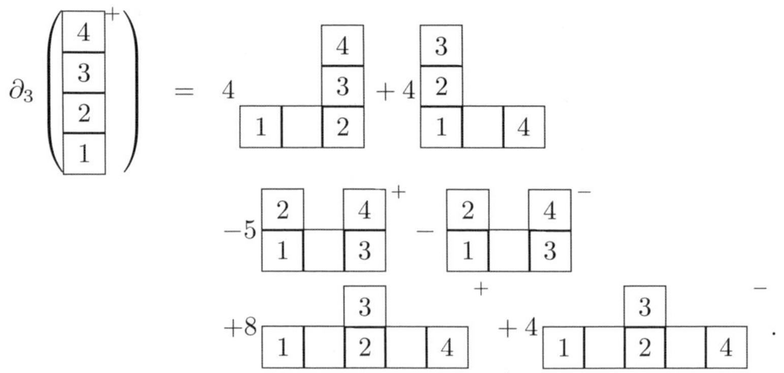

Analogously we have

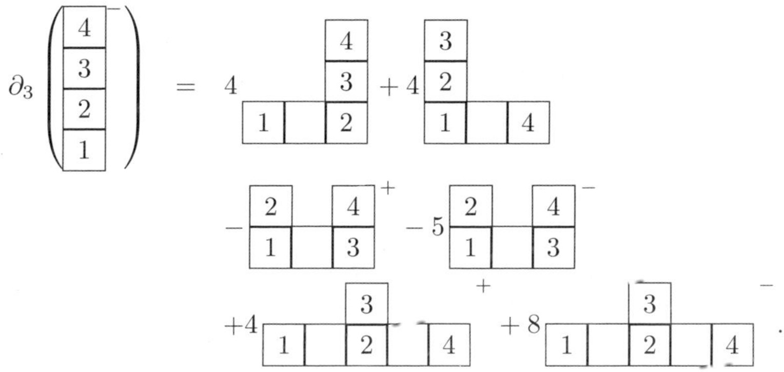

Q.E.D.

4.3. The homology of $\mathcal{C}_3^2$

In this section, we compare $H_*(\mathrm{Sal}(\mathcal{C}_3^2) \otimes_{\Sigma_4} \mathbb{F}_p(\pm 1))$ and $H_*(\mathrm{Sal}(\mathcal{A}_3) \otimes_{\Sigma_4} \mathbb{F}_p(\pm 1))$ for p a prime, following the strategy described in the beginning of §4.

We have analyzed the cell structure of $\mathrm{Sal}(\mathcal{C}_3^2)$ in the previous section based on the structure of $\mathcal{L}(\mathcal{C}_3^2)/\Sigma_4$ investigated in §4.2.

In this section, we compute the homology of

$$K_*^{4,2} = \mathrm{Ker}(i_4^2 : C_*(\mathrm{Sal}(\mathcal{C}_3^2)) \otimes_{\Sigma_4} \mathbb{F}_p(\pm 1) \to C_*(\mathrm{Sal}(\mathcal{A}_3)) \otimes_{\Sigma_4} \mathbb{F}_p(\pm 1))$$

for an odd prime p. We first need to know generators for $K_*^{4,2}$.

The generators of the cellular chain complex $C_*(\mathrm{Sal}(\mathcal{C}_3^2)) \otimes_{\Sigma_4} \mathbb{F}_p(\pm 1)$ are in one-to-one correspondence with cells in $\mathrm{Sal}(\mathcal{C}_3^2)/\Sigma_4$ even when p is odd. Of course, we have to take the sign representation into account, when we compute the boundary homomorphisms.

Lemma 4.10. *Define*

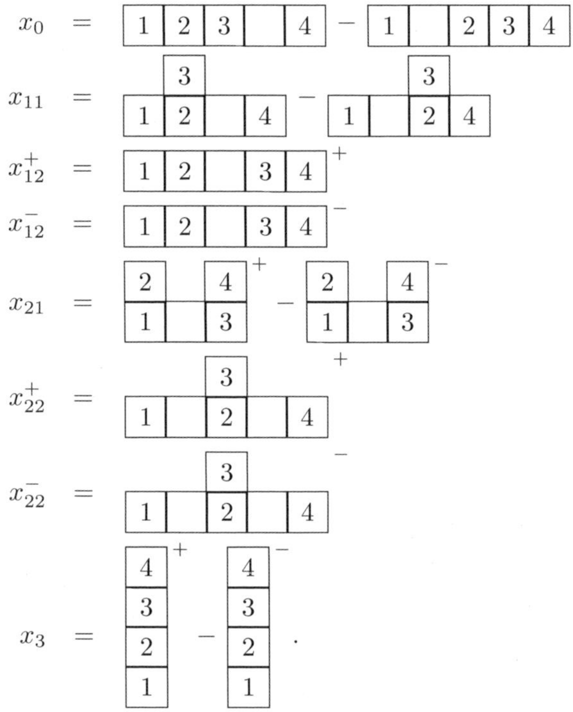

Then these are generators for $K_*^{4,2}$

$$
\begin{aligned}
K_0^{4,2} &= \langle x_0 \rangle, \\
K_1^{4,2} &= \left\langle x_{11}, x_{12}^+, x_{12}^- \right\rangle, \\
K_2^{4,2} &= \left\langle x_{21}, x_{22}^+, x_{22}^- \right\rangle, \\
K_3^{4,2} &= \langle x_3 \rangle.
\end{aligned}
$$

Thanks to the calculations in the previous section, we can easily compute the boundaries on these generators.

Lemma 4.11. *Boundaries are given by*

$$\begin{aligned}
\partial_1(x_{11}) &= 2x_0 \\
\partial_1(x_{12}^+) &= x_0 \\
\partial_1(x_{12}^-) &= x_0 \\
\partial_2(x_{21}) &= 2(x_{12}^+ - x_{12}^-) \\
\partial_2(x_{22}^+) &= x_{11} - 2x_{12}^+ \\
\partial_2(x_{22}^-) &= x_{11} - 2x_{12}^- \\
\partial_3(x_3) &= -4x_{21} + 4(x_{22}^+ - x_{22}^-).
\end{aligned}$$

By an elementary calculation, we obtain the homology of $K_*^{4,2}$.

Proposition 4.12. *When p is odd,*

$$H_i(K_*^{4,2}) \cong 0$$

for all i.

As a corollary, we obtain Theorem 1.3.

Corollary 4.13 (Theorem 1.3). *For an odd prime p, the inclusion $M_2(\mathbb{C}, 4) \hookrightarrow F(\mathbb{C}, 4)$ induces an isomorphism*

$$H_*(S_*(M_2(\mathbb{C}, 4)) \otimes_{\Sigma_4} \mathbb{F}_p(\pm 1)) \cong H_*(S_*(F(\mathbb{C}, 4)) \otimes_{\Sigma_4} \mathbb{F}_p(\pm 1)).$$

When $p = 2$, Lemma 4.11 implies that the boundaries in $K_*^{4,2}$ are given by.

$$\begin{aligned}
\partial_1(x_{11}) &= 0 \\
\partial_1(x_{12}^+) &= x_0 \\
\partial_1(x_{12}^-) &= x_0 \\
\partial_2(x_{21}) &= 0 \\
\partial_2(x_{22}^+) &= x_{11} \\
\partial_2(x_{22}^-) &= x_{11} \\
\partial_3(x_3) &= 0.
\end{aligned}$$

In particular, x_3 represents a nontrivial cycle in $C_3^{4,2}$ that is mapped to 0 under the map i_4^2. Thus we obtain a proof of Theorem 1.2.

Following the suggestion by the referee, we conclude this paper by briefly describing the case of the trivial $\mathbb{F}_p$ coefficients. We have

$$C_*(\mathrm{Sal}(\mathcal{C}_{n-1}^k)) \otimes_{\Sigma_n} \mathbb{F}_p \cong C_*(\mathrm{Sal}(\mathcal{C}_{n-1}^k)/\Sigma_n) \otimes \mathbb{F}_p.$$

By ignoring the changes of signs when we permute labels in the calculations in §4.2, we obtain the following. The details are omitted.

Lemma 4.14. *We have the following formula in* $C_*(\mathrm{Sal}(\mathcal{C}_3^2)/\Sigma_4) \otimes \mathbb{F}_p$.

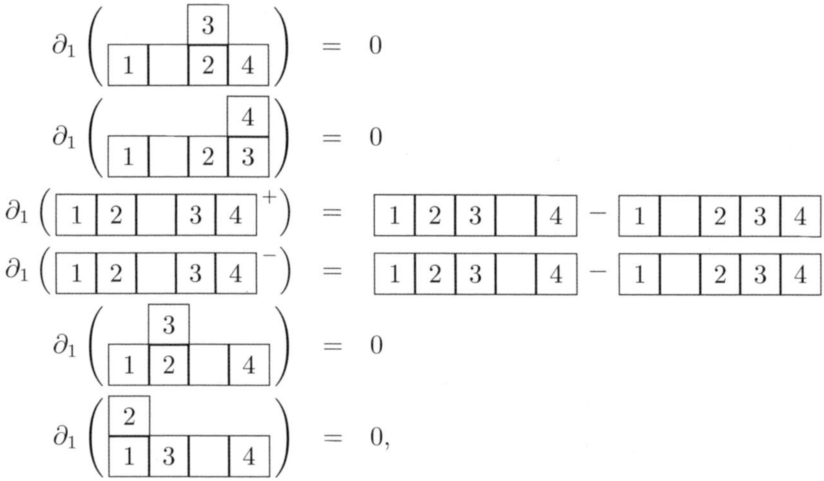

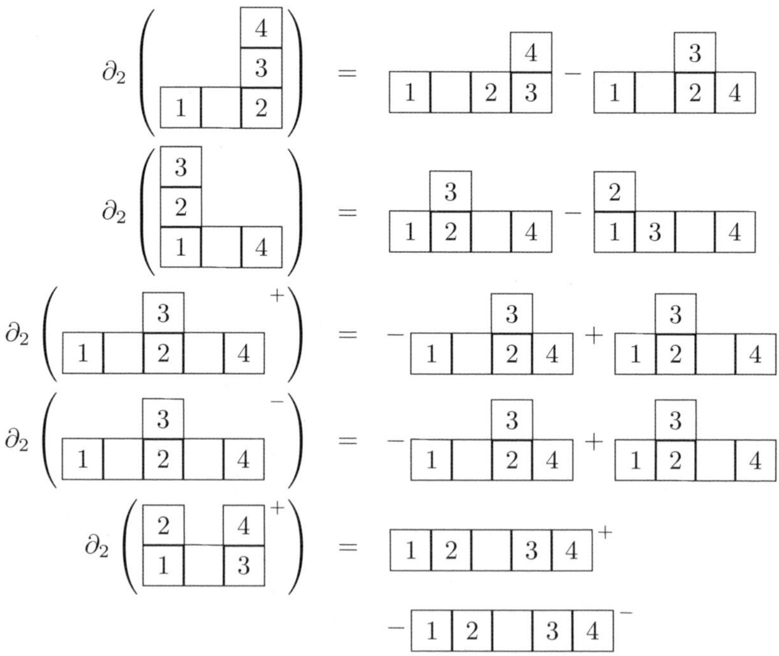

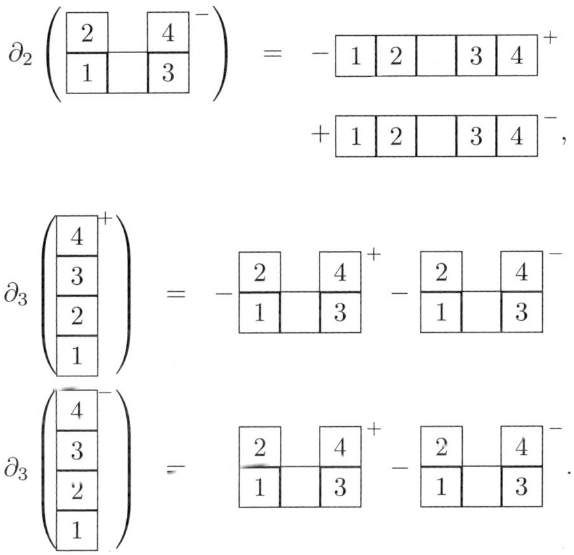

Let $(\bar{K}^{n,k}_*, \bar{\partial})$ denote the chain complex given by the kernel of

$$C_*(\mathrm{Sal}(\mathcal{C}^k_{n-1})/\Sigma_n) \otimes \mathbb{F}_p \longrightarrow C_*(\mathrm{Sal}(\mathcal{C}^{k-1}_{n-1})/\Sigma_n) \otimes \mathbb{F}_p.$$

Then we obtain the homology of $\bar{K}^{4,2}_*$ as follows.

Proposition 4.15. *We have*

$$\begin{aligned}
\bar{\partial}(x_0) &= 0\\
\bar{\partial}(x_{11}) &= 0\\
\bar{\partial}(x_{12}^+) &= x_0\\
\bar{\partial}(x_{12}^-) &= x_0\\
\bar{\partial}(x_{21}) &= 2(x_{12}^+ - x_{12}^-)\\
\bar{\partial}(x_{22}^+) &= x_{11}\\
\bar{\partial}(x_{22}^-) &= x_{11}\\
\bar{\partial}(x_3) &= 0.
\end{aligned}$$

Hence we obtain

$$\begin{aligned}
H_0(\bar{K}^{4,2}_*) &= 0\\
H_1(\bar{K}^{4,2}_*) &= 0\\
H_2(\bar{K}^{4,2}_*) &= \langle [x_{22}^+ - x_{22}^-] \rangle\\
H_3(\bar{K}^{4,2}_*) &= \langle [x_3] \rangle
\end{aligned}$$

when p is odd, and

$$\begin{aligned} H_0(\bar{K}_*^{4,2}) &= 0 \\ H_1(\bar{K}_*^{4,2}) &= \langle [x_{12}^+ - x_{12}^-] \rangle \\ H_2(\bar{K}_*^{4,2}) &= \langle [x_{21}], [x_{22}^+ - x_{22}^-] \rangle \\ H_3(\bar{K}_*^{4,2}) &= \langle [x_3] \rangle \end{aligned}$$

when $p = 2$.

Corollary 4.16. *The map*

$$H_*(M_2(\mathbb{C}, 4)/\Sigma_4; \mathbb{F}_p) \longrightarrow H_*(F(\mathbb{C}, 4)/\Sigma_4; \mathbb{F}_p)$$

induced by the inclusion is not an isomorphism for any prime p.

References

[Arv91] W. A. Arvola, Complexified real arrangements of hyperplanes, Manuscripta Math., **71** (1991), 295–306.

[BLVS+99] A. Björner, M. Las Vergnas, B. Sturmfels, N. White and G. M. Ziegler, Oriented matroids, Encyclopedia Math. Appl., **46**, Cambridge Univ. Press, Cambridge, 1999.

[BZ92] A. Björner and G. M. Ziegler, Combinatorial stratification of complex arrangements, J. Amer. Math. Soc., **5** (1992), 105–149.

[CK07] F. R. Cohen and Y. Kamiyama, Configurations and parallelograms associated to centers of mass, In: Proceedings of the School and Conference in Algebraic Topology, Geom. Topol. Monogr., **11**, Geom. Topol. Publ., Coventry, 2007, pp. 17–32.

[CM82] F. R. Cohen and M. E. Mahowald, Unstable properties of $\Omega^n S^{n+k}$, In: Symposium on Algebraic Topology in honor of José Adem, Oaxtepec, 1981, Contemp. Math., **12**, Amer. Math. Soc., Providence, RI, 1982, pp. 81–90.

[CMN79a] F. R. Cohen, J. C. Moore and J. A. Neisendorfer, The double suspension and exponents of the homotopy groups of spheres, Ann. of Math. (2), **110** (1979), 549–565.

[CMN79b] F. R. Cohen, J. C. Moore and J. A. Neisendorfer, Torsion in homotopy groups, Ann. of Math. (2), **109** (1979), 121–168.

[Coh83] F. R. Cohen, The unstable decomposition of $\Omega^2\Sigma^2 X$ and its applications, Math. Z., **182** (1983), 553–568.

[GR89] I. M. Gel′fand and G. L. Rybnikov, Algebraic and topological invariants of oriented matroids, Dokl. Akad. Nauk SSSR, **307** (1989), 791–795.

[Gra93a] B. Gray, *EHP* spectra and periodicity. I. Geometric constructions, Trans. Amer. Math. Soc., **340** (1993), 595–616.

[Gra93b] B. Gray, *EHP* spectra and periodicity. II. Λ-algebra models, Trans. Amer. Math. Soc., **340** (1993), 617–640.

[Mah77] M. Mahowald, A new infinite family in ${}_2\pi_*{}^s$, Topology, **16** (1977), 249–256.

[Sal87] M. Salvetti, Topology of the complement of real hyperplanes in $\mathbf{C}^N$, Invent. Math., **88** (1987), 603–618.

[Sal94] M. Salvetti, The homotopy type of Artin groups, Math. Res. Lett., **1** (1994), 565–577.

[Sna74] V. P. Snaith, A stable decomposition of $\Omega^n S^n X$, J. London Math. Soc. (2), **7** (1974), 577–583.

[Tam] D. Tamaki, The Salvetti Complex and the Little Cubes.

Dai Tamaki
Department of Mathematical Sciences, Shinshu University,
Matsumoto, 390-8621, Japan
E-mail address: rivulus@math.shinshu-u.ac.jp

Advanced Studies in Pure Mathematics 62, 2012
Arrangements of Hyperplanes—Sapporo 2009
pp. 459–474

Varieties of lines on Fermat hypersurfaces

Tomohide Terasoma

Abstract.

Let X be a hypersurface in the projective space $\mathbf{P}^{n+1}$ and $G(n+1,1)$ be the Grassmann variety $G(n+1,1)$ of lines in $\mathbf{P}^{n+1}$. The subvariety $F(X)$ of $G(n+1,1)$ consisting of lines contained in X is called the Fano variety of X. We study a detailed structure of the Fano variety of the Fermat hypersurface X of degree d for $n \geq d$. More precisely, we show that a certain open subset $F^0(X)$ of $F(X)$ has a fibration structure over a moduli space of marked pointed rational curves and that the fibers are complete intersections of Fermat hypersurfaces introduced in [T]. We also study singularities of $F(X)$.

§1. Introduction

In this paper, all algebraic varieties are considered over the complex number field $\mathbf{C}$ unless otherwise stated. Let n, d be integers such that $n \geq 2$, $d \geq 2$, and $\mathbf{P}^{n+1}$ be the $(n+1)$-dimensional projective space and X a hypersurface of degree d in $\mathbf{P}^{n+1}$. Let $G(n+1,1)$ be the Grassmann variety of lines in $\mathbf{P}^{n+1}$ defined by

$$G(n+1,1) = \{l \mid l \text{ is a line in } \mathbf{P}^{n+1}\}.$$

The variety consisting of lines contained in X is denoted by $F(X)$ and called the Fano variety of lines of X, i.e.

$$F(X) = \{l \in G(n+1,1) \mid l \subset X\}.$$

Let $Sym^d(\mathbf{C}^{n+2})$ be the vector space of homogeneous polynomials of degree d on $\mathbf{C}^{n+2}$ and $\mathbf{P}(Sym^d(\mathbf{C}^{n+2}))$ the projective space associated to $Sym^d(\mathbf{C}^{n+2})$. The class of a non-zero element f of $Sym^d(\mathbf{C}^{n+2})$ in

Received April 23, 2010.
Revised April 1, 2011.
2010 *Mathematics Subject Classification.* Primary 14C05; Secondary 52C35, 14J70.
Key words and phrases. Fano variety, Fermat hypersurface.

$\mathbf{P}(Sym^d(\mathbf{C}^{n+2}))$ is denoted by $[f]$, and the hypersurface in $\mathbf{P}^{n+1}$ defined by the zeros of f is denoted by $V(f)$.

The following theorem is due to Barth-Van de Ven [BV].

Theorem 1 (Generic smoothness). *Assume that $2n > d+1$. There exists a non-empty open set U of $\mathbf{P}(Sym^d(\mathbf{C}^{n+2}))$ such that the Fano variety of lines of $F(V(f))$ is smooth of dimension $2n-d-1$ if $[f] \in U$.*

The open set U in the above theorem is described as follows. Let $\mathcal{U}$ be the universal family of lines contained in the universal family of hypersurfaces defined by

$$\mathcal{U} = \{([f], l) \in \mathbf{P}(Sym^d(\mathbf{C}^{n+2})) \times G(n+1,1) \mid l \subset V(f)\}.$$

Let

$$pr_1 : \mathcal{U} \to \mathbf{P}(Sym^d(\mathbf{C}^{n+2})), \quad pr_2 : \mathcal{U} \to G(n+1,1)$$

be the restriction to $\mathcal{U}$ of the first and second projections of the product $\mathbf{P}(Sym^d(\mathbf{C}^{n+2})) \times G(n+1,1)$. The fiber of pr_2 at a line l is the set of hypersurfaces containing the line l, which is isomorphic to a linear subspace in $\mathbf{P}(Sym^d(\mathbf{C}^{n+2}))$ of codimension $d+1$. Therefore the map pr_1 is smooth and the variety $\mathcal{U}$ is smooth. By the generic smoothness of pr_1, there exists a non-empty open set U of $\mathbf{P}(Sym^d(\mathbf{C}^{n+2}))$ such that the restriction of pr_1 to U is smooth.

On the other hand, there exists a smooth hypersurface X such that the Fano variety $F(X)$ of X is not smooth even if $2n > d+1$. In this paper, we give a description of $F(X)$ of a Fermat hypersurface for $n \geq d$ using a moduli space of marked rational curves. The main theorems are Theorem 2 (the case $n = d$) and Theorem 4 (the case $n \geq d$). In Section 5, we study the singular loci of the Fano varieties $F(X)$ of Fermat hypersurfaces for $n = d$.

§2. Transversal Fano varieties of Fermat hypersurfaces

Let $G(n+1,1)$ be the Grassmann variety of lines in $\mathbf{P}^{n+1}$. The coordinates of the projective space is written as $(X_0 : \cdots : X_{n+1})$. The coordinate hyperplane defined by $\{X_i = 0\}$ is denoted by H_i. We define the subset $G(n+1,1)^0$ of $G(n+1,1)$ by

$$\begin{aligned} &G(n+1,1)^0 \\ =&\{l \in G(n+1,1) \mid \text{ the intersections } l \cap H_0, \dots, l \cap H_{n+1} \text{ are distinct}\}. \end{aligned}$$

The transversal part $F^0(X)$ of $F(X)$ is defined by $F(X) \cap G(n+1,1)^0$.

Definition 1 (Moduli space of rational curves with $(n+2)$ marked points).

(1) *The moduli space of rational curves with distinct* $(n+2)$ *marked points is denoted by* $\mathcal{M}_{0,n+2}$.

(2) *We define the moduli space of rigidified rational curves with* $(n+2)$ *marked points by*

$$\widetilde{\mathcal{M}_{0,n+2}} = (\mathbf{P}^1)^{n+2} - D,$$

where D *is the big diagonal defined by*

$$D = \cup_{0 \le i < j \le n+1} \{(p_0, \ldots, p_{n+1}) \mid p_i = p_j\}.$$

The group $PGL(2)$ acts on $(\mathbf{P}^1)^{n+2}$ diagonally and the open set $\widetilde{\mathcal{M}_{0,n+2}}$ is stable and fixed point free under this action. The variety $\mathcal{M}_{0,n+2}$ is isomorphic to $\widetilde{\mathcal{M}_{0,n+2}}/PGL(2)$. The class of $(p_0, \ldots, p_{n+1})$ in $\mathcal{M}_{0,n+2}$ is denoted by $[p_0, \ldots, p_{n+1}]$. The dimension of $\mathcal{M}_{0,n+2}$ is $n-1$.

Let $l \in F^0(X)$. By the condition of transversality, the intersections $l \cap H_0, \ldots, l \cap H_{n+1}$ are distinct $n+2$ points, and they define a point $conf(l) = [l \cap H_0, \ldots, l \cap H_{n+1}] \in \mathcal{M}_{0,n+2}$. Thus we have a map

$$conf : F^0(X) \to \mathcal{M}_{0,n+2}.$$

We define the Fermat hypersurface X of degree d in $\mathbf{P}^{n+1}$ by

$$X : X_0^d + \cdots + X_{n+1}^d = 0.$$

Let G be the group $\mu_d^{n+2}/\Delta(\mu_d)$, where $\Delta(\mu_d) = \{(\zeta, \cdots, \zeta) \mid \zeta \in \mu_d\} \subset \mu_d^{n+2}$. Then G acts on X by

$$(X_0 : \cdots : X_{n+1}) = (\zeta_0 X_0 : \cdots : \zeta_{n+1} X_{n+1})$$

for $(\zeta_0, \ldots, \zeta_{n+1}) \in G$. The action of G on the Fermat hypersurface X induces that on the Fano variety $F(X)$ of lines on X.

Let l be an element in $F^0(X)$ and $g = (\zeta_0, \ldots, \zeta_{n+1})$ an element in G. Since the image of the point $l \cap H_i$ under the map $g : l \to g(l)$ is equal to $g(l) \cap H_i$, the condition of transversality is stable under the action of G. Therefore G induces an action on $F(X)^0$ and $conf(g(l)) = conf(l)$. Thus we have a map

$$\overline{conf} : F^0(X)/G \to \mathcal{M}_{0,n+2}.$$

Our first main theorem is the following.

Theorem 2. *If* $d = n$, *then the action of* G *on* $F(X)^0$ *is fixed point free and the map* $\overline{conf}$ *is an isomorphism.*

§3. Construction of the family $\mathcal{F}$

3.1. Trivialized Kummer covering

Let k be a field of characteristic zero, S a smooth connected scheme over k, and $\pi : \mathcal{C} \to S$ a geometrically connected smooth projective curve over S. Let $D = \sum_i a_i D_i$ be a divisor on $\mathcal{C}$ such that $\mathcal{O}_{\mathcal{C}}(D) \simeq \mathcal{O}_{\mathcal{C}}$. Let $p_\infty : S \to \mathcal{C}$ be a section of π such that $Im(p_\infty) \cap Supp(D) = \emptyset$.

Definition 2 (Trivialized Kummer covering)**.** *Let d be a positive integer and f a rational function on $\mathcal{C}$ such that*

(1) *The divisor (f) of f is equal to D, and*
(2) $f\,|_{Im(p_\infty)} = 1$.

Then the projective curve D over S defined by $y^d = f$ is called the Kummer covering with the branch divisor D trivialized at p_∞. Since there is a unique rational function f with the properties in Definition 2, the trivialized Kummer covering is uniquely determined by D and p_∞.

3.2. Definition of Kum^q

Let $\widetilde{\mathcal{U}} = \mathbf{P}^1 \times \widetilde{M_{0,n+2}}$ be the universal rational curve over the rigidified moduli space $\widetilde{M_{0,n+2}}$ and $\widetilde{p_i} : \widetilde{\mathcal{M}_{0,n+2}} \to \widetilde{\mathcal{U}}$ be the universal section defined by

$$\widetilde{p_i}(\lambda_0, \ldots, \lambda_{n+1}) = (\lambda_i, (\lambda_0, \ldots, \lambda_{n+1})) \in \widetilde{\mathcal{U}}.$$

The group $PGL(2)$ acts diagonally on $\widetilde{\mathcal{U}}$ freely and the quotient variety is denoted by $\mathcal{U}$. Since the action of $PGL(2)$ on $\widetilde{\mathcal{U}}$ and that on $\widetilde{\mathcal{M}_{0,n+2}}$ are equivariant and the sections $\widetilde{p_i}$ are stable under its action, we have a map $\mathcal{U} \to \mathcal{M}_{0,n+2}$ and sections $p_i : \mathcal{M}_{0,n+2} \to \mathcal{U}$.

We fix $q \in \mathbf{P}^1$. The affine line defined by $\mathbf{P}^1 - q$ is denoted by $\mathbf{A}^1_q$. We choose an inhomogeneous coordinate x of $\mathbf{P}^1$ such that $x(q) = \infty$. The open set $(\mathbf{A}^1_q)^{n+2} \cap \widetilde{\mathcal{M}_{0,n+2}}$ of $\widetilde{\mathcal{M}_{0,n+2}}$ is denoted by $\widetilde{\mathcal{M}_q} = \widetilde{\mathcal{M}_{0,n+2,q}}$ and the restriction of the universal curve $\widetilde{\mathcal{U}}$ to $\widetilde{\mathcal{M}_q}$ is denoted by $\widetilde{\mathcal{U}_q}$. The section of $\widetilde{\mathcal{U}_q} \to \widetilde{\mathcal{M}_q}$ defined by the point q is also denoted by q. The three sections p_i, p_j and q do not intersect to each other. The Kummer covering of $\widetilde{\mathcal{U}_q}$ branching at $(\widetilde{p_i}) - (\widetilde{p_j})$ trivialized at q is denoted by $Kum^q_{i,j}$. Using the above inhomogeneous coordinate x of $\mathbf{P}^1$, we have the coordinates $(x, \lambda_0, \ldots, \lambda_{n+1})$ of $\widetilde{\mathcal{U}}$. Then the covering $Kum^q_{i,j}$ is defined by

$$y_{i,j}^d = \frac{x - \lambda_i}{x - \lambda_j}.$$

We set

$$Kum^q = Kum^q_{0,1} \times_{\widetilde{\mathcal{U}}} Kum^q_{0,2} \times_{\widetilde{\mathcal{U}}} \cdots \times_{\widetilde{\mathcal{U}}} Kum^q_{0,n+1}.$$

An element $(\zeta_0, \dots, \zeta_{n+1})$ in the group G acts on Kum^q by $y_{0,i} \mapsto \dfrac{\zeta_0}{\zeta_i} y_{0,i}$. Under this action of G, $Kum^q \to \widetilde{\mathcal{U}}$ becomes a G-covering.

3.3. Definition of $\widehat{\mathcal{M}_q}$

We define $\Delta_2 = \{(\lambda, \nu) \in \mathbf{A}^1_q \times \mathbf{A}^1_q \mid \lambda \neq \nu\}$. We set

$$\pi : \mathbf{P}^1 \times \Delta_2 \to \Delta_2 : (x, \lambda, \nu) \mapsto (\lambda, \nu)$$

The sections of π defined by $(\lambda, \nu) \mapsto (\lambda, \lambda, \nu)$ and $(\lambda, \nu) \mapsto (\nu, \lambda, \nu)$ are also denoted as λ and ν and the Kummer covering branching at $(\lambda) - (\nu)$ trivialized at q is denoted as $\widetilde{\Delta_3} \to \mathbf{P}^1 \times \Delta_2$. The pull back of $\widetilde{\Delta_3}$ by the map

$$\widetilde{\mathcal{M}_q} \to \mathbf{P}^1 \times \Delta_2 : (\lambda_0, \dots, \lambda_{n+2}) \mapsto (\lambda_i, \lambda_j, \lambda_k)$$

is denoted by $\Delta^i_{j,k}$. Then the μ_d-covering $\Delta^i_{j,k} \to \widetilde{\mathcal{M}_q}$ is defined by $\delta^d_{i,j,k} = \frac{\lambda_i - \lambda_j}{\lambda_i - \lambda_k}$. The Galois group of the covering

$$\prod_{i \neq j,k \, \widetilde{\mathcal{M}_q}} \Delta^i_{j,k} \to \widetilde{\mathcal{M}_q}$$

is isomorphic to μ_d^n and the covering corresponding to the group $\mathrm{Ker}(\mu_d^n \to \mu_d : (\zeta_i)_i \mapsto \sum_i \zeta_i)$ is denoted by $\widehat{\mathcal{M}_{j,k}}$. Here $\prod_{\widetilde{\mathcal{M}_q}}$ is the fiber product over $\widetilde{\mathcal{M}_q}$. The μ_d-covering $\widehat{\mathcal{M}_{j,k}}$ of $\widetilde{\mathcal{M}_q}$ is defined by

$$\delta^d_{j,k} = \prod_{i \neq j,k} \frac{\lambda_i - \lambda_j}{\lambda_i - \lambda_k}.$$

We define a G-covering $\widehat{\mathcal{M}_q} \to \widetilde{\mathcal{M}_q}$ by

$$\widehat{\mathcal{M}_q} = \prod_{i=1 \, \widetilde{\mathcal{M}_q}}^{n+1} \widehat{\mathcal{M}_{0,i}}.$$

3.4. The family $\mathcal{F}_q$

We set $k = n - d$. We define the variety $\widetilde{\mathcal{F}}_q$ by the following fiber product:

$$\widetilde{\mathcal{F}}_q = \underbrace{Kum^q \times_{\widetilde{M}_q} Kum^q \times_{\widetilde{M}_q} \cdots \times_{\widetilde{M}_q} Kum^q}_{k\text{-times}} \times_{\widetilde{M}_q} \widehat{\mathcal{M}_q}.$$

Then $\widetilde{\mathcal{F}}_q$ is a $G^k \times G$ covering of

$$\widetilde{\mathcal{U}}_q^k = \underbrace{\widetilde{\mathcal{U}}_q \times_{\widetilde{\mathcal{M}_q}} \widetilde{\mathcal{U}}_q \times_{\widetilde{\mathcal{M}_q}} \cdots \times_{\widetilde{\mathcal{M}_q}} \widetilde{\mathcal{U}}_q}_{k\text{-times}}.$$

We define the variety $\mathcal{F}_q$ by the G-covering of $\widetilde{\mathcal{U}}_q^k$ corresponding to the subgroup of the Galois group:

$$\operatorname{Ker}\big(G^k \times G \to G : (g_1, \ldots, g_k, g) \mapsto -g + \sum_{i=1}^{k} g_i\big).$$

Let $x_1, \ldots, x_k$ be the coordinates for the fibers of $\widetilde{\mathcal{U}}_q^k \to \widetilde{\mathcal{M}_q}$. Then $\mathcal{F}_q$ is defined by

$$(1) \qquad \xi_j^d = \prod_{i=1}^{k} \frac{(x_i - \lambda_j)}{(x_i - \lambda_0)} \prod_{p \neq 0, j} \frac{(\lambda_0 - \lambda_p)}{(\lambda_j - \lambda_p)} \quad (j = 1, \ldots, n+1).$$

3.5. Patching and an action of $PGL(2)$

We define a fiber space $\mathcal{F}_q \to \widetilde{\mathcal{M}_q}$ of relative dimension k for each $q \in \mathbf{P}^1$. In this subsection, we patch them into a variety $\mathcal{F}$. Let $q_1, \ldots, q_{n+3}$ be distinct elements in $\mathbf{P}^1$. Then we have $\widetilde{\mathcal{M}_{0,n+2}} = \bigcup_{i=1}^{n+3} \widetilde{\mathcal{M}_{q_i}}$.

Proposition 3. *There exists a patching data*

$$(2) \qquad \varphi_{ij} : \mathcal{F}_{q_i} \mid_{\widetilde{\mathcal{M}_{q_i}} \cap \widetilde{\mathcal{M}_{q_j}}} \xrightarrow{\simeq} \mathcal{F}_{q_j} \mid_{\widetilde{\mathcal{M}_{q_i}} \cap \widetilde{\mathcal{M}_{q_j}}}$$

for $\mathcal{F}_{q_i}$ and $\mathcal{F}_{q_j}$ such that the glued variety $\mathcal{F} = \bigcup_{i=1}^{n+3} \mathcal{F}_{q_i}$ admits a lifting of the action of $PGL(2)$ on $\widetilde{\mathcal{M}_{0,n+2}}$.

Proof. We choose coordinates x and x' of $\mathbf{P}^1$ such that $x(q_i) = \infty$ and $x'(q_j) = \infty$. We introduce an isomorphism φ_{ij} in (2). We write

$$x' = \frac{ax + b}{cx + d}.$$

We use these coordinates to express points in $\widetilde{\mathcal{M}_{q_i}}$ and $\widetilde{\mathcal{M}_{q_j}}$. Then a point $(\lambda_0, \dots, \lambda_{n+1}) \in \widetilde{\mathcal{M}_{q_i}}$ corresponds to $(\lambda_0', \dots, \lambda_{n+1}') \in \widetilde{\mathcal{M}_{q_i}}$, where

$$\lambda_i' = \frac{a\lambda_i + b}{c\lambda_i + d}.$$

We write coordinates of fibers for $\mathcal{F}_{q_i} \to \widetilde{\mathcal{M}_{q_i}}$ and $\mathcal{F}_{q_j} \to \widetilde{\mathcal{M}_{q_j}}$ as $x_1, \dots, x_k$ and $x_1', \dots, x_k'$. Then we have

$$x_i' - \lambda_j' = \frac{(ad - bc)(x_i - \lambda_j)}{(cx_i + d)(c\lambda_j + d)}, \quad \lambda_i' - \lambda_j' = \frac{(ad - bc)(\lambda_i - \lambda_j)}{(cx_i + d)(c\lambda_j + d)}.$$

On $\widetilde{\mathcal{M}_{q_i}}$, the equation of $\mathcal{F}_{q_j}$ is given by

$$\begin{aligned} {\xi_j'}^d &= \prod_{i=1}^{k} \frac{(x_i' - \lambda_j')}{(x_i' - \lambda_0')} \prod_{p \neq 0,j} \frac{(\lambda_0' - \lambda_p')}{(\lambda_j' - \lambda_p')} \\ &= \prod_{i=1}^{k} \frac{(x_i - \lambda_j)}{(x_i - \lambda_0)} \prod_{p \neq 0,j} \frac{(\lambda_0 - \lambda_p)}{(\lambda_j - \lambda_p)} \left(\frac{c\lambda_j + d}{c\lambda_0 + d} \right)^d. \end{aligned}$$

By setting

$$\xi_i' = \xi_i \cdot \frac{c\lambda_j + d}{c\lambda_0 + d},$$

we have an isomorphism $\varphi : \mathcal{F}_{q_1} \to \mathcal{F}_{q_2}$. To show that the family of isomorphisms $\{\varphi_{ij}\}$ actually gives a patching data, we check its 1-cocyle condition. Let x, x', x'' be inhomogeneous coordinates of $\mathbf{A}^1_{q_i}, \mathbf{A}^1_{q_j}, \mathbf{A}^1_{q_k}$ and write

$$x' = \frac{ax + b}{cx + d}, \quad x'' = \frac{a'x' + b'}{c'x' + d'}, \quad x'' = \frac{a''x + b''}{c''x + d''}.$$

We set

$$g' = \begin{pmatrix} a & b \\ c & d \end{pmatrix}, \quad g' = \begin{pmatrix} a' & b' \\ c' & d' \end{pmatrix}, \quad g'' = \begin{pmatrix} a'' & b'' \\ c'' & d'' \end{pmatrix},$$

then we have $g'' = g'g$ in $PGL(2)$. Since the rules of the isomorphisms φ_{ij} and φ_{jk} for ξ_p, ξ_p', ξ_p'' are given by

$$\xi_p' = \xi_p \cdot \frac{c\lambda_p + d}{c\lambda_0 + d}, \quad \xi_p'' = \xi_p' \cdot \frac{c'\lambda_p' + d'}{c'\lambda_0' + d'}.$$

Since $\lambda_p' = \dfrac{a\lambda_p + b}{c\lambda_p + d}$, we have $\xi_p'' = \xi_p \cdot \dfrac{c''\lambda_p + d''}{c''\lambda_0 + d''}$. Therefore, we have $\varphi_{ik} = \varphi_{jk} \circ \varphi_{ij}$. This computation also shows that the action of $PGL(2)$ on $\widetilde{\mathcal{U}}^k$ extends to that on $\mathcal{F}$. Q.E.D.

§4. Proof of Theorem 2

We use the same notations as in the last section. We assume that $k = n - d \geq 0$. We have the following sequence of varieties

$$\mathcal{F} \xrightarrow{\alpha} \widetilde{\mathcal{U}}^k \xrightarrow{\beta} \widetilde{\mathcal{M}_{0,n+2}}.$$

We have the followings.

(1) The morphisms α and β are $PGL(2)$ equivariant.
(2) The morphism α is a finite etale Galois covering with Galois group G.
(3) The relative dimension of the morphism β is k.
(4) There are actions of the symmetric group $\mathfrak{S}_k$ of degree k on $\widetilde{\mathcal{U}}^k$ and $\mathcal{F}$ over $\widetilde{\mathcal{M}_{0,n+2}}$ and the morphism α is equivariant. These actions of $\mathfrak{S}_k$ commute with actions of $PGL(2)$.

Therefore we have the following sequence of homomorphisms:

$$\mathcal{F}/(\mathfrak{S}_k \times PGL(2)) \xrightarrow{\overline{\alpha}} \widetilde{\mathcal{U}}^k/(\mathfrak{S}_k \times PGL(2)) \xrightarrow{\overline{\beta}} \widetilde{\mathcal{M}_{0,n+2}}/PGL(2).$$

In this section, we prove the following theorem, which is a generalization of Theorem 2.

Theorem 4. *We assume that $n \geq d$ and set $k = n - d$. Then there is a G-equivariant isomorphism between the variety of transversal lines $F^0(X)$ in the Fermat hypersurface X of degree d in $\mathbf{P}^{n+1}$ and $\mathcal{F}/(\mathfrak{S}_k \times PGL(2))$.*

Remark 5. *Theorem 2 is the special case for $n = d$. In this case, the morphism $\overline{\beta}$ is an isomorphism and $\widetilde{\mathcal{M}_{0,n+2}}$ is isomorphic to $\mathcal{F}$. Therefore $F^0(X) \to \mathcal{M}_{0,n+2}$ is an etale G-covering.*

Let l be an element of the Grassmann variety $G(n+1,1)$. Then by using homogeneous parameter $(t_0 : t_1)$, the line l is expressed as

$$l : (\alpha_0 t_0 - \beta_0 t_1 : \alpha_1 t_0 - \beta_1 t_1 : \cdots \alpha_{n+1} t_0 - \beta_{n+1} t_1). \tag{3}$$

Using this expression and the inhomogeneous coordinate $t = \frac{t_1}{t_0}$, the intersection of l and the coordinate hyperplane H^i is equal to

$$\lambda_i = \frac{\alpha_i}{\beta_i}. \tag{4}$$

The matrix

$$\begin{pmatrix} \alpha_0 & \cdots & \alpha_{n+1} \\ \beta_0 & \cdots & \beta_{n+1} \end{pmatrix}$$

is called a frame of the line l. A $(2 \times (n+2))$-matrix A is a frame of a line in $\mathbf{P}^{n+1}$ if and only if the rank A is equal to 2. The set of $2 \times (n+2)$-matrix of rank 2 is denoted as $M(2,n+2)^0$. Then the space of non-constant maps $Map(\mathbf{P}^1, \mathbf{P}^{n+1}) = \{f : \mathbf{P}^1 \to \mathbf{P}^{n+1} \mid f \text{ is a non-constant map}\}$ is identified with $M(2,n+2)^0/\mathbf{C}^\times$ and the Grassmann variety $G(n+1,1)$ is isomorphic to $M(2,n+2)^0/GL(2)$, where $GL(2)$ acts on $M(2,n+2)^0$ from the right. The line l given by (3) is contained in the Fermat hypersurface if the equality

$$(\alpha_0 t_0 + \beta_0 t_1)^d + \cdots + (\alpha_{n+1} t_0 + \beta_{n+1} t_1)^d = 0$$

is satisfied for all $(t_0 : t_1)$. Therefore the condition of $\alpha_0, \ldots, \alpha_{n+1}$, $\beta_0, \ldots, \beta_{n+1}$ is

$$(5) \quad \begin{cases} \alpha_0^d + \cdots + \alpha_{n+1}^d = 0, \\ \beta_0 \alpha_0^{d-1} + \cdots + \beta_{n+1} \alpha_{n+1}^{d-1} = 0, \\ \vdots \\ \beta_0^{d-1} \alpha_0 + \cdots + \beta_{n+1}^{d-1} \alpha_{n+1} = 0, \\ \beta_0^d + \cdots + \beta_{n+1}^d = 0. \end{cases}$$

We assume that the coordinate $(t_0 : t_1)$ of l satisfies the condition $\beta_i \neq 0$, i.e. $\lambda_i \neq \infty$. Then $(\lambda_0, \ldots, \lambda_{n+1})$ is an element of $\widetilde{\mathcal{M}_{0,n+1,q}}$.

We set

$$\mathcal{F}' = \{A \in M(2,n+2)^0 \mid (1)\ A \text{ satisfies the condition (5), and } (2) \text{ all the } 2 \times 2 \text{ minors of } A \text{ are non-zero}\}/\mathbf{C}^\times.$$

Then we have the following maps

$$\begin{array}{ccc} \{f \in Map(\mathbf{P}^1, \mathbf{P}^{n+1} \mid f(\infty) \notin H_i \text{ for all } i\} & \overset{\psi}{\to} & \widetilde{\mathcal{M}_{0,n+2,q}} \\ \cap & & \cap \\ \mathcal{F}' & \overset{\psi}{\to} & \widetilde{\mathcal{M}_{0,n+2}}. \end{array}$$

Using the relation (4), the fiber of the map ψ at the point $(\lambda_0, \ldots, \lambda_{n+1}) \in \widetilde{\mathcal{M}_q}$ is equal to the subvariety of $\mathbf{P}^{n+1} = \{(\beta_0 : \cdots : \beta_{n+1})\}$ defined by

$$(6) \quad \begin{cases} \lambda_0^d \beta_0^d + \cdots + \lambda_{n+1}^d \beta_{n+1}^d = 0, \\ \lambda_0^{d-1} \beta_0^d + \cdots + \lambda_{n+1}^{d-1} \beta_{n+1}^d = 0, \\ \vdots \\ \lambda_0 \beta_0^d + \cdots + \lambda_{n+1} \beta_{n+1}^d = 0, \\ \beta_0^d + \cdots + \beta_{n+1}^d = 0. \end{cases}$$

Remark 6. *This variety is a special complete intersection of Fermat hypersurfaces defined in* [T].

As a consequence, we have the following proposition.

Proposition 7. *The variety $\mathcal{F}'$ is isomorphic to the variety:*

$$\begin{aligned}\mathcal{F}'' = \{(\beta,\lambda) =&((\beta_0:\cdots:\beta_{n+1}),(\lambda_0,\dots,\lambda_{n+1}))\in \mathbf{P}^{n+1}\times\widetilde{\mathcal{M}_{0,n+2}} \mid \\ &\beta \text{ and } \lambda \text{ satisfies the relation (6)}\}.\end{aligned}$$

The action of $PLG(2)$ on $\mathcal{F}''$ is induced by the action on $\mathcal{F}''$ via this isomorphism. This action is described as follows. Let g be an element of $PGL(2)$ such that $g^*(\lambda_i)=\dfrac{a\lambda_i+b}{c\lambda_i+d}$. Then the action on the variable β_i is given by $g^*(\beta_i)=\beta_i\cdot(c\lambda_i+d)$. We can check that the subvariety $\mathcal{F}''$ is stable under this action. Since $\mathcal{F}'/PGL(2)$ is isomorphic to $F^0(X)$ by the definition of $F(X)$, Proposition 7 implies the following proposition.

Proposition 8. *$\mathcal{F}''/PGL(2)$ is isomorphic to $F^0(X)$.*

Therefore Theorem 4 is a consequence of the following proposition.

Proposition 9. *There exists a $G\times PGL(2)$ equivariant isomorphism between $\mathcal{F}''$ and $\mathcal{F}/\mathfrak{S}_k$.*

Proof. We fix a coordinate t of $\mathbf{P}^1$. Let $x_1,\dots,x_k$ and $\lambda_0,\dots,\lambda_{n+2}$ be coordinates of $\widetilde{\mathcal{U}}^k$ as in §3.4. Then the covering $\mathcal{F}$ of $\widetilde{\mathcal{U}}^k$ is defined by (1). Let ξ_i be the rational function of $\mathcal{F}$ defined in (1). We define the map $\phi:\mathcal{F}\to\mathcal{F}''$ by

$$\begin{aligned}\phi :&(\lambda_0,\dots,\lambda_{n+1},x_1,\dots,x_k,\xi_1,\dots,\xi_{n+1})\\ &\mapsto ((1:\xi_1:\cdots:\xi_{n+1}),\lambda_0,\dots,\lambda_{n+1})\in\mathbf{P}^{n+1}\times\widetilde{\mathcal{M}_{0,n+2}}.\end{aligned}$$

We show that the image of this map ϕ is contained in $\mathcal{F}''$. We set $(\beta_0:\cdots:\beta_{n+1})=(1:\xi_1:\cdots:\xi_{n+1})$. Then by equality (1), we have

$$\begin{aligned}&(\beta_0^d:\cdots:\beta_{n+1}^d)\\ =&\left(\frac{\prod_{i=1}^k(x_i-\lambda_0)}{\prod_{p\neq 0}(\lambda_0-\lambda_p)}:\frac{\prod_{i=1}^k(x_i-\lambda_1)}{\prod_{p\neq 0}(\lambda_1-\lambda_p)}:\cdots:\frac{\prod_{i=1}^k(x_i-\lambda_{n+1})}{\prod_{p\neq 0}(\lambda_{n+1}-\lambda_p)}\right).\end{aligned}$$

Therefore by Lagrange interpolation formula, we have

$$\lambda_0^s\beta_0^d+\cdots\lambda_{n+1}^s\beta_{n+1}^d=0\quad\text{for } s=0,\dots,d.$$

Therefore the image of ψ is contained in $\mathcal{F}''$. Since the map ϕ is invariant under the action of $\mathfrak{S}_k$, it factors through the map $\overline{\phi}:\mathcal{F}/\mathfrak{S}_k\to\mathcal{F}''$. We show that the morphism ϕ is an isomorphism. It is enough to show that each fiber of $\overline{\phi}$ at $(\lambda_0, \dots, \lambda_{n+1})$ is an isomorphism.

The restriction of the morphism $\overline{\phi}$ to the fibers of $\mathcal{F}/\mathfrak{S}_k$ and $\mathcal{F}''$ are G-equivariant and the quotients by the group G are isomorphic to

$$L_1 = \{(\frac{L_0(s_0, \dots, s_k)}{\prod_{j\neq 0}(\lambda_j - \lambda_0)} : \cdots : \frac{L_{n+1}(s_0, \dots, s_k)}{\prod_{j\neq n+1}(\lambda_j - \lambda_{n+1})}) \mid s_0, \dots, s_k \in \mathbf{C}\}$$

and

$$L_2 = \{(b_0 : \cdots : b_{n+1}) \mid \sum_{i=0}^{n+1} \lambda_i^s b_i = 0 \quad \text{for } s = 0, \dots, d\},$$

respectively. Here $L_i(s_0, \dots, s_k) = \sum_{p=0}^{k}(-\lambda_i)^p s_{k-p}$ is a linear form on $s_0, \dots, s_k$ and the quotient map is given by the map

$s_p \mapsto$ the elementary symmetric function of x_i of degree p,

$b_i \mapsto \beta_i^d$.

On the other hand, the two linear spaces L_1 and L_2 are equal. Thus we have the proposition. Q.E.D.

Example 10. *We consider the case $n = d = 3$. In this case the Fano variety $F(X)$ becomes a surface, which is called the Fano surface of the Fermat cubic three fold X. The surface $F(X)^0$ is a $G = (\mathbf{Z}/3\mathbf{Z})^4$-covering of the moduli space $\mathcal{M}_5 = (\mathbf{A}^1 - \{0,1\})^2 - \Delta$, where Δ is the diagonal. In this case, it is known ([CG]) that the Fano variety $F(X)$ is smooth. We refer to [R] for other properties in this case.*

§5. Singularities

In this section, we study the singular locus of $F(X)$ in the case $d = n$. Similar computations are not difficult for the case $n > d$. Let l be an element in $F(X)$. Then the divisor $l \cap (\cup_i H_i)$ defines a partition

$$P(l) = \{P_x = \{i \mid x \in H_i\} \mid x \in l\}$$

of the set $[0, n+1] = \{0, \dots, n+1\}$. The variety $F(X)$ is stratified by the type of partition $P(l)$:

$$F(X) = \coprod_{P:\text{partition of } [0,n+1]} F(X)^P,$$

where $F(X)^P = \{l \in F(X) \mid P(l) = P\}$. For example,

$$F^0(X) = \{\{0\}, \dots, \{n+1\}\}.$$

Let l be an element of $F(X)$ and $i_1, \dots, i_k$ be the cardinalities of the elements in $P(l)$. Then $i_1 + \cdots + i_k = n + 2$. We compute the dimension of the tangent space of $F(X)$ at the point l. By changing the coordinates of the line, we may assume that the frame of l is of the form $A = (A_1, \dots, A_k)$, where

$$A_p = \begin{pmatrix} \lambda_p b_{p1} & \dots & \lambda_p b_{pi_p} \\ b_{p1} & \dots & b_{pi_p} \end{pmatrix}.$$

By changing coordinates of the line l, we may assume that $\lambda_i \neq 0$ and $b_{ij} \neq 0$ for all i and j. We consider a deformation of $A(\epsilon) = (A_1 + E_1, \dots, A_k + E_k)$, where

$$E_p = \begin{pmatrix} \epsilon\xi_{p1} & \dots & \epsilon\xi_{pi_p} \\ \epsilon\eta_{p1} & \dots & \epsilon\eta_{pi_p} \end{pmatrix},$$

with $\epsilon^2 = 0$. Since the line corresponding to the frame $A(\epsilon)$ is contained in the Fermat hyper surface, we have

$$\begin{cases} \sum_{p=1}^k \sum_{i=1}^{i_p} (\lambda_p b_{pi} + \epsilon\xi_{pi})^d = 0, \\ \sum_{p=1}^k \sum_{i=1}^{i_p} (\lambda_p b_{pi} + \epsilon\xi_{pi})^{d-1}(b_{pi} + \epsilon\eta_{pi}) = 0, \\ \vdots \\ \sum_{p=1}^k \sum_{i=1}^{i_p} (\lambda_p b_{pi} + \epsilon\xi_{pi})(b_{pi} + \epsilon\eta_{pi})^{d-1} = 0, \\ \sum_{p=1}^k \sum_{i=1}^{i_p} (b_{pi} + \epsilon\eta_{pi})^d = 0. \end{cases}$$

By looking at the coefficients of ϵ, we have

(7)
$$\begin{cases} \sum_{p=1}^k d\lambda_p^{d-1}(\sum_{i=1}^{i_p} b_{pi}^{d-1}\xi_{pi}) = 0, \\ \sum_{p=1}^k \left[(d-1)\lambda_p^{d-2}(\sum_{i=1}^{i_p} b_{pi}^{d-1}\xi_{pi}) + \lambda_p^{d-1}(\sum_{i=1}^{i_p} b_{pi}^{d-1}\eta_{pi})\right] = 0, \\ \vdots \\ \sum_{p=1}^k \left[(\sum_{i=1}^{i_p} b_{pi}^{d-1}\xi_{pi}) + (d-1)\lambda_p(\sum_{i=1}^{i_p} b_{pi}^{d-1}\eta_{pi})\right] = 0, \\ \sum_{p=1}^k d(\sum_{i=1}^{i_p} b_{pi}^{d-1}\eta_{pi}) = 0. \end{cases}$$

We define linear functions L_p and M_p by

$$L_p = \sum_{i=1}^{i_p} b_{pi}^{d-1}\xi_{pi}, \quad M_p = \sum_{i=1}^{i_p} b_{pi}^{d-1}\eta_{pi}.$$

Then L_p and M_p are independent linear functions for $\xi_0, \dots, \xi_{n+1}, \eta_0, \dots, \eta_{n+1}$. Therefore the linear equation (7) can be written as

$$N\ {}^t\begin{pmatrix} L_1 & \dots & L_k & M_1 & \dots & M_k \end{pmatrix} = 0,$$

where

$$N = \begin{pmatrix} d\lambda_1^{d-1} & \dots & d\lambda_k^{d-1} & 0 & \dots & 0 \\ (d-1)\lambda_1^{d-2} & \dots & (d-1)\lambda_k^{d-2} & \lambda_1^{d-1} & \dots & \lambda_k^{d-1} \\ \vdots & & \vdots & \vdots & & \vdots \\ 1 & \dots & 1 & (d-1)\lambda_1 & \dots & (d-1)\lambda_k \\ 0 & \dots & 0 & d & \dots & d \end{pmatrix}.$$

Lemma 11. *The rank of N is equal to* $\max(n+1, 2k)$.

Proof. We consider Vandermonde matrix for $\lambda_0, \lambda_1, \dots, \lambda_n$ and consider the limit where λ_0 tends to λ_1. We set $\lambda_0 = \lambda_1 + \epsilon$ with $\epsilon^2 = 0$. Then we have

$$\det \begin{pmatrix} 1 & 1 & \dots & 1 \\ \lambda_1+\epsilon & \lambda_1 & \dots & \lambda_n \\ \vdots & \vdots & & \vdots \\ (\lambda_1+\epsilon)^n & \lambda_1^n & \dots & \lambda_n^n \end{pmatrix} = \epsilon \det \begin{pmatrix} 0 & 1 & \dots & 1 \\ 1 & \lambda_1 & \dots & \lambda_n \\ 2\lambda_1 & \lambda_1^2 & \dots & \lambda_n^2 \\ \vdots & \vdots & & \vdots \\ n\lambda_1^{n-1} & \lambda_1^n & \dots & \lambda_n^n \end{pmatrix}.$$

It is equal to

$$-\epsilon \cdot \prod_{1 \le i < j \le n} (\lambda_j - \lambda_i) \cdot \prod_{2 \le i} (\lambda_i - \lambda_1).$$

Using this procedure, we can prove that the determinant

$$\det \begin{pmatrix} 0 & \dots & 0 & 1 & \dots & 1 \\ 1 & \dots & 1 & \lambda_1 & \dots & \lambda_q \\ 2\lambda_1 & \dots & 2\lambda_p & \lambda_1^2 & \dots & \lambda_q^2 \\ \vdots & & \vdots & \vdots & & \vdots \\ n\lambda_1^{n-1} & \dots & n\lambda_p^{n-1} & \lambda_1^n & \dots & \lambda_q^n \end{pmatrix}$$

is non-zero if $p+q = n+1$ and $p \le q$. Thus we have the lemma. Q.E.D.

Definition 3. *Let $P(l)$ be the partition and $i_1, \dots, i_k$ be the cardinalities of the elements in $P(l)$. The number $2k$ is called the rank of l and is denoted by $rk(l)$.*

By the computation of the dimension of the tangent space of $F(X)$ at l, we have the following theorem.

Theorem 12. *Assume that $n = d$. The point $l \in F(X)$ is singular if and only if $rk(l) \le n$.*

Corollary 13. *Assume that $d = n$. The variety $F(X)$ is singular if and only if $4 \le d$.*

§6. Period integrals for Fano varieties

6.1. Arrangement of hyperplanes

We apply Theorem 4 to study period integrals for Fano varieties of Fermat hypersurfaces. We study the case where $n = d$. Then $\widehat{\mathcal{M}_{0,n+2}} \to \widetilde{\mathcal{M}_{0,n+2}}$ is a $(\mathbf{Z}/d\mathbf{Z})^{n+1}$-covering. We take the quotient by $GPL(2)$ and normalize $\lambda_0, \dots, \lambda_{n+1}, \lambda_{n+2}$ so that

$$\lambda_0 = \infty, \lambda_1 = 0, \lambda_2 = 1.$$

The normalized parameter will be written as $\mu_3, \dots, \mu_{n+1}$. Therefore

$$\mu_p = (\lambda_p, \lambda_1; \lambda_2, \lambda_0)$$

for $p = 3, \dots, n+1$. Here we used the notation of the cross ratio

$$(x, y; z, w) = \frac{(x-y)(z-w)}{(x-w)(z-y)}.$$

Therefore $\mathcal{M}_{0,n+2} = \widetilde{\mathcal{M}_{0,n+2}}/PGL(2)$ is isomorphic to

$$\{(\mu_3, \dots, \mu_{n+1}) \mid \mu_i \neq 0, 1, \ \mu_i \neq \mu_j \text{ for } i \neq j\}.$$

We set

$$\Delta_1 = \prod_{i=3}^{n+1} \mu_i = \xi_1^{-d} \left(\frac{\lambda_2 - \lambda_0}{\lambda_2 - \lambda_1} \right)^d,$$

$$\Delta_2 = \prod_{i=3}^{n+1} (1 - \mu_i) = \xi_2^{-d} \left(\frac{\lambda_0 - \lambda_1}{\lambda_2 - \lambda_1} \right)^d,$$

$$\Delta_i = \mu_i(\mu_i - 1) \prod_{j \neq i, 3 \leq j \leq n+1} (\mu_i - \mu_j)$$

$$= \xi_i^{-d} \left(\frac{(\lambda_0 - \lambda_1)(\lambda_2 - \lambda_0)}{(\lambda_j - \lambda_0)(\lambda_2 - \lambda_1)} \right)^d, \quad \text{for } i = 3, \dots, n+1.$$

Then the covering $F^0(X) \simeq \widehat{\mathcal{M}_{0,n+2}}/PGL(2)$ over $\mathcal{M}_{0,n+2}$ is defined by

$$\eta_i^d = \Delta_i \text{ for } i = 1, \dots, n+1. \tag{8}$$

Therefore $F^0(X)$ is an open set of a ramified covering of $\mathbf{A}^{n-1} = \{\mu_3, \dots, \mu_{n+1}\}$ branching at the arrangement of hyperplanes defined by

$$\mathcal{B}_n = \{\mu_i, 1 - \mu_i\}_{3 \leq i \leq n+1} \cup \{\mu_i - \mu_j\}_{3 \leq i < j \leq n+1}.$$

It is called the Selberg arrangement.

6.2. Selberg integral

Using the main theorem, we can show that the period integrals are written down using certain kind of Selberg integrals. The subjects in this subsection are compilations of well known facts, which might be useful for readers. Let $\pi : F^0(X) \to \mathcal{M}_{0,n+2}$ be the covering introduced in the main theorem, G be the Galois group of π, and $K = \mathbf{Q}(\mu_d) \subset \mathbf{C}$. Let

$$G^* = \mathrm{Ker}((\mathbf{Z}/d\mathbf{Z})^{n+2} \xrightarrow{\Sigma} (\mathbf{Z}/d\mathbf{Z}))$$

be the character group of G and $\mathcal{L}(\chi)$ be the χ-part of the direct image sheaf $\pi_*\mathbf{K}$ of $\mathbf{K}$ under the map π. The i-th singular cohomology with the local coefficient $\mathcal{L}(\chi)$ is denoted by $H(\mathcal{M}_{0,n+2}, \mathcal{L}(\chi))$. Then we have a decomposition

$$H^i(F^0(X), \mathbf{K}) \simeq \oplus_{\chi \in G^*} H^i(\mathcal{M}_{0,n+2}, \mathcal{L}(\chi)).$$

Let $j : \mathcal{M}_{0,n+2} \to \mathbf{A}^{n-1}$ be the open immersion defined by $\{\mu_i\}_{3 \le i \le n+1}$ and set $\partial\mathcal{M}_{0,n+2} = \mathbf{A}^{n-1} - \mathcal{M}_{0,n+2}$. Let $j_!\mathcal{L}(\chi)$ be the zero extension of the local system $\mathcal{L}(\chi)$ to $\mathbf{A}^{n-1}$. Then we have a natural homomorphism

$$\varphi^i(\chi) : H^i(\mathbf{A}^{n-1}, j_!\mathcal{L}(\chi)) \to H^i(\mathcal{M}_{0,n+2}, \mathcal{L}(\chi)). \tag{9}$$

Since $F^0(X)$ is an affine variety, the algebraic de Rham cohomology $H^i(F^0(X)/\mathbf{C})$ is the cohomology of the following complex

$$0 \to \Gamma(\mathcal{O}_{F^0(X)}) \to \Gamma(\Omega^1_{F^0(X)}) \to \cdots \to \Gamma(\Omega^{n-1}_{F^0(X)}) \to 0.$$

By de Rham Theorem and taking χ-part, we have an isomorphism

$$H^{n-1}(F^0(X)/\mathbf{C})(\chi) \xrightarrow{\simeq} H^{n-1}(\mathcal{M}_{0,n+2}, \mathcal{L}(\chi)) \otimes_{\mathbf{K}} \mathbf{C}.$$

We construct an element $\widetilde{\Delta_\sigma}$ in the relative homology $H_{n-1}(\mathbf{A}^{n-1}; \partial\mathcal{M}_{0,n+2}, \mathcal{L}(\chi))$ which can be identified with the dual vector space of $H^{n-1}(\mathbf{A}^{n-1}, j_!\mathcal{L}(\chi))$. Let $\mathfrak{S}[3, n+1]$ be the symmetric group of $[3, n+2]$. For an element $\sigma \in \mathfrak{S}[3, n+1]$, we define a simplex $\Delta_\sigma \subset \mathbf{R}^{n-1}$ by

$$\Delta_\sigma = \{0 \le \mu_{\sigma(3)} \le \cdots \le \mu_{\sigma(n+1)} \le 1 \mid \mu_i \in \mathbf{R} \text{ for } 3 \le i \le n+1\}.$$

By choosing a section of $\mathcal{L}(\chi)$ over Δ_σ, we have a chain $\widetilde{\Delta}_\sigma$ with value in $\mathcal{L}(\chi)$, which defines an element in $H_{n-1}(\mathbf{A}^{n-1}; \partial\mathcal{M}_{0,n+2}, \mathcal{L}(\chi))$.

Let $\eta_1, \ldots, \eta_{n+1}$ be rational functions defined in (8). Then a character χ of G can be written as the Kummer character of $f = \prod_{i=1}^{n+1} \eta_i^{\chi_i}$ with $0 \le \chi_i < d$. Then the differential forms

$$\omega = \prod_{i=1}^{n+1} \eta^{\gamma_i} d\mu_3 \wedge \cdots \wedge d\mu_{n+1}, \quad \gamma_i \equiv \chi_i (\mathrm{mod}\ d), \gamma_i \ge 0 \text{ for } i = 1, \ldots, n+1$$

define elements in the image of φ^{n-1} of (9) in the space $H^{n-1}(F^0(X)/\mathbf{C})(\chi)$. By choosing a suitable choice of branch in the integrand, we have

$$(\widetilde{\Delta}_\sigma, \omega) = \int_{0<\mu_{\sigma(3)}<\cdots<\mu_{\sigma(n+1)}<1} \Delta_i^{\frac{\gamma_i}{d}} d\mu_3 \cdots d\mu_{n+1}.$$

The integral in the right hand side is called a Selberg integral.

References

[BV] W. Barth and A. Van de Ven, Fano varieties of lines on hypersurfaces, Arch. Math. (Basel), **31** (1978), 96–104.

[CG] C. H. Clemens and P. Griffiths, Intermediate Jacobian of the cubic threefold, Ann. of Math. (2), **95** (1972), 281–356.

[R] X. Roulleau, Coubes elliptiques, fibrations et automorphismes des surfaces de Fano, C. R. Math. Acad. Sci. Paris, **345** (2007), 209–212.

[T] T. Terasoma, Complete intersections of Fermat hypersurfaces—the Fermat case and the quadric case, Japan. J. Math., **14** (1988), 309–384.

Tomohide Terasoma
Graduate School of Mathematical Science, the University of Tokyo, Komaba, Meguro, 153-8914, Tokyo
E-mail address: terasoma@ms.u-tokyo.ac.jp

Advanced Studies in Pure Mathematics 62, 2012
Arrangements of Hyperplanes—Sapporo 2009
pp. 475–511

Three sides of the geometric Langlands correspondence for $\mathfrak{gl}_N$ Gaudin model and Bethe vector averaging maps

Eugene Mukhin, Vitaly Tarasov and Alexander Varchenko

Abstract.

We consider the $\mathfrak{gl}_N$ Gaudin model of a tensor power of the standard vector representation. The geometric Langlands correspondence in the Gaudin model relates the Bethe algebra of the commuting Gaudin Hamiltonians and the algebra of functions on a suitable space of N-th order differential operators. In this paper we introduce a third side of the correspondence: the algebra of functions on the critical set of a master function. We construct isomorphisms of the third algebra and the first two. Our main technical tool is the Bethe vector averaging maps, which is a new object.

§1. Introduction

We consider the $\mathfrak{gl}_N$ Gaudin model associated with a tensor power of the standard vector representation. The geometric Langlands correspondence identifies the Bethe algebra of the commuting Gaudin Hamiltonians and the algebra of functions on a suitable space of N-th order differential operators. In this paper we introduce a third ingredient of the correspondence: the algebra of functions on the critical set of a master function. We construct isomorphisms of the three algebras.

Master functions were introduced in [16] to construct hypergeometric integral solutions of the KZ equations,

$$\kappa \frac{\partial I}{\partial z_i} = H_i(\boldsymbol{z}) I(\boldsymbol{z}), \ \ i = 1, \dots, n \ , \qquad I(\boldsymbol{z}) \ = \ \int \Phi(\boldsymbol{z}, \boldsymbol{t})^{1/\kappa} \omega(\boldsymbol{z}, \boldsymbol{t}) d\boldsymbol{t} \ ,$$

Received October 25, 2009.
Revised March 14, 2010.
2010 *Mathematics Subject Classification.* Primary 82B23; Secondary 17B80, 32S22.
Key words and phrases. Bethe algebra, Bethe anzats, Bethe vector averaging map, master function, critical points, Wronsky map.

where $H_i(\boldsymbol{z})$ are the Gaudin Hamiltonians, $\Phi(\boldsymbol{z},\boldsymbol{t})$ is a scalar master functions, $\omega(\boldsymbol{z},\boldsymbol{t})$ is a universal weight function, which is a vector-valued function. It was realized almost immediately [2, 14] that the value of the universal weight function at a critical point of the master function is an eigenvector of the Gaudin Hamiltonians. This construction of the eigenvectors is called the Bethe ansatz. The critical point equations for the master function are called the Bethe ansatz equations and the eigenvectors are called the Bethe vectors. The Bethe ansatz gives a relation between the critical points of the master function and the algebra generated by Gaudin Hamiltonians. The algebra of all (in particular, generalized) Gaudin Hamiltonians is called the Bethe algebra. Higher Gaudin Hamiltonians were introduced using different approaches in [6] and [18], see also [9].

In [17, 12], an N-th order differential operator was assigned to every critical point of the master function. The differential operators appearing in that construction form the second component of the geometric Langlands correspondence.

In this paper we introduce the third component of the geometric Langlands correspondence. It is the algebra of functions on the critical set of the master function. We show that all three components of the geometric Langlands correspondence are on equal footing; they are isomorphic.

The main results of the paper are Corollaries 5.4 and 8.6.

The paper is organized as follows. In Section 2 we recall the definition of the Bethe algebra $\mathcal{B}_\mathcal{V}$ of a tensor power of the vector representation of $\mathfrak{gl}_N$ [10]. In Section 3 we introduce the algebra $\mathcal{O}_\mathcal{W}$ of functions on a suitable Schubert cell $\mathcal{W}$. Points of $\mathcal{W}$ are some N-dimensional spaces of polynomials in one variable. Such a space X is characterized by a monic N-th order differential operator with kernel X. The algebra $\mathcal{O}_\mathcal{W}$ can be considered as the algebra of functions on the space of those differential operators. In Section 4 we recall an isomorphism $\zeta : \mathcal{O}_\mathcal{W} \to \mathcal{B}_\mathcal{V}$ constructed in [10]. In Section 5 a master function and its quotient critical set $\mathcal{C}$ are introduced and an isomorphism $\iota^* : \mathcal{O}_\mathcal{W} \to \mathcal{O}_\mathcal{C}$ is constructed. Here $\mathcal{O}_\mathcal{C}$ is the algebra of functions on $\mathcal{C}$. Consequently, we obtain a composition isomorphism $\mathcal{B}_\mathcal{V} \xrightarrow{\zeta^{-1}} \mathcal{O}_\mathcal{W} \xrightarrow{\iota^*} \mathcal{O}_\mathcal{C}$. In Section 6 we introduce the universal weight function $\omega(\boldsymbol{z},\boldsymbol{t})$ and describe the basic facts of the Bethe ansatz. In Section 7 the Bethe vector averaging maps

$$v_F \;:\; \boldsymbol{z} \;\mapsto\; \frac{1}{l_1!\dots l_{N-1}!}\sum_{(\boldsymbol{z},\boldsymbol{p})\in C_{\boldsymbol{z}}} \frac{F(\boldsymbol{z},\boldsymbol{p})\,\omega(\boldsymbol{z},\boldsymbol{p})}{\mathrm{Hess}_{\boldsymbol{t}}\log\Phi(\boldsymbol{z},\boldsymbol{p})}$$

are introduced. Here $\Phi(\boldsymbol{z}, \boldsymbol{t})$ is the master function, $\mathcal{C}_{\boldsymbol{z}}$ the critical set of the function $\Phi(\boldsymbol{z}, \cdot)$, $\omega(\boldsymbol{z}, \boldsymbol{t})$ the Bethe vector, $F(\boldsymbol{z}, \boldsymbol{t})$ an auxiliary polynomial function. Theorem 7.1 says that the Bethe vector averaging maps are polynomial maps. This is the main technical result of the paper. Using the Bethe vector averaging maps, we construct in Section 8 a new (direct) isomorphism $\nu : \mathcal{O}_{\mathcal{C}} \to \mathcal{B}_{\mathcal{V}}$. We prove that the throughout composition $\mathcal{B}_{\mathcal{V}} \xrightarrow{\varsigma^{-1}} \mathcal{O}_{\mathcal{W}} \xrightarrow{\iota^*} \mathcal{O}_{\mathcal{C}} \xrightarrow{\nu} \mathcal{B}_{\mathcal{V}}$ is the identity map. Section 9 contains the proof of Theorem 7.1.

The paper discusses one example: the Gaudin model on a tensor power of the vector representation of $\mathfrak{gl}_N$. But the picture presented here presumably holds for more general representations and more general Lie algebras. All the ingredients of our considerations (the Bethe algebras, master functions, Bethe vector averaging maps) are available in other situations.

Acknowledgments. The authors thank A. Gabrielov for helpful discussions. The work of EM was supported in part by NSF grant DMS-0601005. The work of VT was supported in part by NSF grant DMS-0901616. The work of AV was supported in part by NSF grant DMS-0555327.

§2. Bethe algebra $\mathcal{B}_{\boldsymbol{\lambda}}$

2.1. Lie algebra $\mathfrak{gl}_N$

Let e_{ij}, $i, j = 1, \dots, N$, be the standard generators of the Lie algebra $\mathfrak{gl}_N$ satisfying the relations $[e_{ij}, e_{sk}] = \delta_{js}e_{ik} - \delta_{ik}e_{sj}$. Let $\mathfrak{h} \subset \mathfrak{gl}_N$ be the Cartan subalgebra generated by e_{ii}, $i = 1, \dots, N$.

Let M be a $\mathfrak{gl}_N$-module. A vector $v \in M$ has weight $\boldsymbol{\lambda} = (\lambda_1, \dots, \lambda_N) \in \mathbb{C}^N$ if $e_{ii}v = \lambda_i v$ for $i = 1, \dots, N$. A vector v is singular if $e_{ij}v = 0$ for $1 \leqslant i < j \leqslant N$. Denote by $(M)_{\boldsymbol{\lambda}}$ the subspace of M of weight $\boldsymbol{\lambda}$, by $(M)^{sing}$ the subspace of all singular vectors in M, and by $(M)^{sing}_{\boldsymbol{\lambda}}$ the subspace of all singular vectors of weight $\boldsymbol{\lambda}$.

Denote by $L_{\boldsymbol{\lambda}}$ the irreducible finite-dimensional $\mathfrak{gl}_N$-module with highest weight $\boldsymbol{\lambda}$. The $\mathfrak{gl}_N$-module $L_{(1,0,\dots,0)}$ is the standard N-dimensional vector representation of $\mathfrak{gl}_N$, denoted below by V. We choose a highest weight vector of V and denote it by v_+.

The Shapovalov form on V is the unique symmetric bilinear form S defined by the conditions $S(v_+, v_+) = 1$, $S(e_{ij}u, v) = S(u, e_{ji}v)$, for all $u, v \in V$ and $1 \leqslant i, j \leqslant N$. For a natural number n, the tensor Shapovalov form on $V^{\otimes n}$ is the tensor product of the Shapovalov forms of factors.

A sequence of integers $\boldsymbol{\lambda} = (\lambda_1, \dots, \lambda_N)$ such that $\lambda_1 \geqslant \lambda_2 \geqslant \cdots \geqslant \lambda_N \geqslant 0$ is called a partition with at most N parts. Denote $|\boldsymbol{\lambda}| = \lambda_1 + \cdots + \lambda_N$.

2.2. Current algebra $\mathfrak{gl}_N[t]$

Let $\mathfrak{gl}_N[t] = \mathfrak{gl}_N \otimes \mathbb{C}[t]$ be the complex Lie algebra of $\mathfrak{gl}_N$-valued polynomials with the pointwise commutator. We identify $\mathfrak{gl}_N$ with the subalgebra $\mathfrak{gl}_N \otimes 1$ of constant polynomials in $\mathfrak{gl}_N[t]$. Hence, any $\mathfrak{gl}_N[t]$-module has a canonical structure of a $\mathfrak{gl}_N$-module.

It is convenient to collect elements of $\mathfrak{gl}_N[t]$ in generating series in a formal variable u. For $g \in \mathfrak{gl}_N$, set

$$g(u) = \sum_{s=0}^{\infty} (g \otimes t^s) u^{-s-1}.$$

For each $a \in \mathbb{C}$, there exists an automorphism ρ_a of $\mathfrak{gl}_N[t]$, $\rho_a : g(u) \mapsto g(u-a)$. Given a $\mathfrak{gl}_N[t]$-module M, we denote by $M(a)$ the pull-back of M through the automorphism ρ_a. As $\mathfrak{gl}_N$-modules, M and $M(a)$ are isomorphic by the identity map, since $\rho_a : g \otimes 1 \mapsto g \otimes 1$.

We have the evaluation homomorphism,

$$\mathrm{ev} : \mathfrak{gl}_N[t] \to \mathfrak{gl}_N\,, \qquad \mathrm{ev} : g(u) \mapsto g\,u^{-1}\,,$$

that is, we have $g \otimes 1 \mapsto g \otimes 1$ and $g \otimes t^r \mapsto 0$ for $r > 0$. The restriction of the evaluation homomorphism to the subalgebra $\mathfrak{gl}_N \subset \mathfrak{gl}_N[t]$ is the identity map. For any $\mathfrak{gl}_N$-module M, we denote by the same letter the $\mathfrak{gl}_N[t]$-module, obtained by pulling M back through the evaluation homomorphism.

There is a $\mathbb{Z}_{\geqslant 0}$-grading on $\mathfrak{gl}_N[t]$: for any $g \in \mathfrak{gl}_N$, we have

$$\deg g \otimes t^r = r\,.$$

2.3. The $\mathfrak{gl}_N[t]$-module $\mathcal{V}^S$

Let n be a positive integer. Let $\mathcal{V}$ be the space of polynomials in $z_1, \dots, z_n$ with coefficients in $V^{\otimes n}$, $\mathcal{V} = V^{\otimes n} \otimes_{\mathbb{C}} \mathbb{C}[z_1, \dots, z_n]$. For $v \in V^{\otimes n}$ and $p(z_1, \dots, z_n) \in \mathbb{C}[z_1, \dots, z_n]$, we write $p(z_1, \dots, z_n)\,v$ instead of $v \otimes p(z_1, \dots, z_n)$.

The symmetric group S_n acts on $\mathbb{C}[z_1, \dots, z_n]$ by permuting the variables $z_1, \dots, z_n$ and acts on $\mathcal{V}$ by permuting the factors of $V^{\otimes n}$ and the variables $z_1, \dots, z_n$ simultaneously,

$$\sigma\bigl(p(z_1, \dots, z_n)\, v_1 \otimes \cdots \otimes v_n\bigr) = p(z_{\sigma(1)}, \dots, z_{\sigma(n)})\, v_{\sigma^{-1}(1)} \otimes \cdots \otimes v_{\sigma^{-1}(n)}\,,$$

$\sigma \in S_n$. For an S_n-module M denote by M^S the subspace of S_n-invariants of M. The space $\mathcal{V}^S$ is a free $\mathbb{C}[z_1, \dots, z_n]^S$-module of rank N^n, see [4], cf. [10].

The space $\mathcal{V}$ is a $\mathfrak{gl}_N[t]$-module with a series $g(u)$, $g \in \mathfrak{gl}_N$, acting by

$$g(u)\,\bigl(p(z_1, \dots, z_n)\, v_1 \otimes \dots \otimes v_n\bigr) = \\ = p(z_1, \dots, z_n) \sum_{s=1}^{n} \frac{v_1 \otimes \dots \otimes g v_s \otimes \dots \otimes v_n}{u - z_s}\,.$$

The $\mathfrak{gl}_N[t]$-action on $\mathcal{V}$ commutes with the S_n-action. Hence, $\mathcal{V}^S \subset \mathcal{V}$ is a $\mathfrak{gl}_N[t]$-submodule.

Define a grading on $\mathbb{C}[z_1, \dots, z_n]$ by setting $\deg z_i = 1$ for all i. Define a grading on $\mathcal{V}$ by setting $\deg(v \otimes p) = \deg p$ for any $v \in V^{\otimes n}$ and $p \in \mathbb{C}[z_1, \dots, z_n]$. The grading on $\mathcal{V}$ induces a grading on $\mathcal{V}^S$ and $\operatorname{End}(\mathcal{V}^S)$. The $\mathfrak{gl}_N[t]$-action on $\mathcal{V}^S$ is graded [4].

Let $\boldsymbol{\lambda}$ be a partition of n. The space $(\mathcal{V}^S)^{sing}_{\boldsymbol{\lambda}}$ is a free graded $\mathbb{C}[z_1, \dots, z_n]^S$-module. Its graded character is

$$\operatorname{ch}((\mathcal{V}^S)^{sing}_{\boldsymbol{\lambda}}) = \frac{\prod_{1 \leqslant i < j \leqslant N}(1 - q^{\lambda_i - \lambda_j + j - i})}{\prod_{i=1}^{N} (q)_{\lambda_i + N - i}}\; q^{\sum_{i=1}^{N} (i-1)\lambda_i}, \tag{2.1}$$

where $(q)_a = \prod_{j=1}^{a} (1 - q^j)$, see [4, 3, 10].

2.4. Bethe algebra

Given an $N \times N$ matrix $A = (a_{ij})$, we define its row determinant to be

$$\operatorname{rdet} A = \sum_{\sigma \in S_N} (-1)^\sigma\, a_{1\sigma(1)} a_{2\sigma(2)} \dots a_{N\sigma(N)}\,.$$

Let ∂ be the operator of differentiation in the variable u. Define the *universal differential operator* $\mathcal{D}$ by the formula

$$\mathcal{D} = \operatorname{rdet} \begin{pmatrix} \partial - e_{11}(u) & -e_{21}(u) & \dots & -e_{N1}(u) \\ -e_{12}(u) & \partial - e_{22}(u) & \dots & -e_{N2}(u) \\ \dots & \dots & \dots & \dots \\ -e_{1N}(u) & -e_{2N}(u) & \dots & \partial - e_{NN}(u) \end{pmatrix}.$$

It is a differential operator in u, whose coefficients are formal power series in u^{-1} with coefficients in $U(\mathfrak{gl}_N[t])$,

$$\mathcal{D} = \partial^N + \sum_{i=1}^{N} B_i(u)\, \partial^{N-i}, \qquad B_i(u) = \sum_{j=i}^{\infty} B_{ij}\, u^{-j}\,,$$

and $B_{ij} \in U(\mathfrak{gl}_N[t])$, $i = 1, \dots, N$, $j \in \mathbb{Z}_{\geqslant i}$. The unital subalgebra of $U(\mathfrak{gl}_N[t])$ generated by B_{ij}, $i = 1, \dots, N$, $j \in \mathbb{Z}_{\geqslant 0}$, is called the *Bethe algebra* and denoted by $\mathcal{B}$.

By [18], cf. [9], the algebra $\mathcal{B}$ is commutative, and $\mathcal{B}$ commutes with the subalgebra $U(\mathfrak{gl}_N) \subset U(\mathfrak{gl}_N[t])$.

Let M be a $\mathcal{B}$-module and $v \in M$ an eigenvector of $\mathcal{B}$. For every coefficient $B_i(u)$ we have $B_i(u)v = h_i(u)v$, where $h_i(u)$ is a scalar series. The scalar differential operator

$$\mathcal{D}_v = \partial^N + \sum_{i=1}^{N} h_i(u)\,\partial^{N-i} \tag{2.2}$$

will be called the differential operator associated with an eigenvector v.

As a subalgebra of $U(\mathfrak{gl}_N[t])$, the algebra $\mathcal{B}$ acts on any $\mathfrak{gl}_N[t]$-module M. Since $\mathcal{B}$ commutes with $U(\mathfrak{gl}_N)$, it preserves the weight subspaces of M and the subspaces $(M)^{sing}_{\boldsymbol{\lambda}}$.

For a $\mathcal{B}$-module M, the image of $\mathcal{B}$ in $\mathrm{End}(M)$ is called the Bethe algebra of M.

Let $\boldsymbol{\lambda}$ be a partition of n with at most N parts. The space $(\mathcal{V}^S)^{sing}_{\boldsymbol{\lambda}}$ is a $\mathcal{B}$-module. Set

$$\mathcal{D}^{\mathcal{V}} = \partial^N + \sum_{i=1}^{N} B^{\mathcal{V}}_i(u)\,\partial^{N-i}, \qquad B^{\mathcal{V}}_i(u) = \sum_{j=i}^{\infty} B^{\mathcal{V}}_{ij}\,u^{-j}, \tag{2.3}$$

where $B^{\mathcal{V}}_{ij}$ is the image of B_{ij} in $\mathrm{End}((\mathcal{V}^S)^{sing}_{\boldsymbol{\lambda}})$.

For any (i, j), the element $B^{\mathcal{V}}_{ij}$ is homogeneous of degree $j - i$. For any i the series $B^{\mathcal{V}}_i(u)$ is homogeneous of degree $-i$, see [10].

Denote by $\mathcal{B}_{\mathcal{V}}$ the Bethe algebra of $(\mathcal{V}^S)^{sing}_{\boldsymbol{\lambda}}$. The Bethe algebra $\mathcal{B}_{\mathcal{V}}$ is our first main object.

§3. Algebra of functions $\mathcal{O}_{\mathcal{W}}$

3.1. Cell $\mathcal{W}$ and algebra $\mathcal{O}_{\mathcal{W}}$

Let $N, d \in \mathbb{Z}_{>0}$, $N \leqslant d$. Let $\mathbb{C}_d[u]$ be the space of polynomials in u of degree less than d, $\dim \mathbb{C}_d[u] = d$. Let $\mathrm{Gr}(N, d)$ be the Grassmannian of all N-dimensional vector subspaces of $\mathbb{C}_d[u]$.

Given a partition $\boldsymbol{\lambda} = (\lambda_1, \dots, \lambda_N)$ with $\lambda_1 \leqslant d - N$, introduce a sequence

$$P = \{d_1 > d_2 > \dots > d_N\}, \qquad d_i = \lambda_i + N - i.$$

Denote by $\mathcal{W}$ the subset of $\mathrm{Gr}(N,d)$ consisting of all N-dimensional subspaces $X \subset \mathbb{C}_d[u]$ such that for every $i = 1,\dots,N$, the subspace X contains a polynomial of degree d_i.

In other words, $\mathcal{W}$ consists of subspaces $X \subset \mathbb{C}_d[u]$ with a basis $f_1(u),\dots,f_N(u)$ of the form

$$f_i(u) = u^{d_i} + \sum_{j=1,\ d_i-j\notin P}^{d_i} f_{ij}u^{d_i-j}. \tag{3.1}$$

For a given $X \in \mathcal{W}$, such a basis is unique. The basis $f_1(u),\dots,f_N(u)$ will be called the flag basis of X.

The set $\mathcal{W}$ is a (Schubert) cell isomorphic to an affine space of dimension $|\boldsymbol{\lambda}|$ with coordinate functions f_{ij}. Let $\mathcal{O}_{\mathcal{W}}$ be the algebra of regular functions on $\mathcal{W}$,

$$\mathcal{O}_{\mathcal{W}} \;=\; \mathbb{C}[f_{ij}\,,\ i=1,\dots,N,\ j=1,\dots,d_i,\ d_i-j\notin P]\,.$$

We may regard the polynomials $f_i(u)$, $i=1,\dots,N$, as generating functions for the generators f_{ij} of the algebra $\mathcal{O}_{\mathcal{W}}$.

The algebra $\mathcal{O}_{\mathcal{W}}$ is graded with $\deg f_{ij} = j$. A polynomial $f_i(u)$ is homogeneous of degree d_i. The graded character of $\mathcal{O}_{\mathcal{W}}$ is

$$\begin{aligned}\mathrm{ch}(\mathcal{O}_{\mathcal{W}}) \;&=\; \frac{\prod_{1\leqslant i<j\leqslant N}(1-q^{d_i-d_j})}{\prod_{i=1}^N (q)_{d_i}} \\ &=\; \frac{\prod_{1\leqslant i<j\leqslant N}(1-q^{\lambda_i-\lambda_j+j-i})}{\prod_{i=1}^N (q)_{\lambda_i+N-i}}\,,\end{aligned} \tag{3.2}$$

see [10].

3.2. New generators of $\mathcal{O}_{\mathcal{W}}$

For $g_1,\dots,g_N \in \mathbb{C}[u]$, introduce the Wronskian

$$\mathrm{Wr}(g_1(u),\dots,g_N(u)) \;=\; \det\begin{pmatrix} g_1(u) & g_1'(u) & \dots & g_1^{(N-1)}(u) \\ g_2(u) & g_2'(u) & \dots & g_2^{(N-1)}(u) \\ \dots & \dots & \dots & \dots \\ g_N(u) & g_N'(u) & \dots & g_N^{(N-1)}(u)\end{pmatrix},$$

where an i-th row is formed by derivatives of g_i.

Let $f_i(u)$, $i=1,\dots,N$, be the generating functions in (3.1). We have

$$\mathrm{Wr}(f_1(u),\dots,f_N(u)) \;=\; \prod_{1\leqslant i<j\leqslant N}(d_j-d_i)\,\Big(u^n + \sum_{s=1}^n (-1)^s A_s\,u^{n-s}\Big),$$

where $n = |\boldsymbol{\lambda}|$ and $A_1, \dots, A_n$ are elements of $\mathcal{O}_{\mathcal{W}}$. Define

$$\mathcal{D}^{\mathcal{W}} = \frac{1}{\mathrm{Wr}(f_1(u), \dots, f_N(u))} \,\mathrm{rdet} \begin{pmatrix} f_1(u) & f_1'(u) & \dots & f_1^{(N)}(u) \\ f_2(u) & f_2'(u) & \dots & f_2^{(N)}(u) \\ \dots & \dots & \dots & \dots \\ 1 & \partial & \dots & \partial^N \end{pmatrix}.$$

We have

$$\mathcal{D}^{\mathcal{W}} = \partial^N + \sum_{i=1}^{N} B_i^{\mathcal{W}}(u)\,\partial^{N-i}, \qquad B_i^{\mathcal{W}}(u) = \sum_{j=i}^{\infty} B_{ij}^{\mathcal{W}}\, u^{-j},$$

and $B_{ij}^{\mathcal{W}} \in \mathcal{O}_{\mathcal{W}}$, $i = 1, \dots, N$, $j \in \mathbb{Z}_{\geqslant i}$. For any (i,j), the element $B_{ij}^{\mathcal{W}}$ is homogeneous of degree $j - i$. For any i the series $B_i^{\mathcal{W}}(u)$ is homogeneous of degree $-i$. The elements $B_{ij}^{\mathcal{W}} \in \mathcal{O}_{\mathcal{W}}$, $i = 1, \dots, N$, $j \in \mathbb{Z}_{\geqslant i}$, generate the algebra $\mathcal{O}_{\mathcal{W}}$, see [10].

For $X \in \mathcal{W}$, denote by $\mathcal{D}_X$ the monic scalar differential operator of order N with kernel X. We call $\mathcal{D}_X$ the differential operator associated with X. The operator $\mathcal{D}_X$ is obtained from $\mathcal{D}^{\mathcal{W}}$ by specialization of variables f_{ij} to their values at X.

3.3. Wronski map

Let $X \in \mathcal{W}$. The Wronskian determinant of a basis of the subspace X does not depend on the choice of the basis up to multiplication by a number. The monic polynomial representing the Wronskian determinant of a basis of X is called the Wronskian of X and denoted by $\mathrm{Wr}_X(u)$.

The Wronski map $\mathcal{W} \to \mathbb{C}^n$ sends a point $X \in \mathcal{W}$ to a point $\boldsymbol{a} = (a_1, \dots, a_n)$, if $\mathrm{Wr}_X(u) = u^n + \sum_{s=1}^{n} (-1)^s a_s u^{n-s}$. The Wronski map has finite degree.

§4. Isomorphism of $\mathcal{B}_{\mathcal{V}}$ and $\mathcal{O}_{\mathcal{W}}$

Theorem 4.1 ([10]). *The map*

$$\zeta : \mathcal{O}_{\mathcal{W}} \to \mathcal{B}_{\mathcal{V}}, \qquad B_{ij}^{\mathcal{W}} \mapsto B_{ij}^{\mathcal{V}},$$

is a well-defined isomorphism of graded algebras.

The degrees of elements of $(\mathcal{V}^S)_{\boldsymbol{\lambda}}^{sing}$ are not less than $\sum_{i=1}^{N}(i-1)\lambda_i$ and the homogeneous component of $(\mathcal{V}^S)_{\boldsymbol{\lambda}}^{sing}$ of degree $\sum_{i=1}^{N}(i-1)\lambda_i$ is one-dimensional, see formula (2.1). Let $v_1 \in (\mathcal{V}^S)_{\boldsymbol{\lambda}}^{sing}$ be a nonzero vector of degree $\sum_{i=1}^{N}(i-1)\lambda_i$.

Theorem 4.2 ([10]). *The map*

$$\eta : \mathcal{O}_{\mathcal{W}} \to (\mathcal{V}^S)^{sing}_{\boldsymbol{\lambda}} , \qquad B^{\mathcal{W}}_{ij} \mapsto B^{\mathcal{V}}_{ij} v_1 ,$$

is an isomorphism of degree $\sum_{i=1}^N (i-1)\lambda_i$ of graded vector spaces. The maps ζ and η intertwine the action of the multiplication operators on $\mathcal{O}_{\mathcal{W}}$ and the action of the Bethe algebra $\mathcal{B}_{\mathcal{V}}$ on $(\mathcal{V}^S)^{sing}_{\boldsymbol{\lambda}}$, that is, for any $F, G \in \mathcal{O}_{\mathcal{W}}$, we have

$$\eta(FG) = \zeta(F)\,\eta(G) .$$

§5. Critical points of the master function

5.1. Master function

Let $\boldsymbol{\lambda} = (\lambda_1, \dots, \lambda_N)$ be a partition of n. Set $l_a = \sum_{b=a+1}^N \lambda_b$, $a = 0, \dots, N$, where $l_0 = n$ and $l_N = 0$. Denote $l = l_0 + \dots + l_{N-1}$, $\boldsymbol{l} = (l_0, \dots, l_{N-1})$. Consider a set of l variables

$$\boldsymbol{T} = (t_1^{(0)}, \dots, t_{l_0}^{(0)}, t_1^{(1)}, \dots, t_{l_1}^{(1)}, \dots, t_1^{(N-1)}, \dots, t_{l_{N-1}}^{(N-1)})$$

and its subsets $\boldsymbol{t}^0 = (t_1^{(0)}, \dots, t_{l_0}^{(0)})$ and $\boldsymbol{t} = (t_1^{(1)}, \dots, t_{l_1}^{(1)}, \dots, t_1^{(N-1)}, \dots, t_{l_{N-1}}^{(N-1)})$. Consider the affine space $\mathbb{C}^l_{\boldsymbol{T}} = \mathbb{C}^l$ with coordinates $\boldsymbol{T} = (\boldsymbol{t}^0, \boldsymbol{t})$. The rational function $\Phi : \mathbb{C}^l \to \mathbb{C}$,

$$\Phi(\boldsymbol{T}) = \prod_{a=1}^{N-1} \prod_{1 \leqslant i < j \leqslant l_a} (t_i^{(a)} - t_j^{(a)})^2 \prod_{a=0}^{N-2} \prod_{i=1}^{l_a} \prod_{j=1}^{l_{a+1}} (t_i^{(a)} - t_j^{(a+1)})^{-1}$$

is called a *master function.* The master functions arise in the hypergeometric solutions of the KZ equations, see [8, 16, 19], and in the Bethe ansatz method for the Gaudin model, see [2, 14].

The product of symmetric groups $S_{\boldsymbol{l}} = S_{l_0} \times \dots \times S_{l_{N-1}}$ acts on the coordinates $\boldsymbol{T}$ by permuting the coordinates with the same upper index. The master function is $S_{\boldsymbol{l}}$-invariant.

We consider the master function as a function of $\boldsymbol{t}$ depending on the parameters $\boldsymbol{t}^{(0)}$.

A point $\boldsymbol{T} = (\boldsymbol{t}^0, \boldsymbol{t}) \in \mathbb{C}^l$ is called a *critical point* of $\log \Phi(\,\boldsymbol{t}^0\,, \cdot)$ if

$$\frac{\partial}{\partial t_i^{(a)}} \log \Phi(\boldsymbol{T}) = 0 , \qquad a = 1, \dots, N-1, \quad i = 1, \dots, l_a .$$

That is, a point $\boldsymbol{T}$ is a critical point if the following system of $l-n$ equations is satisfied:

$$(5.1)\quad \sum_{j=1}^{l_{a-1}} \frac{1}{t_i^{(a)}-t_j^{(a-1)}} - \sum_{\substack{j=1\\ j\neq i}}^{l_a} \frac{2}{t_i^{(a)}-t_j^{(a)}} + \sum_{j=1}^{l_{a+1}} \frac{1}{t_i^{(a)}-t_j^{(a+1)}} = 0\,,$$

here $a=1,\dots,N-1$, $j=1,\dots,l_a$. In this definition we assume that all the denominators in (5.1) are nonzero. In the Gaudin model, equations (5.1) are called the Bethe ansatz equations. For a point $\boldsymbol{T}\in\mathbb{C}^l$, denote

$$\mathrm{Hess}_{\boldsymbol{t}} \log\Phi(\boldsymbol{T}) \;=\; \det\left(\frac{\partial^2}{\partial t_i^{(a)}\partial t_j^{(b)}} \log\Phi(\boldsymbol{T})\right)\,,$$

where we take the determinant of the $(l-n)\times(l-n)$ matrix of second derivatives of the function $\log\Phi$ with respect to all of the variables $t_i^{(a)}$ with $a>0$.

For a fixed $\boldsymbol{t}^0$, the function $\log\Phi(\,\boldsymbol{t}^0\,,\cdot)$ has finitely many critical points, see [17, 12, 13].

Theorem 5.1 ([17, 13]). *For generic $\boldsymbol{t}^0\in\mathbb{C}^n$, all critical points of the function $\log\Phi(\,\boldsymbol{t}^0\,,\cdot)$ are nondegenerate. The number of the $S_{l_1}\times\dots\times S_{l_{N-1}}$-orbits of critical points equals $\dim(V^{\otimes n})^{sing}_{\boldsymbol{\lambda}}$.*

Denote by $\widetilde{C}\subset\mathbb{C}^l_{\boldsymbol{T}}$ the union of all critical points of the functions $\log\Phi(\,\boldsymbol{t}^0\,,\cdot)$ for all $\boldsymbol{t}^0\in\mathbb{C}^n$ with distinct coordinates $t_1^{(0)},\dots,t_n^{(0)}$. Denote by $C\subset\mathbb{C}^l_{\boldsymbol{T}}$ the Zariski closure of $\widetilde{C}$. The set C is $S_{\boldsymbol{l}}$-invariant.

5.2. Factorization by $S_{\boldsymbol{l}}$

For $a=0,\dots,N-1$, let $\sigma_1^{(a)},\dots,\sigma_{l_a}^{(a)}$ be the elementary symmetric functions of $t_1^{(a)},\dots,t_{l_a}^{(a)}$. Denote by $\mathbb{C}^l_{\boldsymbol{\Sigma}}=\mathbb{C}^l$ the affine space with coordinates

$$\boldsymbol{\Sigma}=(\sigma_1^{(0)},\dots,\sigma_{l_0}^{(0)},\sigma_1^{(1)},\dots,\sigma_{l_1}^{(1)},\dots,\sigma_1^{(N-1)},\dots,\sigma_{l_{N-1}}^{(N-1)})\;.$$

The space $\mathbb{C}^l_{\boldsymbol{\Sigma}}$ is the quotient of $\mathbb{C}^l_{\boldsymbol{T}}$ by the $S_{\boldsymbol{l}}$-action.

Denote by $\mathcal{C}$ the image of C under the natural projection $\mathbb{C}^l_{\boldsymbol{T}}\to\mathbb{C}^l_{\boldsymbol{\Sigma}}$. The set $\mathcal{C}$ will be called the *quotient critical set* of the master function. Let $\mathcal{O}_{\mathcal{C}}$ be the algebra of regular functions on $\mathcal{C}$, that is, the restriction of $\mathbb{C}[\boldsymbol{\Sigma}]$ to $\mathcal{C}$.

The algebra $\mathbb{C}[\boldsymbol{T}]$ is a graded algebra with $\deg t_i^{(a)}=1$ for all (a,i). The algebra $\mathbb{C}[\boldsymbol{\Sigma}]$ is a graded algebra with $\deg\sigma_i^{(a)}=i$ for all (a,i).

Equations (5.1) are homogeneous. Hence, $\mathcal{C}$ is a quasi-homogeneous algebraic set and the algebra $\mathcal{O}_{\mathcal{C}}$ has a grading with $\deg(\sigma_i^{(a)}|_{\mathcal{C}}) = i$.

5.3. A map $\theta : \mathcal{W} \to \mathbb{C}^l_{\boldsymbol{\Sigma}}$

For $X \in \mathcal{W}$, let $f_{1,X}(u), \dots, f_{N,X}(u)$ be the flag basis of X. Introduce the polynomials $y_{0,X}(u)$, $y_{1,X}(u)$, $\dots$, $y_{N-1,X}(u)$ by the formula

$$y_{a,X}(u) \prod_{a<i<j\leqslant N} (d_i - d_j) \;=\; \mathrm{Wr}(f_{a+1,X}(u), \dots, f_{N,X}(u)),$$

$a = 0, \dots, N-1$. For each a, the polynomial $y_{a,X}(u)$ is a monic polynomial of degree l_a, $y_{a,X}(u) = u^{l_a} + \sum_{i=1}^{l_a} (-1)^i \sigma_{i,X}^{(a)} u^{l_a - i}$. Denote by $t_{1,X}^{(a)}, \dots, t_{l_a,X}^{(a)}$ the roots of $y_{a,X}(u)$. Then $\sigma_{1,X}^{(a)}, \dots, \sigma_{l_a,X}^{(a)}$ are the elementary symmetric functions of $t_{1,X}^{(a)}, \dots, t_{l_a,X}^{(a)}$. The sequence

$$\boldsymbol{T}_X \;=\; (t_{1,X}^{(0)}, \dots, t_{l_0,X}^{(0)}, \; \dots \;, t_{1,X}^{(N-1)}, \dots, t_{l_{N-1},X}^{(N-1)})$$

will be called the *root coordinates* of X. For every a, the numbers $t_{1,X}^{(a)}, \dots, t_{l_a,X}^{(a)}$ are determined up to a permutation. Let $\boldsymbol{\Sigma}_X$ be the image of $\boldsymbol{T}_X$ in $\mathbb{C}^l_{\boldsymbol{\Sigma}}$.

A point $X \in \mathcal{W}$ will be called *nice* if all roots of the polynomials $y_{0,X}(u)$, $y_{1,X}(u)$, $\dots$, $y_{N-1,X}(u)$ are simple and for each $a = 1, \dots, N-1$, the polynomials $y_{a-1,X}(u)$ and $y_{a,X}(u)$ do not have common roots. Nice points form a Zariski open subset of $\mathcal{W}$, see [11]. If X is nice, then the root coordinates $\boldsymbol{T}_X$ satisfy the critical point equations (5.1), see [12].

Define a polynomial map $\theta : \mathcal{W} \to \mathbb{C}^l_{\boldsymbol{\Sigma}}$, $X \mapsto \boldsymbol{\Sigma}_X$. This map induces a graded algebra homomorphism $\mathbb{C}[\boldsymbol{\Sigma}] \to \mathcal{O}_{\mathcal{W}}$.

Lemma 5.2. *We have $\theta(\mathcal{W}) \subset \mathcal{C}$.*

Proof. The lemma follows from the fact that the nice points of $\mathcal{W}$ are mapped to $\mathcal{C}$. Q.E.D.

5.4. Differential operator $\mathcal{D}^{\mathcal{T}}$ and a map $\iota : \mathcal{C} \to \mathcal{W}$

Set

$$\chi^a(u, \boldsymbol{T}) \;=\; \sum_{j=1}^{l_{a-1}} \frac{1}{u - t_j^{(a-1)}} - \sum_{i=1}^{l_a} \frac{1}{u - t_j^{(a)}}, \qquad a = 1, \dots, N,$$

and

$$\mathcal{D}^{\mathcal{T}} \;=\; (\partial - \chi^1(u, \boldsymbol{T})) \dots (\partial - \chi^N(u, \boldsymbol{T})).$$

We have

$$\mathcal{D}^{\mathcal{T}} = \partial^N + \sum_{i=1}^{N} B_i^{\mathcal{T}}(u)\,\partial^{N-i}, \qquad B_i^{\mathcal{T}}(u) = \sum_{j=i}^{\infty} B_{ij}^{\mathcal{T}} u^{-j},$$

and $B_{ij}^{\mathcal{T}} \in \mathbb{C}[\boldsymbol{T}]^{S_l} = \mathbb{C}[\boldsymbol{\Sigma}]$, $i = 1, \dots, N$, $j \in \mathbb{Z}_{\geqslant i}$. For a point $\boldsymbol{T} \in \mathbb{C}^l$, denote by $\mathcal{D}_{\boldsymbol{T}}$ the specialization of $\mathcal{D}^{\mathcal{T}}$ at $\boldsymbol{T}$. We call $\mathcal{D}_{\boldsymbol{T}}$ the differential operator associated with a point $\boldsymbol{T}$.

If $\boldsymbol{T} = (\boldsymbol{t}^0, \boldsymbol{t}) \in \mathbb{C}^l$ is a critical point of $\log \Phi(\boldsymbol{t}^0, \cdot)$, then the kernel $X_{\boldsymbol{T}}$ of $\mathcal{D}_{\boldsymbol{T}}$ consists of polynomials; moreover, $X_{\boldsymbol{T}}$ is a point of $\mathcal{W}$, see [12]. The correspondence $\boldsymbol{T} \mapsto X_{\boldsymbol{T}}$ defines a rational map $\iota : \mathcal{C} \to \mathcal{W}$.

5.5. Quotient critical set is a nonsingular subvariety

Theorem 5.3. *The quotient critical set $\mathcal{C} \subset \mathbb{C}^l_{\boldsymbol{\Sigma}}$ is a nonsingular subvariety. The map $\theta : \mathcal{W} \to \mathbb{C}^l_{\boldsymbol{\Sigma}}$ is an embedding with $\theta(\mathcal{W}) = \mathcal{C}$. The map $\iota : \mathcal{C} \to \mathcal{W}$ is an isomorphism and $\iota\theta = \mathrm{id}_{\mathcal{W}}$.*

Proof. The map θ, considered as a map from $\mathcal{W}$ to $\theta(\mathcal{W})$ is finite. The set $\theta(\mathcal{W})$ is Zariski closed since $\mathcal{W}$ is Zariski closed. We know from [12] that $\theta(\mathcal{W})$ contains the subset $\widetilde{\mathcal{C}} \subset \mathcal{C}$, the image of nondegenerate critical points. We have $\theta(\mathcal{W}) = \mathcal{C}$, since $\mathcal{C}$ is the Zariski closure of $\widetilde{\mathcal{C}}$ and $\theta(\mathcal{W})$ is Zariski closed.

The fact that $\iota\theta = \mathrm{id}_{\mathcal{W}}$ at generic points of $\mathcal{W}$ is proved in [12]. Therefore, $\iota\theta = \mathrm{id}_{\mathcal{W}}$ for all points of $\mathcal{W}$.

Consider the algebra homomorphism $\iota^* : \mathcal{O}_{\mathcal{W}} \to \mathcal{O}_{\mathcal{C}}$ induced by ι. Under the map ι^* the elements $B_{ij}^{\mathcal{W}}$ are mapped to the polynomials $B_{ij}^{\mathcal{T}} \in \mathbb{C}[\boldsymbol{\Sigma}]$ restricted to $\mathcal{C}$. Since the elements $B_{ij}^{\mathcal{W}}$ generate $\mathcal{O}_{\mathcal{W}}$, the map $\theta : \mathcal{W} \to \mathbb{C}^l_{\boldsymbol{\Sigma}}$ is an embedding and the map $\iota : \mathcal{C} \to \mathcal{W}$ is an isomorphism. Q.E.D.

Corollary 5.4. *The map $\iota^* : \mathcal{O}_{\mathcal{W}} \to \mathcal{O}_{\mathcal{C}}$, $B_{ij}^{\mathcal{W}} \mapsto B_{ij}^{\mathcal{T}}|_{\mathcal{C}}$, is an isomorphism of graded algebras. In particular, the elements $B_{ij}^{\mathcal{T}}|_{\mathcal{C}}$ generate $\mathcal{O}_{\mathcal{C}}$.*

§6. Universal weight function and Bethe vectors

We remind a construction of a rational map $\omega : \mathbb{C}^l \to (V^{\otimes n})_{\boldsymbol{\lambda}}$, called the *universal weight function*, see [8, 16], cf. [15].

A basis of $V^{\otimes n}$ is formed by the vectors $e_J v = e_{j_1,1}v_+ \otimes \cdots \otimes e_{j_n,1}v_+$, where $J = (j_1, \dots, j_n)$ and $1 \leqslant j_a \leqslant N$ for $a = 1, \dots, N$. A basis of $(V^{\otimes n})_{\boldsymbol{\lambda}}$ is formed by the vectors $e_J v$ such that $\#\{a \mid j_a > i\} = l_i$ for every $i = 1, \dots, N-1$. Such a multi-index J will be called admissible.

The universal weight function has the form $\omega(\boldsymbol{T}) = \sum_J \omega_J(\boldsymbol{T}) e_J v$ where the sum is over the set of all admissible J, and the functions $\omega_J(\boldsymbol{T})$ are defined below.

For an admissible J and $i = 1, \ldots, N-1$, define $A_i(J) = \{ a \mid 1 \leqslant a \leqslant n, \ 1 \leqslant i < j_a \}$. Then $|A_i(J)| = l_i$.

Let $\Gamma(J)$ be the set of sequences $\boldsymbol{\gamma} = (\gamma_1, \ldots, \gamma_{N-1})$ of bijections $\gamma_i : A_i(J) \to \{1, \ldots, l_i\}$, $i = 1, \ldots, N-1$. Then $|\Gamma(J)| = \prod_{i=1}^{N-1} l_i!$.

For $a \in A_1(J)$ and $\boldsymbol{\gamma} \in \Gamma(J)$, introduce a rational function

$$\omega_{a,\gamma}(\boldsymbol{T}) = \frac{1}{t^{(1)}_{\gamma_1(a)} - t^{(0)}_a} \prod_{i-2}^{j_a-1} \frac{1}{t^{(i)}_{\gamma_i(a)} - t^{(i-1)}_{\gamma_{i-1}(a)}} .$$

Define

$$\omega_J(\boldsymbol{T}) = \sum_{\gamma \in \Gamma(J)} \prod_{a \in A_1(J)} \omega_{a,\gamma} .$$

Theorem 6.1. *Let $\boldsymbol{T} = (\boldsymbol{z}, \boldsymbol{p})$ be a nondegenerate critical point of the function $\log \Phi(\boldsymbol{z}, \cdot)$, here $\boldsymbol{z} = (z_1, \ldots, z_n)$ lies in $\mathbb{C}^n$ and $\boldsymbol{p}$ lies in $\mathbb{C}^{l-n}$ with coordinates $t^{(i)}_j$, $i > 0$. Consider the value $\omega(\boldsymbol{z}, \boldsymbol{p})$ of the universal weight function $\omega : \mathbb{C}^l \to (V^{\otimes n})_{\boldsymbol{\lambda}}$ at $(\boldsymbol{z}, \boldsymbol{p})$. Consider $V^{\otimes n}$ as the $\mathfrak{gl}_N[t]$-module $\otimes_{s=1}^n V(z_s)$. Then*

(i) *The vector $\omega(\boldsymbol{z}, \boldsymbol{p})$ belongs to $(V^{\otimes n})^{sing}_{\boldsymbol{\lambda}}$.*
(ii) *The vector $\omega(\boldsymbol{z}, \boldsymbol{p})$ is an eigenvector of the Bethe algebra $\mathcal{B}$, acting on $\otimes_{s=1}^n V(z_s)$. Moreover, $\mathcal{D}_{\omega(\boldsymbol{z},\boldsymbol{p})} = \mathcal{D}_{(\boldsymbol{z},\boldsymbol{p})}$, where $\mathcal{D}_{\omega(\boldsymbol{z},\boldsymbol{p})}$ and $\mathcal{D}_{(\boldsymbol{z},\boldsymbol{p})}$ are the differential operators associated with the eigenvector $\omega(\boldsymbol{z}, \boldsymbol{p})$ and the point $(\boldsymbol{z}, \boldsymbol{p}) \in \mathbb{C}^l$, respectively, see (2.2) and 5.4.*
(iii) *Let S be the tensor Shapovalov form on $V^{\otimes n}$, then*

$$S(\omega(\boldsymbol{z}, \boldsymbol{p}), \omega(\boldsymbol{z}, \boldsymbol{p})) = \mathrm{Hess}_{\boldsymbol{t}} \log \Phi(\boldsymbol{z}, \boldsymbol{p}) .$$

(iv) *If $(\boldsymbol{z}, \boldsymbol{p})$ and $(\boldsymbol{z}, \boldsymbol{p}')$ are two nondegenerate critical points of the function $\log \Phi(\boldsymbol{z}, \cdot)$, which lie in different $S_{l_1} \times \cdots \times S_{l_{N-1}}$-orbits, then $S(\omega(\boldsymbol{z}, \boldsymbol{p}), \omega(\boldsymbol{z}, \boldsymbol{p}')) = 0$.*

Part (i) is proved in [2] and [14]. Part (i) also follows directly from Theorems 6.16.2 in [16]. Part (ii) is proved in [9]. Part (iii) is proved in [13, 20]. Part (iv) is proved in [20] and also follows from [10].

The vector $\omega(\boldsymbol{z}, \boldsymbol{p})$ is called the *Bethe vector* corresponding to a critical point $(\boldsymbol{z}, \boldsymbol{p})$.

§7. Bethe vector averaging maps

Consider $\mathbb{C}^l$ with coordinates $\boldsymbol{T} = (\boldsymbol{t}^0, \boldsymbol{t})$. We denote the variables $\boldsymbol{t}^0 = (t_1^{(0)}, \dots, t_n^{(0)})$ also by $\boldsymbol{z} = (z_1, \dots, z_n)$.

Let $\boldsymbol{z}$ be a generic point of $\mathbb{C}^n$ with distinct coordinates and such that all the critical points of the function $\log \Phi(\,\boldsymbol{z}\,, \cdot)$ are nondegenerate. The critical set $C_{\boldsymbol{z}}$ of $\log \Phi(\,\boldsymbol{z}\,, \cdot)$ consists of $\dim (V^{\otimes n})^{sing}_{\boldsymbol{\lambda}}$ $S_{l_1} \times \cdots \times S_{l_{N-1}}$-orbits. Each orbit has $l_1! \cdots l_{N-1}!$ points. For any $F \in \mathbb{C}[\boldsymbol{T}]^{S_{\boldsymbol{l}}} = \mathbb{C}[\boldsymbol{\Sigma}]$, let us define

$$v_F(\boldsymbol{z}) \;=\; \frac{1}{l_1! \dots l_{N-1}!} \sum_{(\boldsymbol{z},\boldsymbol{p}) \in C_{\boldsymbol{z}}} \frac{F(\boldsymbol{z},\boldsymbol{p})\, \omega(\boldsymbol{z},\boldsymbol{p})}{\mathrm{Hess}_{\boldsymbol{t}} \log \Phi(\boldsymbol{z},\boldsymbol{p})}\,. \tag{7.1}$$

The term of this sum corresponding to a critical point $(\boldsymbol{z}, \boldsymbol{p})$ can be written as the following integral, see Chapter 5 of [7]. Choose a small neighborhood U of $\boldsymbol{p}$ in $\mathbb{C}^{l-n}$. Define a torus $\Gamma_{\boldsymbol{z},\boldsymbol{p}}$ in U by $l-n$ equations $|\Phi_j^a(\boldsymbol{z}, \boldsymbol{t})| = \epsilon_j^a$ where Φ_j^a are derivatives of $\log \Phi(\,\boldsymbol{z}\,, \cdot)$ with respect to the variables $t_j^{(a)}$, $a > 0$, and where ϵ_i^a are small positive numbers. Then

$$\frac{F(\boldsymbol{z},\boldsymbol{p})\, \omega(\boldsymbol{z},\boldsymbol{p})}{\mathrm{Hess}_{\boldsymbol{t}} \log \Phi(\boldsymbol{z},\boldsymbol{p})} \;=\; \frac{1}{(2\pi i)^{l-n}} \int_{\Gamma_{\boldsymbol{z},\boldsymbol{t}}} \frac{F(\boldsymbol{z},\boldsymbol{t})\, \omega(\boldsymbol{z},\boldsymbol{t})\, d\boldsymbol{t}}{\prod_{a,j} \Phi_j^a(\boldsymbol{z},\boldsymbol{t})}\,. \tag{7.2}$$

The $l_1! \cdots l_{N-1}!$ terms of the sum in (7.1) corresponding to a single $S_{l_1} \times \cdots \times S_{l_{N-1}}$-orbit are all equal due to the $S_{l_1} \times \cdots \times S_{l_{N-1}}$-invariance of Φ, ω and F.

The correspondence $\boldsymbol{z} \mapsto v_F(\boldsymbol{z})$ defines a map $v_F : \mathbb{C}^n \to (V^{\otimes n})^{sing}_{\boldsymbol{\lambda}}$ which will be called *a Bethe vector averaging map*.

The map v_F is a rational map. Indeed, the map is well defined on a Zariski open subset of $\mathbb{C}^n$ and has bounded growth as the argument approached the possible singular points or infinity.

Theorem 7.1. *For any $F \in \mathbb{C}[\boldsymbol{\Sigma}]$, the Bethe vector averaging map v_F is a polynomial map.*

Theorem 7.1 is proved in Section 9.

§8. Quotient critical set and Bethe algebra

8.1. Construction of isomorphisms

Recall that $\mathbb{C}[\boldsymbol{\Sigma}]$ is graded by $\deg \sigma_j^{(a)} = j$. For any $F \in \mathbb{C}[\boldsymbol{\Sigma}]$, consider the Bethe vector averaging map $v_F : \mathbb{C}^n \to (V^{\otimes n})^{sing}_{\boldsymbol{\lambda}}$.

Lemma 8.1. *If F is quasi-homogeneous and $\deg F = d$, then v_F is homogeneous and* $\deg v_F \;=\; d + l - n \;=\; d + \sum_{i=1}^N (i-1)\lambda_i$.

It is clear that the map v_F is an element of $(\mathcal{V}^S)^{sing}_{\boldsymbol{\lambda}}$. Thus, the correspondence $F \mapsto v_F$ defines a graded linear map $\mu : \mathbb{C}[\boldsymbol{\Sigma}] \to (\mathcal{V}^S)^{sing}_{\boldsymbol{\lambda}}$.

Theorem 8.2. *The kernel of $\mu : \mathbb{C}[\boldsymbol{\Sigma}] \to (\mathcal{V}^S)^{sing}_{\boldsymbol{\lambda}}$ is the defining ideal $I_{\mathcal{C}} \subset \mathbb{C}[\boldsymbol{\Sigma}]$ of $\mathcal{C}$. The map μ induces a graded linear isomorphism $\mathcal{O}_{\mathcal{C}} \to (\mathcal{V}^S)^{sing}_{\boldsymbol{\lambda}}$ of degree $\sum_{i=1}^N (i-1)\lambda_i$.*

We shall denote this isomorphism by the same letter μ.

Proof. If $F \in I_{\mathcal{C}}$, then $v_F = 0$ for generic $\boldsymbol{z}$. Hence, $v_F = 0$ as an element of $(\mathcal{V}^S)^{sing}_{\boldsymbol{\lambda}}$. If $v_F = 0$ as an element of $(\mathcal{V}^S)^{sing}_{\boldsymbol{\lambda}}$, then $F = 0$ on a Zariski open subset of $\mathcal{C}$. Hence, $F \in I_{\mathcal{C}}$. Therefore, $\ker \mu = I_{\mathcal{C}}$.

The graded character of $\mathcal{O}_{\mathcal{C}}$ equals the graded character of $\mathcal{O}_{\mathcal{W}}$ by Corollary 5.4. The graded character of $\mathcal{O}_{\mathcal{W}}$ is given by (3.2). The graded character of $(\mathcal{V}^S)^{sing}_{\boldsymbol{\lambda}}$ is given by (2.1). Comparing the characters and using Lemma 8.1, we conclude that the induced map $\mu : \mathcal{O}_{\mathcal{C}} \to (\mathcal{V}^S)^{sing}_{\boldsymbol{\lambda}}$ is an isomorphism. Q.E.D.

Corollary 8.3. *Consider the element $v_1 \in (\mathcal{V}^S)^{sing}_{\boldsymbol{\lambda}}$, corresponding to $F = 1$ under the isomorphism μ. Then v_1 is a generator of the one-dimensional graded component of $(\mathcal{V}^S)^{sing}_{\boldsymbol{\lambda}}$ of degree $\sum_{i=1}^N (i-1)\lambda_i$ (compare this v_1 with the element v_1 in Theorem 4.2).*

Given an element $F \in \mathcal{O}_{\mathcal{C}}$, define a linear map $\nu(F) : (\mathcal{V}^S)^{sing}_{\boldsymbol{\lambda}} \to (\mathcal{V}^S)^{sing}_{\boldsymbol{\lambda}}$, $v_G \mapsto v_{FG}$. By Theorem 8.2, this map is well-defined.

Consider the generators $B^{\mathcal{T}}_{ij}|_{\mathcal{C}}$ of $\mathcal{O}_{\mathcal{C}}$ and generators $B^{\mathcal{V}}_{ij}$ of $\mathcal{B}_{\mathcal{V}}$, see Corollary 5.4 and formula (2.3).

Lemma 8.4. *For any (i,j), the linear map $\nu(B^{\mathcal{T}}_{ij}|_{\mathcal{C}}) : (\mathcal{V}^S)^{sing}_{\boldsymbol{\lambda}} \to (\mathcal{V}^S)^{sing}_{\boldsymbol{\lambda}}$, $v_F \mapsto v_{B^{\mathcal{T}}_{ij}F}$, coincides with the map $B^{\mathcal{V}}_{ij}$.*

Proof. The lemma follows from part (ii) of Theorem 6.1. Q.E.D.

Corollary 8.5. *The map $F \mapsto \nu(F)$ is an algebra isomorphism $\nu : \mathcal{O}_{\mathcal{C}} \to \mathcal{B}_{\mathcal{V}}$.*

Corollary 8.6. *The maps $\mu : \mathcal{O}_{\mathcal{C}} \to (\mathcal{V}^S)^{sing}_{\boldsymbol{\lambda}}$ and $\nu : \mathcal{O}_{\mathcal{C}} \to \mathcal{B}_{\mathcal{V}}$ intertwine the action of the multiplication operators on $\mathcal{O}_{\mathcal{C}}$ and the action of the Bethe algebra $\mathcal{B}_{\mathcal{V}}$ on $(\mathcal{V}^S)^{sing}_{\boldsymbol{\lambda}}$, that is, for any $F, G \in \mathcal{O}_{\mathcal{C}}$, we have*

$$\mu(FG) = \nu(F)\,\mu(G)\,.$$

Corollary 8.7. *Consider the element $v_1 \in (\mathcal{V}^S)^{sing}_{\boldsymbol{\lambda}}$, corresponding to $F = 1$ under the isomorphism μ. Let us use this element in the* *definition of the isomorphism η of Theorem 4.2. Then the throughout compositions*

$$\mathcal{O}_{\mathcal{C}} \xrightarrow{\mu} (\mathcal{V}^S)^{sing}_{\boldsymbol{\lambda}} \xrightarrow{\eta^{-1}} \mathcal{O}_{\mathcal{W}} \xrightarrow{\iota^*} \mathcal{O}_{\mathcal{C}} , \qquad \mathcal{O}_{\mathcal{C}} \xrightarrow{\nu} \mathcal{B}_{\mathcal{V}} \xrightarrow{\zeta^{-1}} \mathcal{O}_{\mathcal{W}} \xrightarrow{\iota^*} \mathcal{O}_{\mathcal{C}}$$

are the identity maps.

8.2. Inverse map to $\nu : \mathcal{O}_{\mathcal{C}} \to (\mathcal{V}^S)^{sing}_{\boldsymbol{\lambda}}$

For $v \in (\mathcal{V}^S)^{sing}_{\boldsymbol{\lambda}}$, define a function f_v on a Zariski open subset of $\mathcal{C}$ as follows. For a generic point $\boldsymbol{\Sigma} \in \mathcal{C}$, let $\boldsymbol{T} = (\boldsymbol{z}, \boldsymbol{t})$ be a point of the critical set $C \subset \mathbb{C}^l$ which projects to $\boldsymbol{\Sigma}$. Let $\omega(\boldsymbol{z}, \boldsymbol{t})$ be the Bethe vector corresponding the point $(\boldsymbol{z}, \boldsymbol{t})$. Set

$$f_v(\boldsymbol{\Sigma}) \;=\; S(v(\boldsymbol{z}), \omega(\boldsymbol{z}, \boldsymbol{t})) \;,$$

where S is the tensor Shapovalov form on $V^{\otimes n}$, cf. Theorem 6.1.

Theorem 8.8. *For any $v \in (\mathcal{V}^S)^{sing}_{\boldsymbol{\lambda}}$, the scalar function f_v is the restriction to $\mathcal{C}$ of a polynomial. Moreover, the map $(\mathcal{V}^S)^{sing}_{\boldsymbol{\lambda}} \to \mathcal{O}_{\mathcal{C}}$, $v \mapsto f_v$, is the inverse map to the isomorphism $\nu : \mathcal{O}_{\mathcal{C}} \to (\mathcal{V}^S)^{sing}_{\boldsymbol{\lambda}}$.*

Proof. Any element of $(\mathcal{V}^S)^{sing}_{\boldsymbol{\lambda}}$ has the form of v_F for a suitable $F \in \mathbb{C}[\boldsymbol{\Sigma}]$, see (7.1). In that case,

$$\begin{aligned} f_{v_F}(\boldsymbol{z}, \boldsymbol{t}) \;&=\; S\left(\frac{1}{l_1! \dots l_{N-1}!} \sum_{(\boldsymbol{z}, \boldsymbol{p}) \in C_{\boldsymbol{z}}} \frac{F(\boldsymbol{z}, \boldsymbol{p})\, \omega(\boldsymbol{z}, \boldsymbol{p})}{\mathrm{Hess}_{\boldsymbol{t}} \log \Phi(\boldsymbol{z}, \boldsymbol{p})}, \omega(\boldsymbol{z}, \boldsymbol{t}))\right) \\ &=\; F(\boldsymbol{z}, \boldsymbol{t}) \end{aligned}$$

by Theorem 6.1. This identity proves the theorem. Q.E.D.

§9. Proof of Theorem 7.1

9.1. The Shapovalov form and asymptotics of v_F

Let $\boldsymbol{T}^0$ be a point of the critical set $C \subset \mathbb{C}^l$, see Section 5.1.

Consider the germ at $0 \in \mathbb{C}$ of a generic analytic curve $\mathbb{C} \to \mathbb{C}^l$, $s \mapsto \boldsymbol{T}(s) = (\boldsymbol{z}(s), \boldsymbol{t}(s))$, with $\boldsymbol{T}(0) = \boldsymbol{T}^0$ such that for any small nonzero s, the point $(\boldsymbol{z}(s), \boldsymbol{t}(s))$ is a nondegenerate critical point of $\log \Phi(\boldsymbol{z}(s), \cdot\,)$, and $\boldsymbol{z}(s)$ has distinct coordinates. The corresponding Bethe vector has the form, $\omega(\boldsymbol{T}(s)) = w_\alpha s^\alpha + o(s^\alpha)$, where α is a rational number and $w_\alpha \in (V^{\otimes n})^{sing}_{\boldsymbol{\lambda}}$ is a nonzero vector.

Let X^0 denote the point of $\mathcal{W}$ corresponding to $\boldsymbol{T}^0$. Namely, we take the image $\boldsymbol{\Sigma}^0$ of $\boldsymbol{T}^0$ in $\mathcal{C}$ under the factorization by the $S_{\boldsymbol{l}}$-action and then set $X^0 = \iota(\boldsymbol{\Sigma}^0)$.

Lemma 9.1. *Assume that X^0 is not a critical point of the Wronski map $\mathcal{W} \to \mathbb{C}^n$. Then $S(w_\alpha, w_\alpha)$ is a nonzero number, where S is the tensor Shapovalov form.*

Proof. For a small nonzero s, the Bethe vectors corresponding to $S_{l_1} \times \cdots \times S_{l_{N-1}}$-orbits of the critical points of $\log \Phi(\boldsymbol{z}(s), \cdot\,)$ form a basis of $(V^{\otimes n})^{sing}_{\boldsymbol{\lambda}}$, see [12]. That basis is orthogonal with respect to the Shapovalov form. The Shapovalov form is nondegenerate on $(V^{\otimes n})^{sing}_{\boldsymbol{\lambda}}$. By assumptions of the lemma, the limit of the direction of the Bethe vector $\omega(\boldsymbol{z}(s), \boldsymbol{t}(s))$ as $s \to 0$ is different from the limits of the directions of the other Bethe vectors of the basis. These remarks imply the lemma. Q.E.D.

Corollary 9.2. *If $\alpha < 0$, then the ratio $\omega(\boldsymbol{T}(s))/\mathrm{Hess}_{\boldsymbol{t}} \log \Phi(\boldsymbol{T}(s))$ has well-defined limit as $s \to 0$.*

Proof. We have

$$\begin{aligned}\mathrm{Hess}_{\boldsymbol{t}} \log \Phi(\boldsymbol{T}(s)) &= S(\omega(\boldsymbol{T}(s)), \omega(\boldsymbol{T}(s))) \\ &= s^{-2\alpha} S(\omega_\alpha, \omega_\alpha) + o(s^{-2\alpha}),\end{aligned}$$

so the ratio $\omega(\boldsymbol{T}(s))/\mathrm{Hess}_{\boldsymbol{t}} \log \Phi(\boldsymbol{T}(s))$ has order $s^{-\alpha}$ as $s \to 0$. Q.E.D.

9.2. Possible places of irregularity of v_F

To prove Theorem 7.1, we need to show that v_F is regular outside of at most a codimension-two algebraic subset of $\mathbb{C}^n$. There are three possible codimension-one irregularity places of v_F:

(9.1) A pole of v_F may occur at a place where $\boldsymbol{z}$ has equal coordinates.

(9.2) A pole of v_F may occur at a place where $\boldsymbol{z}$ has distinct coordinates and the function $\log \Phi(\boldsymbol{z}, \cdot\,)$ has a degenerate critical point.

(9.3) A pole of v_F may occur at a place where $\boldsymbol{z}$ has distinct coordinates and there is a critical point which moved to a position with $t_i^{(1)} = z_j$ for some pair (i, j), or to a position with $t_i^{(a)} = t_j^{(a)}$ for some triple (a, i, j), $a > 0$, $i \neq j$, or to a position with $t_i^{(a)} = t_j^{(a+1)}$ for some triple (a, i, j), $a > 0$.

Problem (9.1) is treated in [13]. By Lemmas 4.3 and 4.4 of [13], the map v_F is regular at generic points of the hyperplanes $z_i = z_j$. (In fact, it is shown in Lemmas 4.3 and 4.4 of [13], that the number α of Corollary 9.2 is negative at generic points of possible irregularity corresponding to such hyperplanes, see [13].)

Problem (9.2) of possible irregularity of v_F at the places, where $\log \Phi(\boldsymbol{z}, \cdot)$ has a degenerate critical point, is treated in a standard way using integral representation (7.2). One replaces the sum in (7.1) by an integral over a cycle which can serve all $\boldsymbol{z}$ that are close to a given one, and then observes that the integral is holomorphic in $\boldsymbol{z}$; see, for example, Sections 5.13, 5.17, 5.18 in [1].

Thus, to prove Theorem 7.1 we need to show that generic points of type (9.3) correspond to the points of $\mathcal{W}$ which are noncritical for the Wronski map and which have $\alpha \leqslant 0$.

9.3. Flag exponents

A point $X \in \mathcal{W}$ is an N-dimensional space of polynomials with a basis $g_1(u), \dots, g_N(u)$ such that $\deg g_i = \lambda_i + N - i$. Each polynomial g_i is defined up to multiplication by a number and addition of a linear combination of $g_{i+1}, \dots, g_N$.

For any $a \in \mathbb{C}$ define distinct integers $\boldsymbol{d}_{X,a} = (d_1, \dots, d_N)$ called the flag exponents of X as follows. Choose a basis $g_1, \dots, g_N$ of X (not changing the degrees of these polynomials) so that $g_1, \dots, g_N$ have different orders at $u = a$ and set d_i to be the order of g_i at $u = a$.

We say that X is of type $\boldsymbol{d}$ if there exists $a \in \mathbb{C}$ such that $\boldsymbol{d}_{X,a} = \boldsymbol{d}$. For every $\boldsymbol{d}$, denote by $\mathcal{W}_{\boldsymbol{d}} \subset \mathcal{W}$ the closure of the subset of points of type $\boldsymbol{d}$. We are interested in the subsets $\mathcal{W}_{\boldsymbol{d}} \subset \mathcal{W}$ which are of codimension one and whose points correspond to Problem (9.3). Such subsets will be called *essential*.

For example, for $N = 2$, the subset $\mathcal{W}_{(0,2)}$ is the only essential subset. For $N = 3$, the only essential subsets are $\mathcal{W}_{(1,3,0)}$, $\mathcal{W}_{(1,0,2)}$ and $\mathcal{W}_{(0,2,1)}$.

Lemma 9.3. *For given N, if $\mathcal{W}_{\boldsymbol{d}}$ is essential, then $\boldsymbol{d}$ is one of the following $2N-3$ indices,*

$$\begin{aligned}
\boldsymbol{d}_{1+} &= (N-2, N, N-3, N-4, \dots, 1, 0)\,, \\
\boldsymbol{d}_{i+} &= (N-1, N-2, \dots, N-i+1, \\
&\qquad N-i-1, N-i-2, N-i, N-i-3, \dots, 1, 0)\,, \\
\boldsymbol{d}_{i-} &= (N-1, N-2, \dots, N-i+1, \\
&\qquad N-i-2, N-i, N-i-1, N-i-3, \dots, 1, 0)\,,
\end{aligned}$$

for $i = 2, \dots, N-1$.

Proof. The lemma is proved by straightforward counting of codimensions. Q.E.D.

If X is a point of $\mathcal{W}_{\boldsymbol{d}_{1+}}$, then for a suitable ordering of its root coordinates we have $z_1 = t_1^{(1)} = t_2^{(1)}$. If X is a point of $\mathcal{W}_{\boldsymbol{d}_{i+}}$, $i > 1$, then for a suitable ordering of its root coordinates we have $t_1^{(i-1)} = t_1^{(i)} = t_2^{(i)}$. If X is a point of $\mathcal{W}_{\boldsymbol{d}_{i-}}$, then for a suitable ordering of its root coordinates we have $t_1^{(i-1)} = t_2^{(i-1)} = t_1^{(i)}$. Each of these properties is a problem of type (9.3).

Lemma 9.4. *Each essential subset is irreducible.*

Proof. It is easy to see that an essential subset is the image of an affine space under a suitable map. Q.E.D.

Lemma 9.5. *Generic points of every essential subset are not critical for the Wronski map.*

Proof. The proof is similar to the proof in Proposition 8 of [5] of the fact that the Jacobian det $\Delta_{\boldsymbol{q}}$ is nonzero. Q.E.D.

9.4. Proof of Theorem 7.1

9.4.1. Let $\mathcal{W}_{\boldsymbol{d}}$ be an arbitrary essential subset. We fix a certain positive integer q. Then for any numbers $\boldsymbol{r} = (r_0, r_1, r_2, \dots, r_q)$, such that $r_0 \in \mathbb{C}$, $r_i \in \mathbb{R}$ for $i > 0$, $0 < r_1 < r_2 < \cdots < r_q$, we choose a point $X_{\boldsymbol{r}}(\epsilon, s) \in \mathcal{W}$ depending on two parameters ϵ, s so that $X_{\boldsymbol{r}}(\epsilon, 0) \in \mathcal{W}_{\boldsymbol{d}}$ and the point $X_{\boldsymbol{r}}(\epsilon, s)$ is nice for small nonzero s. The dependence of $X_{\boldsymbol{r}}(\epsilon, s)$ on $\boldsymbol{r}$ in our construction is generic in the following sense. For any hypersurface $\mathcal{Z} \subset \mathcal{W}_d$ we can fix $\boldsymbol{r}$ so that the curve $X_{\boldsymbol{r}}(\epsilon, 0)$ does not lie in $\mathcal{Z}$.

For any fixed $\boldsymbol{r}$, we choose ordered root coordinates $\boldsymbol{T}_{\boldsymbol{r}}(\epsilon, s)$ of $X_{\boldsymbol{r}}(\epsilon, s)$ and consider the corresponding Bethe vector $\omega(\boldsymbol{T}_{\boldsymbol{r}}(\epsilon, s))$. We choose a suitable coordinate $\omega_J(\boldsymbol{T}_{\boldsymbol{r}}(\epsilon, s))$ of the Bethe vector and show that for small ϵ the coordinate $\omega_J(\boldsymbol{T}_{\boldsymbol{r}}(\epsilon, s))$ has nonzero limit as $s \to 0$. That statement and Corollary 9.2 show that the corresponding summand in (7.1) is regular at $\mathcal{W}_{\boldsymbol{d}}$.

The proof that $\omega_J(\boldsymbol{T}_{\boldsymbol{r}}(\epsilon, s))$ has nonzero limit is lengthy. We present it for the Lie algebras $\mathfrak{gl}_2$ and $\mathfrak{gl}_3$. The proof for the Lie algebra $\mathfrak{gl}_N$ with $N > 3$ is similar, see also the end of Section 9.4.5.

9.4.2. *Proof for $N = 2$.* A point $X \in \mathcal{W}$ is a two-dimensional space of polynomials. The only essential subset is $\mathcal{W}_{(0,2)}$. This essential subset corresponds to the problem $z_{\lambda_1+\lambda_2} = t_{\lambda_2-1}^{(1)} = t_{\lambda_2}^{(1)}$ of type (9.3) (after relabeling the root coordinates).

For any numbers $\boldsymbol{r} = (r_0, r_1, r_2, \dots, r_{\lambda_2+\lambda_1-1})$, such that $r_0 \in \mathbb{C}$, $r_i \in \mathbb{R}$ for $i > 0$, $0 < r_1 < r_2 < \cdots < r_{\lambda_2+\lambda_1-1}$, we choose $X_{\boldsymbol{r}}(\epsilon, s)$ to be the two-dimensional space of polynomials spanned by

$$g_1(u) = (u-r_0)^{\lambda_1+1} + \sum_{i=0}^{\lambda_1} b_i (u-r_0)^i ,$$

$$g_2(u) = (u-r_0)^{\lambda_2} + \sum_{i=2}^{\lambda_2-1} a_i (u-r_0)^i - s^2 a_2 ,$$

where $a_{\lambda_2-1} = \epsilon^{r_1}$, $a_{\lambda_2-i}/a_{\lambda_2-i+1} = \epsilon^{r_i}$, $i = 2,\dots,\lambda_2-2$, $b_{\lambda_1} = \epsilon^{r_{\lambda_2-1}}$, $b_{\lambda_1-i}/b_{\lambda_1-i+1} = \epsilon^{r_{\lambda_2+i-1}}$, $i = 1,\dots,\lambda_1$. We have $X_{\boldsymbol{r}}(\epsilon, 0) \in \mathcal{W}_{(0,2)}$. Clearly, the dependence of $X_{\boldsymbol{r}}(\epsilon, s)$ on $\boldsymbol{r}$ is generic in the sense defined in Section 9.4.1.

We consider the asymptotic zone $1 \gg |\epsilon| \gg |s| > 0$ and describe the asymptotics in that zone of the roots of g_2 and Wronskian $\mathrm{Wr}(g_1, g_2)$. The leading terms of asymptotics are obtained by the Newton polygon method. If the leading term of some root is at least of order s^2, we shall write that this root equals zero.

The roots of g_2 have the form:

$$t_1^{(1)} \sim r_0 - \epsilon^{r_1}, \quad t_2^{(1)} \sim r_0 - \epsilon^{r_2}, \quad \dots , \quad t_{\lambda_2-2}^{(1)} \sim r_0 - \epsilon^{r_{\lambda_2-2}},$$

$$t_{\lambda_2-1}^{(1)} \sim r_0 + s , \quad t_{\lambda_2}^{(1)} \sim r_0 - s .$$

The Wronskian is a polynomial in u, ϵ. Below we present only the monomials corresponding to the line segments of the Newton polygon important for the leading asymptotics of the roots,

$$\begin{aligned} \mathrm{Wr}(g_1, g_2) = {} & (\lambda_1 + 1 - \lambda_2)(u-r_0)^{\lambda_2+\lambda_1} + \\ & + \sum_{i=2}^{\lambda_2-1} (\lambda_1 + 1 - i)\, a_i (u-r_0)^{\lambda_1+i} + \\ & + a_2 \sum_{i=0}^{\lambda_1} (i-2)\, b_i (u-r_0)^{i+1} + \dots . \end{aligned}$$

It follows from this formula that the roots of $\mathrm{Wr}(g_1, g_2)$ have the form:

$$z_1 \sim r_0 - \frac{\lambda_1 - \lambda_2 + 2}{\lambda_1 - \lambda_2 + 1} \epsilon^{r_1}, \quad z_2 \sim r_0 - \frac{\lambda_1 - \lambda_2 + 3}{\lambda_1 - \lambda_2 + 2} \epsilon^{r_2}, \quad \dots ,$$

$$z_{\lambda_2-2} \sim r_0 - \frac{\lambda_1 - 1}{\lambda_1 - 2} \epsilon^{r_{\lambda_2-2}}, \quad z_{\lambda_2-1} \sim r_0 - \frac{\lambda_1 - 2}{\lambda_1 - 1} \epsilon^{r_{\lambda_2-1}}, \quad \dots ,$$

$$z_{\lambda_2+\lambda_1-4} \sim r_0 - \frac{1}{2}\,\epsilon^{r_{\lambda_2+\lambda_1-4}}, \quad z_{\lambda_2+\lambda_1-3} \sim r_0 + \epsilon^{(r_{\lambda_2+\lambda_1-3}+r_{\lambda_2+\lambda_1-2})/2},$$

$$z_{\lambda_2+\lambda_1-2} \sim r_0 - \epsilon^{(r_{\lambda_2+\lambda_1-3}+r_{\lambda_2+\lambda_1-2})/2}, \quad z_{\lambda_2+\lambda_1-1} \sim r_0 - 2\epsilon^{r_{\lambda_2+\lambda_1-1}},$$

$$z_{\lambda_2+\lambda_1} \sim r_0 \,.$$

The point $\boldsymbol{T_r}(\epsilon, s) = (z_1, \dots, z_{\lambda_2+\lambda_1}, t_1^{(1)}, \dots, t_{\lambda_2}^{(1)})$ is a point of root coordinates of $X_{\boldsymbol{r}}(\epsilon, s)$.

Let us call the root coordinates $t_{\lambda_2-1}^{(1)}$, $t_{\lambda_2}^{(1)}$, $z_{\lambda_2+\lambda_1}$ exceptional, and the remaining root coordinates regular. For each regular root coordinate y the leading term of asymptotics of $y - r_0$ as $\epsilon \to 0$ has the form $A\epsilon^B$ for suitable numbers $A \neq 0$, B.

Lemma 9.6. *The pairs (A, B) are different for different regular root coordinates.*

Proof. A proof is by inspection of the list. Q.E.D.

For each exceptional coordinate y the absolute value of the difference $y - r_0$ is much smaller as $\epsilon \to 0$ than for any regular coordinate.

The Bethe vector is the vector $\omega(\boldsymbol{T_r}(\epsilon, s)) = \sum_J \omega_J(\boldsymbol{T_r}(\epsilon, s))\, e_J v$, where the sum is over all admissible J, see Section 6. An admissible $J = (j_1, \dots, j_{\lambda_1+\lambda_2})$ consists of ones and twos with exactly λ_2 twos. Choose J with $j_i = 2$ for $i = 1, \dots, \lambda_2$. Then

$$\omega_J(\boldsymbol{T_r}(\epsilon, s)) = \sum_{\sigma \in S_{\lambda_2}} \prod_{i=1}^{\lambda_2} \frac{1}{t_{\sigma(i)}^{(1)} - z_i}\,. \tag{9.4}$$

Lemma 9.7. *For small ϵ, the function $\omega_J(\boldsymbol{T_r}(\epsilon, s))$ has well-defined limit as $s \to 0$.*

Proof. By Lemma 9.6, each summand in (9.4) has well-defined limit. Q.E.D.

Our goal is to show that $\bar\omega_J(\epsilon) = \lim_{s\to 0} \omega_J(\boldsymbol{T_r}(\epsilon, s))$ is nonzero for small ϵ.

If f is a function of ϵ and $f \sim A(f)\epsilon^{B(f)}$ for some numbers $A(f) \neq 0$, $B(f)$ as $\epsilon \to 0$, then we call f acceptable, $B(f)$ the order of f and $A(f)$ the leading coefficient of f. If the absolute value of f is smaller than any positive power of ϵ or is the zero function, then we set $B(f) = \infty$.

For every σ, the limit $q_\sigma = \lim_{s\to 0}(\prod_{i=1}^{\lambda_2} \frac{1}{t_{\sigma(i)}^{(1)} - z_i})$ is an acceptable function of order $B(q_\sigma) = -\sum_{i=1}^{\lambda_2} \min(B(t_{\sigma(i)}^{(1)} - r_0), B(z_i - r_0))$. In particular, $B(q_\sigma) \geqslant -B(z_{\lambda_2-1} - r_0) - B(z_{\lambda_2} - r_0) - \sum_{i=1}^{\lambda_2-2} B(t_i^{(1)} - r_0)$.

Lemma 9.8. *The function $\bar\omega_J(\epsilon)$ is acceptable. Its order and leading coefficient are given by the formulas*

$$B(\bar\omega_J(\epsilon)) = -B(z_{\lambda_2-1}-r_0) - B(z_{\lambda_2}-r_0) - \sum_{i=1}^{\lambda_2-2} B(t_i^{(1)}-r_0)\,,$$

$$A(\bar\omega_J(\epsilon)) = \frac{2}{A(z_{\lambda_2-1}-r_0)A(z_{\lambda_2}-r_0)} \prod_{i=1}^{\lambda_2-2} \frac{1}{A(t_i^{(1)}-r_0)-A(z_i-r_0)}\,.$$

By Lemma 9.8, $\bar\omega_J(\epsilon)$ is nonzero for small ϵ and therefore, $\alpha \leqslant 0$ for generic points of $\mathcal{W}_{(0,2)}$. Theorem 7.1 is proved for $N=2$.

9.4.3. *Proof for $N = 3$ and $\mathcal{W}_{(1,3,0)}$.* A point $X \in \mathcal{W}$ is a three-dimensional space of polynomials. We study the problem $z_{\lambda_1+\lambda_2+\lambda_3} = t^{(1)}_{\lambda_2+\lambda_3-1} = t^{(1)}_{\lambda_2+\lambda_3}$ of type (9.3) (after relabeling the root coordinates).

For any numbers $\boldsymbol{r} = (r_0, r_1, r_2, \dots, r_{\lambda_3+\lambda_2+\lambda_1-1})$, such that $r_0 \in \mathbb{C}$, $r_i \in \mathbb{R}$ for $i > 0$, $0 < r_1 < r_2 < \dots < r_{\lambda_3+\lambda_2+\lambda_1-1}$, we choose $X_{\boldsymbol{r}}(\epsilon, s)$ to be the three-dimensional space of polynomials spanned by

$$g_1(u) = (u-r_0)^{\lambda_1+2} + \sum_{i=1}^{\lambda_1+1} c_i(u-r_0)^i\,,$$

$$g_2(u) = (u-r_0)^{\lambda_2+1} + \sum_{i=3}^{\lambda_2} b_i(u-r_0)^i - 3s^2b_3(u-r_0)\,,$$

$$g_3(u) = (u-r_0)^{\lambda_3} + \sum_{i=0}^{\lambda_3-1} a_i(u-r_0)^i\,,$$

where

$$\begin{aligned}
&a_{\lambda_3-1} = \epsilon^{r_1}, && a_{\lambda_3-i}/a_{\lambda_3-i+1} = \epsilon^{r_i}, && i = 2,\dots,\lambda_3\,,\\
&b_{\lambda_2} = \epsilon^{r_{\lambda_3+1}}, && b_{\lambda_2-i}/b_{\lambda_2-i+1} = \epsilon^{r_{\lambda_3+i+1}}, && i = 1,\dots,\lambda_2-3\,,\\
&c_{\lambda_1+1} = \epsilon^{r_{\lambda_3+\lambda_2-1}}, && c_{\lambda_1-i}/c_{\lambda_1-i+1} = \epsilon^{r_{\lambda_3+\lambda_2+i}}, && i = 0,\dots,\lambda_1-1\,.
\end{aligned}$$

We have $X_{\boldsymbol{r}}(\epsilon,0) \in \mathcal{W}_{(1,3,0)}$. Clearly, the dependence of $X_{\boldsymbol{r}}(\epsilon,s)$ on $\boldsymbol{r}$ is generic in the sense defined in Section 9.4.1.

We consider the asymptotic zone $1 \gg |\epsilon| \gg |s| > 0$ and describe the asymptotics in that zone of the roots of the polynomials g_3, $\mathrm{Wr}(g_2, g_3)$, $\mathrm{Wr}(g_1, g_2, g_3)$. We obtain the leading terms of asymptotics by the Newton polygon method. If the leading term of some root is at least of order s^2, we shall write that this root equals zero.

The roots of g_3 are of the form:

$$t_1^{(2)} \sim r_0 - \epsilon^{r_1}, \quad t_2^{(2)} \sim r_0 - \epsilon^{r_2}, \ \dots \ , \quad t_{\lambda_3}^{(2)} \sim r_0 - \epsilon^{r_{\lambda_3}} \ .$$

We have

$$\begin{aligned}
\mathrm{Wr}(g_2, g_3) &= (\lambda_2 + 1 - \lambda_3)(u - r_0)^{\lambda_2+\lambda_3} + \\
&+ \sum_{i=0}^{\lambda_3-1} (\lambda_2 + 1 - i)\, a_i (u - r_0)^{\lambda_2+i-1} + \\
&+ a_0 \sum_{i=3}^{\lambda_2} i b_i (u - r_0)^{i-1} - 3s^2 a_0 b_3 + \ \dots \ ,
\end{aligned}$$

where the dots denote the monomials which are not important for the leading asymptotics of the roots. The roots of $\mathrm{Wr}(g_2, g_3)$ are of the form

$$t_1^{(1)} \sim r_0 - \frac{\lambda_2 - \lambda_3 + 2}{\lambda_2 - \lambda_3 + 1}\, \epsilon^{r_1}, \ \dots \ , \ t_{\lambda_3}^{(1)} \sim r_0 - \frac{\lambda_2 + 1}{\lambda_2}\, \epsilon^{r_{\lambda_3}},$$

$$t_{\lambda_3+1}^{(1)} \sim r_0 - \frac{\lambda_2}{\lambda_2 + 1}\, \epsilon^{r_{\lambda_3+1}}, \ \dots \ , \ t_{\lambda_3+\lambda_2-2}^{(1)} \sim r_0 - \frac{3}{4}\, \epsilon^{r_{\lambda_3+\lambda_2-2}},$$

$$t_{\lambda_3+\lambda_2-1}^{(1)} \sim r_0 + s, \quad t_{\lambda_3+\lambda_2}^{(1)} \sim r_0 - s.$$

We have

$$\begin{aligned}
&\mathrm{Wr}\,(g_1, g_2, g_3) = \\
&= (\lambda_1 + 1 - \lambda_2)(\lambda_1 + 2 - \lambda_3)(\lambda_2 + 1 - \lambda_3)(u - r_0)^{\lambda_3+\lambda_2+\lambda_1} + \\
&+ \sum_{i=0}^{\lambda_3-1} (\lambda_1 + 1 - \lambda_2)(\lambda_1 + 2 - i)(\lambda_2 + 1 - i)\, a_i (u - r_0)^{i+\lambda_2+\lambda_1} + \\
&+ a_0 \sum_{i=3}^{\lambda_2} (\lambda_1 + 2 - i)(\lambda_1 + 2)\, i\, b_i (u - r_0)^{\lambda_1+i-1} + \\
&+ a_0 b_3 \sum_{i=1}^{\lambda_1+1} 3i(i-3)\, c_i (u - r_0)^i + \ \dots \ .
\end{aligned}$$

The roots of $\mathrm{Wr}(g_1, g_2, g_3)$ are of the form

$$z_1 \sim r_0 - \frac{(\lambda_1 + 3 - \lambda_3)(\lambda_2 + 2 - \lambda_3)}{(\lambda_1 + 2 - \lambda_3)(\lambda_2 + 1 - \lambda_3)}\, \epsilon^{r_1}, \ \dots \ ,$$

$$z_{\lambda_3} \sim r_0 - \frac{(\lambda_1 + 2)(\lambda_2 + 1)}{(\lambda_1 + 1)\lambda_2}\, \epsilon^{r_{\lambda_3}},$$

$$z_{\lambda_3+1} \sim r_0 - \frac{(\lambda_1+2-\lambda_2)\lambda_2}{(\lambda_1+1-\lambda_2)(\lambda_2+1)}\,\epsilon^{r_{\lambda_3+1}}, \ \dots\ ,$$

$$z_{\lambda_3+\lambda_2-2} \sim r_0 - \frac{3(\lambda_1-1)}{4(\lambda_1-2)}\,\epsilon^{r_{\lambda_3+\lambda_2-2}},$$

$$z_{\lambda_3+\lambda_2-1} \sim r_0 - \frac{(\lambda_1+1)(\lambda_1-2)}{(\lambda_1+2)(\lambda_1-1)}\,\epsilon^{r_{\lambda_3+\lambda_2-1}}, \ \dots\ ,$$

$$z_{\lambda_3+\lambda_2+\lambda_1-4} \sim r_0 - \frac{2}{5}\,\epsilon^{r_{\lambda_3+\lambda_2+\lambda_1-4}},$$

$$z_{\lambda_3+\lambda_2+\lambda_1-3} \sim r_0 + \frac{1}{\sqrt{2}}\,\epsilon^{(r_{\lambda_3+\lambda_2+\lambda_1-3}+r_{\lambda_3+\lambda_2+\lambda_1-2})/2},$$

$$z_{\lambda_3+\lambda_2+\lambda_1-2} \sim r_0 - \frac{1}{\sqrt{2}}\,\epsilon^{(r_{\lambda_3+\lambda_2+\lambda_1-3}+r_{\lambda_3+\lambda_2+\lambda_1-2})/2},$$

$$z_{\lambda_3+\lambda_2+\lambda_1-1} \sim r_0 - \epsilon^{r_{\lambda_3+\lambda_2+\lambda_1-1}}, \quad z_{\lambda_3+\lambda_2+\lambda_1} \sim r_0\,.$$

The point $\boldsymbol{T}_{\boldsymbol{r}}(\epsilon, s) = (z_1, \dots, z_{\lambda_3+\lambda_2+\lambda_1}, t_1^{(1)}, \dots, t^{(1)}_{\lambda_3+\lambda_2}, t_1^{(2)}, \dots, t^{(2)}_{\lambda_3})$ is a point of root coordinates of $X_{\boldsymbol{r}}(\epsilon, s)$.

Let us call the root coordinates $t^{(1)}_{\lambda_3+\lambda_2-1}$, $t^{(1)}_{\lambda_3+\lambda_2}$, $z_{\lambda_3+\lambda_2+\lambda_1}$ exceptional, and the remaining root coordinates regular. For each regular root coordinate y the leading term of asymptotics of $y - r_0$ as $\epsilon \to 0$ has the form $A\epsilon^B$ for suitable numbers $A \neq 0$, B.

Lemma 9.9. *The pairs (A, B) are different for different regular root coordinates.*

Proof. A proof is by inspection of the list. Q.E.D.

For each exceptional coordinate y the absolute value of the difference $y - r_0$ is much smaller as $\epsilon \to 0$ than for any regular coordinate.

The Bethe vector has the form $\omega(\boldsymbol{T}_{\boldsymbol{r}}(\epsilon, s)) = \sum_J \omega_J(\boldsymbol{T}_{\boldsymbol{r}}(\epsilon, s))\, e_J v$, where the sum is over all admissible J, see Section 6. An admissible $J = (j_1, \dots, j_{\lambda_3+\lambda_2+\lambda_1})$ consists of ones, twos and threes with exactly λ_3 threes and λ_2 twos. Choose J with $j_i = 3$ for $i = 1, \dots, \lambda_3$ and $j_i = 2$ for $i = \lambda_3+\lambda_2-1, \lambda_3+\lambda_2, \dots, \lambda_3+2\lambda_2-2$. Then $\omega_J(\boldsymbol{T}_{\boldsymbol{r}}(\epsilon, s))$ is given by the formula

$$\omega_J(\boldsymbol{T}_{\boldsymbol{r}}(\epsilon, s)) = \sum_{\sigma\in S_{\lambda_3+\lambda_2}} \sum_{\tau \in S_{\lambda_3}} \prod_{i=1}^{\lambda_3} \frac{1}{(t^{(2)}_{\tau(i)} - t^{(1)}_{\sigma(i)})(t^{(1)}_{\sigma(i)} - z_i)} \times \prod_{i=\lambda_3+1}^{\lambda_3+\lambda_2} \frac{1}{t^{(1)}_{\sigma(i)} - z_{\lambda_2+i-2}}\,. \tag{9.5}$$

Lemma 9.10. *For small ϵ, the function $\omega_J(\boldsymbol{T_r}(\epsilon, s))$ has well-defined limit as $s \to 0$.*

Proof. By Lemma 9.9, each summand in (9.5) has well-defined limit. Q.E.D.

Our goal is to show that $\bar{\omega}_J(\epsilon) = \lim_{s\to 0} \omega_J(\boldsymbol{T_r}(\epsilon, s))$ is nonzero for small ϵ.

For every σ the second product in (9.5), has well-defined limit

$$q_\sigma = \lim_{s\to 0} \left(\prod_{i=\lambda_3+1}^{\lambda_3+\lambda_2} \frac{1}{t^{(1)}_{\sigma(i)} - z_{\lambda_2+i-2}} \right).$$

That limit is an acceptable function of order

$$B(q_\sigma) = -\sum_{i=\lambda_3+1}^{\lambda_3+\lambda_2} \min(B(t^{(1)}_{\sigma(i)} - r_0), B(z_{\lambda_2+i-2} - r_0)).$$

In particular,

$$B(q_\sigma) \geqslant -B(z_{\lambda_3+2\lambda_2-3} - r_0) - B(z_{\lambda_3+2\lambda_2-2} - r_0) - \sum_{i=\lambda_3+1}^{\lambda_3+\lambda_2-2} B(t^{(1)}_i - r_0).$$

The largest second products are those with

$$B(q_\sigma) = -B(z_{\lambda_3+2\lambda_2-3} - r_0) - B(z_{\lambda_3+2\lambda_2-2} - r_0) - \sum_{i=\lambda_3+1}^{\lambda_3+\lambda_2-2} B(t^{(1)}_i - r_0).$$

For every σ, τ, the first product in (9.5) has well-defined limit

$$p_{\sigma\tau} = \lim_{s\to 0} \left(\prod_{i=1}^{\lambda_3} \frac{1}{(t^{(2)}_{\tau(i)} - t^{(1)}_{\sigma(i)})(t^{(1)}_{\sigma(i)} - z_i)} \right).$$

That limit is an acceptable function of order

$$B(p_{\sigma\tau}) = -\sum_{i=1}^{\lambda_3} \big(\min(B(t^{(2)}_{\tau(i)} - r_0), B(t^{(1)}_{\sigma(i)} - r_0)) + \\ + \min(B(t^{(1)}_{\sigma(i)} - r_0), B(z_i - r_0))\big).$$

In particular, $B(p_{\sigma\tau}) \geqslant -\sum_{i=1}^{\lambda_3}(B(t_i^{(2)} - r_0) + B(z_i - r_0))$. The largest first products are those with

$$B(p_{\sigma\tau}) = -\sum_{i=1}^{\lambda_3} (B(t_i^{(2)} - r_0) + B(z_i - r_0)).$$

Lemma 9.11. *The function $\bar\omega_J(\epsilon)$ is acceptable. Its order and leading coefficient are given by the formulas*

$$B(\bar\omega_J(\epsilon)) = -\sum_{i=1}^{\lambda_3} (B(t_i^{(2)} - r_0) + B(z_i - r_0)) - B(z_{\lambda_3+2\lambda_2-3} - r_0) - \\ - B(z_{\lambda_3+2\lambda_2-2} - r_0) - \sum_{i=\lambda_3+1}^{\lambda_3+\lambda_2-2} B(t_i^{(1)} - r_0)\,.$$

$$A(\bar\omega_J(\epsilon)) = 2\,(\lambda_2 - 2)!\,\frac{1}{A(z_{\lambda_3+2\lambda_2-3} - r_0)A(z_{\lambda_3+2\lambda_2-2} - r_0)} \times \\ \times \prod_{i=1}^{\lambda_3} \frac{1}{(A(t_i^{(2)} - r_0) - A(t_i^{(1)} - r_0))(A(t_i^{(1)} - r_0) - A(z_i - r_0))} \times \\ \times \prod_{i=\lambda_3+1}^{\lambda_3+\lambda_2-2} \frac{1}{A(t_i^{(1)} - r_0)}\,.$$

Proof. It is easy to see that if σ, τ are such that the second product in (9.5) has order

$$-B(z_{\lambda_3+2\lambda_2-3} - r_0) - B(z_{\lambda_3+2\lambda_2-2} - r_0) - \sum_{i=\lambda_3+1}^{\lambda_3+\lambda_2-2} B(t_i^{(1)} - r_0)\,,$$

then the first product has order $-\sum_{i=1}^{\lambda_3}(B(t_i^{(2)} - r_0) + B(z_i - r_0))$ only if it equals $\prod_{i=1}^{\lambda_3}(t_i^{(2)} - t_i^{(1)})^{-1}(t_i^{(1)} - z_i)^{-1}$. This implies the lemma. Q.E.D.

9.4.4. *Proof for $N = 3$ and $\mathcal{W}_{(0,2,1)}$.* We study the problem $t^{(1)}_{\lambda_2+\lambda_3} = t^{(1)}_{\lambda_2+\lambda_3-1} = t^{(2)}_{\lambda_3}$ of type (9.3) (after relabeling the root coordinates).

For any numbers $\boldsymbol{r} = (r_0, r_1, r_2, \ldots, r_{\lambda_3+\lambda_2+\lambda_1})$, such that $r_0 \in \mathbb{C}$, $r_i \in \mathbb{R}$ for $i > 0$, $0 < r_1 < r_2 < \cdots < r_{\lambda_3+\lambda_2+\lambda_1}$, we choose $X_{\boldsymbol{r}}(\epsilon, s)$ to be the three-dimensional space of polynomials spanned by

$$g_1(u) = (u-r_0)^{\lambda_1+2} + \sum_{i=0}^{\lambda_1+1} c_i(u-r_0)^i ,$$

$$g_2(u) = (u-r_0)^{\lambda_2+1} + \sum_{i=2}^{\lambda_2} b_i(u-r_0)^i + s^2 b_2 ,$$

$$g_3(u) = (u-r_0)^{\lambda_3} + \sum_{i=1}^{\lambda_3-1} a_i(u-r_0)^i ,$$

where

$$\begin{aligned} &a_{\lambda_3-1} = \epsilon^{r_1}, && a_{\lambda_3-i}/a_{\lambda_3-i+1} = \epsilon^{r_i}, && i = 2,\dots,\lambda_3-1,\\ &b_{\lambda_2} = \epsilon^{r_{\lambda_3}}, && b_{\lambda_2-i}/b_{\lambda_2-i+1} = \epsilon^{r_{\lambda_3+i}}, && i = 1,\dots,\lambda_2-2,\\ &c_{\lambda_1+1} = \epsilon^{r_{\lambda_3+\lambda_2-1}}, && c_{\lambda_1-i}/c_{\lambda_1-i+1} = \epsilon^{r_{\lambda_3+\lambda_2+i}} && i = 0,\dots,\lambda_1 . \end{aligned}$$

We have $X_{\boldsymbol{r}}(\epsilon,0) \in \mathcal{W}_{(0,2,1)}$. Clearly, the dependence of $X_{\boldsymbol{r}}(\epsilon,s)$ on $\boldsymbol{r}$ is generic in the sense defined in Section 9.4.1.

We consider the same asymptotic zone $1 \gg |\epsilon| \gg |s| > 0$. The roots of g_3 are of the form:

$$t_1^{(2)} \sim r_0 - \epsilon^{r_1},\ \ t_2^{(2)} \sim r_0 - \epsilon^{r_2},\ \dots\ ,\ t_{\lambda_3-1}^{(2)} \sim r_0 - \epsilon^{r_{\lambda_3-1}},\ \ t_{\lambda_3}^{(2)} = r_0 .$$

We have

$$\begin{aligned} \mathrm{Wr}(g_2,g_3) &= (\lambda_2+1-\lambda_3)(u-r_0)^{\lambda_3+\lambda_2} + \\ &\quad + \sum_{i=1}^{\lambda_3-1} (\lambda_2+1-i)\, a_i(u-r_0)^{\lambda_2+i} + \\ &\quad + a_1 \sum_{i=2}^{\lambda_2} (i-1)\, b_i(u-r_0)^i - s^2 a_1 b_2 + \dots . \end{aligned}$$

The roots of $\mathrm{Wr}(g_2,g_3)$ are of the form

$$t_1^{(2)} \sim r_0 - \frac{\lambda_2-\lambda_3+2}{\lambda_2-\lambda_3+1}\,\epsilon^{r_1},\ \dots\ ,\ t_{\lambda_3-1}^{(2)} \sim r_0 - \frac{\lambda_2}{\lambda_2-1}\,\epsilon^{r_{\lambda_3-1}},$$

$$t_{\lambda_3}^{(2)} \sim r_0 - \frac{\lambda_2-1}{\lambda_2}\,\epsilon^{r_{\lambda_3}},\ \dots\ ,\ t_{\lambda_3+\lambda_2-2}^{(2)} \sim r_0 - \frac{1}{2}\,\epsilon^{r_{\lambda_3+\lambda_2-2}},$$

$$t_{\lambda_3+\lambda_2-1}^{(2)} \sim r_0 + s,\ \ t_{\lambda_3+\lambda_2}^{(2)} \sim r_0 - s .$$

We have

$$\begin{aligned}
&\mathrm{Wr}\,(g_1, g_2, g_3) = \\
&= (\lambda_1 + 1 - \lambda_2)(\lambda_1 + 2 - \lambda_3)(\lambda_2 + 1 - \lambda_3)(u - r_0)^{\lambda_3+\lambda_2+\lambda_1} + \\
&+ \sum_{i=1}^{\lambda_3-1} (\lambda_1 + 1 - \lambda_2)(\lambda_1 + 2 - i)(\lambda_2 + 1 - i)\, a_i (u - r_0)^{i+\lambda_2+\lambda_1} + \\
&+ a_1 \sum_{i=2}^{\lambda_2} (\lambda_1 + 2 - i)(\lambda_1 + 1)(i - 1)\, b_i (u - r_0)^{\lambda_1+i} + \\
&+ a_1 b_2 \sum_{i=0}^{\lambda_1+1} (i - 2)(i - 1)\, c_i (u - r_0)^i + \dots\,.
\end{aligned}$$

The roots of $\mathrm{Wr}(g_1, g_2, g_3)$ are of the form

$$\begin{aligned}
&z_1 \sim r_0 - \frac{(\lambda_1 + 3 - \lambda_3)(\lambda_2 + 2 - \lambda_3)}{(\lambda_1 + 2 - \lambda_3)(\lambda_2 + 1 - \lambda_3)}\, \epsilon^{r_1}, \ \dots\,, \\
&z_{\lambda_3-1} \sim r_0 - \frac{(\lambda_1 + 1)\lambda_2}{\lambda_1(\lambda_2 - 1)}\, \epsilon^{r_{\lambda_3-1}}, \\
&z_{\lambda_3} \sim r_0 - \frac{(\lambda_1 + 2 - \lambda_2)(\lambda_2 - 1)}{(\lambda_1 + 1 - \lambda_2)\lambda_2}\, \epsilon^{r_{\lambda_3}}, \ \dots\,, \\
&z_{\lambda_3+\lambda_2-3} \sim r_0 - \frac{\lambda_1}{2(\lambda_1 + 1)}\, \epsilon^{r_{\lambda_3+\lambda_2-3}}, \\
&z_{\lambda_3+\lambda_2-2} \sim r_0 - \frac{(\lambda_1 - 1)}{(\lambda_1 + 1)}\, \epsilon^{r_{\lambda_3+\lambda_2-2}}, \ \dots\,, \\
&z_{\lambda_3+\lambda_2+\lambda_1-3} \sim r_0 - \frac{1}{3}\, \epsilon^{r_{\lambda_3+\lambda_2+\lambda_1-3}}, \\
&z_{\lambda_3+\lambda_2+\lambda_1-2} \sim r_0 + x_1 \epsilon^m, \quad z_{\lambda_3+\lambda_2+\lambda_1-1} \sim r_0 + x_2 \epsilon^m, \\
&z_{\lambda_3+\lambda_2+\lambda_1} \sim r_0 + x_3 \epsilon^m,
\end{aligned}$$

where x_1, x_2, x_3 are distinct roots of the equation $x^3 + 1 = 0$ and

$$m = \frac{1}{3}\left(r_{\lambda_3+\lambda_2+\lambda_1-2} + r_{\lambda_3+\lambda_2+\lambda_1-1} + r_{\lambda_3+\lambda_2+\lambda_1}\right).$$

The point $\boldsymbol{T_r}(\epsilon, s) = (z_1, \dots, z_{\lambda_3+\lambda_2+\lambda_1}, t_1^{(1)}, \dots, t_{\lambda_3+\lambda_2}^{(1)}, t_1^{(2)}, \dots, t_{\lambda_3}^{(2)})$ is a point of root coordinates of $\boldsymbol{X_r}(\epsilon, s)$.

Let us call the root coordinates $t^{(2)}_{\lambda_3}$, $t^{(1)}_{\lambda_3+\lambda_2-1}$, $t^{(1)}_{\lambda_3+\lambda_2}$ exceptional, and the remaining root coordinates regular. For each regular root coordinate y the leading term of asymptotics of $y - r_0$ as $\epsilon \to 0$ has the form $A\epsilon^B$ for suitable numbers $A \neq 0$, B.

Lemma 9.12. *The pairs (A, B) are different for different regular root coordinates.*

Proof. A proof is by inspection of the list. Q.E.D.

For each exceptional coordinate y the absolute value of the difference $y - r_0$ is much smaller as $\epsilon \to 0$ than for any regular coordinate.

The Bethe vector has the form $\omega(\boldsymbol{T_r}(\epsilon, s)) = \sum_J \omega_J(\boldsymbol{T_r}(\epsilon, s))\, e_J v$, where the sum is over all admissible J, see Section 6. An admissible $J = (j_1, \dots, j_{\lambda_3+\lambda_2+\lambda_1})$ consists of ones, twos and threes with exactly λ_3 threes and λ_2 twos. Choose J with $j_i = 3$ for $i = 1, 2, \dots, \lambda_3 - 1$, $\lambda_3 + \lambda_2 - 2$ and $j_i = 2$ for $i = \lambda_3, \lambda_3 + 1, \dots, \lambda_3 + \lambda_2 - 3, \lambda_3 + \lambda_2 - 1$, $\lambda_3 + \lambda_2$. Then $\omega_J(\boldsymbol{T_r}(\epsilon, s))$ is given by the formula

$$\omega_J(\boldsymbol{T_r}(\epsilon, s)) = \sum_{\sigma \in S_{\lambda_3+\lambda_2}} \sum_{\tau \in S_{\lambda_3}} \prod_{i=1}^{\lambda_3-1} \frac{1}{(t^{(2)}_{\tau(i)} - t^{(1)}_{\sigma(i)})(t^{(1)}_{\sigma(i)} - z_i)} \times \tag{9.6}$$
$$\times \prod_{i=\lambda_3+1}^{\lambda_3+\lambda_2-2} \frac{1}{t^{(1)}_{\sigma(i)} - z_{i-1}} \times$$
$$\times \frac{1}{(t^{(2)}_{\tau(\lambda_3)} - t^{(1)}_{\sigma(\lambda_3)})(t^{(1)}_{\sigma(\lambda_3)} - z_{\lambda_3+\lambda_2-2})} \prod_{i=\lambda_3+\lambda_2-1}^{\lambda_3+\lambda_2} \frac{1}{t^{(1)}_{\sigma(i)} - z_i}.$$

Lemma 9.13. *For small ϵ, the function $\omega_J(\boldsymbol{T_r}(\epsilon, s))$ has well-defined limit as $s \to 0$.*

Proof. Divergent summands in (9.6) are the summands with factors $(t^{(2)}_{\lambda_3} - t^{(1)}_{\lambda_3+\lambda_2-1})^{-1}$ or $(t^{(2)}_{\lambda_3} - t^{(1)}_{\lambda_3+\lambda_2})^{-1}$. The divergent summands come in pairs. There are two types of such pairs. The first type has the form

$$p_1^{Ckij} = \frac{C}{(t^{(2)}_{\lambda_3} - t^{(1)}_{\lambda_3+\lambda_2-1})(t^{(1)}_{\lambda_3+\lambda_2-1} - z_i)(t^{(2)}_k - t^{(1)}_{\lambda_3+\lambda_2})(t^{(1)}_{\lambda_3+\lambda_2} - z_j)},$$
$$p_2^{Ckij} = \frac{C}{(t^{(2)}_{\lambda_3} - t^{(1)}_{\lambda_3+\lambda_2})(t^{(1)}_{\lambda_3+\lambda_2} - z_i)(t^{(2)}_k - t^{(1)}_{\lambda_3+\lambda_2-1})(t^{(1)}_{\lambda_3+\lambda_2-1} - z_j)},$$

where C is a common factor, regular as $s \to 0$. The second type has the form

$$q_1^{Cij} = \frac{C}{(t^{(2)}_{\lambda_3} - t^{(1)}_{\lambda_3+\lambda_2-1})(t^{(1)}_{\lambda_3+\lambda_2-1} - z_i)(t^{(1)}_{\lambda_3+\lambda_2} - z_j)}\,,$$

$$q_2^{Cij} = \frac{C}{(t^{(2)}_{\lambda_3} - t^{(1)}_{\lambda_3+\lambda_2})(t^{(1)}_{\lambda_3+\lambda_2} - z_i)(t^{(1)}_{\lambda_3+\lambda_2-1} - z_j)}\,,$$

where C is a common factor, regular as $s \to 0$. Each pair has well-defined limit as $s \to 0$,

$$\lim_{s\to 0}\,(p_1^{Ckij} + p_2^{Ckij}) =$$

$$= \lim_{s\to 0}\,\frac{C}{(t^{(2)}_k - r_0)(z_i - r_0)(z_j - r_0)}\left(\frac{2}{t^{(2)}_k - r_0} + \frac{2}{z_j - r_0} - \frac{2}{z_i - r_0}\right),$$

$$\lim_{s\to 0}\,(q_1^{Cij} + q_2^{Cij}) =$$

$$= \lim_{s\to 0}\,\frac{C}{(z_i - r_0)(z_j - r_0)}\left(\frac{2}{z_j - r_0} - \frac{2}{z_i - r_0}\right).$$

These limits will be called resonant pairs. Q.E.D.

It is easy to see that $\bar\omega_J(\epsilon) = \lim_{s\to 0}\omega_J(\boldsymbol{T}_{\boldsymbol{r}}(\epsilon, s))$ is an acceptable function and its order equals

$$b = -B(z_{\lambda_3+\lambda_2} - r_0) - \sum_{i=1}^{\lambda_3-1} B(t^{(2)}_i - r_0) - \sum_{i=1}^{\lambda_3+\lambda_2} B(z_i - r_0)\,.$$

Indeed, the order of the limit of any convergent summand in (9.6) is greater than b. The order of any resonant pair is not less than b. There is exactly one resonant pair of order b. That pair is of the second type and corresponds to

$$q_1 = \prod_{i=1}^{\lambda_3-1}\frac{1}{(t^{(2)}_i - t^{(1)}_i)(t^{(1)}_i - z_i)}\prod_{i=\lambda_3}^{\lambda_3+\lambda_2-3}\frac{1}{t^{(1)}_i - z_i}\times$$

$$\times\,\frac{1}{(0 - s)(s - z_{\lambda_3+\lambda_2-2})(t^{(1)}_{\lambda_3+\lambda_2-2} - z_{\lambda_3+\lambda_2-1})(-s - z_{\lambda_3+\lambda_2})}\,,$$

$$q_2 = \prod_{i=1}^{\lambda_3-1} \frac{1}{(t_i^{(2)} - t_i^{(1)})(t_i^{(1)} - z_i)} \prod_{i=\lambda_3}^{\lambda_3+\lambda_2-3} \frac{1}{t_i^{(1)} - z_i} \times$$

$$\times \frac{1}{(0+s)(-s - z_{\lambda_3+\lambda_2-2})(t_{\lambda_3+\lambda_2-2}^{(1)} - z_{\lambda_3+\lambda_2-1})(s - z_{\lambda_3+\lambda_2})} .$$

Thus, $\bar{\omega}_J(\epsilon)$ is nonzero for small ϵ.

9.4.5. *Proof for $N = 3$ and $\mathcal{W}_{(1,0,2)}$.* We study the problem $t_{\lambda_2+\lambda_3}^{(1)} = t_{\lambda_3-1}^{(2)} = t_{\lambda_3}^{(2)}$ of type (9.3) (after relabeling the root coordinates).

For any numbers $\boldsymbol{r} = (r_0, r_1, r_2, \ldots, r_{\lambda_3+\lambda_2+\lambda_1})$, such that $r_0 \subset \mathbb{C}$, $r_i \in \mathbb{R}$ for $i > 0$, $0 < r_1 < r_2 < \cdots < r_{\lambda_3+\lambda_2+\lambda_1}$, we choose $X_{\boldsymbol{r}}(\epsilon, s) \in \mathcal{W}$ to be the three-dimensional space of polynomials spanned by

$$g_1(u) = (u - r_0)^{\lambda_1+2} + \sum_{i=1}^{\lambda_1+1} c_i (u - r_0)^i ,$$

$$g_2(u) = (u - r_0)^{\lambda_2+1} + \sum_{i=0}^{\lambda_2} b_i (u - r_0)^i ,$$

$$g_3(u) = (u - r_0)^{\lambda_3} + \sum_{i=2}^{\lambda_3-1} a_i (u - r_0)^i - a_2 s^2 ,$$

where

$$a_{\lambda_3-1} = \epsilon^{r_1}, \qquad a_{\lambda_3-i}/a_{\lambda_3-i+1} = \epsilon^{r_i}, \qquad i = 2, \ldots, \lambda_3 - 2 ,$$

$$b_{\lambda_2} = \epsilon^{r_{\lambda_3-1}}, \qquad b_{\lambda_2-i}/b_{\lambda_2-i+1} = \epsilon^{r_{\lambda_3+i-1}}, \qquad i = 1, \ldots, \lambda_2 ,$$

$$c_{\lambda_1+1} = \epsilon^{r_{\lambda_3+\lambda_2}}, \quad c_{\lambda_1-i}/c_{\lambda_1-i+1} = \epsilon^{r_{\lambda_3+\lambda_2+i+1}}, \qquad i = 0, \ldots, \lambda_1 - 1 .$$

We have $X_{\boldsymbol{r}}(\epsilon, 0) \in \mathcal{W}_{(1,0,2)}$. Clearly, the dependence of $X_{\boldsymbol{r}}(\epsilon, s)$ on $\boldsymbol{r}$ is such that the corresponding curve $X_{\boldsymbol{r}}(\epsilon, 0)$ is generic in $\mathcal{W}_{\boldsymbol{d}}$ in the sense defined in Section 9.4.1.

We consider the same asymptotic zone $1 \gg |\epsilon| \gg |s| > 0$. The roots of g_3 are of the form:

$$t_1^{(2)} \sim r_0 - \epsilon^{r_1}, \; \ldots \; , \; t_{\lambda_3-2}^{(2)} \sim r_0 - \epsilon^{r_{\lambda_3-2}}, \; t_{\lambda_3-1}^{(2)} \sim r_0 + s , \; t_{\lambda_3}^{(1)} \sim r_0 - s .$$

We have

$$\begin{aligned}\mathrm{Wr}(g_2, g_3) &= (\lambda_2 + 1 - \lambda_3)(u - r_0)^{\lambda_3+\lambda_2} + \\ &+ \sum_{i=2}^{\lambda_3-1} (\lambda_2 + 1 - i)\, a_i (u - r_0)^{\lambda_3+i-1} + \\ &+ a_2 \sum_{i=0}^{\lambda_2} (i-2)\, b_i (u - r_0)^{i+1} + \dots\,.\end{aligned}$$

The roots of $\mathrm{Wr}(g_2, g_3)$ are of the form

$$t_1^{(1)} \sim r_0 - \frac{\lambda_2 - \lambda_3 + 2}{\lambda_2 - \lambda_3 + 1}\, \epsilon^{r_1}, \ \dots\ ,\ t_{\lambda_3-2}^{(1)} \sim r_0 - \frac{\lambda_2 - 1}{\lambda_2 - 2}\, \epsilon^{r_{\lambda_3-2}},$$

$$t_{\lambda_3-1}^{(1)} \sim r_0 - \frac{\lambda_2 - 2}{\lambda_2 - 1}\, \epsilon^{r_{\lambda_3-1}}, \ \dots\ ,\ t_{\lambda_3+\lambda_2-4}^{(1)} \sim r_0 - \frac{1}{2}\, \epsilon^{r_{\lambda_3+\lambda_2-4}},$$

$$t_{\lambda_3+\lambda_2-3}^{(1)} \sim \epsilon^{(r_{\lambda_3+\lambda_2-3} + r_{\lambda_3+\lambda_2-2})/2},\ \ t_{\lambda_3+\lambda_2-2}^{(1)} \sim -\epsilon^{(r_{\lambda_3+\lambda_2-3} + r_{\lambda_3+\lambda_2-2})/2},$$

$$t_{\lambda_3+\lambda_2-1}^{(1)} \sim r_0 - 2\epsilon^{r_{\lambda_3+\lambda_2-1}},\ \ t_{\lambda_3+\lambda_2}^{(1)} \sim r_0\,.$$

We have

$$\begin{aligned}&\mathrm{Wr}\,(g_1, g_2, g_3) = \\ &+ (\lambda_1 + 1 - \lambda_2)(\lambda_1 + 2 - \lambda_3)(\lambda_2 + 1 - \lambda_3)(u - r_0)^{\lambda_3+\lambda_2+\lambda_1} + \\ &+ \sum_{i=2}^{\lambda_3-1} (\lambda_1 + 1 - \lambda_2)(\lambda_1 + 2 - i)(\lambda_2 + 1 - i)\, a_i (u - r_0)^{i+\lambda_2+\lambda_1} + \\ &+ a_2 \sum_{i=0}^{\lambda_2} (\lambda_1 + 2 - i)\, \lambda_1 (i-2)\, b_i (u - r_0)^{\lambda_1+i+1} - \\ &- a_2 b_0 \sum_{i=1}^{\lambda_1+1} 2i\,(i-2)\, c_i (u - r_0)^{i-1} + \dots\,.\end{aligned}$$

The roots of $\mathrm{Wr}(g_1, g_2, g_3)$ are of the form

$$z_1 \sim r_0 - \frac{(\lambda_1 + 3 - \lambda_3)(\lambda_2 + 2 - \lambda_3)}{(\lambda_1 + 2 - \lambda_3)(\lambda_2 + 1 - \lambda_3)}\, \epsilon^{r_1}, \ \dots\ ,$$

$$z_{\lambda_3-2} \sim r_0 - \frac{\lambda_1(\lambda_2 - 1)}{(\lambda_1 - 1)(\lambda_2 - 2)}\, \epsilon^{r_{\lambda_3-2}},$$

$$z_{\lambda_3-1} \sim r_0 - \frac{(\lambda_1+2-\lambda_2)(\lambda_2-2)}{(\lambda_1+1-\lambda_2)(\lambda_2-1)}\, \epsilon^{r_{\lambda_3+1}}, \;\ldots\;,$$

$$z_{\lambda_3+\lambda_2-4} \sim r_0 - \frac{\lambda_1-1}{2(\lambda_1-2)}\, \epsilon^{r_{\lambda_3+\lambda_2-4}},$$

$$z_{\lambda_3+\lambda_2-3} \sim r_0 + \sqrt{\frac{\lambda_1+1}{\lambda_1-1}}\, \epsilon^{(r_{\lambda_3+\lambda_2-3}+r_{\lambda_3+\lambda_2-2})/2},$$

$$z_{\lambda_3+\lambda_2-2} \sim r_0 - \sqrt{\frac{\lambda_1+1}{\lambda_1-1}}\, \epsilon^{(r_{\lambda_3+\lambda_2-3}+r_{\lambda_3+\lambda_2-2})/2},$$

$$z_{\lambda_3+\lambda_2-1} \sim r_0 - \frac{2(\lambda_1+2)}{\lambda_1+1}\, \epsilon^{r_{\lambda_3+\lambda_2-1}},$$

$$z_{\lambda_3+\lambda_2} \sim r_0 \quad \frac{(\lambda_1+1)(\lambda_1-1)}{(\lambda_1+2)\lambda_1}\, \epsilon^{r_{\lambda_3+\lambda_2}}, \;\ldots\;,$$

$$z_{\lambda_3+\lambda_2+\lambda_1-2} \sim r_0 - \frac{3}{8}\, \epsilon^{r_{\lambda_3+\lambda_2+\lambda_1-2}},$$

$$z_{\lambda_3+\lambda_2+\lambda_1-1} \sim r_0 + \frac{1}{\sqrt{3}}\, \epsilon^{(r_{\lambda_3+\lambda_2+\lambda_1-1}+r_{\lambda_3+\lambda_2+\lambda_1})/2},$$

$$z_{\lambda_3+\lambda_2+\lambda_1} \sim r_0 - \frac{1}{\sqrt{3}}\, \epsilon^{(r_{\lambda_3+\lambda_2+\lambda_1-1}+r_{\lambda_3+\lambda_2+\lambda_1})/2}.$$

The point $\boldsymbol{T_r}(\epsilon, s) = (z_1, \ldots, z_{\lambda_3+\lambda_2+\lambda_1}, t^{(1)}_1, \ldots, t^{(1)}_{\lambda_3+\lambda_2}, t^{(2)}_1, \ldots, t^{(2)}_{\lambda_3})$ is a point of root coordinates of $X_{\boldsymbol{r}}(\epsilon, s)$.

Let us call the root coordinates $t^{(2)}_{\lambda_3-1}, t^{(2)}_{\lambda_3}, t^{(1)}_{\lambda_3+\lambda_2}$ exceptional, and the remaining root coordinates regular. For each regular root coordinate y the leading term of asymptotics of $y - r_0$ as $\epsilon \to 0$ has the form $A\epsilon^B$ for suitable numbers $A \neq 0$, B.

Lemma 9.14. *The pairs (A, B) are different for different regular root coordinates.*

Proof. A proof is by inspection of the list. Q.E.D.

For each exceptional coordinate y the absolute value of the difference $y - r_0$ is much smaller as $\epsilon \to 0$ than for any regular coordinate.

The Bethe vector has the form $\omega(\boldsymbol{T_r}(\epsilon, s)) = \sum_J \omega_J(\boldsymbol{T_r}(\epsilon, s))\, e_J v$, where the sum is over all admissible J, see Section 6. An admissible $J = (j_1, \ldots, j_{\lambda_3+\lambda_2+\lambda_1})$ consists of ones, twos and threes with exactly λ_3 threes and λ_2 twos. Choose J with $j_i = 3$ for $i = 1, 2, \ldots, \lambda_3 - 3, \lambda_3 - 2, \lambda_3 + \lambda_2 - 1, \lambda_3 + \lambda_2$ and $j_i = 2$ for $i = \lambda_3 - 1, \lambda_3, \ldots, \lambda_3 + \lambda_2 - 2$. Then

$\omega_J(\boldsymbol{T_r}(\epsilon, s))$ is given by the formula

$$(9.7)\quad \omega_J(\boldsymbol{T_r}(\epsilon, s)) = \sum_{\sigma\in S_{\lambda_3+\lambda_2}} \sum_{\tau\in S_{\lambda_3}} \prod_{i=1}^{\lambda_3-2} \frac{1}{(t^{(2)}_{\tau(i)} - t^{(1)}_{\sigma(i)})(t^{(1)}_{\sigma(i)} - z_i)} \times$$

$$\times \prod_{i=\lambda_3-1}^{\lambda_3} \frac{1}{(t^{(2)}_{\tau(i)} - t^{(1)}_{\sigma(i)})(t^{(1)}_{\sigma(i)} - z_{\lambda_2+i})} \prod_{i=\lambda_3+1}^{\lambda_3+\lambda_2} \frac{1}{t^{(1)}_{\sigma(i)} - z_{i-2}}\,.$$

It is easy to see that $\bar{\omega}_J(\epsilon) = \lim_{s\to 0} \omega_J(\boldsymbol{T_r}(\epsilon, s))$ is an acceptable function and its order equals

$$b = -\sum_{i=1}^{\lambda_3-2} B(t^{(2)}_i - r_0) - \sum_{i=1}^{\lambda_3+\lambda_2} B(z_i - r_0) - 2B(t^{(1)}_{\lambda_3+\lambda_2-1} - r_0)\,.$$

Namely, consider the following four summands in (9.7):

$$q = \prod_{i=1}^{\lambda_3-2} \frac{1}{t^{(2)}_i - t^{(1)}_i} \prod_{i=1}^{\lambda_3+\lambda_2-4} \frac{1}{t^{(1)}_i - z_i} \prod_{\lambda_3+\lambda_2-1}^{\lambda_3+\lambda_2} \frac{1}{t^{(1)}_i - z_i} \times$$

$$\times \left(\frac{1}{(t^{(2)}_{\lambda_3-1} - t^{(1)}_{\lambda_3+\lambda_2-1})\,(t^{(2)}_{\lambda_3} - t^{(1)}_{\lambda_3+\lambda_2})} + \right.$$

$$\left. + \frac{1}{(t^{(2)}_{\lambda_3-1} - t^{(1)}_{\lambda_3+\lambda_2})\,(t^{(2)}_{\lambda_3} - t^{(1)}_{\lambda_3+\lambda_2-1})} \right) \times$$

$$\times \left(\frac{1}{(t^{(1)}_{\lambda_3+\lambda_2-3} - z_{\lambda_3+\lambda_2-3})\,(t^{(1)}_{\lambda_3+\lambda_2-2} - z_{\lambda_3+\lambda_2-2})} + \right.$$

$$\left. + \frac{1}{(t^{(1)}_{\lambda_3+\lambda_2-3} - z_{\lambda_3+\lambda_2-2})\,(t^{(1)}_{\lambda_3+\lambda_2-2} - z_{\lambda_3+\lambda_2-3})} \right).$$

Then the order of $\lim_{s\to 0} q$ equals b and the order of

$$\lim_{s\to 0} (\omega_J((\boldsymbol{T_r}(\epsilon, s)) - q))$$

is greater than b. Therefore, $\lim_{s\to 0} \omega_J(\boldsymbol{T_r}(\epsilon, s))$ is nonzero for small ϵ.

For $N = 3$ and every essential subset $\mathcal{W}_{\boldsymbol{d}}$, we proved that the Bethe vector is nonzero at generic points of $\mathcal{W}_{\boldsymbol{d}}$ and, hence, the number α of Corollary 9.2 is nonpositive. Thus, Theorem 7.1 is proved for $N = 3$.

For $N > 3$, Theorem 7.1 is proved similarly. Namely, let $\mathcal{W}_{\boldsymbol{d}}$, $\boldsymbol{d} = (d_1, \dots, d_N)$, be an arbitrary essential subset. A point $X \in \mathcal{W}$ is an N-dimensional space of polynomials. For any numbers $\boldsymbol{r} = (r_0, r_1, r_2, \dots)$, such that $r_0 \in \mathbb{C}$, $r_i \in \mathbb{R}$ for $i > 0$, $0 < r_1 < r_2 < \dots$, we choose $X_{\boldsymbol{r}}(\epsilon, s)$ to be the N-dimensional space of polynomials spanned by the polynomials $g_1, \dots, g_N$, where

$$g_j(u) = (u - r_0)^{\lambda_j + N - j} + \sum_{i=d_j}^{\lambda_j+N-j-1} a_{j,i}(u - r_0)^i ,$$

if $j = 1$ or $j > 1$ and $d_j - d_{j-1} \neq 2$, and

$$g_j(u) = (u - r_0)^{\lambda_j + N - j} + \sum_{i=d_j}^{\lambda_j+N-j-1} a_{j,i}(u - r_0)^i - s^2 a_{j,d_j}(u - r_0)^{d_j - 2} ,$$

if $j > 1$ and $d_j - d_{j-1} = 2$. We order the coefficients $a_{j,i}$ by defining $(j, i) > (j', i')$ if either $j > j'$, or $j = j'$ and $i > i'$. According to that order we define the coefficients $a_{j,i}$ by setting

$$a_{N,\lambda_N+N-2} = \epsilon^{r_1}, \quad a_{N,\lambda_N+N-3}/a_{N,\lambda_N+N-2} = \epsilon^{r_2}, \quad \dots \quad ,$$

see the cases $N = 2, 3$. We have $X_{\boldsymbol{r}}(\epsilon, 0) \in \mathcal{W}_{\boldsymbol{d}}$ and the point $X_{\boldsymbol{r}}(\epsilon, s)$ is nice for small nonzero s.

We consider the asymptotic zone $1 \gg |\epsilon| \gg |s| > 0$ and determine the asymptotics of the roots of the Wronskians $\mathrm{Wr}(g_N, g_{N-1}, \dots, g_i)$. That gives the asymptotics of coordinates of the corresponding critical point of the master function. Among the coordinates there will be exactly three exceptional coordinates with the same ϵ-asymptotics, see examples for $N = 2, 3$. Depending on the indices of the exceptional coordinates we choose a coordinate of the associated Bethe vector and show that the chosen coordinate has a well-defined limit as $s \to 0$. That statement, proved for any exceptional subset $\mathcal{W}_{\boldsymbol{d}}$, implies Theorem 7.1.

References

[1] V. I. Arnol'd, S. M. Gusein-Zade and A. N. Varchenko, Singularities of Differentiable Maps. Vol. I, The Classification of Critical Points, Caustics and Wave Fronts, Monogr. Math., **82**, Birkhäuser Boston, Inc., Boston, MA, 1985.

[2] H. M. Babujian, Off-shell Bethe ansatz equation and N-point correlators in the $SU(2)$, WZNW Theory, J. Phys. A, **26** (1993), 6981–6990.

[3] V. Chari and S. Loktev, Weyl, Fusion and Demazure modules for the current algebra of sl_{r+1}, Adv. Math., **207** (2006), 928–960.

[4] V. Chari and A. Pressley, Weyl modules for classical and quantum affine algebras, Represent. Theory, **5** (2001), 191–223 (electronic).

[5] A. Eremenko and A. Gabrielov, Degrees of real Wronski maps, Discrete Comput. Geom., **28** (2002), 331–347.

[6] B. Feigin, E. Frenkel and N. Reshetikhin, Gaudin model, Bethe ansatz and critical level, Comm. Math. Phys., **166** (1994), 27–62.

[7] Ph. Griffiths and J. Harris, Principles of Algebraic Geometry, Wiley, 1994.

[8] A. Matsuo, An application of Aomoto–Gelfand hypergeometric functions to the SU(n) Knizhnik–Zamolodchikov equation, Comm. Math. Phys., **134** (1990), 65–77.

[9] E. Mukhin, V. Tarasov and A. Varchenko, Bethe eigenvectors of higher transfer matrices, J. Stat. Mech., **2006**, no. 8, P08002, 1–44.

[10] E. Mukhin, V. Tarasov and A. Varchenko, Schubert calculus and representations of a general linear group, J. Amer. Math. Soc., **22** (2009), 909–940.

[11] E. Mukhin, V. Tarasov and A. Varchenko, On separation of variables and completeness of the Bethe ansatz for quantum $\mathfrak{gl}_N$ Gaudin model, Glasg. Math. J., **51** (2009), no. A, 137–145.

[12] E. Mukhin and A. Varchenko, Critical points of master functions and flag varieties, Commun. Contemp. Math., **6** (2004), 111–163.

[13] E. Mukhin and A. Varchenko, Norm of a Bethe vector and the Hessian of the master function, Compos. Math., **141** (2005), 1012–1028.

[14] N. Reshetikhin and A. Varchenko, Quasiclassical asymptotics of solutions to the KZ equations, In: Geometry, Topology and Physics for R. Bott, Intern. Press, 1995, pp. 293–322, arXiv:hep-th/9402126.

[15] R. Rimanyi, L. Stevens and A. Varchenko, Combinatorics of rational functions and Poincaré–Birkhoff–Witt expansions of the canonical $U(\mathfrak{n}_-)$-valued differential form, Ann. Comb., **9** (2005), 57–74.

[16] V. Schechtman and A. Varchenko, Arrangements of hyperplanes and Lie algebra homology, Invent. Math., **106** (1991), 139–194.

[17] I. Scherbak and A. Varchenko, Critical points of functions, $\mathfrak{sl}_2$-representations, and Fuchsian differential equations with only univalued solutions, Dedicated to Vladimir I. Arnold on the occasion of his 65th birthday, Mosc. Math. J., **3** (2003), no. 2, 621–645, 745.

[18] D. Talalaev, Quantization of the Gaudin system, preprint, arXiv:hep-th/0404153.

[19] A. Varchenko, Hypergeometric Functions and Representation Theory of Lie Algebras and Quantum Groups, Adv. Ser. Math. Phys., **21**, World Scientific, 1995.

[20] A. Varchenko, Bethe ansatz for arrangements of hyperplanes and the Gaudin model, Mosc. Math. J., **6** (2006), no. 1, 195–210, 223–224.

Eugene Mukhin
Department of Mathematical Sciences,
Indiana University – Purdue University Indianapolis,
402 North Blackford St.
Indianapolis IN 46202-3216, USA
E-mail address: mukhin@math.iupui.edu

Vitaly Tarasov
Department of Mathematical Sciences,
Indiana University – Purdue University Indianapolis,
402 North Blackford St.
Indianapolis IN 46202-3216, USA
and
St. Petersburg Branch of Steklov Mathematical Institute,
Fontanka 27, St. Petersburg 191023, Russia
E-mail address: vt@math.iupui.edu, vt@pdmi.ras.ru

Alexander Varchenko
Department of Mathematics,
University of North Carolina at Chapel Hill,
Chapel Hill NC 27599-3250, USA
E-mail address: anv@email.unc.edu

Advanced Studies in Pure Mathematics 62, 2012
Arrangements of Hyperplanes—Sapporo 2009
pp. 513–521

A note on Bockstein homomorphisms in local cohomology

Anurag K. Singh and Uli Walther

Abstract.

This note is an exposition on the results from [SiW] related to Lyubeznik's conjecture on local cohomology. We use the method of Bockstein homomorphisms, which to our knowledge has not been employed before in commutative algebra. As an example, we strengthen the classical Stanley–Reisner correspondence by relating Bockstein homomorphisms on local cohomology in monomial rings to the topological Bockstein homomorphisms of simplicial complexes.

We introduce the necessary language in Section 1, then present the motivation with our main theorem in Section 2, and finally illustrate our results in the case of Stanley–Reisner ideals in Section 3.

§1. Basic notions

Definition 1.1. Let R be a commutative Noetherian ring, and M an R-module. Let p be an element of R which is a non-zero divisor on M. Let $\mathrm{F}^\bullet$ be a covariant δ-functor on the category of R-modules. The exact sequence

$$0 \longrightarrow M \xrightarrow{\ p\ } M \longrightarrow M/pM \longrightarrow 0$$

then induces an exact sequence

$$\mathrm{F}^k(M/pM) \xrightarrow{\ \delta_p^k\ } \mathrm{F}^{k+1}(M) \xrightarrow{\ \mathrm{F}^{k+1}(p)\ } \mathrm{F}^{k+1}(M) \xrightarrow{\ \pi_p^{k+1}\ } \mathrm{F}^{k+1}(M/pM)\,.$$

Received April 16, 2010.

2010 *Mathematics Subject Classification.* Primary 13D45; Secondary 13F20, 13F55.

Key words and phrases. Bockstein, local cohomology, Lyubeznik conjecture, Stanley-Reisner.

A.K.S. was supported by NSF grants DMS 0600819 and DMS 0856044.

U.W. was supported by NSF grants DMS 0555319 and 0901123 and by NSA grant H98230-06-1-0012.

The composition

$$\pi_p^{k+1} \circ \delta_p^k \colon \mathrm{F}^k(M/pM) \to \mathrm{F}^{k+1}(M/pM)$$

is the *Bockstein homomorphism* $\beta_p^k = \beta_p^k(F(M))$.

An elementary check shows that β_p^k is the connecting homomorphism that results from applying $\mathrm{F}^\bullet$ to the short exact sequence

$$0 \longrightarrow M/pM \xrightarrow{\ p\ } M/p^2M \longrightarrow M/pM \longrightarrow 0.$$

Let $\mathfrak{a}$ be an ideal of R, generated by elements $f_1, \dots, f_t$. The covariant δ-functors of interest to us are local cohomology $H^\bullet_{\mathfrak{a}}(-)$, Koszul cohomology $H^\bullet(f_1, \dots, f_t; -)$, and $\mathrm{Ext}^\bullet_R(R/\mathfrak{a}, -)$ discussed briefly below.

Setting $\boldsymbol{f}^e = f_1^e, \dots, f_t^e$, the Koszul cohomology modules $H^\bullet(\boldsymbol{f}^e; M)$ are the cohomology modules of the Koszul complex $K^\bullet(\boldsymbol{f}^e; M)$. For each $e \geq 1$, one has a map of complexes

$$K^\bullet(\boldsymbol{f}^e; M) \to K^\bullet(\boldsymbol{f}^{e+1}; M)\,,$$

and thus a filtered directed system $\{K^\bullet(\boldsymbol{f}^e; M)\}_{e \geq 1}$. The direct limit of this system can be identified with the Čech complex $\check{C}^\bullet(\boldsymbol{f}; M)$ given by

$$0 \to R \to \bigoplus_i R_{f_i} \to \bigoplus_{i<j} R_{f_i f_j} \to \cdots \to R_{f_1 \cdots f_t} \to 0\,.$$

The local cohomology modules $H^\bullet_{\mathfrak{a}}(M)$ may be computed as the cohomology modules of $\check{C}^\bullet(\boldsymbol{f}; M)$, or equivalently as the direct limit of the Koszul cohomology modules $H^\bullet(\boldsymbol{f}^e; M)$: the filter condition yields an isomorphism of functors

$$H^k_{\mathfrak{a}}(-) \;\cong\; \varinjlim_e H^k(\boldsymbol{f}^e; -)\,,$$

and each element of $H^k_{\mathfrak{a}}(M)$ lifts to an element of $H^k(\boldsymbol{f}^e; M)$ for $e \gg 0$.

Remark 1.2. Let $\mathrm{F}^\bullet$ and $G^\bullet$ be covariant δ-functors on the category of R-modules, and let $\tau \colon \mathrm{F}^\bullet \to G^\bullet$ be a natural transformation. The Bockstein homomorphism is functorial: one has a commutative diagram

$$\begin{array}{ccc} \mathrm{F}^k(M/pM) & \longrightarrow & \mathrm{F}^{k+1}(M/pM) \\ {\scriptstyle\tau}\big\downarrow & & \big\downarrow{\scriptstyle\tau} \\ G^k(M/pM) & \longrightarrow & G^{k+1}(M/pM)\,, \end{array}$$

where the horizontal maps are the respective Bockstein homomorphisms. With $\mathfrak{a} = (f_1, \dots, f_k)$, the natural transformations of interest to us are

$$\mathrm{Ext}^\bullet(R/\mathfrak{a}^e, -) \to H^\bullet_{\mathfrak{a}}(-)\,,$$

$$H^\bullet(\boldsymbol{f}^e; -) \to H^\bullet(\boldsymbol{f}^{e+1}; -)\,,$$

and

$$H^\bullet(\boldsymbol{f}^e; -) \to H^\bullet_{\mathfrak{a}}(-)\,.$$

Let M be an R-module, let p be an element of R which is a non-zero divisor on M, and let $\boldsymbol{f} = f_1, \dots, f_t$ and $\boldsymbol{g} = g_1, \dots, g_t$ be elements of R such that $f_i \equiv g_i \mod p$ for each i. One then has isomorphisms

$$H^\bullet(\boldsymbol{f}; M/pM) \cong H^\bullet(\boldsymbol{g}; M/pM)\,.$$

Example 1.3. Let p be a non-zero divisor on R. Let x be an element of R. The Bockstein homomorphism on Koszul cohomology $H^\bullet(x; R/pR)$ is

$$\begin{aligned} (0 :_{R/pR} x) = H^0(x; R/pR) &\to H^1(x; R/pR) = R/(p,x)R \\ r \mod (p) &\mapsto rx/p \mod (p,x)\,. \end{aligned}$$

Let y be an element of R with $x \equiv y \mod p$. Comparing the Bockstein homomorphisms β, β' on $H^\bullet(x; R/pR)$ and $H^\bullet(y; R/pR)$ respectively, one sees that the diagram

$$\begin{array}{ccc} (0 :_{R/pR} x) & \xrightarrow{\beta} & R/(p,x)R \\ \| & & \| \\ (0 :_{R/pR} y) & \xrightarrow{\beta'} & R/(p,y)R \end{array}$$

commutes if and only if $(x - y)/p \in (p, y)R$.

A key point in the proof of Theorem 2.2 is Lemma 1.4, which states that upon passing to the direct limits $\varinjlim_e H^\bullet(\boldsymbol{f}^e; M/pM)$ and $\varinjlim_e H^\bullet(\boldsymbol{g}^e; M/pM)$, the phenomenon of Example 1.3 disappears: the Bockstein homomorphisms always commute with the induced isomorphisms

$$\varinjlim_e H^\bullet(\boldsymbol{f}^e; M/pM) \cong \varinjlim_e H^\bullet(\boldsymbol{g}^e; M/pM)\,.$$

Lemma 1.4 ([SiW]). *Let M be an R-module, and let p be an element of R which is M-regular. Suppose $\mathfrak{a}$ and $\mathfrak{b}$ are ideals of R with $\sqrt{(\mathfrak{a}+pR)} = \sqrt{(\mathfrak{b}+pR)}$. Then there exists a commutative diagram*

$$\begin{array}{ccccccc} \cdots & \longrightarrow & H^k_{\mathfrak{a}}(M/pM) & \longrightarrow & H^{k+1}_{\mathfrak{a}}(M/pM) & \longrightarrow & \cdots \\ & & \downarrow & & \downarrow & & \\ \cdots & \longrightarrow & H^k_{\mathfrak{b}}(M/pM) & \longrightarrow & H^{k+1}_{\mathfrak{b}}(M/pM) & \longrightarrow & \cdots \end{array}$$

where the horizontal maps are the respective Bockstein homomorphisms, and the vertical maps are isomorphisms.

§2. Motivation and main result

Huneke [H, Problem 4] asked whether local cohomology $H^\bullet_{\mathfrak{a}}(R)$ modules of Noetherian rings R have finitely many associated prime ideals.

Suppose p is a prime number such that the Bockstein homomorphism

$$\beta^k_p \colon H^k_{\mathfrak{a}}(R/pR) \to H^{k+1}_{\mathfrak{a}}(R/pR)$$

is nonzero. Clearly, $H^{k+1}_{\mathfrak{a}}(R)$ must then have an associated prime containing p. But different prime numbers cannot be in the same associated prime ideal, so Huneke's conjecture implies the vanishing of all but finitely many β^k_p with p a varying prime number.

Example 2.1. Consider the hypersurface

$$R = \mathbb{Z}[u, v, w, x, y, z]/(ux + vy + wz)$$

and ideal $\mathfrak{a} = (x, y, z)R$. A variation of the argument given in [Si2] shows that

$$\beta^2_p \colon H^2_{\mathfrak{a}}(R/pR) \to H^3_{\mathfrak{a}}(R/pR)$$

is nonzero for each prime integer p. Hence, $H^3_{\mathfrak{a}}(R)$ has infinitely many associated primes.

Lyubeznik conjectured in [L1, Remark 3.7] that for regular rings R, each local cohomology module $H^k_{\mathfrak{a}}(R)$ has finitely many associated prime ideals. This conjecture has been verified for regular rings of positive characteristic by Huneke and Sharp [HS], and for regular local rings of characteristic zero as well as unramified regular local rings of mixed characteristic by Lyubeznik [L1, L2]. Surprisingly, it remains unresolved for polynomial rings over $\mathbb{Z}$ where the issue of p-torsion appears to be central in studying Lyubeznik's conjecture. The following theorem provides supporting evidence for Lyubeznik's conjecture.

Theorem 2.2 ([SiW]). *Let $R = \mathbb{Z}[x_1, \ldots, x_n]$ be a polynomial ring in finitely many variables over the ring of integers. Let $\mathfrak{a} = (f_1, \ldots, f_t)$ be an ideal of R, and let p be a prime integer.*

If p is not a zero divisor on the Koszul cohomology module $H^{k+1}(\boldsymbol{f}; R)$, then the Bockstein homomorphism $\beta_p^k \colon H_{\mathfrak{a}}^k(R/pR) \to H_{\mathfrak{a}}^{k+1}(R/pR)$ is zero.

We outline the proof of Theorem 2.2. One investigates the endomorphism φ of R with $\varphi(x_i) = x_i^p$ for each i. Since

$$H^{k+1}(\boldsymbol{f}; R) \xrightarrow{\ p\ } H^{k+1}(\boldsymbol{f}; R)$$

is injective and φ is flat, it follows that

$$H^{k+1}(\varphi^e(\boldsymbol{f}); R) \xrightarrow{\ p\ } H^{k+1}(\varphi^e(\boldsymbol{f}); R)$$

is injective for each $e \geq 0$. Thus, the Bockstein map on Koszul cohomology

$$H^k(\varphi^e(\boldsymbol{f}); R/pR) \longrightarrow H^{k+1}(\varphi^e(\boldsymbol{f}); R/pR)$$

must be the zero map. While $\{\varphi^e(\boldsymbol{f})R\}$ and $\{\boldsymbol{f}^e R\}$ are not cofinal families of ideals, Lemma 1.4 allows to replace one by the other for the purpose of studying Bockstein homomorphisms on local cohomology.

Suppose $\eta \in H_{\mathfrak{a}}^k(R/pR)$. Then η has a lift in $H^k(\varphi^e(\boldsymbol{f}); R/pR)$ for large e, and the commutativity of the diagram

$$\begin{array}{ccc} H^k(\varphi^e(\boldsymbol{f}); R/pR) & \longrightarrow & H^{k+1}(\varphi^e(\boldsymbol{f}); R/pR) \\ \downarrow & & \downarrow \\ H^k_{(\varphi^e(\boldsymbol{f}))}(R/pR) & \longrightarrow & H^{k+1}_{(\varphi^e(\boldsymbol{f}))}(R/pR) \\ \downarrow & & \downarrow \\ H^k_{(\boldsymbol{f})}(R/pR) & \longrightarrow & H^{k+1}_{(\boldsymbol{f})}(R/pR), \end{array}$$

(each horizontal map is a Bockstein homomorphism) implies that η maps to zero in $H_{\mathfrak{a}}^{k+1}(R/pR)$.

§3. Stanley–Reisner theory

Definition 3.1. Let Δ_n be the n-simplex, viewed as a simplicial complex (i.e., with its collection of faces), and fix a subcomplex Δ. Then the associated *Stanley–Reisner ideal* is the ideal

$$\mathfrak{a}_\Delta = (\boldsymbol{x}^\sigma \mid \sigma \notin \Delta) \subseteq R$$

generated by the monomials $\boldsymbol{x}^\sigma = \prod_{i=0}^n x_i^{\sigma_i}$ for which σ is not a face of Δ. The quotient $S_\Delta = R/\mathfrak{a}_\Delta$ is the *Stanley–Reisner ring* of Δ.

For $\boldsymbol{a} \in \mathbb{Z}^{n+1}$ write $\boldsymbol{a}_+$ and $\boldsymbol{a}_-$ for the sets $\{i \mid a_i > 0\}$ and $\{i \mid a_i < 0\}$ respectively. Both can be viewed as subsimplices of Δ_n. Hochster's theorem, [BH, Section 5.3] states the following, with notation explained below:

$$\left(H^i_{\mathfrak{m}}(S_\Delta/pS_\Delta)\right)_{\boldsymbol{a}} = \tilde{H}^{i-|\boldsymbol{a}_-|-1}(\mathrm{link}_{\star_\Delta(\boldsymbol{a}_+)}(\boldsymbol{a}_-); \mathbb{Z}/p\mathbb{Z}).$$

Definition 3.2. Let Δ be a simplicial complex and suppose τ is a subset of its vertex set (but not necessarily a simplex). The *link*, $\mathrm{link}_\Delta(\tau)$, is the set

$$\mathrm{link}_\Delta(\tau) = \{\tau' \mid \tau \cap \tau' = \emptyset, \quad \tau \cup \tau' \in \Delta\},$$

while its *star*, $\star(\tau)$, is simply

$$\star_\Delta(\tau) = \{\tau' \mid \tau \cup \tau' \in \Delta\}.$$

The Bockstein construction respects the identifications of Hochster's result:

Theorem 3.3 ([SiW]). *Let $\Delta \subseteq \Delta_n$ be a simplicial complex on $n+1$ vertices, and let $S_\Delta = R/\mathfrak{a}_\Delta$ be its integral Stanley–Reisner ring. If p is a prime number then the following are equivalent:*

(1) *For some multi-degree $\boldsymbol{a}$, the topological Bockstein morphism on the singular reduced $(i - |\boldsymbol{a}_-| - 1)$-th $\mathbb{Z}/p\mathbb{Z}$-cohomology of $\mathrm{link}_{\star_\Delta(\boldsymbol{a}_+)}(\boldsymbol{a}_-)$ is nonzero.*

(2) *The Bockstein morphism $\beta_p^{n+1-i}(H_{\mathfrak{a}_\Delta}(R))$ on the local cohomology group $H^{n+1-i}_{\mathfrak{a}_\Delta}(R/pR)$ is nonzero.*

An ingredient for this is that Bocksteins commute with local duality:

Theorem 3.4 ([SiW]). *Let $(S, \mathfrak{m}, \mathbb{K})$ be a local Gorenstein ring of dimension at least one. Let $p \in S$ be a regular element on both ω_S and on the finitely generated S-module M.*

The Bockstein morphism on $\mathrm{Ext}_S^{\dim(S)-i}(M, \omega_S/p\omega_S)$ induced by

$$0 \to \omega_S \xrightarrow{p} \omega_S \to \omega_{S/pS} = \omega_S/p\omega_S \to 0$$

is Matlis dual to the Bockstein morphism on $H^i_{\mathfrak{m}}(M/pM)$ induced by

$$0 \to M \xrightarrow{p} M \to M/pM \to 0.$$

Example 3.5. Let Λ_m be the *m-fold dunce cap*, i.e., the quotient of the unit disk obtained by identifying each point on the boundary circle with its translates under rotation by $2\pi/m$; the 2-fold dunce cap Λ_2 is the real projective plane.

Suppose m is the product of distinct primes $p_1, \ldots, p_r$. It is readily computed that the Bockstein homomorphisms

$$\widetilde{H}^1(\Lambda_m; \mathbb{Z}/p_i) \to \widetilde{H}^2(\Lambda_m; \mathbb{Z}/p_i)$$

are nonzero. Let Δ be the simplicial complex corresponding to a triangulation of Λ_m, and let $\mathfrak{a}$ in $R = \mathbb{Z}[x_1, \ldots, x_n]$ be the corresponding Stanley–Reisner ideal. The theorem then implies that the Bockstein homomorphisms

$$H^{n-3}_{\mathfrak{a}}(R/p_iR) \to H^{n-2}_{\mathfrak{a}}(R/p_iR)$$

are nonzero for each p_i. It follows that the local cohomology module $H^{n-2}_{\mathfrak{a}}(R)$ has a p_i-torsion element for each $i = 1, \ldots, r$.

If $m = 2$ (with standard triangulation shown below), Δ has singular cohomology

$$H^i(\Delta; \mathbb{Z}) = \begin{cases} \mathbb{Z} & \text{if } i = 0; \\ \mathbb{Z}/2\mathbb{Z} & \text{if } i = 1, 2; \\ 0 & \text{else.} \end{cases}$$

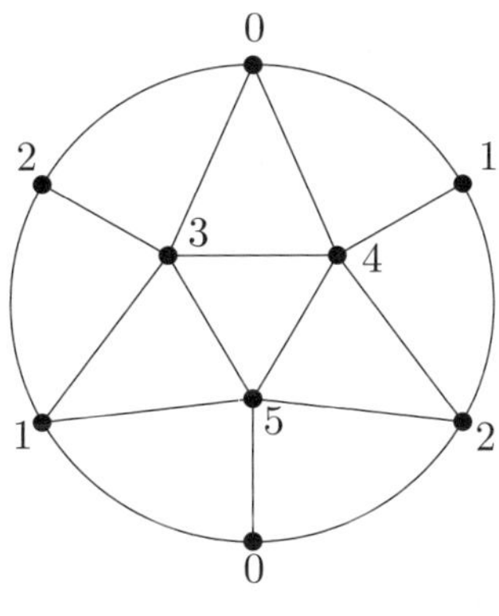

Fig. 1. The minimal triangulation of the real projective plane

The associated Stanley–Reisner ideal is

$$\begin{aligned} \mathfrak{a}_{\mathbb{RP}^2} &= (x_1x_2x_5,\ x_1x_2x_0,\ x_1x_3x_4,\ x_1x_3x_0,\ x_1x_4x_5, \\ &\quad x_2x_3x_4,\ x_2x_3x_5,\ x_2x_4x_0,\ x_3x_5x_0,\ x_4x_5x_0) \\ &\quad \subseteq \mathbb{Z}[x_0,\dots,x_5]; \end{aligned} \tag{1}$$

the minimal free resolution is of the form

$$R \to R^7 \to R^{15} \to R^{10} \to R$$

where the top degree map is the transpose of the the peculiar matrix

$$R^7 \xrightarrow{(x_1,x_2,x_3,x_4,x_5,x_6,2)} R.$$

One may show that the last entry in this map carries topological meaning and is "responsible" for β_p^2 being nonzero on $H^3_{\mathfrak{a}}(R/pR)$.

Remark 3.6. For a δ-functor F, vanishing of all p-Bocksteins on M is certain if $\mathrm{F}^i(M)$ is torsionfree for all i. However, it is possible that (for fixed p) the p-Bockstein homomorphisms are all zero and yet $\mathrm{F}^i(M)$ has p-torsion. For example, the 4-fold dunce cap Λ_4 has non-vanishing p-Bocksteins ($p \in \mathbb{N}_+$) if and only if $p = 4u$, u odd. (In this case, $H^1(\Lambda_4;\mathbb{Z}) = \mathbb{Z}/4\mathbb{Z}$ while $H^2(\Lambda_4;\mathbb{Z}) = 0$. So $H^1(\Lambda_4,\mathbb{Z}/p\mathbb{Z}) = H^2(\Lambda_4;\mathbb{Z}/p\mathbb{Z}) = \mathbb{Z}/\gcd(4,p)\mathbb{Z}$. The p-Bockstein homomorphism linking these two, when interpreted as endomorphism of $\mathbb{Z}/\gcd(4,p)\mathbb{Z}$ is multiplication by $4/\gcd(4,p)$.)

In general, there is a spectral sequence relating Bockstein homomorphisms to varying powers of p. It should be noted that it is possible that the Bockstein homomorphisms to all powers of p vanish and yet $\mathrm{F}(M)$ has p-torsion. For example, $R = \mathbb{Z}$, p a prime number, $\mathrm{F}^i = H^i_p(-)$ has for $M = R$ only one nonzero $\mathrm{F}^i(M/pM)$, namely when $i = 0$. So nonzero Bocksteins cannot exist but clearly $\mathrm{F}^1(M)$ is nonzero and p-torsion.

References

[BH] W. Bruns and J. Herzog, Cohen–Macaulay rings, revised ed., Cambridge Stud. Adv. Math., **39**, Cambridge Univ. Press, Cambridge, 1998.

[H] C. Huneke, Problems on local cohomology, In: Free Resolutions in Commutative Algebra and Algebraic Geometry, Sundance, Utah, 1990, Res. Notes Math., **2**, Jones and Bartlett, Boston, MA, 1992, pp. 93–108.

[HS] C. Huneke and R. Sharp, Bass numbers of local cohomology modules, Trans. Amer. Math. Soc., **339** (1993), 765–779.

[L1] G. Lyubeznik, Finiteness properties of local cohomology modules (an application of D-modules to commutative algebra), Invent. Math., **113** (1993), 41–55.

[L2] G. Lyubeznik, Finiteness properties of local cohomology modules for regular local rings of mixed characteristic: the unramified case, Comm. Alg., **28** (2000), 5867–5882.

[Si2] A. K. Singh, p-torsion elements in local cohomology modules, Math. Res. Lett., **7** (2000), 165–176.

[SiW] A. K. Singh and U. Walther, Bockstein homomorphisms in local cohomology, J. Reine Angew. Math., **655** (2011), 147–164.

[24h] S. B. Iyengar, G. J. Leuschke, A. Leykin, C. Miller, E. Miller, A. K. Singh and U. Walther, Twenty-Four Hours of Local Cohomology, Grad. Stud. Math., **87**, Amer. Math. Soc., Providence, RI, 2007, xviii+282 pp.

Anurag K. Singh
Department of Mathematics, University of Utah,
155 South 1400 East, Salt Lake City, UT 84112, USA
E-mail address: `singh@math.utah.edu`

Uli Walther
Department of Mathematics, Purdue University,
150 N. University Street, West Lafayette, IN 47907, USA
E-mail address: `walther@math.purdue.edu`

Advanced Studies in Pure Mathematics 62, 2012
Arrangements of Hyperplanes—Sapporo 2009
pp. 523–552

Arrangements, multiderivations, and adjoint quotient map of type ADE

Masahiko Yoshinaga

Abstract.

The first part of this paper is a survey on algebro-geometric aspects of sheaves of logarithmic vector fields of hyperplane arrangements. In the second part we prove that the relative de Rham cohomology (of degree two) of ADE-type adjoint quotient map is naturally isomorphic to the module of certain multiderivations. The isomorphism is obtained by the Gauss–Manin derivative of the Kostant–Kirillov form.

§1. Introduction

We begin with an example to illustrate how the structure of the module of logarithmic vector fields $D(\mathcal{A})$ is related to combinatorial problems of a hyperplane arrangement $\mathcal{A}$. Let $\mathcal{A}$ be a collection $\{H_{ij} \mid 1 \leq i < j \leq n\}$ of hyperplanes $H_{ij} = \{(x_1, \dots, x_n) \in \mathbb{K}^n \mid x_i - x_j = 0\} \subset \mathbb{K}^n$, where $\mathbb{K}$ is a fixed field. According to the field $\mathbb{K}$, several enumerative problems appear for the complement $M(\mathcal{A}) = \mathbb{K}^n \setminus \bigcup H_{ij}$.

(i) If $\mathbb{K} = \mathbb{F}_q$ is a finite filed, then the complement $M(\mathcal{A})$ is a finite set, of cardinality $|M(\mathcal{A})| = q(q-1)(q-2)\dots(q-n+1)$.

(ii) If $\mathbb{K} = \mathbb{R}$ is the real numbers, then each connected component of $M(\mathcal{A})$ (the chamber) is expressed by the inequality $x_{i_1} < x_{i_2} < \cdots < x_{i_n}$, where $(i_1, \dots, i_n)$ is a permutation of $(1, \dots, n)$. There are $n!$ chambers.

(iii) If $\mathbb{K} = \mathbb{C}$ is the complex numbers, then $M(\mathcal{A})$ is an affine complex variety of $\dim_{\mathbb{C}} = n$. Using the fibration $(x_1, \dots, x_n) \longmapsto (x_1, \dots, x_{n-1})$ and the Leray–Hirsch theorem, the Poincaré

Received April 30, 2010.
Revised October 25, 2010.
2010 *Mathematics Subject Classification.* Primary 32S22; Secondary 05E14, 20F55.
Key words and phrases. Arrangements, logarithmic vector fields, adjoint quotient.

polynomial is computed as $\sum_i b_i(M(\mathcal{A}))t^i = (1+t)(1+2t)\dots(1+(n-1)t)$.

The formulas in (i)–(iii) are similar in appearance. The general theory of arrangements [23] tells us that these invariants are combinatorial. Namely, they are determined from the poset $L(\mathcal{A})$ of subspaces obtained as intersections. Computations of the characteristic polynomial $\chi(\mathcal{A},t) \in \mathbb{Z}[t]$ unify these enumerative problems.

We also consider derivations

$$\delta_p = \sum_{i=1}^{n} x_i^p \partial_i,$$

($\partial_i = \frac{\partial}{\partial x_i}$) with $p = 0, 1, \dots, n-1$. These satisfy

$$\delta_p(x_i - x_j) = x_i^p - x_j^p \in (x_i - x_j), \tag{1}$$

for all i, j and the determinant of coefficients

$$\det \begin{pmatrix} 1 & x_1 & x_1^2 & \dots & x_1^{n-1} \\ 1 & x_2 & x_2^2 & \dots & x_2^{n-1} \\ \vdots & \vdots & \vdots & \ddots & \vdots \\ 1 & x_n & x_n^2 & \dots & x_n^{n-1} \end{pmatrix} = \prod_{1 \le i < j \le n} (x_j - x_i) \tag{2}$$

is the product of defining equations. The properties (1) and (2) guarantee that the module

$$D(\mathcal{A}) = \{\delta \in \mathrm{Der}_S \mid \delta(x_i - x_j) \in (x_i - x_j), \forall i, j\}$$

is a free module over $S = \mathbb{K}[x_1, \dots, x_n]$ with basis $\delta_0, \dots, \delta_{n-1}$ (this is Saito's criterion [25]). Remarkably, the decomposition of $D(\mathcal{A})$ into a direct sum of rank one free modules implies the product formulas (i)–(iii) above (Terao's factorization theorem [38]). More generally, the algebraic structure of $D(\mathcal{A})$ determines the characteristic polynomial $\chi(\mathcal{A}, t)$ by Solomon–Terao's formula [35] (see also §2.2 below).

The graded S-module $D(\mathcal{A})$ can also be considered as a coherent sheaf $\widetilde{D}(\mathcal{A})$ on projective space $\mathbb{P}^{n-1}$. This fact enables us to employ algebro-geometric methods to study $\mathcal{A}$. The structures of these sheaves contain information on $\mathcal{A}$.

The purpose of this paper is to survey algebro-geometric aspects of $D(\mathcal{A})$ and give some related results. The paper is organized as follows. In §2, we start with recalling basic notions on logarithmic vector fields for a Cartier divisor. We also introduce the module $D(\mathcal{A}, \mathbf{m})$ of logarithmic vector fields for an arrangement with multiplicity (multiarrangements) in §2.2. In general, the logarithmic vector fields for multiarrangements are much more difficult to analyze than simple arrangements. However, freeness of rank ℓ simple arrangements is closely related to freeness of rank $\ell-1$ multiarrangements. We will describe freeness criteria for these objects in §2.3–§2.4. In §2.5, we give a new necessary condition for a 3-dimensional arrangement to be free, in terms of plane curves. In §3, we will review results on freeness of Coxeter multiarrangements. Coxeter arrangements are the best understood class of multiarrangements. In §4, we will give two applications of freeness of Coxeter multiarrangements. The first concerns the adjoint quotient map $\chi : \mathfrak{g} \to \mathfrak{g}//\operatorname{ad}(G)$ of a simple Lie algebra $\mathfrak{g}$ of ADE-type. To describe the relative de Rham cohomology of χ, the module $D(\mathcal{A}, \mathbf{m})$ naturally appears. In the second application, we will give another proof for the freeness of A_n-Catalan arrangements, which was first proved by Edelman and Reiner [15].

Acknowledgements. The author deeply thanks Professor Kyoji Saito. Parts of this article (especially §4.1) were done under his supervision. This work was supported by JSPS Grant-in-Aid for Young Scientists (B) 20740038.

§2. Algebraic geometry of logarithmic vector fields

2.1. Sheaf of logarithmic vector fields

Let X be a smooth complex variety and $D \subset X$ a Cartier divisor. Let $U \subset X$ be an open subset of X. Suppose that there exists $h \in \Gamma(U, \mathcal{O}_X)$ such that $U \cap D = \{h = 0\}$. Let $\delta \in \Gamma(U, \mathcal{T}_X)$ be a section of the tangent sheaf on an open subset $U \subset X$ (i.e., a holomorphic vector field on X). The section δ is said to be *logarithmic tangent to* D if $\delta h \in h \cdot \mathcal{O}_U$. The sheaf of vector fields logarithmic tangent to D is denoted by $\mathcal{T}_X(-\log D)$. The sheaves of logarithmic forms are also similarly defined as

$$\Omega_X^p(\log D) = \{\omega \in \frac{1}{h}\Omega_X^p \mid \omega \wedge dh \text{ is holomorphic}\}.$$

They were introduced by K. Saito in [25]. He proved that they are reflexive sheaves and if $\mathcal{T}_X(-\log D)$ (or $\Omega_X^1(\log D)$) is locally free then $\Omega_X^p(\log D) = \bigwedge^p \Omega_X^1(\log D)$. We also note that if $\dim X = 2$, $\mathcal{T}_X(-\log D)$ is a locally free sheaf.

Example 2.1. *Let* $X = \mathbb{P}^2_{\mathbb{C}} = \operatorname{Proj}\mathbb{C}[z_0, z_1, z_2]$. *Using the Euler sequence ([21])*

$$0 \longrightarrow \mathcal{O}_{\mathbb{P}^2} \longrightarrow \mathcal{O}_{\mathbb{P}^2}(1)^3 \longrightarrow \mathcal{T}_{\mathbb{P}^2} \longrightarrow 0,$$

we have the following.

(1) *If* $D_0 = \{z_0 = 0\} \subset \mathbb{P}^2$. *Then* $\mathcal{T}_{\mathbb{P}^2}(-\log D_0) \cong \mathcal{O}_{\mathbb{P}^2}(1)^2$.
(2) *If* $D_1 = \{z_0 z_1 = 0\} \subset \mathbb{P}^2$. *Then* $\mathcal{T}_{\mathbb{P}^2}(-\log D_1) \cong \mathcal{O}_{\mathbb{P}^2}(1) \oplus \mathcal{O}_{\mathbb{P}^2}$.
(3) *If* $D_2 = \{z_0 z_1 z_2 = 0\} \subset \mathbb{P}^2$. *Then* $\mathcal{T}_{\mathbb{P}^2}(-\log D_2) \cong \mathcal{O}_{\mathbb{P}^2}^2$.
(4) *If* $D_3 = \{z_0^2 + z_1^2 + z_2^2 = 0\} \subset \mathbb{P}^2$. *Then* $\mathcal{T}_{\mathbb{P}^2}(-\log D_3) \cong \mathcal{T}_{\mathbb{P}^2}(-1)$. *(Sketch:* $\bigoplus_d \Gamma(\mathcal{T}_{\mathbb{P}^2}(-\log D_3)(d))$ *is generated by* $\delta_0 = z_2\partial_1 - z_1\partial_2, \delta_1 = -z_2\partial_0 + z_0\partial_2$ *and* $\delta_2 = z_1\partial_0 - z_0\partial_1$ *with a relation* $z_0\delta_0 + z_1\delta_1 + z_2\delta_2 = 0$ *this induces a resolution which is isomorphic to shifted Euler sequence.)*

The examples above, $\mathcal{T}_X(-\log D)$ is always a uniform sheaf. However for "generic" divisors of higher degrees, we obtain "generic" sheaves. We can sometimes recover the original divisor D from the sheaf $\mathcal{T}_{\mathbb{P}^n}(-\log D)$. (Dolgachev and Kapranov called this type of result a "Torelli-type" theorem in [14].) Let us recall two results in this direction. First one is due to Dolgachev and Kapranov, concerning the case of a union of generic hyperplanes.

Theorem 2.2. [14] *Let* $m \geq 2n+3$ *and* $\mathcal{A}_i = \{H_{i1}, H_{i2}, \dots, H_{im}\}$ $(i = 1, 2)$ *be arrangements of generic* m *hyperplanes* $H_{ik} \subset \mathbb{P}^n_{\mathbb{C}}$ *in* n*-dimensional projective space. We denote the union by* $\cup\mathcal{A} = \bigcup_{H\in\mathcal{A}} H$. *If*

$$\mathcal{T}_{\mathbb{P}^n}(-\log(\cup\mathcal{A}_1)) \cong \mathcal{T}_{\mathbb{P}^n}(-\log(\cup\mathcal{A}_2)),$$

then $\cup\mathcal{A}_1 = \cup\mathcal{A}_2$.

For smooth divisors $D \subset \mathbb{P}^n$ defined by a homogeneous polynomial $\{f(z_0, \dots, z_n) = 0\}$, Torelli-type results are related to the following property.

Definition 2.3. *The homogeneous polynomial* $f(z_0, \dots, z_n)$ *is said to be* Thom–Sebastiani type *if there exists a linear transformation* $A : \mathbb{C}^{n+1} \to \mathbb{C}^{n+1}$ *such that* $f(A(z_0, \dots, z_n)) = g(z_0, \dots, z_k) + h(z_{k+1}, \dots, z_n)$ *for some* $0 \leq k \leq n-1$.

Theorem 2.4. [42, 43]

(i) *Let* $D_1, D_2 \subset \mathbb{P}^n$ *be degree* d *smooth divisors which are not Thom–Sebastiani type. Then* $\mathcal{T}_{\mathbb{P}^n}(-\log D_1) \cong \mathcal{T}_{\mathbb{P}^n}(-\log D_2)$ *if and only if* $D_1 = D_2$.
(ii) *Let* $D_1, D_2 \subset \mathbb{P}^2$ *be smooth cubic curves with non-zero* j*-invariant* $j(D_i) \neq 0$. *Then* $\mathcal{T}_{\mathbb{P}^n}(-\log D_1) \cong \mathcal{T}_{\mathbb{P}^n}(-\log D_2)$ *if and only if* $D_1 = D_2$.

2.2. Log vector fields for multiarrangements

The main theme of this paper is logarithmic vector fields for arrangements of hyperplanes. Freeness is one of the important properties for arrangements. Ziegler [55] showed that a free arrangement of rank ℓ induces several free multiarrangements of rank $\ell - 1$ (see §2.4). This means that freeness of multiarrangements will be necessary for that of simple arrangements. Recently, several results on free simple arrangements have been generalized to free multiarrangements.

Let $V = \mathbb{C}^\ell$ be a complex vector space with coordinate $(x_1, \cdots, x_\ell)$, $\mathcal{A} = \{H_1, \ldots, H_n\}$ be an arrangement of hyperplanes. Let us denote by $S = \mathbb{C}[x_1, \ldots, x_\ell]$ the polynomial ring and fix $\alpha_i \in V^*$ a defining equation of H_i, i.e., $H_i = \alpha_i^{-1}(0)$. We also put $Q(\mathcal{A}, \mathbf{m}) = \prod_{i=1}^n \alpha_i^{\mathbf{m}(H_i)}$ and $|\mathbf{m}| = \sum_i \mathbf{m}(H_i)$

Definition 2.5. *A multiarrangement is a pair $(\mathcal{A}, \mathbf{m})$ of an arrangement $\mathcal{A}$ with a map $\mathbf{m} : \mathcal{A} \to \mathbb{Z}_{\geq 0}$, called the multiplicity.*

An arrangement $\mathcal{A}$ can be identified with a multiarrangement with constant multiplicity $m \equiv 1$, which is sometimes called a *simple arrangement.* With this notation, the main object is the following module of S-derivations which has contact to each hyperplane of order m.

Definition 2.6. *Let $(\mathcal{A}, \mathbf{m})$ be a multiarrangement, and define*

$$D(\mathcal{A}, \mathbf{m}) = \{\delta \in \mathrm{Der}_S \,|\delta\alpha_i \in (\alpha_i)^{\mathbf{m}(H_i)}, \forall i\},$$

and

$$\Omega^p(\mathcal{A}, \mathbf{m}) = \left\{ \omega \in \frac{1}{Q}\Omega^p_V \,\middle|\, d\alpha_i \wedge \omega \text{ does not have pole along } H_i, \forall i \right\},$$

The module $D(\mathcal{A}, \mathbf{m})$ is obviously a graded S-module. A multiarrangement $(\mathcal{A}, \mathbf{m})$ is said to be *free* with exponents $(e_1, \ldots, e_\ell)$ if and only if $D(\mathcal{A}, \mathbf{m})$ is an S-free module and there exists a basis $\delta_1, \ldots, \delta_\ell \in D(\mathcal{A}, \mathbf{m})$ such that $\det \delta_i = e_i$. Here note that the degree $\deg \delta$ of a derivation δ is the polynomial degree, in other words, $\deg(\delta f) = \deg \delta + \deg f - 1$ for a homogeneous polynomial f. An arrangement $\mathcal{A}$ is said to be free if $(\mathcal{A}, 1)$ is free.

Let $\delta_1, \ldots, \delta_\ell \in D(\mathcal{A}, \mathbf{m})$. Then $\delta_1, \cdots, \delta_\ell$ form a S-basis of $D(\mathcal{A}, \mathbf{m})$ if and only if

$$\delta_1 \wedge \cdots \wedge \delta_\ell = c \cdot Q(\mathcal{A}, \mathbf{m}) \cdot \frac{\partial}{\partial z_1} \wedge \cdots \wedge \frac{\partial}{\partial z_\ell},$$

where $c \in \mathbb{C}^*$ is a non-zero constant (Saito's criterion [25]). From Saito's criterion, we also obtain that if a multiarrangement $(\mathcal{A}, \mathbf{m})$ is free with exponents $(e_1, \ldots, e_\ell)$, then $|\mathbf{m}| = \sum_{i=1}^\ell e_i$.

The Euler vector field $\theta_E = \sum_{i=1}^{\ell} x_i \partial_i$ is always contained in $D(\mathcal{A})$. Thus it is natural to define $D_0(\mathcal{A}) := D(\mathcal{A})/S \cdot \theta_E$. Since $D_0(\mathcal{A})$ is a graded S-module, it determines a coherent sheaf $\widetilde{D_0}(\mathcal{A})$ on $\mathbb{P}^{\ell-1}$. On the other hand, an arrangement $\mathcal{A}$ defines a Cartier divisor $\cup\overline{\mathcal{A}} = \bigcup \overline{H} \subset \mathbb{P}^{\ell-1}$. The logarithmic sheaf $\mathcal{T}_{\mathbb{P}^{\ell-1}}(-\log(\cup\overline{\mathcal{A}}))$ determined by the divisor $(\cup\overline{\mathcal{A}})$ is related to $\widetilde{D_0}(\mathcal{A})$ by the following formula.

$$\mathcal{T}_{\mathbb{P}^{\ell-1}}(-\log(\cup\overline{\mathcal{A}})) \cong \widetilde{D_0}(\mathcal{A})[+1].$$

From the sheaf $\widetilde{D}(\mathcal{A})$, we can reconstruct the graded module $D(\mathcal{A})$ as global sections $\Gamma_*(\mathbb{P}^{\ell-1}, \widetilde{D}(\mathcal{A})) := \bigoplus_{d\in\mathbb{Z}} \Gamma(\mathbb{P}^{\ell-1}, \widetilde{D}(\mathcal{A}))$. More generally, we have the following.

Proposition 2.7. *Let $(\mathcal{A}, \mathbf{m})$ be a multiarrangement. Then the natural map*

$$\Omega^p(\mathcal{A}, \mathbf{m}) \longrightarrow \Gamma_*(\mathbb{P}^{\ell-1}, \widetilde{\Omega}^p(\mathcal{A}, \mathbf{m}))$$

is an isomorphism.

Proof. We prove the surjectivity. Since $\bigcup_{i=1}^{\ell} U_i = \mathbb{P}^{\ell-1}$, where $U_i = \{z_i \neq 0\} \subset \mathbb{P}^{\ell-1}$, is an affine open covering, any element of the right hand side $\frac{\omega}{Q} \in \Gamma_*(\mathbb{P}^{\ell-1}, \widetilde{\Omega}^p(\mathcal{A}, \mathbf{m}))$ can be expressed as

$$\frac{\omega}{Q} = \frac{\omega_1}{z_1^{d_1} Q} = \frac{\omega_2}{z_2^{d_2} Q} = \cdots = \frac{\omega_\ell}{z_\ell^{d_\ell} Q},$$

with $\frac{\omega_i}{Q} \in \Omega^p(\mathcal{A}, \mathbf{m})$. Using the fact that S is a UFD, it is easily seen that ω is a regular differential form. Assume that z_i and α_H are linearly independent. Taking the wedge with $d\alpha_H$, $d\alpha_H \wedge \frac{\omega_i}{Q}$ does not have pole along H. So $d\alpha_H \wedge \frac{\omega}{Q} = d\alpha_H \wedge \frac{\omega_i}{z_i^{d_i} Q}$. Hence $d\alpha_H \wedge \frac{\omega}{Q} \in \Omega^p(\mathcal{A}, \mathbf{m})$.

Q.E.D.

Combining the above proposition with a sheaf theoretic property of reflexive sheaves, we can prove that $\Omega^p(\mathcal{A}, \mathbf{m})$ is determined by $\Omega^1(\mathcal{A}, \mathbf{m})$ in general.

Proposition 2.8. *Assume that $\Omega^1(\mathcal{A}_1, m_1) \cong \Omega^1(\mathcal{A}_2, m_2)$. Then $\Omega^p(\mathcal{A}_1, m_1) \cong \Omega^p(\mathcal{A}_2, m_2)$.*

Proof. Let us denote $\mathcal{E}^p = \widetilde{\Omega}^p(\mathcal{A}, \mathbf{m})$. Let $U := \mathbb{P}^{\ell-1} \setminus \bigcup_{H,H'\in\mathcal{A}, H\neq H'} (\overline{H} \cap \overline{H'})$ be the complement to the union of codimension ≥ 2 strata. We denote the inclusion map $i : U \hookrightarrow \mathbb{P}^{\ell-1}$. The restriction $i^*\mathcal{E}^p$ is locally free. Hence we have $i^*\mathcal{E}^p = \bigwedge^p i^*\mathcal{E}^1$. Since $\mathcal{E}^p$ is reflexive, hence normal, we have $\mathcal{E}^p = i_*(i^*(\mathcal{E}^p))$. Thus $\mathcal{E}^p = i_*(\wedge^p i^*(\mathcal{E}^1))$. (See [17, §1] for basic properties of reflexive sheaves.)

Thus $\mathcal{E}^p$ is determined by $\mathcal{E}^1$. Then by Proposition 2.7, we obtain the graded module $\Omega^p(\mathcal{A}, \mathbf{m})$ from its sheafification. Q.E.D.

The following result shows that $D(\mathcal{A})$ determines the characteristic polynomial $\chi(\mathcal{A}, t)$. Corollary 2.10 is known as Terao's factorization theorem.

Theorem 2.9. [35] *Denote by* $H(\Omega^p(\mathcal{A}), x) \in \mathbb{Z}[[x]][x^{-1}]$ *the Hilbert series of the graded module* $\Omega^p(\mathcal{A})$. *Define*

$$\Phi(\mathcal{A}; x, y) = \sum_{p=0}^{\ell} H(\Omega^p(\mathcal{A}), x) y^p. \tag{3}$$

Then

$$\chi(\mathcal{A}, t) = \lim_{x \to 1} \Phi(\mathcal{A}; x, t(1-x) - 1). \tag{4}$$

Corollary 2.10. [38] *Suppose that* $\mathcal{A}$ *is a free arrangement with exponents* $(e_1, \dots, e_\ell)$. *Then*

$$\chi(\mathcal{A}, t) = \prod_{i=1}^{\ell} (t - e_i). \tag{5}$$

The notion of freeness has a geometric interpretation. It is equivalent to a splitting $\widetilde{D}(\mathcal{A}) = \bigoplus_{i=1}^{\ell} \mathcal{O}(-e_i)$ of the sheaf $\widetilde{D}(\mathcal{A})$. Then the formula (5) indicates that the characteristic polynomial is related to the Chern polynomial $c_t(\mathcal{E}) = c_0(\mathcal{E}) + c_1(\mathcal{E})t + \dots$ (recall that $c_t(\mathcal{O}(-e)) = 1 - et$) [32]. Indeed, for locally free arrangements, Mustaţă and Schenck gave a beautiful formula connecting $\chi(\mathcal{A}, t)$ and the Chern polynomial.

Theorem 2.11. [20] *If* $\widetilde{D}(\mathcal{A})$ *is a locally free sheaf on* $\mathbb{P}^{\ell-1}$, *then*

$$c_t(\widetilde{D_0}(\mathcal{A})) = t^{\ell-1} \chi_0(\mathcal{A}, 1/t), \tag{6}$$

where $\chi_0(\mathcal{A}, t) = \chi(\mathcal{A}, t)/(t-1)$.

Note that in the case $\ell \le 3$, the local freeness is always satisfied. Thus the Chern polynomial is essentially equivalent to $\chi(\mathcal{A}, t)$ and combinatorially computable [28].

2.3. Characterizing freeness

A vector bundle on $\mathbb{P}^1$ is always a direct sum of line bundles (Grothendieck). The splitting of vector bundles on $\mathbb{P}^n (n \ge 2)$ is also a well studied subject, e.g., see [21]. There are several criterion to be split. The next result is known as Horrocks' criterion.

Theorem 2.12. *Let E be a rank r holomorphic vector bundle on $\mathbb{P}^n (n \geq 2)$. The following conditions are equivalent.*

(i) $E = \bigoplus_{i=1}^{r} \mathcal{O}_{\mathbb{P}^n}(d_i)$ *for some* $d_1, \dots, d_r \in \mathbb{Z}$.

(ii) $H^i(\mathbb{P}^n, E(d)) = 0$ *for* $\forall 1 \leq i \leq n-1$ *and* $\forall d \in \mathbb{Z}$.

(iii) *(If $n \geq 3$) $\exists H \subset \mathbb{P}^n$ a hyperplane such that the restriction splits as* $E|_H = \bigoplus_{i=1}^{r} \mathcal{O}_H(d_i)$.

Yuzvinsky [52, 53, 54] developed sheaf theory on the intersection lattice $L(\mathcal{A})$ and gave a cohomological criterion for an arrangement $\mathcal{A}$ to be free which is similar to Theorem 2.12 (ii). As an application he proved that the set of free arrangements form a Zariski open subset in the moduli space of all arrangements having the fixed combinatorial type.

Here we describe a criterion similar to Theorem 2.12 (iii). We begin with recalling Ziegler's restriction [55].

Choose a hyperplane $H \in \mathcal{A}$ and coordinate $(z_1, \dots, z_\ell)$ such that $H = \{z_\ell = 0\}$. Define a submodule $D_0^H(\mathcal{A})$ of $D(\mathcal{A})$ as follows:

$$D_0^H(\mathcal{A}) := \{\delta \in D(\mathcal{A}) \mid \delta z_\ell = 0\}.$$

Lemma 2.13. $D(\mathcal{A}) = S \cdot \theta_E \oplus D_0^H(\mathcal{A})$.

Proof. Let $\delta \in D(\mathcal{A})$. The assertion is obvious from $\delta = \left(\frac{\delta z_\ell}{z_\ell}\right)\theta_E + \left(\delta - \frac{\delta z_\ell}{z_\ell}\theta_E\right)$. Q.E.D.

The arrangement $\mathcal{A}$ determines the restricted arrangement $\mathcal{A}^H = \{H \cap H' \mid H' \in \mathcal{A}, H' \neq H\}$ on H. The restricted arrangement $\mathcal{A}^H$ possesses a natural multiplicity

$$\begin{array}{rccl} m^H: & \mathcal{A}^H & \longrightarrow & \mathbb{Z} \\ & X & \longmapsto & \sharp\{H' \in \mathcal{A} \mid X = H \cap H'\}. \end{array}$$

Ziegler [55] proved that the freeness of $\mathcal{A}$ implies that of $(\mathcal{A}^H, m^H)$.

Theorem 2.14. [55]

(1) *If* $\delta \in D_0^H(\mathcal{A})$, *then* $\delta|_{z_\ell = 0} \in D(\mathcal{A}^H, m^H)$.

(2) *If $\mathcal{A}$ is free with exponents $(1, e_2, \dots, e_\ell)$, then $(\mathcal{A}^H, m^H)$ is free with exponents $(e_2, \dots, e_\ell)$.*

Corollary 2.15. *$\mathcal{A}$ is free with exponents $(1, e_2, \dots, e_\ell)$ if and only if the following are satisfied.*

- *$(\mathcal{A}^H, m^H)$ is free with exponents $(e_2, \dots, e_\ell)$.*
- *The restriction induces the surjection $D_0^H(\mathcal{A}) \longrightarrow D(\mathcal{A}^H, m^H)$.*

Using Corollary 2.15, we can establish a Horrocks' type criterion for freeness. Namely, we will characterize freeness by using the freeness of the restriction $D(\mathcal{A}^H, m^H)$. We first consider the case $\ell = 3$. By analyzing the Hilbert series of these graded modules using the restriction map and Solomon–Terao's formula (Theorem 2.9), we have the following.

Theorem 2.16. [51] *If $\ell = 3$, then the cokernel of the restriction map is finite dimensional. Furthermore, suppose that $\exp(\mathcal{A}^H, m^H) = (e_1, e_2)$, then*

$$\dim_{\mathbb{C}} \mathrm{Coker} = b_3(\mathbb{C}^3 \setminus \bigcup_i H_i) - e_1 e_2.$$

Corollary 2.17. [51] *Suppose $\ell = 3$. Then the following conditions are equivalent.*

- *$\mathcal{A}$ is free with exponents $(1, e_2, e_3)$.*
- *$\chi(\mathcal{A}, t) = (t-1)(t-e_2)(t-e_3)$ and there exists $H \in \mathcal{A}$ such that $\exp(\mathcal{A}^H, m^H) = (e_2, e_3)$.*

Remark 2.18. *Recently a higher dimensional version of Corollary 2.17 has been obtained by Schulze* [31].

The characterization in the case $\ell \geq 4$ is the following.

Theorem 2.19. [50] *Suppose $\ell \geq 4$. Then an arrangement $\mathcal{A}$ is free with exponents $(1, e_2, \ldots, e_\ell)$ if and only if there exists $H \in \mathcal{A}$ such that*

(a) *$(\mathcal{A}^H, m^H)$ is free with exponents $(e_2, \ldots, e_\ell)$, and*
(b) *the localization $\mathcal{A}_x = \{H \in \mathcal{A} \mid x \in H\}$ is free for any $x \in H \setminus \{0\}$.*

2.4. Freeness for multiarrangements

The notion of multiarrangement is a natural generalization of simple arrangement. For a 2-dimensional simple arrangement $\mathcal{A}$, it is easy to construct explicit basis of $D(\mathcal{A})$. However, for the case of multiarrangements, describing an explicit basis for $D(\mathcal{A}, \mathbf{m})$ is difficult even for $\ell = 2$. Wakamiko [44] gave an explicit basis for $D(\mathcal{A}, \mathbf{m})$ with $\ell = 2, |\mathcal{A}| = 3$. Wakefield and Yuzvinsky [46] computed the exponents for $\ell = 2$ and generic $\mathcal{A}$. Both results show that the exponents tend to $(\lfloor \frac{|\mathbf{m}|}{2} \rfloor, \lceil \frac{|\mathbf{m}|}{2} \rceil)$, where $|\mathbf{m}| = \sum_{H \in \mathcal{A}} \mathbf{m}(H)$.

Remark 2.20. *The above mentioned results remind the author results of Dolgachev and Kapranov (cf. §2.1) and Schenck* [29] *on stability of $\mathcal{T}_{\mathbb{P}^n}(-\log(\cup\mathcal{A}))$. It seems natural to ask whether for generic $\mathcal{A}$ with $\ell = 3$, $\widetilde{D}(\mathcal{A}, \mathbf{m})$ is a stable rank 3 vector bundle on $\mathbb{P}^2$.*

Recently several results on $D(\mathcal{A})$ has been generalized to multiarrangements. Abe, Terao and Wakefield [3] proved that Solomon–Terao's formula (4) (and (3)) gives a polynomial $\chi((\mathcal{A},\mathbf{m}),t)$ for any multiarrangement $(\mathcal{A},\mathbf{m})$. The polynomial $\chi((\mathcal{A},\mathbf{m}),t)$ is called the characteristic polynomial of a multiarrangement $(\mathcal{A},\mathbf{m})$, which is a basic tool for proving non-freeness for multiarrangements.

Another important result on free multiarrangements is the Addition-Deletion Theorem [4]. Let $(\mathcal{A},\mathbf{m})$ be a multiarrangement. Choose a hyperplane $H_0\in\mathcal{A}$ with $\mathbf{m}(H_0)>0$. One can associate two multiarrangements to $(\mathcal{A},\mathbf{m},H_0)$ as follows.

- The deletion $(\mathcal{A}',m')$: $\mathcal{A}'=\mathcal{A}$ and the multiplicity $m':\mathcal{A}'\to\mathbb{Z}_{\geq 0}$ is defined by
$$m'(H)=\begin{cases}\mathbf{m}(H) & \text{if } H\neq H_0,\\ \mathbf{m}(H)-1 & \text{if } H=H_0.\end{cases}$$
- The restriction $(\mathcal{A}'',\mathbf{m}^*)$: $\mathcal{A}''=\{H\cap H_0\mid H\in\mathcal{A}, H\neq H_0\}$. Let $X\in\mathcal{A}''$. Then X has codimension two. Thus the multiarrangement $\mathcal{A}_X=\{H\in\mathcal{A}\mid H\supset X\}$ with the multiplicity $\mathbf{m}|_{\mathcal{A}_X}$ is free. We can choose the basis $\theta_X,\psi_X,\partial_3,\dots,\partial_\ell$ with $\theta_X\notin\alpha_{H_0}\cdot\mathrm{Der}_S$ and $\psi_X\in\alpha_{H_0}\cdot\mathrm{Der}_S$. Define the multiplicity $\mathbf{m}^*:\mathcal{A}''\to\mathbb{Z}_{\geq 0}$ by $\mathbf{m}^*(X)=\deg\theta_X$.

The following theorem generalizes the classical Addition-Deletion Theorem [37] to multiarrangements.

Theorem 2.21. [4] *With the notations above, any two of the following statements imply the third:*

(i) $(\mathcal{A},\mathbf{m})$ *is free with exponents* $(d_1,\dots,d_\ell)$.
(ii) $(\mathcal{A}',\mathbf{m}')$ *is free with exponents* $(d_1,\dots,d_\ell-1)$.
(iii) $(\mathcal{A}'',\mathbf{m}^*)$ *is free with exponents* $(d_1,\dots,d_{\ell-1})$.

Using Theorem 2.21, one can construct a lot of free multiarrangements inductively.

2.5. Free arrangements and intersection of plane curves

In this section we consider a 3-dimensional free arrangement $\mathcal{A}$ with exponents $(1,e_1,e_2)$. Choose $H_0\in\mathcal{A}$. Then the deconing $\mathbf{d}_{H_0}\mathcal{A}$ is an affine line arrangement in $\mathbb{C}^2$. Freeness of $\mathcal{A}$ imposes strong conditions on the positions of intersections $L_2(\mathcal{A}):=\{L\cap L'\in\mathbb{C}^2\mid L,L'\in\mathbf{d}_{H_0}\mathcal{A}, L\neq L'\}$ and their multiplicities. Let $\mu(p):=\sharp\{L\in\mathbf{d}_{H_0}\mathcal{A}\mid L\ni p\}-1$.

Theorem 2.22. *Assume that $\mathcal{A}$ is free. With notation as above, there exist plane curves $C_1,C_2\subset\mathbb{C}^2$ with degrees e_1 and e_2 respectively* *such that* $C_1 \cap C_2 = L_2(\mathcal{A})$ *and the intersection multiplicity is*

$$\mathrm{mult}_p(C_1, C_2) = \mu(p).$$

Remark 2.23. *If* $\mathcal{A}$ *is a fiber-type arrangement, we can find easily such* C_1 *and* C_2 *as union of lines.*

Proof. Choose coordinates (z_0, z_1, z_2) so that $H_0 = \{z_0 = 0\}$. We can choose a basis $\theta_E, \delta_1, \delta_2 \in D(\mathcal{A})$ such that $\delta_1 z_0 = \delta_2 z_0 = 0$ (see Lemma 2.13). Let $\alpha = a_1 z_1 + a_2 z_2$ be a linear form such that the line $\{\alpha = 0\} \subset \mathbb{C}^2$ is not parallel to any line $L \in \mathbf{d}_{H_0}\mathcal{A}$. By definition $\delta_i \alpha \in \mathbb{C}[z_1, z_2]$ is a polynomial of degree e_i. Note that

$$\delta_i \alpha = 0 \text{ at } p \in \mathbb{C}^2 \iff \begin{cases} \delta_i(p) = 0 \text{ or} \\ \delta_i(p) \text{ is parallel to } \{\alpha = 0\}. \end{cases}$$

If $p \notin \bigcup_{L \in \mathbf{d}_{H_0}\mathcal{A}} L$, $\delta_1(p)$ and $\delta_2(p)$ are linearly independent. If $p \in L$ and $p \notin L_2(\mathcal{A})$, $\delta_1(p)$ and $\delta_2(p)$ span the tangent space T_pL. In any case, either $\delta_1(p)\alpha \neq 0$ or $\delta_2(p)\alpha \neq 0$. Hence $\delta_1(p)\alpha = \delta_2(p)\alpha = 0$ precisely when $p \in L_2(\mathbf{d}_{H_0}\mathcal{A})$. Fix $p \in L_2(\mathcal{A})$ and choose the coordinate (z_1, z_2) such that $p = (0, 0)$ and $\{z_1 = 0\} \in \mathbf{d}_{H_0}\mathcal{A}$. Let Q be the product of defining equations which contain p. Then

$$\eta_1 = z_1 \partial_{z_1} + z_2 \partial_{z_2} \text{ and } \eta_2 = \frac{Q}{z_1}\partial_{z_2}$$

form a basis of $D(\mathbf{d}_{H_0}\mathcal{A}_p)$. It is easily seen that $\mathrm{mult}_p(\eta_1\alpha, \eta_2\alpha) = \dim \mathbb{C}[[z_1, z_2]]/(\eta_1\alpha, \eta_2\alpha) = \mu(p)$. Germs of δ_i at p is expressed as $\delta_i = f_{i1}\eta_1 + f_{i2}\eta_2$ with $\det = f_{11}f_{22} - f_{12}f_{21}$ is contained in the unit $\mathbb{C}[[z_1, z_2]]^{\times}$. Thus intersection multiplicity is

$$\mathrm{mult}_p(C_1, C_2) = \dim \mathbb{C}[[z_1, z_2]]/(\delta_1\alpha, \delta_2\alpha) = \mu(p).$$

Q.E.D.

Remark 2.24. *Although there exists a free arrangement which has non-vanishing homotopy group* $\pi_2(M(\mathcal{A}))$, [16], *it is challenging to see the homotopy types of free arrangements.*

2.6. An example of a non-free arrangement

Factorization of the characteristic polynomial (Corollary 2.10) is a necessary combinatorial condition for an arrangement to be free. However the converse is not true. Indeed, there are non-free arrangements which have factored characteristic polynomials.

Example 2.25. *(Stanley's example* [23]*) Let* $\mathcal{A} = \{H_0, H_1, \dots, H_6\}$ *be an arrangement of* 7 *planes in* $\mathbb{R}^3$ *defined as Figure 1 (real lines). The characteristic polynomial is* $\chi(\mathcal{A}, t) = (t-1)(t-3)^2$*. However* $\mathcal{A}$ *is not free. We shall give three proofs.*

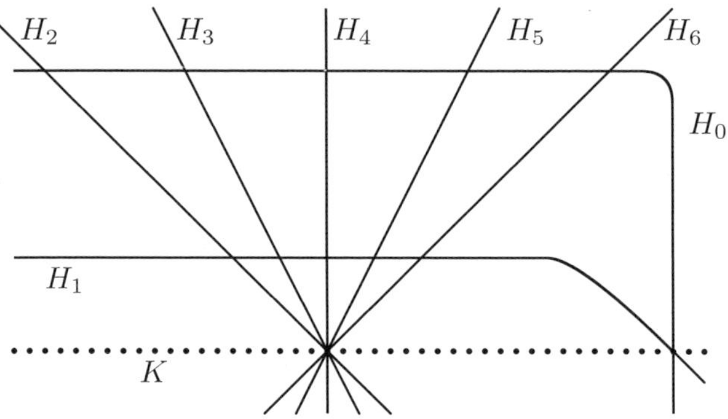

Fig. 1. $\mathcal{A} = \{H_0, \dots, H_6\}$

First note that by 2.10, if $\mathcal{A}$ is free, then the exponents should be $(1, 3, 3)$.

(1) Consider another hyperplane K (dotted line). The extended arrangement $\mathcal{A} \cup \{K\}$ is of fiber-type and hence free with exponents $(1, 2, 5)$ (also easily proved by using Addition-Deletion Theorem 2.21). Hence $D(\mathcal{A} \cup \{K\})$ has degree 2 element δ which is linearly independent from the Euler vector field θ_E. By definition, $\delta \in D(\mathcal{A})$. However this contradicts the fact that $D(\mathcal{A})$ does not have a basis element of degree ≤ 2 other than θ_E.

(2) Consider the restriction to H_0. Then $(\mathcal{A}^{H_0}, m^{H_0})$ is free with exponents $(1, 5)$. From Corollary 2.17, $\mathcal{A}$ is not free.

(3) Consider the deconing $\mathbf{d}_{H_0}\mathcal{A}$ with respect to H_0. If $\mathcal{A}$ is free, then by Theorem 2.22, the intersections satisfy $L(\mathbf{d}_{H_0}\mathcal{A}) = C_1 \cap C_2$, where C_i is a cubic curve. We may assume that C_1 does not have H_1 as a component. Then $H_1 \cap C_1$ consists of five points. This contradicts Bezout theorem.

§3. Coxeter multiarrangements

Coxeter multiarrangements are a well-studied class of multiarrangements. Using the notion of primitive derivation, we can construct a basis for several Coxeter multiarrangements. Here we give a brief review.

The importance of the primitive derivation was first realized by K. Saito [27] in the context of singularity theory. K. Saito's theory of primitive forms reveals that the parameter space B of semi-universal deformation $X \to B$ of an isolated singularity $0 \in X_0$ possesses rich geometric structures [26, 19]. On the other hand, Grothendieck–Brieskorn–Slodowy's theory [11] shows that for simple singularities, the semi-universal family can be described in terms of Lie theory. In particular, the parameter space B can be canonically identified with the Weyl group quotient $\mathfrak{h}/W$ of an ADE-type Cartan subalgebra $\mathfrak{h}$ (see also §4.1). In [27], Saito describes the flat structure for any finite reflection group $W \curvearrowright V$ in purely invariant theoretic way by using the primitive derivation. Later Terao [40, 41] pulled back the theory to V via the natural projection $\pi : V \to V/W$ and proved freeness of Coxeter multiarrangements with constant multiplicity.

In this section, we will describe the structure of $D(\mathcal{A}, \mathbf{m})$ for a Coxeter arrangement $\mathcal{A}$ based on [40, 41, 49, 6].

Let V be an ℓ-dimensional Euclidean space over $\mathbb{R}$ with inner product $I : V \times V \to \mathbb{R}$. Fix a coordinate $(x_1, \cdots, x_\ell)$ and put $S = S(V^*) \otimes_{\mathbb{R}} \mathbb{C} = \mathbb{C}[x_1, \ldots, x_\ell]$. Let $W \subset O(V, I)$ be a finite irreducible reflection group with the Coxeter number h. Let $\mathcal{A}$ be the corresponding Coxeter arrangement, i.e., the collection of all reflecting hyperplanes of W. Fix a defining linear form $\alpha_H \in V^*$ for each hyperplane $H \in \mathcal{A}$.

It is proved by Chevalley [13] that the invariant ring S^W is a polynomial ring $S^W = \mathbb{C}[P_1, \ldots, P_\ell]$ with $P_1, \ldots, P_\ell$ are homogeneous generators. Suppose that $\deg P_1 \leq \cdots \leq \deg P_\ell$. Then it is known that $\deg P_1 = 2 < \deg P_2 \leq \cdots \leq \deg P_{\ell-1} < \deg P_\ell = h$. Note that we may choose $P_1(x) = I(x, x)$. Then $\frac{\partial}{\partial P_i}$ $(i = 1, \ldots, \ell)$ can be considered as a rational vector field on V with order one poles along $H \in \mathcal{A}$. Indeed by using the fact

$$\Delta := \det \left(\frac{\partial P_i}{\partial x_j} \right)_{i,j=1,\ldots,\ell} \doteq \prod_{H \in \mathcal{A}} \alpha_H,$$

we may define the action of the differential operator $\frac{\partial}{\partial P_i}$ to $f \in S$ by

$$\frac{\partial f}{\partial P_i} = \frac{1}{\Delta} \det \begin{pmatrix} \frac{\partial P_1}{\partial x_1} & \cdots & \frac{\partial f}{\partial x_1} & \cdots & \frac{\partial P_\ell}{\partial x_1} \\ \frac{\partial P_1}{\partial x_2} & \cdots & \frac{\partial f}{\partial x_2} & \cdots & \frac{\partial P_\ell}{\partial x_2} \\ \vdots & \ddots & \vdots & \ddots & \vdots \\ \frac{\partial P_1}{\partial x_\ell} & \cdots & \frac{\partial f}{\partial x_\ell} & \cdots & \frac{\partial P_\ell}{\partial x_\ell} \end{pmatrix}.$$

Obviously, we have $\frac{\partial P_i}{\partial P_i} = 1$ and $\frac{\partial P_j}{\partial P_i} = 0$ for $i \neq j$.

Definition 3.1. *We denote* $D = \frac{\partial}{\partial P_\ell}$ *and call it the* primitive derivation.

Since $\deg P_i < \deg P_\ell$ for $i \leq \ell - 1$, the primitive derivation D is uniquely determined up to nonzero constant multiple independent of the choice of the generators $P_1, \dots, P_\ell$.

Next we define the affine connection ∇.

Definition 3.2. *For a given rational vector field* $\delta = \sum_{i=1}^{\ell} f_i \frac{\partial}{\partial x_i}$ *and a rational differential k-form* $\omega = \sum_{i_1,\dots,i_k} g_{i_1,\dots,i_k} dx_{i_1,\dots,i_k}$ *(where* $dx_{i_1,\dots,i_k} = dx_{i_1} \wedge \cdots \wedge dx_{i_k}$*), define* $\nabla_\delta \omega$ *by*

$$\nabla_\delta \omega = \sum_{i_1,\dots,i_k} \delta(g_{i_1,\dots,i_k}) dx_{i_1,\dots,i_k}.$$

Let $\mathbf{m} : \mathcal{A} \longrightarrow \{0, 1\}$ be a map. The differentiation ∇_D by the primitive derivation changes the degree by h. This action connects $D(\mathcal{A}, \mathbf{m})$ with $D(\mathcal{A}, 2k + \mathbf{m})$ and $\Omega^1(\mathcal{A}, 2k - \mathbf{m})$.

Theorem 3.3. *Fix notation as above, and let k be a positive integer.*

(1) *The map*

$$\begin{array}{rccc} \Phi_k : & D(\mathcal{A}, \mathbf{m})(kh) & \longrightarrow & \Omega^1(\mathcal{A}, 2k - \mathbf{m}) \\ & \delta & \longmapsto & \nabla_\delta \nabla_D^k dP_1 \end{array}$$

gives an S-isomorphism of graded modules.

(2) *The map*

$$\begin{array}{rccc} \Psi_k : & D(\mathcal{A}, \mathbf{m})(-kh) & \longrightarrow & D(\mathcal{A}, 2k + \mathbf{m}) \\ & \delta & \longmapsto & \nabla_\delta \nabla_D^{-k} E \end{array}$$

gives an S-isomorphism of graded modules.

Corollary 3.4. *For a* $\{0, 1\}$*-valued multiplicity* $\mathbf{m} : \mathcal{A} \to \{0, 1\}$ *and an integer* $k > 0$*, the following conditions are equivalent.*

- $(\mathcal{A}, \mathbf{m})$ *is free with exponents* $(e_1, \dots, e_\ell)$.
- $(\mathcal{A}, 2k + \mathbf{m})$ *is free with exponents* $(kh + e_1, \dots, kh + e_\ell)$.
- $(\mathcal{A}, 2k - \mathbf{m})$ *is free with exponents* $(kh - e_1, \dots, kh - e_\ell)$.

If $\mathbf{m} \equiv 0$, then $(\mathcal{A}, \mathbf{m})$ is free with exponents $(0, \dots, 0)$. Hence $(\mathcal{A}, 2k)$ is free with exponents $(kh, kh, \dots, kh)$. If $\mathbf{m} \equiv 1$, then $(\mathcal{A}, \mathbf{m})$ is free with exponents $(e_1, \dots, e_\ell)$, where $e_i = \deg P_i - 1$ (by [25, 27]). Hence $(\mathcal{A}, 2k + 1)$ is free with exponents $(e_1 + kh, \dots, e_\ell + kh)$. In particular, Coxeter multiarrangements with constant multiplicities are free [40].

The primitive derivation acts on W-invariant forms. The following will be used in the next section.

Theorem 3.5. [41] *With notation as above, the set of W-invariant derivations $D(\mathcal{A}, 2k+1)^W$ is a free S^W-module. Furthermore, if $k > 0$,*

$$\nabla_{\frac{\partial}{\partial P_i}} D(\mathcal{A}, 2k+1)^W \subset D(\mathcal{A}, 2k-1)^W,$$

$$\nabla_{\frac{\partial}{\partial P_\ell}} D(\mathcal{A}, 2k+1)^W = D(\mathcal{A}, 2k-1)^W.$$

Remark 3.6. *Recently Theorem 3.3 and Corollary 3.4 are generalized for $\mathbf{m} : \mathcal{A} \longrightarrow \{-1, 0, +1\}$ by Abe* [1].

§4. Applications of freeness of $D(\mathcal{A}, \mathbf{m})$

In this section we will describe two applications of freeness of $D(\mathcal{A}, \mathbf{m})$.

4.1. Relative de Rham cohomology of adjoint quotient maps

Let $\mathfrak{g}$ be a simple Lie algebra of type ADE over $\mathbb{C}$. The categorical quotient map $\chi : \mathfrak{g} \to B := \mathfrak{g}//G$ of the adjoint group action on $\mathfrak{g}$ is called the adjoint quotient map. The purpose of this section is to investigate the $\mathcal{O}_B$-module structure of the relative de Rham cohomology $H^2(\Omega^\bullet_\chi)$ (see below the definition of $\Omega^\bullet_\chi$) of $\chi : \mathfrak{g} \to B$ through an action of vector fields Der_B on B (the Gauß–Manin connection).

The study of the relative de Rham cohomology for an affine morphism goes back to E. Brieskorn [10] who proved the coherence of relative de Rham cohomology for any polynomial map $f : \mathbb{C}^n \to \mathbb{C}$ with isolated critical point $0 \in \mathbb{C}^n$ (and M. Sebastiani proved $\mathcal{O}_\mathbb{C}$-freeness of rank μ, where μ is the Milnor number of f). Further, K. Saito proved the freeness for the semi-universal deformation $F : X \to B := \mathbb{C}^\mu$ of an isolated hypersurface singularity defined by f. More precisely, he gave an isomorphism between a certain submodule of vector fields Der_B on B and $H^n(\Omega^\bullet_{X/B})$. The isomorphism is given by the following correspondence, we first fix a special cohomology class ζ called a primitive form, then for given vector field $\delta \in \mathrm{Der}_B$ take a lift up $\tilde{\delta} \in \mathrm{Der}_X$ of the vector field on the total space X, and differentiate ζ by $\tilde{\delta}$, we have a new cohomology class $\mathcal{L}_{\tilde{\delta}}\zeta$, where $\mathcal{L}$ is the Lie derivative. On the other hand, the semi-universal deformation of a simple singularity is constructed by using the adjoint quotient map χ of type ADE [11, 33].

Indeed, if we restrict the map χ to a certain affine subspace $X \subset \mathfrak{g}$, we have the semi-universal deformation of a simple singularity. In this case H. Yamada [48] showed that the restriction of the Kostant–Kirillov form ζ to X becomes the primitive form which generates the relative de Rham cohomology $H^2(\Omega^\bullet_{X/B})$ by differentiation by means of vector fields $\delta \in \mathrm{Der}_B$.

Let us introduce some notation. Let $\mathfrak{g}$ a simple Lie algebra over $\mathbb{C}$ (later we will restrict $\mathfrak{g}$ to ADE-type). Let $\mathfrak{g} = \mathfrak{h} \oplus \bigoplus_{\alpha\in\Phi} \mathfrak{g}_\alpha$, $(\Phi \subset \mathfrak{h}^*)$ a Cartan decomposition with respect to a Cartan algebra $\mathfrak{h}$ with $\ell = \dim \mathfrak{h}$, G the adjoint group of $\mathfrak{g}$ and T the maximal torus of G with Lie algebra $\mathfrak{h}$. We denote by W the Weyl group $N_G(T)/T$. The classical Chevalley's restriction theorem states that the restriction $\rho : \mathbb{C}[\mathfrak{g}] \to \mathbb{C}[\mathfrak{h}]$ of polynomial functions induces an isomorphism

$$\mathbb{C}[\mathfrak{g}]^G \xrightarrow{\cong} \mathbb{C}[\mathfrak{h}]^W \tag{7}$$

of algebras of invariants. We also denote $\mathbb{C}[\mathfrak{h}] = S$ and $S^W = \mathbb{C}[P_1, \ldots, P_\ell]$ as in §3. The categorical quotient of the adjoint action is $B = \mathfrak{g}//G \cong \mathfrak{h}//W \cong \mathrm{Spec}\, S^W$. We call the quotient map $\chi : \mathfrak{g} \to B$ the adjoint quotient map as mentioned above. The construction is summarized in the following diagram.

(8)

$$\begin{array}{ccccccc} & & \mathfrak{g} & & & & \mathbb{C}[\mathfrak{g}] \\ & & \downarrow & & & & \uparrow \\ V = \mathfrak{h} & \xrightarrow{\pi} & \mathfrak{h}/W = B = \mathfrak{g}//G & , & S & \hookleftarrow & S^W = \mathbb{C}[B] = \mathbb{C}[\mathfrak{g}]^G. \end{array}$$

Definition 4.1. *Define the relative de Rham complex $\Omega^\bullet_\chi$ for the adjoint quotient map $\chi : \mathfrak{g} \to B$ by*

$$\Omega^\bullet_\chi = \frac{\Omega^\bullet_\mathfrak{g}}{\chi^*\Omega^1_B \wedge \Omega^{\bullet-1}_\mathfrak{g}} = \frac{\Omega^\bullet_\mathfrak{g}}{\sum_{i=1}^\ell dP_i \wedge \Omega^{\bullet-1}_\mathfrak{g}}.$$

By the formula $d(P\cdot\omega) = dP \wedge \omega + P \cdot d\omega$, the differential $d_\chi : \Omega^\bullet_\chi \to \Omega^{\bullet+1}_\chi$ is a S^W-module homomorphism. Hence the cohomology group $H^k(\Omega^\bullet_\chi)$ possesses S^W-module structure.

Let $D \subset B$ be the set of critical points of the quotient map $\pi : \mathfrak{h} \to B$. It is proved in [25] that π induces an isomorphism

$$D(\mathcal{A})^W \xrightarrow{\cong} \mathrm{Der}_B(-\log D).$$

Thus for $\delta \in D(\mathcal{A})^W$, we may differentiate the Kostant-Kirillov form ζ by δ and obtain a relative 2-form $\nabla_\delta\zeta$ (which has poles along D in general).

After Yamada's result, it was naturally conjectured that $H^2(\Omega_\chi^\bullet)$ is a free S^W-module of rank ℓ.

Theorem 4.2. *Let $\mathfrak{g}$ a simple Lie algebra of type ADE, with a Cartan subalgebra $\mathfrak{h}$ and the Weyl group W. Let $\mathcal{A}$ be the corresponding Weyl arrangement on $\mathfrak{h}$. The map $D(\mathcal{A})^W \ni \delta \longmapsto \nabla_\delta \zeta$ induces a natural isomorphism*

$$H^2(\Omega_\chi^\bullet) \cong D(\mathcal{A}, 5)^W,$$

of S^W-modules.

The rest of this section is devoted to a proof of this theorem.

We first recall a result due to J. Vey [47], which is an analogue of Weyl's unitary trick.

Theorem 4.3. *Let G be a connected reductive algebraic group over $\mathbb{C}$ with a linear action on a finite dimensional $\mathbb{C}$-vector space E. Let $\Omega_E^\bullet$ be the de Rham complex of holomorphic differential forms on E and $\mathcal{I}^\bullet$ the ideal of $\Omega^\bullet$ generated by differentials $df_1, df_2, \cdots, df_r$, where $f_1, f_2, \cdots, f_r$ are G-invariant homogeneous polynomials on E. Then the morphism*

$$(\Omega^\bullet)^G/(\mathcal{I}^\bullet)^G \to \Omega^\bullet/\mathcal{I}^\bullet$$

is a quasi-isomorphism.

By this, we can compute cohomology of $\Omega_\chi^\bullet$ by using the complex $\Omega_{\mathfrak{g}}^{G,\bullet}/(\sum_i dP_i \wedge \Omega_{\mathfrak{g}}^{\bullet-1})^G$ of G-invariant relative forms. Next we shall describe $\Omega_{\mathfrak{g}}^{G,\bullet}$.

Broer [12] considered a generalization of Chevalley's restriction theorem (7) in the following setting. Let M be a finite dimensional G-module and $\mathrm{Mor}(\mathfrak{g}, M)$ (resp. $\mathrm{Mor}_G(\mathfrak{g}, M)$) the space of polynomial (resp. G-equivariant polynomial) morphisms of $\mathfrak{g}$ into M. It is isomorphic to $\mathbb{C}[\mathfrak{g}] \otimes M$ (resp. $(\mathbb{C}[\mathfrak{g}] \otimes M)^G$). For any G-module M the restriction map ρ induces a homomorphism

$$\rho_M : \mathrm{Mor}_G(\mathfrak{g}, M) \longrightarrow \mathrm{Mor}_W(\mathfrak{h}, M^T).$$

Since the union of all Cartan subalgebras is Zariski dense in $\mathfrak{g}$, ρ_M is injective for all M. If $M = \mathbb{C}$ is a trivial G-module, ρ_M is bijective because of Chevalley's theorem. However it is not necessarily bijective in general. Broer [12] proved that

Theorem 4.4. *Let M be a G-module. Restriction induces an isomorphism*

$$\rho_M : \mathrm{Mor}_G\left(\mathfrak{g}, M\right) \xrightarrow{\cong} \mathrm{Mor}_W\left(\mathfrak{h}, M^T\right)$$

if and only if the weights 2α ($\alpha \in \Phi$ is a root of $\mathfrak{g}$) do not occur as T-weights in M. (We shall call M small *if it satisfies this assumption.)*

We need this theorem to describe the set of G-invariant differential forms $\Omega_{\mathfrak{g}}^{\bullet,G}$ on $\mathfrak{g}$ below. By definition the set of all differential p-forms on $\mathfrak{g}$ is $\Omega_{\mathfrak{g}}^{p} = \mathbb{C}[\mathfrak{g}] \otimes \overset{p}{\wedge} \mathfrak{g}^*$. Thus we apply Theorem 4.4 for $M = \overset{p}{\wedge} \mathfrak{g}^*$. If $p = 1$, since $\mathfrak{g} \cong \mathfrak{g}^*$ by Killing form, the T-weights of $\mathfrak{g}^*$ are nothing but the roots of $\mathfrak{g}$, so $\mathfrak{g}^*$ is small. Thus we have an isomorphism

$$\rho_1 : \Omega_{\mathfrak{g}}^{1,G} \cong \left(\mathbb{C}[\mathfrak{h}] \otimes \mathfrak{h}^*\right)^W \cong \Omega_{\mathfrak{h}}^{1,W}.$$

It follows from a result of Solomon [34] that

$$\pi^* : \Omega_B^p \xrightarrow{\cong} \Omega_{\mathfrak{h}}^{p,W}. \tag{9}$$

Thus we conclude that G-invariant 1-forms $\Omega_{\mathfrak{g}}^{1,G}$ on $\mathfrak{g}$ are nothing but the pull back $\chi^*\Omega_B^1$ of 1-forms on B. In particular, $\Omega_{\chi}^{1,G} = 0$, we have

$$H^2(\Omega_{\chi}^{\bullet,G}) \cong \operatorname{Ker}(d_\chi : \Omega_{\chi}^{2,G} \to \Omega_{\chi}^{3,G}). \tag{10}$$

From the classification of simple root systems, it is easily seen that $M = \overset{2}{\wedge} \mathfrak{g}^*$ is small if and only if $\mathfrak{g}$ is of type ADE, since the set of weights of $\overset{2}{\wedge} \mathfrak{g}^*$ is

$$\{0\} \cup \Phi \cup \{\alpha + \beta | \alpha, \beta \in \Phi, \alpha \neq \beta\}.$$

Furthermore, $(\overset{2}{\wedge} \mathfrak{g}^*)^T \cong \overset{2}{\wedge} \mathfrak{h}^* \otimes (\overset{2}{\wedge} \mathfrak{h}^{\perp})^T$ is a direct sum decomposition of W-submodules. From (9), we obtain,

Proposition 4.5. *Let $\mathfrak{g}$ be a simple Lie algebra of type ADE. Then*

$$\Omega_{\mathfrak{g}}^{2,G} \cong \chi^*\Omega_B^2 \oplus \Omega_{\chi}^{2,G}. \tag{11}$$

$$\rho_2 : \Omega_{\chi}^{2,G} \xrightarrow{\cong} \left(\mathbb{C}[\mathfrak{h}] \otimes (\overset{2}{\wedge} \mathfrak{h}^{\perp})^T\right)^W. \tag{12}$$

By Proposition 4.5, we can identify G-invariant relative 2-forms $\Omega_{\chi}^{2,G}$ with the submodule $(\mathbb{C}[\mathfrak{h}] \otimes (\overset{2}{\wedge} \mathfrak{h}^{\perp})^T)^W$ of $\Omega_{\mathfrak{g}}^{2,G}$. Now we define two submodules $\mathcal{H}_\chi \subset \mathcal{H}'_\chi \subset \Omega_{\mathfrak{g}}^{2,G}$ which are related to $H^2(\Omega_{\chi}^{\bullet})$.

Definition 4.6.

$$\mathcal{H}_\chi := \{\omega \in \Omega_{\chi}^{2,G} \mid d_{\mathfrak{g}}\omega \in \sum_i dP_i \wedge \Omega_{\mathfrak{g}}^{2,G}\}, \tag{13}$$

$$(14) \qquad \mathcal{H}'_\chi := \{\omega \in \Omega_\chi^{2,G} \mid d_{\mathfrak{g}}\omega \wedge dP_1 \wedge \cdots \wedge dP_\ell = 0 \ in \ \Omega_{\mathfrak{g}}^{3+\ell}\},$$

where $\mathbb{C}[\mathfrak{g}]^G = \mathbb{C}[P_1, \cdots, P_\ell]$.

By (10) and definition above,

$$(15) \qquad \mathcal{H}_\chi \cong H^2(\Omega_\chi^{\bullet,G})$$

and obviously $\mathcal{H}_\chi \subset \mathcal{H}'_\chi$. Later it will be proved that $\mathcal{H}_\chi \subsetneqq \mathcal{H}'_\chi$.

Let $e_\alpha \in \mathfrak{g}_\alpha$ $(\alpha \in \Phi)$ be non-zero root vectors such that $I([e_\alpha, e_{-\alpha}], h) = \alpha(h)$ (for all $\alpha \in \Phi,\ h \in \mathfrak{h}$), where $I(\bullet, \bullet)$ is the Killing form, and $e_\alpha^* \in \mathfrak{g}_\alpha^*$ be the dual basis. Then each element of $(\mathbb{C}[\mathfrak{h}] \otimes (\stackrel{2}{\wedge} \mathfrak{h}^\perp)^T)^W$ can be expressed in the form

$$\omega = \sum_{\alpha \in \Phi^+} f_\alpha \ e_\alpha^* \wedge e_{-\alpha}^* \in \left(\mathbb{C}[\mathfrak{h}^*] \otimes (\stackrel{2}{\wedge} \mathfrak{h}^\perp)^T\right)^W.$$

Since ω is W-invariant, if we apply the simple reflection $s_\alpha \in W$ with respect to a root $\alpha \in \Phi^+$ to ω we have $s_\alpha f_\alpha = -f_\alpha$. Hence f_α is divisible by α.

Next let us recall the definition of the Kostant–Kirillov form. The Kostant–Kirillov form ζ is a symplectic form on the (co)adjoint orbit $G \cdot x \subset \mathfrak{g}$ of $x \in \mathfrak{g}$. Let $Y, Z \in \mathfrak{g}$. Then $[Y, x] = \frac{d}{dt}\big|_{t=0} \mathrm{ad}(e^{tX})x \in T_x(G \cdot x)$. For two tangent vectors $[Y, x], [Z, x] \in T_x(G \cdot x)$, the 2-form ζ is given by the formula

$$\zeta([Y, x], [Z, x]) = I(x, [Y, Z]),$$

where $[Y, Z]$ is the bracket in $\mathfrak{g}$.

Proposition 4.7. *By restricting the Kostant–Kirillov form ζ to $\mathfrak{h}$, we have the following expression*

$$(16) \qquad \rho_2(\zeta) = -\sum_{\alpha \in \Phi^+} \frac{e_\alpha^* \wedge e_{-\alpha}^*}{\alpha}.$$

Proof. Let $h \in \mathfrak{h} \backslash \bigcup H_\alpha$. We compute $\zeta([e_\alpha, h], [e_\beta, h])$ in two ways. First, using the property $[h, e_\alpha] = \alpha(h)e_\alpha$, we have $\zeta([e_\alpha, h], [e_\beta, h]) = \alpha(h)\beta(h)\zeta(e_\alpha, e_\beta)$. On the other hand, using the definition of ζ, we have $\zeta([e_\alpha, h], [e_\beta, h]) = I(h, [e_\alpha, e_\beta])$. Note that it is non-zero only if $\beta = -\alpha$, and in this case, we have $I(h, [e_\alpha, e_{-\alpha}]) = \alpha(h)$. Hence we have $\zeta(e_\alpha, e_{-\alpha}) = \frac{-1}{\alpha(h)}$, which implies (16). Q.E.D.

The generic fiber of $\chi : \mathfrak{g} \to B$ is isomorphic to G/T, which is homotopy equivalent to the flag manifold of G. We recall the Borel–Hirzebruch description of the de Rham cohomology of G/T in degree 2 [9]. A G-invariant differential form on G/T can be seen as a G-invariant section of the vector bundle $\overset{\bullet}{\wedge} T^*(G/T)$. Hence the evaluation at the base point $[T] \in G/T$ induces an isomorphism

$$\lambda : \Omega_{G/T}^{\bullet,G} \xrightarrow{\cong} \left(\overset{\bullet}{\wedge} T^*_{[T]}(G/T)\right)^T \cong \left(\overset{\bullet}{\wedge} \mathfrak{h}^{\perp}\right)^T. \tag{17}$$

For degree 2, the above map induces the isomorphism $\Omega_{G/T}^{\bullet,G} \cong (\wedge^2\mathfrak{h}^{\perp})^T = \bigoplus_{\alpha\in\Phi^+}(\mathfrak{g}_\alpha \oplus \mathfrak{g}_{-\alpha})$. Using the map λ, we can show that $H^2(G/T, \mathbb{C}) \cong \mathfrak{h}$.

Theorem 4.8. [9]

$$\begin{array}{rccc} \omega : & \mathfrak{h} & \longrightarrow & \Omega^{2,G}_{(G/T)} \\ & h & \longmapsto & \sum_{\alpha\in\Phi} \alpha(h) \cdot \lambda^{-1}(e^*_\alpha \wedge e^*_{-\alpha}) \end{array}$$

induces an isomorphism of $\mathbb{C}$*-vector spaces* $\mathfrak{h} \xrightarrow{\cong} H^2(G/T, \mathbb{C})$. *In particular,* $\omega(h)$ *is closed,*

$$\sum_{\alpha\in\Phi} \alpha(h) \cdot d\lambda^{-1}(e^*_\alpha \wedge e^*_{-\alpha}) = 0. \tag{18}$$

Next we consider the relative de Rham cohomology for the projection

$$\mathrm{pr} : \mathfrak{h} \times G/T \to \mathfrak{h}.$$

We may consider G acts on each fiber of pr from the left. Since $\Omega^2_{\mathrm{pr}} \cong \mathbb{C}[\mathfrak{h}] \otimes \Omega^2_{G/T}$, the set of G-invariant relative 2-forms Ω^2_{pr} is described by the following isomorphism

$$1 \otimes \lambda : \Omega^2_{\mathrm{pr}} \xrightarrow{\cong} \mathbb{C}[\mathfrak{h}] \otimes (\overset{2}{\wedge} \mathfrak{h}^{\perp})^T. \tag{19}$$

By definition and Theorem 4.8,

$$H^2(\Omega^\bullet_{\mathrm{pr}}) \cong \mathbb{C}[\mathfrak{h}] \otimes H^2(\Omega^\bullet_{(G/T)}) \cong \mathbb{C}[\mathfrak{h}] \otimes \mathfrak{h} \cong \mathrm{Der}_{\mathfrak{h}}.$$

The isomorphism is given by

Theorem 4.9.

$$\begin{array}{rccc} 1 \otimes \omega : & \mathrm{Der}_{\mathfrak{h}} & \longrightarrow & \mathbb{C}[\mathfrak{h}] \otimes \Omega^{2,G}_{(G/T)} \cong \Omega^{2,G}_{\mathrm{pr}} \\ & \delta & \longmapsto & (1\otimes\omega)\delta = \sum_{\alpha\in\Phi} (\delta\alpha) \cdot \lambda^{-1}(e^*_\alpha \wedge e^*_{-\alpha}) \end{array} \tag{20}$$

induces a $\mathbb{C}[\mathfrak{h}]$*-module isomorphism* $\mathrm{Der}_{\mathfrak{h}} \cong H^2(\Omega^\bullet_{\mathrm{pr}})$*, where* $\delta\alpha$ *is the differentiation of a function* α *by a vector field* δ*.*

We define a submodule $\mathcal{H}_{\mathrm{pr}} \subset \Omega^{2,G}_{(G/T)\times\mathfrak{h}}$ as

$$\mathcal{H}_{\mathrm{pr}} := \left\{ \sum_{\alpha\in\Phi} (\delta\alpha)\cdot\lambda^{-1}(e^*_\alpha \wedge e^*_{-\alpha}) \in \Omega^{2,G}_{(G/T)\times\mathfrak{h}} \;\middle|\; \delta \in \mathrm{Der}_{\mathfrak{h}} \right\}.$$

From the decomposition $\Omega^{2,G}_{(G/T)\times\mathfrak{h}} = \bigoplus_{i+j=2} \Omega^{i,G}_{G/T} \wedge \Omega^j_{\mathfrak{h}}$ and $\Omega^{2,G}_{\mathrm{pr}} \cong \mathbb{C}[\mathfrak{h}] \otimes \Omega^{2,G}_{G/T}$, we may consider

$$\Omega^{2,G}_{(G/T)\times\mathfrak{h}} \supset \Omega^{2,G}_{\mathrm{pr}} \supset \mathcal{H}_{\mathrm{pr}} \;\left(\cong H^2(\Omega^\bullet_{\mathrm{pr}})\right). \tag{21}$$

Let $x_1, \cdots, x_\ell$ be a coordinate system of $\mathfrak{h}$, then $\mathcal{H}_{\mathrm{pr}}$ has another expression as in (14):

$$\mathcal{H}_{\mathrm{pr}} = \left\{ \omega \in \Omega^{2,G}_{\mathrm{pr}} \;\middle|\; d\omega \wedge dx_1 \wedge \cdots \wedge dx_\ell = 0 \text{ in } \Omega^{3+\ell,G}_{(G/T\times\mathfrak{h})} \right\}. \tag{22}$$

To study the relative de Rham cohomology of $\chi : \mathfrak{g} \to B$, we consider the following diagram

$$\begin{array}{ccc} (G/T)\times\mathfrak{h} & \xrightarrow{\widetilde{\pi}} & \mathfrak{g} \\ \mathrm{pr}\downarrow & & \downarrow\chi \\ \mathfrak{h} & \xrightarrow{\pi} & B \end{array} \quad \left(\begin{array}{ccc} (g[T],h) & \longmapsto & \mathrm{ad}(g)h \\ \downarrow & & \downarrow \\ h & \longmapsto & \overline{h} \end{array} \right). \tag{23}$$

More precisely, from diagram (23), there is a natural homomorphism

$$\widetilde{\pi}^* : H^2\left(\Omega^\bullet_\chi\right) \hookrightarrow H^2\left(\Omega^\bullet_{\mathrm{pr}}\right),$$

which is injective because we have realized these cohomology groups as subspaces of absolute differential forms (see (15) and (21)). We consider the image of $H^2(\Omega^\bullet_\chi)$ in $H^2(\Omega^\bullet_{\mathrm{pr}}) \cong \mathrm{Der}_{\mathfrak{h}}$. Note that if we define a W action on $G/T \times \mathfrak{h}$ by

$$w \cdot (g[T], h) = (gn_w^{-1}, \mathrm{ad}(n_w)h),$$

where $n_w \in N_G(T)$ is a representative of $w \in W = N_G(T)/T$, then obviously $\widetilde{\pi}$ is a W-invariant map and the pull back of differential form (resp. relative cohomology class) on $\mathfrak{g}$ by $\widetilde{\pi}$ becomes a W-invariant differential form (resp. W-invariant cohomology class).

$$\widetilde{\pi}^*\mathcal{H}_\chi \subset \mathcal{H}^W_{\mathrm{pr}}. \tag{24}$$

Now recall two expressions of relative 2-forms (12) in Proposition 4.5 and (19), we have a diagram:

(25)

$$\begin{array}{ccccc} \Omega^{2,G}_{\mathcal{X}} & \overset{\widetilde{\pi}^*}{\hookrightarrow} & \Omega^{2,G\times W}_{\mathrm{pr}} & \subset & \Omega^{2,G}_{\mathrm{pr}} \\ \rho_2 \downarrow \wr & & 1\otimes\lambda \downarrow \wr & & \downarrow \wr \\ \left(\mathbb{C}[\mathfrak{h}^*]\otimes(\bigwedge^2 \mathfrak{h}^\perp)^T\right)^W & \overset{(1\otimes\lambda)\circ\widetilde{\pi}^*\circ\rho_2^{-1}}{\longrightarrow} & \left(\mathbb{C}[\mathfrak{h}^*]\otimes(\bigwedge^2 \mathfrak{h}^\perp)^T\right)^W & \subset & \mathbb{C}[\mathfrak{h}^*]\otimes(\bigwedge^2 \mathfrak{h}^\perp)^T. \end{array}$$

We compute the map $(1\otimes\lambda)\circ\widetilde{\pi}^*\circ\rho_2^{-1}$.

Lemma 4.10. *The map* $(1\otimes\lambda)\circ\widetilde{\pi}^*\circ\rho_2^{-1} : \left(\mathbb{C}[\mathfrak{h}^*]\otimes(\bigwedge^2 \mathfrak{h}^\perp)^T\right)^W \to \left(\mathbb{C}[\mathfrak{h}^*]\otimes(\bigwedge^2 \mathfrak{h}^\perp)^T\right)^W$ *can be expressed as*

$$\sum_{\alpha\in\Phi+} \alpha f_\alpha\cdot e^*_\alpha\wedge e^*_{-\alpha} \longmapsto -\sum_{\alpha\in\Phi+}\alpha^3 f_\alpha\cdot e^*_\alpha\wedge e^*_{-\alpha}.$$

Proof. The derivation of $\widetilde{\pi}$ is given by

$$\begin{array}{cccc} (d\widetilde{\pi})_{([T],h)}: & T_{([T],h)}(G/T\times\mathfrak{h}) & \longrightarrow & T_h\mathfrak{g} \\ & \|\wr & & \|\wr \\ & \mathfrak{g}/\mathfrak{h}\oplus\mathfrak{h} & & \mathfrak{g} \\ & & & \\ & (\bar{X}_1, X_2) & \longmapsto & [X_1,h]+X_2. \end{array}$$

Indeed

$$(d\widetilde{\pi})_{([T],h)}(\bar{X}_1,X_2) = \left.\frac{d}{dt}\right|_{t=0} \mathrm{ad}(\exp(tX_1))(h+tX_2) = [X_1,h]+X_2.$$

In particular $\mathfrak{g}/\mathfrak{h}\oplus\mathfrak{h}\ni(\bar{e}_\alpha,0)\longmapsto -\alpha(h)e_\alpha\in\mathfrak{g}_\alpha$. Hence we have

$$\widetilde{\pi}^*(e^*_\alpha) = -\alpha(h)e^*_\alpha\ ,\ \widetilde{\pi}^*(e^*_{-\alpha}) = \alpha(h)e^*_{-\alpha}.$$

Q.E.D.

Example 4.11. *From Proposition 4.7 and the preceding lemma, the pull back of the Kostant–Kirillov form* ζ *is*

$$\widetilde{\pi}^*(\zeta) = \sum_{\alpha\in\Phi+}\alpha\cdot\lambda^{-1}(e^*_\alpha\wedge e^*_{-\alpha}).$$

Using the Euler vector field $\theta_E := \sum_{i=1}^{\ell} x_i\frac{\partial}{\partial x_i}$, *it is expressed as* $\widetilde{\pi}^*(\zeta) = (1\otimes\omega)(\theta_E)$.

As a corollary of Lemma 4.10, we can characterize the image of the map $\widetilde{\pi}^* : \Omega_X^{2,G} \to \Omega_{\mathrm{pr}}^{2,G\times W}$.

Corollary 4.12.

$$(1\otimes\lambda)\circ\widetilde{\pi}^*(\Omega_X^{2,G}) = \left\{ \sum_{\alpha\in\Phi^+} F_\alpha \cdot e_\alpha^* \wedge e_{-\alpha}^* \;\middle|\; F_\alpha \text{ can be divisible by } \alpha^3 \right\}^W .$$

Proof. If $\sum_{\alpha\in\Phi^+} F_\alpha \cdot e_\alpha^* \wedge e_{-\alpha}^*$ is contained in the left hand side above, F_α have to be divisible by α^3 from the preceding lemma. Conversely, if F_α is divisible by α^3 for all $\alpha \in \Phi^+$, it is the image of

$$\rho_2^{-1}\left(\sum_{\alpha\in\Phi^+} \frac{F_\alpha}{\alpha^2} \cdot e_\alpha^* \wedge e_{-\alpha}^*\right) \in \Omega_X^{2,G}.$$

Q.E.D.

Here it is possible to characterize the image of $\widetilde{\pi}^*$. We have a diagram deduced from (25),

$$\text{(26)} \qquad \begin{array}{ccccc} \mathcal{H}_X \subset \mathcal{H}'_X & \overset{\widetilde{\pi}^*}{\hookrightarrow} & \mathcal{H}_{\mathrm{pr}}^W & \subset & \mathcal{H}_{\mathrm{pr}} \\ & & (1\otimes\omega)^{-1} \downarrow\wr & & \downarrow\wr \\ & & \mathrm{Der}_{\mathfrak{h}}^W & \subset & \mathrm{Der}_{\mathfrak{h}} . \end{array}$$

Combining (20) and Corollary 4.12, we have

Theorem 4.13. $(1\otimes\omega)^{-1}\circ\widetilde{\pi}^*$ *induces an isomorphism*

$$\mathcal{H}'_X \cong D(\mathcal{A},3)^W .$$

Proof. For any $\delta \in \mathrm{Der}_{\mathfrak{h}}^W$, since

$$\begin{aligned} d\left((1\otimes\omega)\delta\right) \wedge dP_1 \wedge \cdots \wedge dP_\ell &= Q \cdot d\left((1\otimes\omega)\delta\right) \wedge dx_1 \wedge \cdots \wedge dx_\ell \\ &= 0, \end{aligned}$$

$(1\otimes\omega)\delta \in \widetilde{\pi}^*\mathcal{H}'_X$ if and only if $\delta\alpha$ is divisible by α^3 for all $\alpha \in \Phi^+$.

Q.E.D.

Now we are in a position to prove our main result.

Theorem 4.14. $(1\otimes\omega)^{-1}\circ\widetilde{\pi}^*$ *induces an isomorphism*

$$\mathcal{H}_X \cong D(\mathcal{A},5)^W .$$

Hence $H^2(\Omega_X^\bullet)$ is a free $\mathbb{C}[B]$-module of rank ℓ.

Proof. Suppose $\eta \in \mathcal{H}_\chi$ and put $\widetilde{\pi}^*\eta = \sum_{\alpha\in\Phi}(\delta\alpha)\cdot\lambda^{-1}(e^*_\alpha \wedge e^*_{-\alpha})$ for $\delta \in D(\mathcal{A},3)^W$, then by definition there exist $\eta_1,\cdots,\eta_\ell \in \Omega^{2,G}_{\mathfrak{g}}$ such that

$$d\eta = \sum_{i=1}^{\ell} dP_i \wedge \eta_i.$$

Applying the operator d and multiplying by $dP_1\wedge\cdots\wedge\widehat{dP_i}\wedge\cdots\wedge dP_\ell$,

$$dP_1\wedge\cdots\wedge dP_i\wedge\cdots\wedge dP_\ell\wedge d\eta_i = 0.$$

Hence $\eta_i \in \mathcal{H}'_\chi$ for all $i=1,\cdots,\ell$ and $\widetilde{\pi}^*\eta_i = \sum_{\alpha\in\Phi}(\delta_i\alpha)\cdot\lambda^{-1}(e^*_\alpha\wedge e^*_{-\alpha})$ for some $\delta_i \in D(\mathcal{A},3)^W$. Using (18), we have

$$\begin{aligned} d\widetilde{\pi}^*\eta &= \sum_{\alpha\in\Phi} d(\delta\alpha)\wedge\lambda^{-1}(e^*_\alpha\wedge e^*_{-\alpha}) \\ &= \sum_{\alpha\in\Phi}\sum_i \frac{\partial}{\partial P_i}(\delta\alpha)dP_i\wedge\lambda^{-1}(e^*_\alpha\wedge e^*_{-\alpha}) \\ &= \sum_{\alpha\in\Phi}\sum_i \left((\nabla_{\frac{\partial}{\partial P_i}}\delta)\alpha\right)dP_i\wedge\lambda^{-1}(e^*_\alpha\wedge e^*_{-\alpha}), \end{aligned}$$

and $\delta_i = \nabla_{\frac{\partial}{\partial P_i}}\delta$. By Theorem 3.5, $\eta\in\mathcal{H}_\chi$ if and only if $\delta \in D(\mathcal{A},5)^W$.

Q.E.D.

4.2. Freeness of A_n-Catalan arrangements

As another application, we prove that Catalan arrangements of type A are free.

Let $(x_1,\ldots,x_n,z)$ be a coordinate of $\mathbb{C}^{n+1}$. (The cone of) Catalan arrangement Cat_n is defined by

$$z\times\prod_{1\le i<j\le n}((x_i-x_j)(x_i-x_j-z)(x_i-x_j+z)) = 0.$$

The terminology "Catalan arrangement" comes from the fact that the number of chambers divided by $2n!$ is equal to n-th Catalan number. See [7] for more combinatorial aspects of Cat_n. (Note that the definition of Cat_n in this article is the coning of that of [7].)

Theorem 4.15. *The Catalan arrangement* Cat_n *is free with exponents* $(0,1,n+1,n+2,\ldots,2n-1)$.

Remark 4.16. *This result was first proved by Edelman and Reiner* [15] *using Addition-Deletion Theorem 2.21. It can be also proved by* *using the freeness of $D(\mathcal{A},3)$ and Theorem 2.19. We give another proof which is also based on the freeness of $D(\mathcal{A},3)$. However, instead of using Theorem 2.19, we will directly show the existence of basis of $D(\mathrm{Cat}_n)$ by invariant theoretic arguments.*

Lemma 4.17. *For non-negative integers $p, i \geq 0$, define a symmetric polynomial $F_{p,i}(x_1, x_2)$ of two variables by $F_{p,i} := \frac{x_1^{p+1} - x_2^{p+1}}{x_1 - x_2} \cdot (x_1 - x_2)^{2i}$. Then*

$$\mathbb{C}[x_1, x_2]^{\mathfrak{S}_2} = \bigoplus_{p,i \geq 0} \mathbb{C} \cdot F_{p,i}.$$

Proof. Let $F(x_1, x_2)$ be a homogeneous symmetric polynomial. We will prove F is expressed a linear combination $\{F_{p,i}\}$ by induction on $\deg F$. If $\deg F = 1$, then $F = x_1 + x_2 = F_{1,0}$. Consider the case $\deg F \geq 2$. If $F(x, x) = 0$, then we have $F = (x_1 - x_2)^2 \cdot G(x_1, x_2)$ with G symmetric. Thus by inductive hypothesis, G is a linear combination of $\{F_{p,i}\}$, and so is F. Suppose $F(x, x) = ax^n \neq 0$. Consider $G(x_1, x_2) = F - \frac{a}{n+1} F_{n,0}$. Then $G(x, x) = 0$, and is reduced to the previous case. Thus $\mathbb{C}[x_1, x_2]^{\mathfrak{S}_2}$ is spanned by $\{F_{p,i}\}$. By computing the Hilbert series, we have (note that $\deg F_{p,i} = p + 2i$)

$$H(\mathbb{C}[x_1, x_2]^{\mathfrak{S}_2}, t) = \frac{1}{(1-t)(1-t^2)} \leq H(\sum \mathbb{C} \cdot F_{p,i}, t) \leq \frac{1}{(1-t)(1-t^2)}.$$

We conclude that $\{F_{p,i}\}$ forms a basis. Q.E.D.

Let $S = \mathbb{C}[x_1, \ldots, x_n]$. We consider a subgroup $\mathfrak{S}_2 \times \mathfrak{S}_{n-2} \subset \mathfrak{S}_n$ which acts $\{x_1, x_2\}$ and $\{x_3, \ldots, x_n\}$ respectively.

Lemma 4.18.

$$S^{\mathfrak{S}_2 \times \mathfrak{S}_{n-2}} = \mathbb{C}[x_1, x_2]^{\mathfrak{S}_2} \cdot S^{\mathfrak{S}_n}.$$

Proof. The inclusion $\supseteq$ is clear. For the reverse inclusion, we first note that

$$S^{\mathfrak{S}_2 \times \mathfrak{S}_{n-2}} = \mathbb{C}[x_1, x_2]^{\mathfrak{S}_2} \otimes \mathbb{C}[x_3, \ldots, x_n]^{\mathfrak{S}_{n-2}}.$$

As is well known, $S^{\mathfrak{S}_n}$ is generated by $x_1^k + \cdots + x_n^k$ $(k \geq 0)$ as a $\mathbb{C}$-algebra. Thus

$$x_3^k + \cdots + x_n^k = (x_1^k + \cdots + x_n^k) - (x_1^k + x_2^k) \in \mathbb{C}[x_1, x_2]^{\mathfrak{S}_2} \cdot S^{\mathfrak{S}_n}.$$

Q.E.D.

Combining Lemma 4.17 and 4.18, we have.

Lemma 4.19.

$$S^{\mathfrak{S}_2\times\mathfrak{S}_{n-2}} = \sum_{p,i\geq 0} S^{\mathfrak{S}_n}\cdot F_{p,i}.$$

Now we prove Theorem 4.15. Denote $H_0 = \{z=0\}$ and $\alpha_{ij} = x_i - x_j$. The restriction $(\mathcal{A}^{H_0}, m^{H_0})$ is equal to $(A_{n-1}, 3)$. From the result of §3, $(\mathcal{A}^{H_0}, m^{H_0})$ is free with exponents $(0, n+1, \ldots, 2n-1)$. We can choose basis $\eta_1, \ldots, \eta_n \in D(\mathcal{A}^{H_0}, m^{H_0})^{\mathfrak{S}_n}$ from $\mathfrak{S}_n$-invariant vector fields. By definition, $\eta_i(x_1-x_2) = (x_1-x_2)^3 G_i$, and $G_i \in S^{\mathfrak{S}_2\times\mathfrak{S}_{n-2}}$. By Lemma 4.19, there exist symmetric polynomials $B_i^{p,r} \in S^{\mathfrak{S}_n}$ such that

$$G_i = \sum_{p,r\geq 0} B_i^{p,r}\cdot F_{p,r}. \tag{27}$$

With $B_i^{p,r}$ and $\mathfrak{S}_n$-invariant vector field $\delta_p = \sum_{i=1}^n x_i^p \partial_i$, let us define

$$\widetilde{\eta}_i := \eta_i - \sum_{p,r\geq 0} z^{2r+2} B_i^{p,r}\delta_{p+1}. \tag{28}$$

Then $\widetilde{\eta}_1, \ldots, \widetilde{\eta}_n$ and $\theta_E = \sum_{i=1}^n x_i\partial_i + z\partial_z$ form a basis of $D(\mathrm{Cat}_n)$. Indeed, they are linearly independent over $\mathbb{C}[x_1, \ldots, x_n, z]$ (since η_i's are independent), and

$$\begin{aligned}
\widetilde{\eta}_i(\alpha_{12}\pm z) &= \widetilde{\eta}_i\alpha_{12}\\
&= \eta(x_1-x_2) - \sum_{p,r\geq 0} z^{2r+2}B_i^{p,r}(x_1^{p+1}-x_2^{p+1})\\
&= (x_1-x_2)^3 G_i - \sum_{p,r\geq 0} z^{2r+2}B_i^{p,r}(x_1^{p+1}-x_2^{p+1}).
\end{aligned}$$

The last polynomial is divisible by $z^2-\alpha_{12}^2$. Indeed, put $z=\pm\alpha_{12}$ in the last formula, we have from (27)

$$\begin{aligned}
&= (x_1-x_2)^3 G_i - \sum_{p,r\geq 0}(x_1-x_2)^{2r+2}B_i^{p,r}(x_1^{p+1}-x_2^{p+1})\\
&= (x_1-x_2)^3\left(G_i - \sum_{p,r\geq 0}(x_1-x_2)^{2r}B_i^{p,r}\frac{x_1^{p+1}-x_2^{p+1}}{x_1-x_2}\right)\\
&= (x_1-x_2)^3\left(G_i - \sum_{p,r\geq 0}B_i^{p,r}F_{p,r}\right)\\
&= 0.
\end{aligned}$$

Since $\widetilde{\eta}_i$ is $\mathfrak{S}_n$-invariant, $\widetilde{\eta}_i\alpha_{jk}$ is divisible by $\alpha_{jk}(\alpha_{jk}^2 - z^2)$ for any j, k. This completes the proof.

§5. Concluding remarks and open problems

One of the central problems in the theory of hyperplane arrangements is to decide to what extent the structure of an arrangement is determined by the combinatorics of the arrangement.

Problem 5.1. *Let $\mathcal{A}_1$ and $\mathcal{A}_2$ be central arrangements in $\mathbb{K}^\ell$. Assume that $L(\mathcal{A}_1) \simeq L(\mathcal{A}_2)$. Does the freeness of $\mathcal{A}_1$ imply the freeness of $\mathcal{A}_2$?*

The above (by Terao [39]) is a long standing problem, even for the case $\ell = 3$, since the beginning of this area. Note that several variants of this problem are known to have counter examples:

- $\mathcal{A}_1$ and $\mathcal{A}_2$ are multiarrangements ([55]),
- $\mathcal{A}_1$ and $\mathcal{A}_2$ are defined over different fields ([56]),
- $\mathcal{A}_1$ and $\mathcal{A}_2$ are line-conic arrangements ([30]).

There are several characterizations for 3-arrangements to be free via Ziegler's restriction map (§2.3 and see [2] for recent developments). However the author does not know the answer to the following.

Problem 5.2. *Does the converse to Theorem 2.22 hold?*

As in §4.1 the modules of multiderivations naturally appear in the study of relative de Rham cohomology groups for the adjoint quotient map $\chi : \mathfrak{g} \to \mathfrak{h}/W$. However the idea of the proof of Theorem 4.2 works only for the type ADE and the cohomology of degree 2.

Problem 5.3. *Study the structure of the relative de Rham cohomology group $H^k(\Omega_\chi)$ as S^W-module.*

The structures of $H^2(\Omega_\chi)$ for non simply laced cases are expected to be related with the module of derivations with a non constant multiplicity. Similarly higher degree cases are expected to be related with the module of higher derivations $D^k(\mathcal{A}, m)$ defined in [3].

Postnikov and Stanley [24] observed curious properties of the characteristic polynomials that for some truncated affine Weyl arrangements, the all roots of the characteristic polynomial have the same real part ("Riemann hypothesis"). By Solomon–Terao's formula (Theorem 2.9), the characteristic polynomial is determined by the module $D(\mathcal{A})$. It would be natural to expect curious behaviours of the characteristic polynomial reflect the structure of $D(\mathcal{A})$. The following question posed by Athanasiadis ([7, Question 6.5]) is still challenging.

Problem 5.4. *Are there any natural algebraic structures of $D(\mathcal{A})$ which cause Riemann hypothesis?*

For instance, is the "functional equation" ([24, (9.12)]) deduced from the self duality (up to degree shift) $D_0(\mathcal{A}) \simeq D_0(\mathcal{A})[-d]^\vee$ of certain module?

References

[1] T. Abe, A generalized logarithmic module and duality of Coxeter multiarrangements, arXiv:0807.2552.

[2] T. Abe, Exponents of 2-multiarrangements and freeness of 3-arrangements, arXiv:1005.5276.

[3] T. Abe, H. Terao, and M. Wakefield, The characteristic polynomial of a multiarrangement, Adv. in Math., **215** (2007), 825–838.

[4] T. Abe, H. Terao and M. Wakefield, The Euler multiplicity and the addition-deletion theorems for multiarrangements, J. Lond. Math. Soc. (2), **77** (2008), 335–348.

[5] T. Abe, H. Terao and M. Yoshinaga, Totally free arrangements of hyperplanes, Proc. Amer. Math. Soc., **137** (2009), 1405–1410.

[6] T. Abe and M. Yoshinaga, Coxeter multiarrangements with quasi-constant multiplicities, J. Algebra, **322** (2009), 2839–2847.

[7] C. A. Athanasiadis, Deformations of Coxeter hyperplane arrangements and their characteristic polynomials, In: Arrangements—Tokyo 1998, Adv. Stud. Pure Math., **27**, Kinokuniya, Tokyo, 2000, pp. 1–26.

[8] C. A. Athanasiadis, Extended Linial hyperplane arrangements for root systems and a conjecture of Postnikov and Stanley, J. Algebraic Combin., **10** (1999), 207–225.

[9] A. Borel and F. Hirzebruch, Characteristic classes and homogeneous spaces. I, Amer. J. Math., **80** (1958), 458–538.

[10] E. Brieskorn, Die Monodromie der isolierten Singularitäten von Hyperflächen, Manuscripta Math., **2** (1970), 103–161.

[11] E. Brieskorn, Singular elements of semi-simple algebraic groups, In: Actes du Congrès International des Mathématiciens, Nice, 1970, Tome 2, pp. 279–284.

[12] A. Broer, The sum of generalized exponents and Chevalley's restriction theorem for modules of covariants, Indag. Math. (N.S.), **6** (1995), 385–396.

[13] C. Chevalley, Invariants of finite groups generated by reflections, Amer. J. Math., **77** (1955), 778–782.

[14] I. Dolgachev and M. Kapranov, Arrangements of hyperplanes and vector bundles on $\mathbf{P}^n$, Duke Math. J., **71** (1993), 633–664.

[15] P. H. Edelman and V. Reiner, Free arrangements and rhombic tilings, Discrete Comput. Geom., **15** (1996), 307–340.

[16] P. H. Edelman and V. Reiner, Not all free arrangements are $K(\pi, 1)$, Bull. Amer. Math. Soc. (N.S.), **32** (1995), 61–65.
[17] R. Hartshorne, Stable reflexive sheaves, Math. Ann., **254** (1980), 121–176.
[18] M. Lehn, Y. Namikawa and Ch. Sorger, Slodowy slices and universal Poisson deformations, preprint, arXiv:1002.4107.
[19] A. Matsuo, Summary of the theory of primitive forms, In: Topological Field Theory, Primitive Forms and Related Topics, Kyoto, 1996, Progr. Math., **160**, Birkhäuser Boston, Boston, MA, 1998, pp. 337–363.
[20] M. Mustaţă and H. Schenck, The module of logarithmic p-forms of a locally free arrangement, J. Algebra, **241** (2001), 699–719.
[21] C. Okonek, M. Schneider and H. Spindler, Vector bundles on complex projective spaces, Progr. Math., **3**, Birkhäuser Boston, Boston, MA, 1980.
[22] P. Orlik and L. Solomon, Combinatorics and topology of complements of hyperplanes, Invent. Math., **56** (1980), 167–189.
[23] P. Orlik and H. Terao, Arrangements of Hyperplanes, Grundlehren Math. Wiss., **300**, Springer-Verlag, 1992. xviii+325 pp.
[24] A. Postnikov and R. Stanley, Deformations of Coxeter hyperplane arrangements, J. Combin. Theory Ser. A, **91** (2000), 544–597.
[25] K. Saito, Theory of logarithmic differential forms and logarithmic vector fields, J. Fac. Sci. Univ. Tokyo Sect. IA Math., **27** (1980), 265–291.
[26] K. Saito, Period mapping associated to a primitive form, Publ. Res. Inst. Math. Sci., **19** (1983), 1231–1264.
[27] K. Saito, On a linear structure of the quotient variety by a finite reflexion group, Publ. Res. Inst. Math. Sci., **29** (1993), 535–579.
[28] H. Schenck, A rank two vector bundle associated to a three arrangement, and its Chern polynomial, Adv. Math., **149** (2000), 214–229.
[29] H. Schenck, Elementary modifications and line configurations in $\mathbb{P}^2$, Comment. Math. Helv., **78** (2003), 447–462.
[30] H. Schenck and S. Tohăneanu, Freeness of conic-line arrangements in $\mathbb{P}^2$, Comment. Math. Helv., **84** (2009), 235–258.
[31] M. Schulze, Freeness and multirestriction of hyperplane arrangements, arXiv:1003.0917.
[32] R. Silvotti, On the Poincaré polynomial of a complement of hyperplanes, Math. Res. Lett., **4** (1997), 645–661.
[33] P. Slodowy, Simple Singularities and Simple Algebraic Groups, Lecture Notes in Math., **815**, Springer-Verlag, 1980.
[34] L. Solomon, Invariants of finite reflection groups, Nagoya Math. J., **22** (1963), 57–64.
[35] L. Solomon and H. Terao, A formula for the characteristic polynomial of an arrangement, Adv. in Math., **64** (1987), 305–325.
[36] L. Solomon and H. Terao, The double Coxeter arrangement, Comm. Math. Helv., **73** (1998), 237–258.
[37] H. Terao, Arrangements of hyperplanes and their freeness. I, II, J. Fac. Sci. Univ. Tokyo Sect. IA Math., **27** (1980), 293–320.

[38] H. Terao, Generalized exponents of a free arrangement of hyperplanes and Shepherd–Todd–Brieskorn formula, Invent. Math., **63** (1981), 159–179.

[39] H. Terao, The exponents of a free hypersurface, In: Singularities, Part 2, Arcata, CA, 1981, Proc. Sympos. Pure Math., **40**, Amer. Math. Soc., Providence, RI, 1983, pp. 561–566.

[40] H. Terao, Multiderivations of Coxeter arrangements, Invent. Math., **148** (2002), 659–674.

[41] H. Terao, The Hodge filtration and the contact-order filtration of derivations of Coxeter arrangements, Manuscripta Math., **118** (2005), 1–9.

[42] K. Ueda and M. Yoshinaga, Logarithmic vector fields along smooth plane cubic curves, Kumamoto J. Math., **21** (2008), 11–20.

[43] K. Ueda and M. Yoshinaga, Logarithmic vector fields along smooth divisors in projective spaces, Hokkaido Math. J., **38** (2009), 409–415.

[44] A. Wakamiko, On the exponents of 2-multiarrangements, Tokyo J. Math., **30** (2007), 99–116.

[45] M. Wakefield and M. Yoshinaga, The Jacobian ideal of a hyperplane arrangement, Math. Res. Letters, **15** (2008), 795–799.

[46] M. Wakefield and S. Yuzvinsky, Derivations of an effective divisor on the complex projective line, Trans. Amer. Math. Ssoc., **359** (2007), 4389–4403.

[47] J. Vey, Un problème de cohomologie relative, Ark. Mat., **15** (1977), 21–31.

[48] H. Yamada, Lie group theoretical construction of period mapping, Math. Z., **220** (1995), 231–255.

[49] M. Yoshinaga, The primitive derivation and freeness of multi-Coxeter arrangements, Proc. Japan Acad. Ser. A Math. Sci., **78** (2002), 116–119.

[50] M. Yoshinaga, Characterization of a free arrangement and conjecture of Edelman and Reiner, Invent. Math., **157** (2004), 449–454.

[51] M. Yoshinaga, On the freeness of 3-arrangements, Bull. London Math. Soc., **37** (2005), 126–134.

[52] S. Yuzvinsky, Cohomology of local sheaves on arrangement lattices, Proc. Amer. Math. Soc., **112** (1991), 1207–1217.

[53] S. Yuzvinsky, The first two obstructions to the freeness of arrangements, Trans. Amer. Math. Soc., **335** (1993), 231–244.

[54] S. Yuzvinsky, Free and locally free arrangements with a given intersection lattice, Proc. Amer. Math. Soc., **118** (1993), 745–752.

[55] G. Ziegler, Multiarrangements of hyperplanes and their freeness, In: Singularities, Iowa City, IA, 1986, Contemp. Math., **90**, Amer. Math. Soc., Providence, RI, 1989, pp. 345–359.

[56] G. Ziegler, Matroid representations and free arrangements, Trans. Amer. Math. Soc., **320** (1990), 525–541.

Masahiko Yoshinaga
Department of Mathematics, Kyoto University,
Kyoto 606-8502, Japan
E-mail address: `mhyo@math.kyoto-u.ac.jp`

Advanced Studies in Pure Mathematics 62, 2012
Arrangements of Hyperplanes—Sapporo 2009
pp. 553–570

Resonance varieties of arrangement complements

Sergey Yuzvinsky

§1. Introduction

This paper is a survey based on the lecture given by the author in the conference "Arrangements of Hyperplanes -Sapporo, 2009" at the University of Hokkaido. The main topic of the survey is the resonance varieties of the complement of an arrangement of several linear complex hyperplanes. These varieties can be defined for an arbitrary topological space M as the jumping loci for the kind of 'secondary cohomology' of M. More precisely this is the cohomology of the graded commutative algebra $H^*(M)$ provided with the differential given by the multiplication by an element of $H^1(M)$. This cohomology has appeared first in topology as the first sheet of the Farber–Novikov spectral system (see, for example [25]) which converges to cohomology with local coefficients for compact manifolds. For arrangement complements first results for this cohomology were vanishing theorems from [30] and [35] and comparison theorems [14]. The jumping loci for this cohomology were first considered explicitly in [15] and called resonance varieties. In [1], this cohomology was considered for an arbitrary graded commutative algebra as the measure of its complexity. For initial results about resonance varieties over arbitrary fields see [16].

At the beginning, the resonance varieties were mainly considered due to their connections with the jumping loci for the cohomology with local coefficients; the most recent results about these connections can be found in [10, 11]. Now resonance varieties appear in many areas of arrangement theory. The most recent appearance is in [5, 6] where these varieties have been used for results on the Milnor fiber cohomology and roots of b-functions. There are also several recent papers (see, for example

Received April 26, 2010.
Revised September 14, 2010.
2010 *Mathematics Subject Classification*. Primary: 52C35, 05B35; Secondary: 16E05, 15A75.

[12, 33]) where some properties of resonance varieties for arrangement complements were analyzed from the point of view which properties of the spaces imply them. This has allowed the authors to generalize them to other topological spaces. See also the survey by A. Suciu in the Proceedings.

The outline of the paper is as follows. In Section 2, we give the main definitions and list the properties of the resonance varieties with just hints of proofs. Then in Section 3, we discuss the structure of the first resonance variety, that is the jumping locus for the first cohomology of the algebra $H^*(M)$. The components of this variety can be characterized equivalently in terms of classical projective geometry and in terms of pencils of plane curves. This theorem has been proved in [17] but here we give a proof that fills a small gap and uses some new ideas and shortcuts. In Section 3, we survey the results on the dimension of the first resonance variety. In Section 4, we list several open questions.

§2. Resonance varieties and their properties

Definition 2.1. *Let M be a topological space and $A = H^*(M)$ its graded cohomology algebra (the coefficients can be apriori from an arbitrary field but for later use we assume they are in $\mathbb{C}$). We write $A^p = H^p(M)$. For every $x \in A^1$ we have $x^2 = 0$ since A is graded-commutative. Thus the multiplication by x defines the differential $A \to A$ of degree +1, i.e., converts A into a cochain complex that we denote by (A, x).*

The main object of this paper is the *cohomology* $H^p(A, x)$ as a function of x.

For the next definition we assume that the linear spaces A^p are finite dimensional.

Definition 2.2. *The p-th resonance variety $R^p = R^p(M)$ is the (determinantal) subvariety of A^1 defined as $R^p = \{x \in A^1 | H^p(A, x) \neq 0\}$.*

The rest of the paper will be devoted to the case where M is the complement in a finite dimensional linear space V ($V \approx \mathbb{C}^\ell$) to an arrangement $\mathcal{A}$ of several (linear) hyperplanes. We will always assume that $\mathcal{A}$ is *essential*, i.e., $\bigcap_{H \in \mathcal{A}} H = 0$. If we consider several arrangements at the same time we will use the symbol $A(\mathcal{A})$ for A.

The cohomology of such an M is determined by theorems of Arnold, Brieskorn and Orlik–Solomon, [3, 4, 26]. For each hyperplane $H \in \mathcal{A}$ fix a linear form α_H with $\ker \alpha_H = H$. Then A can be identified with the subalgbera of the algebra of all the (holomorphic) differential forms on M generated by the logarithmic forms $\frac{d\alpha_H}{\alpha_H}$ $(H \in \mathcal{A})$. The classes e_H of these forms form a canonical basis of A^1 whence for every $x \in A^1$ we have $x = \sum_{H\in\mathcal{A}} x_H e_H$ for some $x_H \in \mathbb{C}$. Relations for the generators are well-known and can be found in [27].

Properties of R^p.

There are several properties of varieties R^p for arrangement complements that hold for all p, $0 \leq p \leq \ell$.

(i) *(linearity of components)* The irreducible components of R^p are linear subspaces of A^1.

(ii) *(sufficient condition for vanishing)* If $\sum_{H\in\mathcal{A}} x_H \neq 0$ then $R^p = 0$ for all p (we will write $R^* = 0$).

Remark 2.3. *If $\mathcal{A}_i$ $(i = 1, 2)$ are arrangements in linear spaces V_i then their product is the arrangement $\mathcal{A}_1 \times \mathcal{A}_2$ in $V_1 \oplus V_2$ of the hyperplanes $H \oplus V_2$ $(H \in \mathcal{A}_1)$ and $V_1 \oplus H$ $(H \in \mathcal{A}_2)$, see* [27]. *One easily checks that $A(\mathcal{A}_1 \times \mathcal{A}_2) = A(\mathcal{A}_1) \otimes_{\mathbb{C}} A(\mathcal{A}_2)$. An non-empty arrangement $\mathcal{A}$ is irreducible if it is not isomorphic to the product of non-empty arrangements. Every non-empty arrangement $\mathcal{A}$ is the product of several uniquely defined irreducible arrangements, irreducible components of $\mathcal{A}$.*

(iii) *(equivalent condition for vanishing)* $R^* \neq 0$ if and only if $\sum_{H\in\mathcal{A}_j} x_H = 0$ for every $j = 1, \ldots, r$ where $\mathcal{A}_j$ is an irreducible component of $\mathcal{A}$.

Remark 2.4. *The subset $sing(\mathcal{A}) = sing(A)$ of A^1 consisting of all x satisfying $H^*(A, x) \neq 0$ is called the singular variety of A. It is usually viewed as a module over the exterior algebra E generated in degree 1 by $\{e_H | H \in \mathcal{A}\}$. See* [1] *for another approach to it and its applications.*

(iv)*(propagation of cohomology)* If $H^p(A, x) \neq 0$ for some p then $H^q(A, x) \neq 0$ (i.e., $R^p \subset R^q$) for every q, $p \leq q \leq \ell$ (i.e., $R^* \neq 0$ is equivalent to $R^\ell \neq 0$).

The known proofs of the properties are of different difficulties. The easiest property is (ii). It is a consequence of the existence of a differential $\partial : A \to A$ of degree -1 that satisfies the signed Leibniz formula and is normed by the condition $\partial(e_H) = 1$ for every $H \in \mathcal{A}$. Then $\frac{1}{\sum_H x_H}\partial$ is a contracting homotopy for (A, x).

If $\mathcal{A}$ is not a product of non-empty arrangements than the converse of (ii) follows from the non-vanishing of the Euler characteristic of $\ker \partial$

(cf. [9]). This and the fact $sing(\mathcal{A}_1 \times \mathcal{A}_2) = sing(\mathcal{A}_1) \times sing(\mathcal{A}_2)$ implies (iii).

Property (i) has been proved for an arbitrary p in [7] and [22]. The idea is to view the cohomology of (A, x) as a linearization of cohomology of another complex.

Finally, property (iv) has been proved in [13] as a corollary of the linearity of the minimal injective resolution for A viewed as an E-module.

A direct linear algebra proof of the propagation for $p = 1$ has been given by J. V. Pereira and goes as follows.

Proof. Assume that $a \in A^1$ satisfies $H^1(A, a) \neq 0$ and $H^p(A, a) = 0$, for some $1 < p \leq \ell$. Let $b \in A^1 \setminus \mathbb{C}a$ with $ab = 0$. Since a and b can be identified with some rational one-forms on $\mathbb{C}^\ell$, there exists a rational function h such that $b = ha$. This immediately implies that the cocycle spaces for a and b in all degrees r coincide: $Z_a^r = Z_b^r$. In particular, this is true for $r = p - 1$ and $r = p$. Since by assumption $B_a^p = Z_a^p$, where B_a^p is the space of coboundaries of degree p for a, we obtain that $\dim B_b^p = \dim B_a^p = \dim Z_a^p = \dim Z_b^p$ whence $B_b^p = B_a^p$.

Now consider the isomorphisms $\phi_a, \phi_b \colon C = A^{p-1}/Z_a^{p-1} \to B_a^p$, given by multiplication by a and b, respectively, and the automorphism $\phi = \phi_a^{-1}\phi_b$ of C. If $c \in A^{p-1}$ is such that its projection to C is an eigenvector of ϕ, then we have $bc = \lambda ac$ for some $\lambda \in \mathbb{C}^*$. But since also $bc = hac$, we have $h = \lambda$, which contradicts the choice of b. Q.E.D.

No proof of this kind is known for arbitrary p.

§3. The first resonance variety

For the first resonance variety R^1 the situation can be simplified.

First, we projectivize the linear space and study an arrangement of projective hyperplanes in the complex projective space. The cohomology algebra of the projectivized complement (that we still denote by A in this section) is the graded subalgebra of the cohomology algebra of M generated by $\bar{e}_{ij} = e_i - e_j$ for $i \neq j$. For these generators the basic relations are $\prod_{j=2,3,\ldots,k} \bar{e}_{i_1 i_j}$ where $\{\alpha_{i_1}, \alpha_{i_2}, \ldots, \alpha_{i_k}\}$ runs through all minimal linearly dependent sets. Second, it suffices to consider arrangements of lines in $\mathbb{P}^2$. Indeed intersect arbitrary $\mathcal{A}$ with a generic plane and either apply the weak Lefschetz theorem or the description of A in terms of the intersection lattice of $\mathcal{A}$. An arrangement of projective lines will be denoted by L.

3.1. Local components

Here we describe the class of the simplest components of R^1 (in some sense trivial).

Suppose $P \in \mathbb{P}^2$ is a point where k lines of L intersect, $k \geq 3$. Define the linear subspace

$$V_P = \{x = \sum_{P \in \ell} x_\ell e_\ell | \sum x_\ell = 0\}$$

of A^1. Using the relations above it is easy to show that $xy = 0$ for every $x, y \in V_P$ and for every $x \in V_P$ the equality $xy = 0$ implies $y \in V_P$. Thus V_P is a component (of dimension $k-1$) of R^1. This component is called a *local component of* R^1.

Notice that there is no uniform upper bound for the dimension of local components of arrangements. As we will see later, this makes local components different from all others.

3.2. Nets in $\mathbb{P}^2$

Now we want to study the non-local components. Our goal is to state the theorem that gives at least two different characterizations of them. For that we need to define the terms to be used.

First we discuss some special configurations of lines and points in $\mathbb{P}^2$.

Definition 3.1. *A finite set L of lines partitioned in k $(k > 2)$ blocks $L = \cup_{j=1}^k L_j$ is a k-net if for every point P which is the intersection of lines from different blocks there is a precisely one line from each block passing through P.*

The following numerical equalities for a k-net are almost obvious. In order to state them we denote by $\mathcal{X}$ the set of all points of intersection of lines from different blocks.

(i) For arbitrary $1 \leq i, j \leq k$ and $\ell \in L$ we have $|L_i| = |L_j| = |\mathcal{X} \cap \ell|$. The latter integer is denoted by d and the net is called (k, d)-*net.*
(ii) $|L| = kd$.
(iii) $|\mathcal{X}| = d^2$.

Nets can be defined purely combinatorially using an incidence relation. Then after identifying two blocks of a (k, d)-net with each other, every other block gives a Latin square of size d and these $k-2$ squares are pairwise orthogonal. If $k = 3$ identifying all blocks gives a multiplication table of a quasi-group.

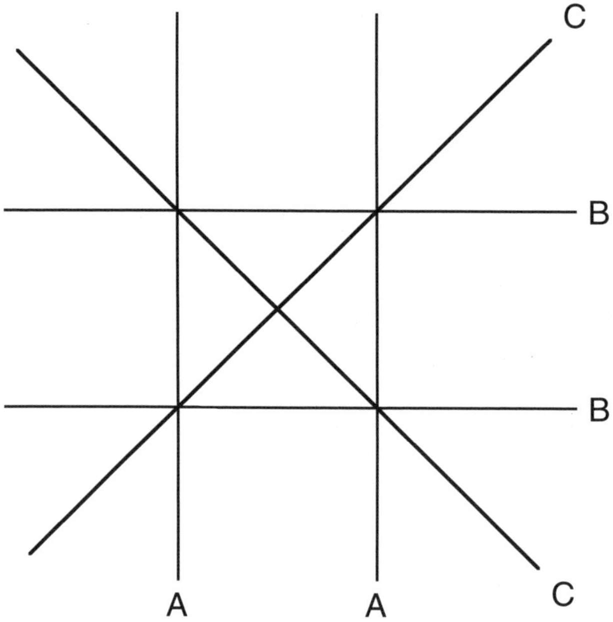

Fig. 1. The (3,2) net

A $(k,1)$-net consists of k lines passing through a point with each block consisting of one line. Clearly maximal $(k,1)$ nets in an arrangements correspond to local components of R^1. Thus from now on we will assume that $d > 1$.

The combinatorial nets that can be realized in $\mathbb{P}^2$ form a very restricted class (e.g., see the restrictions on k below). However there are plenty of examples of 3-nets. The simplest nontrivial one is given by all the reflection lines of the Coxeter group of type A_3. In appropriate coordinates the blocks can be described by the equations

$$(x^2 - y^2) = (x^2 - z^2) = (y^2 - z^2) = 0,$$

see Fig. 1 for the picture of it in the affine plane $z = 1$. This is the only example of a $(3,2)$-net up to a projective isomorphism. As a classical example of a $(3,3)$-net one can use the generic Pappus configuration taking for $\mathcal{X}$ all the triple points.

The first attempt to study 3-nets in $\mathbb{P}^2$ was taken in [36]. In particular, a general way to construct examples of $(3, d)$-nets for every d was found there.

If H is a group of order d then we say that a $(3, d)$-net *realizes* H if there is a way to identify all the blocks of the net to obtain the multiplication table of H.

Theorem 3.2. *Let H be a finite subgroup of a two dimensional torus. Then there exists a 3-net in $\mathbb{P}^2$ realizing H.*

Proof. Fix a non-singular plane cubic curve C and define the additive group operation on it (converting it to the two dimensional torus) by choosing one of the flexes of C as the neutral element. Then three points p_1, p_2, and p_3 from C are collinear if and only if $p_1 + p_2 + p_3 = 0$. We can identify H with a subset of C and put $d = |H|$. Now choose $\alpha, \beta \in C/H$ such that α, β, and $-\alpha - \beta$ are distinct. The union of these three cosets is a set of $3d$ points partitioned in three blocks. It is clear that the lines in the projective plane dual to these points form a 3-net and this net realizes H. Q.E.D.

The construction used in that proof can be extended to reducible cubics. We say that a 3-net is *algebraic* if there exists a cubic curve $C \subset (P^2)^*$ containing the points dual to the lines of the net in the set $\mathbb{C}^0$ of its regular points. Then the following partial converse of the previous theorem has been proved in [36].

Theorem 3.3. *Let H be a finite Abelian group that is either cyclic or it has at least one element of order greater than 9. Then every realization of H by a 3-net in $\mathbb{P}^2$ is algebraic.*

Remark 3.4. *It was also proved in* [36] *that under the extra condition that all the lines in one class of the net are concurrent then* 6 *instead of* 9 *suffices for the conclusion of Theorem. In the case where each of three classes consists of concurrent lines,* 2 *can be substituted for* 9. *In fact, the conclusion of the theorem holds for $\mathbb{Z}_2$ and is false for $\mathbb{Z}_2^k$ for $k > 2$. The group $\mathbb{Z}_2^2$ is an unknown case.*

Further examples of nets and attempts to classify some of them can be found in [31, 34]. In particular, J. Stipins found in [31] example of a $(3, 5)$-net that does not represent a group whence is not algebraic.

3.3. Multinets in $\mathbb{P}^2$.

The example of the B_3-arrangement (i.e., the arrangement of all reflection hyperplanes of the Coxeter group of type B_3) shows that there are arrangements that do not support nets but would support very similar configurations if provided with some multiplicities.

Definition 3.5. *(1) A multi-arrangement of lines in $\mathbb{P}^2$ is an arrangement L of lines together with a multiplicity function $m : L \to \mathbb{Z}_{>o}$. For every $\ell \in L$ the integer $m(\ell)$ is called the multiplicity of ℓ.*

(2) Let k be an integer, $k \geq 3$. A k-multinet is a multi-arrangement (L, m) with L partitioned into k blocks $L_1, \ldots, L_k$ subject to the following conditions:

(i) As for the nets, let $\mathcal{X}$ be the set of the intersections of lines from different blocks; then for each $P \in \mathcal{X}$ the number $n(P) = \sum_{\ell \in L_i, P \in \ell} m(\ell)$ is independent of i;

This number is called the multiplicity of P;

(ii) For every $1 \leq i \leq k$ and $\ell, \ell' \in L_i$ there exists a sequence $\ell_0 = \ell, \ell_1, \ldots, \ell_r = \ell'$ such that $\ell_{j-1} \cap \ell_j \notin \mathcal{X}$, for all $j = 1, 2, \ldots, r$.

Remark 3.6. *(1) The axiom (i) is a direct generalization of the definition of nets.*

(2) The more technical axiom (ii) guarantees that the partition is the finest for given (L, m) and $\mathcal{X}$. Due to this axiom, the blocks L_i are the equivalent classes of the transitive closure of the relation $\ell \sim \ell'$ if $\ell \cap \ell' \notin \mathcal{X}$.

(3) If $n(P) = 1$ for all $P \in \mathcal{X}$ then clearly $m(\ell) = 1$ for all $\ell \in L$ and we have a k-net. The converse is false—there are multinets with $m(\ell) = 1$ for all lines ℓ but $n(P) \neq 1$ for some P (see Fig. 3).

Multinets satisfy the numerical equalities which generalize those for nets.

(1) $\sum_{\ell \in L_i} m(\ell)$ does not depend on i;

This integer is denoted by d and the multinet is called (k, d)-multinet.

(2) $\sum_{\ell \in L} m(\ell) = dk$;

(3) $\sum_{P \in \mathcal{X} \cap \ell} n(P) = d$ for every $\ell \in L$;

(4) $\sum_{P \in \mathcal{X}} n(P)^2 = d^2$.

Also dividing all $m(\ell)$ by their common divisor does not spoil the properties. Thus we can and always will assume that the numbers $\{m(\ell) | \ell \in L\}$ are mutually relatively prime.

As stated above, the simplest and motivating example of a multinet that is not a net is supported by the Coxeter arrangement of the type B_3, see Fig. 2. It can be given (up to projective isomorphism) by the

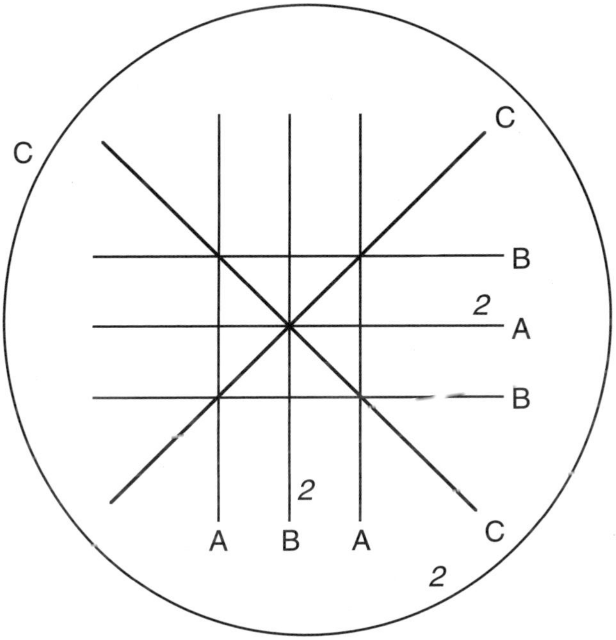

Fig. 2. A (3,4) multinet

equation

$$[x^2(y^2-z^2)][y^2(x^2-y^2)][z^2(x^2-y^2)]=0$$

where the exponents signify that the coordinate axes are taken each with multiplicity 2. The brackets show the partition into 3 classes. It is a $(3,4)$-multinet with 9 lines (12 counting with multiplicities) and 7 points (16 counting with squares of multiplicities).

3.4. Pencils of plane algebraic curves

Here we briefly discuss the second ingredient for the characterization of R^1.

We will identify homogeneous polynomials in three variables that differ by a non-zero constant multiplier. Thus we do not make a distinction between a homogeneous polynomial and the projective plane curve (perhaps non-reduced) defined by it, and treat it either as a polynomial or as a curve. A *pencil* of plane curves is a line in the projective space of homogeneous polynomials of some fixed degree. Thus any two distinct curves of the same degree generate a pencil, and conversely a pencil is determined by any two of its curves C_1, C_2. An arbitrary curve in the

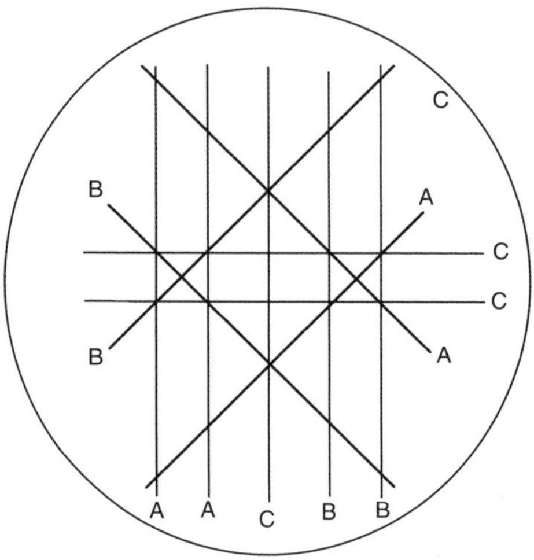

Fig. 3. A multinet with $m(\ell) = 1$ for all $\ell \in L$, which is not a net

pencil (called a *fiber*) is then $aC_1 + bC_2$, $[a : b] \in \mathbb{P}^1$. A pencil π with generators C_1 and C_2 can be equivalently defined as a rational morphism $\mathbb{P}^2 \to \mathbb{P}^1$ via $x \mapsto [C_2(x) : -C_1(x)]$. We denote this morphism also by π.

Every two fibers in a pencil intersect in the same set of points $\mathcal{X} = C_1 \cap C_2$, called the *base* of the pencil. If fibers do not have a common component then the base is a finite set of points. The base $\mathcal{X}$ coincides with the indeterminacy locus of the rational morphism π.

We call a pencil *connected* if the proper transform of each fiber after the blowup at all the points of the base stays connected. Notice that if generic fibers of a pencil π are irreducible then π is connected. A curve of the form $\prod_{i=1}^{q} \alpha_i^{m_i}$, where α_i are different linear forms and m_i are positive integer for $1 \leq i \leq q$ will be called *completely reducible*. We are interested in connected pencils with at least three completely reducible fibers. Following [17], we say that such a pencil is of *Ceva type*.

3.5. Characterizations of R^1

Now we can give a characterization of the resonance variety R^1 by characterizing its components. Let $\mathcal{V}$ be a component of R^1. The *support* U *of* $\mathcal{V}$ is the smallest subarrangement $U = supp\mathcal{V}$ of L such that for every $a = \sum_{\ell \in L} a_\ell e_\ell \in \mathcal{V}$ if $a_\ell \neq 0$ then $\ell \in U$. If $supp\mathcal{V}$ is a set of several concurrent lines from L then $\mathcal{V}$ is local. For every component $\mathcal{V}$

we can restrict our consideration to *supp*$\mathcal{V}$ whence we can always assume that $\mathcal{V}$ is supported on the whole L.

Theorem 3.7. *Let (L, m) be a multi-arrangement of lines in $\mathbb{P}^2$ not passing all through a point. The following conditions are equivalent:*

(i) There exists a partition $\pi = \{L_1, L_2, \dots, L_k\}$ $(k \geq 3)$ of (L, m) that forms a (k, d)-multinet $(d > 1)$;

(ii) The curves $C_i = \prod_{\ell \in L_i} \alpha_\ell^{m(\ell)}$ are all the completely reducible fibers of a pencil of Ceva type of curves of degree $d > 1$;

(iii) There is an irreducible component of R^1 of dimension $k - 1$ supported on the whole L. More precisely, the cohomology classes in A^1 of the logarithmic forms $\frac{dC_i}{C_i} - \frac{dC_1}{C_1}$, $i = 2, 3, \dots, k$, form a basis of the component.

Partial results in this direction have been obtained in [24] in a completely different way.

Proof. The implication (ii)$\Longrightarrow$(iii) can be checked straightforwardly.

The implications (i)$\Longrightarrow$(ii) and (iii)$\Longrightarrow$(i) have been proved in [17]. Here we give the proofs that are somewhat different with several new ideas and short cuts. Also the proof of the former implication fills a small gap in [17].

Proof of implication (i)$\Longrightarrow$(ii). It suffices to prove that the pencil π generated by C_1 and C_2 is connected and C_i is a fiber of π for every $i > 2$. Fix $i > 2$ and put $D = D_i = \prod_{\ell \in L_i} \alpha_\ell$.

(a) **Claim 1**. There exists a fiber $F = F_i$ of π such that D divides F. Indeed by the part (ii) of the definition of multinets we can order $\ell_1, \ell_2, \dots, \ell_k$ of L_i so that $\ell_j \cap \ell_{j+1} \notin \mathcal{X}$ for all j, $1 \leq j \leq k - 1$. Choose a point $P \in \ell_1 \setminus \mathcal{X}$ and put $F = C_2(P)C_1 - C_1(P)C_2$. Then by Bèzout's Theorem ℓ_1 is a component of F. Now substituting points $\ell_j \cap \ell_{j+1}$ for P, claim 1 follows by induction on j.

(b) For each $\ell \in L_i$ denote by $k(\ell)$ the largest integer such that $\alpha_\ell^{k(\ell)}$ divides F and put $\bar{D} = \prod_{\ell \in L_i} \alpha_\ell^{k(\ell)}$. Notice that $1 \leq k(\ell) \leq m(\ell)$. We have $F = \bar{D}E$ where E is a polynomial relatively prime with D of degree

$$(1) \qquad \deg E = d - \sum_{\ell \in L_i} k(\ell) = \sum_{\ell \in L_i} (m(\ell) - k(\ell)).$$

In particular, using the same idea as in (a) we have $E \cap D \subset \mathcal{X}$. Thus, due again to Bézout's Theorem, we have for every $\ell_0 \in L_i$

$$(2) \qquad \deg E = \sum_{\ell \in L_i'} (m(\ell) - k(\ell)) + s(m(\ell_0) - k(\ell_0))$$

where $L_i' = \{\ell \in L_i | \ell \cap \ell_0 \in \mathcal{X}\}$ and $s = |L_i'|$. Also by properties of multinets we have

$$sm(\ell_0) = \sum_{\ell \in L_i \setminus L_i'} m(\ell) \tag{3}$$

for every $\ell_0 \in L_i$.

Comparing (1) and (2) and using (3) we obtain

$$sk(\ell_0) = \sum_{\ell \in L_i \setminus L_i'} k(\ell) \tag{4}$$

for every $\ell_0 \in L_i$.

(c) **Claim 2.** The function $\rho : L_i \to \mathbb{Q}$ given by $\ell \mapsto \frac{m(\ell)}{k(\ell)}$ is constant.

Indeed choose $\ell_0 \in L_i$ such that $\rho(\ell_0)$ is the greatest value on L_i. Then multiplying the equality (3) by $\frac{1}{\rho(\ell_0)}$ we have

$$sk(\ell_0) = \sum_{\ell \in L_i \setminus L_i'} m(\ell)\frac{k(\ell_0)}{m(\ell_0)} \leq \sum_{\ell \in L_i \setminus L_i'} k(\ell) \tag{5}$$

which has to be an equality. Thus $\rho(\ell) = \rho(\ell_0)$ for every $\ell \in L_i \setminus L_i'$, i.e., ρ is constant on $L_i \setminus L_i'$. Applying again part (ii) of the definition of multinets, ρ is constant on L_i.

(d) Here we conclude the proof using Stein Factorization Theorem (SFT) ([19, Cor. III.11.5], see also [18, p. 556]).

As it proved in (c), for every $i \geq 1$ there is a constant ρ_i such that $m(\ell)/k(\ell) = \rho_i$ for all $\ell \in L_i$. Suppose π is not connected which is definitely the case if at least one $\rho_i \neq 1$. The rational map $\pi : \mathbb{P}^2 \to \mathbb{P}^1$ lifts to a regular map $\tilde{\pi} : \mathbb{S} \to \mathbb{P}^1$ where $\varphi : \mathbb{S} \to \mathbb{P}^2$ is the blowup at $\mathcal{X}$. By SFT we can write $\tilde{\pi} = f \circ \tilde{\pi}_0$ where $\tilde{\pi}_0 : \mathbb{S} \to \mathcal{C}$ is a regular map with connected fibers to a curve $\mathcal{C}$, and $f : \mathcal{C} \to \mathbb{P}^1$ is a finite regular map. Since φ is birational, $\tilde{\pi}_0$ can be pushed down to a rational map $\pi_0 : \mathbb{P}^2 \to \mathcal{C}$. Since π is not connected the degree e of $f : \mathcal{C} \to \mathbb{P}^1$ is greater than 1. The fibers of π_0 are curves of degree d' where $d = ed'$ for some integer $e > 1$.

Now, again by condition (ii) of definition of multinets, the proper transform $\tilde{D}_i$ of $\bar{D}_i$ is connected. Since D_i and E_i are disjoint away from the base locus, $\tilde{D}_i$ is in fact a connected component of the proper transform of $F_i = \bar{D}_i E_i$. Then $\bar{D}_i$ is a (perhaps multiple) fiber of π_0 whence there exists a curve D_{0i} such that $D_{0i}^{\mu} = \bar{D}_i$ for an integer μ and $D_{0i}^{e} = C_i$. Since the proper transforms of C_1 and C_2 are also connected by definition of multinets the respective fibers of f are points of multiplicity e each. This implies that all the multiplicities $m(\ell)$ $(\ell \in L)$ are divisible by e contradicting the condition that the line multiplicities $m(\ell)$ are mutually relatively prime. We conclude that $e = 1$, π is connected, and C_i is its fiber for every i.

Remark 3.8. *Notice that unlike in the proof of the same implication in* [17] *the condition (ii) from definition of multinets is used for all C_i. In fact, the Example 3.9 from* [17] *shows that if this condition fails even for one C_i then the implication may not hold.*

Proof of the implication (iii)$\Longrightarrow$(i).

For the proof of this implication [17] refers to [23]. Here we give details of a simplified version of the proof in [23].

Fix an irreducible component $\mathcal{V}$ of R^1 supported on a the whole L. For every $a \in \mathcal{V}$ put $Z(a) = \{b \in A^1 | ab = 0\}$. Recall from the end of section 2 that for every $b \in Z(a)$ we have $Z(b) = Z(a)$. This implies that $Z(c) \cap Z(a) = \{0\}$ for each $c \subset R^1 \setminus Z(a)$. Since $\mathcal{V}$ is an irreducible component of R^1 it coincides with $Z(a)$ and in particular is a linear subspace of A^1. Now without any loss we can assume that a is in general position in $\mathcal{V}$ in the sense that $a_\ell \neq 0$ for every $\ell \in L$.

Now we define the subset $\mathcal{X}$ of the set of all multiple points of intersections of lines from L: $\mathcal{X} = \{P | \sum_{P \in \ell} a_\ell = 0\}$. Denote by J the incidence matrix (of the size $|\mathcal{X}| \times |L|$) of points from $\mathcal{X}$ and lines from L and define the square $|L| \times |L|$-matrix $Q = J^T J - E$ where J^T is the transpose of J and E is the matrix with all entries equal to 1. The importance of Q for the component $\mathcal{V}$ is that

$$\mathcal{V} = Z(a) = \ker J \cap \ker E = \ker Q \cap \ker E.$$

This follows from the description of local components. Besides Q is a symmetric integer matrix with the diagonal entries equal to $t_\ell - 1$, where t_ℓ is the number of points in $\mathcal{X}$ which are contained in ℓ. The assumption that the support of $\mathcal{V}$ is equal to the whole L implies that $Q_{\ell,\ell} > 0$ for $\ell \in L$. An off-diagonal entry $Q_{\ell,\ell'}$ is equal to zero when $\ell \cap \ell' \in \mathcal{X}$, and is equal to -1 otherwise. In particular, Q is a *(generalized) Cartan matrix*. These matrices were classified by E. Vinberg, see [20].

In order to apply Vinberg's classification (and construct the partition for the multinet) we represent Q as the direct sum of indecomposable matrices, as follows. Define the equivalence relation on L as the transitive hull of the relation $\ell \approx \ell'$ if $\ell \cap \ell' \notin \mathcal{X}$ and denote by $\{L_1, L_2, \dots, L_k\}$ the partition of L into the equivalence classes. Due to properties of Q we have a direct sum decomposition $Q = \oplus_{i=1}^k Q_i$ where the matrices Q_i are again generalized Cartan and besides are indecomposable.

According to Vinberg's classification, each of these matrices is of one of three possible types. We denote by u elements of the linear space over $\mathbb{Q}$ with L as a basis and write $u > 0$ if all coordinates $u(\ell)$ are positive.

(1) Q_i is *finite* if it is positive definite; equivalently $Q_i u > 0$ for every $u > 0$;
(2) Q_i is *affine* if it is positive semidefinite (i.e., $Q_i u \geq 0$ for every $u > 0$) and $\ker(Q_i) = \mathbb{Q}u_i$ for some $u_i > 0$;
(3) Q_i is *indefinite* if $Q_i u < 0$ for some $u > 0$.

Since $\mathcal{V}$ is supported on the whole L the partition that we defined does not have finite blocks. Since matrix $J^T J$ is positive Q can have at most one indefinite block and in that case no affine blocks, in particular $k = 1$. Indeed otherwise it would be easy to find a positive vector u with $\sum_{\ell \in L} u(\ell) = 0$ and $(J^T J u, u) = (Qu, u) < 0$. On the other hand, by definition of the partition, the restriction of $\mathcal{V}$ to every L_i is at most one-dmensional and $\dim \mathcal{V} \geq 2$. Thus $k \neq 1$ whence all the blocks are affine. We see that $\mathcal{V} = \ker Q$ is generated by the vectors $u_i - u_j$ where u_i is a positive integer vector generating $\ker Q_i$ such that $\sum_{\ell \in L_i} u_i(\ell)$ does not depend on i. Now we choose for each $i = 1, 2, \ldots, k$ such vectors u_i and put $m(\ell) = u_i(\ell)$ if $\ell \in L_i$. Then it follows from the construction that the partition of the multiarrangement (L, m) defines a (k, d) multinet. Q.E.D.

Examples

We start with the example of a multinet of type B_3 and change the exponent 2 to an arbitrary $d \geq 2$. What we obtain are examples of $(3, d)$-multinets that we call N_d ($N_2 = B_3$). This can be seen easier if we consider first the respective pencil of Ceva type. Indeed the curves $z^d(x^d - y^d)$, and $x^d(y^d - z^d)$, generate the pencil (the *Fermat pencil*) of Ceva type with the third completely rerducible fiber $y^d(x^d - z^d)$.

The other classical example of a pencil of Ceva type is the *Hesse pencil* of cubics generated by $x^3 + y^3 + z^3$ and xyz. This pencil has four completely reducible fibers, each of which is the product of three distinct lines. The resulting $(4, 3)$-net has 12 lines and $|\mathcal{X}| = 9$. In this example each of the four blocks is in general position (i.e., lines of each block intersect at three double points). The set $\mathcal{X}$ can be realized as the the set of all inflection points of a smooth irreducible cubic. It is the only known example of a $(4, d)$-net for any d (see below).

§4. Upper bound on k

Theorem 4.1. *A Ceva pencil of degree $d > 1$ cannot have more than four completely reducible fibers.*

Due to the main theorem this result can be also formulated in at least two other equivalent ways.

(i) For a (k,d)-multinet in $\mathbb{P}^2$ if $d > 1$ then $k < 5$.

(ii) Every non-local irreducible resonance component has dimension either two or three.

(iii) $\dim H^1(A,x) \leq 2$ if x is not supported on several lines passing through a point.

These inequalities have resulted from work by different authors. The first result $k \leq 5$ was proved in [23] for nets (not explicitly defined there) using the pencil corresponding to a components of R^1 and computing the Euler characteristic of the blowup at the base in two different ways. This proof was generalized in [17] to the multinets with all $m(\ell) = 1$. Then this inequality was proved in general case in [29] with a help of foliations of $\mathbb{P}^2$ generated by pencils of curves. The stronger inequality $k < 4$ was proved in the dissertation of J. Stipins [31] for nets. He used the pencil of Hessians of the fibers of π. Finally this method was generalized in [37] to the general case.

Let us recall that while there are plenty of examples with $k = 3$, the Hesse pencil is the only example of a Ceva pencil with 4 completely reducible fibers whence the only known $(4,d)$-multinet. $(4,d)$-multinets that are not nets do not exist (see [37]). The only possible examples would be nets with $d \geq 7$.

Open problems

1. Give a direct proof of the propagation property of the resonance varieties R^p.

2. Give a more constructive description of irreducible components of R^p for $p > 1$. In particular, how are they related to linear systems of subsurfaces?

3. Describe all the groups that can be represented by a 3-net in $\mathbb{P}^2$.

Conjecture 4.2. *These are all the finite subgroups of* $PGL(2,\mathbb{C})$.

4. Find other 4-nets besides the Hesse configuration.

Conjecture 4.3. *They do not exist.*

The conjecture has been proved for d=4,5,6.

5. Is every multinet the limit of a family of nets (in $\mathbb{P}^2$)?

For instance, $(3,2d)$-multinet N_d $(d = 2,3,\ldots)$ considered above has this property. The easiest way to see this is to consider an arrangement of planes in $\mathbb{P}^3$ given by the polynomial

$$(x_0^d - x_1^d)(x_2^d - x_3^d)(x_0^d - x_2^d)(x_1^d - x_3^d)(x_0^d - x_3^d)(x_1^d - x_2^d).$$

Intersection of this arrangement with a generic (projective) plane gives a $(3, 2d)$-net in $\mathbb{P}^2$. Moving this plane to $x_3 = 0$ makes the intersection approach the multinet N_d.

Acknowledgements. The author is grateful to Professor Terao and other organizers for organizing the very useful and pleasant conference and for the invitation to give a lecture in it. He also thanks the referee for careful reading of the paper.

References

[1] A. Aramova, L. Avramov and J. Herzog, Resolutions of monomial ideals and cohomology over exterior algebras, Trans. Amer. Math. Soc., **352** (1999), 579–594.

[2] D. Arapura, Geometry of cohomology support loci for local systems I, J. Alg. Geom., **6** (1997), 563–597.

[3] V. I. Arnol'd, The cohomology ring of the group of dyed braids, Math. Notes, **5** (1969), 138–140.

[4] E. Brieskorn, Sur les groupes de tresses, In: Séminaire Bourbaki, 1971/72, Lect. Notes in Math., **317**, Springer-Verlag, 1973, pp. 21–44.

[5] N. Budur, A. Dimca and M. Saito, First Milnor cohomology of hyperplane arrangements, Contemp. Math., **538** (2011), 279–292.

[6] N. Budur, M. Saito and S. Yuzvinsky, On the local zeta functions and the b-functions of certain hyperplane arrangements, J. London Math. Soc. (2), **84** (2011), 631–648.

[7] D.C. Cohen and P. Orlik, Arrangements and local systems, Math. Res. Lett., **7** (2000), 299–316.

[8] D.C. Cohen and A. Suciu, Characteristic varieties of arrangements, Math. Proc. Cambridge Phil. Soc., **127** (1999), 33–53.

[9] H. Crapo, A higher invariant for matroids, J. Combinatorial Theory, **2** (1967), 406–417.

[10] A. Dimca, On the irreducible components of characteristic varieties, An. Ştiinţ. Univ. Ovidius Constanţa Ser. Mat., **15** (2007), 67–74.

[11] A. Dimca, On admissible rank one local systems, J. Algebra, **321** (2009), arXiv:0707.4646.

[12] A. Dimca, S. Papadima and A. Suciu, Topology and geometry of cohomology jump loci, Duke Math. J., **148** (2009), 405–457.

[13] D. Eisenbud, S. Popescu and S. Yuzvinsky, Hyperplane arrangement cohomology and monomials in the exterior algebra, Trans. Amer. Math. Soc., **355** (2003), 4365–4383.

[14] H. Esnault, V. Schechtman and E. Viehweg, Cohomology of local systems on the complement of hyperplanes, Invent. Math., **109** (1992), 557–561.
[15] M. Falk, Arrangements and cohomology, Ann. Combin., **1** (1997), 135–157.
[16] M. Falk, Resonance varieties over fields of positive characteristic, Int. Math. Res. Not. IMRN, **2007** (2007), no. 3, Art. ID rnm009.
[17] M. Falk and S. Yuzvinsky, Multinets, resonance varieties, and pencils of plane curves, Compositio Math., **143** (2007), 1069–1088.
[18] P. Griffiths and J. Harris, Principles of Algebraic Geometry, Wiley Interscience, New York, 1978.
[19] R. Hartshorne, Algebraic Geometry, Grad. Texts in Math., **52**, Springer-Verlag, 1977.
[20] V. Kac, Infinite Dimensional Lie Algebras, Cambridge Univ. Press, Cambridge, 1990.
[21] T. Kohno, Homology of a local system on the complement of hyperplanes, Proc. Japan Acad. Ser. A Math. Sci., **62** (1986), 144–147.
[22] A. Libgober, First order deformations for rank one local systems with non vanishing cohomology, Topology Appl., **118** (2002), 159–168.
[23] A. Libgober and S. Yuzvinsky, Cohomology of the Orlik–Solomon algebras and local systems, Compositio Math., **21** (2000), 337–361.
[24] M. Marco-Buzunáriz, Resonance varieties, admissible line combinatorics and combinatorial pencils, arXiv:CO/0505435.
[25] S. P. Novikov, Bloch homology, critical points of functions, and closed 1-forms, Soviet Math. Dokl., **33** (1986), 551–555.
[26] P. Orlik and L. Solomon, Combinatorics and topology of complements of hyperplanes, Invent. Math., **56** (1980), 167–189.
[27] P. Orlik and H. Terao, Arrangements of hyperplanes, Grundlehren Math. Wiss., **300**, Springer-Verlag, 1992.
[28] P. Orlik and H. Terao, The number of critical points of a product of powers of linear functions, Invent. Math., **120** (1995), 1–14.
[29] J. Pereira and S. Yuzvinsky, On the dimension of resonance and characteristic varieties, Adv. in Math., **219** (2008), 672–688.
[30] V. Schechtman and A. Varchenko, Arrangements of hyperplanes and Lie algebra homology, Invent. Math., **106** (1991), 139–194.
[31] J. Stipins, On finite nets in the complex projective plane, Ph.D. thesis, Univ. of Michigan, 2007.
[32] A. Suciu, Translated tori in the characteristic varieties of complex hyperplane arrangements, Topology Appl., **118** (2002), 209–223.
[33] A. Suciu, Fundamental groups, Alexander invariants, and cohomology jumping loci, Contemp. Math., **538** (2011), 179–223.
[34] G. Urzua, On line arrangements with applications to 3-nets, Ph. D. thesis, Univ. of Michigan, 2007.
[35] S. Yuzvinsky, Cohomology of Brieskorn–Orlik–Solomon algebras, Comm. Algebra, **23** (1995), 5339–5354.

[36] S. Yuzvinsky, Realization of finite abelian groups by nets in $\mathbb{P}^2$, Compos. Math., **140** (2004), 1614–1624.
[37] S. Yuzvinsky, A new bound on the number of special fibers in a pencil of curves, Proc. Amer. Math. Soc., **137** (2009), 1641–1648.

Sergey Yuzvinsky
Department of Mathematics, University of Oregon,
Eugene OR 97403-1222, USA
E-mail address: yuz@uoregon.edu

Advanced Studies in Pure Mathematics

A SERIES OF UP-TO-DATE GUIDES OF LASTING INTEREST TO ADVANCED MATHEMATICS

Volume 1 Algebraic Varieties and Analytic Varieties.
Edited by S. Iitaka. February, 1983

Volume 2 Galois Groups and their Representations.
Edited by Y. Ihara. December, 1983

Volume 3 Geometry of Geodesics and Related Topics.
Edited by K. Shiohama. June, 1984

Volume 4 Group Representations and Systems of Differential Equations.
Edited by K. Okamoto. March, 1985

Volume 5 Foliations.
Edited by I. Tamura. February, 1986

Volume 6 Algebraic Groups and Related Topics.
Edited by R. Hotta. March, 1985

Volume 7 Automorphic Forms and Number Theory.
Edited by I. Satake. February, 1986

Volume 8 Complex Analytic Singularities.
Edited by T. Suwa and P. Wagreich. February, 1987

Volume 9 Homotopy Theory and Related Topics.
Edited by H. Toda. February, 1987

Volume 10 Algebraic Geometry, Sendai, 1985.
Edited by T. Oda. July, 1987

Volume 11 Commutative Algebra and Combinatorics.
Edited by M. Nagata and H. Matsumura.
October, 1987

Volume 12 Galois Representations and Arithmetic Algebraic Geometry.
Edited by Y. Ihara. November, 1987

Volume 13 Investigations in Number Theory.
Edited by T. Kubota. March, 1988

Volume 14 Representations of Lie Groups, Kyoto, Hiroshima, 1986.
Edited by K. Okamoto and T. Oshima. February, 1989

Volume 15 Automorphic Forms and Geometry of Arithmetic Varieties.
Edited by K. Hashimoto and Y. Namikawa. July, 1989

Volume 16 Conformal Field Theory and Solvable Lattice Models.
Edited by M. Jimbo, T. Miwa and A. Tsuchiya.
June, 1988

Volume 17 Algebraic Number Theory—in honor of K. Iwasawa.
Edited by J. Coates, R. Greenberg, B. Mazur and I. Satake. August, 1989

Volume 18 I—Recent Topics in Differential and Analytic Geometry.
Edited by T. Ochiai. November, 1990
II—Kähler Metric and Moduli Spaces.
Edited by Edited by T. Ochiai. November, 1990

Volume 19 Integrable Systems in Quantum Field Theory and Statistical Mechanics.
Edited by M. Jimbo, T. Miwa and A. Tsuchiya.
November, 1989

Volume 20 Aspects of Low Dimensional Manifolds.
Edited by Y. Matsumoto and S. Morita.
December, 1992

Volume 21 Zeta Functions in Geometry.
Edited by N. Kurokawa and T. Sunada.
December, 1992

Volume 22 Progress in Differential Geometry.
Edited by K. Shiohama. September, 1993

Volume 23 Spectral and Scattering Theory and Applications.
Edited by K. Yajima. February, 1994

Volume 24 Progress in Algebraic Combinatorics.
Edited by E. Bannai and A. Munemasa. May, 1996

Volume 25 CR-Geometry and Overdetermined Systems.
Edited by T. Akahori, G. Komatsu, K. Miyajima, M. Namba and K. Yamaguchi. July, 1997

Volume 26 Analysis on Homogeneous Spaces and Representation Theory of Lie Groups, Okayama-Kyoto.
Edited by M. Kashiwara, T. Kobayashi, T. Matsuki, K. Nishiyama and T. Oshima. April, 2000

Volume 27 Arrangements – Tokyo 1998.
Edited by M. Falk and H. Terao. August, 2000

Volume 28 Combinatorial Methods in Representation Theory.
Edited by M. Kashiwara, K. Koike, S. Okada, I. Terada, and H.-F. Yamada. December, 2000

Volume 29 Singularities – Sapporo 1998.
Edited by J.-P. Brasselet and T. Suwa.
December, 2000

Volume 30 Class Field Theory – Its Centenary and Prospect.
Edited by K. Miyake. April, 2001

Volume 31 Taniguchi Conference on Mathematics Nara '98.
Edited by M. Maruyama and T. Sunada. May, 2001

Volume 32 Groups and Combinatorics — in memory of Michio Suzuki.
Edited by E. Bannai, H. Suzuki, H. Yamaki and T. Yoshida. October, 2001

Volume 33 Computational Commutative Algebra and Combinatorics.
Edited by T. Hibi. February, 2002

Volume 34 Minimal Surfaces, Geometric Analysis and Symplectic Geometry.
Edited by K. Fukaya, S. Nishikawa and J. Spruck.
June, 2002

Volume 35 Higher Dimensional Birational Geometry.
Edited by S. Mori and Y. Miyaoka. July, 2002

Volume 36 Algebraic Geometry 2000, Azumino.
Edited by S. Usui, M. Green, L. Illusie, K. Kato, E. Looijenga, S. Mukai and S. Saito. November, 2002

Volume 37 Lie Groups, Geometric Structures and Differential Equations — One Hundred Years after Sophus Lie —.
Edited by T. Morimoto, H. Sato and K. Yamaguchi.
December, 2002

Volume 38 Operator Algebras and Applications.
Edited by H. Kosaki. January, 2004

Volume 39 Stochastic Analysis on Large Scale Interacting Systems.
Edited by T. Funaki and H. Osada. March, 2004

Volume 40 Representation Theory of Algebraic Groups and Quantum Groups.
Edited by T. Shoji, M. Kashiwara, N. Kawanaka, G. Lusztig and K. Shinoda. March, 2004

Volume 41 Stochastic Analysis and Related Topics in Kyoto
In honour of Kiyosi Itô.
Edited by H. Kunita, S. Watanabe and Y. Takahashi.
March, 2004

Volume 42 Complex Analysis in Several Variables
– Memorial Conference of Kiyoshi Oka's Centennial Birthday, Kyoto/Nara 2001.
Edited by K. Miyajima, M. Furushima, H. Kazama, A. Kodama, J. Noguchi, T. Ohsawa, H. Tsuji and T. Ueda. June, 2004

Volume 43 Singularity Theory and Its Applications.
Edited by S. Izumiya, G. Ishikawa, H. Tokunaga, I. Shimada and T. Sano. January, 2007

Volume 44 Potential Theory in Matsue.
Edited by H. Aikawa, T. Kumagai, Y. Mizuta and N. Suzuki. August, 2006

Volume 45 Moduli Spaces and Arithmetic Geometry (Kyoto, 2004).
Edited by S. Mukai, Y. Miyaoka, S. Mori, A. Moriwaki and I. Nakamura. January, 2007

Volume 46 Singularities in Geometry and Topology 2004.
Edited by J.-P. Brassele and T. Suwa. February, 2007

Volume 47 Asymptotic Analysis and Singularities.
1—Hyperbolic and dispersive PDEs and fluid mechanics.
2—Elliptic and parabolic PDEs and related problems.
Edited by H. Kozono, T. Ogawa, K. Tanaka, Y. Tsutsumi and E. Yanagida. October, 2007

Volume 48 Finsler Geometry, Sapporo 2005
— In Memory of Makoto Matsumoto.
Edited by S. V. Sabau and H. Shimada.
November, 2007

Volume 49 Probability and Number Theory — Kanazawa 2005.
Edited by S. Akiyama, K. Matsumoto, L. Murata and H. Sugita. December, 2007

Volume 50 Algebraic Geometry in East Asia — Hanoi 2005.
Edited by K. Konno and V. Nguyen-Khac.
February, 2008

Volume 51 Surveys on Geometry and Integrable Systems.
Edited by M. Guest, R. Miyaoka and Y. Ohnita.
November, 2008

Volume 52 Groups of Diffeomorphisms
in honor of Shigeyuki Morita
on the occasion of his 60th birthday.
Edited by R. Penner, D. Kotschick, T. Tsuboi, N. Kawazumi, T. Kitano and Y. Mitsumatsu. November, 2008

Volume 53 Advances in Discrete Dynamical Systems.
Edited by S. Elaydi, K. Nishimura, M. Shishikura and N. Tose. June, 2009

Volume 54 Algebraic Analysis and Around
in honor of Professor Masaki Kashiwara's 60th birthday.
Edited by T. Miwa, A. Matsuo, T. Nakashima and Y. Saito. February, 2009

Volume 55 Noncommutativity and Singularities
Proceedings of French–Japanese symposia held at IHÉS in 2006.
Edited by J.-P. Bourguignon, M. Kotani, Y. Maeda and N. Tose. July, 2009

Volume 56 Singularities — Niigata–Toyama 2007.
Edited by J.-P. Brasselet, S. Ishii, T. Suwa and M. Vaquie. November, 2009

Volume 57 Probabilistic Approach to Geometry.
Edited by M. Kotani, M. Hino and T. Kumagai. March, 2010

Volume 58 Algebraic and Arithmetic Structures of Moduli Spaces (Sapporo 2007).
Edited by I. Nakamura and L. Weng. June, 2010

Volume 59 New Developments in Algebraic Geometry, Integrable Systems and Mirror Symmetry (RIMS, Kyoto, 2008).
Edited by M.-H. Saito, S. Hosono and K. Yoshioka. August, 2010

Volume 60 Algebraic Geometry in East Asia — Seoul 2008.
Edited by J. Keum, S. Kondō, K. Konno and K. Oguiso. September, 2010

Volume 61 Exploring New Structures and Natural Constructions in Mathematical Physics.
Edited by K. Hasegawa, T. Hayashi, S. Hosono and Y. Yamada. October, 2011

Volume 62 Arrangements of Hyperplanes—Sapporo 2009.
Edited by H. Terao and S. Yuzvinsky. March, 2012

To be continued